IMPORTANT:

HERE IS YOUR REGISTRATION CODE TO ACCESS
YOUR PREMIUM McGRAW-HILL ONLINE RESOURCES.

For key premium online resources you need THIS CODE to gain access. Once the code is entered, you will be able to use the Web resources for the length of your course.

If your course is using **WebCT** or **Blackboard**, you'll be able to use this code to access the McGraw-Hill content within your instructor's online course.

Access is provided if you have purchased a new book. If the registration code is missing from this book, the registration screen on our Website, and within your WebCT or Blackboard course, will tell you how to obtain your new code.

Registering for McGraw-Hill Online Resources

TO gain access to your MCGraw-Hill web resources simply follow the steps below:

1. USE YOUR WEB BROWSER TO GO TO: **www.mhhe.com/fellmann8e**
2. CLICK ON **FIRST TIME USER**.
3. ENTER THE REGISTRATION CODE* PRINTED ON THE TEAR-OFF BOOKMARK ON THE RIGHT.
4. AFTER YOU HAVE ENTERED YOUR REGISTRATION CODE, CLICK **REGISTER**.
5. FOLLOW THE INSTRUCTIONS TO SET-UP YOUR PERSONAL UserID AND PASSWORD.
6. WRITE YOUR UserID AND PASSWORD DOWN FOR FUTURE REFERENCE. KEEP IT IN A SAFE PLACE.

TO GAIN ACCESS to the McGraw-Hill content in your instructor's **WebCT** or **Blackboard** course simply log in to the course with the UserID and Password provided by your instructor. Enter the registration code exactly as it appears in the box to the right when prompted by the system. You will only need to use the code the first time you click on McGraw-Hill content.

Thank you, and welcome to your MCGraw-Hill online Resources!

2UIQ-SNVW-79HK-IOIN-F5R9

REGISTRATION CODE

* YOUR REGISTRATION CODE CAN BE USED ONLY ONCE TO ESTABLISH ACCESS. IT IS NOT TRANSFERABLE.

0-07-293965-6 T/A FELLMANN/GETIS/GETIS/MALINOWSKI: HUMAN GEOGRAPHY, 8E

Inset map (top — Europe):

NORWAY
NORTH SEA
SWEDEN
BALTIC SEA
ESTONIA
100 Miles
100 Kilometers
DENMARK
LATVIA
RUSSIA
LITHUANIA
RUSSIA
NETHERLANDS
BELARUS
GERMANY
POLAND
BELGIUM
CZECH REPUBLIC
UKRAINE
LUXEMBOURG
SLOVAKIA
FRANCE
LIECHTENSTEIN
MOLDOVA
SWITZERLAND
AUSTRIA
HUNGARY
ROMANIA
SLOVENIA
CROATIA
YUGOSLAVIA (SERBIA-MONTENEGRO)
MONACO
SAN MARINO
BOSNIA-HERZEGOVINA
BLACK SEA
ITALY
BULGARIA
MEDITERRANEAN
ALBANIA
MACEDONIA
SEA
GREECE
TURKEY
MALTA

Main map (world):

ARCTIC OCEAN
ircle
AND
NORWAY
SWEDEN
FINLAND
RUSSIA
RANCE
A
SPAIN
TURKEY
TUNISIA
CYPRUS
LEBANON
SYRIA
IRAQ
KAZAKSTAN
MONGOLIA
NORTH PACIFIC OCEAN
UZBEKISTAN
KYRGYZSTAN
TURKMENISTAN
TAJIKISTAN
NORTH KOREA
JAPAN
SOUTH KOREA
ALGERIA
LIBYA
ISRAEL
EGYPT
JORDAN
BAHRAIN
SAUDI ARABIA
IRAN
KUWAIT
QATAR
AFGHANISTAN
PAKISTAN
NEPAL
BHUTAN
CHINA
MALI
NIGER
CHAD
SUDAN
ERITREA
YEMEN
OMAN
UNITED ARAB EMIRATES
INDIA
MYANMAR (BURMA)
TAIWAN
Tropic of Cancer
NIGERIA
BANGLADESH
LAOS
THAILAND
VIETNAM
PHILIPPINES
MARSHALL ISLANDS
EROON
EPUBLIC
NCIPE
UINEA
GABON
UBLIC
UGANDA
ETHIOPIA
DJIBOUTI
SOMALIA
Equator
SRI LANKA
MALDIVES
CAMBODIA (KAMPUCHEA)
BRUNEI
MALAYSIA
PALAU
MICRONESIA
NAURU
KIRIBATI
RWANDA
DEM. REP. OF THE CONGO
KENYA
BURUNDI
TANZANIA
MALAWI
SEYCHELLES
SINGAPORE
INDONESIA
PAPUA NEW GUINEA
SOLOMON ISLANDS
TUVALU
ANGOLA
ZAMBIA
COMOROS
MOZAMBIQUE
INDIAN OCEAN
EAST TIMOR
VANUATU
FIJI
MADAGASCAR
MAURITIUS
Tropic of Capricorn
NAMIBIA
BOTSWANA
ZIMBABWE
SWAZILAND
LESOTHO
SOUTH AFRICA
AUSTRALIA
NEW ZEALAND

Inset map (bottom left — West Africa):

MAURITANIA
SENEGAL
MALI
NIGER
IA
NEA-
SSAU
BURKINA FASO
GUINEA
BENIN
SIERRA LEONE
GHANA
NIGERIA
ANTIC
CEAN
IVORY COAST
LIBERIA
TOGO
150
300 Miles
300 Kilometers

Inset map (bottom right — Caucasus):

100 Miles
100 Kilometers
CASPIAN SEA
RUSSIA
BLACK SEA
GEORGIA
AZERBAIJAN
ARMENIA
TURKEY
AZERBAIJAN
IRAN

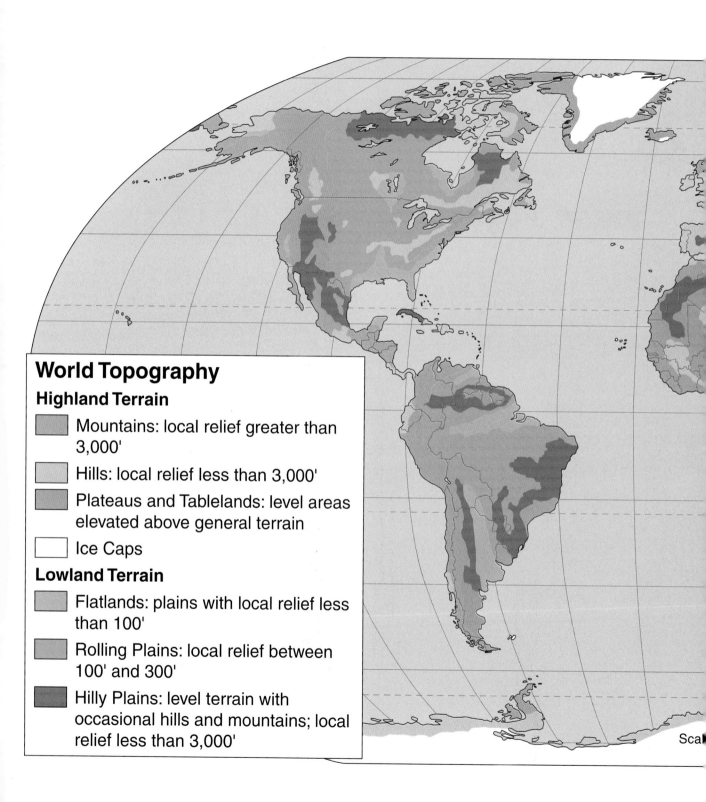

World Topography

Highland Terrain

Mountains: local relief greater than 3,000'

Hills: local relief less than 3,000'

Plateaus and Tablelands: level areas elevated above general terrain

Ice Caps

Lowland Terrain

Flatlands: plains with local relief less than 100'

Rolling Plains: local relief between 100' and 300'

Hilly Plains: level terrain with occasional hills and mountains; local relief less than 3,000'

Sca

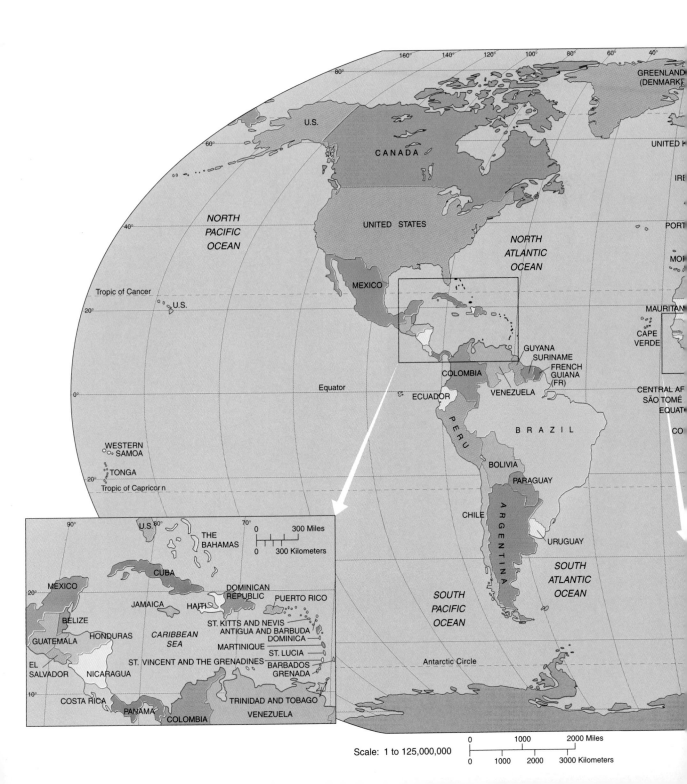

160°   140°   120°   100°   80°   60°   40°

GREENLAND
(DENMARK)

80°

U.S.

60°

CANADA

UNITED K

IRE

NORTH
PACIFIC
OCEAN

40°

UNITED STATES

NORTH
ATLANTIC
OCEAN

PORT

MOF

Tropic of Cancer

MEXICO

MAURITAN

20°

U.S.

CAPE
VERDE

GUYANA
SURINAME
FRENCH
GUIANA
(FR)

COLOMBIA

VENEZUELA

CENTRAL AF
SÃO TOMÉ
EQUAT

Equator

0°

ECUADOR

CO

P
E
R
U

B R A Z I L

WESTERN
SAMOA

BOLIVIA

PARAGUAY

20°

TONGA

Tropic of Capricorn

CHILE

A
R
G
E
N
T
I
N
A

URUGUAY

SOUTH
ATLANTIC
OCEAN

SOUTH
PACIFIC
OCEAN

SOUTH
PACIFIC
OCEAN

Antarctic Circle

90°

U.S. 80°

THE
BAHAMAS

70°

0   300 Miles

0   300 Kilometers

CUBA

20°

MEXICO

DOMINICAN
REPUBLIC

PUERTO RICO

JAMAICA

HAITI

BELIZE

ST. KITTS AND NEVIS
ANTIGUA AND BARBUDA
DOMINICA

GUATEMALA

HONDURAS

CARIBBEAN
SEA

MARTINIQUE

ST. LUCIA

EL
SALVADOR

NICARAGUA

ST. VINCENT AND THE GRENADINES

BARBADOS
GRENADA

10°

COSTA RICA

PANAMA

TRINIDAD AND TOBAGO

COLOMBIA

VENEZUELA

Scale: 1 to 125,000,000

0   1000   2000 Miles

0   1000   2000   3000 Kilometers

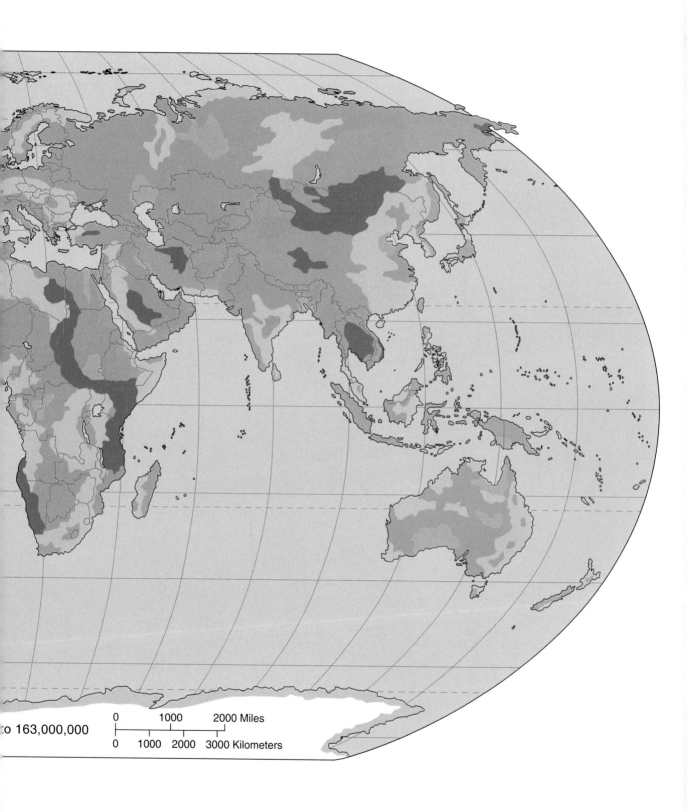

o 163,000,000

0	1000		2000 Miles
0	1000	2000	3000 Kilometers

Eighth Edition

Human
GEOGRAPHY

Landscapes of Human Activities

Jerome D. Fellmann
University of Illinois, Urbana-Champaign

Arthur Getis
San Diego State University

Judith Getis

with contributions by
Jon C. Malinowski
United States Military Academy

Boston   Burr Ridge, IL   Dubuque, IA   Madison, WI   New York   San Francisco   St. Louis
Bangkok   Bogotá   Caracas   Kuala Lumpur   Lisbon   London   Madrid   Mexico City
Milan   Montreal   New Delhi   Santiago   Seoul   Singapore   Sydney   Taipei   Toronto

Higher Education

HUMAN GEOGRAPHY: LANDSCAPES OF HUMAN ACTIVITIES, EIGHTH EDITION

Published by McGraw-Hill, a business unit of The McGraw-Hill Companies, Inc., 1221 Avenue of the Americas, New York, NY 10020. Copyright © 2005, 2003, 2001, 1999, 1997 by The McGraw-Hill Companies, Inc. All rights reserved. No part of this publication may be reproduced or distributed in any form or by any means, or stored in a database or retrieval system, without the prior written consent of The McGraw-Hill Companies, Inc., including, but not limited to, in any network or other electronic storage or transmission, or broadcast for distance learning.

Some ancillaries, including electronic and print components, may not be available to customers outside the United States.

♲ This book is printed on recycled, acid-free paper containing 10% postconsumer waste.

International   1 2 3 4 5 6 7 8 9 0 VNH/VNH 0 9 8 7 6 5 4 3
Domestic   1 2 3 4 5 6 7 8 9 0 VNH/VNH 0 9 8 7 6 5 4 3

ISBN 0-07-282688-6
ISBN 0-07-111233-2 (ISE)

Publisher: *Margaret J. Kemp*
Developmental editor: *Lisa A. Bruflodt*
Executive marketing manager: *Lisa L. Gottschalk*
Senior project manager: *Kay J. Brimeyer*
Production supervisor: *Sherry L. Kane*
Lead media project manager: *Judi David*
Senior media technology producer: *Jeffry Schmitt*
Senior coordinator of freelance design: *Michelle D. Whitaker*
Cover/interior designer: *Jamie E. O'Neal*
Cover image: *D. Boschung/Zefa Masterfile*
Senior photo research coordinator: *Lori Hancock*
Photo research: *Toni Michaels/PhotoFind, LLC*
Supplement producer: *Brenda A. Ernzen*
Compositor: *Shepherd, Inc.*
Typeface: *10/12 Times Roman*
Printer: *Von Hoffmann Corporation*

The credits section for this book begins on page 553 and is considered an extension of the copyright page.

The views and opinions of authors expressed herein do not necessarily state or reflect those of the United States Government or any agency thereof.

Library of Congress Cataloging-in-Publication Data

Human geography : landscapes of human activities / Jerome D. Fellmann . . . [et al.].—8th ed.
    p. cm.
Rev. ed. of: Human geography / Jerome D. Fellmann, Arthur Getis, Judith Getis. 7th ed. © 2003.
Includes bibliographical references and index.
ISBN 0-07-282688-6 (hard copy : alk. paper)
 1. Human geography. I. Fellmann, Jerome Donald, 1926–. II. Fellmann, Jerome Donald, 1926–.
Human geography.

GF41.H893   2005
604.2—dc22   2003022660
                                                  CIP

INTERNATIONAL EDITION ISBN 0-07-111233-2
Copyright © 2005. Exclusive rights by The McGraw-Hill Companies, Inc., for manufacture and export. This book cannot be re-exported from the country to which it is sold by McGraw-Hill. The International Edition is not available in North America.

www.mhhe.com

Brief Contents

Contents

PART Two
Patterns of Diversity and Unity 140

PART *Three*
Dynamic Patterns of the Space Economy 266

C H A P T E R

Eight

Livelihood and Economy:
Primary Activities 269

C H A P T E R

Nine

Livelihood and Economy:
From Blue Collar to Gold
Collar 315

PART Four
Landscapes of Functional Organization 388

PART *Five*
Human Actions
and Environmental
Impacts 480

———— C H A P T E R ————
Thirteen

Human Impacts on Natural
Systems 483

Focus Preview 483

List of Boxes

Chapter 12

Chapter 13

Preface

This eighth edition of *Human Geography* retains the organization and structure of its earlier versions. Like them, it seeks to introduce its users to the scope and excitement of geography and its relevance to their daily lives and roles as informed citizens. We recognize that for many students, human geography may be their first or only work in geography and this their first or only textbook in the discipline. For these students particularly, we seek to convey the richness and breadth of human geography and to give insight into the nature and intellectual challenges of the field of geography itself. Our goals are to be inclusive in content, current in data, and relevant in interpretations. These goals are elusive. Because of the time lapse between world events and the publication of a book, inevitably events outpace analysis. We therefore depend on a continuing partnership with classroom instructors to provide the currency of information and the interpretation of new patterns of human geographic substance that changing conditions demand.

Organization

The text can easily be read in a one-semester or one-quarter course. The emphasis on human geographic current events and interpretations builds on our initial obligation to set the stage in Chapter 1 by briefly introducing students to the scope, methods, and background basics of geography as a discipline and to the tools—especially maps—that all geographers employ. It is supplemented by Appendix A giving a more detailed treatment of map projections than is appropriate in a general introductory chapter. Both are designed to be helpful, with content supportive of, not essential to, the later chapters of the text.

The arrangement of those chapters reflects our own sense of logic and teaching experiences. The chapters are unevenly divided among five parts, each with a brief orienting introduction. Those of Part One, "Themes and Fundamentals," examine the basis of culture, culture change, and cultural regionalism, review the concepts of spatial interaction and spatial behavior, and consider population structures, patterns, and change. Parts Two through Four (Chapters 5 through 12) discuss the landscapes of cultural distinction and social organization resulting from human occupance of the earth. These include linguistic, religious, ethnic, folk, and popular differentiation of peoples and societies and the economic, urban, and political organization of space. Chapter 13—Part Five—draws together in sharper focus selected aspects of the human impact on the natural landscape to make clear to students the relevance of the earlier-studied human geographic concepts and patterns to matters of current national and world environmental concern.

Among those concepts is the centrality of gender issues that underlie all facets of human geographic inquiry. Because they are so pervasive and significant, we felt it unwise to relegate their consideration to a single separate chapter, thus artificially isolating women and women's concerns from all the topics of human geography for which gender distinctions and interests are relevant. Instead, we have incorporated significant gender/female issues within the several chapters where those issues apply—either within the running text of the chapter or, very often, highlighted in boxed discussions.

We hope by means of these chapter clusters and sequence to convey to students the logic and integration we recognize in the broad field of human geography. We realize that our sense of organization and continuity is not necessarily that of instructors using this text and have designed each chapter to be reasonably self-contained, able to be assigned in any sequence that satisfies the arrangement preferred by the instructor.

New to this Edition

- We note with pride and pleasure the addition of Jon C. Malinowski to the original author team. He brings special interests and background in human geography in general and in environmental perception and Asian studies in particular.
- Although the text's established framework of presentation has been retained in this eighth edition, every chapter contains at least brief text additions or modifications to reflect current data, and many chapters contain new or revised illustrations, maps, and photos.

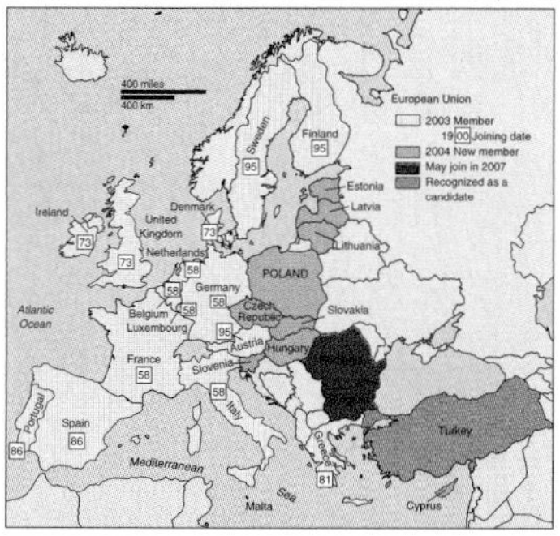

Figure 12.26 **The 15 members of the European Union (EU)** as of January, 2001, when 13 additional states were applicants for membership. The EU has stipulated that in order to join, a country must have stable institutions guaranteeing democracy, the rule of law, human rights and protection of minorities; a functioning market economy; and the ability to accept the obligations of membership, including the aims of political, economic, and monetary union. Ten applicants accepted and met those conditions and, subject to final negotiations, were admitted to the Union by 2004. The EU now spreads from the Mediterranean to the Arctic. In addition, some 70 states in Africa, the Caribbean, and the Pacific have been affiliated with the EU by the Lomé Convention, which provides for developmental aid and favored trade access to EU markets.

the Association of Southeast Asian Nations (ASEAN), formed in 1967. A similar, but much less wealthy African example is ECOWAS, the Economic Community of West African States. The Asia Pacific Economic Cooperation (APEC) forum includes China, Japan, Australia, Canada, and the United States among its 18 members and has a grand plan for "free trade in the Pacific" by 2020. More restricted bilateral and regional preferential trade arrangements have also proliferated, numbering over 400 early in this century and creating a maze of rules, tariffs, and commodity agreements that result in trade restrictions and preferences contrary to the free trade intent of the World Trade Organization.

Some supranational alliances, of course, are more cultural and political in orientation than these cited agencies. The League of Arab States, for example, was established in 1945 primarily to promote social, political, military, and foreign policy cooperation among its 22 members. In the Western Hemisphere, the Organization of American States (OAS) founded in 1948 concerns itself largely with social, cultural, human rights, and security matters affecting the hemisphere. A similar concern with peace and security underlay the Organization of African Unity (OAU) formed in 1963 by 32 African countries and, by 2001, expanded to 53 members and renamed the African Union.

Economic interests, therefore, may motivate the establishment of most international alliances, but political, social, and cultural objectives also figure largely or exclusively in many. Although the alliances themselves may change, the idea of single- and multiple-purpose supranational associations has been permanently added to the national political and global realities of the 21st century. The world map pattern those alliances create must be recognized to understand the current international order.

Three further points about regional international alliances are worth noting. The first is that the formation of a coalition in one area often stimulates the creation of another alliance by countries left out of the first. Thus, the union of the Inner Six gave rise to the treaty among the Outer Seven. Similarly, a counterpart of the Common Market was the Council of Mutual Economic Assistance (CMEA), also known as Comecon, which linked the former communist countries of Eastern Europe and the USSR through trade agreements.

Second, the new supranational unions tend to be composed of contiguous states (Figure 12.27). This was not the case with the recently dissolved empires, which included far-flung territories. Contiguity facilitates the movement of people and goods. Communication and transportation are simpler and more effective among adjoining countries than among those far removed from one another, and common cultural, linguistic, historical, and political traits and interests are more to be expected in spatially proximate countries.

The Political Ordering of Space 469

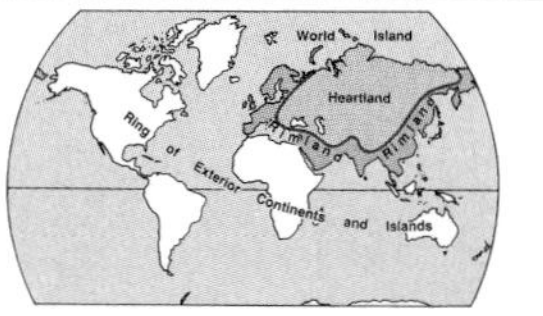

Figure 12.21 **Geopolitical viewpoints.** Both Mackinder and Spykman believed that Eurasia possessed strategic advantages, but they disagreed on whether its heartland or rimland provided the most likely base for world domination. Mahan recognized sea power as the key to national strength, advocating American occupation of the Hawaiian Islands, control of the Caribbean, and construction of an interocean canal through Central America.

based on the notion of **containment,** or confining the USSR within its borders by means of a string of regional alliances in the Rimland: The North Atlantic Treaty Organization (NATO) in Western Europe, the Central Treaty Organization (CENTO) in West Asia, and the Southeast Asia Treaty Organization (SEATO). Military intervention was deemed necessary where communist expansion, whether Soviet or Chinese, was a threat—in Berlin, the Middle East, and Korea, for example.

A simple spatial model, the **domino theory,** was used as an adjunct to the policy of containment. According to this analogy, adjacent countries are lined up like dominoes; if one topples, the rest will fall. In the early 1960s, the domino theory was invoked to explain and justify U.S. intervention in Vietnam, and in the 1980s, the theory was applied to involvement in Central America. The fear that war among the Serbs, Croatians, and Bosnians in Bosnia-Herzegovina would lead to the destruction of that state and spread into other parts of the former Yugoslavia led in 1995 to NATO airstrikes against the Serbs, a peace agreement forged with American help in Dayton, Ohio, and stationing of United Nations peacekeeping forces in Bosnia.

These (and other) models aimed at realistic assessments of national power and foreign policy stand in contrast to "organic state theory" based on the 19th-century idea of German geographer Friedrich Ratzel (1844–1904) that the state acted as if it were an organism conforming to natural laws and forced to grow and expand into new territories (*Lebensraum*) in order to secure the resources needed for survival. Without that growth, the state would wither and die. These ideas, later expanded in the 1920s by the German Karl Haushofer (1869–1946) as *Geopolitik*, were used by the Nazi party as the presumed intellectual basis for wartime Germany's theories of race superiority and need for territorial conquest. Repudiated by events and Germany's defeat, *Geopolitik* for many years gave bad odor to any study of geopolitics, which only recently has again become a serious subfield of political geography.

In a rapidly changing world, many analysts believe the older geopolitical concepts and ideas of geostrategy no longer apply. A number of developments have rendered them obsolete: the dissolution of the USSR and the presumed end of the Cold War; the proliferation of nuclear technology; and the rise of Japan, China, and Western Europe to world power status. Geopolitical reality is now seen less in terms of military advantage and confrontation—the East-West rivalry of the Cold War era—and more as a reflection of other forms of competition.

One of those forms expresses itself in violent assaults carried out by individuals and groups motivated by their total rejection of an established order they despise. Traditional geopolitical theories and assessments have dealt with projections of state power and influence embodied in military and economic strength. As a series of destructive incidents throughout the world before and after the tragic September 11, 2001 assaults on the World Trade Center and the Pentagon—and the aroused response to them from the United States and other governments—made clear, terrorism, including state-sponsored terrorism, must also be recognized as a substantially different form of global geopolitical activity. **Terrorism,** defined as the use or threat of violence to advance a political cause, was usually thought of as dissident individual or small group assault on civilian populations to weaken state authority; force change in governmental policy; or erode the social, cultural, or political organization of a society. The truck bombing of the Oklahoma City Murrah Federal Building in 1995 was of that individual terrorist nature. It caused 168 deaths and led to the conviction and execution of Timothy McVeigh, who attributed his action to rage over lethal federal agency actions against a family at Ruby Ridge, Idaho, in 1992 and a religious group in Waço, Texas, in 1993. But other, broader forms of organized terrorism exist.

On the wider global scene, the State Department in late 2002 listed 28 international terrorist organizations with goals as varied as eliminating Israel; replacing secular Muslim state governments with strict Islamic rule; securing regional separatism or unification in Ireland, India, Spain, Sri Lanka, the Philippines, Indonesia, and elsewhere; and the like. The State Department's annually updated terrorist list cites groups dedicated to changing—by violent action if necessary—current conditions they deem intolerable. Many are subnational ethnic groups (Sri Lanka's Tamil Liberation Tigers, for example) or those with specific ideological objectives, such as the Marxist Shining Path in Peru.

Religious or faith-based groups, dominantly though not exclusively Muslim, are particularly prominent on the late 20th and early 21st centuries' terrorist rosters. In their home countries, the common objective of the Muslim organizations is to replace existing secular or westernized governments with regimes committed to strict enforcement of fundamentalist religious law. On the international scene, however, they have primarily targeted the United States and its citizens. The bomb destruction of Pan Am flight over Scotland in 1988, car bombing of the World Trade Center in New York in 1993, bombing of American embassies in

462 Landscapes of Functional Organization

- Many new and updated topics, including the following:
  - HIV/AIDS
  - fertility rates
  - agricultural density
  - population momentum and aging
  - "Amalgamation Theory"
  - Hispanic and Asian Americans
  - immigrant gateways and clusters
  - globalization of popular culture and reactions to it
  - nomadic herding
  - urban agriculture
  - Green Revolution
  - trade in primary products
  - fishing
  - fur trapping and trade
  - containerization and intermodal transportation
  - transnational corporations
  - adverse impacts of gentrification
  - terrorism
- New boxed readings include "Militant Fundamentalism," "World Englishes," and "The Gated Community."
- Two new maps inside the front cover of the book, "World Political Map" and "Topographic Regions of the World," replace earlier versions and are accessible from any chapter in the book.

The Gated Community

Approximately one in six Americans—some 47 million people—lives in a master-planned community. Particularly characteristic of the fastest growing parts of the country, most of these communities are in the South and West, but they are increasingly common everywhere. In many regions, more than half of all new houses are being built in private developments. Master-planned communities in the United States trace their modern start back to the 1960s, when Irvine, California, and Sun City, Arizona, were built, but their roots can be found much earlier. Tuxedo Park, New York, for example, was planned and built in 1886 as a fully protected, socially exclusive community, and in the 1920s Kansas City's Country Club District was established as a restricted residential development with land use controlled by planning and deed restrictions and a self-governing homeowners association providing a variety of governmental, cultural, and recreational services.

A subset of the master-planned community is the **gated community,** a fenced or walled residential area with checkpoints staffed by guards and access limited to designated individuals and identified guests. By 2002, 9 million Americans were living in these middle- and high-income gated communities within communities. With private security forces, surveillance systems monitoring common recreational areas such as community swimming pools, tennis courts, and health clubs and—often—with individual home security systems, the walled enclaves provide a sense of refuge from high crime rates, drug abuse, and other social problems of urban America.

Gated and sheltered communities are not just an American phenomenon but are increasingly found in all parts of the world. More and more guarded residential enclaves have been sited in such stable Western European states as Spain, Portugal, and France.

Elsewhere, as in Argentina or Venezuela in South America or Lebanon in the Near East, with little urban planning, unstable city administration, and inadequate police protection, not only rich but also middle-class citizens are opting for protected residential districts. In China and Russia, the sudden boom in private and guarded settlements reflects in part a new form of post-communist social class distinction, while in South Africa gated communities serve as effective racial barriers.

Within the United States, the typical developer-created gated community is conceived as a unit and built following a master plan. To preserve the upscale nature of the development and protect land values, community associations enact and enforce a range of conditions and restrictions on private property use. Pervasive and detailed, they may specify such things as the size, construction materials, and design of houses, the color of walls and fences, the size and permitted uses of rear and side yards, even the design of exterior lights and mailboxes and the display of flags and outdoor decorations. Some go so far as to tell residents what trees they can plant and what pets they may have. The comfort of protected living may carry a high price in loss of individuality.

Urban Systems and Urban Structures 421

Features

- The "Focus Preview" alerts students to the main themes of the chapter.

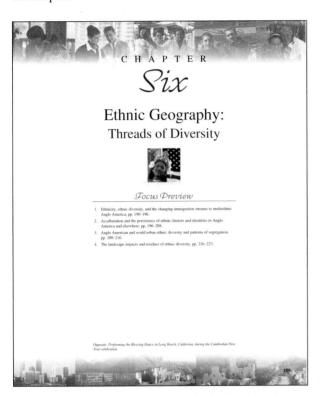

- Chapter introductions take the form of interest-arousing vignettes to focus student attention on the subject matter that follows.

- The boxed inserts that are part of each chapter expand on ideas included within the text or introduce related examples of chapter concepts and conclusions, often in gender-related contexts.

- Almost every chapter contains at least one special-purpose box labeled "Geography and Public Policy," introducing a discussion of a topic of current national or international interest and concluding with a set of questions designed to induce thought and class discussion of the topic viewed against the background of human geographic insights students have mastered.

- New terms and special usages of common words and phrases are identified by boldface or italic type. The boldface terms are included in the "Key Words" list at the end of each chapter and defined in an inclusive cross-referenced glossary at the end of the text.

- The "Focus Follow-up" section in the end-of-chapter material summarizes the main points of the chapter and conveys additional information and explanation as integral parts of the text.

- Each chapter also includes other repeated pedagogical aids. The "Summary" reiterates the main points of the chapter and provides a bridge to the chapter that follows. "For Review" contains questions that direct student attention to important concepts developed within the chapter. "Selected References" suggests a number of book and journal articles that expand on topics presented within the chapter.

- Appendix B is a modified version of the Population Reference Bureau's *2003 World Population Data Sheet* containing economic and demographic data and projections for countries, regions, and continents. These provide a wealth of useful comparative statistics for student projects and study of world patterns.
- Appendix C, a single-page "Anglo American Reference Map," provides name identification of all U.S. states and Canadian provinces and shows the location of principal cities.

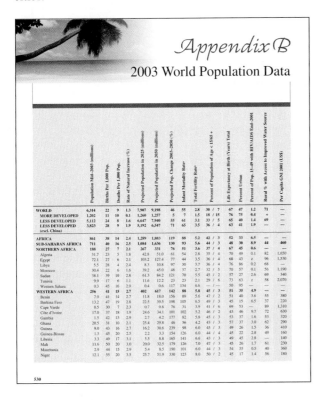

Appendix B

2003 World Population Data

	Population Mid-2003 (millions)	Births Per 1,000 Pop.	Deaths Per 1,000 Pop.	Rate of Natural Increase (%)	Projected Population in 2025 (millions)	Projected Population in 2050 (millions)	Projected Pop. Change 2003–2050 (%)	Infant Mortality Rate	Total Fertility Rate	Percent of Population of Age <15/65+	Life Expectancy at Birth (Years) Total	Percent Urban	Percent with HIV/AIDS End-2001	Rural % with Access to Improved Water Source	Per Capita GNI 2001 (US$)
WORLD	6,314	22	9	1.3	7,907	9,198	46	55	2.8	30 / 7	67	47	1.2	71	—
MORE DEVELOPED	1,202	11	10	0.1	1,260	1,257	5	7	1.5	18 / 15	76	75	0.4	+	—
LESS DEVELOPED	5,112	24	8	1.6	6,647	7,940	55	61	3.1	33 / 5	65	40	1.4	69	—
LESS DEVELOPED (excl. China)	3,823	28	9	1.9	5,192	6,547	71	66	3.5	36 / 4	63	41	1.9	—	—
AFRICA	861	38	14	2.4	1,289	1,803	119	88	5.2	42 / 3	52	33	6.5	—	—
SUB-SAHARAN AFRICA	711	40	16	2.5	1,084	1,636	130	93	5.6	44 / 3	48	30	8.9	44	460
NORTHERN AFRICA	188	27	7	2.1	267	331	76	53	3.6	37 / 4	67	45	0.6	—	—
Algeria	31.7	23	5	1.8	42.8	51.0	61	54	2.8	35 / 4	70	49	0.1	82	1,650
Egypt	72.1	27	6	2.1	103.2	127.4	77	44	3.5	36 / 4	68	43	z	96	1,530
Libya	5.5	28	4	2.4	8.3	10.8	97	30	3.7	36 / 4	76	86	0.2	68	—
Morocco	30.4	22	6	1.6	39.2	45.0	48	37	2.7	32 / 5	70	57	0.1	56	1,190
Sudan	38.1	39	10	2.8	61.3	84.2	121	70	5.5	45 / 2	57	27	2.6	69	340
Tunisia	9.9	17	6	1.1	11.6	12.2	23	23	2.1	29 / 6	73	63	z	58	2,070
Western Sahara	0.3	45	16	2.9	0.4	0.6	117	134	6.6	— / —	50	95	—	—	—
WESTERN AFRICA	256	41	15	2.7	402	617	142	88	5.8	45 / 3	51	35	4.9	—	—
Benin	7.0	41	14	2.7	11.8	18.0	156	89	5.6	47 / 2	51	40	3.6	55	380
Burkina Faso	13.2	47	19	2.8	22.5	39.5	198	105	6.5	49 / 3	45	15	6.5	37	220
Cape Verde	0.5	30	7	2.3	0.7	0.8	76	31	3.9	41 / 6	69	53	—	89	1,310
Côte d'Ivoire	17.0	37	18	1.9	24.6	34.1	101	102	5.2	46 / 2	43	46	9.7	72	630
Gambia	1.5	42	13	2.9	2.7	4.2	177	82	5.8	45 / 3	53	37	1.6	53	320
Ghana	20.5	31	10	2.1	25.4	29.8	46	56	4.2	43 / 3	57	37	3.0	62	290
Guinea	9.0	43	16	2.7	16.2	30.6	239	98	6.0	45 / 3	49	26	1.5	36	410
Guinea-Bissau	1.3	45	20	2.5	2.2	3.3	154	126	6.0	44 / 4	45	22	2.8	49	160
Liberia	3.3	49	17	3.1	5.5	8.8	165	141	6.6	43 / 3	49	45	2.8	—	140
Mali	11.6	50	20	3.0	20.0	32.5	179	126	7.0	47 / 3	45	26	1.7	61	230
Mauritania	2.9	44	15	2.9	5.4	8.5	190	101	6.0	44 / 3	54	55	0.5	40	360
Niger	12.1	55	20	3.5	25.7	51.9	330	123	8.0	50 / 2	45	17	1.4	56	180

530

Supplements

The eighth edition provides a complete human geography program for the student and instructor.

For the Student

On-line Learning Center at http://www.mhhe.com/fellmann8e/

This site gives you the opportunity to further explore topics presented in the book using the Internet. The site contains interactive quizzes with immediate feedback, a student study guide, flashcards, crossword puzzles, base maps, critical-thinking questions, Internet exercises, and a career center. We've integrated *Power Web: Geography*'s information and timely world news, Web links, and much more into the site to make these valuable resources easily accessible to students.

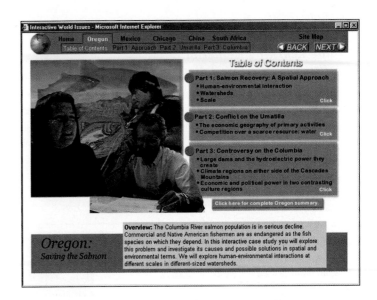

Interactive World Issues CD-ROM

Your instructor may require the *Interactive World Issues* CD-ROM. This CD-ROM allows you to have hands-on exercises and to see videos of different case studies. The five case studies include Chicago, Oregon, Mexico, China, and South Africa. Since most of us are unable to visit different world regions, this is a good way to understand the issues facing different parts of the world.

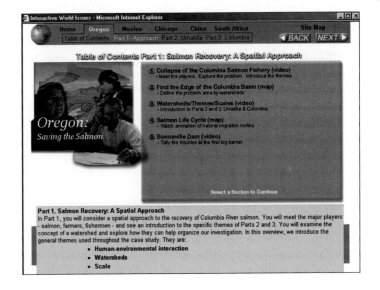

For the Instructor

On-line Learning Center at http://www.mhhe.com/fellmann8e/

Included in the password-protected section of the On-line Learning Center is an Instructor's Manual that includes a chapter summary, key words, teaching strategies, and active learning tips. The On-line Learning Center also contains PowerPoint lecture outlines and a correlation guide to the *Interactive World Issues* CD-ROM.

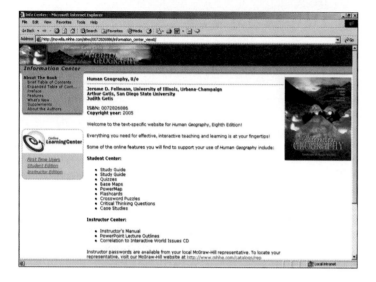

Digital Content Manager CD-ROM

This CD-ROM contains all of the illustrations, photographs, and tables from the text. The software makes customizing your multimedia presentation easy. You can organize figures in any order you want; add labels, lines, and your own artwork; integrate materials from other sources; edit and annotate lecture notes; and then have the option of placing your multimedia lecture into another presentation program such as PowerPoint.

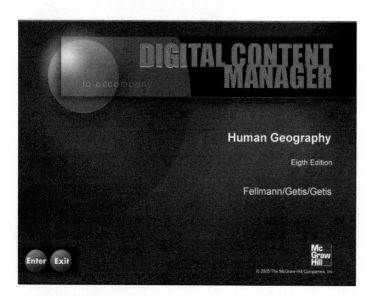

Transparencies

Included are 150 illustrations from the text, all enlarged for excellent visibility in the classroom.

Instructor's Testing and Resource CD-ROM

This cross-platform CD-ROM provides a wealth of resources for the instructor. Supplements featured on this CD-ROM include a computerized test bank utilizing Brownstone Diploma testing software to quickly create customized exams. This user-friendly program allows instructors to search for questions by topic, format, or difficulty level; edit existing questions or add new ones; and scramble questions and answer keys for multiple versions of the same test.

Other assets on the Instructor's Testing and Resource CD-ROM are grouped within easy-to-use folders. The Instructor's Manual and Test Item File are available in both Word and PDF formats. Word files of the test bank are included for those instructors who prefer to work outside of the test-generator software.

Videotape Library

An extensive array of videotapes is available to qualified adopters. Check with your representative for details.

Interactive World Issues CD-ROM

This CD allows you to have hands-on exercises and to see videos of different case studies. The five case studies include Chicago, Oregon, Mexico, China, and South Africa.

Course Management Systems

On-line course content is available for a variety of course management systems, including

- Blackboard
- WebCT
- eCollege
- PageOut

Acknowledgements

It is with great pleasure that we again acknowledge our debts of gratitude to both departmental colleagues—at the University of Illinois, Urbana-Champaign, and at both San Diego State University and the University of California, Santa Barbara—and all others who have given generously of their time and knowledge in response to our requests. These have been identified in earlier editions, and although their names are not repeated here, they know of our continuing appreciation.

We specifically, however, wish to recognize with gratitude the advice, suggestions, corrections, and general assistance in matters of content and emphasis provided by the following reviewers of the manuscript for this edition.

C. Murray Austin, *University of Northern Iowa*

Lisa Benton-Short, *The George Washington University*

Dylan Clark, *University of Colorado—Boulder*

Fiona Davidson, *University of Arkansas—Fayetteville*

James A. Davis, *Brigham Young University*

James E. DiLisio, *Towson University*

Owen Dwyer, *Indiana University-Purdue University, Indianapolis*

Michael Hopps, *Central Lakes College*

Peter Hossler, *Indiana University—Bloomington*

Artimus Keiffer, *Wittenberg University*

Naresh Kumar, *University of Toledo*

Amy Lilienfeld, *Central Michigan University*

James D. Lowry, Jr., *East Central University*

Robin Lyons, *San Joaquin Delta College*

Cindy Miller, *Minnesota State University*

Douglas Munski, *University of North Dakota*

James C. Saku, *Frostburg State University*

We appreciate their invaluable help, as do we that of the many other previous reviewers recognized in earlier editions of this book. None except the authors, of course, is responsible for final decisions on content or for errors of fact or interpretation the reader may detect.

A final note of thanks is reserved for the publisher's "book team" members separately named on the copyright page. It is a privilege to emphasize here their professional competence, unflagging interest, and always courteous helpfulness.

Jerome D. Fellmann

Arthur Getis

Judith Getis

Jon C. Malinowski

Meet the Authors

Jerome D. Fellmann

Jerome Fellmann received his B.S., M.S., and Ph.D. degrees from the University of Chicago. Except for visiting professorships at Wayne State University, the University of British Columbia, and California State University/Northridge, his professional career has been spent at the University of Illinois at Urbana-Champaign. His teaching and research interests have been concentrated in the areas of human geography in general and urban and economic geography in particular, in geographic bibliography, the geography of Russia and the CIS, and geographic education. His varied interests have been reflected in articles published in the *Annals of the Association of American Geographers, Professional Geographer, Journal of Geography,* the *Geographical Review,* and elsewhere. He is the coauthor of McGraw-Hill's *Introduction to Geography.* In addition to teaching and research, he has held administrative appointments at the University of Illinois and served as a consultant to private corporations on matters of economic and community development.

Arthur Getis

Arthur Getis received his B.S. and M.S. degrees from Pennsylvania State University and his Ph.D. from the University of Washington. He is the coauthor of several geography textbooks as well as two books dealing with map pattern analysis. He has also published widely in the areas of urban geography, spatial analysis, and geographical information systems. He is coeditor of *Journal of Geographical Systems* and for many years served on the editorial boards of *Geographical Analysis* and *Papers in Regional Science.* He has held administrative appointments at Rutgers University, the University of Illinois, and San Diego State University (SDSU) and currently holds the Birch Chair of Geographical Studies at SDSU. In 2002 he received the Association of American Geographers Distinguished Scholarship Award. Professor Getis is a member of many professional organizations and has served as an officer in, among others, the Western Regional Science Association and the University Consortium for Geographic Information Science.

Judith Getis

Judith Getis earned her B.A. and a teaching credential from the University of Michigan and her M.A. from Michigan State University. She has coauthored several geography textbooks and written the environmental handbook *You Can Make a Difference.* In addition to numerous articles in the fields of urban geography and geography education, she has written technical reports on topics such as solar power and coal gasification. She and her husband, Arthur Getis, were among the original unit authors of the High School Geography Project, sponsored by the National Science Foundation and the Association of American Geographers. In addition, Mrs. Getis was employed by the Urban Studies Center at Rutgers University; taught at Rutgers; was a social science examiner at Educational Testing Service, Princeton, New Jersey; developed educational materials for Edcom Systems, Princeton, New Jersey; and was a professional associate in the Office of Energy Research, University of Illinois.

Jon C. Malinowski

Jon Malinowski received his B.S. from Georgetown University's Edmund A. Walsh School of Foreign Service and his M.A. and PhD. from the University of North Carolina—Chapel Hill, where he taught five semesters as a teaching fellow. He currently holds the position of Associate Professor of Geography at the United States Military Academy, West Point, New York. His teaching interests focus on human geography, including the geography of Asia, and research methods. Dr. Malinowski's research interests are concentrated in the field of behavioral geography, primarily human navigation and childrens' geographies. In addition to several book chapters, his research has been published in journals such as *Environment and Behavior, Journal of Environmental Psychology,* and *Perceptual and Motor Skills.* He is a member of the Association of American Geographers and the National Council for Geographic Education. In addition, he is the coauthor of two trade books, *The Summer Camp Handbook* and *The Spirit of West Point: Celebrating 200 Years.* In addition to his work as a geographer, Dr Malinowski is a recognized speaker and consultant on leadership in the summer-camp industry.

Human
GEOGRAPHY

CHAPTER

One

Introduction:
Some Background Basics

Focus Preview

1. The nature of geography and the role of human geography, pp. 4–7.

2. Seven fundamental geographic observations and the basic concepts that underlie them, pp. 7–17.

3. The regional concept and the characteristics of regions, pp. 17–20.

4. Why geographers use maps and how maps show spatial information, pp. 20–26.

5. Other means of visualizing and analyzing spatial data: mental maps, systems, and models, pp. 26–29.

Opposite: A graceful pavilion carries the human imprint high onto Hua Shan mountain, Shaanxi Province, China.

The fundamental question asked by geographers is "Does it make a difference where things are located?" If for any one item or group of objects the answer is "You bet it does!", the geographer's interest is aroused and geographic investigation is appropriate. For example, it matters a great deal that languages of a certain kind are spoken in certain places. But knowledge of the location of a specific language group is not of itself particularly significant. Geographic study of a language requires that we try to answer questions about why and how the language shows different characteristics in different locations and how the present distribution of its speakers came about. In the course of our study, we would logically discuss such concepts as migration, acculturation, the diffusion of innovation, the effect of physical barriers on communication, and the relationship of language to other aspects of culture. As geographers, we are interested in how things are interrelated in different regions and give evidence of the existence of "spatial systems."

Geography is often referred to as the *spatial* science, that is, the discipline concerned with the use of earth space. In fact, *geography* literally means "description of the earth," but that task is really the responsibility of nearly all the sciences. Geography might better be defined as the study of spatial variation, of how—and why—physical and cultural items differ from place to place on the surface of the earth. It is, further, the study of how observable spatial patterns evolved through time. If things were everywhere the same, if there were no spatial variation, the kind of human curiosity that we call "geographic" simply would not exist. Without the certain conviction that in some interesting and important way landscapes, peoples, and opportunities differ from place to place, there would be no discipline of geography.

But we do not have to deal in such abstract terms. You consciously or subconsciously display geographic awareness in your own daily life. You are where you are, doing what you are doing, because of locational choices you faced and spatial decisions you made. You cannot be here reading this book and simultaneously be somewhere else—working, perhaps, or at the gym. And should you now want to go to work or take an exercise break, the time involved in going from here to there (wherever "there" is) is time not available for other activities in other locations. Of course, the act of going implies knowing where you are now, where "there" is in relation to "here," and the paths or routes you can take to cover the distance.

These are simple examples of the observation that "space matters" in a very personal way. You cannot avoid the implications of geography in your everyday affairs. Your understanding of your hometown, your neighborhood, or your college campus is essentially a geographic understanding. It is based on your awareness of where things are, of their spatial relationships, and of the varying content of the different areas and places you frequent. You carry out your routine activities in particular places and move on your daily rounds within defined geographic space, following logical paths of connection between different locations.

Just as geography matters in your personal life, so it matters on the larger stage as well. Decisions made by corporations about the locations of manufacturing plants or warehouses in relation to transportation routes and markets are spatially rooted. So, too, are those made by shopping center developers and locators of parks and grade schools. On an even grander scale, judgments about the projection of national power or the claim and recognition of "spheres of influence and interest" among rival countries are related to the implications of distance and area.

Geography, therefore, is about space and the content of space. We think of and respond to places from the standpoint not only of where they are but, rather more importantly, of what they contain or what we think they contain. Reference to a place or an area usually calls up images about its physical nature or what people do there and often suggests, without conscious thought, how those physical objects and human activities are related. "Colorado," "mountains," and "skiing" might be a simple example. The content of area, that is, has both physical and cultural aspects, and geography is always concerned with understanding both (Figure 1.1).

Evolution of the Discipline

Geography's combination of interests was apparent even in the work of the early Greek geographers who first gave structure to the discipline. Geography's name was reputedly coined by the Greek scientist Eratosthenes over 2200 years ago from the words *geo,* "the earth," and *graphein,* "to write." From the beginning, that writing focused both on the physical structure of the earth and on the nature and activities of the people who inhabited the different lands of the known world. To Strabo (*ca.* 64 B.C.–A.D. 20) the task of geography was to "describe the several parts of the inhabited world . . . to write the assessment of the countries of the world [and] to treat the differences between countries." Even earlier, Herodotus (*ca.* 484–425 B.C.) had found it necessary to devote much of his book to the lands, peoples, economies, and customs of the various parts of the Persian Empire as necessary background to an understanding of the causes and course of the Persian wars.

Greek (and, later, Roman) geographers measured the earth, devised the global grid of parallels and meridians (marking latitudes and longitudes—see page 20), and drew upon that grid surprisingly sophisticated maps of their known world (Figure 1.2). They explored the apparent latitudinal variations in climate and described in numerous works the familiar Mediterranean basin and the more remote, partly rumored lands of northern Europe, Asia, and equatorial Africa. Employing nearly modern concepts, they described river systems, explored causes of erosion, cited the dangers of deforestation, described areal variations in the natural landscape, and discussed patterns and processes of climates, vegetation, and landforms. Against that physical backdrop, they

Figure 1.1 The ski development at Whistler Mountain, British Columbia, Canada, clearly shows the interaction of physical environment and human activity. Climate and terrain have made specialized human use attractive and possible. Human exploitation has placed a cultural landscape on the natural environment, thereby altering it.

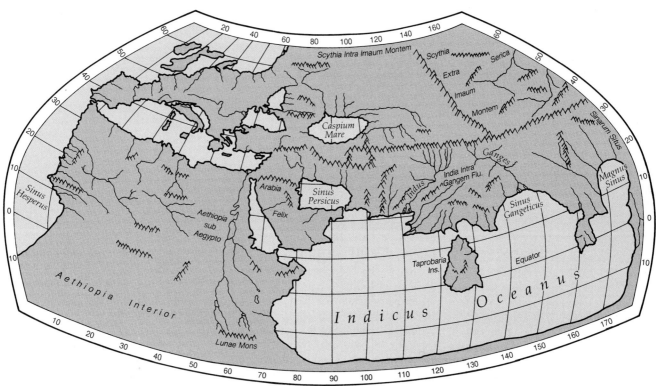

Figure 1.2 **World map of the 2d-century A.D. Greco–Egyptian geographer-astronomer Ptolemy.** Ptolemy (Claudius Ptolemaeus) adopted a previously developed map grid of latitude and longitude based on the division of the circle into 360°, permitting a precise mathematical location for every recorded place. Unfortunately, errors of assumption and measurement rendered both the map and its accompanying six-volume gazetteer inaccurate. Ptolemy's map, accepted in Europe as authoritative for nearly 1500 years, was published in many variants in the 15th and 16th centuries. The version shown here summarizes the extent and content of the original. Its underestimation of the earth's size convinced Columbus a short westward voyage would carry him to Asia.

focused their attention on what humans did in home and distant areas—how they lived; what their distinctive similarities and differences were in language, religion, and custom; and how they used, altered, and perhaps destroyed the lands they inhabited.

These are enduring and universal interests. The ancient Chinese, for example, were as involved in geography as an explanatory viewpoint as were Westerners, though there was no exchange between them. Further, as Christian Europe entered its Middle Ages between A.D. 500 and 1400 and lost its knowledge of Greek and Roman geographical work, Muslim scholars—who retained that knowledge—undertook to describe and analyze their known world in its physical, cultural, and regional variation (see "Roger's Book").

Modern geography had its origins in the surge of scholarly inquiry that, beginning in the 17th century, gave rise to many of the traditional academic disciplines we know today. In its European rebirth, geography from the outset was recognized—as it always had been—as a broadly based integrative study. Patterns and processes of the physical landscape were early interests, as was concern with humans as part of the earth's variation from place to place. The rapid development of geology, botany, zoology, and other natural sciences by the end of the 18th century strengthened regional geographic investigation and increased scholarly and popular awareness of the intricate interconnections of items in space and between places. By that same time, accurate determination of latitude and longitude and scientific mapping of the earth made assignment of place information more reliable and comprehensive.

During the 19th century, national censuses, trade statistics, and ethnographic studies gave firmer foundation to human geographic investigation. By the end of the 19th century, geography had become a distinctive and respected discipline in universities throughout Europe and in other regions of the world where European academic examples were followed. The proliferation of professional geographers and geography programs resulted in the development of a whole series of increasingly specialized disciplinary subdivisions.

Geography and Human Geography

Geography's specialized subfields are not divisive but are interrelated. Geography in all its subdivisions is characterized by three dominating interests. The first is in the areal variation of physical and human phenomena on the surface of the earth. Geography examines relationships between human societies and the natural environments that they occupy and modify. The second is a focus on the spatial systems[1] that link physical phenomena and human activities in one area of the earth with other areas. Together, these interests lead to a third enduring theme, that of regional analysis: geography studies human–environmental—"ecological"—relationships and spatial systems in specific locational settings. This areal orientation pursued by some geographers is called *regional geography* (see also page 17). For some, the regions of interest may be large: Southeast Asia or Latin America, for example; others may focus on smaller areas differently defined, such as Alpine France or the United States Corn Belt.

---

[1] A "system" is simply a group of elements organized in a way that every element is to some degree directly or indirectly interdependent with every other element. For geographers, the systems of interest are those that distinguish or characterize different regions or areas of the earth.

Roger's Book

The Arab geographer Idrisi, or Edrisi (*ca.* A.D. 1099–1154), a descendant of the Prophet Mohammed, was directed by Roger II, the Christian king of Sicily in whose court he served, to collect all known geographical information and assemble it in a truly accurate representation of the world. An academy of geographers and scholars was gathered to assist Idrisi in the project. Books and maps of classical and Islamic origins were consulted, mariners and travelers interviewed, and scientific expeditions dispatched to foreign lands to observe and record. Data collection took 15 years before the final world map was fabricated on a silver disc some 200 centimeters (80 inches) in diameter and weighing over 135 kilograms (300 pounds).

Lost to looters in 1160, the map is survived by "Roger's Book," containing the information amassed by Idrisi's academy and including a world map, 71 part maps, and 70 sectional itinerary maps.

Idrisi's "inhabited earth" is divided into the seven "climates" of Greek geographers, beginning at the equator and stretching northward to the limit at which, it was supposed, the earth was too cold to be inhabited. Each climate was then subdivided by perpendicular lines into 11 equal parts beginning with the west coast of Africa on the west and ending with the east coast of Asia. Each of the resulting 77 square compartments was then discussed in sequence in "Roger's Book."

Though Idrisi worked in one of the most prestigious courts of Europe, there is little evidence that his work had any impact on European geographic thought. He was strongly influenced by Ptolemy's work and misconceptions and shared the then common Muslim fear of the unknown western ocean. Yet Idrisi's clear understanding of such scientific truths as the roundness of the earth, his grasp of the scholarly writings of his Greek and Muslim predecessors, and the faithful recording of information on little-known portions of Europe, the Near East, and North Africa set his work far above the mediocre standards of contemporary Christian geography.

Other geographers choose to identify particular classes of things, rather than segments of the earth's surface, for specialized study. These *systematic geographers* may focus their attention on one or a few related aspects of the physical environment or of human populations and societies. In each case, the topic selected for study is examined in its interrelationships with other spatial systems and areal patterns. *Physical geography* directs its attention to the natural environmental side of the human–environment structure. Its concerns are with landforms and their distribution, with atmospheric conditions and climatic patterns, with soils or vegetation associations, and the like. The other systematic branch of geography—and the subject of this book—is *human geography*.

Human Geography

Human geography deals with the world as it is and with the world as it might be made to be. Its emphasis is on people: where they are, what they are like, how they interact over space, and what kinds of landscapes of human use they erect on the natural landscapes they occupy. It encompasses all those interests and topics of geography that are not directly concerned with the physical environment or, like cartography, are technical in orientation. Its content provides integration for all of the social sciences, for it gives to those sciences the necessary spatial and systems viewpoint that they otherwise lack. For example, economists are generally concerned with trends and patterns over time, not space, and psychology rarely considers the interaction between space and behavior. At the same time, human geography draws on other social sciences in the analyses identified with its subfields, such as *behavioral, political, economic,* or *social geography* (Figure 1.3).

Human geography admirably serves the objectives of a liberal education. It helps us to understand the world we occupy and to appreciate the circumstances affecting peoples and countries other than our own. It clarifies the contrasts in societies and cultures and in the human landscapes they have created in different regions of the earth. Its models and explanations of how things are interrelated in earth space give us a clearer understanding of the economic, social, and political systems within which we live

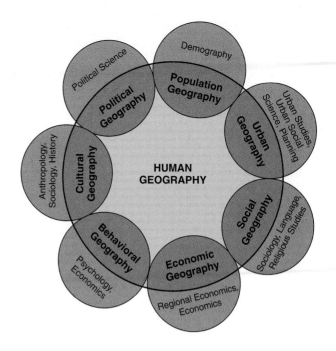

Figure 1.3 Some of the subdivisions of human geography and the allied fields to which they are related. Geography, "the mother of sciences," initiated in antiquity the lines of inquiry that later led to the development of these and other separate disciplines. That geography retains its ties to them and shares their insights and data reinforces its role as an essential synthesizer of all data, concepts, and models that have integrative regional and spatial implications.

and operate. Its analyses of those spatial systems make us more aware of the realities and prospects of our own society in an increasingly connected and competitive world. Our study of human geography, therefore, can help make us better-informed citizens, more able to understand the important issues facing our communities and our countries and better prepared to contribute to their solutions. Importantly, it can also help open the way to wonderfully rewarding and diversified careers as professional geographers (see "Careers in Geography").

 *Background Basics*

Core Geographic Concepts

The topics included in human geography are diverse, but that very diversity emphasizes the reality that all geographers—whatever their particular topical or regional specialties—are united by the similar questions they ask and the common set of basic concepts they employ to consider their answers. Of either a physical or cultural phenomenon they will inquire: What is it? Where is it? How did it come to be what and where it is? Where is it in relation to other things that affect it or are affected by it? How is it part of a functioning whole? How does its location

affect people's lives and the content of the area in which it is found?

These questions are spatial in focus and systems analytical in approach and are derived from enduring central themes in geography.[2] In answering them, geographers draw upon a common

[2] Five fundamental themes of geography—basic concepts and topics that are essential elements in all geographic inquiry and at all levels of instruction—have been recognized by a joint committee of the National Council for Geographic Education and the Association of American Geographers. They are (1) the significance of absolute and relative *location;* (2) the distinctive physical and human characteristics of *place;* (3) *relationships* including human-environmental relationships, within places; (4) *movements* expressing patterns and change in human spatial interaction; and (5) how *regions* form and change.

Careers in Geography

Geography admirably serves the objectives of a liberal education. It can make us better informed citizens, more able to understand the important issues facing our communities, our country, and our world and better prepared to contribute solutions.

Can it, as well, be a pathway to employment for those who wish to specialize in the discipline? The answer is "Yes," in a number of different types of jobs. One broad cluster is concerned with supporting the field itself through teaching and research. Teaching opportunities exist at all levels, from elementary to university postgraduate. Teachers with some training in geography are increasingly in demand in elementary and high schools throughout the United States, reflecting geography's inclusion as a core subject in the federally adopted *Educate America Act* (Public Law 103-227) and the national determination to create a geographically literate society (see "The National Standards," p. 10). At the college level, specialized teaching and research in all branches of geography have long been established, and geographically trained scholars are prominently associated with urban, community, and environmental studies, regional science, locational economics, and other interdisciplinary programs.

Because of the breadth and diversity of the field, training in geography involves the acquisition of techniques and approaches applicable to a wide variety of jobs outside the academic world. Modern geography is both a physical and social science and fosters a wealth of technical skills. The employment possibilities it presents are as many and varied as are the agencies and enterprises dealing with the natural environment and human activities and with the acquisition and analysis of spatial data.

Many professional geographers work in government, either at the state or local level or in a variety of federal agencies and international organizations. Although many positions do not carry a geography title, physical geographers serve as water, mineral, and other natural resource analysts; weather and climate experts; soil scientists; and the like. An area of recent high demand is for environmental managers and technicians. Geographers who have specialized in environmental studies find jobs in both public and private agencies. Their work may include assessing the environmental impact of proposed development projects on such things as air and water quality and endangered species, as well as preparing the environmental impact statements required before construction can begin.

Human geographers work in many different roles in the public sector. Jobs include data acquisition and analysis in health care, transportation, population studies, economic development, and international economics. Many geography graduates find positions as planners in local and state governmental agencies concerned with housing and community development, park and recreation planning, and urban and regional planning. They map and analyze land use plans and transportation systems, monitor urban land development, make informed recommendations about the location of public facilities, and engage in basic social science research.

Most of these same specializations are also found in the private sector. Geographic training is ideal for such tasks as business planning and market analysis; factory, store, and shopping-center site selection; community and economic development programs for banks, public utilities, and railroads, and similar applications. Publishers of maps, atlases, news and travel magazines, and the like employ geographers as writers, editors, and mapmakers.

The combination of a traditional, broadly based liberal arts perspective with the technical skills required in geographic research and analysis gives geography graduates a

store of concepts, terms, and methods of study that together form the basic structure and vocabulary of geography. Collectively, they reflect the fundamental truths addressed by geography: that things are rationally organized on the earth's surface and that recognizing spatial patterns is an essential starting point for understanding how people live on and shape the earth's surface. That understanding is not just the task and interest of the professional geographer; it should be, as well, part of the mental framework of all informed persons. As the publication *Geography for Life* summarizes, "There is now a widespread acceptance . . . that being literate in geography is essential . . . to earn a decent living, enjoy the richness of life, and participate responsibly in local, national, and international affairs." (See "The National Standards.")

Geographers use the word *spatial* as an essential modifier in framing their questions and forming their concepts. Geography, they say, is a *spatial* science. It is concerned with *spatial behavior* of people, with the *spatial relationships* that are observed between places on the earth's surface, and with the *spatial processes* that create or maintain those behaviors and relation-

ships. The word *spatial* comes, of course, from *space,* and to geographers, it always carries the idea of the way items are distributed, the way movements occur, and the way processes operate over the whole or a part of the surface of the earth. The geographer's space, then, is earth space, the surface area occupied or available to be occupied by humans. Spatial phenomena have locations on that surface, and spatial interactions occur between places, things, and people within the earth area available to them. The need to understand those relationships, interactions, and processes helps frame the questions that geographers ask.

Those questions have their starting point in basic observations about the location and nature of places and about how places are similar to or different from one another. Such observations, though simply stated, are profoundly important to our comprehension of the world we occupy.

- Places have location, direction, and distance with respect to other places.
- A place has size; it may be large or small. Scale is important.
- A place has both physical structure and cultural content.

competitive edge in the labor market. These field-based skills include familiarity with geographic information systems (GIS), cartography and computer mapping, remote sensing and photogrammetry, and competence in data analysis and problem solving. In particular, students with expertise in GIS, who are knowledgeable about data sources, hardware, and software, are finding that they have ready access to employment opportunities. The following table, based on the booklet "Careers in Geography,"* summarizes some of the professional opportunities open to students who have specialized in one (or more) of the various subfields of geography. Also, be sure to read the informative discussions under the "Careers in Geography" option on the home page of the Association of American Geographers at **www.aag.org/.**

Geographic Field of Concentration	Employment Opportunities
Cartography and geographic information systems	Cartographer for federal government (agencies such as Defense Mapping Agency, U.S. Geological Survey, or Environmental Protection Agency) or private sector (e.g., Environmental Systems Research Institute, ERDAS, Intergraph, or Bentley); map librarian; GIS specialist for planners, land developers, real estate agencies, utility companies, local government; remote-sensing analyst; surveyor
Physical geography	Weather forecaster; outdoor guide; coastal zone manager; hydrologist; soil conservation/agricultural extension agent
Environmental studies	Environmental manager; forestry technician; park ranger; hazardous waste planner
Cultural geography	Community developer; Peace Corps volunteer; health care analyst
Economic geography	Site selection analyst for business and industry; market researcher; traffic/route delivery manager; real estate agent/broker/appraiser; economic development researcher
Urban and regional planning	Urban and community planner; transportation planner; housing, park, and recreation planner; health services planner
Regional geography	Area specialist for federal government; international business representative; travel agent; travel writer
Geographic education or general geography	Elementary/secondary school teacher; college professor; overseas teacher

*"Careers in Geography," by Richard G. Boehm. Washington, D.C.: National Geographic Society, 1996. Previously published by Peterson's Guides, Inc.

- The attributes of places develop and change over time.
- The elements of places interrelate with other places.
- The content of places is rationally structured.
- Places may be generalized into regions of similarities and differences.

These are basic notions understandable to everyone. They also are the means by which geographers express fundamental observations about the earth spaces they examine and put those observations into a common framework of reference. Each of the concepts is worth further discussion, for they are not quite as simple as they at first seem.

Location, Direction, and Distance

Location, direction, and *distance* are everyday ways of assessing the space around us and identifying our position in relation to other items and places of interest. They are also essential in understanding the processes of spatial interaction that figure so importantly in the study of human geography.

Location

The location of places and objects is the starting point of all geographic study as well as of all our personal movements and spatial actions in everyday life. We think of and refer to location in at least two different senses, *absolute* and *relative.*

Absolute location is the identification of place by some precise and accepted system of coordinates; it therefore is sometimes called *mathematical location.* We have several such accepted systems of pinpointing positions. One of them is the global grid of parallels and meridians (discussed later on page 20). With it the absolute location of any point on the earth can be accurately described by reference to its degrees, minutes, and seconds of *latitude* and *longitude* (Figure 1.4).

Other coordinate systems are also in use. Survey systems such as the township, range, and section description of property in much of the United States give mathematical locations on a regional level, while street address precisely defines a building according to the reference system of an individual town. For convenience or special purposes, locational grid references may be superimposed on the basic global grid. The Universal Transverse Mercator

Geography is a core subject in the national *Educate America Act.* Its inclusion reflects a national conviction that a grasp of the skills and understandings of geography are essential in an American educational system "tailored to the needs of productive and responsible citizenship in the global economy." The National Geography Standards 1994 were developed to help achieve that goal. They specify the essential subject matter, skills, and perspectives that students who have gone through the U.S. public school system should acquire and use. Although not all of the standards are relevant to our study of human geography, together they help frame the kinds of understanding we will seek in the following pages and suggest the purpose and benefit of further study of geography.

The 18 standards from *Geography for Life* tell us

The geographically informed person knows and understands:

The World in Spatial Terms

1. How to use maps and other geographic tools and technologies to acquire, process, and report information from a spatial perspective.

2. How to use mental maps to organize information about people, places, and environments in a spatial context.

3. How to analyze the spatial organization of people, places, and environments on Earth's surface.

Places and Regions

4. The physical and human characteristics of places.

5. That people create regions to interpret Earth's complexity.

6. How culture and experience influence people's perceptions of places and regions.

Physical Systems

7. The physical processes that shape the patterns of Earth's surface.

8. The characteristics and spatial distribution of ecosystems on Earth's surface.

Human Systems

9. The characteristics, distribution, and migration of human populations on Earth's surface.

10. The characteristics, distribution, and complexity of Earth's cultural mosaics.

11. The patterns and networks of economic interdependence on Earth's surface.

12. The processes, patterns, and functions of human settlement.

13. How the forces of cooperation and conflict among people influence the division and control of Earth's surface.

Environment and Society

14. How human actions modify the physical environment.

15. How physical systems affect human systems.

16. The changes that occur in the meaning, use, distribution, and importance of resources.

The Uses of Geography

17. How to apply geography to interpret the past.

18. How to apply geography to interpret the present and plan for the future.

Source: *Geography for Life: National Geography Standards 1994.* Washington, D.C.: National Geographic Research and Exploration, 1994.

(UTM) system, for example, based on a set of 60 longitude zones, is widely used in geographic information system (GIS) applications and, with different notations, as a military grid reference system. Absolute location is unique to each described place, is independent of any other characteristic or observation about that place, and has obvious value in the legal description of places, in measuring the distance separating places, or in finding directions between places on the earth's surface.

When geographers—or real estate agents—remark that "location matters," however, their reference is usually not to absolute but to **relative location**—the position of a place in relation to that of other places or activities (Figure 1.5). Relative location expresses spatial interconnection and interdependence and may carry social (neighborhood character) and economic (assessed valuations of vacant land) implications. On an immediate and personal level, we think of the location of the school library not in terms of its street address or room number but where it is relative to our classrooms, or the cafeteria, or some other reference point. On the larger scene, relative location tells us that people, things, and places exist not in a spatial vacuum but in a world of physical and cultural characteristics that differ from place to place.

New York City, for example, may in absolute terms be described as located at (approximately) latitude 40° 43′ N and longitude 73° 58′ W. We have a better understanding of the *meaning* of its location, however, when reference is made to its spatial relationships: to the continental interior through the Hudson–Mohawk lowland corridor or to its position on the eastern seaboard of the United States. Within the city, we gain understanding of the locational significance of Central Park or the Lower East Side not solely by reference to the street addresses or city blocks they occupy, but by their spatial and functional relationships to the total land use, activity, and population patterns of New York City.

In view of these different ways of looking at location, geographers make a distinction between the *site* and the *situation* of a place. **Site,** an absolute location concept, refers to the physical and cultural characteristics and attributes of the place itself. It is more than mathematical location, for it tells us something about the internal features of that place. The site of Philadelphia, for example, is an area bordering and west of the Delaware River north of its intersection with the Schuylkill River in southeast Pennsylvania (Figure 1.6). **Situation,** on the other hand, refers to the

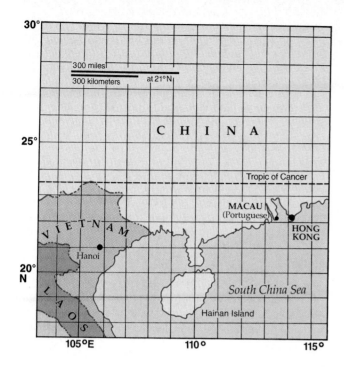

Figure 1.4 The latitude and longitude of Hong Kong is 22° 15′ N, 114° 10′ E (read as 22 degrees, 15 minutes north; 114 degrees, 10 minutes east). The circumference of the earth measures 360 degrees; each degree contains 60 minutes and each minute has 60 seconds of latitude or longitude. What are the coordinates of Hanoi?

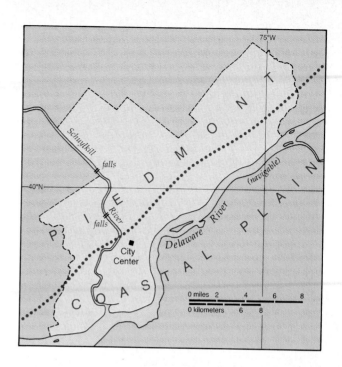

Figure 1.6 **The *site* of Philadelphia.**

external relations of a locale. It is an expression of relative location with particular reference to items of significance to the place in question. The situation of Chicago might be described as at the deepest penetration of the Great Lakes system into the interior of the United States, astride the Great Lakes–Mississippi waterways, and near the western margin of the manufacturing belt, the northern boundary of the Corn Belt, and the southeastern reaches of a major dairy region. Reference to railroads, coal deposits, and ore fields would amplify its situational characteristics (Figure 1.7).

Direction

Direction is a second universal spatial concept. Like location, it has more than one meaning and can be expressed in absolute or relative terms. **Absolute direction** is based on the cardinal points of north, south, east, and west. These appear uniformly and independently in all cultures, derived from the obvious "givens" of nature: the rising and setting of the sun for east and west, the sky location of the noontime sun and of certain fixed stars for north and south.

We also commonly use **relative** or *relational* **directions.** In the United States we go "out West," "back East," or "down South"; we worry about conflict in the "Near East" or economic competition from the "Far Eastern countries." These directional references are culturally based and locationally variable, despite their reference to cardinal compass points. The Near and the Far East locate parts of Asia from the European perspective; they are retained in the Americas by custom and usage, even though one would normally travel westward across the Pacific, for example, to reach the "Far East" from California, British Columbia, or Chile. For many Americans, "back East" and "out West" are reflections of the migration paths of earlier generations for whom home was in the

Figure 1.5 **The reality of *relative location*** on the globe may be strikingly different from the impressions we form from flat maps. The position of Russia with respect to North America when viewed from a polar perspective emphasizes that relative location properly viewed is important to our understanding of spatial relationships and interactions between the two world areas.

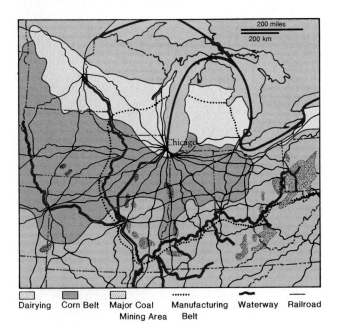

Figure 1.7 **The *situation* of Chicago** helps suggest the reasons for its functional diversity.

eastern part of the country, to which they might look back. "Up North" and "down South" reflect our accepted custom of putting north at the top and south at the bottom of our maps.

Distance

Distance joins location and direction as a commonly understood term that has dual meanings for geographers. Like its two companion spatial concepts, distance may be viewed in both an absolute and a relative sense.

Absolute distance refers to the spatial separation between two points on the earth's surface measured by some accepted standard unit such as miles or kilometers for widely separated locales, feet or meters for more closely spaced points. **Relative distance** transforms those linear measurements into other units more meaningful for the space relationship in question.

To know that two competing malls are about equidistant in miles from your residence is perhaps less important in planning your shopping trip than is knowing that because of street conditions or traffic congestion one is 5 minutes and the other 15 minutes away (Figure 1.8). Most people, in fact, think of time distance rather than linear distance in their daily activities; downtown is 20 minutes by bus, the library is a 5-minute walk. In some instances, money rather than time may be the distance transformation. An urban destination might be estimated to be a $10 cab ride away, information that may affect either the decision to make the trip at all or the choice of travel mode to get there. As a college student, you already know that rooms and apartments are less expensive at a greater distance from campus.

A *psychological* transformation of linear distance is also frequent. The solitary late-night walk back to the car through an unfamiliar or dangerous neighborhood seems far longer than a daytime stroll of the same distance through familiar and friendly territory. A first-time trip to a new destination frequently seems much longer than the return trip over the same path. Distance relationships, their measurement, and their meaning for human spatial interaction are fundamental to our understanding of human geography. They are a subject of Chapter 3, and reference to them recurs throughout this book.

Size and Scale

When we say that a place may be large or small, we speak both of the nature of the place itself and of the generalizations that can be made about it. In either instance, geographers are concerned with **scale,** though we may use that term in different ways. We can, for example, study a problem—say, population or agriculture—at the local scale, the regional scale, or on a global scale. Here the reference is purely to the size of unit studied. More technically, scale tells us the mathematical relationship between the size of an area on a map and the actual size of the mapped area on the surface of the earth (see page 20). In this sense, scale is a feature of every map and essential to recognizing the areal meaning of what is shown on that map.

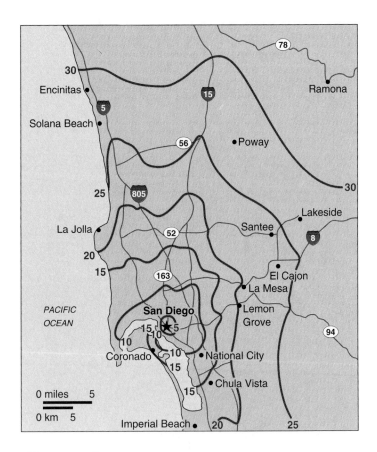

Figure 1.8 **Lines of equal travel time** (*isochrones:* from Greek, *isos,* equal, and *chronos,* time) mark off the different linear distances accessible within given spans of time from a starting point. The fingerlike outlines of isochrone boundaries reflect variations in road conditions, terrain, traffic congestion, and other aids or impediments to movement. On this map, the areas within 30 minutes' travel time from downtown San Diego are recorded for the year 2002. Note the effect of freeways on travel time.

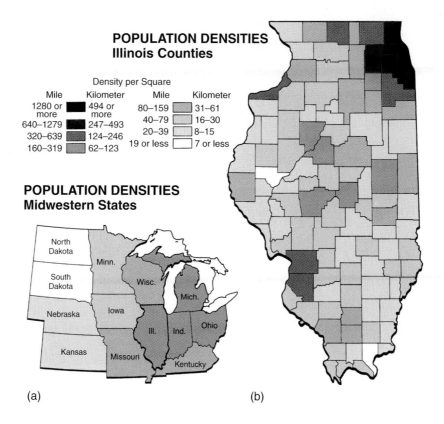

POPULATION DENSITIES
Illinois Counties

Density per Square

Mile	Kilometer		Mile	Kilometer
1280 or more	494 or more		80–159	31–61
640–1279	247–493		40–79	16–30
320–639	124–246		20–39	8–15
160–319	62–123		19 or less	7 or less

POPULATION DENSITIES
Midwestern States

(a)   (b)

Figure 1.9 **Population density and map scale.** "Truth" depends on one's scale of inquiry. Map (*a*) reveals that the maximum year 2000 population density of Midwestern states was no more than 123 people per square kilometer (319 per sq mi). From map (*b*), however, we see that population densities in three Illinois counties exceeded 494 people per square kilometer (1280 per sq mi) in 2000. If we were to reduce our scale of inquiry even further, examining individual city blocks in Chicago, we would find densities reaching 2500 or more people per square kilometer (10,000 per sq mi). Scale matters!

In both senses of the word, *scale* implies the degree of generalization represented (Figure 1.9). Geographic inquiry may be broad or narrow; it occurs at many different size scales. Climate may be an object of study, but research and generalization focused on climates of the world will differ in degree and kind from study of the microclimates of a city. Awareness of scale is very important. In geographic work, concepts, relationships, and understandings that have meaning at one scale may not be applicable at another.

For example, the study of world agricultural patterns may refer to global climatic regimes, cultural food preferences, levels of economic development, and patterns of world trade. These large-scale relationships are of little concern in the study of crop patterns within single counties of the United States, where topography, soil and drainage conditions, farm size, ownership, and capitalization, or even personal management preferences may be of greater explanatory significance.

Physical and Cultural Attributes

All places have physical and cultural attributes that distinguish them from other places and give them character, potential, and meaning. Geographers are concerned with identifying and analyzing the details of those attributes and, particularly, with recognizing the interrelationship between the physical and cultural components of area: the human–environmental interface.

Physical characteristics refer to such natural aspects of a locale as its climate and soil, the presence or absence of water supplies and mineral resources, its terrain features, and the like. These **natural landscape** attributes provide the setting within which human action occurs. They help shape—but do not dictate—how people live. The resource base, for example, is physically determined, though how resources are perceived and utilized is culturally conditioned.

People modify the environmental conditions of a given place simply by occupying it. The existence of the U.S. Environmental Protection Agency (and its counterparts elsewhere) is a reminder that humans are the active and frequently harmful agents in the continuing interplay between the cultural and physical worlds (Figure 1.10). Virtually every human activity leaves its imprint on an area's soils, water, vegetation, animal life, and other resources and on the atmosphere common to all earth space. The impact of humans has been so universal and so long exerted that essentially no "natural landscape" any longer exists.

The visible expression of that human activity is the **cultural landscape.** It, too, exists at different scales and different levels of visibility. Differences in agricultural practices and land use between Mexico and southern California are evident in Figure 1.11, while the signs, structures, and people of, for instance, Los Angeles's Chinatown leave a smaller, more confined imprint within the larger cultural landscape of the metropolitan area itself.

Figure 1.10 Sites (and sights) such as this devastation of ruptured barrels and petrochemical contamination near Texas City, Texas, are all-too-frequent reminders of the adverse environmental impacts of humans and their waste products. Many of those impacts are more hidden in the form of soil erosion, water pollution, increased stream sedimentation, plant and animal extinctions, deforestation, and the like.

Although the focus of this book is on the human characteristics of places, geographers are ever aware that the physical content of an area is also important in understanding the activity patterns of people and the interconnections between people and the environments they occupy and modify. Those interconnections and modifications are not static or permanent, however, but are subject to continual change. For example, marshes and wetlands, when drained, may be transformed into productive, densely settled farmland, while the threat or occurrence of eruption of a long-dormant volcano may quickly and drastically alter established patterns of farming, housing, and transportation on or near its flanks.

The Changing Attributes of Place

The physical environment surrounding us seems eternal and unchanging but, of course, it is not. In the framework of geologic time, change is both continuous and pronounced. Islands form and disappear; mountains rise and are worn low to swampy plains; vast continental glaciers form, move, and melt away, and sea levels fall and rise in response. Geologic time is long, but the forces that give shape to the land are timeless and relentless.

Even within the short period of time since the most recent retreat of continental glaciers—some 11,000 or 12,000 years ago—the environments occupied by humans have been subject to change. Glacial retreat itself marked a period of climatic alteration, extending the area habitable by humans to include vast

reaches of northern Eurasia and North America formerly covered by thousands of feet of ice. With moderating climatic conditions came associated changes in vegetation and fauna. On the global scale, these were natural environmental changes; humans were as yet too few in numbers and too limited in technology to alter materially the course of physical events. On the regional scale, however, even early human societies exerted an impact on the environments they occupied. Fire was used to clear forest undergrowth, to maintain or extend grassland for grazing animals and to drive them in the hunt, and, later, to clear openings for rudimentary agriculture.

With the dawn of civilizations and the invention and spread of agricultural technologies, humans accelerated their management and alteration of the now no longer "natural" environment. Even the classical Greeks noted how the landscape they occupied differed—for the worse—from its former condition. With growing numbers of people and particularly with industrialization and the spread of European exploitative technologies throughout the world, the pace of change in the content of area accelerated. The built landscape—the product of human effort—increasingly replaced the natural landscape. Each new settlement or city, each agricultural assault on forests, each new mine, dam,

Figure 1.11 This Landsat image reveals contrasting cultural landscapes along the Mexico-California border. Move your eyes from the Salton Sea (the dark patch at the top of the image) southward to the agricultural land extending to the edge of the image. Notice how the regularity of the fields and the bright colors (representing growing vegetation) give way to a marked break, where irregularly shaped fields and less prosperous agriculture are evident. Above the break is the Imperial Valley of California; below the border is Mexico.
© *NASA.*

or factory changed the content of regions and altered the temporarily established spatial interconnections between humans and the environment.

Characteristics of places today, therefore, are the result of constantly changing past conditions. They are, as well, the forerunners of differing human–environmental balances yet to be struck. Geographers are concerned with places at given moments of time. But to understand fully the nature and development of places, to appreciate the significance of their relative locations, and to comprehend the interplay of their physical and cultural characteristics, geographers must view places as the present result of the past operation of distinctive physical and cultural processes (Figure 1.12).

You will recall that one of the questions geographers ask about a place or thing is: How did it come to be what and where it is? This is an inquiry about process and about becoming. The forces and events shaping the physical and explaining the cultural environment of places today are an important focus of geography. They are, particularly in their human context, the subjects of most of the separate chapters of this book. To understand them is to appreciate more fully the changing human spatial order of our world.

Interrelations between Places

The concepts of relative location and distance that we earlier introduced lead directly to a fundamental spatial reality: Places interact with other places in structured and comprehensible ways. In describing the processes and patterns of that **spatial interac-** tion, geographers add *accessibility* and *connectivity* to the ideas of location and distance.

A basic law of geography tells us that in a spatial sense everything is related to everything else but that relationships are stronger when items are near one another. Our observation, therefore, is that interaction between places diminishes in intensity and frequency as distance between them increases—a statement of the idea of *distance decay,* which we explore in Chapter 3.

Consideration of distance implies assessment of **accessibility.** How easy or difficult is it to overcome the "friction of distance"? That is, how easy or difficult is it to surmount the barrier of the time and space separation of places? Distance isolated North America from Europe until the development of ships (and aircraft) that reduced the effective distance between the continents. All parts of the ancient and medieval city were accessible by walking; they were "pedestrian cities," a status lost as cities expanded in area and population with industrialization. Accessibility between city districts could be maintained only by the development of public transit systems whose fixed lines of travel increased ease of movement between connected points and reduced it between areas not on the transit lines themselves.

Accessibility therefore suggests the idea of **connectivity,** a broader concept implying all the tangible and intangible ways in which places are connected: by physical telephone lines, street and road systems, pipelines and sewers; by unrestrained walking across open countryside; by radio and TV broadcasts beamed outward uniformly from a central source. Where routes are fixed and flow is channelized, *networks*—the patterns of routes

Figure 1.12 The process of change in a cultural landscape. Before the advent of the freeway, this portion of suburban Long Island, New York, was largely devoted to agriculture (left). The construction of the freeway and cloverleaf interchange ramps altered nearby land use patterns (right) to replace farming with housing developments and new commercial and light industrial activities.

connecting sets of places—determine the efficiency of movement and the connectedness of points. Demand for universal instantaneous accessibility and connectivity is common and unquestioned in today's advanced societies. Technologies and devices to achieve it proliferate, as our own lifestyles show. Cell phones, e-mail, broadband wireless Internet access, instant messaging, and more have erased time and distance barriers formerly separating and isolating individuals and groups and have reduced our dependence on physical movement and on networks fixed in the landscape. The realities of accessibility and connectivity, that is, clearly change over time (Figure 1.13).

There is, inevitably, interchange between connected places. **Spatial diffusion** is the process of dispersion of an idea or an item from a center of origin to more distant points with which it is directly or indirectly connected. The rate and extent of that diffusion are affected by the distance separating the originating center of, say, a new idea or technology and other places where it is eventually adopted. Diffusion rates are also affected by population densities, means of communication, obvious advantages of the innovation, and importance or prestige of the originating *node*. These ideas of diffusion are further explored in Chapter 2.

Geographers study the dynamics of spatial relationships. Movement, connection, and interaction are part of the social and economic processes that give character to places and regions. Geography's study of those relationships recognizes that spatial interaction is not just an awkward necessity but a fundamental organizing principle of human life on earth. That recognition has become universal, repeatedly expressed in the term *globalization*. **Globalization** implies the increasing interconnection of all peoples and societies in all parts of the world as the full range of social, cultural, political, economic, and environmental processes becomes international in scale and effect. Promoted by continuing advances in worldwide accessibility and connectivity, globalization encompasses other core geographic concepts of spatial interaction, accessibility, connectivity, and diffusion. More detailed implications of globalization will be touched on in later chapters of this text.

The Structured Content of Place

A starting point for geographic inquiry is how objects are distributed in area—for example, the placement of churches or supermarkets within a town. That interest distinguishes geography from other sciences, physical or social, and underlies many of the questions geographers ask: Where is a thing located? How is that location related to other items? How did the location we observe come to exist? Such questions carry the conviction that the contents of an area are comprehensibly arranged or structured. The arrangement of items on the earth's surface is called **spatial distribution** and may be analyzed by the elements common to all spatial distributions: *density, dispersion,* and *pattern.*

Density

The measure of the number or quantity of anything within a defined unit of area is its **density.** It is therefore not simply a count

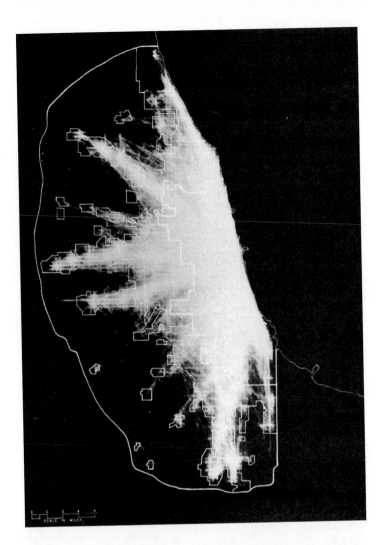

Figure 1.13 **The routes of the 5 million automobile trips** made each day in Chicago during the late 1950s are recorded on this light-display map. The boundaries of the region of interaction that they created are clearly marked and document the centrality of Chicago at that time as the employment destination of city-fringe and suburban residents. Those boundaries (and the dynamic region they defined) were subject to change as residential neighborhoods expanded or developed, as population relocations occurred, and as the road pattern was altered over time. If made today, the light-display would show a much more complex commuting pattern, with most trips between suburbs and not from suburbs to the central city.

From Chicago Area Transportation Study, Final Report. *1959, Vol. 1, p. 44, figure 22 "Desire Lines of Internal Automobile Driver Trips."*

of items but of items in relation to the space in which they are found. When the relationship is absolute, as in population per square kilometer, for example, or dwelling units per acre, we are defining *arithmetic density* (see Figure 1.9). Sometimes it is more meaningful to relate item numbers to a specific kind of area. *Physiological density,* for example, is a measure of the number of persons per unit area of arable land. Density defined in population terms is discussed in Chapter 4.

A density figure is a statement of fact but not necessarily one useful in itself. Densities are normally employed comparatively, relative to one another. High or low density implies a comparison with a known standard, with an average, or with a different area. Ohio, with (2000) 107 persons per square kilometer (277 per sq mi), might be thought to have a high density compared to neighboring Michigan at 68 per square kilometer (175 per sq mi), and a low one in relation to New Jersey at 438 (1134 per sq mi).

Dispersion

Dispersion (or its opposite, **concentration**) is a statement of the amount of *spread* of a phenomenon over an area. It tells us not how many or how much but how far things are spread out. If they are close together spatially, they are considered *clustered* or *agglomerated*. If they are spread out, they are *dispersed* or *scattered* (Figure 1.14).

If the entire population of a metropolitan county were all located within a confined central city, we might say the population was clustered. If, however, that same population redistributed itself, with many city residents moving to the suburbs and occupying a larger portion of the county's territory, it would become more dispersed. In both cases, the *density* of population (numbers in relation to area of the county) would be the same, but the distribution would have changed. Since dispersion deals with separation of things one from another, a distribution that might be described as *clustered* (closely spaced) at one scale of reference might equally well be considered *dispersed* (widely spread) at another scale.

Pattern

The geometric arrangement of objects in space is called **pattern.** Like dispersion, pattern refers to distribution, but that reference emphasizes design rather than spacing (Figure 1.15). The distribution of towns along a railroad or houses along a street may be seen as *linear*. A *centralized* pattern may involve items concen-

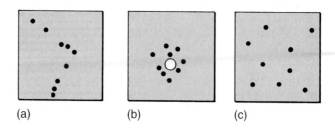

(a)　　　　　(b)　　　　　(c)

Figure 1.15 *Pattern* describes spatial arrangement and design. The *linear* pattern of towns in (*a*) perhaps traces the route of a road or railroad or the course of a river. The central city in (*b*) with its nearby suburbs represents a *centralized* pattern, while the dots in (*c*) are *randomly* distributed.

trated around a single node. A *random* pattern may be the best description of an unstructured irregular distribution.

The rectangular system of land survey adopted in much of the United States under the Ordinance of 1785 creates a checkerboard rural pattern of "sections" and "quarter-sections" of farmland (see Figure 6.26). As a result, in most American cities, streets display a *grid* or *rectilinear* pattern. The same is true of cities in Canada, Australia, New Zealand, and South Africa, which adopted similar geometric survey systems. The *hexagonal* pattern of service areas of farm towns is a mainstay of central place theory discussed in Chapter 11. These references to the geometry of distribution patterns help us visualize and describe the structured arrangement of items in space. They help us make informed comparisons between areas and use the patterns we discern to ask further questions about the interrelationship of things.

Place Similarity and Regions

The distinctive characteristics of places in content and structure immediately suggest two geographically important ideas. The first is that no two places on the surface of the earth can be *exactly* the same. Not only do they have different absolute locations, but—as in the features of the human face—the precise mix of physical and cultural characteristics of a place is never exactly duplicated.

Because geography is a spatial science, the inevitable uniqueness of place would seem to impose impossible problems of generalizing spatial information. That this is not the case results from the second important idea: The physical and cultural content of an area and the dynamic interconnections of people and places show patterns of spatial similarity. Often the similarities are striking enough for us to conclude that spatial regularities exist. They permit us to recognize and define **regions**—earth areas that display significant elements of internal uniformity and external difference from surrounding territories. Places are, therefore, both unlike and like other places, creating patterns of areal differences and of coherent spatial similarity.

The problem of the historian and the geographer is similar. Each must generalize about items of study that are essentially unique. The historian creates arbitrary but meaningful and useful historical periods for reference and study. The "Roaring

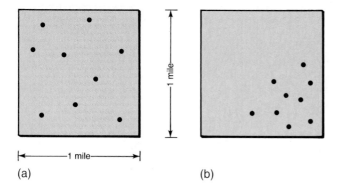

(a)　　　　　　　　(b)

Figure 1.14 Density and dispersion each tell us something different about how items are distributed in an area. *Density* is simply the number of items or observations within a defined area; it remains the same no matter how the items are distributed. The density of houses per square mile, for example, is the same in both (*a*) and (*b*). *Dispersion* is a statement about nearness or separation. The houses in (*a*) are more *dispersed* than those shown *clustered* in (*b*).

Twenties" and the "Victorian Era" are shorthand summary names for specific time spans, internally quite complex and varied but significantly distinct from what went before or followed after. The region is the geographer's equivalent of the historian's epoch. It is a device of areal generalization that segregates into component parts the complex reality of the earth's surface. In both the time and the space need for generalization, attention is focused on key unifying elements or similarities of the era or area selected for study. In both the historical and geographical cases, the names assigned to those times and places serve to identify the time span or region and to convey between speaker and listener a complex set of interrelated attributes.

All of us have a general idea of the meaning of region, and all of us refer to regions in everyday speech and action. We visit "the old neighborhood" or "go downtown"; we plan to vacation or retire in the "Sunbelt"; or we speculate about the effects of weather conditions in the "Corn Belt" on next year's food prices. In each instance, we have mental images of the areas mentioned, and in each, we have engaged in an informal place classification to pass along quite complex spatial, organizational, or content ideas. We have applied the **regional concept** to bring order to the immense diversity of the earth's surface.

Regions are not "given" in nature any more than "eras" are given in the course of human events. Regions are devised; they are spatial summaries designed to bring order to the infinite diversity of the earth's surface. At their root, they are based on the recognition and mapping of *spatial distributions*—the territorial occurrence of environmental, human, or organizational features selected for study. As many spatial distributions exist as there are imaginable physical, cultural, or connectivity elements of area to examine. Since regions are mental constructs, different observers employing different criteria may bestow the same regional identity on differently bounded areal units (Figure 1.16). In each case, however, the key characteristics that are selected for study are those that contribute to the understanding of a specific topic or problem.

Types of Regions

Regions may be either *formal, functional,* or *perceptual.* **Formal or uniform regions** are areas of essential uniformity in one or a limited combination of physical or cultural features. Your home state is a precisely bounded formal political region within which uniformity of law and administration is found. Later in this book we will encounter formal (homogeneous) cultural regions in which standardized characteristics of language, religion, ethnicity, or economy exist. The frontpaper foldout maps of landform regions and country units show other formal regional patterns. Whatever the basis of its definition, the formal region is the largest area over which a valid generalization of attribute uniformity may be made. Whatever is stated about one part of it holds true for its remainder.

The **functional** or **nodal region,** in contrast, may be visualized as a spatial system. Its parts are interdependent, and throughout its extent the functional region operates as a dynamic, organizational unit. A functional region has unity not in the sense of static content but in the manner of its operational connectivity. It has a *core* area in which its characterizing features are most clearly defined; they lessen in prominence toward the region's margins or *periphery.* As the degree and extent of areal control and interaction change, the boundaries of the functional region change in response. Trade areas of towns, national "spheres of influence," and the territories subordinate to the financial, administrative, wholesaling, or retailing centrality exercised by such regional capitals as Chicago, Atlanta, or Minneapolis are cases in point (Figure 1.17).

Perceptual regions are less rigorously structured than the formal and functional regions geographers devise. They reflect feelings and images rather than objective data and because of that may be more meaningful in the lives and actions of those who recognize them than are the more abstract regions of geographers.

Ordinary people have a clear idea of spatial variation and employ the regional concept to distinguish between territorial

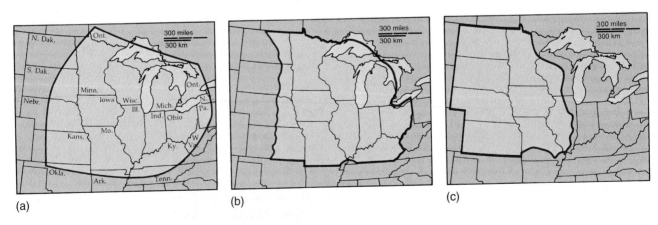

(a)   (b)   (c)

Figure 1.16 **The Middle West as seen by different professional geographers.** Agreement on the need to recognize spatial order and to define regional units does not imply unanimity in the selection of boundary criteria. All the sources concur in the significance of the Middle West as a regional entity in the spatial structure of the United States and agree on its core area. These sources differ, however, in their assessment of its limiting characteristics.

Sources: (a) John H. Garland, ed., The North American Midwest *(New York: John Wiley & Sons, 1955); (b) John R. Borchert and Jane McGuigan,* Geography of the New World *(Chicago: Rand McNally, 1961); and (c) Otis P. Starkey and J. Lewis Robinson,* The Anglo-American Realm *(New York: McGraw-Hill, 1969).*

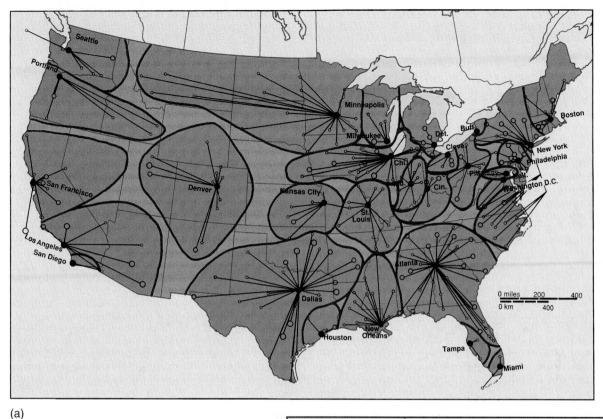

(a)

Figure 1.17 (a) The *functional* (or *nodal*) *regions* shown on this map were based on linkages between large banks of major central cities and the "correspondent" banks they served in smaller towns in the 1970s, before the advent of electronic banking and bank consolidation. (b) A different form of *connectivity* is suggested by the "desire line" map recording the volume of daily work trips within the San Francisco Bay area to the Silicon Valley employment node. The outer *periphery* of a dynamic functional region is marked by the farthest extent of the commuting lines. The intensity of interchange and the strength of regional identity increases toward the center, or *core*. The region changed in size and shape over time as the network was enlarged and improved, the Valley employment base expanded, and the commuting range of workers increased. The map, of course, gives no indication of the global reach of the Valley's *accessibility* and interaction through other means of communication and exchanges.

(a) Redrawn by permission from Annals of the Association of American Geographers, *John R. Borchert, vol. 62, p. 358, Association of American Geographers, 1972. (b) Map "1981 Desired Lines and Recorded Volumes of Daily Work Trips from the Entire Bay Area of the Silicon Valley" from Robert Cervero,* Suburban Gridlock, *p. 188. Copyright © 1986 Rutgers, The State University of New Jersey, Center for Urban Policy Research. Reprinted by permission.*

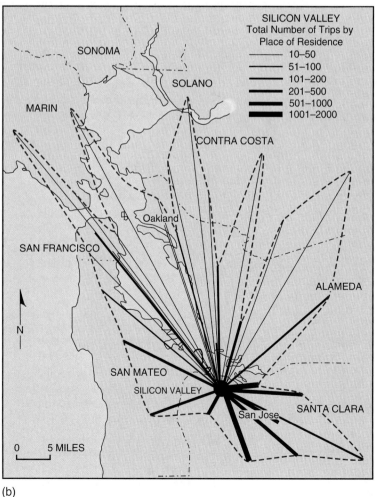

(b)

entities. People individually and collectively agree on where they live. The *vernacular regions* they recognize have reality in their minds and are reflected in regionally based names employed in businesses, by sports teams, or in advertising slogans. The frequency of references to "Dixie" in the southeastern United States represents that kind of regional consensus and awareness. Such vernacular regions reflect the way people view space, assign their loyalties, and interpret their world. At a different scale, such urban ethnic enclaves (see Chapter 6) as "Little Italy" or "Chinatown" have comparable regional identity in the minds of their inhabitants. Less clearly perceived by outsiders but unmistakable to their inhabitants are the "turfs" of urban clubs or gangs. Their boundaries are sharp, and the perceived distinctions between them are paramount in the daily lives and activities of their occupants. What perceptual regions do you have clearly in mind?

Maps

Maps are tools to identify regions and to analyze their content. The spatial distributions, patterns, and relations of interest to geographers usually cannot easily be observed or interpreted in the landscape itself. Many, such as landform or agricultural regions or major cities, are so extensive spatially that they cannot be seen or studied in their totality from one or a few vantage points. Others, such as regions of language usage or religious belief, are spatial phenomena, but are not tangible or visible. Various interactions, flows, and exchanges imparting the dynamic quality to spatial interaction may not be directly observable at all. And even if all matters of geographic interest could be seen and measured through field examination, the infinite variety of tangible and intangible content of area would make it nearly impossible to isolate for study and interpretation the few items of regional interest selected for special investigation.

Therefore, the map has become the essential and distinctive tool of geographers. Only through the map can spatial distributions and interactions of whatever nature be reduced to an observable scale, isolated for individual study, and combined or recombined to reveal relationships not directly measurable in the landscape itself. But maps can serve their purpose only if their users have a clear idea of their strengths, limitations, and diversity and of the conventions observed in their preparation and interpretation.

Map Scale

We have already seen that scale (page 12) is a vital element of every map. Because it is a much reduced version of the reality it summarizes, a map generalizes the data it displays. *Scale*—the relationship between size or length of a feature on the map and the same item on the earth's surface—determines the amount of that generalization. The smaller the scale of the map, the larger is the area it covers and the more generalized are the data it portrays. The larger the scale, the smaller is the depicted area and the more accurately can its content be represented (Figure 1.18). It may seem backward, but large-scale maps show small areas, and small-scale maps show large areas.

Map scale is selected according to the amount of generalization of data that is acceptable and the size of area that must be depicted. The user must consider map scale in evaluating the reliability of the spatial data that are presented. Regional boundary lines drawn on the world maps in this and other books or atlases would cover many kilometers or miles on the earth's surface. They obviously distort the reality they are meant to define, and on small-scale maps major distortion is inevitable. In fact, a general rule of thumb is that the larger the earth area depicted on a map, the greater is the distortion built into the map.

This is so because a map has to depict the curved surface of the three-dimensional earth on a two-dimensional sheet of paper. The term **projection** designates the method chosen to represent the earth's curved surface as a flat map. Since absolutely accurate representation is impossible, all projections inevitably distort. Specific projections may be selected, however, to minimize the distortion of at least one of the four main map properties—area, shape, distance, and direction.[3]

The Globe Grid

Maps are geographers' primary tools of spatial analysis. All spatial analysis starts with locations, and all locations are related to the global grid of latitude and longitude. Since these lines of reference are drawn on the spherical earth, their projection onto a map distorts their grid relationships. The extent of variance between the globe grid and a map grid helps tell us the kind and degree of distortion the map will contain.

The key reference points in the *grid system* are the North and South poles and the equator, which are given in nature, and the *prime meridian,* which is agreed on by cartographers. Because a circle contains 360 degrees, the distance between the poles is 180 degrees and between the equator and each pole, 90 degrees (Figure 1.19). *Latitude* measures distance north and south of the equator (0°), and *parallels* of latitude run due east-west. *Longitude* is the angular distance east or west of the prime meridian and is depicted by north-south lines called *meridians,* which converge at the poles. The properties of the globe grid the mapmaker tries to retain and the map user should look for are as follows:

1. All meridians are of equal length; each is one-half the length of the equator.
2. All meridians converge at the poles and are true north–south lines.
3. All lines of latitude (parallels) are parallel to the equator and to each other.
4. Parallels decrease in length as one nears the poles.
5. Meridians and parallels intersect at right angles.
6. The scale on the surface of the globe is the same in every direction.

Only the globe grid itself retains all of these characteristics. To project it onto a surface that can be laid flat is to distort some or all of these properties and consequently to distort the reality the map attempts to portray.

---

[3]A more detailed discussion of map projections, including examples of their different types and purposes, may be found in Appendix A, beginning on page 521.

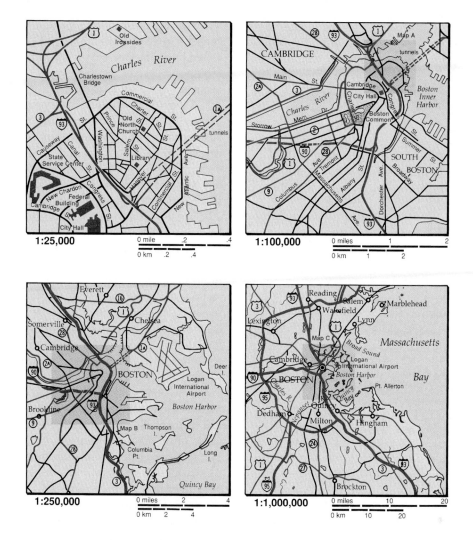

Figure 1.18 **The effect of scale on area and detail.** The larger the scale, the greater the number and kinds of features that can be included. Scale may be reported to the map user in one (or more) of three ways. A *verbal* scale is given in words ("1 centimeter to 1 kilometer" or "1 inch to 1 mile"). A *representative fraction* (such as that placed at the left, below each of the four maps shown here) is a statement of how many linear units on the earth's surface are represented by one unit on the map. In the upper left map, for example, one map inch represents 25,000 inches on the ground. A *graphic* scale (such as that placed at the right and below each of these maps) is a line or bar marked off in map units but labeled in ground units.

How Maps Show Data

The properties of the globe grid and of various projections are the concern of the cartographer. Geographers are more interested in the depiction of spatial data and in the analysis of the patterns and interrelationships those data present. Out of the myriad items comprising the content of an area, the geographer must, first, select those that are of concern to the problem at hand and, second, decide on how best to display them for study or demonstration. In that effort, geographers can choose between different types of maps and different systems of symbolization.

General-purpose, reference, or *location maps* make up one major class of maps familiar to everyone. Their purpose is simply to show without analysis or interpretation a variety of natural or human-made features of an area or of the world as a whole. Familiar examples are highway maps, city street maps, topographic maps (Figure 1.20), atlas maps, and the like. Until about the mid-

dle of the 18th century, the general-purpose or reference map was the dominant map form, for the primary function of the mapmaker (and the explorer who supplied the new data) was to "fill in" the world's unknown areas with reliable locational information. With the passage of time, scholars saw the possibilities to use the accumulating locational information to display and study the spatial patterns of social and physical data. The maps they made of climate, vegetation, soil, population, and other distributions introduced the thematic map, the second major class of maps.

Thematic map is the general term applied to a map of any scale that presents a specific spatial distribution or a single category of data—that is, presents a graphic theme. The way the information is shown on such a map may vary according to the type of information to be conveyed, the level of generalization that is desired, and the symbolization selected. Thematic maps may be either *qualitative* or *quantitative*. The principal purpose of the

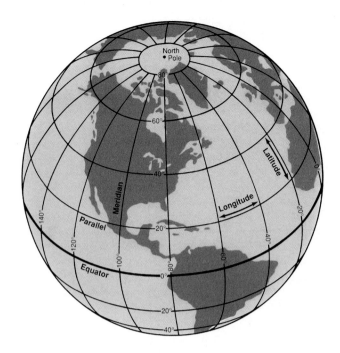

Figure 1.19 **The grid system of parallels of latitude and meridians of longitude.** Since the meridians converge at the poles, parallels become increasingly shorter away from the equator. On the globe, the 60th parallel is only one-half as long as the equator, and a degree of longitude along it measures only about 55½ kilometers (about 34½ miles) compared to about 111 kilometers (about 69 miles) at the equator (0°).

qualitative map is to show the distribution of a particular class of information. The world location of producing oil fields, the distribution of U.S. national parks, or the pattern of areas of agricultural specialization within a state or country are examples. The interest is in where things are and nothing is reported about—in the examples cited—barrels of oil extracted or in reserve, number of park visitors, or value or volume of crops or livestock produced.

In contrast, quantitative thematic maps show the spatial characteristic of numerical data. Usually, a single variable such as population, income, wheat, or land value is chosen, and the map displays the variation from place to place in that feature. Important types of quantitative thematic maps include graduated circle, dot, isometric and isopleth, and choropleth maps (Figure 1.21).

Graduated circle maps use circles of different size to show the frequency of occurrence of a topic in different places; the larger the circle, the more frequent the incidence. On *dot maps,* a single or specified number of occurrences of the item studied is recorded by a single dot. The dot map serves not only to record data but to suggest their spatial pattern, distribution, and dispersion.

An *isometric map* features lines (*isolines*) that connect points registering equal values of the item mapped (*iso* means "equal"). The *isotherms* shown on the daily weather map connect points recording the same temperature at the same moment of time or the same average temperature during the day. Identical elevations above sea level may be shown by a form of isoline called a *contour line.* On *isopleth maps,* the calculation refers not to a point but to an areal statistic—for example, persons per square kilome-

ter or average percentage of cropland in corn—and the isoline connects average values for unit areas. For emphasis, the area enclosed by isolines may be shaded to indicate approximately uniform occurrence of the thing mapped, and the isoline itself may be treated as the boundary of a uniform region.

A *choropleth map* presents average value of the data studied per preexisting areal unit—dwelling unit rents or assessed values by city block, for example, or (in the United States) population densities by individual townships within countries. Each unit area on the map is then shaded or colored to suggest the magnitude of the event or item found within its borders. Where the choropleth map is based on the absolute number of items within the unit area, as it is in Figure 1.21d, rather than on areal averaging (total numbers, that is, instead of, for example, numbers per square kilometer), a misleading statement about density may be conveyed.

A *statistical map* records the actual numbers or occurrences of the mapped item per established unit area or location. The actual count of each state's colleges and universities shown on an outline map of the United States or the number of traffic accidents at each street intersection within a city are examples of statistical maps. A *cartogram* uses such statistical data to transform territorial space so that the largest areal unit on the map is the one showing the greatest statistical value (Figure 1.22).

Maps communicate information but, as in all forms of communication, the message conveyed by a map reflects the intent and, perhaps, the biases of its author. Maps are persuasive because of the implied precision of their lines, scales, color and symbol placement, and information content. But maps, as communication devices, can subtly or blatantly manipulate the message they impart or contain intentionally false information (Figure 1.23). Maps, then, can distort and lie as readily as they can convey verifiable spatial data or scientifically valid analyses. The more map users are aware of those possibilities and the more understanding of map projections, symbolization, and common forms of thematic and reference mapping standards they possess, the more likely are they to reasonably question and clearly understand the messages maps communicate.

Geographic Information Systems (GIS)

Increasingly, digital computers, mapping software, and computer-based display units and printers are employed in the design and production of maps and in the development of databases used in map production. In computer-assisted cartography, the content of standard maps—locational and thematic—is digitized and stored in computers. The use of computers and printers in map production permits increases in the speed, flexibility, and accuracy of many steps in the mapmaking process but in no way reduces the obligation of the mapmaker to employ sound judgment in the design of the map or the communication of its content.

Geographic information systems (GIS) extend the use of digitized data and computer manipulation to investigate and display spatial information. A GIS is both an integrated software package for handling, processing, and analyzing geographical data and a computer database in which every item of information is tied to a precise geographic location. In the *raster approach,* that tie involves dividing the study area into a set of rectangular

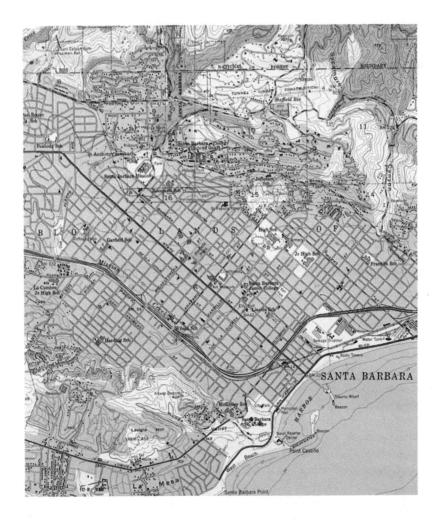

Figure 1.20 **A portion of the Santa Barbara, California, topographic quadrangle** of the U.S. Geological Survey 1:24,000 series. Topographic maps portray the natural landscape features of relatively small areas. Elevations and shapes of landforms, streams, and other water bodies, vegetation, and coastal features are recorded, often with great accuracy. Because cultural items that people have added to the physical landscape, such as roads, railroads, buildings, political boundaries, and the like, are also frequently depicted on them, topographic maps are classed as general purpose or reference maps by the International Cartographic Association. The scale of the original map no longer applies to this photographic reduction.

Source: U.S. Geological Survey.

cells and describing the content of each cell. In the *vector approach,* the precise location of each object—point, line, or area—in a distribution is described. In either approach, a vast amount of different spatial information can be stored, accessed, compared, processed, analyzed, and displayed.

A GIS database, then, can be envisioned as a set of discrete informational overlays linked by reference to a basic locational grid of latitude and longitude (Figure 1.24). The system then permits the separate display of the spatial information contained in the database. It allows the user to overlay maps of different themes, analyze the relations revealed, and compute spatial relationships. It shows aspects of spatial associations otherwise difficult to display on conventional maps, such as flows, interactions, and three-dimensional characteristics. In short, a GIS database, as a structured set of spatial information, has become a powerful tool for automating geographical analysis and synthesis.

A GIS data set may contain the great amount of place-specific information collected and published by the U.S. Census Bureau, in-

cluding population distribution, race, ethnicity, income, housing, employment, industry, farming, and so on. It may also hold environmental information downloaded from satellite imagery or taken from Geological Survey maps and other governmental and private sources. In human geography, the vast and growing array of spatial data has encouraged the use of GIS to explore models of regional economic and social structure, to examine transportation systems and urban growth patterns, to study patterns of voting behavior, disease incidence, the accessibility of public services, and a vast array of other topics. For physical geographers, the analytic and modeling capabilities of GIS are fundamental to the understanding of processes and interrelations in the natural environment.

Because of the growing importance of GIS in all manner of public and private spatial inquiries, demand in the job market is growing for those skilled in its techniques. Most university courses in GIS are taught in Geography departments, and "GIS/remote sensing" is a primary occupational specialty for which many geography undergraduate and graduate majors seek preparation.

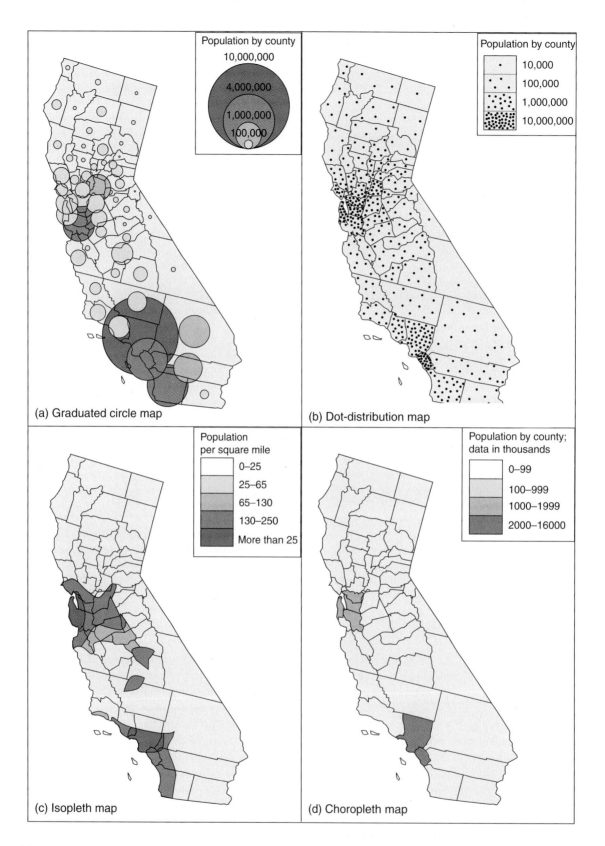

(a) Graduated circle map

Population by county
10,000,000
4,000,000
1,000,000
100,000

(b) Dot-distribution map

Population by county
· 10,000
100,000
1,000,000
10,000,000

(c) Isopleth map

Population per square mile
0–25
25–65
65–130
130–250
More than 25

(d) Choropleth map

Population by county; data in thousands
0–99
100–999
1000–1999
2000–16000

Figure 1.21 **Types of thematic maps.** Although population is the theme of each, these different California maps present their information in strikingly different ways. (*a*) In the graduated circle map, the area of the circle is approximately proportional to the absolute number of people within each county. (*b*) In a dot-distribution map where large numbers of items are involved, the value of each dot is identical and stated in the map legend. The placement of dots on this map does not indicate precise locations of people within the county, but simply their total number. (*c*) Population density is recorded by the isopleth map, while the choropleth map (*d*) may show absolute values as here or, more usually, ratio values such as population per square kilometer.

Source: From Fred M. Shelley and Audrey E. Clarke, Human and Cultural Geography, © 1994. *Reproduced by permission of The McGraw-Hill Companies.*

24 Introduction: Some Background Basics

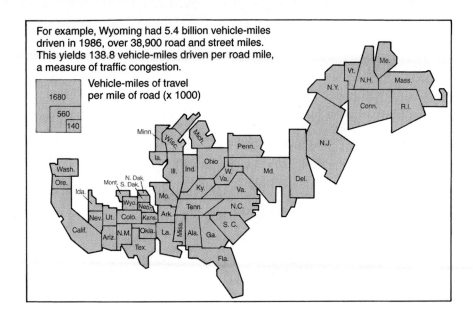

For example, Wyoming had 5.4 billion vehicle-miles driven in 1986, over 38,900 road and street miles. This yields 138.8 vehicle-miles driven per road mile, a measure of traffic congestion.

Vehicle-miles of travel per mile of road (x 1000)

1680
560
140

Figure 1.22 **Relative traffic congestion.** A typical value-by-area cartogram with states drawn in proportion to the number of vehicle-miles driven per road mile.

Source: Borden D. Dent, Cartography: Thematic Map Design, *4th ed., © 1996. Reproduced by permission of Times Mirror Higher Education Group, Dubuque, Iowa.*

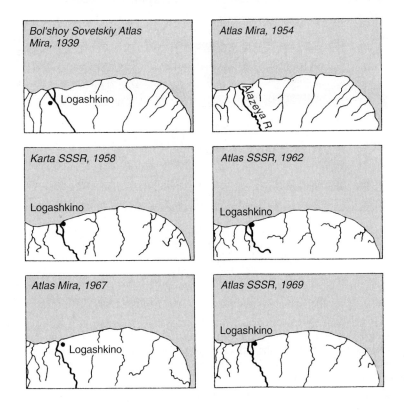

Figure 1.23 **The wandering town of Logashkino,** as traced in various Soviet atlases by Mark Monmonier. Deliberate, extensive cartographic "disinformation" and locational falsification, he reports, became a Cold War tactic of the Soviet Union. We usually use—and trust—maps to tell us exactly where things are located. On the maps shown, however, Logashkino migrates from west of the river away from the coast to east of the river on the coast, while the river itself gains and loses a distributary and, in 1954, the town itself disappears. The changing misinformation, Monmonier suggests, was intended to obscure from potential enemies the precise location of possible military targets.

Source: Mark Monmonier, How to Lie with Maps, *2d ed. © 1996. Reproduced by permission of the University of Chicago Press.*

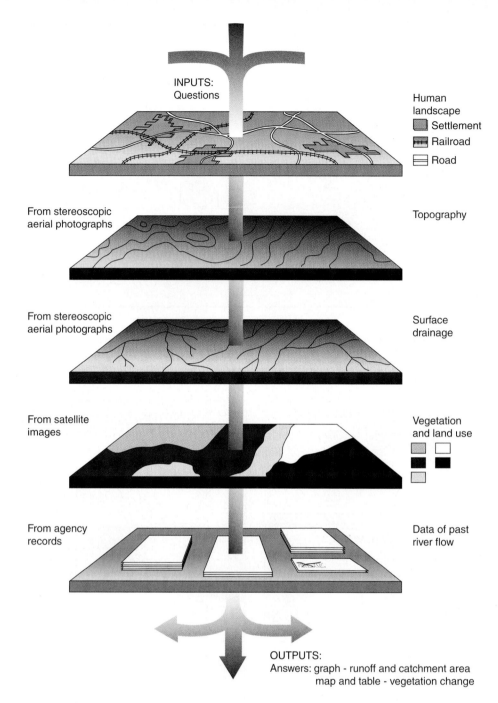

INPUTS:
Questions

Human landscape
▨ Settlement
▦ Railroad
⊟ Road

From stereoscopic aerial photographs

Topography

From stereoscopic aerial photographs

Surface drainage

From satellite images

Vegetation and land use

From agency records

Data of past river flow

OUTPUTS:
Answers: graph - runoff and catchment area
map and table - vegetation change

Figure 1.24 **A model of a geographic information system.** A GIS incorporates three primary components: data storage capability, computer graphics programs, and statistical packages. In this example, the different layers of information held are important in monitoring a river system. Different data sets, all selected for applicability to the questions asked, may be developed and used in human geography, economic geography, transportation planning, industrial location work, and similar applications.

Source: From Michael Bradshaw and Ruth Weaver, Foundations of Physical Geography. *© 1995. Reprinted by permission of The McGraw-Hill Companies.*

Mental Maps

Maps that shape our understanding of distributions and locations or influence our perception of the world around us are not always drawn on paper. We carry with us *mental maps* that in some ways are more accurate in reflecting our view of spatial reality than the formal maps created by geographers or cartographers. **Mental maps** are images about an area or an environment developed by an individual on the basis of information or impressions received, interpreted, and stored. What are believed to be unnecessary details are left out, and only the important elements are incorporated. We use this information—this mental map—in organizing our daily activities: selecting our destinations and the sequence in which they will be visited, deciding on our routes of travel, recognizing where we are in relation to where we wish to be. A mental route map may also include reference points to be encountered on the chosen path of connection or on alternate lines of travel.

Such mental maps are every bit as real to their creators (and we all have them) as are the street maps or highway maps commercially available, and they are a great deal more immediate in their impact on our spatial decisions. We may choose routes or avoid neighborhoods not on objective grounds but on emotional or perceptual ones. Whole sections of a community may be voids on our mental maps, as unknown as the interiors of Africa and South America were to Western Europeans two centuries ago. Our areas of awareness generally increase with the increasing mobility that comes with age (Figure 1.25), affluence, and education and may be enlarged or restricted for different social groups within the city (Figure 1.26).

Systems, Maps, and Models

The content of area is interrelated and constitutes a **spatial system** that, in common with all systems, functions as a unit because its component parts are interdependent. Only rarely do individual elements of area operate in isolation, and to treat them as if they do is to lose touch with spatial reality. The systems of geographic concern are those in which the functionally important variables are spatial: location, distance, direction, density, and the other basic concepts we have reviewed. The systems that they define are not the same as regions, though spatial systems may be the basis for regional identification.

Systems have components, and the analysis of the role of components helps reveal the operation of the system as a whole. To conduct that analysis, individual system elements must be isolated for separate identification and, perhaps, manipulated to see their function within the structure of the system or subsystem. Maps and models are the devices geographers use to achieve that isolation and separate study.

Maps, as we have seen, are effective to the degree that they can segregate at an appropriate level of generalization those system elements selected for examination. By compressing, simplifying, and abstracting reality, maps record in manageable dimension the real-world conditions of interest. A **model** is a simplified abstraction of reality, structured to clarify causal relationships. Maps are a kind of model. They represent reality in an idealized form so that certain aspects of its properties may be more clearly seen. They are a special form of model, of course. Their abstractions are rendered visually and at a reduced scale so they may be displayed, for example, on the pages of this book.

The complexities of spatial systems analysis—and the opportunities for quantitative analysis of systems made possible by computers and sophisticated statistical techniques—have led geographers to use other kinds of models in their work. Model building is the technique social scientists use to simplify complex situations, to eliminate (as does the map) unimportant details, and to isolate for special study and analysis the role of one or more interacting elements in a total system.

An interaction model discussed in Chapter 3, for instance, suggests that the amount of exchange expected between two

(a)

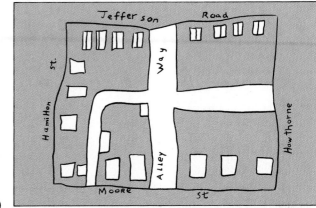

(b)

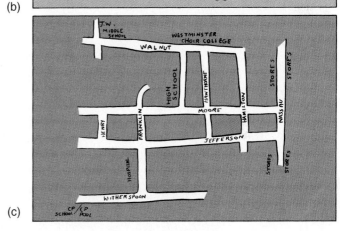

(c)

Figure 1.25 Three children, aged 6, 10, and 13, who lived in the same house, were asked to draw maps of their neighborhood. They received no further instructions. Notice how perspectives broaden and neighborhoods expand with age. (*a*) For the 6-year-old, the "neighborhood" consists of the houses on either side of her own. (*b*) The square block on which she lives is the neighborhood for the 10-year-old. (*c*) The wider horizons of the 13-year-old are reflected in her drawing. The square block that the 10-year-old drew is shaded in this sketch.

places depends on the distance separating them and on their population size. The model indicates that the larger the places and the closer their distance, the greater is the amount of interaction. Such a model helps us to isolate the important components of the spatial system, to manipulate them separately, and to reach conclusions concerning their relative importance. When a model satisfactorily predicts the volume of intercity interaction in the

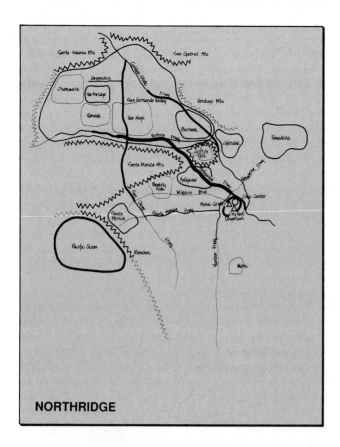

NORTHRIDGE

BOYLE HEIGHTS

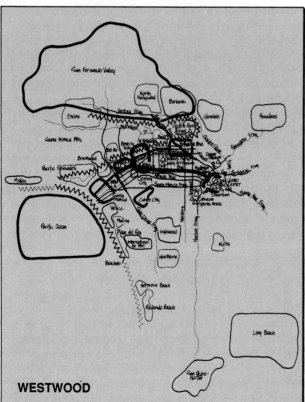

WESTWOOD

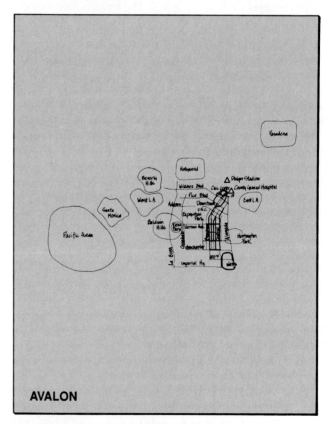

AVALON

Figure 1.26 **Four mental maps of Los Angeles.** The upper-middle-income residents of Northridge and Westwood have expansive views of the metropolis, reflecting their mobility and area of travel. Residents of Boyle Heights and Avalon, both minority districts, have a much more restricted and incomplete mental image of the city. Their limited mental maps reflect and reinforce their spatial isolation within the metropolitan area.

From Department of City Planning, City of Los Angeles, The Visual Environment of Los Angeles, *1971. Reprinted by permission.*

majority of cases, the lack of agreement between what is observed and what is expected in a particular case leads to an examination of the circumstances contributing to the disparity. The quality of connecting roads, political barriers, or other variables may affect the specific places examined, and these causative elements may be isolated for further study.

Indeed, the steady pursuit of more refined and definitive analysis of human geographic questions—the "further study" that continues to add to our understanding of how people occupy and utilize the earth, interact with each other, and organize and alter earth space—has led to the remarkably diversified yet coherent field of modern human geography. With the content of this introductory chapter as background to the nature, traditions, and tools of geography, we are ready to begin its exploration.

The Structure of This Book

By way of getting started, it is useful for you to know how the organization and topics of this text have been structured to help you reach the kinds of understandings we seek.

We begin by exploring the roots and meaning of culture (Chapter 2), establishing the observed ground rules of spatial interaction and spatial behavior (Chapter 3), and examining the areal variations in patterns of population distribution and change (Chapter 4). These set the stage for following separate discussions of spatial patterns of language and religion, ethnic distinctions, and folk and popular culture (Chapters 5–7). These are the principal expressions of unity and diversity and of areal differentiation among the peoples and societies of the earth. Understanding their spatial patterns and interrelations goes far toward providing the world view that is our objective.

Beginning with Chapter 8, our focus shifts more to the economic and organizational landscapes humans have created. In turn, we look at economic geography and economic development (Chapters 8–10), urban systems and structures (Chapter 11), and patterns of the political ordering of space (Chapter 12). Finally, in Chapter 13, dealing with human impacts, we return to the underlying concern of all geographic study: the relationship between human geographic patterns and processes and both the present conditions and the future prospects of the physical and cultural environments we occupy, create, or modify. To help clarify the connections between the various topics of human geography, the chapters of this book are grouped by common theme and separately introduced.

Summary

Geography is about earth space and its physical and cultural content. Throughout its long history, geography has remained consistent in its focus on human–environmental interactions, the interrelatedness of places, and the likenesses and differences in physical and cultural content of area that exist from place to place. The collective interests of geographers are summarized by the spatial and systems analytical questions they ask. The responses to those questions are interpreted through basic concepts of location, distance, direction, content evolution, spatial interaction, and regional organization.

Geographers employ maps and models to abstract the complex reality of space and to isolate its components for separate study. Maps are imperfect renderings of the three-dimensional earth and its parts on a two-dimensional surface. In that rendering, some or all of the characteristics of the globe grid are distorted, but convenience and data manageability are gained. Spatial information may be depicted visually in a number of ways, each designed to simplify and to clarify the infinite complexity of spatial content. Geographers also use verbal and mathematical models for the same purpose, to abstract and analyze.

In their study of the earth's surface as the occupied and altered space within which humans operate, geographers may concentrate on the integration of physical and cultural phenomena in a specific earth area (regional geography). They may, instead, emphasize systematic geography through study of the earth's physical systems of spatial and human concern or, as here, devote primary attention to people. This is a text in *human geography*. Its focus is on human interactions both with the physical environments people occupy and alter and with the cultural environments they have created. We are concerned with the ways people perceive the landscapes and regions they occupy, act within and between them, make choices about them, and organize them according to the varying cultural, political, and economic interests of human societies. This is a text clearly within the social sciences, but like all geography, its background is the physical earth as the home of humans. As a human geography, its concern is with how that home has been altered by societies and cultures. Culture is the starting point, and in the next chapter we begin with an inquiry about the roots and nature of culture.

 *Key Words* _____

 For Review _____

1. In what two meanings and for what different purposes do we refer to *location?*

2. Describe the *site* and the *situation* of the town where you live, work, or go to school.

3. What kinds of distance transformations are suggested by the term *relative distance?* How is the concept of *psychological distance* related to relative distance?

4. What are the common elements of *spatial distribution?* What different aspects of the spatial arrangement of things do they address?

5. What are the common characteristics of *regions?* How are *formal* and *functional* regions different in concept and definition? What is a *perceptual region?*

6. List at least four properties of the globe grid. Why are globe grid properties apt to be distorted on maps?

7. What does *prime meridian* mean? What happens to the length of a degree of longitude as one approaches the poles?

8. What different ways of displaying statistical data on maps can you name and describe?

 Focus Follow-up _____

1. **What is the nature of geography and the role of human geography?** pp. 4–7.

   Geography is a *spatial* science concerned with how the content of earth areas differs from place to place. It is the study of spatial variation in the world's physical and cultural (human) features. The emphasis of human geography is on the spatial variations in characteristics of peoples and cultures, on the way humans interact over space, and the ways they utilize and alter the natural landscapes they occupy.

2. **What are the fundamental geographic observations and their underlying concepts?** pp. 7–17.

   Basic geographic observations all concern the characteristics, content, and interactions of places. Their underlying concepts involve such place specifics as location, direction, distance, size, scale, physical and cultural attributes, interrelationships, and regional similarities and differences.

3. **What are the regional concept and the generalized characteristics of regions?** pp. 17–20.

   The regional concept tells us that physical and cultural features of the earth's surface are rationally arranged by understandable processes. All recognized regions are characterized by location, spatial extent, defined boundaries, and position within a hierarchy of regions. Regions may be "formal" (uniform) or "functional" (nodal) in nature.

4. **Why do geographers use maps, and how do maps show spatial information?** pp. 20–26.

   Maps are tools geographers use to identify and delimit regions and to analyze their content. They permit the study of areas and areal features too extensive to be completely viewed or understood on the earth's surface itself. Thematic (single category) maps may be either qualitative or quantitative. Their data may be shown in graduated circle, dot distribution, isometric, choropleth, statistical, or cartogram form.

5. **In what ways in addition to maps may spatial data be visualized or analyzed?** pp. 26–29.

Informally, we all create "mental maps" reflecting highly personalized impressions and information about the spatial arrangement of things (for example, buildings, streets, landscape features). More formally, geographers recognize the content of area as forming a spatial system to which techniques of spatial systems analysis and model building are applicable.

 Selected References

Agnew, John, David N. Livingstone, and Alisdair Rogers, eds. *Human Geography: An Essential Anthology.* Cambridge, Mass.: Blackwell, 1996.

Demko, George J., with Jerome Agel and Eugene Boe. *Why in the World: Adventures in Geography.* New York: Anchor Books/Doubleday, 1992.

Dent, Borden D. *Cartography: Thematic Map Design.* 5th ed. Dubuque, Iowa: WCB/McGraw-Hill, 1999.

Gersmehl, Phil. *The Language of Maps.* 15th ed. Indiana, Pa.: National Council for Geographic Education, 1996.

Gould, Peter, and Rodney White. *Mental Maps.* 2d. ed. New York: Routledge, 1986.

Gritzner, Charles F., Jr. "The Scope of Cultural Geography." *Journal of Geography* 65 (1966): 4–11.

Holt-Jensen, Arild. *Geography: Its History and Concepts.* 3d ed. Thousand Oaks, Calif.: Sage Publications, 1999.

Johnston, Ronald J., Derek Gregory, Geraldine Pratt, and Michael Watts. *The Dictionary of Human Geography.*

4th ed. Oxford, England: Blackwell Publishers, 2000.

Lanegran, David A., and Risa Palm. *An Invitation to Geography.* 2d ed. New York: McGraw-Hill, 1978.

Ley, David. "Cultural/Humanistic Geography." *Progress in Human Geography* 5 (1981): 249–257; 7 (1983): 267–275.

Livingstone, David N. *The Geographical Tradition.* Cambridge, Mass.: Blackwell, 1992.

Lobeck, Armin K. *Things Maps Don't Tell Us: An Adventure into Map Interpretation.* Chicago: University of Chicago Press, 1993.

Martin, Geoffrey J., and Preston E. James. *All Possible Worlds: A History of Geographical Ideas.* 3d ed. New York: Wiley, 1993.

Monmonier, Mark. *How to Lie with Maps.* 2d ed. Chicago: University of Chicago Press, 1996.

Morrill, Richard L. "The Nature, Unity and Value of Geography." *Professional Geographer* 35 (1983): 1–9.

Muehrcke, Phillip C., and Juliana O. Muehrcke. *Map Use: Reading, Analysis, and Interpretation.* 4th ed. Madison, Wis.: J.P. Publications, 1998.

Pattison, William D. "The Four Traditions of Geography." *Journal of Geography* 63 (1964): 211–216.

Rogers, Alisdair, and Heather Viles. *The Student's Companion to Geography.* 2d ed. Malden, Mass.: Blackwell, 2003.

White, Gilbert F. "Geographers in a Perilously Changing World." *Annals of the Association of American Geographers* 75 (1985): 10–15.

Wood, Tim F. "Thinking in Geography." *Geography* 72 (1987): 289–299.

Websites: The World Wide Web has a tremendous number and variety of sites pertaining to geography. Websites relevant to the subject matter of this chapter appear in the "Web Links" section of the On-line Learning Center associated with this book. Access it at **www.mhhe.com/fellmann8e**

PART One

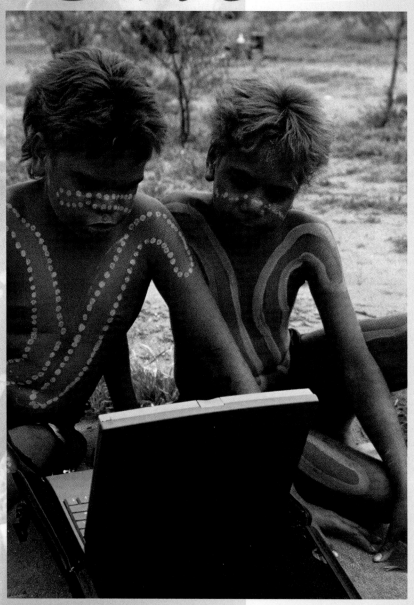

Cultural convergence is increasingly a worldwide reality as these Native Australian boys with their laptop computer demonstrate.

Themes and Fundamentals
of Human Geography

Human geography studies the ways in which people and societies are regionally different in their distinguishing characteristics. Additionally, it examines the ways that different societies perceive, use, and alter the landscapes they occupy. These interests would seem to imply an unmanageable range and variety of topics. The implication is misleading, however, for the diversified subject matter of human geography can be accommodated within two general themes connected by one continuing and unifying thread of concern.

One theme considers the traits of *culture* that characterize different social groups and comprise the individual pieces of the human geographic mosaic. These are matters of learned behaviors, attitudes, and group beliefs that are fundamental and identifying features of specific social groups and larger societies. They are cultural identifiers that are transmitted within the group by tradition, example, and instruction. A second theme has to do with the *systems* of production, livelihood, spatial organization, and administration—and the institutions appropriate to those systems—that a society erects in response to opportunity, technology, resources, conflict, or the need to adapt and change. This second theme recognizes what French geographers early in the 20th century called *genre de vie*—the way of life—of a population that might be adopted or pursued no matter what the other intangible cultural traits of that social group might be. Interwoven with and unifying these primary themes is the continuing background concern of geographers: humans and environments in interaction.

We shall pursue each of these themes in separate sections of this book and address the unifying interest in human impact on the earth surface both as an integral part of each chapter and as the topic of our concluding chapter. Throughout, we shall keep returning to a small number of basic observations that underlie all of human geographic study: (1) People and the societies they form are differentiated by a limited set of identifying cultural characteristics and organizational structures; (2) without regard to those cultural and organizational differences, human spatial behavior has common and recurring motivations and patterns; (3) cultural variations and spatial actions are rooted in the distribution, numbers, and movements of people.

These observations are the concerns of the following three chapters, which together make up this first part of our study of human geography. We begin by setting the stage in Chapter 2 with a review of the meaning, components, and structure of *culture* and of the processes of cultural change, diffusion, and divergence. Those processes underlie an observable world mosaic of culture regions and realms. Despite regional distinctiveness, however, common characteristics of *spatial behavior* affect and unify all peoples and social systems. These are the topic of Chapter 3 and might be called the "ground rules" of spatial interaction. They are physical and behavioral constants whose recognition is a necessary first step in understanding the world, regional, and local patterns of people and social systems that are so central to human geography.

To conclude the first section of our study, Chapter 4 focuses on *population:* people in their numbers, movements, distributions, and growth trends. Part of our understanding of those matters in both their world and regional expressions is based on the examination of cultural origins and diffusions and the principles of spatial interaction conducted in Chapters 2 and 3.

The first phase of our exploration of human geography, then, expresses a unifying concern with the cultural processes and spatial interactions of an unevenly distributed and expanding world population.

C H A P T E R

Two

Roots and Meaning of Culture

Focus Preview

Opposite: The once universal hunting-gathering economy and lifestyle is still practiced by the San people of Namibia.

They buried him there in the cave where they were working, less than 6 kilometers (4 miles) from the edge of the ice sheet. Outside stretched the tundra, summer feeding grounds for the mammoths whose ivory they had come so far to collect. Inside, near where they dug his grave, were stacked the tusks they had gathered and were cutting and shaping. They prepared the body carefully and dusted it with red ochre, then buried it in an elaborate grave with tundra flowers and offerings of food, a bracelet on its arm, a pendant about its throat, and 40 to 50 polished rods of ivory by its side. It rested there, in modern Wales, undisturbed for some 18,000 years until discovered early in the 19th century. The 25-year-old hunter had died far from the group's home some 650 kilometers (400 miles) away near present-day Paris, France. He had been part of a routine annual summer expedition overland from the forested south across the as-yet-unflooded English Channel to the mammoths' grazing grounds at the edge of the glacier.

As always, they were well prepared for the trip. Their boots were carefully made. Their sewn skin leggings and tunics served well for travel and work; heavier fur parkas warded off the evening chill. They carried emergency food, fire-making equipment, and braided cord that they could fashion into nets, fishing lines, ropes, or thread. They traveled by reference to sun and stars, recognizing landmarks from past journeys and occasionally consulting a crude map etched on bone.

Although the hunters returned bearing the sad news of their companion's death, they also brought the ivory to be carved and traded among the scattered peoples of Europe from the Atlantic Ocean to the Ural Mountains.

As shown by their tools and equipment, their behaviors and beliefs, these Stone Age travelers displayed highly developed and distinctive characteristics, primitive only from the vantage point of our own different technologies and customs. They represented the culmination of a long history of development of skills, of invention of tools, and of creation of lifestyles that set them apart from peoples elsewhere in Europe, Asia, and Africa who possessed still different cultural heritages.

To writers in newspapers and the popular press, "culture" means the arts (literature, painting, music, and the like). To a social scientist, **culture** is the specialized behavioral patterns, understandings, adaptations, and social systems that summarize a group of people's learned way of life. In this broader sense, culture is an ever-present part of the regional differences that are the essence of human geography. The visible and invisible evidences of culture—buildings and farming patterns, language, political organization, and ways of earning a living, for example—are all parts of the spatial diversity human geographers study. Cultural differences over time may present contrasts as great as those between the Stone Age ivory hunters and modern urban Americans. Cultural differences in area result in human landscapes with variations as subtle as the differing "feel" of urban Paris, Moscow, or New York or as obvious as the sharp contrasts of rural Zimbabwe and the U.S. Midwest (Figure 2.1).

Since such tangible and intangible cultural differences exist and have existed in various forms for thousands of years, human geography addresses the question, Why? Why, since humankind constitutes a single species, are cultures so varied? What and where were the origins of the different culture regions we now observe? How, from whatever limited areas individual culture traits developed, were they diffused over a wider portion of the globe? How did people who had roughly similar origins come to display significant areal differences in technology, social structure, ideology, and the innumerable other expressions of human geographic diversity? In what ways and why are there distinctive cultural variations even in presumed "melting pot" societies such as the United States and Canada or in the historically homogeneous, long-established countries of Europe? Part of the answer to these questions is to be found in the way separate human groups developed techniques to solve regionally varied problems of securing food, clothing, and shelter and, in the process, created areally distinctive customs and ways of life.

Components of Culture

Culture is transmitted within a society to succeeding generations by imitation, instruction, and example. In short, it is learned, not biological. It has nothing to do with instinct or with genes. As members of a social group, individuals acquire integrated sets of behavioral patterns, environmental and social perceptions, and knowledge of existing technologies. Of necessity, each of us learns the culture in which we are born and reared. But we need not—indeed, cannot—learn its totality. Age, sex, status, or occupation may dictate the aspects of the cultural whole in which an individual becomes indoctrinated.

A culture, that is, despite overall generalized and identifying characteristics and even an outward appearance of uniformity, displays a social structure—a framework of roles and interrelationships of individuals and established groups. Each individual learns and is expected to adhere to the rules and conventions not only of the culture as a whole but also of those specific to the subculture to which he or she belongs. And that subgroup may have its own recognized social structure (Figure 2.2). Think back to the different subgroups and aspects of your own national culture that you became part of (and left) as you progressed from

(a)

(b)

Figure 2.1 Cultural contrasts are clearly evident between (*a*) a subsistence maize plot in Zimbabwe and (*b*) the extensive fields and mechanized farming of the U.S. Midwest.

Figure 2.2 Both the traditional rice farmer of rural Japan and the harried commuter of Tokyo are part of a common Japanese culture. They occupy, however, vastly different positions in its social structure.

childhood through high school and on to college-age adulthood and, perhaps, to first employment.

Many different cultures, then, can coexist within a given area, each with its own influence on the thoughts and behaviors of their separate members. Within the United States, for example, we can readily recognize masculine and feminine; majority white and minority black, Hispanic, Asian American, or other ethnic groups; gay and straight, urban and rural; and many other subcultures. All, of course, are simultaneously part of a larger "American" culture marked by commonalities of traditions, behaviors, loyalties, and beliefs. Human geography increasingly recognizes the plurality of cultures within regions. In addition to examining the separate content and influence of those subcultures, it attempts to record and analyze the varieties of contested cultural in-

teractions between them, including those of political and economic nature.

Culture is a complexly interlocked web of behaviors and attitudes. Realistically, its full and diverse content cannot be appreciated, and in fact may be wholly misunderstood, if we concentrate our attention only on limited, obvious traits. Distinctive eating utensils, the use of gestures, or the ritual of religious ceremony may summarize and characterize a culture for the casual observer. These are, however, individually insignificant parts of a much more complex structure that can be appreciated only when the whole is experienced.

Out of the richness and intricacy of human life we seek to isolate for special study those more fundamental cultural variables that give structure and spatial order to societies. We begin with *culture traits,* the smallest distinctive items of culture. **Culture traits** are units of learned behavior ranging from the language spoken to the tools used or the games played. A trait may be an object (a fishhook, for example), a technique (weaving and knotting of a fishnet), a belief (in the spirits resident in water bodies), or an attitude (a conviction that fish is superior to other animal protein). Such traits are the most elementary expression of culture, the building blocks of the complex behavioral patterns of distinctive groups of peoples.

Individual cultural traits that are functionally interrelated comprise a **culture complex.** The existence of such complexes is universal. Keeping cattle was a *culture trait* of the Maasai of Kenya and Tanzania. Related traits included the measurement of personal wealth by the number of cattle owned, a diet containing milk and the blood of cattle, and disdain for labor unrelated to herding. The assemblage of these and other related traits yielded a culture complex descriptive of one aspect of Maasai society (Figure 2.3). In exactly analogous ways, religious complexes, business behavior complexes, sports complexes, and others can easily be recognized in any society. In the United States, for example, a culture complex exists around the automobile. Americans often buy car brands and models to reflect their income, employment, or status in society. Cinema, television, and sports often have autos at their heart; movies such as *American Graffiti* and the mass popularity of NASCAR races are familiar examples. Even rites of passage may focus on the automobile: driver education in high school, passing the driving exam and, perhaps, getting a car of one's own as a teenager, or the common practice of decorating automobiles at the end of a wedding ceremony.

Culture traits and complexes have areal extent. When they are plotted on maps, the regional character of the components of culture is revealed. Although human geographers are interested in the spatial distribution of these individual elements of culture, their usual concern is with the **culture region,** a portion of the earth's surface occupied by populations sharing recognizable and distinctive cultural characteristics. Examples include the political organizations societies devise, the religions they espouse, the form of economy they pursue, and even the type of clothing they wear, eating utensils they use, or kind of housing they occupy. There are as many such conceptual culture regions as there are culture traits and complexes recognized for population groups. Their recognition will be particularly important in discussions of ethnic, folk, and popular cultures in later chapters of this book. In those later reviews as within the present chapter, we must keep in mind that within any one recognized culture region, groups united by the specific mapped characteristics may be competing and distinctive in other important cultural traits.

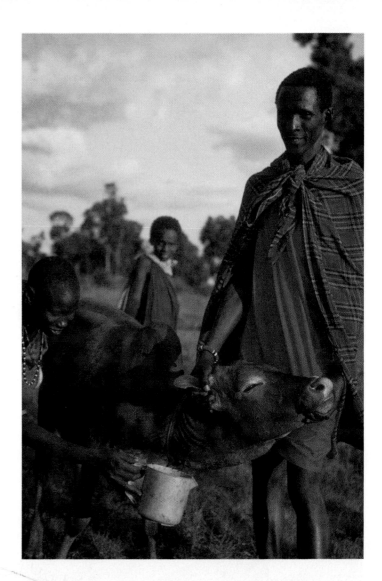

Figure 2.3 The formerly migratory Maasai of Kenya are now largely sedentary, partially urbanized, and frequently owners of fenced farms. Cattle formed the traditional basis of Maasai culture and were the evidence of wealth and social status. They provided, as well, the milk and blood important in the Maasai diet. Here, a herdsman catches blood released from a small neck incision he has just made.

Finally, a set of culture regions showing related culture complexes and landscapes may be grouped to form a **culture realm.** The term recognizes a large segment of the earth's surface having an assumed fundamental uniformity in its cultural characteristics and showing a significant difference in them from adjacent realms. Culture realms are, in a sense, culture regions at the broadest scale of recognition. In fact, the scale is so broad and the diversity within the recognized realms so great that the very concept of realm may mislead more than it informs.

Indeed, the current validity of distinctive culture realms has been questioned in light of an assumed globalization of all aspects of human society and economy. The result of that globalization, it has been suggested, is a homogenization of cultures as economies are integrated and uniform consumer demands are satisfied by standardized commodities produced by international corporations.

Certainly, the increasing mobility of people, goods, and information has reduced the rigidly compartmentalized ethnicities, languages, and religions of earlier periods. Cultural flows and exchanges have increased over the recent decades and with them has come a growing worldwide intermixture of peoples and customs. Despite that growing globalism in all facets of life and economy, however, the world is far from homogenized. Although an increased sameness of commodities and experiences is encountered in distant places, even common and standardized items of everyday life—branded soft drinks, for example, or American fast-food franchises—take on unique regional meanings and roles, conditioned by the total cultural mix they enter. Those multiple regional cultural mixes are often defiantly distinctive and separatist as recurring incidents of ethnic conflict, civil war, and strident regionalism attest.

If a global culture can be discerned, it may best be seen as a combination of multiple territorial cultures, rather than a completely standardized uniformity. It is those territorially different cultural mixtures that are recognized by the culture realms suggested on Figure 2.4, which itself is only one of many such possible divisions. The spatial pattern and characteristics of these generalized realms will help us place the discussions and examples of human geography of later chapters in their regional context.

Interaction of People and Environment

Culture develops in a physical environment that, in its way, contributes to differences among people. In premodern subsistence societies, the acquisition of food, shelter, and clothing, all parts of culture, depends on the utilization of the natural resources at hand. The interrelations of people to the environment of a given area, their perceptions and utilization of it, and their impact on it are interwoven themes of **cultural ecology**—the study of the relationship between a culture group and the natural environment it occupies.

Environments as Controls

Geographers have long dismissed as intellectually limiting and demonstrably invalid the ideas of **environmental determinism,** the belief that the physical environment exclusively shapes humans, their actions, and their thoughts. Environmental factors alone cannot account for the cultural variations that occur around the world. Levels of technology, systems of organization, and ideas about what is true and right have no obvious relationship to environmental circumstances.

The environment does place certain limitations on the human use of territory. Such limitations, however, must be seen not as absolute, enduring restrictions but as relative to technologies, cost

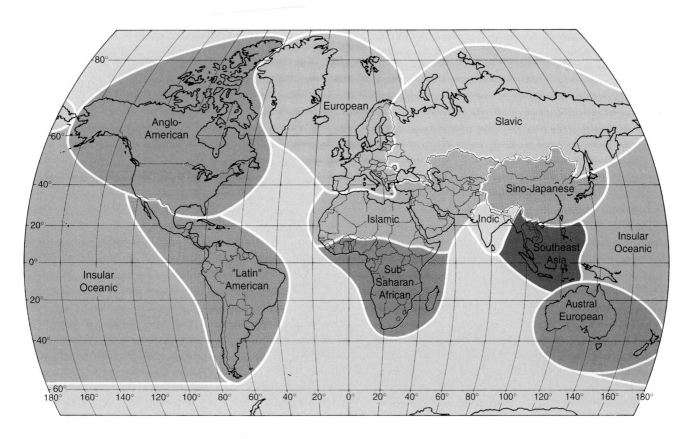

Figure 2.4 **Culture realms of the modern world.** This is just one of many possible subdivisions of the world into multifactor cultural regions.

considerations, national aspirations, and linkages with the larger world. Human choices in the use of landscapes are affected by group perception of the feasibility and desirability of their settlement and exploitation. These are not circumstances inherent in the land. Mines, factories, and cities have been created in the formerly nearly unpopulated tundra and forests of Siberia as a reflection of Russian developmental programs, not in response to recent environmental improvement.

Possibilism is the viewpoint that people, not environments, are the dynamic forces of cultural development. The needs, traditions, and level of technology of a culture affect how that culture assesses the possibilities of an area and shape what choices the culture makes regarding them. Each society uses natural resources in accordance with its circumstances. Changes in a group's technical abilities or objectives bring about changes in its perceptions of the usefulness of the land. Simply put, the impact of the environment appears inversely related to the level of development of a culture, while perception of environmental opportunities increases directly with growth in economic and cultural development.

Map evidence suggests the nature of some environmental limitations on use of area. The vast majority of the world's population is differentially concentrated on less than one-half of the earth's land surface, as Figure 4.22 suggests. Areas with relatively mild climates that offer a supply of fresh water, fertile soil, and abundant mineral resources are densely settled, reflecting in part the different potentials of the land under earlier technologies to support population. Even today, the polar regions, high and rugged mountains, deserts, and some hot and humid lowland areas contain very few people. If resources for feeding, clothing, or housing ourselves within an area are lacking or if we do not recognize them there, there is no inducement for people to occupy the territory.

Environments that do contain such resources provide the framework within which a culture operates. Coal, oil, and natural gas have been in their present locations throughout human history, but they were rarely of use to preindustrial cultures and did not impart any recognized advantage to their sites of occurrence. Not until the Industrial Revolution did coal deposits gain importance and come to influence the location of such great industrial complexes as the Midlands in England, the Ruhr in Germany, and the steel-making districts formerly so important in parts of northeastern United States. Native Americans made one use of the environment around Pittsburgh, while 19th-century industrialists made quite another.

Human Impacts

People are also able to modify their environment, and this is the other half of the human–environment relationship of geographic concern. Geography, including cultural geography, examines both the reactions of people to the physical environment and their impact on that environment. By using it, we modify our environment—in part, through the material objects we place on the landscape: cities, farms, roads, and so on (Figure 2.5). The form these take is the product of the kind of culture group in which we live. The **cultural landscape,** the earth's surface as modified by human

action, is the tangible physical record of a given culture. House types, transportation networks, parks and cemeteries, and the size and distribution of settlements are among the indicators of the use that humans have made of the land.

Human actions, both deliberate and inadvertent, modifying or even destroying the environment are perhaps as old as humankind itself. People have used, altered, and replaced the vegetation in wide areas of the tropics and midlatitudes. They have hunted to extinction vast herds and whole species of animals. They have, through overuse and abuse of the earth and its resources, rendered sterile and unpopulated formerly productive and attractive regions.

Fire has been called the first great tool of humans, and the impact of its early and continuing use is found on nearly every continent. Poleward of the great rain forests of equatorial South America, Africa, and South Asia lies the *tropical savanna* of extensive grassy vegetation separating scattered trees and forest groves (Figure 2.6). The trees appear to be the remnants of naturally occurring tropical dry forests, thorn forests, and scrub now largely obliterated by the use, over many millennia, of fire to remove the unwanted and unproductive trees and to clear off old grasses for more nutritious new growth. The grasses supported the immense herds of grazing animals that were the basis of hunting societies. After independence, the government of Kenya in East Africa sought to protect its national game preserves by prohibiting the periodic use of fire. It quickly found that the immense herds of gazelles, zebras, antelope, and other grazers (and the lions and other predators that fed on them) that tourists came to see were being replaced by less-appealing browsing species—rhinos, hippos, and elephants. With fire prohibited, the forests began to reclaim their natural habitat and the grassland fauna was replaced.

The same form of vegetation replacement occurred in midlatitudes. The grasslands of North America were greatly extended by Native Americans who burned the forest margin to extend grazing areas and to drive animals in the hunt. The control of fire in modern times has resulted in the advance of the forest once again in formerly grassy areas ("parks") of Colorado, northern Arizona, and other parts of the United States West.

Examples of adverse human impact abound. The *Pleistocene overkill*—the Stone Age loss of whole species of large animals on all inhabited continents—is often ascribed to the unrestricted hunting to extinction carried on by societies familiar with fire to drive animals and hafted (with handles) weapons to slaughter them. With the use of these, according to one estimate, about 40% of African large-animal genera passed to extinction. The majority of large animal, reptile, and flightless bird species had disappeared from Australia around 46,000 years ago; in North America, some two-thirds of original large mammals had succumbed by 11,000 years ago under pressure from the hunters migrating to and spreading across the continent. Although some have suggested that climatic changes or pathogens carried by dogs, rats, and other camp followers were at least partially responsible, human action is the more generally accepted explanation for the abrupt faunal changes. No uncertainty exists in the record of faunal destruction by the Maoris of New Zealand or of Polynesians who had exterminated some 80% to 90% of South

Figure 2.5 The physical and cultural landscapes in juxtaposition. Advanced societies are capable of so altering the circumstances of nature that the cultural landscapes they create become the controlling environment. The city of Cape Town, South Africa, is a "built environment" largely unrelated to its physical surroundings.

Pacific bird species—as many as 2000 in all—by the time Captain Cook arrived in the 18th century. Similar destruction of key marine species—Caribbean sea turtles, sea cows off the coast of Australia, sea otters near Alaska, and others elsewhere—as early as 10,000 years ago resulted in environmental damage whose effects continue to the present.

Not only destruction of animals but of the life-supporting environment itself has been a frequent consequence of human misuse of area (see "Chaco Canyon Desolation"). North Africa, the "granary of Rome" during the empire, became wasted and sterile in part because of mismanagement. Roman roads standing high above the surrounding barren wastes give testimony to the erosive power of wind and water when natural vegetation is unwisely removed and farming techniques are inappropriate. Easter Island in the South Pacific was covered lushly with palms and

Figure 2.6 The parklike landscape of grasses and trees characteristic of the tropical savanna is seen in this view from Kenya, Africa.

It is not certain when they first came, but by A.D. 1000, the Anasazi people were building a flourishing civilization in present-day Arizona and New Mexico. In the Chaco Canyon alone, they erected as many as 75 towns, all centered around pueblos, huge stone-and-adobe apartment buildings as tall as five stories and with as many as 800 rooms. These were the largest and tallest buildings of North America prior to the construction of iron-framed "cloudscrapers" in major cities at the end of the 19th century. An elaborate network of roads and irrigation canals connected and supported the pueblos. About A.D.1200, the settlements were abruptly abandoned. The Anasazi, advanced in their skills of agriculture and communal dwelling, were—according to some scholars—forced to move on by the ecological disaster their pressures had brought to a fragile environment.

They needed forests for fuel and for the hundreds of thousands of logs used as beams and bulwarks in their dwellings. The pinyon-juniper woodland of the canyon was quickly depleted. For larger timbers needed for construction, the Anasazi first harvested stands of ponderosa pine found some 40 kilometers (25 miles) away. As early as A.D. 1030 these, too, were exhausted, and the community switched to spruce and Douglas fir from mountaintops surrounding the canyon. When they were gone by 1200, the Anasazi fate was sealed—not only by the loss of forest

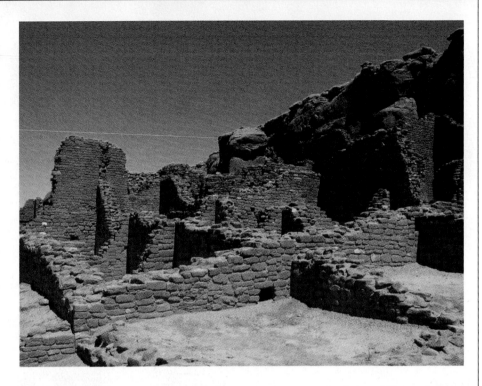

but by the irreversible ecological changes deforestation and agriculture had occasioned. With forest loss came erosion that destroyed the topsoil. The surface water channels that had been built for irrigation were deepened by accelerated erosion, converting them into enlarging arroyos useless for agriculture.

The material roots of their culture destroyed, the Anasazi turned upon themselves; warfare convulsed the region and, compelling evidence suggests, cannibalism was practiced. Smaller groups sought refuge elsewhere, re-creating on reduced scale their pueblo way of life but now in nearly inaccessible, highly defensible mesa and cliff locations. The destruction they had wrought destroyed the Anasazi in turn.

---

other trees when Polynesians settled there about A.D. 400. By the beginning of the 18th century, Easter Island had become the barren wasteland it remains today. Deforestation increased soil erosion, removed the supply of timbers needed for the vital dugout fishing canoes, and made it impossible to move the massive stone statues that were significant in the islanders' religion (Figure 2.7). With the loss of livelihood resources and the collapse of religion, warfare broke out and the population was decimated. A similar tragic sequence is occurring on Madagascar in the Indian Ocean today. Despite current romantic notions, not all early societies lived in harmony with their environment.

The more technologically advanced and complex the culture, the more apparent is its impact on the natural landscape. In sprawling urban-industrial societies, the cultural landscape has come to outweigh the natural physical environment in its impact on people's daily lives. It interposes itself between "nature" and humans, and residents of the cities of such societies—living and

working in climate-controlled buildings, driving to enclosed shopping malls—can go through life with very little contact with or concern about the physical environment.

Roots of Culture

Earlier humans found the physical environment more immediate and controlling than we do today. Some 11,000 years ago, the massive glaciers—moving ice sheets of great depth—that had covered much of the land and water of the Northern Hemisphere (Figure 2.8) began to retreat. Animal, plant, and human populations that had been spatially confined by both the ice margin and the harsh climates of middle-latitude regions began to spread, colonizing newly opened territories. The name *Paleolithic* (Old Stone Age) is used to describe the period near the end of glaciation during which small and scattered groups like the ivory

Figure 2.7 Now treeless, Easter Island once was lushly forested. The statues (some weighing up to 85 tons) dotting the island were rolled to their locations and lifted into place with logs.

hunters at this chapter's start began to develop regional variations in their ways of life and livelihood.

All were **hunter-gatherers,** preagricultural people dependent on the year-round availability of plant and animal foodstuffs they could secure with the limited variety of rudimentary stone tools and weapons at their disposal. Even during the height of the Ice Age, the unglaciated sections of western, central, and northeastern Europe (the continent with the best-documented evidence of Paleolithic culture) were covered with tundra vegetation, the mosses, lichens, and low shrubs typical of areas too cold to support forests. Southeastern Europe and southern Russia had forest, tundra, and grasslands, and the Mediterranean areas had forest cover (Figure 2.9). Gigantic herds of herbivores—reindeer, bison, mammoth, and horses—browsed, bred, and migrated throughout the tundra and the grasslands. An abundant animal life filled the forests.

Human migration northward into present-day Sweden, Finland, and Russia demanded a much more elaborate set of tools and provision for shelter and clothing than had previously been required. It necessitated the crossing of a number of ecological barriers and the occupation of previously avoided difficult environments. By the end of the Paleolithic period, humans had spread to all the continents but Antarctica, carrying with them their common hunting-gathering culture and social organization. The settlement of the lands bordering the Pacific Ocean is suggested in Figure 2.10.

While spreading, the total population also increased. But hunting and foraging bands require considerable territory to support a relatively small number of individuals. There were contacts between groups and, apparently, even planned gatherings for trade, socializing, and selecting mates from outside the home group. Nevertheless, the bands tended to live in isolation. Estimates place the Paleolithic population of the entire island of Great Britain, which was on the northern margin of habitation, at only some 400–500 persons living in widely separated families of 20–40 people. Total world population at about 9000 B.C. proba-

bly ranged from 5 to 10 million. Variations in the types of tools characteristic of different population groups steadily increased as people migrated and encountered new environmental problems.

Improved tool technology greatly extended the range of possibilities in the use of locally available materials. The result was more efficient and extensive exploitation of the physical environment than earlier had been possible. At the same time, regional contrasts in plant and animal life and in environmental conditions accelerated the differentiation of culture between isolated groups who under earlier, less varied conditions had shared common characteristics.

Within many environments, even harsh ones, the hunting and foraging process was not particularly demanding of either time or energy. Recent studies of South African San people (Bushmen), for example, indicate that such bands survive well on the equivalent of a 2½-day workweek. Time was available for developing skills in working flint and bone for tools, in developing regionally distinctive art and sculpture, and in making decorative beads and shells for personal adornment and trade. By the end of the Ice Age (about 11,000 to 12,000 years ago), language, religion, long-distance trade, permanent settlements, and social stratification within groups appear to have been well developed in many European culture areas.

What was learned and created was transmitted within the cultural group. The increasing variety of adaptive strategies and technologies and the diversity of noneconomic creations in art, religion, language, and custom meant an inevitable cultural variation of humankind. That diversification began to replace the rough cultural uniformity among hunting and gathering people

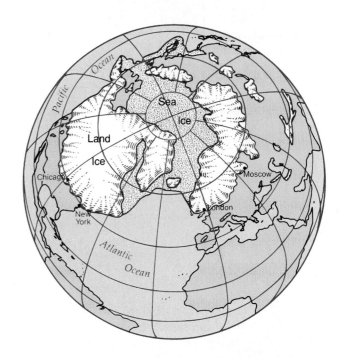

Figure 2.8 **Maximum extent of glaciation.** In their fullest development, glaciers of the most recent Ice Age covered large parts of Eurasia and North America. Even areas not covered by ice were affected as ocean levels dropped and rose and climate and vegetation regions changed with glacial advance and retreat.

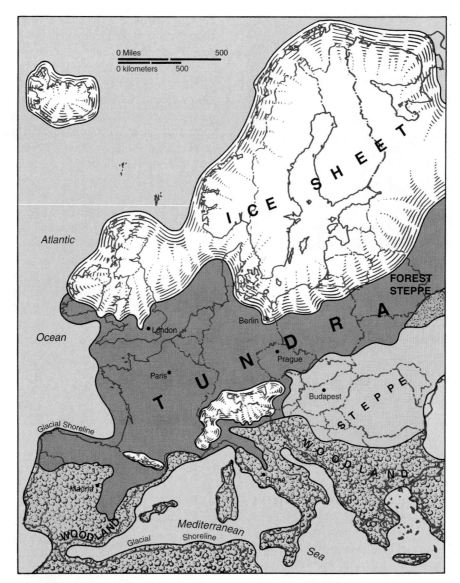

Figure 2.9 **Late Paleolithic environments of Europe.** During the late Paleolithic period, new food-gathering, shelter, and clothing strategies were developed to cope with harsh and changing environments, so different from those in Europe today.

that had been based on their similar livelihood patterns, informal leadership structures, small-band kinship groups, and the like (Figure 2.11).

Seeds of Change

The retreat of the last glaciers marked the end of the Paleolithic era and the beginning of successive periods of cultural evolution leading from basic hunting and gathering economies at the outset through the development of agriculture and animal husbandry to, ultimately, the urbanization and industrialization of modern societies and economies. Since not all cultures passed through all stages at the same time, or even at all, **cultural divergence** between human groups became evident.

Glacial recession brought new ecological conditions to which people had to adapt. The weather became warmer and forests began to appear on the open plains and tundras of Europe and northern China. In the Middle East, where much plant and animal domestication would later occur, savanna (grassland) vegetation replaced more arid landscapes. Populations grew and through hunting depleted the large herds of grazing animals already retiring northward with the retreating glacial front.

Further population growth demanded new food bases and production techniques, for the **carrying capacity**—the number of persons supportable within a given area by the technologies at their disposal—of the earth for hunter-gatherers is low. The *Mesolithic* (Middle Stone Age) period, from about 11,000 to 5000 B.C. in Europe, marked the transition from the collection of food to its production. These stages of the Stone Age—occurring

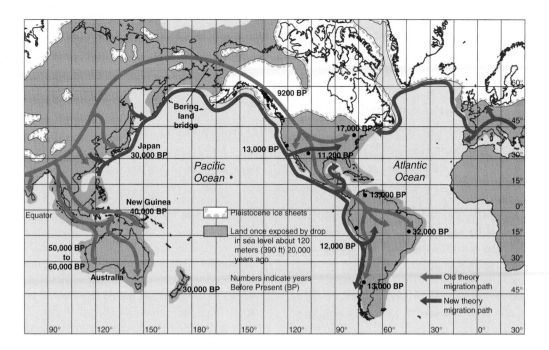

Figure 2.10 **Settlement of the Americas and the Pacific basin.** Genetic studies suggest humans spread around the globe from their Old World origins beginning some 100,000 years ago. Their time of arrival in the Western Hemisphere, however, is uncertain. The older view claimed earliest migrants to the Americas, the ancestors of modern Amerindian groups, crossed the Bering land bridge in three different waves beginning 11,500 years ago. Recent evidence suggests those North Asian land migrants encountered (and conquered or absorbed) earlier occupants who had arrived from Europe, Polynesia, and coastal East Asia by boat traveling along frozen or open shorelines. Although genetic and linguistic research yields mixed conclusions, physical evidence considered solid by some investigators indicates first Asian arrivals came at least 22,000 and more likely 30,000 or more years ago. Eastern United States artifacts that have been assigned dates of 17,000 to 30,000 years ago hint at European arrivals as early as those of coastal Asians. Other researchers, however, caution that any New World population dates earlier than 11,500 to 12,000 years ago are questionable and first migrants within those dates probably were most closely related to prehistoric Jomon and later Ainu groups of Japan who crossed over the Bering land bridge.

Figure 2.11 Hunter-gatherers practiced the most enduring lifestyle in human history, trading it for the more arduous life of farmers under the necessity to provide larger quantities of less diversified foodstuffs for a growing population. For hunter-gatherers (unlike their settled farmer rivals and successors), age and sex differences, not caste or economic status, were and are the primary basis for the division of labor and of interpersonal relations. Here a San (Bushman) hunter of Botswana, Africa, stalks his prey. Men also help collect the gathered food that constitutes 80% of the San diet.

during different time spans in different world areas—mark distinctive changes in tools, tasks, and social complexities of the cultures that experience the transition from "Old" to "Middle" to "New."

Agricultural Origins and Spread

The population of hunter-gatherers rose slowly at the end of the glacial period. As rapid climatic fluctuation adversely affected their established plant and animal food sources, people independently in more than one world area experimented with the *domestication* of plants and animals. There is no agreement on whether the domestication of animals preceded or followed that of plants. The sequence may well have been different in different areas. What appears certain is that animal domestication—the successful breeding of species that are dependent on human beings—began during the Mesolithic, not as a conscious economic effort by humans but as outgrowths of the keeping of small or young wild animals as pets and the attraction of scavenger animals to the refuse of human settlements. The assignment of religious significance to certain animals and the docility of others to herding by hunters all strengthened the human-animal connections that ultimately led to full domestication.

Radiocarbon dates suggest the domestication of pigs in southeastern Turkey and of goats in the Near East as early as 8000 to 8400 B.C., of sheep in Turkey by about 7500 B.C., and of cattle and pigs in both Greece and the Near East about 7000 B.C. North Africa, India, and southeastern Asia were other Old World domestication sources, as were—less successfully—Meso-America and the Andean Uplands. Although there is evidence that the concept of animal domestication diffused from limited source regions, once its advantages were learned numerous additional domestications were accomplished elsewhere. The widespread natural occurrence of species able to be domesticated made that certain. Cattle of different varieties, for example, were domesticated in India, north-central Eurasia, Southeast Asia, and Africa. Pigs and various domestic fowl are other examples.

The domestication of plants, like that of animals, appears to have occurred independently in more than one world region over a time span of between 10,000 and perhaps as long as 20,000 years ago. A strong case can be made that most widespread Eurasian food crops were first cultivated in the Near East beginning some 10,000 years ago and dispersed rapidly from there across the midlatitudes of the Old World. However, clear evidence also exists that African peoples were raising crops of wheat, barley, dates, lentils, and chickpeas on the floodplains of the Nile River as early as 18,500 years ago. In other world regions, farming began more recently; the first true farmers in the Americas appeared in Mexico no more than 5000 years ago.

Familiarity with plants of desirable characteristics is universal among hunter-gatherers. In those societies, females were assigned the primary food-gathering role and thus developed the greatest familiarity with nutritive plants. Their fundamental role in initiating crop production to replace less reliable food gathering seems certain. Indeed, women's major contributions as innovators of technology—in food preparation and clothing production, for example—or as inventors of such useful and important items as baskets and other containers, baby slings, yokes for carrying burdens, and the like are unquestioned.

Agriculture itself, however, seems most likely to have been not an "invention" but the logical extension to food species of plant selection and nurturing habits developed for nonfood varieties. Plant poisons applied to hunting arrows or spread on lakes or streams to stun fish made food gathering easier and more certain. Plant dyes and pigments were universally collected or prepared for personal adornment or article decoration. Medicinal and mood-altering plants and derivatives were known, gathered, protected, and cultivated by all early cultures. Indeed, persuasive evidence exists to suggest that early gathering and cultivation of grains was not for grinding and baking as bread but for brewing as beer, a beverage that became so important in some cultures for religious and nutritional reasons that it may well have been a first and continuing reason for sedentary agricultural activities.

Nevertheless, full-scale domestication of food plants, like that of animals, can be traced to a limited number of origin areas from which its techniques spread (Figure 2.12). Although there were several source regions, certain uniformities united them. In each, domestication focused on plant species selected apparently for their capability of providing large quantities of storable calories or protein. In each, there was a population already well fed and able to devote time to the selection, propagation, and improvement of plants available from a diversified vegetation. Some speculate, however, that grain domestication in the Near East may have been a forced inventive response, starting some 12,000 years ago, to food shortages reflecting abrupt increases in summertime temperatures and aridity in the Jordan Valley. That environmental stress—reducing summer food supplies and destroying habitats of wild game—favored selection and cultivation of short-season annual grains and legumes whose seeds could be stored and planted during cooler, wetter winter growing seasons.

In the tropics and humid subtropics, selected plants were apt to be those that reproduced vegetatively—from roots, tubers, or cuttings. Outside of those regions, wild plants reproducing from seeds were more common and the objects of domestication. Although there was some duplication, each of the origin areas developed crop complexes characteristic of itself alone, as Figure 2.12 summarizes. From each, there was dispersion of crop plants to other areas, slowly at first under primitive systems of population movement and communication (Figure 2.13), more rapidly and extensively with the onset of European exploration and colonization after A.D. 1500.

While adapting wild plant stock to agricultural purposes, the human cultivators, too, adapted. They assumed sedentary residence to protect the planted areas from animal, insect, and human predators. They developed labor specializations and created more formalized and expansive religious structures in which fertility and harvest rites became important elements. The regional contrasts between hunter-gatherer and sedentary agricultural societies increased. Where the two groups came in contact, farmers were the victors and hunter-gatherers the losers in competition for territorial control.

The contest continued into modern times. During the past 200 years, European expansion totally dominated the hunting and gathering cultures it encountered in large parts of the world such

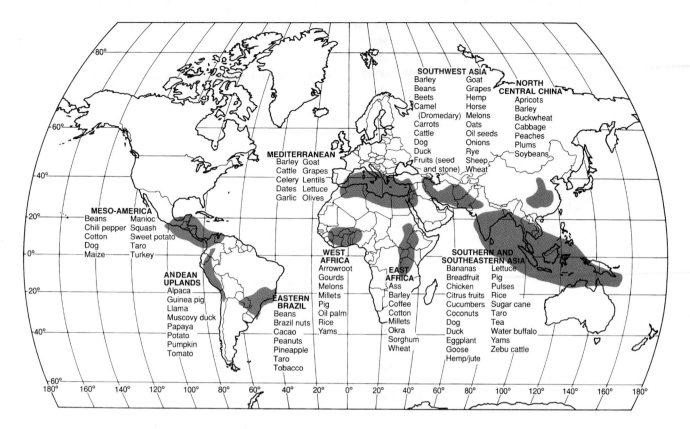

Figure 2.12 **Chief centers of plant and animal domestication.** The southern and southeastern Asian center was characterized by the domestication of plants such as taro, which are propagated by the division and replanting of existing plants (vegetative reproduction). Reproduction by the planting of seeds (e.g., maize and wheat) was more characteristic of Meso-America and the Near East. The African and Andean areas developed crops reproduced by both methods. The lists of crops and livestock associated with the separate origin areas are selective, not exhaustive.

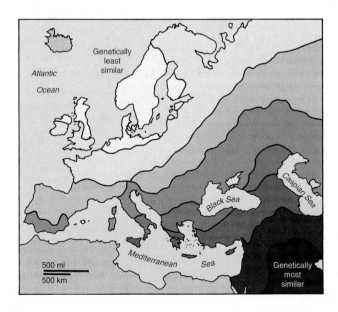

Figure 2.13 **The migration of first farmers** out of the Middle East into Europe starting about 10,000 years ago is presumably traced by blood and gene markers. If the gene evidence interpretation is valid, the migrants spread at a rate of about 1 kilometer (five-eighths of a mile) per year, gradually interbreeding with and replacing the indigenous European hunter-gatherers throughout that continent.

Source: L. Luca Cavalli-Sforza, Paolo Menozzi, and Alberto Piazza. The History and Geography of Human Genes. Copyright © 1994 Princeton University Press, Princeton, N.J.

as North America and Australia. Even today, in the rain forests of central Africa, Bantu farmers put continuing pressure on hunting and gathering Pygmies and in southern Africa, Hottentot herders and Bantu farmers constantly advance on the territories of the San (Bushmen) hunter-gatherer bands. The contrast and conflict between the hunter-gatherers and agriculturalists provide dramatic evidence of cultural divergence.

Neolithic Innovations

The domestication of plants and animals began during the Mesolithic period, but in its refined form it marked the onset of the *Neolithic* (New Stone Age). Like other Stone Age levels, the Neolithic was more a stage of cultural development than a specific span of time. The term implies the creation of an advanced set of tools and technologies to deal with the conditions and needs encountered by an expanding, sedentary population whose economy was based on the agricultural management of the environment (Figure 2.14).

Not all peoples in all areas of the earth made the same cultural transition at the same time. In the Near East, from which most of our knowledge of this late prehistoric period comes, the Neolithic lasted from approximately 8000 to 3500 B.C. There, as elsewhere, it brought complex and revolutionary changes in human life. Culture began to alter at an accelerating pace, and change itself became a way of life. In an interconnected adaptive

(a)

(b)

Figure 2.14 (a) The Mediterranean scratch plow, the earliest form of this basic agricultural tool, was essentially an enlarged digging stick dragged by an ass, an ox, or—as here in the mountains of Peru—by a pair of oxen. The scratch plow represented a significant technological breakthrough in human use of tools and animal power in food production. (b) Its earliest evidence is found in Egyptian tomb drawings and in art preserved from the ancient Middle East but it was elsewhere either independently invented or introduced by those familiar with its use. See also Figure 2.17a.

web, technological and social innovations came with a speed and genius surpassing all previous periods.

Humans learned the arts of spinning and weaving plant and animal fibers. They learned to use the potter's wheel and to fire clay and make utensils. They developed techniques of brick making, mortaring, and construction, and they discovered the skills of mining, smelting, and casting metals. On the foundation of such technical advancements, a more complex exploitative culture appeared and a more formal economy emerged. A stratified society based on labor and role specialization replaced the rough equality of adults in hunting and gathering economies. Special local advantages in resources or products promoted the development of

long-distance trading connections, which the invention of the sailboat helped to maintain.

By the end of the Neolithic period, certain spatially restricted groups, having created a food-producing rather than a foraging society, undertook the purposeful restructuring of their environment. They began to modify plant and animal species; to manage soil, terrain, water, and mineral resources; and to utilize animal energy to supplement that of humans. They used metal to make refined tools and superior weapons—first pure copper and later the alloy of tin and copper that produced the harder, more durable bronze. Humans had moved from adopting and shaping to the art of creating.

As people gathered together in larger communities, new and more formalized rules of conduct and control emerged, especially important where the use of land was involved. We see the beginnings of governments to enforce laws and specify punishments for wrongdoers. The protection of private property, so much greater in amount and variety than that carried by the nomad, demanded more complex legal codes, as did the enforcement of the rules of societies increasingly stratified by social privileges and economic status.

Religions became more formalized. For the hunter, religion could be individualistic, and his worship was concerned with personal health and safety. The collective concerns of farmers were based on the calendar: the cycle of rainfall, the seasons of planting and harvesting, the rise and fall of waters to irrigate the crops. Religions responsive to those concerns developed rituals appropriate to seasons of planting, irrigation, harvesting, and thanksgiving. An established priesthood was required, one that stood not only as intermediary between people and the forces of nature but also as authenticator of the timing and structure of the needed rituals.

In daily life, occupations became increasingly specialized. Metalworkers, potters, sailors, priests, merchants, scribes, and in some areas, warriors complemented the work of farmers and hunters.

Culture Hearths

The social and technical revolutions that began in and characterized the Neolithic period were initially spatially confined. The new technologies, the new ways of life, and the new social structures diffused from those points of origin and were selectively adopted by people who were not a party to their creation. The term **culture hearth** is used to describe such centers of innovation and invention from which key culture traits and elements moved to exert an influence on surrounding regions.

The hearth may be viewed as the "cradle" of any culture group whose developed systems of livelihood and life created a distinctive cultural landscape. Most of the thousands of hearths that evolved across the world in all regions and natural settings remained at low levels of social and technical development. Only a few developed the trappings of *civilizations*. The definition of that term is not precise, but indicators of its achievement are commonly assumed to be writing, metallurgy, long-distance trade connections, astronomy and mathematics, social stratification and labor specialization, formalized governmental systems, and a structured urban culture.

Several major culture hearths emerged in the Neolithic period. Prominent centers of early creativity were found in Egypt, Crete, Mesopotamia, the Indus Valley of the Indian subcontinent, northern China, southeastern Asia, several locations in sub-Saharan Africa, in the Americas, and elsewhere (Figure 2.15). They arose in widely separated areas of the world, at different times, and under differing ecological circumstances. Each displayed its own unique mix of culture traits and amalgams.

All were urban centered, the indisputable mark of civilization first encountered in the Near East 5500–6000 years ago, but the urbanization of each was differently arrived at and expressed (Figure 2.16). In some hearth areas, such as Mesopotamia and Egypt, the transition from settled agricultural village to urban form was gradual and prolonged. In Minoan Crete, urban life was less explicitly developed than in the Indus Valley, where early trade contacts with the Near East suggest the importance of exchange in fostering urban growth (see "Cities Brought Low"). Trade seems particularly important in the development of West African culture hearths, such as Ghana and Kanem. Coming later (from the 8th to the 10th centuries) than the Nile or Mesopotamian centers, their numerous stone-built towns seem to have been supported both by an extensive agriculture whose origins were probably as early as those of the Middle East and, par-

ticularly, by long-distance trade across the Sahara. The Shang kingdom on the middle course of the Huang He (Yellow River) on the North China Plain had walled cities containing wattle-and-daub buildings but no monumental architecture.

Each culture hearth showed a rigorous organization of agriculture resulting in local productivity sufficient to enable a significant number of people to engage in nonfarm activities. Therefore, each hearth region saw the creation of a stratified society that included artisans, warriors, merchants, scholars, priests, and administrators. Each also developed or adopted astronomy, mathematics, and the all-essential calendar. Each, while advancing in cultural diversity and complexity, exported technologies, skills, and learned behaviors far beyond its own boundaries.

Writing appeared first in Mesopotamia and Egypt at least 5000 years ago, as cuneiform in the former and as hieroglyphics in the latter. The separate forms of writing have suggested to some that they arose independently in separate hearths. Others maintain that the idea of writing originated in Mesopotamia and spread outward to Egypt, to the Indus Valley, to Crete, and perhaps even to China, though independent development of Chinese ideographic writing is usually assumed. The systems of record keeping developed in New World hearths were not related to those of the Old, but once created they spread widely in areas under the influence of Andean and Meso-American hearths. Skill in working iron, so important in Near Eastern kingdoms, was an export of sub-Saharan African hearths.

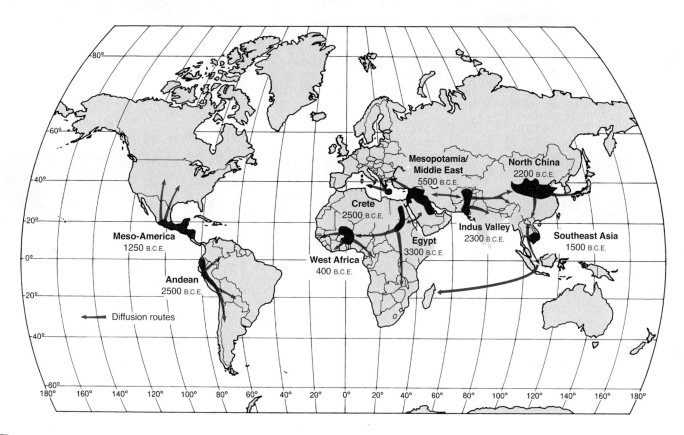

Figure 2.15 **Early culture hearths of the Old World and the Americas.** The B.C.E. (Before the Common Era) dates approximate times when the hearths developed complex social, intellectual, and technological bases and served as cultural diffusion centers.

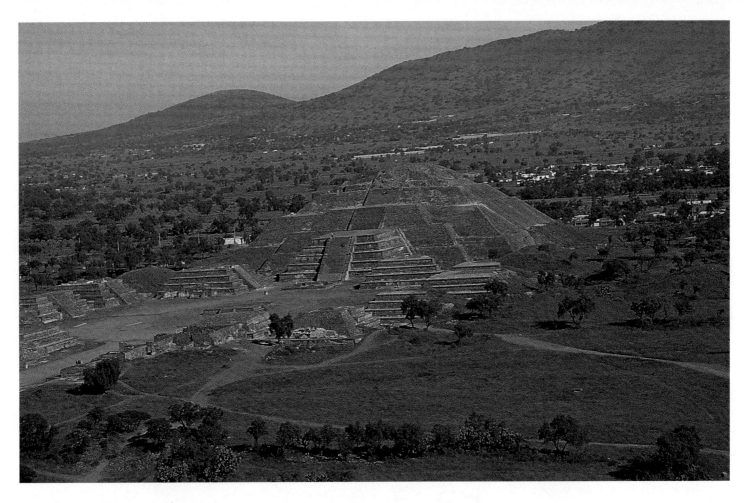

Figure 2.16 Urbanization was invariably a characteristic of culture hearths of both the Old and the New Worlds. Pictured is the Pyramid of the Moon and Avenue of the Dead at Teotihuacán, a city that at its height between A.D. 300 and 700 spread over nearly 18 square kilometers (7 square miles). Located some 50 kilometers (30 miles) northeast of Mexico City in the Valley of Mexico, the planned city of Teotihuacán featured broad, straight avenues and an enormous pyramid complex. The Avenue of the Dead, bordered with low stone-faced buildings, was some 3 kilometers (nearly 2 miles) in length.

Cities Brought Low

Sustainable development requires a long-term balance between human actions and environmental conditions. When either poor management of resources by an exploiting culture or natural environmental alteration unrelated to human actions destroys that balance, a society's use of a region is no longer "sustainable" in the form previously established. Recent research shows that over 4000 years ago an unmanageable natural disaster spelled the death of half a dozen ancient civilizations from the Mediterranean Sea on the west to the Indus Valley on the east.

That disaster took the form of an intense 300-year drought that destroyed the rain-based agriculture on which many of the early civilizations were dependent. Although they prospered through trade, urban societies were sustained by the efforts of farmers. When, about 2200 B.C., fields dried and crops failed through lack of rain, urban and rural inhabitants alike were forced to flee the dust storms and famine of intolerable environmental deterioration.

Evidence of the killer drought that destroyed so many Bronze Age cultures—for example, those of Mesopotamia, early Minoan Crete, and the Old Kingdom in Egypt—includes cities abandoned in 2200 B.C. and not reoccupied for over 300 years; deep accumulations (20–25 cm, or 8–10 in.) of windblown sand over farmlands during the same three centuries; abrupt declines in lake water levels; and thick lake- and seabed deposits of windblown debris.

Similar, but differently timed drought periods—such as the catastrophic aridity between A.D. 800 and 1000 that destroyed Mayan culture in Meso-America—have been blamed for the collapse of advanced societies in the New World as well. Not even the most thriving of early urban cultures were immune to restrictions arbitrarily imposed by nature.

The anthropologist Julian Steward (1902–1972) proposed the concept of **multilinear evolution** to explain the common characteristics of widely separated cultures developed under similar ecological circumstances. He suggested that each major environmental zone—arid, high altitude, midlatitude steppe, tropical forest, and so on—tends to induce common adaptive traits in the cultures of those who exploit it. Those traits were, at base, founded on the development of agriculture and the emergence of similar cultural and administrative structures in the several culture hearths. But *similar* does not imply *identical*. Steward simply suggested that since comparable sequences of developmental events cannot always or even often be explained on the basis of borrowing or exporting of ideas and techniques (because of time and space differences in cultures sharing them), they must be regarded as evidence of parallel creations based on similar ecologies. From similar origins, but through separate adaptations and innovations, distinctive cultures emerged.

Diffusionism is the belief that cultural similarities occur primarily—perhaps even solely—by spatial spread (diffusion) from one or, at most, a very few common origin sites. Cultural advancement and civilizations, that is, are passed on along trade routes and through group contact rather than being the result of separate and independent creation. Although long out of favor, diffusionism has recently received renewed support from archaeological discoveries apparently documenting very long-distance transfer of ideas, technologies, and language by migrating peoples.

In any event, the common characteristics deriving from multilinear evolution and the spread of specific culture traits and complexes contained the roots of **cultural convergence.** That term describes the sharing of technologies, organizational structures, and even cultural traits and artifacts that is so evident among widely separated societies in a modern world united by instantaneous communication and efficient transportation. Convergence in those worldwide terms is, for many observers, proof of the pervasive globalization of culture.

The Structure of Culture

Understanding a culture fully is, perhaps, impossible for one who is not part of it. For analytical purposes, however, the traits and complexes of culture—its building blocks and expressions—may be grouped and examined as subsets of the whole. The anthropologist Leslie White (1900–1975) suggested that for analytical purposes, a culture could be viewed as a three-part structure composed of subsystems that he termed *ideological, technological,* and *sociological.* In a similar classification, the biologist Julian Huxley (1887–1975) identified three components of culture: *mentifacts, artifacts,* and *sociofacts.* Together, according to these interpretations, the subsystems—identified by their separate components—comprise the system of culture as a whole. But they are integrated; each reacts on the others and is affected by them in turn.

The **ideological subsystem** consists of ideas, beliefs, and knowledge of a culture and of the ways in which these things are expressed in speech or other forms of communication. Mythologies and theologies, legend, literature, philosophy, and folk wis-

dom make up this category. Passed on from generation to generation, these abstract belief systems, or **mentifacts,** tell us what we ought to believe, what we should value, and how we ought to act. Beliefs form the basis of the socialization process. Often we know—or think we know—what the beliefs of a group are from their oral or written statements. Sometimes, however, we must depend on the actions or objectives of a group to tell us what its true ideas and values are. "Actions speak louder than words" or "Do as I say, not as I do" are commonplace recognitions of the fact that actions, values, and words do not always coincide. Two basic strands of the ideological subsystem— language and religion—are the subject of Chapter 5.

The **technological subsystem** is composed of the material objects, together with the techniques of their use, by means of which people are able to live. The objects are the tools and other instruments that enable us to feed, clothe, house, defend, transport, and amuse ourselves. We must have food, we must be protected from the elements, and we must be able to defend ourselves. Huxley termed the material objects we use to fill these basic needs **artifacts** (Figure 2.17). In Chapter 10 we will examine the relationship between technological subsystems and regional patterns of economic development.

The **sociological subsystem** of a culture is the sum of those expected and accepted patterns of interpersonal relations that find their outlet in economic, political, military, religious, kinship, and other associations. These **sociofacts** define the social organization of a culture. They regulate how the individual functions relative to the group—whether it be family, church, or state. There are no "givens" as far as the patterns of interaction in any of these associations are concerned, except that most cultures possess a variety of formal and informal ways of structuring behavior. Differing patterns of behavior are learned and are transmitted from one generation to the next (Figure 2.18).

Classifications are of necessity arbitrary, and these classifications of the subsystems and components of culture are no exception. The three-part categorization of culture, while helping us to appreciate its structure and complexity, can simultaneously obscure the many-sided nature of individual elements of culture. A dwelling, for example, is an artifact providing shelter for its occupants. It is, simultaneously, a sociofact reflecting the nature of the family or kinship group it is designed to house, and a mentifact summarizing a culture group's convictions about appropriate design, orientation, and building materials of dwelling units. In the same vein, clothing serves as an artifact of bodily protection appropriate to climatic conditions, available materials and techniques, or the activity in which the wearer is engaged. But garments also may be sociofacts, identifying an individual's role in the social structure of the community or culture, and mentifacts, evoking larger community value systems (Figure 2.19).

Nothing in a culture stands totally alone. Changes in the ideas that a society holds may affect the sociological and technological systems just as changes in technology force adjustments in the social system. The abrupt alteration of the ideological structure of Russia following the 1917 communist revolution from a monarchical, agrarian, capitalistic system to an industrialized, communistic society involved sudden, interrelated alteration of all facets of that country's culture system. The equally abrupt

(a)

(b)

Figure 2.17 (*a*) This Balinese farmer working with draft animals uses tools typical of the lower technological levels of subsistence economies. (*b*) Cultures with advanced technological subsystems use complex machinery to harness inanimate energy for productive use.

disintegration of Russian communism in the early 1990s was similarly disruptive of all its established economic, social, and administrative structures. The interlocking nature of all aspects of a culture is termed **cultural integration.**

Culture Change

The recurring theme of cultural geography is change. No culture is, or has been, characterized by a permanently fixed set of material objects, systems of organization, or even ideologies. Admittedly, all of these may be long-enduring within a stable, isolated society at equilibrium with its resource base. Such isolation and stability have always been rare. On the whole, while cultures are essentially conservative, they are always in a state of flux. Some changes are major and pervasive. The transition from hunter-gatherer to sedentary farmer, as we have seen, affected markedly every facet of the cultures experiencing that change. Profound, too, has been the impact of the Industrial Revolution and its associated urbanization on all societies it has touched.

Not all change is so extensive as that following the introduction of agriculture or the Industrial Revolution. Many changes are so slight individually as to go almost unnoticed at their inception, though cumulatively they may substantially alter the affected culture. Think of how the culture of the United States differs today from what you know it to have been in 1940—not in essentials, perhaps, but in the innumerable electrical, electronic, and transportational devices that have been introduced and in the social, behavioral, and recreational changes they and other technological changes have wrought. Among these latter have been shifts in employment patterns to include greater participation by women

in the waged workforce and associated adjustments in attitudes toward the role of women in the society at large. Such cumulative changes occur because the cultural traits of any group are not independent; they are clustered in a coherent and integrated pattern. Change on a small scale will have wide repercussions as associated traits arrive at accommodation with the adopted adjustment. Change, both major and minor, within cultures is induced by *innovation, diffusion,* and *acculturation.*

Innovation

Innovation implies changes to a culture that result from ideas created within the social group itself and adopted by the culture. The novelty may be an invented improvement in material technology, like the bow and arrow or the jet engine. It may involve the development of nonmaterial forms of social structure and interaction: feudalism, for example, or Christianity.

Many innovations are of little consequence by themselves, but sometimes the widespread adoption of seemingly inconsequential innovations may bring about large changes when viewed over a period of time. A new musical tune, "adopted" by a few people, may lead many individuals to fancy that tune plus others of similar sound. This, in turn, may have a bearing on dance routines, which may then bear on clothing selection, which, in turn, may affect retailers' advertising campaigns and consumers' patterns of expenditures. Eventually, a new cultural form will be identified that may have an important impact on the thinking processes of the adopters and on those who come into contact with the adopters. Notice that a broad definition of innovation is used, but notice also that what is important is whether or not innovations are accepted and adopted.

Figure 2.18 All societies prepare their children for membership in the culture group. In each of these settings, certain values, beliefs, skills, and proper ways of acting are being transmitted to the youngsters.

Premodern and traditional societies characteristically are not innovative. In societies at equilibrium with their environment and with no unmet needs, change has no adaptive value and no reason to occur. Indeed, all societies have an innate resistance to change. Complaints about youthful fads or the glorification of times past are familiar cases in point. However, when a social group is inappropriately unresponsive—mentally, psychologically, or economically—to changing circumstances and to innovation, it is said to exhibit

cultural lag. For example, for at least 1500 years, most California Indians were in contact with cultures utilizing both maize and pottery, yet they failed to accept either innovation.

Innovation—invention—frequently under stress, has marked the history of humankind. As we have seen, growing populations at the end of the Ice Age necessitated an expanded food base. In response, domestication of plants and animals appears to have occurred independently in more than one world area. Indeed, a most

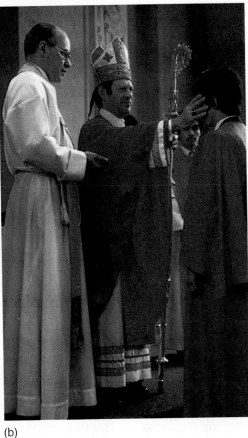

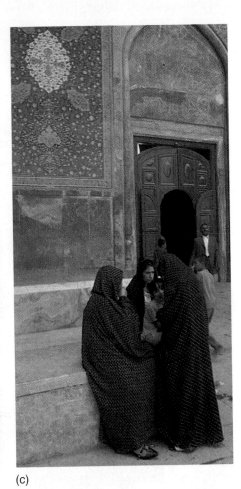

(a)　　　　　　　　　　　　　(b)　　　　　　　　　　　　　(c)

Figure 2.19 (*a*) When clothing serves primarily to cover, protect, or assist in activities, it is an *artifact*. (*b*) Some garments are *sociofacts*, identifying a role or position within the social structure: the distinctive "uniforms" of the soldier, the cleric, or the beribboned ambassador immediately proclaim their respective roles in a culture's social organizations. (*c*) The mandatory chadors of Iranian females are *mentifacts*, indicative not specifically of the role of the wearer but of the values of the culture the wearer represents.

striking fact about early agriculture is the universality of its development or adoption within a very short span of human history. In 10,000 B.C., the world population of no more than 10 million was exclusively hunter-gatherers. By A.D. 1500, only 1% of the world's 350 million people still followed that way of life. The revolution in food production affected every facet of the threefold subsystems of culture of every society accepting it. All innovation has a radiating impact on the web of culture; the more basic the innovation, the more pervasive its consequences.

In most modern societies, innovative change has become common, expected, and inevitable. The rate of invention, at least as measured by the number of patents granted, has steadily increased, and the period between idea conception and product availability has been decreasing. A general axiom is that the more ideas available and the more minds able to exploit and combine them, the greater the rate of innovation. The spatial implication is that larger urban centers of advanced technologies tend to be centers of innovation. This is not just because of their size but because of the number of ideas interchanged. Indeed, ideas not only stimulate new thoughts and viewpoints but also create circumstances in which the society must develop new solutions to maintain its forward momentum (Figure 2.20).

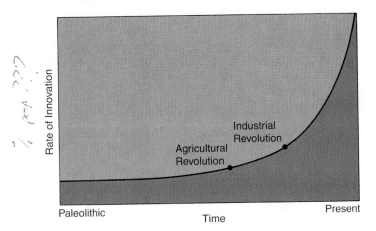

Figure 2.20 **The rate of innovation through human history.** Hunter-gatherers, living in easy equilibrium with their environment and their resource base, had little need for innovation and no necessity for cultural change. Increased population pressures led to the development of agriculture and the diffusion of the ideas and techniques of domestication, urbanization, and trade. With the Industrial Revolution, dramatic increases in innovation began to alter cultures throughout the world.

Diffusion

Diffusion is the process by which an idea or innovation is transmitted from one individual or group to another across space. Diffusion may assume a variety of forms, each different in its impact on social groups. Basically, however, two processes are involved: (1) People move, for any of a number of reasons, to a new area and take their culture with them. For example, immigrants to the American colonies brought along crops and farming techniques, building styles, or concepts of government alien to their new home. (2) Information about an innovation (e.g., hybrid corn or compact discs) may spread throughout a society, perhaps aided by local or mass media advertising; or new adopters of an ideology or way of life—for example, a new religious creed—may be inspired or recruited by immigrant or native converts. The former is known as *relocation diffusion,* the latter as *expansion diffusion* (Figure 2.21).

Expansion diffusion involves the spread of an item or idea from one place to others. In the process, the thing diffused also remains—and is frequently intensified—in the origin area. Islam, for example, expanded from its Arabian Peninsula origin locale across much of Asia and North Africa. At the same time, it strengthened its hold over its Near Eastern birthplace by displacing pagan, Christian, and Jewish populations. When expansion diffusion affects nearly uniformly all individuals and areas outward from the source region, it is termed *contagious diffusion.* The term implies the importance of direct contact between those who developed or have adopted the innovation and those who newly encounter it, and is reminiscent of the course of infectious diseases (Figure 2.22).

If an idea has merit in the eyes of potential adopters and they themselves become adopters, the number of contacts of adopters with potential adopters will compound. Consequently, the innovation will spread slowly at first and then more and more rapidly until saturation occurs or a barrier is reached. The incidence of adoption under contagious diffusion is represented by the S-shaped curve in Figure 2.23. The rate of diffusion of a trait or idea may be influenced by *time-distance decay,* which simply tells us that the spread or acceptance of an idea is usually delayed as distance from the source of the innovation increases.

In some instances, however, geographic distance is less important in the transfer of ideas than is communication between major centers or important people. News of new clothing styles, for example, quickly spreads internationally between major cities and only later filters down irregularly to smaller towns and rural areas. The process of transferring ideas first between larger places or prominent people and only later to smaller or less important points or people is known as *hierarchical diffusion.* The Christian faith in Europe, for example, spread from Rome as the principal center to provincial capitals and thence to smaller Roman settlements in largely pagan occupied territories (see Figure 5.22). Today, new discoveries are shared among scientists at leading universities before they appear in textbooks or become general knowledge through the public press. The process works because, for many things, distance is relative to the communication network involved. Big cities or leading scientists, connected by strong information flows, are "closer" than their simple distance separation suggests.

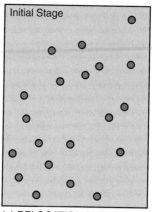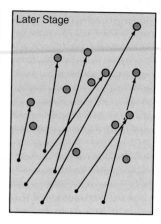

(a) RELOCATION DIFFUSION

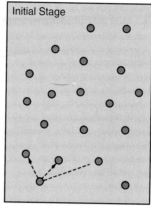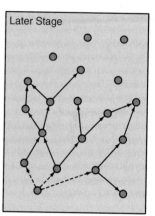

(b) EXPANSION DIFFUSION

Figure 2.21 **Patterns of diffusion.** (*a*) In *relocation diffusion,* innovations or ideas are transported to new areas by carriers who permanently leave the home locale. The "Pennsylvania Dutch" barn (Figure 6.24) was brought to Pennsylvania by German immigrants and spread to other groups and areas southward through Appalachia and westward into Ohio, Indiana, Illinois, and Missouri, Not all farmers or farm districts in the path of advancement adopted the new barn design. (*b*) In *expansion diffusion,* a phenomenon spreads from one place to neighboring locations, but in the process remains and is often intensified in the place of origin (see Figure 5.28).

Source: Redrawn by permission from Spatial Diffusion, *by Peter R. Gould, Resource Paper no. 4, page 4, Association of American Geographers, 1969.*

While the diffusion of ideas may be slowed by time-distance decay, their speed of spread may be increased to the point of becoming instantaneous through the *space-time compression* made possible by modern communication. Given access to radios; telephones; worldwide transmission of television news, sports, and entertainment programs; and—perhaps most importantly—to computers and the Internet, people and areas distantly separated can immediately share in a common fund of thought and innovation. Modern communication technology, that is, has encouraged and facilitated the globalization of culture.

Stimulus diffusion is a third form of expansion diffusion. The term summarizes situations in which a fundamental idea, though not the specific trait itself, stimulates imitative behavior within a receptive population. A documented case in point involves the

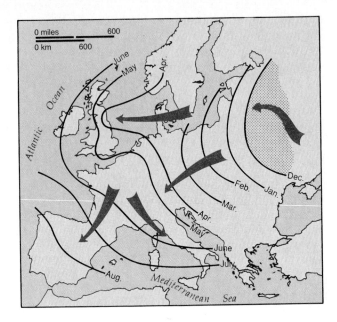

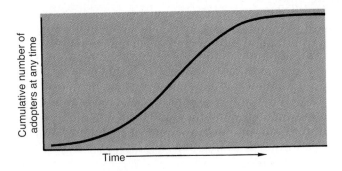

Figure 2.23 **The diffusion of innovations over time.** The number of adopters of an innovation rises at an increasing rate until the point at which about one-half the total who ultimately decide to adopt the innovation have made the decision. At this point, the number of adopters increases at a decreasing rate.

Figure 2.22 **The process of *contagious diffusion*** is sensitive to both time and distance, as suggested by the diffusion pathways of the European influenza pandemic of 1781. The pattern there was a wavelike radiation from a Russian nodal origin area.

Source: Based on Gerald F. Pyle and K. David Patterson, Ecology of Disease 2, no. 3 (1984): 179.

spread of the concept but not of a specific system of writing from European American settlers to at least one Native American culture group. Observing that white people could make marks on pieces of paper to record agreements and repeat lengthy speeches, Sequoyah, a Cherokee who could neither read nor write any language, around 1820 devised a system for writing the Cherokee language, eventually refining his initially complex pictorial system to a set of 86 syllabic signs. With time, literacy in the new system spread to others and *The Cherokee Phoenix,* a Cherokee language newspaper, was established in 1828. There was no transfer between cultures or groups of a specific technique of writing, but there was a clear-cut case of the *idea* of writing diffusing by stimulating imitative behavior.

In **relocation diffusion,** the innovation or idea is physically carried to new areas by migrating individuals or populations that possess it (Figure 2.21a). Mentifacts or artifacts are therefore introduced into new locales by new settlers who become part of populations not themselves associated or in contact with the origin area of the innovation. The spread of religions by settlers or conquerors is a clear example of relocation diffusion, as was the diffusion of agriculture to Europe from the Middle East (Figure 2.13). Christian Europeans brought their faiths to areas of colonization or economic penetration throughout the world. At the world scale, massive relocation diffusion resulted from the European colonization and economic penetration that began in the 16th century. More localized relocation diffusion continues today as Asian refugees or foreign "guest workers" bring their cultural traits to their new areas of settlement in Europe or North America.

For either expansion or relocation diffusion, innovations in the technological or ideological subsystems may be relatively readily diffused to, and accepted by, cultures that have basic similarities and compatibilities. Continental Europe and North America, for example, could easily and quickly adopt the innovations of the Industrial Revolution diffused from England with which they shared a common economic and technological background. Industrialization was not quickly accepted in Asian and African societies of totally different cultural conditioning. On the ideological level, too, successful diffusion depends on acceptability of the innovations. The Shah of Iran's attempt at rapid westernization of traditional Iranian, Islamic culture after World War II provoked a traditionalist backlash and revolution that deposed the Shah and reestablished clerical control of the state.

The conclusion must be, therefore, that diffusion cannot be viewed solely as the outcome of knowledge dispersed. The acceptance of new traits, articles, or ways of doing or thinking by a potential receiving population depends not just on information flow to that population but also upon its entire cultural and economic structure. Innovation may be rejected not because of lack of knowledge but because the new trait violates the established cultural norms of the culture to which it is introduced. For example, cash crop specialization recommended to a peasant agricultural society may be rejected not because it is not understood, but because it unacceptably disrupts the knowledge base and culture complex devoted to assured food security in a subsistence farming economy. Similarly, less disruptive new production ideas—chemical fertilizers, deep-well irrigation, hybrid seeds, and the like—may be rejected simply because, though understood, they are not affordable. Culture is a complex organized system and culture change involves alteration of the system's established structure in ways that may be rejected even after knowledge of an innovation is received and understood.

It is not always possible, of course, to determine the precise point of origin or the routes of diffusion of innovations now widely adopted (see "Documenting Diffusion"). Nor is it always certain whether the existence of a cultural trait in two different areas is the result of diffusion or of **independent** (or *parallel*) **invention.** Cultural similarities do not necessarily prove that diffusion has occurred. The pyramids of Egypt and of the Central American Maya civilization most likely were separately

The places of origin of many ideas, items, and technologies important in contemporary cultures are only dimly known or supposed, and their routes of diffusion are speculative at best. Gunpowder, printing, and spaghetti are presumed to be the products of Chinese inventiveness; the lateen sail has been traced to the Near Eastern culture world. The mold-board plow is ascribed to 6th-century Slavs of northeastern Europe. The sequence and routes of the diffusion of these innovations has not been documented.

In other cases, such documentation exists, and the process of diffusion is open to analysis. Clearly marked is the diffusion path of the custom of smoking tobacco, a practice that originated among Amerindians. Sir Walter Raleigh's Virginia colonists, returning home in 1586, introduced smoking in English court circles, and the habit very quickly spread among the general populace. England became the source region of the new custom for northern Europe; smoking was introduced to Holland by English medical students in 1590. Dutch and English together spread the habit by sea to the Baltic and Scandinavian areas and overland through Germany to Russia. The innovation continued its eastward diffusion, and within a hundred years tobacco had spread across Siberia and was, in the 1740s, reintroduced to the American continent at Alaska by Russian fur traders. A second route of diffusion for to-

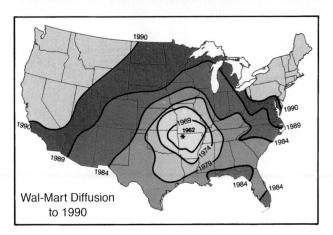

Wal-Mart Diffusion to 1990

Source: Map based on data from Thomas O. Graff and Dub Ashton, "Spatial Diffusion of Wal-Mart: Contagious and Reverse Hierarchical Elements." Professional Geographer *46, no. 1 (1994): 19–29.*

bacco smoking can be traced from Spain, where the custom was introduced in 1558, and from which it spread more slowly through the Mediterranean area into Africa, the Near East, and Southeast Asia.

In more recent times, hybrid corn was first adopted by imaginative farmers of northern Illinois and eastern Iowa in the mid-1930s. By the late 1930s and early 1940s, the new seeds were being planted as far east as Ohio and north to Minnesota, Wisconsin, and northern Michigan. By the late 1940s, all commercial corn-growing districts of the

United States and southern Canada were cultivating hybrid corn varieties.

A similar pattern of diffusion marked the expansion of the Wal-Mart stores chain. From its origin in northwest Arkansas in 1962, the discount chain had dispersed throughout the United States by the 1990s to become the country's largest retailer in sales volume. In its expansion, Wal-Mart displayed a "reverse hierarchical" diffusion, initially spreading by way of small towns before opening its first stores in larger cities and metropolitan areas (see map).

conceived and are not necessarily evidence, as some have proposed, of pre-Columbian voyages from the Mediterranean to the Americas. A monument-building culture, after all, has only a limited number of shapes from which to choose.

Historical examples of independent, parallel invention are numerous: logarithms by Napier (1614) and Burgi (1620), the calculus by Newton (1672) and Leibnitz (1675), the telephone by Elisha Gray and Alexander Graham Bell (1876) are commonly cited. It appears beyond doubt that agriculture was independently developed not only in both the New World and the Old but also in more than one culture hearth in each of the hemispheres.

Acculturation and Cultural Modification

A culture group may undergo major modifications in its own identifying traits by adopting some or all of the characteristics of another, dominant culture group. Such is the case in **acculturation**—

discussed at greater length in Chapter 6 (pp. 196 to 198)—as immigrant populations take on the values, attitudes, customs, and speech of the receiving society, which itself undergoes change from absorption of the arriving group. A different form of contact and subsequent cultural alteration may occur in a conquered or colonized region where the subordinate or subject population is either forced to adopt the culture of the new ruling group, introduced through relocation diffusion, or does so voluntarily, overwhelmed by the superiority in numbers or the technical level of the conqueror. Tribal Europeans in areas of Roman conquest, native populations in the wake of Slavic occupation of Siberia, and Native Americans stripped of their lands following European settlement of North America experienced this kind of cultural modification or adoption.

In extreme cases, of course, small and, particularly, primitive indigenous groups brought into contact with conquering or absorbing societies may simply cease to exist as separate cultural entities. Although presumably such cultural loss has been part of all of human history, its occurrence has been noted and its pace quickened over the past 500 years. By one informed estimate, at least one-third of the world's inventory of human cultures has totally disappeared since A.D. 1500, along with their languages, traditions, ways of life, and, indeed, with their very identity or remembrance.

In many instances, close contact between two different groups may involve adjustments of the original cultural patterns of both rather than disappearance of either. For example, changes in Japanese political organization and philosophy were imposed by occupying Americans after World War II, and the Japanese voluntarily adopted some more frivolous aspects of American life (Figure 2.24). In turn, American society was enriched by the selective importation of Japanese cuisine, architecture, and philosophy, demonstrating the two-way nature of cultural diffusion. Where that two-way flow reflects a more equal exchange of cultural outlooks and ways of life, a process of *transculturation* has occurred. That process is observable within the United States as massive South and Central American immigration begins to intertwine formerly contrasting cultures, altering both.

Contact between Regions

All cultures are amalgams of innumerable innovations spread spatially from their points of origin and integrated into the structure of the receiving societies. It has been estimated that no more than 10% of the cultural items of any society are traceable to innovations created by its members and that the other 90% come to the society through diffusion (see "A Homemade Culture"). Since, as we have seen, the pace of innovation is affected strongly by the mixing of ideas among alert, responsive people and is increased by exposure to a variety of cultures, the most active and innovative historical hearths of culture were those at crossroads locations and those deeply involved in distant trade and colonization. Ancient Mesopotamia and classical Greece and Rome had such locations and involvements, as did the West African culture hearth after the 5th century and, much later, England during the Industrial Revolution and the spread of its empire.

Recent changes in technology permit us to travel farther than ever before, with greater safety and speed, and to communicate without physical contact more easily and completely than was previously possible. This intensification of contact has resulted in an acceleration of innovation and in the rapid spread of goods and ideas. Several millennia ago, innovations such as smelting of metals took hundreds of years to diffuse. Today, worldwide diffusion—through Internet interest groups, for example—may be almost instantaneous.

Obstacles do exist, of course. **Diffusion barriers** are any conditions that hinder either the flow of information or the movement of people and thus retard or prevent the acceptance of an innovation. Because of the *friction of distance,* generally the farther two areas are from each other, the less likely is interaction to occur, an observation earlier (p. 55) summarized by the term *time-distance*

Figure 2.24 Baseball, an import from America, is one of the most popular sports in Japan, attracting millions of spectators annually.

A Homemade Culture

Reflecting on an average morning in the life of a "100% American," Ralph Linton noted:

> Our solid American citizen awakens in a bed built on a pattern which originated in the Near East but which was modified in Northern Europe before it was transmitted to America. He throws back covers made from cotton, domesticated in India, or linen, domesticated in the Near East, or wool from sheep, also domesticated in the Near East, or silk, the use of which was discovered in China. All of these materials have been spun and woven by processes invented in the Near East. . . . He takes off his pajamas, a garment invented in India, and washes with soap invented by the ancient Gauls. . . .
>
> Returning to the bedroom, . . . he puts on garments whose form originally derived from the skin clothing of the nomads of the Asiatic steppes [and] puts on shoes made from skins tanned by a process invented in ancient Egypt and cut to a pattern derived from the classical civilizations of the Mediterranean. . . . Before going out for breakfast he glances through the window, made of glass invented in Egypt, and if it is raining puts on overshoes made of rubber discovered by the Central American Indians and takes an umbrella invented in southeastern Asia. . . .
>
> [At breakfast] a whole new series of borrowed elements confronts him. His plate is made of a form of pottery invented in China. His knife is of steel, an alloy first made in southern India, his fork a medieval Italian invention, and his spoon a derivative of a Roman original. He begins breakfast with an orange, from the eastern Mediterranean, a cantaloupe from Persia, or perhaps a piece of African watermelon. With this he has coffee, an Abyssinian plant. . . . [H]e may have the egg of a species of bird domesticated in Indo-China, or thin strips of flesh of an animal domesticated in Eastern Asia which have been salted and smoked by a process developed in northern Europe.
>
> When our friend has finished eating . . . he reads the news of the day, imprinted in characters invented by the ancient Semites upon a material invented in China by a process invented in Germany. As he absorbs the accounts of foreign troubles he will, if he is a good conservative citizen, thank a Hebrew deity in an Indo-European language that he is 100% American.

Ralph Linton, *The Study of Man: An Introduction.* © 1936, renewed 1964, pp. 326–327. Reprinted by permission of Prentice Hall, Inc., Upper Saddle River, N.J.

decay. Distance as a factor in spatial interaction is further explored in Chapter 3. For now it is sufficient to note that distance may be an *absorbing barrier,* halting the spread of an innovation.

Interregional contact can also be hindered by the physical environment and by a lack of receptivity by a contacted culture. Oceans and rugged terrain can and have acted as physical *interrupting barriers,* delaying or deflecting the path of diffusion. Cultural obstacles that are equally impenetrable may also exist. Should reluctant adopters or nonadopters of innovations (the California natives who rejected acceptance of maize and pottery, for example) intervene between hearths and receptive cultures, the spread of an innovation can be slowed. It can also be delayed when cultural contact is overtly impeded by governments that interfere with radio reception, control the flow of foreign literature, and discourage contact between their citizens and foreign nationals.

More commonly, barriers are at least partially *permeable;* they permit passage (acceptance) of at least some innovations encountering them. The more similar two cultural areas are to each other, the greater is the likelihood of the adoption of an innovation, for diffusion is a selective process. The receiver culture may adopt some goods or ideas from the donor society and reject others. The decision to adopt is governed by the receiving group's own culture. Political restrictions, religious taboos, and other so-cial customs are cultural barriers to diffusion. The French Canadians, although close geographically to many centers of diffusion such as Toronto, New York, and Boston, strive to be only minimally influenced by such centers. Both their language and culture complex govern their selective acceptance of Anglo influences, and restrictive French-only language regulations are enforced to preserve the integrity of French culture. Traditional groups, perhaps controlled by firm religious conviction, may very largely reject culture traits and technologies of the larger society in whose midst they live (see Figure 7.2).

Adopting cultures do not usually accept intact items originating outside the receiving society. Diffused ideas and artifacts commonly undergo some alteration of meaning or form that makes them acceptable to a borrowing group. The process of the fusion of the old and new is called **syncretism** and is a major feature of culture change. It can be seen in alterations to religious ritual and dogma made by convert societies seeking acceptable conformity between old and new beliefs; the mixture of Catholic rites and theological and magical elements taken from West African religions that produced voodooism in Haiti is an example. On a more familiar level, syncretism is reflected in subtle or blatant alterations of imported cuisines to make them conform to the demands of America's palate and its fast-food franchises (Figure 2.25).

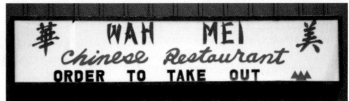

Figure 2.25 Foreign foods modified for American tastes and American palates growing accustomed to dishes from all cultures together represent *syncretism* in action.

 ## Summary

The web of culture is composed of many strands. Together, culture traits and complexes in their spatial patterns create human landscapes, define culture regions, and distinguish culture groups. Those landscapes, regions, and group characteristics change through time as human societies interact with their environment, develop for themselves new solutions to collective needs, or are altered through innovations adopted from outside the group itself. The cultural uniformity of a preagricultural world composed solely of hunter-gatherers was lost as domestication of plants and animals in many world areas led to the emergence of culture hearths of wide-ranging innovation and to a cultural divergence between farmers and gatherers. Innovations spread outward from their origin points, carried by migrants through relocation diffusion or adopted by others through a variety of expansion diffusion and acculturation processes. Although diffusion barriers exist, most successful or advantageous innovations find adopters, and both cultural modification and cultural convergence of different societies result. The details of the technological, sociological, and ideological subsystems of culture define the differences that still exist between world areas.

The ivory hunters who opened our chapter showed how varied and complex the culture of even a primitive group can be. Their artifacts of clothing, fire making, hunting, and fishing displayed diversity and ingenuity. They were part of a structured kinship system and engaged in organized production and trade. Their artistic efforts and ritual burial customs speak of a sophisticated set of abstract beliefs and philosophies. Their culture complex did not develop in isolation; it reflected at least in part their contacts with other groups, even those far distant from their Paris Basin homeland. As have culture groups always and everywhere, the hunters carried on their own pursuits and interacted with others in spatial settings. They exhibited and benefitted from structured *spatial behavior,* the topic to which we next turn our attention.

Key Words

acculturation 57
artifact 51
carrying capacity 44
cultural convergence 51
cultural divergence 44
cultural ecology 39
cultural integration 52
cultural lag 53
cultural landscape 40
culture 36
culture complex 38

culture hearth 48
culture realm 38
culture region 38
culture trait 38
diffusion 55
diffusion barrier 58
environmental determinism 39
expansion diffusion 55
hunter-gatherer 43
ideological subsystem 51

independent invention 56
innovation 52
mentifact 51
multilinear evolution 51
possibilism 40
relocation diffusion 56
sociofact 51
sociological subsystem 51
syncretism 59
technological subsystem 51

For Review

1. What is included in the concept of *culture?* How is culture transmitted? What personal characteristics affect the aspects of culture that any single individual acquires or fully masters?

2. What do we mean by *domestication?* When and where did the domestication of plants and animals occur? What impact on culture and population numbers did plant domestication have?

3. What is a *culture hearth?* What new traits of culture characterized the early hearths? Identify and locate some of the major culture hearths that emerged at the close of the Neolithic period.

4. What do we mean by *innovation?* By *diffusion?* What different patterns of diffusion can you describe? Discuss the role played by innovation and diffusion in altering the cultural structure in which you are a

participant from that experienced by your great-grandparents.

5. Differentiate between *culture traits* and *culture complexes.* Between *environmental determinism* and *possibilism.*

6. What are the components or subsystems of the three-part system of culture? What characteristics are included in each of the subsystems?

Focus Follow-up

1. **What are the components of culture and nature of culture–environment interactions?** pp. 36–42.

   Culture traits and complexes may be grouped into culture regions and realms. Differing developmental levels color human perceptions of environmental opportunities. In general, as the active agents in the relationship, humans exert adverse impacts on the natural environment.

2. **How did cultures develop and diverge** (pp. 42–44), **and where did cultural advances originate?** pp. 44–51.

   From Paleolithic hunting and gathering to Neolithic farming and then to city civilizations, different groups made differently timed cultural transitions. All early cultural advances had their origins in a few areally distinct "hearths."

3. **What are the structures of culture and forms of culture change?** pp. 51–59.

   All cultures contain ideological, technological, and sociological components that work together to create cultural integration. Cultures change through innovations they themselves invent or that diffuse from other areas and are accepted or adapted.

Selected References

Blaut, James M. "Two Views of Diffusion." *Annals of the Association of American Geographers* 67, no. 3 (1977): 345–349.

Brown, Lawrence A. *Innovation Diffusion: A New Perspective.* London and New York: Methuen, 1981.

Coe, Michael, Dean Snow, and Elizabeth Benson. *Atlas of Ancient America.* New York: Facts on File Incorporated, 1986.

Cowan, C. Wesley, and Patty Jo Watson, eds. *The Origins of Agriculture: An International Perspective.* Washington, D.C.: Smithsonian Institution Press, 1992.

Denevan, William M. "The Pristine Myth: The Landscape of the Americas in 1492." *Annals of the Association of American Geographers* 82, no. 3 (1992): 369–385.

Diamond, Jared. *Guns, Germs, and Steel: The Fates of Human Societies.* New York: Norton, 1997.

Gebauer, Anne B., and T. Douglas Price, eds. *Transitions to Agriculture in Prehistory.* Monographs in World Archeology, no. 4. Madison, Wis.: Prehistory Press, 1992

Gore, Rick. "The Most Ancient Americans." *National Geographic* (October 1997): 92–99.

Gould, Peter. *Spatial Diffusion.* Association of American Geographers, Commission on College Geography. *Resource Paper* No. 4. Washington, D.C.: Association of American Geographers, 1969.

Hägerstrand, Torsten. *Innovation Diffusion as a Spatial Process.* University of Chicago Press, 1967.

Haggett, Peter. "Geographical Aspects of the Emergence of Infectious Diseases." *Geografiska Annaler* 76B, no. 2 (1994): 91–104.

Isaac, Erich. *Geography of Domestication.* Englewood Cliffs, N.J.: Prentice Hall, 1970.

Kroeber, Alfred L., and Clyde Kluckhohn. "Culture: A Critical Review of Concepts and Definitions," Harvard University. *Papers of the Peabody Museum of American Archaeology and Ethnology* 47, no. 2 (1952).

Lamb, H. H. *Climate, History, and the Modern World.* New York: Routledge, 1995.

MacNeish, Richard S. *The Origins of Agriculture and Settled Life.* Norman: University of Oklahoma Press, 1991.

Morrill, Richard, Gary L. Gaile, and Grant Ian Thrall. *Spatial Diffusion.* Scientific Geography Series vol. 10. Newbury Park, Calif.: SAGE Publications, 1988.

Parfit, Michael. "Hunt for the First Americans." *National Geographic* (December 2000): 41–67.

"The Peopling of the Earth." *National Geographic* (October 1988): 434–503.

Rodrique, Christine M. "Can Religion Account for Early Animal Domestications . . . ?" *Professional Geographer* 44, no. 4 (1992): 417–430.

Rogers, Alisdair, ed. *Peoples and Cultures.* The Illustrated Encyclopedia of World Geography. New York: Oxford University Press, 1992.

Runnels, Curtis N. "Environmental Degradation in Ancient Greece." *Scientific American* (March 1995): 96–99.

Sauer, Carl. *Agricultural Origins and Dispersals.* New York: American Geographical Society, 1952.

Sebastian, Lynne. *The Chaco Anasazi: Sociopolitical Evolution in the Prehistoric Southwest.* New York: Cambridge University Press, 1992.

Sjoberg, Gideon. "The Origin and Evolution of Cities." *Scientific American* 213 (1965): 54–63.

Steward, Julian H. *Theory of Culture Change.* Urbana: University of Illinois Press, 1955.

Thomas, William L., Jr., ed. *Man's Role in Changing the Face of the Earth.* Chicago: University of Chicago Press, 1956.

White, Leslie A. *The Science of Culture: A Study of Man and Civilization.* New York: Farrar, Straus and Giroux, 1969.

White, Randall. *Dark Caves, Bright Visions: Life in Ice Age Europe.* New York: American Museum of Natural History in Association with W. W. Norton & Company, 1986.

Zohary, Daniel, and Mari Hopf. *Domestication of Plants in the Old World.* 2d ed. Oxford, England: Clarendon Press, 1993.

Websites: The World Wide Web has a tremendous number and variety of sites pertaining to geography. Websites relevant to the subject matter of this chapter appear in the "Web Links" section of the On-line Learning Center associated with this book. Access it at **www.mhhe.com/fellmann8e**

CHAPTER
Three

Spatial Interaction and Spatial Behavior

Focus Preview

1. The three bases for all spatial interaction, pp. 66–68.
2. How the probability of aggregate spatial interaction is measured, pp. 68–71.
3. The special forms and nature of human spatial behavior, pp. 71–75.
4. The roles of information and perception in human spatial behavior, pp. 75–84.
5. Migration patterns, types, and controls, pp. 84–94.

Opposite: Spatial interaction in the motorized world: commuter traffic on Gardiner Expressway, Toronto, Ontario, Canada.

65

arly in January of 1849 we first thought of migrating to California. It was a period of National hard times . . . and we longed to go to the new El Dorado and "pick up" gold enough with which to return and pay off our debts.

Our discontent and restlessness were enhanced by the fact that my health was not good. . . . The physician advised an entire change of climate thus to avoid the intense cold of Iowa, and recommended a sea voyage, but finally approved of our contemplated trip across the plains in a "prairie schooner."

Full of the energy and enthusiasm of youth, the prospects of so hazardous an undertaking had no terror for us, indeed, as we had been married but a few months, it appealed to us as a romantic wedding tour.[1]

So begins Catherine Haun's account of their 9-month journey from Iowa to California, just two of the quarter-million people who traveled across the continent on the Overland Trail in one of the world's great migrations. The migrants faced months of grueling struggle over badly marked routes that crossed swollen rivers, deserts, and mountains. The weather was often foul, with hailstorms, drenching rains, and burning summer temperatures. Graves along the route were a silent testimony to the lives claimed by buffalo stampedes, Indian skirmishes, cholera epidemics, and other disasters.

What inducements were so great as to make emigrants leave behind all that was familiar and risk their lives on an uncertain venture? Catherine Haun alludes to economic hard times gripping the country and to their hope for riches to be found in California. Like other migrants, the Hauns were attracted by the climate in the West, which was said to be always sunny and free of disease. Finally, like most who undertook the perilous journey West, the Hauns were young, moved by restlessness, a sense of adventure, and a perception of greater opportunities in a new land. They, like their predecessors back to the beginnings of humankind, were acting in space and across space on the basis of acquired information and anticipation of opportunity—prepared to pay the price in time, money, and hardship costs of overcoming distance.

A fundamental question in human geography is: What considerations influence how individual human beings use space and act within it? Related queries include: Are there discernible controls on human spatial behavior? How does distance affect human interaction? How do our perceptions of places influence our spatial activities? How do we overcome the consequences of distance in the exchange of commodities and information? How are movement and migration decisions (like that of the Hauns) reached? These are questions addressing geography's concern with understanding spatial interaction.

Spatial interaction means the movement of peoples, ideas, and commodities within and between areas. The Hauns were engaging in spatial interaction (Figure 3.1). International trade, the movement of semitrailers on the expressways, radio broadcasts, and business or personal telephone calls are more familiar examples. Such movements and exchanges are designed to achieve effective integration between different points of human activity. Movement of whatever nature satisfies some felt need or desire. It represents the attempt to smooth out the spatially differing

Figure 3.1 Cross-country movement was slow, arduous, and dangerous early in the 19th century, and the price of long-distance spatial interaction was far higher in time and risks than a comparable journey today.

availability of required resources, commodities, information, or opportunities. Whatever the particular purpose of a movement, there is inevitably some manner of trade-off balancing the benefit of the interaction with the costs that are incurred in overcoming spatial separation. Because commodity movements represent simple demonstrations of the principles underlying all spatial interactions, let us turn to them first.

Bases for Interaction

Neither the world's resources nor the products of people's efforts are uniformly distributed. Commodity flows are responses to these differences; they are links between points of supply and

---

[1]From Catherine Haun, "A Woman's Trip Across the Plains in 1849," in Lillian Schlissel, *Women's Diaries of the Westward Journey.* (New York: Schocken Books, 1982).

locales of demand. Such response may not be immediate or even direct. Matters of awareness of supplies or markets, the presence or absence of transportation connections, costs of movement, ability to pay for things wanted and needed—all and more are factors in the structure of trade. Underlying even these, however, is a set of controlling principles governing spatial interaction.

A Summarizing Model

The conviction that spatial interaction reflects areal differences led the geographer Edward Ullman (1912–1976) to speculate on the essential conditions affecting such interactions and to propose an explanatory model. He observed that spatial interaction is effectively controlled by three flow-determining factors that he called *complementarity, transferability,* and *intervening opportunity.* Although Ullman's model deals with commodity flows, it has—as we shall see—applicability to informational transfers and patterns of human movements as well.

Complementarity

For two places to interact, one place must have what another place wants and can secure. That is, one place must have a supply of an item for which there is an effective demand in the other, as evidenced by desire for the item, purchasing power to acquire it,

and means to transport it. The word describing this circumstance is **complementarity.** *Effective* supply and demand are important considerations; mere differences from place to place in commodity surplus or deficit are not enough to initiate exchange. Greenland and the Amazon basin are notably unlike in their natural resources and economies, but their amount of interaction is minimal. Supply and market must come together, as they do in the flow of seasonal fruits and vegetables from California's Imperial Valley to the urban markets of the American Midwest and East or in the movement of manganese from Ukraine to the steel mills of Western Europe. The massive movement of crude and refined petroleum between spatially separated effective supplies and markets clearly demonstrates complementarity in international trade (Figure 3.2). More generalized patterns of complementarity underlie the exchanges of the raw materials and agricultural goods of less developed countries for the industrial commodities of the developed states.

Transferability

Even when complementarity exists, spatial interaction occurs only when conditions of **transferability**—acceptable costs of an exchange—are met. Spatial movement responds not just to availability and demand but to considerations of time and cost. Transferability is an expression of the mobility of a commodity and is a function of three interrelated conditions: (1) the characteristics

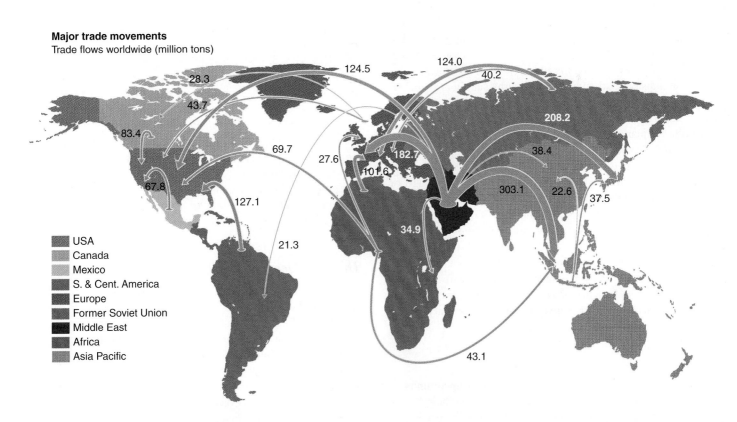

Major trade movements
Trade flows worldwide (million tons)

124.5   124.0
28.3   40.2
43.7
83.4   208.2
69.7   182.7   38.4
27.6
101.6
67.8   303.1   22.6
127.1   34.9   37.5
21.3

- ■ USA
- ■ Canada
- ■ Mexico
- ■ S. & Cent. America
- ■ Europe
- ■ Former Soviet Union
- ■ Middle East
- ■ Africa
- ■ Asia Pacific

43.1

Figure 3.2 **Interregional trade in oil.** Complementarity is so basic in initiating interaction that even relatively low-value bulk commodities such as coal, fertilizer, and grain move in trade over long distances. For many years, despite fluctuating prices, petroleum has been the most important commodity in international trade, moving long distances in response to effective supply and demand considerations.

Source: The BP Amoco Statistical Review of World Energy, 2002. Reprinted by permission.

and value of the product; (2) the distance, measured in time and money penalties, over which it must be moved; and (3) the ability of the commodity to bear the costs of movement. If the time and money costs of traversing a distance are too great, exchange does not occur. That is, mobility is not just a physical matter but an economic one as well. If a given commodity is not affordable upon delivery to an otherwise willing buyer, it will not move in trade, and the potential buyer must seek a substitute or go without.

Transferability is not a constant condition. It differs between places, over time, and in relation to what is being transferred and how it is to be moved. The opening of a logging road will connect a sawmill with stands of timber formerly inaccessible (nontransferable). An increasing scarcity of high-quality ores will enhance the transferability of lower-quality mine outputs by increasing their value. Low-cost bulk commodities not economically moved by air may be fully transferable by rail or water. Poorly developed and costly transportation may inhibit exchanges even at short distance between otherwise willing traders. In short, transferability expresses the changing relationships between the costs of transportation and the value of the product to be shipped.

Intervening Opportunity

Complementarity can be effective only in the absence of more attractive alternative sources of supply or demand closer at hand or cheaper. **Intervening opportunities** serve to reduce supply/demand interactions that otherwise might develop between distant complementary areas. A supply of Saharan sand is not enough to assure its flow to sand-deficient Manhattan Island because supplies of sand are more easily and cheaply available within the New York metropolitan region. For reasons of cost and convenience, a purchaser is unlikely to buy identical commodities at a distance when a suitable nearby supply is available. When it is, the intervening opportunity demonstrates complementarity at a shorter distance.

Similarly, markets and destinations are sought, if possible, close at hand. Growing metropolitan demand in California reduces the importance of midwestern markets for western fruit growers. The intervening opportunities offered by Chicago or Philadelphia reduce the number of job seekers from Iowa searching for employment in New York City. People from New England are more likely to take winter vacations in Florida, which is relatively near and accessible, than in Southern California, which is not. That is, opportunities that are discerned closer at hand reduce the pull of opportunities offered by a distant destination (Figure 3.3). Patterns of spatial interaction are dynamic, reflecting the changeable structure of apparent opportunity.

Measuring Interaction

Complementarity, transferability, and intervening opportunity—the controlling conditions of commodity movement—help us understand all forms of spatial interaction, including the placing of long-distance phone calls, the residential locational decisions of commuters, and the once-in-a-lifetime transcontinental adventure of the Hauns. Interaction of whatever nature between places is

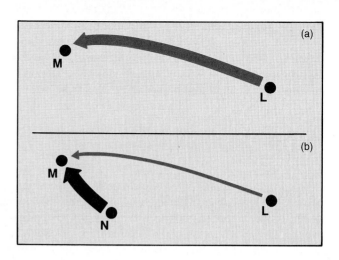

Figure 3.3 (*a*) The volume of expected flow of a good between centers L and M, based solely on their complementarity and distance apart, may be (*b*) materially reduced if an alternate supplier is introduced as an intervening opportunity nearer to the market. A sand and gravel dealer of City L, for example, may find its role as sole supplier to the City M market reduced or eliminated if a new supplier with a shorter haul and lower delivery costs locates at City N.

not, of course, meaningfully described by the movement of a single commodity, by the habits of an individual commuter, or the once-only decision of a migrant. The discovery of an Inuit (Eskimo) ivory carving in a Miami gift shop does not establish significant interaction between the Arctic coast and a Florida resort.

The study of unique events is suggestive but not particularly informative. We seek general principles that govern the frequency and intensity of interaction both to validate the three preconditions of spatial exchange and to establish the probability that any given potential interaction will actually occur. Our interest is similar to that of the physical scientist investigating, for example, the response of a gas to variations in temperature and pressure. The concern there is with *all* of the gas molecules and the probability of their collective reactions; the responses of any particular molecule are of little interest. Similarly, we are concerned here with the probability of aggregate, not individual, behavior.

Distance Decay

In all manner of ways, our lives and activities are influenced by the **friction of distance.** That phrase reminds us that distance has a retarding effect on human interaction because there are increasing penalties in time and cost associated with longer-distance, more expensive interchanges. We visit nearby friends more often than distant relatives; we go more frequently to the neighborhood convenience store cluster than to the farther regional shopping center. Telephone calls or mail deliveries between nearby towns are greater in volume than those to more distant locations. An informal study showed that college students are more likely to order out for food if they are close to the delivery drop-off point; students farther away do not order out as often.

Our common experience, clearly supported by maps and statistics tracing all kinds of flows, is that most interactions occur over short distances. That is, interchange decreases as distance increases, a reflection of the fact that transferability costs increase with distance. More generally stated, **distance decay** describes the decline of an activity or function with increasing distance from its point of origin. As the examples in Figure 3.4 demonstrate, near destinations have a disproportionate pull over more distant points in commodity movements. However, it is also evident that the rate of distance decay varies with the type of activity.

Study of all manner of spatial interconnections has led to the very general conclusion that interaction between places is inversely related to the square of the distance separating them. That is, volume of flow between two points 80 kilometers (50 miles) apart would probably be only one-quarter of that between centers at 40 kilometers (25 miles) separation. Such a rigid *inverse-square* relationship is well documented in the physical sciences. For social, cultural, and economic relations, however, it is at best a useful approximation. In human interaction, linear distance is only one aspect of transferability; cost and time are often more meaningful measures of separation.

When the friction of distance is reduced by lowered costs or increased ease of flow, the slope of the distance decay curve is flattened and more total area is effectively united than when those costs are high. When telephone calls are charged by uniform area rates rather than strictly by distance, more calls are placed to the outer margins of the rate area than expected. Expressways extend commuting travel ranges to central cities and expand the total area conveniently accessible for weekend recreation. Figure 3.4 shows that shipping distances for high-cost truck transport are, on the average, shorter than for lower-cost rail hauls.

The Gravity Concept

Interaction decisions are not based on distance or distance/cost considerations alone. The large regional shopping center attracts customers from a wide radius because of the variety of shops and goods its very size promises. We go to distant big cities "to seek our fortune" rather than to the nearer small town. We are, that is, attracted by the expectation of opportunity that we associate with larger rather than smaller places. That expectation is summarized by another model of spatial interaction, the **gravity model,** also drawn from the physical sciences.

In the 1850s, Henry C. Carey (1793–1879), in his *Principles of Social Science,* observed that the law of "molecular gravitation" is an essential condition in human existence and that the attractive force existing between areas is akin to the force of gravity. According to Carey, the physical laws of gravity and motion developed by Sir Isaac Newton (1642–1727) have applicability to the aggregate actions of humans.

Newton's *law of universal gravitation* states that any two objects attract each other with a force that is proportional to the product of their masses and inversely proportional to the square of the distance between them. Thus, the force of attraction, F, between two masses M_i and M_j separated by distance, d, is

$$F = g \frac{M_i M_j}{d_{ij}^2}$$

where g is the "gravitational constant."

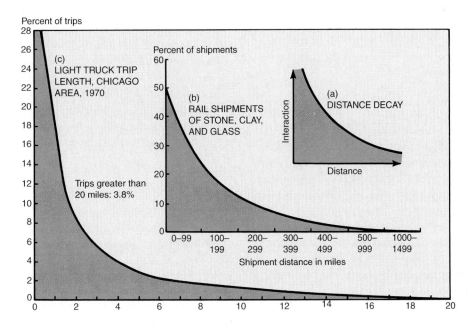

Figure 3.4 **The shape of distance decay.** The geographer W. Tobler summarized the concept of distance decay in proposing his "first law of geography: everything is related to everything else, but near things are more related than distant things." Distance decay curves vary with the type of flow. (*a*) is a generalized statement of distance decay, (*b*) summarizes United States data for a single year, and (*c*) suggests the primary use of light trucks as short haul pickup and delivery vehicles.

Source: (c) Data from Chicago Area Transportation Study, A Summary of Travel Characteristics, *1977.*

Carey's interests were in the interaction between urban centers and in the observation that a large city is more likely to attract an individual than is a small hamlet. His first interest could be quickly satisfied by simple analogy. The expected interaction (*I*) between two places, *i* and *j*, can be calculated by converting physical mass in the gravity model to population size (*P*), so that

$$I_{ij} = \frac{P_i P_j}{d_{ij}^2}$$

Exchanges between any set of two cities, *A* and *B*, can therefore be quickly estimated:

$$I_{AB} = \frac{\text{population of } A \times \text{ population of } B}{(\text{distance between } A \text{ and } B)^2}$$

In social—rather than physical—science applications of the gravity model, distance may be calculated by travel time or travel cost modifications rather than by straight line separation. Whatever the unit of measure, however, the model assures us that although spatial interaction always tends to decrease with increasing distance between places, at a given distance it tends to expand with increases in their size.

Carey's second observation—that large cities have greater drawing power for individuals than small ones—was subsequently addressed by the *law of retail gravitation,* proposed by William J. Reilly (1899–1970) in 1931. Using the population and distance inputs of the gravity model, Reilly concluded that the *breaking point* (*BP*), or boundary, marking the outer edge of either of the cities' trade area could be located by the expression:

$$BP = \frac{d_{ij}}{1 + \sqrt{\dfrac{P_2}{P_1}}}$$

where

BP = distance from city 1 to the breaking point (or boundary)
d_{ij} = distance between city 1 and city 2
P_1 = population of city 1
P_2 = population of city 2

Any farm or small-town resident located between the two cities would be inclined to shop in one or the other of them according to that resident's position relative to the calculated breaking point. Since the breaking point between cities of unequal size will lie farther from the larger of the two, its spatially greater drawing power is assured (Figure 3.5).

Later studies in location theory, city systems, trade area analysis, and other social topics all suggest that the gravity model can be used to account for a wide variety of flow patterns in human geography, including population migration, commodity flows, journeys to work or to shop, telephone call volumes, and the like. Each such flow pattern suggests that size as well as distance influences spatial interaction. Carey's observation made nearly 150 years ago initiated a type of analysis that has continuing relevance. In modified form it is used today for a variety of practical studies that help us better understand the "friction of distance."

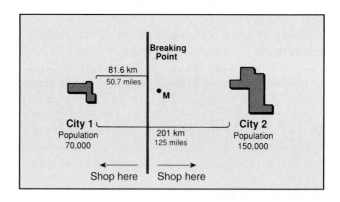

$\mathcal{F}igure$ 3.5 The *law of retail gravitation* provides a quick determination of the trade boundary (or breaking point) between two cities. In the diagram, cities 1 and 2 are 201 kilometers (125 mi) apart. Reilly's law tells us that the breaking point between them lies 81.6 kilometers (50.7 mi) distant from City 1. A potential customer located at *M,* midway (100.5 km or 62.5 mi) between the cities, would lie well within the trade zone of City 2. A series of such calculations would define the "trade area" of any single city.

Interaction Potential

Spatial interaction models of distance decay and gravitational pull deal with only two places at a time. The world of reality is rather more complex. All cities, not just city pairs, within a regional system of cities have the possibility of interacting with each other. Indeed, the more specialized the goods produced in each separate center—that is, the greater their collective complementarity—the more likely it is that such multiple interactions will occur.

A **potential model,** also based on Newtonian physics, provides an estimate of the interaction opportunities available to a center in such a multicentered network. It tells us the relative position of each point in relation to all other places within a region. It does so by summing the size and distance relationships between all points of potential interaction within an area. The concept of potential is applicable whenever the measurement of the intensity of spatial interaction is of concern—as it is in studies of marketing, land values, broadcasting, commuting patterns, and the like.

Movement Biases

Distance decay and the gravity and potential models help us understand the bases for interaction in an idealized area without natural or cultural barriers to movement or restrictions on routes followed, and in which only rational interaction decisions are made. Even under those model conditions, the pattern of spatial interaction that develops for whatever reason inevitably affects the conditions under which future interactions will occur. An initial structure of centers and connecting flows will tend to freeze into the landscape a mutually reinforcing continuation of that same pattern. The predictable flows of shoppers to existing shopping centers make those centers attractive to other merchants. New store openings increase customer flow; increased flow strengthens the developed pattern of spatial interaction. And increased road traffic calls for the highway improvement that encourages additional traffic volume.

Such an aggregate regularity of flow is called a **movement bias.** We have already noted a *distance bias* favoring short movements over long ones. There is also *direction bias,* in which of all possible directions of movement, actual flows are restricted to only one or a few. Direction bias is simply a statement that from a given origin, flows are not random (Figure 3.6); rather, certain places have a greater attraction than do others. The movement patterns from an isolated farmstead are likely oriented to a favored shopping town. On a larger scale, in North America or Siberia, long-distance freight movements are directionally biased in favor of east-west flows. Direction bias reflects not just the orientation but also the intensity of flow. Movements from a single point—from Novosibirsk in Siberia, for example, or from Winnipeg, Canada, or Kansas City in the United States—may occur in all directions; they are in reality more intense along the east-west axis.

Such directional biases are in part a reflection of *network bias,* a shorthand way of saying that the presence or absence of connecting channels strongly affects the likelihood that spatial interaction will occur. A set of routes and the set of places that they connect are collectively called a **network.** Flows cannot occur between all points if not all points are linked. In Figure 3.6a, the interchange between A and X, though not necessarily impossible, is unlikely because the routeway between them is indirect and circuitous. In information flows, a worker on the assembly line is less likely to know of company production plans than is a secretary in the executive offices; these two workers are tied into quite different information networks.

A recognition of movement biases helps to refine the coarser generalizations of spatial interaction based solely on complementarity, transferability, and intervening opportunity. Other modifying statements have been developed, but each further refinement moves us away from aggregate behavior toward less predictable individual movements and responses. The spatial interaction questions we ask and the degree of refinement of the answers we require determine the modifications we must introduce into the models we employ.

Human Spatial Behavior

Humans are not commodities and individually do not necessarily respond predictably to the impersonal dictates of spatial interaction constraints. Yet, to survive, people must be mobile and collectively do react to distance, time, and cost considerations of movement in space and to the implications of complementarity, transferability, and intervening opportunity. Indeed, an exciting line of geographic inquiry involves how individuals make spatial behavioral decisions and how those separate decisions may be summarized by models and generalizations to explain collective actions.

Mobility is the general term applied to all types of human territorial movement. Two aspects of that mobility behavior concern us. The first is the daily or temporary use of space—the journeys to stores, to work, or to school, or for longer periods on vacation or college students' relocation between home and school dormitory. These types of mobility are often designated as *circulation* and have no suggestion of relocation of residence (Figure 3.7). The second type of mobility is the longer-term commitment related to decisions to permanently leave the home territory and find residence in a new location. This second form of spatial behavior is termed *migration.*

Both aspects imply a time dimension. Humans' spatial actions are not instantaneous. They operate over time, frequently imparting a rhythm to individual and group activity patterns and imposing choices among time-consuming behaviors. Elements of both aspects of human spatial behavior are also embodied in how individuals perceive space and act within it and how they respond to information affecting their space-behavioral decisions. The nature of those perceptions and responses affect us all in our daily movements. The more permanent movement embodied in migration involves additional and less common decisions and behaviors, as we shall see later in this chapter.

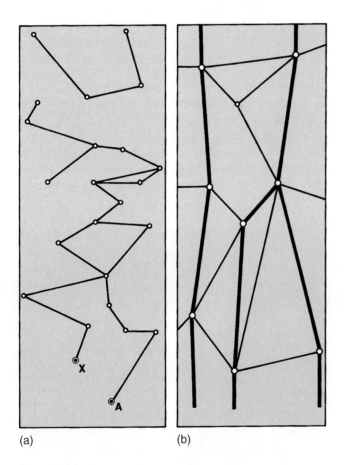

(a)   (b)

Figure 3.6 **Direction bias.** (*a*) When direction bias is absent, movements tend to be almost random, occurring in all possible directions, but less likely between points, such as A and X, not directly connected. (*b*) Direction bias indicating predominantly north-south movements. Direction bias implies greatest intensity of movement within a restricted number of directions.

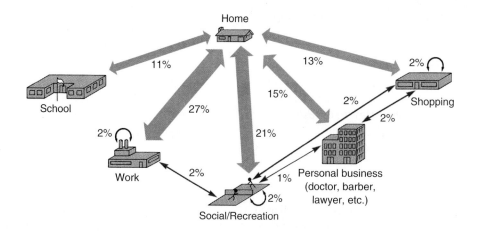

Figure 3.7 **Southern California travel patterns.** The numbers are the percentages of all urban trips taken in Southern California on a typical weekday. The greatest single movement is the journey to and from work. Notice that 12% of all trips are for more than one of the purposes shown.

Source of data: Association of Governments Survey.

Individual Activity Space

One of the realities of life is that groups and countries draw boundaries around themselves and divide space into territories that are, if necessary, defended. Some see the concept of **territoriality**—the emotional attachment to and the defense of home ground—as a root explanation of much of human action and response. It is true that some individual and collective activity appears to be governed by territorial defense responses: the conflict between street groups in claiming and protecting their "turf" (and their fear for their lives when venturing beyond it) and the sometimes violent rejection by ethnic urban neighborhoods of any different advancing population group it considers threatening. On a more individualized basis, each of us claims as

personal space the zone of privacy and separation from others our culture or our physical circumstances require or permit. Anglo Americans demand greater face-to-face separation in conversations than do Latin Americans. Personal space on a crowded beach or in a department store is acceptably more limited than it is in our homes or when we are studying in a library (Figure 3.8).

For most of us, our personal sense of territoriality is a tempered one. We regard our homes and property as defensible private domains but open them to innocent visitors, known and unknown, or to those on private or official business. Nor do we confine our activities so exclusively within controlled home territories as street-gang members do within theirs. Rather, we have a more or less extended home range, an **activity space** or area

(a)

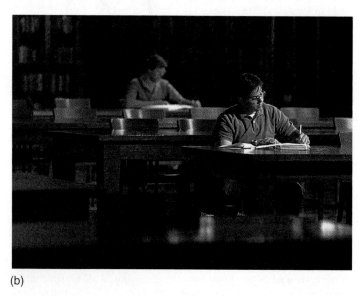

(b)

Figure 3.8 Our demanded *personal space* is not necessarily uniform in shape or constant in size. We tolerate strangers closer to our sides than directly in front of us; we accept more crowding in an elevator than in a store. We accept the press of the crowd on a popular beach—as do these students on spring break in the Florida Keys (*a*), but tend to distance ourselves from others in a library (*b*).

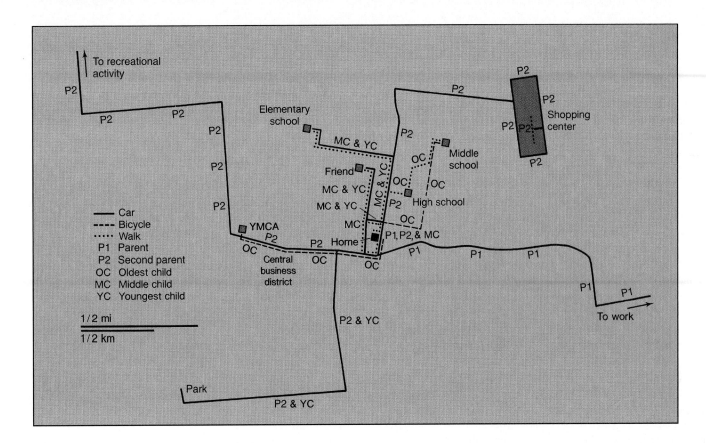

Figure 3.9 Activity space for each member of one author's family of five for a typical weekday. Routes of regular movement and areas recurrently visited help to foster a sense of territoriality and to color one's perceptions of space.

within which we move freely on our rounds of regular activity, sharing that space with others who are also about their daily affairs. Figure 3.9 suggests probable activity spaces for a suburban family of five for a day. Note that the activity space is different and for the mapped day rather limited for each individual, even though two members of the family use automobiles. If 1 week's activity were shown, more paths would be added to the map, and in a year's time, one or more long trips would probably have to be noted.

The types of trips that individuals make and thus the extent of their activity space depend on at least three interrelated variables: their stage in life course (age); the means of mobility at their command; and the demands or opportunities implicit in their daily activities. The first variable, *stage in life,* refers to membership in specific age groups. School-age children usually travel short distances to lower schools and longer distances to upper-level schools. After-school activities tend to be limited to walking or bicycle trips to nearby locations. Greater mobility is characteristic of high-school students. Adults responsible for household duties make shopping trips and trips related to child care as well as journeys away from home for social, cultural, or recreational purposes. Wage-earning adults usually travel farther from home than other family members. Elderly people may, through infirmity or interests, have less extensive activity spaces.

The second variable that affects the extent of activity space is *mobility,* or the ability to travel. An informal consideration of the cost and effort required to overcome the friction of distance is implicit. Where incomes are high, automobiles are available, and the cost of fuel is reckoned minor in the family budget, mobility may be great and individual activity space large. In societies or neighborhoods where cars are not a standard means of conveyance, the daily nonemergency activity space may be limited to walking, bicycling, or taking infrequent trips on public transportation. Wealthy suburbanites are far more mobile than are residents of inner-city slums, a circumstance that affects ability to learn about, seek, or retain work and to have access to medical care, educational facilities, and social services.

A third factor limiting activity space is the individual assessment of the existence of possible activities or *opportunities.* In subsistence economies where the needs of daily life are satisfied at home, the impetus for journeys away from home is minimal. If there are no stores, schools, factories, or even roads, expectations and opportunities are limited. Not only are activities spatially restricted, but **awareness space**—knowledge of opportunity locations beyond normal activity space—is minimal, distorted, or absent. In low-income neighborhoods of modern cities in any country, poverty and isolation limit the inducements, opportunities, destinations, and necessity of travel (Figure 1.26). Opportunities plus mobility conditioned by life stage bear heavily on the amount of spatial interaction in which individuals engage.

The Tyranny of Time

The daily activities of humans—eating, sleeping, traveling between home and destination, working or attending classes—all consume time as well as involve space. An individual's spatial reach is restricted because one cannot be in two different places at the same moment or engage simultaneously in activities that are spatially separate. Further, since there is a finite amount of time within a day and each of us is biologically bound to a daily rhythm of day and night, sleeping and eating, time tyrannically limits the spatial choices we can make and the activity space we can command.

Our daily space-time constraints—our *time-geography*—may be represented by a **space-time prism,** the volume of space and length of time within which our activities must be confined. Its size and shape are determined by our mobility; its boundaries define what we can or cannot accomplish spatially or temporally (Figure 3.10). If our circumstances demand that we walk to work or school (Figure 3.10b), the sides of our prism are steep and the space available for our activities is narrow. We cannot use time spent in transit for other activities, and the area reasonably accessible to the pedestrian is limited. The space-time prism for the driver (Figure 3.10c) has angled sides and the individual's spatial range is wide. The dimensions of the prism determine what spatially defined activities are possible, for no activity can exceed the bounds of the prism (see "Space, Time, and Women"). Since most activities have their own time constraints, the choices of

things you can do and the places you can do them are strictly limited. Defined class hours, travel time from residence to campus, and dining hall location and opening and closing hours, for example, may be the constraints on your *space-time path* (Figure 3.11). If you also need part-time work, your choice of jobs is restricted by their respective locations and work hours, for the job, too, must fit within your daily space-time prism.

Distance and Human Interaction

People make many more short-distance trips than long ones, a statement in human behavioral terms of the concept of *distance decay.* If we drew a boundary line around our activity space, it would be evident that trips to the boundary are taken much less often than short-distance trips around the home. The tendency is for the frequency of trips to fall off very rapidly beyond an individual's **critical distance**—the distance beyond which cost, effort, and means strongly influence our willingness to travel. Figure 3.12 illustrates the point with regard to journeys from the homesite.

Regular movements defining our individual activity space are undertaken for different purposes and are differently influenced by time and distance considerations. The kinds of activities individuals engage in can be classified according to type of trip: journeys to work, to school, to shop, for recreation, and so on. People in nearly all parts of the world make these same types of journeys, though the spatially variable requirements of culture, economy,

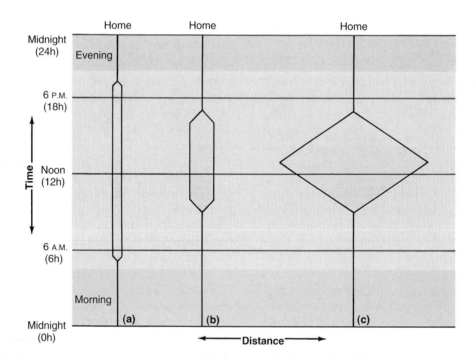

Figure 3.10 **The space-time prism.** An individual's daily prism has both geographical limits and totally surrounding space-time walls. The *time* (vertical axis) involved in movement affects the space that is accessible, along with the time and space available for other than travel purposes. (*a*) When collecting firewood for household use may take an entire day, as it does in some deforested developing countries, no time or space is left for other activities, and the gatherer's space-time prism may be represented by a straight line. (*b*) Walking to and from work or school and spending the required number of hours there leave little time to broaden one's area of activity. (*c*) The automobile permits an extension of the geographical boundaries of the driver's space-time prism; the range of activity possibilities and locations is expanded for the highly mobile.

Space, Time, and Women

From a time-geographic perspective, it is apparent that many of the limitations women face in their choices of employment or other activities outside the home reflect the restrictions that women's time budgets and travel paths place on their individual daily activities.

Consider the case* of the unmarried working woman with one or more preschool-age children. The location and operating hours of available child-care facilities may have more of an influence on her choice of job than do her labor skills or the relative merits of alternative employment opportunities. For example, the woman may not be able to leave her home base before a given hour because the only available full-day child-care service is not open earlier. She

*Suggested by Risa Palm and Allan Pred, *A Time-Geographic Perspective on Problems of Inequality for Women.* Institute of Urban and Regional Development, Working Paper no. 236. University of California, Berkeley, 1974.

must return at the specified child pickup time and arrive home to prepare food at a reasonable (for the child) dinner time. Her travel mode and speed determine the outer limits of her daily space-time prism.

Suppose both of two solid job offers have the same working hours and fall within her possible activity space. She cannot accept the preferred, better paying job because drop-off time at the child-care center would make her late for work, and work hours would make her miss the center's closing time. On the other hand, although the other job is acceptable from a child-care standpoint, it leaves no time (or store options) for shopping or errands except during the lunch break. Job choice and shopping opportunities are thus determined not by the woman's labor skills or awareness of store price comparisons but by her time-geographic constraints. Other women in other job skill, parenthood, locational, or mobility circumstances experience different but comparable space-path restrictions.

Mobility is a key to activity mix, time-budget, and activity space configurations. Again, research indicates that women are frequently disadvantaged. Because of their multiple work, child-care, and home maintenance tasks, women on average make more—though shorter—trips than men, leaving less time for alternate activities.

Although the automobile reduces those time demands, women have less access to cars than do males, in part because in many cities they are less likely to have a driver's license and because they typically cede use of a single family car to husbands. The lower income level of many single women with or without children limits their ability to own cars and leads them to use public transit disproportionately to their numbers—to the detriment of both their money and time-space budgets. They are, it has been observed, "transportation deprived and transit dependent."

---

and personal circumstance dictate their frequency, duration, and significance to an individual (Figure 3.13). A small child, for example, will make many trips up and down the block but is inhibited by parental admonitions from crossing the street. Different but equally effective distance constraints control adult behavior.

The journey to work plays a decisive role in defining the activity space of most adults. Formerly restricted by walking distance or by the routes and schedules of mass transit systems, the critical distances of work trips have steadily increased in European and Anglo American cities as the private automobile figures more importantly in the movement of workers (Figure 3.14). Daily or weekly shopping may be within the critical distance of an individual, and little thought may be given to the cost or the effort involved. That same individual, however, may relegate shopping for special goods to infrequent trips and carefully consider their cost and effort. The majority of our social contacts tend to be at short distance within our own neighborhoods or with friends who live relatively close at hand; longer social trips to visit relatives are less frequent. In all such trips, however, the distance decay function is clearly at work (Figure 3.15).

Spatial Interaction and the Accumulation of Information

Critical distances, even for the same activity, are different for each person. The variables of life stage, mobility, and opportunity, together with an individual's interests and demands, help

define how often and how far a person will travel. On the basis of these variables, we can make inferences about the amount of information a person is likely to acquire about his or her activity space and the area beyond. The accumulation of information about the opportunities and rewards of spatial interaction helps increase and justify movement decisions.

For information flows, however, space has a different meaning than it does for the movement of commodities. Communication, for example, does not necessarily imply the time-consuming physical relocations of freight transportation (though in the case of letters and print media it usually does). Indeed, in modern telecommunications, the process of information flow may be instantaneous regardless of distance. The result is space-time convergence to the point of the obliteration of space. A Bell System report tells us that in 1920, putting through a transcontinental telephone call took 14 minutes and eight operators and cost more than $15.00 for a 3-minute call. By 1940, the call completion time was reduced to less than $1\frac{1}{2}$ minutes, and the cost fell to $4.00. In the 1960s, direct distance dialing allowed a transcontinental connection in less than 30 seconds, and electronic switching has now reduced the completion time to that involved in dialing a number and answering a phone. The price of long-distance conversation essentially disappeared with the advent of voice communication over the Internet in the late 1990s.

The Internet and communication satellites have made worldwide personal and mass communication immediate and data transfers instantaneous. The same technologies that have led to

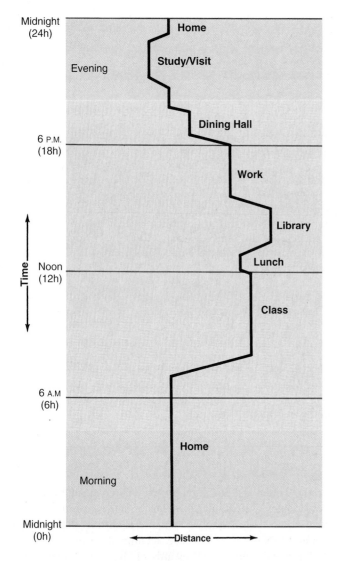

Figure 3.11 School-day space-time path for a hypothetical college student.

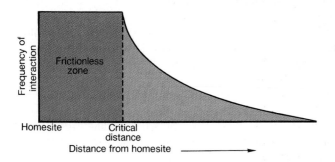

Figure 3.12 **Critical distance.** This general diagram indicates how most people observe distance. For each activity, there is a distance beyond which the intensity of contact declines. This is called the *critical distance* if distance alone is being considered, or the *critical isochrone* (from Greek *isos,* "equal," and *chronos,* "time") if time is the measuring rod. The distance up to the critical distance is identified as a *frictionless zone,* in which time or distance considerations do not effectively figure in the trip decision.

communication space-time convergence have tended toward a space-cost convergence (Figure 3.16). Domestic mail, which once charged a distance-based postage, is now carried nationwide or across town for the same price. In the modern world, transferability is no longer a consideration in information flows.

A speculative view of the future suggests that as distance ceases to be a determinant of the cost or speed of communication, the spatial structure of economic and social decision making may be fundamentally altered. Determinations about where people live and work, the role of cities and other existing command centers, flows of domestic and international trade, constraints on human mobility, and even the concepts and impacts of national boundaries may fundamentally change with new and unanticipated consequences for patterns of spatial interaction.

Information Flows

Spatially significant information flows are of two types: individual (person-to-person) exchanges and mass (source-to-area) communication. A further subdivision into formal and informal interchange recognizes, in the former, the need for an interposed channel (radio, press, postal service, or telephone, for example) to convey messages. Informal communication requires no such institutionalized message carrier.

Short-range informal *individual communication* is as old as humankind itself. Contacts and exchanges between individuals and within small groups tend to increase as the complexity of social organization increases, as the size and importance of the population center grow, and as the range of interests and associations of the communicating person expands. Each individual develops a **personal communication field,** the informational counterpart of that person's activity space. Its size and shape are defined by the individual's contacts in work, recreation, shopping, school, or other regular activities. Those activities, as we have seen, are functions of the age, sex, education, employment, income, and so on of each person. An idealized personal communication field is suggested in Figure 3.17.

Each interpersonal exchange constitutes a link in the individual's personal communication field. Each person, in turn, is a node in the communication field of those with whom he or she makes or maintains contact. The total number of such separate informal networks equals the total count of people alive. Despite the number of those networks, all people, in theory, are interconnected by multiple shared nodes (Figure 3.18). One debated experiment suggested that through such interconnections no person in the United States is more than six links removed from any other person, no matter where located or how unlikely the association.

Mass communication is the formal, structured transmission of information in essentially a one-way flow between single points of origin and broad areas of reception. There are few transmitters and many receivers. The mass media are by nature "space filling." From single origin points, they address their messages by print, radio, or television to potential receivers within a defined area. The number and location of disseminating points, therefore, are related to their spatial coverage characteristics, to the minimum size of area and population necessary for their support, and to the capability of the potential audiences to receive their message. The

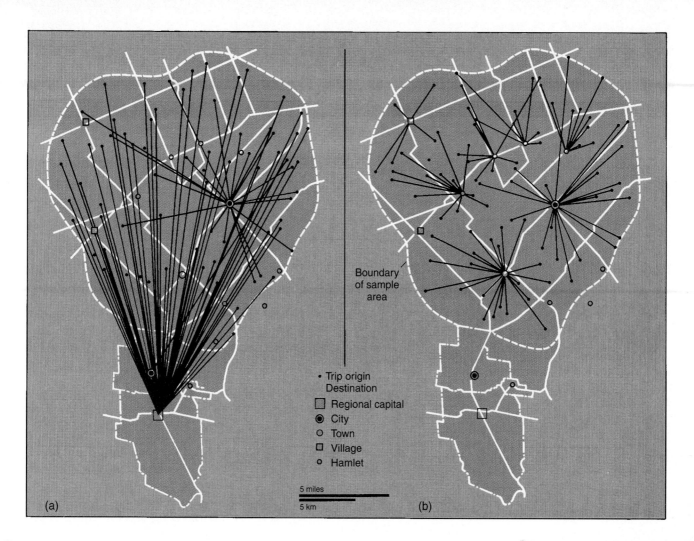

Trip origin
Destination
- ☐ Regional capital
- ◉ City
- ○ Town
- ▢ Village
- ○ Hamlet

5 miles
5 km

(a)   (b)

Boundary of sample area

Figure 3.13 **Travel patterns for purchases of clothing and yard goods** of (*a*) rural cash-economy Canadians and (*b*) Canadians of the Old Order Mennonite sect. These strikingly different travel behaviors mapped many years ago in midwestern Canada demonstrate the great differences that may exist in the action spaces of different culture groups occupying the same territory. At that time, "modern" rural Canadians, owning cars and wishing to take advantage of the variety of goods offered in the more distant regional capital, were willing and able to travel longer distances than were neighboring people of a traditionalist culture who had different mobility and whose different demands in clothing and other consumer goods were, by preference or necessity, satisfied in nearby small settlements. Unpublished studies suggest similar contrasts in mobility and purchase travel patterns currently exist between buggy-using Old Order Amish (see Figure 7.2) and their car-driving neighbors.

Source: Robert A. Murdie, "Cultural Differences in Consumer Travel," Economic Geography *41, no. 3 (Worcester, Mass.: Clark University, 1965). Redrawn by permission.*

coverage area is determined both by the nature of the medium and by the corporate intent of the agency.

There are no inherent spatial restrictions on the dissemination of printed materials. In the United States, much book and national magazine publishing has localized in metropolitan New York City, as have the services supplying news and features for sale to the print media located there and elsewhere in the country. Paris, Buenos Aires, Moscow, London—indeed, the major metropolises and/or capital cities of other countries—show the same spatial concentration. Regional journals emanate from regional capitals, and major metropolitan newspapers, though serving primarily their home markets, are distributed over (or produce special editions for distribution within) tributary areas whose size and shape depend on the intensity of competition from other metropolises. A spatial information hierarchy has thus emerged.

Hierarchies are also reflected in the market-size requirements for different levels of media offerings. National and international organizations are required to expedite information flows (and, perhaps, to control their content), but market demand is heavily weighted in favor of regional and local coverage. In the electronic media, the result has been national networks with local affiliates acting as the gatekeepers of network offerings and adding to them locally originating programs and news content. A similar market subdivision is represented by the regional editions of national newspapers and magazines.

The technological ability to fill space with messages from different mass media is unavailing if receiving audiences do not exist. In illiterate societies, publications cannot inform or influence. Unless the appropriate receivers are widely available, television and radio broadcasts are a waste of resources. Perhaps no

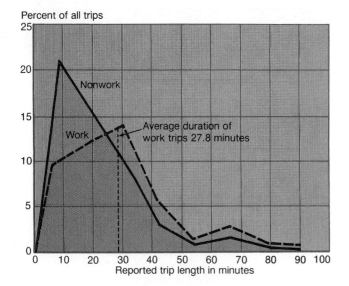

Percent of all trips

Figure 3.14 **The frequency distribution of work and nonwork trip lengths in minutes in Toronto.** More recent studies in different metropolitan areas support the conclusions documented by this graph: work trips are usually longer than other recurring journeys. In the United States in the early 1990s, the average work trip covered 17.1 kilometers (10.6 mi) and half of all trips to work took under 22 minutes; for suburbanites commuting to the central business district, the journey to work involved between 30 and 45 minutes. By 2000, increasing sprawl had lengthened average commuting distances and, because of growing traffic congestion, had increased the average work trip commuting time to 25 minutes; 15% of workers had commutes of more than 45 minutes. The situation is similar elsewhere; in the middle 1990s, the average British commuting distance was 12.5 kilometers. Most nonwork trips in all countries are relatively short.

Source: Maurice Yeates, Metropolitan Toronto and Region Transportation Study, *figure 42, The Queen's Printer, Toronto: 1966.*

invention in history has done more to weld isolated individuals and purely person-to-person communicators into national societies exposed to centralized information flows than has the low-cost transistor radio. Its battery-powered transportability converts the remotest village and the most isolated individual into a receiving node of entertainment, information, and political messages. The direct satellite broadcast of television programs to community antennae or communal sets brings that mass medium to remote areas of Arctic Canada, India, Indonesia, and other world areas able to invest in the technology but as yet unserved by ground stations.

Information and Perception

Human spatial interaction, as we have seen, is conditioned by a number of factors. Complementarity, transferability, and intervening opportunities help pattern the movement of commodities and peoples. Flows between points and over area are influenced by distance decay and partially explained by gravity and potential models. Individuals in their daily affairs operate in activity spaces that are partly determined by stage in life, mobility, and a variety

of socioeconomic characteristics. In every instance of spatial interaction, however, decisions are based on information about opportunity or feasibility of movement, exchange, or want satisfaction.

More precisely, actions and decisions are based on **place perception**—the awareness we have, as individuals, of home and distant places and the beliefs we hold about them. Place perception involves our feelings and understandings, reasoned or irrational, about the natural and cultural characteristics of an area and about its opportunity structure. Whether our view accords with that of others or truly reflects the "real" world seen in abstract descriptive terms is not the major concern. Our perceptions are the important thing, for the decisions people make about the use of their lives or about their actions in space are based not necessarily on reality but on their assumptions and impressions of reality.

Perception of Environment

Psychologists and geographers are interested in determining how we arrive at our perceptions of place and environment both within and beyond our normal activity space. The images we form firsthand of our home territory have been in part reviewed in the discussion of mental maps in Chapter 1. The perceptions we have of more distant places are less directly derived (Figure 3.19). In technologically advanced societies, television and radio, magazines and newspapers, books and lectures, travel brochures and hearsay all combine to help us develop a mental picture of unfamiliar places and of the interaction opportunities

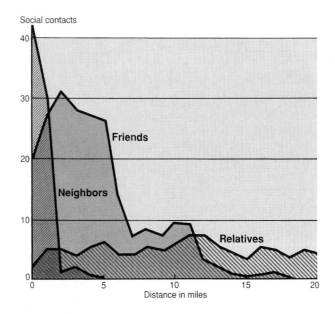

Figure 3.15 **Social interaction as a function of distance.** Visits with neighbors on the same street are frequent; they are less common with neighbors around the corner and diminish quickly to the vanishing point after a residential relocation. Friends exert a greater spatial pull, though the distance decay factor is clearly evident. Visits with relatives offer the greatest incentive for longer distance (though relatively infrequent) journeys.

Source: Frederick P. Stutz, "Distance and Network Effects on Urban Social Travel Fields," Economic Geography *49, no. 2 (Worcester, Mass.: Clark University, 1973), p. 139. Redrawn by permission.*

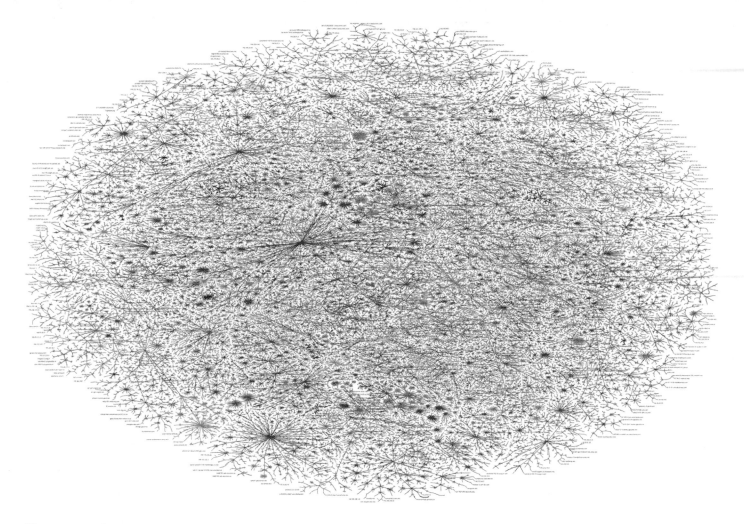

Figure 2.16 **The Internet.** This map was created on May 3, 1999, by sending test data from a networked computer in Murray Hill, New Jersey, to each of the 95,842 networks in 137 countries then listed in the global Internet register kept by Merit Network, Inc., and graphing the resulting connections. Colors show the Internet domains where network switches (routers) were registered.

Source: Mercator's World, *Nov./Dec. 1999, p. 80. Courtesy of Peacock Maps, Washington, D.C.*

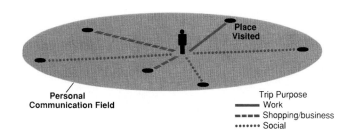

Figure 2.17 **A personal communication field** is determined by individual spatial patterns of communication related to work, shopping, business trips, social visits, and so on.

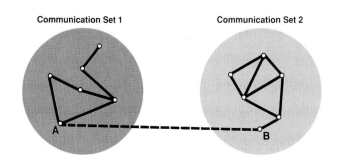

Figure 2.18 Separate population sets are interconnected by the links between individuals. If link A–B exists, everyone in the two sets is linked.

they may contain. Again, however, the most effectively transmitted information seems to come from word-of-mouth reports. These may be in the form of letters or visits from relatives, friends, and associates who supply information that helps us develop lines of attachment to relatively unknown areas.

There are, of course, barriers to the flow of information, including that of distance decay. Our knowledge of close places is greater than our knowledge of distant points; our contacts with nearby persons theoretically yield more information than we receive from afar. Yet in crowded areas with maximum interaction

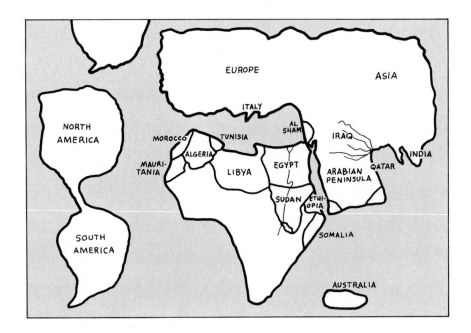

Figure 3.19 **A Palestinian student's view of the world.** The map was drawn by a Palestinian high-school student from Gaza. The map reflects the instruction and classroom impressions the student has received. The Gaza curriculum conforms to the Egyptian national standards and thus is influenced by the importance of the Nile River and pan-Arabism. Al Sham is the old, but still used, name for the area including Syria, Lebanon, and Palestine. The map might be quite different in emphasis if the Gaza school curriculum were designed by Palestinians or if it had been drawn by an Israeli student.

potential, people commonly set psychological barriers around themselves so that only a limited number of those possible interactions and information exchanges actually occur. We raise barriers against information overload and to preserve a sense of privacy that permits the filtering out of information that does not directly affect us. There are obvious barriers to long-distance information flows as well, such as time and money costs, mountains, oceans, rivers, and differing religions, languages, ideologies, and political systems.

Barriers to information flow give rise to what we earlier (p. 71) called *direction bias*. In the present usage, this implies a tendency to have greater knowledge of places in some directions than in others. Not having friends or relatives in one part of a country may represent a barrier to individuals, so interest in and knowledge of the area beyond the "unknown" region are low. In the United States, both northerners and southerners tend to be less well informed about each other's areas than about the western part of the country. Traditional communication lines in the United States follow an east-west rather than a north-south direction, the result of early migration patterns, business connections, and the pattern of the development of major cities. In Russia, directional bias favors a north-south information flow within the European part of the country and less familiarity with areas far to the east. Within Siberia, however, east-west flows dominate.

When information about a place is sketchy, blurred pictures develop. These influence the impression—the perception—we have of places and cannot be discounted. Many important decisions are made on the basis of incomplete information or biased reports, such as decisions to visit or not, to migrate or not, to hate or not, even to make war or not. Awareness of places is usually accompanied by opinions about them, but there is no necessary relationship between the depth of knowledge and the perceptions held. In general, the more familiar we are with a locale, the more sound the factual basis of our mental image of it will be. But individuals form firm impressions of places totally unknown to them personally, and these may color interaction decisions.

One way to determine how individuals envisage home or distant places is to ask them what they think of different locales. For instance, they may be asked to rate places according to desirability—perhaps residential desirability—or to make a list of the 10 best and the 10 worst cities in their country of residence. Certain regularities appear in such inquiries. Figure 3.20 presents some residential desirability data elicited from college students in three provinces of Canada. These and comparable mental maps derived from studies conducted by researchers in many countries suggest that near places are preferred to far places unless much information is available about the far places. Places of similar culture are favored, as are places with high standards of living. Individuals tend to be indifferent to unfamiliar places and areas and to dislike those that have competing interests (such as distasteful political and military activities or conflicting economic concerns) or a physical environment perceived to be unpleasant.

On the other hand, places perceived to have superior climates or landscape amenities are rated highly in mental map studies and favored in tourism and migration decisions. The southern and southwestern coast of England is attractive to citizens of generally wet and cloudy Britain, and holiday tours to Spain, the south of France, and the Mediterranean islands are heavily booked by the English. A U.S. Census Bureau study indicates that "climate" is, after work and family proximity, the most often reported reason

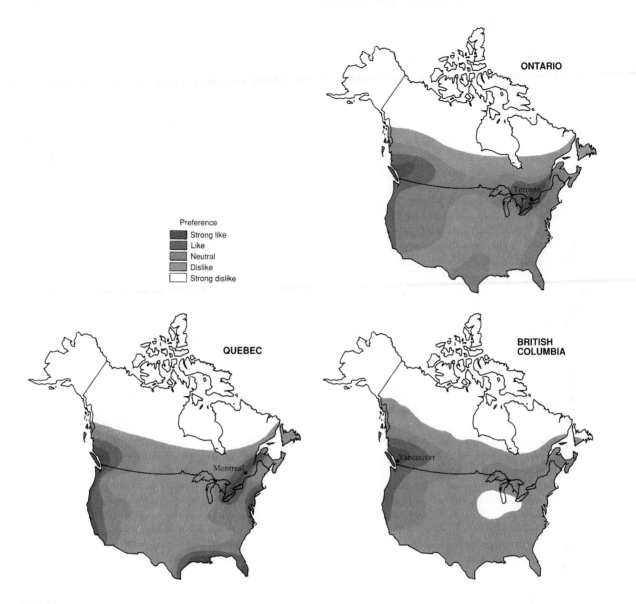

Preference
- Strong like
- Like
- Neutral
- Dislike
- Strong dislike

𝒻igure 3.20 **Residential preferences of Canadians.** Each of these maps shows the residential preference of a sampled group of Canadians from the Provinces of British Columbia, Ontario, and Quebec, respectively. Note that each group of respondents prefers its own area, but all like the Canadian and U.S. west coasts.

Source: Herbert A. Whitney, "Preferred Locations in North America: Canadians, Clues, and Conjectures," Journal of Geography 83, no. 5, p. 222. (Indiana, Pa.: National Council for Geographic Education, 1984). Redrawn by permission.

for interstate moves by adults of all ages. International studies reveal a similar migration motivation based not only on climate but also on concepts of natural beauty and amenities.

Perception of Natural Hazards

Less certain is the negative impact on spatial interaction or relocation decisions of assessments of *natural hazards,* processes or events in the physical environment that are not caused by humans but that have consequences harmful to them. Distinction is made between chronic, low-level hazards (health-affecting mineral content of drinking water, for example) and high-consequence/low-probability events such as hurricanes, earthquakes, landslides, and the like. Remedial low-level hazards do

not appear to create negative space perceptions, though highly publicized chronic natural conditions, such as suspected cancer-related radon emissions (Figure 3.21) may be an exception. Space perception studies do reveal, however, a small but measurable adverse assessment of locales deemed "dangerous," no matter what the statistical probability of the hazard occurring.

Mental images of home areas do not generally include as an overriding concern an acknowledgment of potential natural dangers. The cyclone that struck the delta area of Bangladesh on November 12, 1970, left at least 500,000 people dead, yet after the disaster the movement of people into the area swelled population above precyclone levels—a resettlement repeated after other, more recent cyclones. The July 28, 1976, earthquake in the Tangshan

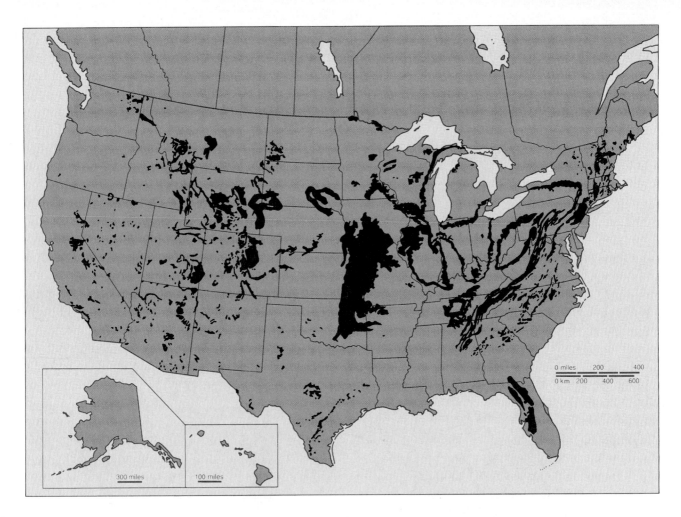

Figure 3.21 **Areas with potentially high radon levels.** The radon "scare" began in 1984 with the discovery that a Pennsylvania family was being exposed in its home to the equivalent of 455,000 chest X rays per year. With the estimate that as many as 20% of the nation's annual lung cancer deaths may be attributable to radon, homeowners and seekers were made aware of a presumed new but localized environmental hazard. More recent reassessments suggest the earlier warnings of radon danger were partially or largely unwarranted.

Source: U.S. Environmental Protection Agency, August 1987.

area of China devastated a major urban industrial complex, with casualties estimated at about a quarter-million, and between 50,000 and 100,000 city dwellers and villagers reportedly perished during and after the January, 2001, quake in Gujarat state of western India. In both cases, rebuilding began almost immediately, as it usually does following earthquake damage (Figure 3.22). The human response to even such major and exceptional natural hazards is duplicated by a general tendency to discount dangers from more common hazard occurrences. Johnstown, Pennsylvania, has suffered recurrent floods, and yet its residents rebuild; violent storms strike the Gulf and East coasts of the United States, and people remain or return. Californians may be concerned about Kansas tornadoes if contemplating a move there but be unconcerned about earthquake dangers at home.

Why do people choose to settle in areas of high-consequence hazards in spite of the potential threat to their lives and property? Why do hundreds of thousands of people live along the San Andreas Fault in California, build houses in Pacific coastal areas known to experience severe erosion during storms, return to flood-

prone river valleys in Europe or Asia, or avalanche-threatened Andean valleys? What is it that makes the risk worth taking? Ignorance of natural hazard danger is not necessarily a consideration. People in seismically active regions of the United States and Europe, at least, do believe that damaging earthquakes are a possibility in their districts but, research indicates, are reluctant to do anything about the risk. Similar awareness and reticence accompanies other low-incidence/high-consequence natural dangers. Less than one-tenth of 1% of respondents to a federal survey gave "natural disaster" as the reason for their interstate residential move.

There are many reasons why natural hazard risk does not deter settlement or adversely affect space-behavioral decisions. Of importance, of course, is the persistent belief that the likelihood of an earthquake or a flood or other natural calamity is sufficiently remote so that it is not reasonable or pressing to modify behavior because of it. People are influenced by their innate optimism and the predictive uncertainty about timing or severity of a calamitous event and by their past experiences in high-hazard

Figure 3.22 Destruction from the San Francisco earthquake and fire. The first shock struck San Francisco early on the morning of April 18, 1906, damaging the city's water system. Fire broke out and raged for 3 days. It was finally stopped by dynamiting buildings in its path. When it was over, some 700 people were dead or missing, and 25,000 buildings had been destroyed. Locally, the event is usually referred to as the Great Fire of 1906, suggesting a denial of the natural hazard in favor of assigning blame to correctable human error. Post-destruction reconstruction began at once. Rebuilding following earthquake damage is the rule, though the immediate return of population to northern Italian areas after a major quake in 1976 was followed by an abrupt longer-term exodus after a subsequent, much weaker shock.

areas. If they have not suffered much damage in the past, they may be optimistic about the future. If, on the other hand, past damage has been great, they may think that the probability of repetition in the future is low (Table 3.1).

Perception of place as attractive or desirable may be quite divorced from any understanding of its hazard potential. Attachment to locale or region may be an expression of emotion and economic or cultural attraction, not just a rational assessment of risk. The culture hearths of antiquity discussed in Chapter 2 and shown on Figure 2.15 were for the most part sited in flood-prone river valleys; their enduring attraction was undiminished by that potential danger. The home area, whatever disadvantages an outside observer may discern, exerts a force not easily dismissed or ignored.

Indeed, high-hazard areas are often sought out because they possess desirable topography or scenic views, as do, for instance, coastal areas subject to storm damage. Once people have purchased property in a known hazard area, they may be unable to sell it for a reasonable price even if they so desire. They think that they have no choice but to remain and protect their investment. The cultural hazard—loss of livelihood and investment—appears more serious than whatever natural hazards there may be.

Carried further, it has been observed that spatial adjustment to perceived natural hazards is a luxury not affordable to impoverished people in general or to the urban and rural poor of Third World

Table 3.1
Common Responses to the Uncertainty of Natural Hazards

Eliminate the Hazard

Deny or Denigrate Its Existence	Deny or Denigrate Its Recurrence
"We have no floods here, only high water."	"Lightning never strikes twice in the same place."
"It can't happen here."	"It's a freak of nature."

Eliminate the Uncertainty

Make It Determinate and Knowable	Transfer Uncertainty to a Higher Power
"Seven years of great plenty. . . . After them seven years of famine."	"It's in the hands of God."
"Floods come every five years."	"The government is taking care of it."

Burton and Kates, The Perception of Natural Hazards in Resource Management, 3 *Natural Resources Journal* 435 (1964). Used by permission of the University of New Mexico School of Law, Albuquerque, N. M.

countries in particular. Forced by population growth and economic necessity to exert ever-greater pressures upon fragile environments or to occupy at higher densities hazardous hillside and floodplain slums, their margin of safety in the face of both chronic and low-probability hazards is minimal to nonexistent (Figure 3.23).

Migration

When continental glaciers began their retreat some 11,000 years ago, the activity space and awareness space of Stone Age humans were limited. As a result of pressures of numbers, need for food, changes in climate, and other inducements, those spaces were collectively enlarged to encompass the world. **Migration**—the permanent relocation of residential place and activity space—has been one of the enduring themes of human history. It has contributed to the evolution of separate cultures, to the diffusion of those cultures and their components by interchange and commu-

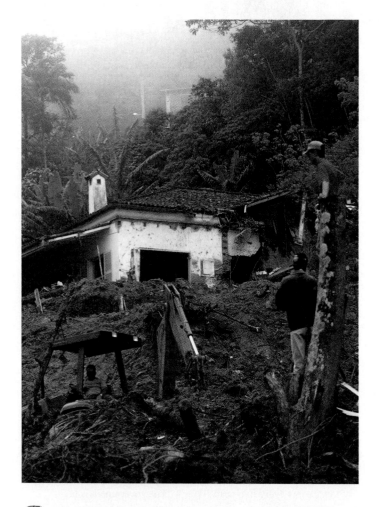

Figure 3.23 Many of the poor of Rio de Janeiro, Brazil, occupy steep hillside locations above the reach of sewer, water, and power lines that hold the more affluent at lower elevations. Frequent heavy rains cause mudflows from the saturated hillsides that wipe away the shacks and shelters that insecurely cling to them, and deposit the homes and hopes of the poor in richer neighborhoods below.

nication, and to the frequently complex mix of peoples and cultures found in different areas of the world.

Massive movements of people within countries, across national borders, and between continents have emerged as a pressing concern of recent decades. They affect national economic structures, determine population density and distribution patterns, alter traditional ethnic, linguistic, and religious mixtures, and inflame national debates and international tensions. Because migration patterns and conflicts touch so many aspects of social and economic relations and have become so important a part of current human geographic realities, their specific impact is a significant aspect of several of our topical concerns. Portions of the story of migration have been touched on already in Chapter 2; other elements of it are part of later discussions of population (Chapter 4), ethnicity (Chapter 6), economic development (Chapter 10), urbanization (Chapter 11), and international political relations (Chapter12). Because migration is above all the result of individual and family decisions, our interest here is with migration as an unmistakable, recurring, and near-universal expression of human spatial behavior. Reviewing that behavioral basis of migration now will give us common ground for understanding its impacts in other contexts later.

Migration embodies all the principles of spatial interaction and space relations we have already discussed. Complementarity, transferability, and intervening opportunities and barriers all play a role. Space information and perception are important, as are the sociocultural and economic characteristics of the migrants and the distance relationships between their original and prospective locations of settlement. In less abstract terms, mass and individual migration decisions may express real-life responses to poverty, rapid population growth, environmental deterioration, or international and civil conflict or war. In its current troubling dimensions, migration may be as much a strategy for survival as an unforced but reasoned response to economic and social opportunity.

Naturally, the length of a specific move and its degree of disruption of established activity space patterns raise distinctions important in the study of migration. A change of residence from the central city to the suburbs certainly changes both residence and activity space of schoolchildren and of adults in many of their nonworking activities, but the working adults may still retain the city—indeed, the same place of employment there—as an action space. On the other hand, immigration from Europe to the United States and the massive farm-to-city movements of rural Americans late in the 19th and early in the 20th centuries clearly meant a total change of all aspects of behavioral patterns.

Principal Migration Patterns

Migration flows may be discussed at different scales, from massive intercontinental torrents to individual decisions to move to a new house or apartment within the same metropolitan area. At each level, although the underlying controls on spatial behavior remain constant, the immediate motivating factors influencing the spatial interaction are different, with differing impacts on population patterns and cultural landscapes.

At the broadest scale, *intercontinental* movements range from the earliest peopling of the habitable world to the most recent flight of Asian or African refugees to countries of Europe or the Western Hemisphere. The population structure of the United States, Canada, Australia and New Zealand, Argentina, Brazil, and other South American countries—as Chapter 4 suggests—is a reflection and result of massive intercontinental flows of immigrants that began as a trickle during the 16th and 17th centuries and reached a flood during the 19th and early 20th (Figure 4.21). Later in the 20th century, World War II (1939–1945) and its immediate aftermath involved more than 25 million permanent population relocations, all of them international but not all intercontinental.

Intracontinental and *interregional* migrations involve movements between countries and within countries, most commonly in response to individual and group assessments of improved economic prospects, but often reflecting flight from difficult or dangerous environmental, military, economic, or political conditions. The millions of refugees leaving their homelands following the dissolution of Eastern European communist states, including the former USSR and Yugoslavia, exemplify that kind of flight. Between 1980 and 2000, Europe received some 20 million newcomers, often refugees, who joined the 15 million labor migrants ("guest workers") already in West European countries by the early 1990s (Figure 3.24).

North America has its counterparts in the hundreds of thousands of immigrants coming (many illegally) to the United States each year from Mexico, Central America, and the Caribbean region. The Hauns, whose westward trek opened this chapter, were part of a massive 19th-century regional shift of Americans that continues today (Figure 3.25). Russia experienced a similar, though eastward, flow of people in the 20th century. Some 175 million people—3% of world population—lived in a country other than the country of their birth in the early 2000s, and migration had become a world social, economic, and political issue of first priority.

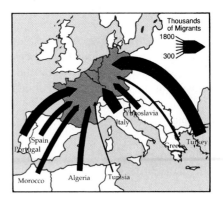

Figure 3.24 **International "guest worker" flows to Western Europe.** Labor shortages in expanding Western European economies beginning in the 1960s offered job opportunities to workers immigrating under labor contract from Eastern and Southern Europe and North Africa. Economic stagnation and domestic unemployment halted foreign worker contracting in Germany, France, Belgium, Netherlands, and Switzerland in the later 1980s and 1990s, but continuing immigration raised the share of foreign workers in the labor force to 20% in Switzerland, 10% in Austria, and 9.5% in Germany by 2000.

Source: Data from Gunther Glebe and John O'Loughlin, eds., "Foreign Minorities in Continental European Cities," Erdkundliches Wissen 84 (Wiesbaden, Germany: Franz Steiner Verlag, 1987).

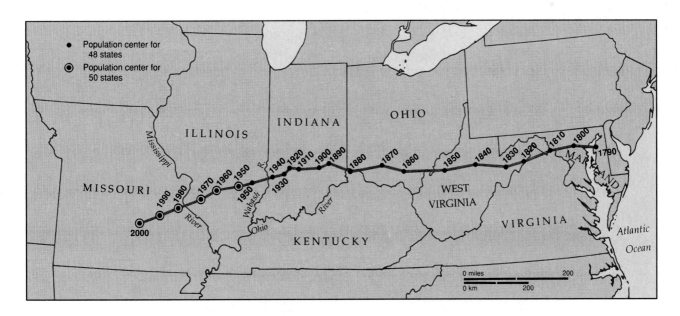

Figure 3.25 **Westward shift of population, 1790–2000.** More than 200 years of western migration and population growth are recorded by the changing U.S. center of population. (The "center of population" is that point at which a rigid map of the United States would balance, reflecting the identical weights of all residents in their location on the census date.) The westward movement was rapid for the first 100 years of census history and slowed between 1890 and 1950. Some of the post-1950 acceleration reflects population growth in the "Sunbelt." However, the two different locations for the population center in 1950 and the symbol change indicate the geographic pull on the center of population exerted by the admission of Alaska and Hawaii to statehood.

In the 20th century, nearly all countries experienced a great movement of peoples from agricultural areas to the cities, continuing a pattern of *rural-to-urban* migration that first became prominent during the 18th- and 19th-century Industrial Revolution in advanced economies and now is even more massive than international migrant flows. Rapid increases in impoverished rural populations of developing countries put increasing and unsustainable pressures on land, fuel, and water in the countryside. Landlessness and hunger as well as the loss of social cohesion that growing competition for declining resources induces help force migration to cities. As a result, while the rate of urban growth is decreasing in the more developed countries, urbanization in the developing world continues apace, as will be discussed more fully in Chapter 11.

Types of Migration

Migrations may be forced or voluntary or, in many instances, reluctant relocations imposed on the migrants by circumstances.

In *forced migrations*, the relocation decision is made solely by people other than the migrants themselves (Figure 3.26). Perhaps 10 to 12 million Africans were forcibly transferred as slaves to the Western Hemisphere from the late 16th to early 19th centuries. Half or more were destined for the Caribbean and most of the remainder for Central and South America, though nearly a million arrived in the United States. Australia owed its earliest European settlement to convicts transported after the 1780s to the British penal colony established in southeastern Australia (New South Wales). More recent involuntary migrants include millions of Soviet citizens forcibly relocated from countryside to cities and from the western areas to labor camps in Siberia and the Russian Far East beginning in the late 1920s. During the 1980s and 90s, many refugee destination countries in Africa, Europe, and Asia expelled immigrants or encouraged or forced the repatriation of foreign nationals within their borders.

Less than fully voluntary migration—*reluctant relocation*—of some 8 million Indonesians has taken place under an aggressive governmental campaign begun in 1969 to move people from densely settled Java (roughly 775 per square kilometer or 2000 people per square mile) to other islands and territories of the country in what has been called the "biggest colonization program in history." International refugees from war and political turmoil or repression numbered some 15 million in 2003, according to the World Refugee Survey—one out of every 415 people on the planet. In the past, refugees sought asylum mainly in Europe and other developed areas. More recently, the flight of people is primarily from developing countries to other developing regions, and many countries with the largest refugee populations are among the world's poorest. Sub-Saharan Africa alone housed over 3 million refugees (Figure 3.27). Worldwide, an additional 22 million persons were "internally displaced," effectively internal refugees within their own countries. In a search for security or sustenance, they have left their home areas but not crossed an international boundary.

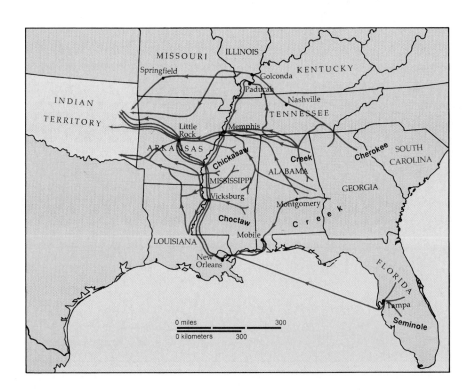

Figure 3.26 **Forced migrations: The Five Civilized Tribes.** Between 1825 and 1840, some 100,000 southeastern Amerindians were removed from their homelands and transferred by the Army across the Mississippi River to "Indian Territory" in present-day Oklahoma. By far, the largest number were members of the Five Civilized Tribes of the South: Cherokees, Choctaws, Chickasaws, Creeks, and Seminoles. Settled, Christianized, literate small-farmers, their forced eviction and arduous journey—particularly along what the Cherokees named their "Trail of Tears" in the harsh winter of 1837–1838—resulted in much suffering and death.

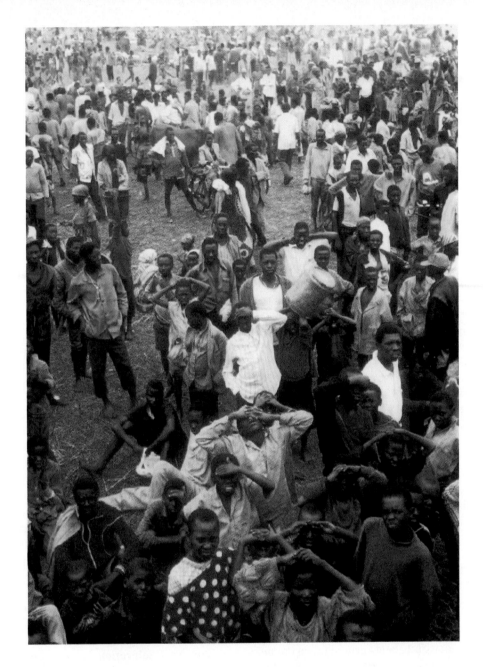

Figure 2.27 Rwandan refugees near the border of Rwanda and Tanzania. More than 1 million Rwandans fled into neighboring Zaire (now, the Democratic Republic of the Congo), Tanzania, Uganda, and Burundi in 1994 to escape an interethnic civil war in their home country and the genocide that killed at least 750,000 people. Early in the 21st century, more than 14 million Africans remained uprooted (that is, internally displaced and refugees combined). Fleeing war, repression, and famine, millions of people in developing nations have become reluctant migrants from their homelands.

The great majority of migratory movements, however, are *voluntary* (volitional), representing individual response to the factors influencing all spatial interaction decisions. At root, migrations take place because the migrants believe that their opportunities and life circumstances will be better at their destination than they are at their present location.

Poverty is the great motivator. Some 30% of the world's population—nearly 2 billion persons—have less than $1.00 per day income. Many additionally are victims of drought, floods, other natural catastrophes or of wars and terrorism. Poverty in de-

veloping countries is greatest in the countryside; rural areas are home to around 750 million of the world's poorest people. Of these, some 20 to 30 million move each year to towns and cities, many as "environmental refugees" abandoning land so eroded or exhausted it can no longer support them. In the cities, they join the 40% or more of the labor force that is unemployed or underemployed in their home country and seek legal or illegal entry into more promising economies of the developed world. All, rural or urban, respond to the same basic forces—the push of poverty and the pull of perceived or hoped-for opportunity.

Controls on Migration

Economic considerations crystallize most migration decisions, though nomads fleeing the famine and spreading deserts of the Sahel obviously are impelled by different economic imperatives than is the executive considering a job transfer to Montreal or the resident of Appalachia seeking factory employment in the city. Among the aging, affluent populations of highly developed countries, retirement amenities figure importantly in perceptions of residential attractiveness of areas. Educational opportunities, changes in life cycle, and environmental attractions or repulsions are but a few other possible migration motivations.

Migration theorists attribute international economic migrations to a series of often overlapping mechanisms. Differentials in wages and job opportunities between home and destination countries are perhaps the major driving force in such individual migration decisions. Those differentials are in part rooted in a built-in demand for workers at the bottom of the labor hierarchy in more prosperous developed countries whose own workers disdain low-income, menial jobs. Migrants are available to fill those jobs, some argue, because advanced economies make industrial investment in developing or colonial economies to take advantage of lower labor costs there. New factories inevitably disturb existing peasant economies, employ primarily short-term female workers, and leave a residue of unemployed males available and prone to migrate in search of opportunity. If successful, international economic migrants, male or female, help diversify sources of family income through their remittances from abroad, a form of household security that in itself helps motivate some international economic migration.

Negative home conditions that impel the decision to migrate are called **push factors.** They might include loss of job, lack of professional opportunity, overcrowding or slum clearance, or a variety of other influences including poverty, war, and famine. The presumed positive attractions of the migration destination are known as **pull factors.** They include all the attractive attributes perceived to exist at the new location—safety and food, perhaps, or job opportunities, better climate, lower taxes, more room, and so forth. Very often migration is a result of both perceived push and pull factors. It is *perception* of the areal pattern of opportunities and want satisfaction that is important here, whether or not perceptions are supported by objective reality. In China, for example, a "floating" population of more than 100 million surplus workers has flooded into cities from the countryside, seeking urban employment that exists primarily in their anticipation.

The concept of *place utility* helps us to understand the decision-making process that potential voluntary migrants undergo. **Place utility** is the measure of an individual's satisfaction with a given residential location. The decision to migrate is a reflection of the appraisal—the perception—by the prospective migrant of the current homesite as opposed to other sites of which something is known or hoped for. In the evaluation of comparative place utility, the decision maker considers not only perceived value of the present location, but also expected place utility of potential destinations.

Those evaluations are matched with the individual's *aspiration level,* that is, the level of accomplishment or ambition that the person sees for herself or himself. Aspirations tend to be adjusted to what one considers attainable. If one finds present circumstances satisfactory, then **spatial search** behavior—the process by which locational alternatives are evaluated—is not initiated. If, on the other hand, dissatisfaction with the home location is felt, then a utility is assigned to each of the possible migration locations. The utility is based on past or expected future rewards at various sites. Because new places are unfamiliar to the searcher, the information received about them acts as a substitute for the personal experience of the homesite. Decision makers can do no more than sample information about place alternatives and, of course, there may be errors in both information and interpretation. Ultimately, they depend on their image—perhaps a mental map—of the place being considered and on the motivations that impel them to consider long distance migration or even local area relocation of residence. In the latter instance, of course, the spatial search usually involves actual site visits in evaluating the potential move (Figure 3.28).

One goal of the potential migrant is to avoid physically dangerous or economically unprofitable outcomes in the final migration decision. Place utility evaluation, therefore, requires assessments not only of hoped-for pull factors of new sites but also of the potentially negative economic and social reception the migrant might experience at those sites. An example of that observation can be seen in the case of the large numbers of young Mexicans and Central Americans who have migrated both legally and illegally to the United States (Figure 3.29). Faced with poverty and overpopulation at home, they regard the place utility in Mexico as minimal. With a willingness to work, they learn from friends and relatives of job opportunities north of the border and, hoping for success or even wealth, quickly place high utility on relocation to the United States. Many know that dangerous risks are involved in entering the country illegally, but even legal immigrants face legal restrictions or rejections that are advocated or designed to reduce the pull attractions of the United States (see "Backlash").

Another migrant goal is to reduce uncertainty. That objective may be achieved either through a series of transitional relocation stages or when the migrant follows the example of known predecessors. **Step migration** involves the place transition from, for example, rural to central city residence through a series of less extreme locational changes—from farm to small town to suburb and, finally, to the major central city itself. **Chain migration** assures that the mover is part of an established migrant flow from a common origin to a prepared destination. An advance group of migrants, having established itself in a new home area, is followed by second and subsequent migrations originating in the same home district and frequently united by kinship or friendship ties. Public and private services for legal migrants and informal service networks for undocumented or illegal migrants become established and contribute to the continuation or expansion of the chain migration flow. Ethnic and foreign-born enclaves in major cities and rural areas in a number of countries are the immediate result, as we shall see more fully in Chapter 6.

Sometimes the chain migration is specific to occupational groups. For example, nearly all newspaper vendors in New Delhi, in the north of India, come from one small district in Tamil Nadu,

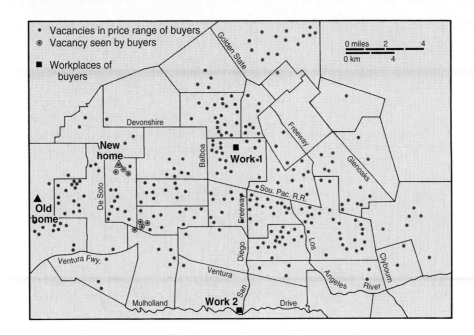

Figure 3.28 **An example of a residential spatial search.** The dots represent the house vacancies in the price range of a sample family. Note (1) the relationship of the new house location to the workplaces of the married couple; (2) the relationship of the old house location to the chosen new home site; and (3) the limited total area of the spatial search. This example from the San Fernando Valley area of Los Angeles is typical of intraurban moves.

Redrawn by permission from J. O. Huff, Annals of the Association of American Geographers, *Vol. 76, pp. 217–221. Association of American Geographers, 1986.*

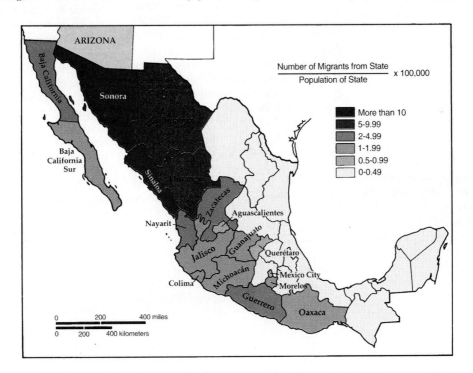

Figure 3.29 **Undocumented migration rate to Arizona.** The Arizona region and nearby Mexican states have long-standing ties that reach back to the early 1800s. In many respects, the international border cuts through a cultural region. Note that distance plays a large role in the decision to migrate to the United States. Over half of the migrants come from three nearby Mexican states: Sonora, Sinaloa, and Chihuahua.

Redrawn from John Harner, "Continuity Amidst Change," The Professional Geographer *47, no. 4. Fig. 2, p. 403. Association of American Geographers, 1995.*

Geography and Public Policy

Backlash

Migrants can enter a country legally—with a passport, visa, work permit, or other authorization—or illegally. Some aliens initially enter a country legally but on a temporary basis (as a student or tourist, for example), but then remain after their departure date. Others may arrive claiming the right of political asylum but actually seeking economic opportunity. Recent years have seen a rising tide of emotion against the estimated 5 million people who reside illegally in the United States, a sentiment that has been reflected in a number of actions.

- Security fears since the September 11, 2001, World Trade Center and Pentagon assaults have led to more stringent visa applicant background checks, greater restrictions on admitting refugees and asylum seekers, tighter border controls, and stricter enforcement of Immigration and Naturalization Service (INS) rules on alien residency reports and visa time restrictions.

- Greater efforts are being made to deter illegal crossings along the Mexican border by increasing the number of Border Patrol agents and by building steel fences near El Paso, Texas, Nogales, Arizona, San Ysidro, California, and elsewhere.

- Four states—Florida, Texas, Arizona, and California—are suing the federal government to win reimbursement for their costs of illegal immigration.

- The U.S.–Mexico Border Counties Coalition, composed of representatives from the 24 counties in the United States that abut Mexico, is demanding that the federal government reimburse local administrations for money spent on legal and medical services for undocumented aliens. These services include the detention, prosecution, and defense of immigrants, emergency medical care, ambulance service, even autopsies and burials for those who die while trying to cross the border.

- The governor of California, which is home to an estimated 2.2 million illegal immigrants, proposed an amendment to the Constitution to deny citizenship to children born on American soil if their parents are not legal residents of the United States.

In the last several years, California voters have approved a trio of ballot initiatives aimed at curbing what their proponents see as unwarranted privileges for immigrants. Proposition 187, passed in November, 1994, prohibited state and local government agencies from providing publicly funded education, nonemergency health care, welfare benefits, and social services to any person they could not verify as either a U.S. citizen or a person legally admitted to the country. The measure also required state and local agencies to report suspected illegal immigrants to the Immigration and Naturalization Service and to certain state officials.

Proponents of Proposition 187 argued that California could no longer support the burden of high levels of immigration, especially if the immigrants cannot enter the more skilled professions. They contended that welfare, medical, and educational benefits are magnets that draw illegal aliens into the state. These unauthorized immigrants were estimated to cost California taxpayers more than $3.5 billion per year and result in overcrowded schools and public health clinics, and the reduction of services to legal residents. Why should the latter pay for benefits for people who are breaking the law? 187-supporters asked.

Those opposed to 187 contended that projected savings would be illusory because the proposition collided with federal laws that guarantee access to public education for all children in the United States. It also, they said, violated federal Medicaid laws, so California would be in danger of losing all regular Medicaid funding. Forcing an estimated 300,000 children out of school and onto the streets would increase the risk of juvenile crime. Forbidding doctors from giving immunizations or basic medical care to anyone suspected of being an illegal immigrant would encourage the spread of communicable diseases throughout the state, putting everyone at risk. Educators, doctors, and other public service officials would be turned into immigration officers, a task for which they are ill-suited. Finally, opponents argued that the proposition would not stop the flow of illegal aliens because it did nothing to increase enforcement at the border or to punish employers who hire undocumented workers.

A week after the passage of Proposition 187, a federal judge issued a temporary restraining order blocking enforcement of most of its provisions pending the resolution of legal issues. Shortly thereafter, the U.S. District Court struck down portions of the proposition, declaring them unconstitutional. "The state is powerless to enact its own scheme to regulate immigration or to devise immigration regulations," the court wrote. Ultimately, 5 years after its passage, the U.S. Court of Appeals for the Ninth Circuit in 1999 permanently voided the core provisions of Proposition 187, including those that prevented illegal immigrants from attending public schools and receiving social services and

in the south of India. Most construction workers in New Delhi come either from Orissa, in the east of India, or Rajasthan, in the northwest. The diamond trade of Bombay, India, is dominated by a network of about 250 related families who come from a small town several hundred miles to the north.

Certainly, not all immigrants stay permanently at their first destination. Of the some 80 million newcomers to the United States between 1900 and 1980, some 10 million returned to their homelands or moved to another country. Estimates for Canada indicate that perhaps 40 of each 100 immigrants eventually leave,

and about 25% of newcomers to Australia also depart permanently. A corollary of all out-migration flows is, therefore, **counter** (or **return) migration,** the likelihood that as many as 25% of all migrants will return to their place of origin (Figure 3.30).

Within the United States, return migration—defined as moving back to one's state of birth—makes up about 20% of all domestic moves. That figure varies dramatically between states. More than a third of recent in-migrants to West Virginia, for example, were returnees—as were over 25% of those moving to Pennsylvania, Alabama, Iowa, and a few other states. Such

[Continued]

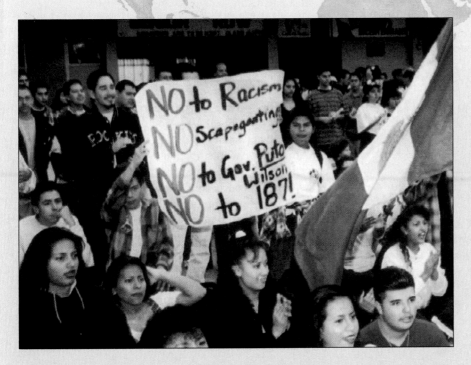

receive 1 year of instruction in English. While opponents of the measure called it immigrant-bashing, its supporters argued that bilingual education has been a failure; few children graduate into English-speaking classes each year, and many leave school unable to speak, read, or write well in the language of their adopted country. Whatever the final fate of these or other state initiatives, the immigration backlash they express remains as a local and national issue.

Questions to Consider

1. What do you think are the magnets that draw immigrants across the border: jobs or benefits? Would a denial of services likely lessen the perceived place utility of the United States and thus reduce illegal immigration?

2. People who believe that states should receive full federal payment for all costs associated with illegal immigrants argue that "State taxpayers should not bear the burden of the federal government's failure to control the border." Do you believe the federal government has an obligation to fully or partially reimburse states for the costs of education, medical care, and incarceration for unauthorized immigrants? Why or why not?

3. Should the United States require citizens to have a national identification card? Why or why not?

4. If you had been able to vote on Proposition 187, how would you have voted? Why?

5. Is it good policy not to educate or give basic medical care to any persons, even those not legally in the country? If so, under what circumstance?

health care. It also voided the requirement that local law-enforcement authorities, school administrators, and social and medical workers turn in suspected illegal immigrants to federal and state authorities. "Today's announcement is the final shovel of dirt on the grave of Proposition 187," said the director of the American Civil Liberties Union of Southern California. "Hopefully, it brings to a close what has been a very ugly chapter in California politics," added a spokesman for the state Attorney General's office. The original sponsors of Proposition 187, however, warned that "the will of the people has been frustrated," and predicted that "the battle may not be over."

California voters also approved Proposition 209 in November, 1996, banning state and local government preferences based on race and gender in hiring and school admissions. No "positive" discrimination for racial minorities is allowed, and affirmative action programs are to be discontinued. The U.S. Supreme Court in 1997 allowed the ban on racial and gender preferences to stand.

Next, in November, 1998, California voters overwhelmingly approved Proposition 227, characterized by a spokesman for the Mexican American Legal Defense and Education Fund as the third in a row of anti-Latino measures. The proposition scraps the system of bilingual education, in which non-English-speaking children were taught in their native language until they learned English well enough to be mainstreamed into regular classrooms. Instead, students would

widely different states as New Hampshire, Maryland, California, Florida, Wyoming, and Alaska were among the several that found returnees were fewer than 10% of their in-migrants. Interviews suggest that states deemed attractive draw new migrants in large numbers, while those with high proportions of returnees in the migrant stream are not perceived as desirable destinations by other than former residents.

Once established, origin and destination pairs of places tend to persist. Areas that dominate a locale's in- and out-migration patterns make up the *migration fields* of the place in question. As

we would expect, areas near the point of origin comprise the largest part of the migration field (Figure 3.31), though larger cities more distantly located may also be prominent as the ultimate destination of hierarchical step migration. Some migration fields reveal a distinctly *channelized* pattern of flow. The channels link areas that are in some way tied to one another by past migrations, by economic trade considerations, or some other affinity. The flow along them is greater than otherwise would be the case but does not necessarily involve individuals with personal or family ties. The former streams of southern blacks and

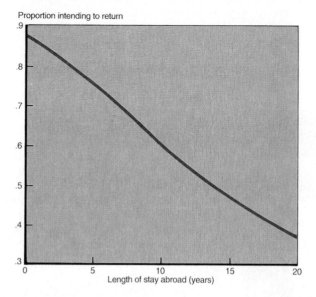

Proportion intending to return

Figure 3.30 **Intended return migration of Yugoslavs from Germany.** As the length of stay in Germany increased, the proportion of Yugoslavs intending to return decreased, but even after 10 years abroad, more than half intended to leave.

Source: Brigitte Waldorf, "Determinants of International Return Migration Intentions," Professional Geographer 47, no. 2 (1995), Fig. 2, p. 132.

whites to northern cities, of Scandinavians to Minnesota and Wisconsin, and of U.S. retirees to Florida and Arizona or their European counterparts to Iberia or the Mediterranean coast are all examples of **channelized migration.**

Voluntary migration is responsive to other controls that influence all forms of spatial interaction. Push-pull factors may be equated with *complementarity;* costs (emotional and financial) of a residence relocation are expressions of *transferability.* Other things being equal, large cities exert a stronger migrant pull than do small towns, a reflection of the impact of the *gravity model.* The *distance decay* effect has often been noted in migration studies. Movers seek to minimize the *friction of distance.* In selecting between two potential destinations of equal merit, a migrant tends to choose the nearer as involving less effort and expense. And since information about distant areas is less complete and satisfying than awareness of nearer localities, short moves are favored over long ones. Research indicates that determined migrants with specific destinations in mind are unlikely to be deterred by distance considerations. However, groups for whom push factors are more determining than specific destination pulls are likely to limit their migration distance in response to encountered apparent opportunities. For them, *intervening opportunity* affects locational decisions.

Observations such as these were summarized in the 1870s and 1880s as a series of "laws of migration" by E. G. Ravenstein (1834–1913). Among those that remain relevant are the following:

1. Most migrants go only a short distance.
2. Longer-distance migration favors big city destinations.
3. Most migration proceeds step-by-step.
4. Most migration is rural to urban.
5. Each migration flow produces a counterflow.
6. Most migrants are adults; families are less likely to make international moves.
7. Most international migrants are young males.

The latter two "laws" introduce the role of personal attributes (and attitudes) of migrants: their age, sex, education, and economic status. Migrants do not represent a cross section of the populace from which they come. Selectivity of movers is evident, and the selection shows some regional differences. In most societies, young adults are the most mobile (Figure 3.32). In the United States, mobility peaks among those in their twenties, especially the later twenties, and tends to decline thereafter. Among West African cross-border migrants, a World Bank study reveals, the age group 15–39 predominated.

Ravenstein's conclusion that young adult males are dominant in economically pushed international movement is less valid today than when first proposed. In reality, women and girls now comprise 40% to 60% of all international migrants worldwide (see "Gender and Migration"). It is true that legal and illegal migrants to the United States from Mexico and Central America are primarily young men, as were first generation "guest workers" in European cities. But population projections for West European countries suggest that women will shortly make up the largest part of their foreign-born population, and in one-third of the countries of sub-Saharan Africa, including Burkina Faso, Swaziland, and Togo, the female share of foreign-born populations was as large as the male. Further, among rural to urban migrants in Latin America since the 1960s, women have been in the majority.

Female migrants are motivated primarily by economic pushes and pulls. Surveys of women migrants in southeast Asia and Latin America indicate that 50% to 70% moved in search of employment and commonly first moved while in their teens. The proportion of young, single women is particularly high in rural-to-urban migration flows, reflecting their limited opportunities in increasingly overcrowded agricultural areas. To the push and pull factors normally associated with migration decisions are sometimes added family pressures that encourage young women with few employment opportunities to migrate as part of a household's survival strategy. In Latin America, the Philippines, and parts of Asia, emigration of young girls from large, landless families is more common than from smaller families or those with land rights. Their remittances of foreign earnings help maintain their parents and siblings at home.

An eighth internationally relevant observation may be added to those cited in Ravenstein's list: On average, emigrants tend to be relatively well-educated. A British government study reveals three-quarters of Africa's emigrants have higher (beyond high-school) education, as do about half of Asia's and South America's. Of the more than 1 million Asian Indians living in the United States, more than three-quarters of those of working age have at least a bachelor's degree. The loss to home countries can be draining; about 30% of all highly educated West African Ghanaians and Sierra Leoneans live abroad. Outward migration of the educated affects developed countries as well as poorer

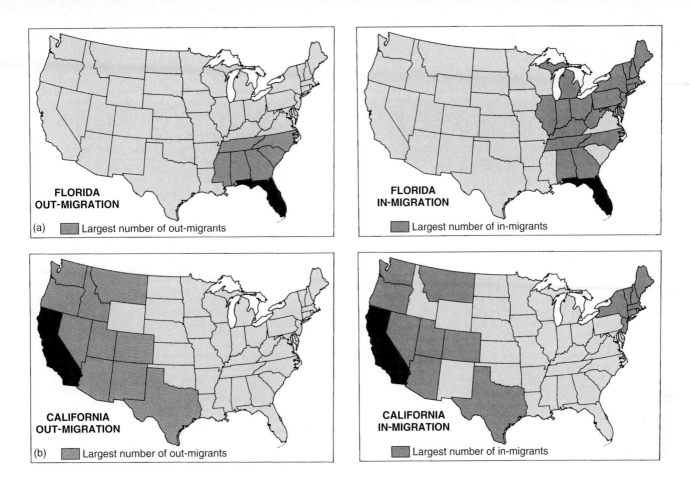

Figure 3.31 **The migration fields of Florida and California.** (*a*) For Florida, nearby southern states received most out-migrants, but in-migrants, especially retirees, originated from much of the eastern United States. (*b*) For California, the out-migration areas were the western states; the in-migration origins included both western and heavily populated northeastern states.

Source: Kavita Pandit, "Differentiating Between Subsystems and Typologies in the Analysis of Migration Regions: A U.S. Example," Professional Geographer 46, no. 3 (1994), figures 5 and 6, pp. 342–343.

Figure 3.32 **Percentage of 2000 population over 5 years of age with a different residence than in 1999.** Young adults figure most prominently in both short- and long-distance moves in the United States, an age-related pattern of mobility that has remained constant over time. For the sample year shown, 33% of people in their twenties moved while fewer than 5% of those 65 and older did so. Short-distance moves predominate; 56% of the 43 million U.S. movers between March, 1999 and March, 2000, relocated within the same county and another 20% moved to another county in the same state. Some two-thirds of intracounty (mobility) moves were made for housing-related reasons; long-distance moves (migration), a Census Bureau survey reveals, are likely to be made for work-related reasons.

Source: U.S. Bureau of the Census.

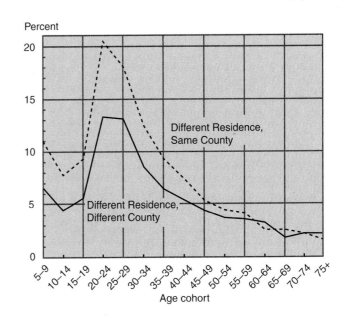

Gender and Migration

Gender is involved in migration at every level. In a household or family, women and men are likely to play different roles regarding decisions or responsibilities for activities such as child care. These differences, and the inequalities that underlie them, help determine who decides whether the household moves, which household members migrate, and the destination for the move. Outside the household, societal norms about women's mobility and independence often restrict their ability to migrate.

The economies of sending and receiving areas play a role as well. If jobs are available for women in the receiving area, women have an incentive to migrate, and families are more likely to encourage the migration of women as necessary and beneficial. Thousands of women from East and Southeast Asia have migrated to the oil-rich countries of the Middle East, for example, to take service jobs.

The impact of migration is also likely to be different for women and men. Moving to a new economic or social setting can affect the regular relationships and processes that occur within a household or family. In some cases, women might remain subordinate to the men in their families. A study of Greek-Cypriot immigrant women in London and of Turkish immigrant women in the Netherlands found that although these women were working for wages in their new societies, these new economic roles did not affect their subordinate standing in the family in any fundamental way.

In other situations, however, migration can give women more power in the family. In former Zaire [now the Democratic Republic of the Congo], women in rural areas moved to towns to take advantage of job opportunities there, and gain independence from men in the process.

One of the keys to understanding the role of gender in migration is to disentangle household decision-making processes. Many researchers see migration as a family decision or strategy, but some members will benefit more than others from those decisions.

For many years, men predominated in the migration streams flowing from Mexico to the United States. Women played an important role in this migration stream, even when they remained in Mexico. Mexican women influenced the migration decisions of other family members; they married migrants to gain the benefits from and opportunity for migration; and they resisted or accepted the new roles in their families that migration created.

In the 1980s, Mexican women began to migrate to the United States in increasing numbers. Economic crises in Mexico and an increase in the number of jobs available for women in the United States, especially in factories, domestic service, and service industries, have changed the backdrop of individual migration decisions. Now, women often initiate family moves or resettlement efforts.

Mexican women have begun to build their own migration networks, which are key to successful migration and resettlement in the United States. Networks provide migrants with information about jobs and places to live and have enabled many Mexican women to make independent decisions about migrating.

In immigrant communities in the United States, women are often the vital links to social institution services and to other immigrants. Thus, women have been instrumental in the way that Mexican immigrants have settled and become integrated into new communities.

From "Gender, Power, and Population Change" by Nancy E. Riley in *Population Bulletin*, Vol. 52, No. 1, May 1997, pp. 32–33. Reprinted by permission of Population Reference Bureau.

developing states. Between 1997 and 2002, it is claimed, between 15% and 40% of each year's Canadian colleges' graduating class emigrated to the United States, while in Europe, for one example, half the mid-1990s' graduating physics classes of Bucharest University left the country.

For modern Americans, the decisions to migrate are more ordinary but individually just as compelling. They appear to involve (1) changes in life course (e.g., getting married, having children, getting a divorce); (2) changes in the career course (getting a first job or a promotion, receiving a career transfer, seeking work in a new location, retiring); (3) changes of residence associated with individual personality (Figure 3.33). Work-related relocations are most important in U.S. long-distance (intercounty) migrations, and in both intra- and interstate relocations, more migrants move down the urban hierarchy—that is, from larger to smaller centers—than vice versa. Some observers suggest that pattern of deconcentration reflects modern transportation and communication technologies, more and younger retirees, and the attractions of amenity-rich smaller places. Some people, of course, simply seem to move often for no discernible reason, whereas others, *stayers,* settle into a community permanently. For other developed countries, a different set of summary migration factors may be present.

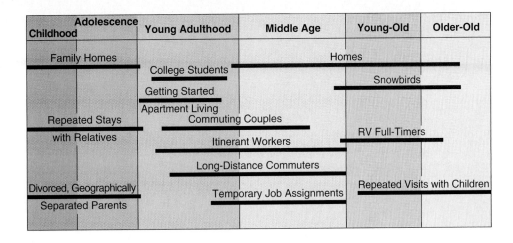

Childhood	Adolescence	Young Adulthood	Middle Age	Young-Old	Older-Old
Family Homes			Homes		
	College Students				
	Getting Started			Snowbirds	
	Apartment Living				
Repeated Stays	Commuting Couples				
with Relatives		Itinerant Workers		RV Full-Timers	
	Long-Distance Commuters				
Divorced, Geographically		Temporary Job Assignments		Repeated Visits with Children	
Separated Parents					

Figure 3.33 **Examples of multiple residences by stage in life.** Each horizontal line represents a period of time in a possible new residence.

From K. McHugh, T. Hogan, and S. Happel, "Multiple Residence and Cyclical Migration." The Professional Geographer 47, no. 3. Figure 1, p. 253. Association of American Geographers, 1995.

 *Summary*

Spatial interaction is the dynamic evidence of the areal differentiation of the earth's surface and of the interdependence between geographic locations. The term refers to the movement of goods, information, people, ideas—indeed, of every facet of economy and society—between one place and another. It includes the daily spatial activities of individuals and the collective patterns of their short- and long-distance behavior in space. The principles and constraints that unite, define, and control spatial behavior in this sense constitute an essential organizing focus for the study of human geographic patterns of the earth.

We have seen that whatever the type of spatial behavior or flow, a limited number of recurring mechanisms of guidance and control are encountered. Three underlying bases for spatial interaction are *complementarity,* which encourages flows between areas by balancing supply with demand or satisfying need with opportunity; *transferability,* which affects movement decisions by introducing cost, effort, and time considerations; and *intervening opportunities,* which suggests that costs of overcoming distance may be reduced by finding closer alternate points where needs can be satisfied. The flows of commodities, ideas, or people governed by these interaction factors are interdependent and additive. Flows of commodities establish and reinforce traffic patterns, for example, and also channelize the movement of information and people.

Those flows and interactions may further be understood by the application of uniform models to all forms of spatial interaction from interregional commodity exchanges to an individual's daily pattern of movement. Distance decay tells us of the inevitable decline of interaction with increasing distance. The gravity model suggests that major centers of activity can exert interaction pulls that partly compensate for distance decay. Recognition of movement biases explains why spatial interaction in the objective world may deviate from that proposed by abstract models.

Humans in their individual and collective short- and long-distance movements are responsive to these impersonal spatial controls. Their spatial behaviors are also influenced by their separate circumstances. Each has an activity and awareness space reflective of individual socioeconomic and life-cycle conditions. Each differs in mobility. Each has unique wants and needs and perceptions of their satisfaction. Human response to distance decay is expressed in a controlling critical distance beyond which the frequency of interaction quickly declines. That decline is partly conditioned by unfamiliarity with distant points outside normal activity space. Perceptions of home and distant territory therefore color interaction flows and space evaluations. In turn, those perceptions, well or poorly based, underlie travel and migration decisions, part of the continuing spatial diffusion and interaction of people. It is to people and their patterns of distribution and regional growth and change that we turn our attention in the following chapter.

Key Words

For Review

1. What is meant by *spatial interaction?* What are the three fundamental conditions governing all forms of spatial interaction? What is the distinctive impact or importance of each of the conditions?

2. What variations in *distance decay* curves might you expect if you were to plot shipments of ready-mixed concrete, potato chips, and computer parts? What do these respective curves tell us about transferability?

3. What is *activity space?* What factors affect the areal extent of an individual's activity space?

4. On a piece of paper, and following the model of Figure 3.11, plot your *space-time path* for your movements on a typical class day. What alterations in your established movement habits might be necessary (or become possible) if (a) Instead of walking, you rode a bike? (b) Instead of biking, you drove a car? (c) Instead of driving, you had to use the bus or go by bike or afoot?

5. What does the thought that transportation and communication are *space-adjusting* imply? In what ways has technology affected the "space adjustment" in commodity flows? In information flows?

6. Recall the places you have visited in the past week. In your movements, were the rules of *distance decay* and *critical distance* operative? What variables affect *your* critical distances?

7. What considerations appear to influence the decision to migrate? How do perceptions of *place utility* induce or inhibit migration?

8. What is a *migration field?* Some migration fields show a *channelized* flow of people. Select a particular channelized migration flow (such as the movement of Scandinavians to Michigan, Wisconsin, and Minnesota, or people from the Great Plains to California, or southern blacks to the North) and speculate why a channelized flow developed.

Focus Follow-up

1. **What are the three bases for all spatial interaction?** pp. 66–68.

Spatial interaction reflects areal differences and is controlled by three "flow-determining" factors. *Complementarity* implies a local supply of an item for which effective demand exists elsewhere. *Transferability* expresses the costs of movement from source of supply to locale of demand. An *intervening opportunity* serves to reduce flows of goods between two points by presenting nearer or cheaper sources.

2. **How is spatial interaction probability measured?** pp. 68–71.

The probability of aggregate spatial movements and interactions may be assessed by the application of established models. *Distance decay* reports the decline of interaction with increase in separation; the *gravity model* tells us that distance decay can in part be overcome by the enhanced attraction of larger centers of activity; and *movement bias* helps explain interaction flows contrary to model predictions.

3. **What are the special forms, attributes, and controls of human spatial behavior?** pp. 71–75.

While humans react to distance, time, and cost considerations of spatial movement, their spatial behavior is also affected by separate conditions of activity and awareness space, of individual economic and life-cycle circumstances, by degree of mobility, and by unique perceptions of wants and needs.

4. **What roles do information and perception play in conditioning human spatial actions?** pp. 75–84.

Humans base decisions about the opportunity or feasibility of spatial movements, exchanges, or want satisfactions on *place perceptions.* These condition the feelings we have about physical and cultural characteristics of areas, the opportunities they possess, and their degree of attractiveness. Those perceptions may not be based on reality or supported by balanced information. Distant places are less known than nearby ones, for example, and real natural hazards of areas may be mentally minimized through familiarity or rationalization.

5. **What kinds of migration movements can be recognized and what influences their occurrence?** pp. 84–94.

Migration means the permanent relocation of residence and activity space. It is subject to all the principles of spatial interaction and behavior and represents both a survival strategy for threatened people and a reasoned response to perceptions of opportunity. Migration has been enduring throughout human history and occurs at separate scales from intercontinental to regional, and includes flights of refugees and relocations of retirees. Negative home conditions (push factors) coupled with perceived positive destination attractions (pull factors) are important, as are age and sex of migrants and the spatial search they conduct. Step and chain migration and return migratory flows all affect patterns and volume of flows.

 Selected References

Boyle, Paul, and Keith Halfacre, eds. *Migration and Gender in the Developed World.* New York: Routledge, 1999.

Brunn, Stanley, and Thomas Leinbach. *Collapsing Space and Time: Geographic Aspects of Communication and Information.* Winchester, Mass.: Unwin Hyman, 1991.

Castles, Stephen, and Mark J. Miller. *The Age of Migration: International Population Movements in the Modern World.* 3d ed. New York: Guilford Publications, 2003.

Clark, W. A. V. *Human Migration.* Vol. 7, *Scientific Geography Series.* Newbury Park, Calif.: Sage, 1986.

Cohen, Robin, ed. *The Cambridge Survey of World Migration.* Cambridge, England: Cambridge University Press, 1995.

Gober, Patricia. "Americans on the Move." *Population Bulletin* 48, no. 3. Washington, D.C.: Population Reference Bureau, 1993.

Golledge, Reginald G., and Robert J. Stimson. *Spatial Behavior: A Geographic Perspective.* New York: Guilford Publications, 1996.

Haggett, Peter, *Geography: A Global Synthesis.* Harlow, England: Pearson Education, 2001.

Hanson, Susan, and Geraldine Pratt. "Geographic Perspectives on the Occupational Segregation of Women." *National Geographic Research* 6, no. 4 (1990): 376–399.

Janelle, Don G., and David C. Hodge, eds. *Information, Place, and Cyberspace: Issues in Accessibility.* Berlin: Springer Verlag, 2000.

Kane, Hal. *The Hour of Departure: Forces that Create Refugees and Migrants.* Worldwatch Paper 125. Washington, D.C.: Worldwatch Institute, 1995.

Kellerman, Aharon. *Telecommunications and Geography.* New York: Halsted, 1993.

King, Russell, ed. *The New Geography of European Migrations.* New York: Belhaven Press, 1993.

Manson, Gary A., and Richard E. Groop. "U.S. Intercounty Migration in the 1990s: People and Income Move Down the Urban Hierarchy." *Professional Geographer* 52, no. 3 (2000): 493–504.

Martin, Philip, and Jonas Widgren. "International Migration: Facing the Challenge." *Population Bulletin* 57, no. 1. Washington, D.C.: Population Reference Bureau, 2002.

Massey, Douglas S., et al. "Theories of International Migration: A Review and Appraisal." *Population and Development Review* 19, no. 3 (1993): 431–466.

Michelson, William. *From Sun to Sun: Daily Obligations and Community Structure in the Lives of Employed Women and Their Families.* Totowa, N.J.: Rowman & Allanheld, 1985.

Newbold, K. Bruce. "Race and Primary, Return, and Onward Interstate Migration." *Professional Geographer* 49, no. 1 (1997): 1–14.

Palm, Risa. *Natural Hazards.* Baltimore, Md.: Johns Hopkins University Press, 1990.

Plane, David A. "Age-Composition Change and the Geographical Dynamics of Interregional Migration in the U.S." *Annals of the Association of American Geographers* 82, no. 1 (1992): 64–85.

Pooley, Colin G., and Ian D. Whyte, eds. *Migrants, Emigrants and Immigrants: A Social History of Migration.* New York: Routledge, 1991.

Ravenstein, E. G. "The Laws of Migration." *Journal of the Royal Statistical Society* 48 (1885): 167–227; 52 (1889): 241–301.

Rogers, Andrei, and Stuart Sweeney. "Measuring the Spatial Focus of Migration Patterns." *Professional Geographer* 50, no. 2 (1998): 232–242.

Roseman, Curtis C. "Channelization of Migration Flows from the Rural South to the Industrial Midwest." *Proceedings of the Association of*

American Geographers 3 (1971): 140–146.

Simon, Rita James, and Caroline B. Brettell, eds. *International Migration: The Female Experience.* Totowa, N.J.: Rowman & Allenheld, 1986.

Ullman, Edward L. "The Role of Transportation and the Basis for Interaction." In *Man's Role in Changing the Face of the Earth*, edited by William E. Thomas, Jr., pp. 862–880. Chicago: University of Chicago Press, 1956.

United Nations. High Commissioner for Refugees. *The State of the World's Refugees.* New York: Oxford University Press, annual.

Wood, William B. "Forced Migration: Local Conflicts and International Dilemmas." *Annals of the Association of American Geographers* 84, no. 4 (1994): 607–634.

Websites: The World Wide Web has a tremendous number and variety of sites pertaining to geography. Websites relevant to the subject matter of this chapter appear in the "Web links" section of the On-line Learning Center associated with this book. Access it at **www.mhhe.com/fellmann8e**

CHAPTER

Four

Population:
World Patterns, Regional Trends

Focus Preview

1. Data and measures used by population geographers: the meaning and purpose of population cohorts, rates, and other measurements, pp. 102–118.

2. What we are told by the demographic transition model and the demographic equation, pp. 118–124.

3. World population distributions, densities, and urban components, pp. 125–130.

4. Population projections, controls, and prospects: estimating the future, pp. 130–135.

Opposite: The Friday market in Dhaka, Bangladesh, attracts a throng of shoppers in the largest city of the most densely populated mainland Asian country.

*"*ero, possibly even negative [population] growth*"* was the 1972 slogan proposed by the prime minister of Singapore, an island country in Southeast Asia. His nation's population, which stood at 1 million at the end of World War II (1945), had doubled by the mid-1960s. To avoid the overpopulation he foresaw, the government decreed *"Boy or girl, two is enough"* and refused maternity leaves and access to health insurance for third or subsequent births. Abortion and sterilization were legalized, and children born fourth or later in a family were to be discriminated against in school admissions policy. In response, birth rates by the mid-1980s fell to below the level necessary to replace the population, and abortions were terminating more than one-third of all pregnancies.

"At least two. Better three. Four if you can afford it" was the national slogan proposed by that same prime minister in 1986, reflecting fears that the stringencies of the earlier campaign had gone too far. From concern that overpopulation would doom the country to perpetual Third World poverty, Prime Minister Lee Kuan Yew was moved to worry that population limitation would deprive it of the growth potential and national strength implicit in a youthful, educated workforce adequate to replace and support the present aging population. His 1990 national budget provided for sizable long-term tax rebates for second children born to mothers under 28. Not certain that financial inducements alone would suffice to increase population, the Singapore government annually renewed its offer to take 100,000 Hong Kong Chinese who might choose to leave when China took over that territory in 1997.

The policy reversal in Singapore reflects an inflexible population reality: The structure of the present controls the content of the future. The size, characteristics, growth trends, and migrations of today's populations help shape the well-being of peoples yet unborn but whose numbers and distributions are now being determined. The numbers, age, and sex distribution of people; patterns and trends in their fertility and mortality; their density of settlement and rate of growth all affect and are affected by the social, political, and economic organization of a society. Through population data, we begin to understand how the people in a given area live, how they may interact with one another, how they use the land, what pressure on resources exists, and what the future may bring.

Population geography provides the background tools and understandings of those interests. It focuses on the number, composition, and distribution of human beings in relation to variations in the conditions of earth space. It differs from **demography,** the statistical study of human population, in its concern with *spatial* analysis—the relationship of numbers to area. Regional circumstances of resource base, type of economic development, level of living, food supply, and conditions of health and well-being are basic to geography's population concerns. They are, as well, fundamental expressions of the human–environmental relationships that are the substance of all human geographic inquiry.

Population Growth

Sometime in early 2004, a human birth raised the earth's population to about 6.4 billion people. At the start of 1991, the count was nearly 5.4 billion. That is, over the 13 years between those dates, the world population grew on average by about 77 million people annually, or some 211,000 per day. The average, however, conceals the reality that annual increases have been declining over the years. During the early 1990s, the U.S. Census Bureau and the United Nations Population Division regularly reported yearly growth at well over 85 million. Even with the slower pace of estimated increase, the United Nations early this century still projected that the world would likely contain nearly 9 billion inhabitants in 2050 and grow to perhaps 9.5 billion around the year 2100. Many demographers, however, impressed by dramatic birth rate reductions reported by 2002 for many developing and populous countries—India, importantly—lowered their estimates to predict end-of-century world totals peaking at between 8 and 9 billion, followed by numerical decline, not stability. All do agree, however, that essentially all of any future growth will occur in countries now considered "developing" (Figure 4.1). We will return to these projections and to the difficulties and disagreements inherent in making them later in this chapter.

Just what is implied by numbers in the millions and billions? With what can we compare the 2003 population of Guinea–Bissau in Africa (about 1.3 million) or of China (about 1.3 billion)? Unless we have some grasp of their scale and meaning, our understanding of the data and data manipulations of the population geographer can at best be superficial. It is difficult to appreciate a number as vast as 1 million or 1 billion, and the great distinction between them. Some examples offered by the Population Reference Bureau may help in visualizing their immensity and implications.

- A 2.5-centimeter (1-inch) stack of U.S. paper currency contains 233 bills. If you had a *million* dollars in thousand-dollar bills, the stack would be 11 centimeters (4.3 inches) high. If you had a *billion* dollars in thousand-dollar bills, your pile of money would reach 109 meters (358 feet)—about the length of a football field.

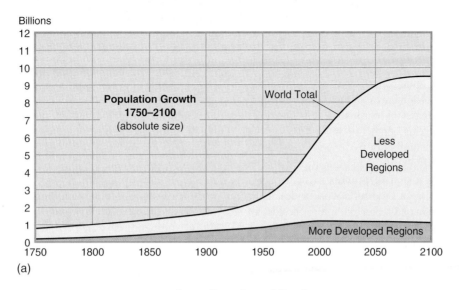

(a)

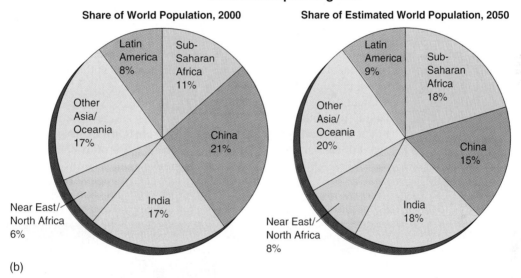

Less Developed Regions

Share of World Population, 2000

Latin America 8%
Sub-Saharan Africa 11%
Other Asia/Oceania 17%
China 21%
Near East/North Africa 6%
India 17%

Share of Estimated World Population, 2050

Latin America 9%
Sub-Saharan Africa 18%
Other Asia/Oceania 20%
China 15%
Near East/North Africa 8%
India 18%

(b)

Figure 4.1 **World population numbers and projections.** (*a*) After two centuries of slow growth, world poulation began explosive expansion after World War II. United Nations demographers project a global population of nearly 9 billion in 2050. Declining growth rates in much of the developing world have lowered earlier year 2100 estimates of global population from 10 billion to no more than 9.5 billion; some demographers argue for further reducing it to between 8 and 9 billion. Numbers in more developed regions at mid-century will be the same or lower than at its start thanks to anticipated population loss in Europe. However, higher fertility rates and immigration are projected to increase U.S. population by nearly 45% between 2000 and 2050, and large volume immigration into Europe could alter its population decline projections. In contrast, the populations of the less developed regions may increase by almost 60% between 2000 and 2050. (*b*) While only a little more than 80% of world population was found in regions considered "less developed" in 2000 (left diagram), nearly 9 out of 10 of a larger total will be located there in 2050 (right diagram).

Sources: (a) Estimates from Population Reference Bureau and United Nations Population Fund; (b) Based on United Nations and U.S. Bureau of the Census projections.

- You had lived a *million* seconds when you were 11.6 days old. You won't be a *billion* seconds old until you are 31.7 years of age.

- The supersonic airplane, the Concorde, could theoretically circle the globe in only 18.5 hours at its cruising speed of 2150 kilometers (1336 mi) per hour. It would take 31 days for a passenger to journey a *million* miles on the Concorde, while a trip of a *billion* miles would last 85 years.

The implications of the present numbers and the potential increases in population are of vital current social, political, and ecological concern. Population numbers were much smaller some 12,000 years ago when continental glaciers began their retreat, people spread to formerly unoccupied portions of the globe, and human experimentation with food sources initiated the Agricultural Revolution. The 5 or 10 million people who then constituted all of humanity obviously had considerable potential to expand their numbers. In retrospect, we see that the natural resource base of the earth had a population-supporting capacity far in excess of the pressures exerted on it by early hunting and gathering groups.

Some observers maintain that despite present numbers or even those we can reasonably anticipate for the future, the adaptive and exploitive ingenuity of humans is in no danger of being taxed. Others, however, compare the earth to a self-contained spaceship and declare with chilling conviction that a finite vessel cannot bear an ever-increasing number of passengers. They point to recurring problems of malnutrition and starvation (though these are realistically more a matter of failures of distribution than of inability to produce enough foodstuffs worldwide). They cite dangerous conditions of air and water pollution, the loss of forest and farmland, the apparent nearing exhaustion of many minerals and fossil fuels, and other evidences of strains on world resources as foretelling the discernible outer limits of population growth.

On a worldwide basis, populations grow only one way: The number of births in a given period exceeds the number of deaths. Ignoring for the moment regional population changes resulting from migration, we can conclude that observed and projected increases in population must result from the failure of natural controls to limit the number of births or to increase the number of deaths, or from the success of human ingenuity in circumventing such controls when they exist. In contrast, current estimates of slowing population growth and eventual stability or decline in world totals clearly indicate that humans by their individual and collective decisions may effectively limit growth and control global population numbers. The implications of these observa-

tions will become clearer after we define some terms important in the study of world population and explore their significance.

Some Population Definitions

Demographers employ a wide range of measures of population composition and trends, though all their calculations start with a count of events: of individuals in the population, of births, deaths, marriages, and so on. To those basic counts, demographers bring refinements that make the figures more meaningful and useful in population analysis. Among them are *rates* and *cohort* measures.

Rates simply record the frequency of occurrence of an event during a given time frame for a designated population—for example, the marriage rate as the number of marriages performed per 1000 population in the United States last year. **Cohort** measures refer data to a population group unified by a specified common characteristic—the age cohort of 1–5 years, perhaps, or the college class of 2007 (Figure 4.2). Basic numbers and rates useful in the analysis of world population and population trends have been reprinted with the permission of the Population Reference Bureau as Appendix B to this book. Examination of them will document the discussion that follows.

Birth Rates

The **crude birth rate (CBR),** often referred to simply as the *birth rate,* is the annual number of live births per 1000 population. It is "crude" because it relates births to total population without

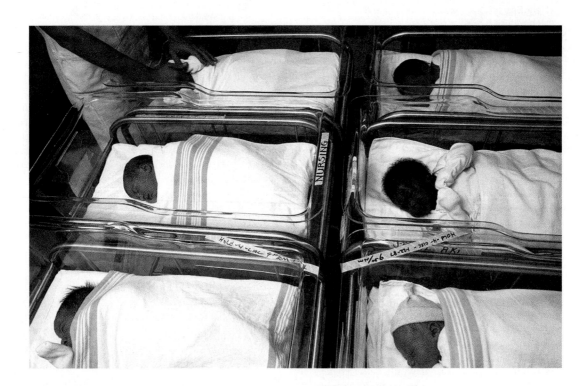

Figure 4.2 Whatever their differences may be by race, sex, or ethnicity, these babies will forever be clustered demographically into a single *birth cohort.*

regard to the age or sex composition of that population. A country with a population of 2 million and with 40,000 births a year would have a crude birth rate of 20 per 1000.

$$\frac{40,000}{2,000,000} = \frac{20}{1000} = 20 \text{ per thousand}$$

The birth rate of a country is, of course, strongly influenced by the age and sex structure of its population, by the customs and family size expectations of its inhabitants, and by its adopted population policies. Because these conditions vary widely, recorded national birth rates vary—in the early 21st century, from a high of 45 to 50 or more in some West African states to lows of 9 or 10 per 1000 in 20 or more European countries. Although birth rates of 30 or above per 1000 are considered *high,* one-sixth of the world's people live in countries with rates that are that high or higher (Figure 4.3). In these countries—found chiefly in Africa, western and southern Asia, and Latin America—the population is predominantly agricultural and rural, and a high proportion of the female population is young. In many of them, birth rates may be significantly higher than official records indicate. Available data suggest that every year around 50 million births go unregistered and therefore uncounted.

Birth rates of less than 18 per 1000 are reckoned *low* and are characteristic of industrialized, urbanized countries. All European countries including Russia, as well as Anglo America, Japan, Australia, and New Zealand, have low rates as, importantly, do an increasing number of developing states such as China (see "China's Way—and Others") that have adopted effective family planning programs. *Transitional* birth rates (between 18 and 30 per 1000) characterize some, mainly smaller, "developing" countries, though giant India entered that group in 1994.

As the recent population histories of Singapore and China indicate, birth rates are subject to change. The decline to current low birth rates of European countries and of some of the areas that they colonized is usually ascribed to industrialization, urbanization, and, in recent years, maturing populations. While restrictive family planning policies in China rapidly reduced the birth rate from over 33 per 1000 in 1970 to 18 per 1000 in 1986, industrializing Japan experienced a comparable 15-point decline in the decade 1948–1958 with little governmental intervention. Indeed, the stage of economic development appears closely related to variations in birth rates among countries, although rigorous testing of this relationship proves it to be imperfect (Figure 4.3). As a group, the more developed states of the world showed a crude birth rate of 11 per 1000 at the start of the 21st century; less developed countries (excluding China) registered almost 30 per 1000.

Religious and political beliefs can also affect birth rates. The convictions of many Roman Catholics and Muslims that their religion forbids the use of artificial birth control techniques often lead to high birth rates among believers. However, dominantly

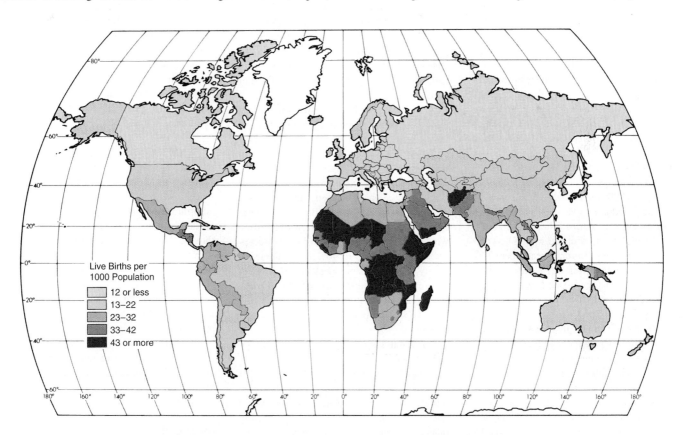

Figure 4.3 **Crude birth rates.** The map suggests a degree of precision that is misleading in the absence of reliable, universal registration of births. The pattern shown serves, however, as a generally useful summary of comparative reproduction patterns if class divisions are not taken too literally. Reported or estimated population data vary annually, so this and other population maps may not agree in all details with the figures recorded in Appendix B.

Source: Data from Population Reference Bureau.

An ever larger population is "a good thing," Chairman Mao announced in 1965 when China's birth rate was 37 per 1000 and population totaled 540 million. At Mao's death in 1976, numbers reached 852 million, though the birth rate then had dropped to 25. During the 1970s, when it became evident that population growth was consuming more than half of the annual increase in the country's gross domestic product, China introduced a well-publicized campaign advocating the "two-child family" and providing services, including abortions, supporting that program. In response, China's birth rate dropped to 19.5 per 1000 by the late 1970s.

"One couple, one child" became the slogan of a new and more vigorous population control drive launched in 1979, backed by both incentives and penalties to assure its success in China's tightly controlled society. Late marriages were encouraged; free contraceptives, cash awards, abortions, and sterilizations were provided to families limited to a single child. Penalties, including steep fines, were levied for second births. At the campaign's height in 1983, the government ordered the sterilization of either husband or wife for couples with more than one child. Infanticide—particularly the exposure or murder of female babies—was a reported means both of conforming to a one-child limit and of increasing the chances that the one child would be male. By 1986, China's officially reported crude birth rate had fallen to 18 per 1000, far below the 37 per 1000 then registered among the rest of the world's less developed countries. The one-child policy was effectively dropped in 1984 to permit two-child limits in rural areas where 70% of Chinese population still resides, but in 2002 it was reinstated as nationwide law following documentation of extensive underreporting of rural births.

In contrast, newly prosperous urbanites have voluntarily reduced their fertility to well below replacement levels, with childless couples increasingly common. Nationally, past and current population controls have been so successful that by 2001 serious concerns were being expressed by demographers and government officials that population decrease, not increase, is the problem next to be confronted. Projections suggest that by 2042, because of lowered fertility rates, China's

FAMILY PLANNING—A BASIS NATIONAL POLICY OF CHINA

population numbers will actually start falling. The country is already beginning to face a pressing social problem: a declining proportion of working-age persons and an absence of an adequate welfare network to care for a rapidly growing number of senior citizens.

Concerned with their own increasing numbers, many developing countries have introduced their own less extreme programs of family planning, stressing access to contraception and sterilization. International agencies have encouraged these programs, buoyed by such presumed success as the 21% fall in fertility rates in Bangladesh from 1970 to 1990 as the proportion of married women of reproductive age using contraceptives rose from 3% to 40% under intensive family planning encouragement and frequent adviser visits. The costs per birth averted, however, were reckoned at more than the country's $160 per capita gross domestic product.

Research suggests that fertility falls because women decide they want smaller families, not because they have unmet needs for contraceptive advice and devices. Nineteenth-century northern Europeans without the aid of science had lower fertility rates than their counterparts today in middle-income countries. With some convincing evidence, improved women's education has been proposed as a surer way to reduce births

than either encouraged contraception or China's coercive efforts. Studies from individual countries indicate that 1 year of female schooling can reduce the fertility rate by between 5% and 10%. Yet the fertility rate of uneducated Thai women is only two-thirds that of Ugandan women with secondary education. Obviously, the demand for children is not absolutely related to educational levels.

Instead, that demand seems closely tied to the use value placed on children by poor families in some parts of the developing world. Where those families share in such communal resources as firewood, animal fodder, grazing land, and fish, the more of those collective resources that can be converted to private family property and use, the better off is the family. Indeed, the more communal resources that are available for "capture," the greater are the incentives for a household to have more children to appropriate them. Some population economists conclude that only when population numbers increase to the point of total conversion of communal resources to private property—and children have to be supported and educated rather than employed—will poor families in developing countries want fewer children. If so, coercion, contraception, and education may be less effective as checks on births than the economic consequence of population increase itself.

Catholic Italy has nearly the world's lowest birth rate, and Islam itself does not prohibit contraception. Similarly, some European governments—concerned about birth rates too low to sustain present population levels—subsidize births in an attempt to raise those rates. Regional variations in projected percentage contributions to world population growth are summarized in Figure 4.4.

Fertility Rates

Crude birth rates may display such regional variability because of differences in age and sex composition or disparities in births among the reproductive-age, rather than total, population. **Total fertility rate (TFR)** is a more accurate statement than the birth rate in showing the amount of reproduction in the population (Figure 4.5). The TFR tells us the average number of children that would be born to each woman if, during her childbearing years, she bore children at the current year's rate for women that age. The fertility rate minimizes the effects of fluctuation in the population structure and is thus a more reliable figure for regional comparative and predictive purposes than the crude birth rate.

Although a TFR of 2.0 would seem sufficient exactly to replace present population (one baby to replace each parent), in reality replacement levels are reached only with TFRs of 2.1 to 2.3. The fractions over 2.0 are required to compensate for infant and childhood mortalities, childless women, and unexpected deaths in the general population. On a worldwide basis, the TFR early in the 21st century was 2.7 to 2.8; in the mid-1980s it was 3.7. The

more developed countries recorded a 1.5 rate at the start of the present century, down from a near-replacement 2.0 in 1985. That decrease has been dwarfed in amount and significance by the rapid changes in reproductive behavior in much of the developing world. Since 1960, the average TFR in the less developed world has fallen by half from the traditional 6.0 or more to below 3.0 today. Between the early 1960s and the end of the 20th century, the largest fertility declines occurred in Latin America and Asia (down by 55% and 52%, respectively). The smallest decrease—of 15%—came in sub-Saharan Africa.

The recent fertility declines in developing states, however, have been more rapid and widespread than anyone expected. Indeed, the TFRs for so many less developed countries have dropped so dramatically since the early 1960s (Figure 4.6), that earlier widely believed world population projections anticipating 10 billion or more at the end of this century are now generally discounted and rejected. Indeed, worldwide in 2003, 65 countries and territories containing some 45% of global population had fertility rates less than 2.1, with more poised to join their ranks. China's decrease from a TFR of 5.9 births per woman in the period 1960–1965 to (officially) about 1.8 in 2000 and comparable drops in TFRs of Bangladesh, Brazil, Mozambique, and other states demonstrate that fertility reflects cultural values, not biological imperatives. If those values now favor fewer children than formerly, population projections based on earlier, higher TFR rates must be adjusted.

In fact, demographers have long assumed that recently observed developing country—and therefore global—fertility rate declines to the replacement level would continue and in the long run lead to stable population numbers. However, nothing in logic or history requires population stability at any level. Indeed, rather than assume, as in the past, a fertility decline to a constant continuing rate of 2.1, the 2002 revision of the United Nations' world population projections predicts a long-term world fertility rate of 1.9—*below* the replacement rate. Should the UN's new assessment of global fertility prove correct, world population would not just stop growing as past UN projections envisioned; it would inevitably decline (see "A Population Implosion?"). Of course, should cultural values change to again favor children, growth would resume. Different TFR estimates imply conflicting population projections and vastly different regional and world population concerns.

Individual country projections based on current fertility rates, it should be noted, may not accurately anticipate population levels even in the near future. As we saw in Chapter 3, massive international population movements are occurring in response to political instabilities and, particularly, to differentials in perceived economic opportunities. Resulting migration flows may cause otherwise declining national populations to stabilize or even grow. For example, the European Union in recent years has had a negative rate of natural increase, yet since 2000 has experienced essentially a constant population solely because of immigrant influx from Eastern Europe, Asia, and Africa.

World regional and national fertility rates reported in Appendix B and other sources are summaries that conceal significant variations between population groups. The Caribbean region, for example, showed a total fertility rate of 2.6 in 2002, but the TFRs

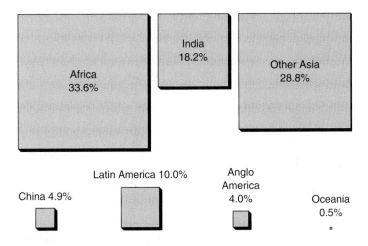

Figure 4.4 **Projected percentage contributions to world population growth, by region, 2000–2050.** Birth rate changes recorded by differently sized regional populations with differing age structures are altering the world pattern of population increase. Africa, containing 13% of world population in 2000, will probably account for more than one-third of total world increase between 2000 and 2050. Between 1965 and 1975, China's contribution to world growth was 2.5 times that of Africa; between 2000 and 2050, Africa's numerical growth will be over 8 times that of China. India, which reached the 1 billion level in 2000, is projected to grow by more than 50% over the first half of the 21st century and have by far the world's largest population. In contrast to the growth within the world regions shown, Europe's population is projected to decrease by 70 million over the same half-century period.

Sources: Projections based on World Bank and United Nations figures.

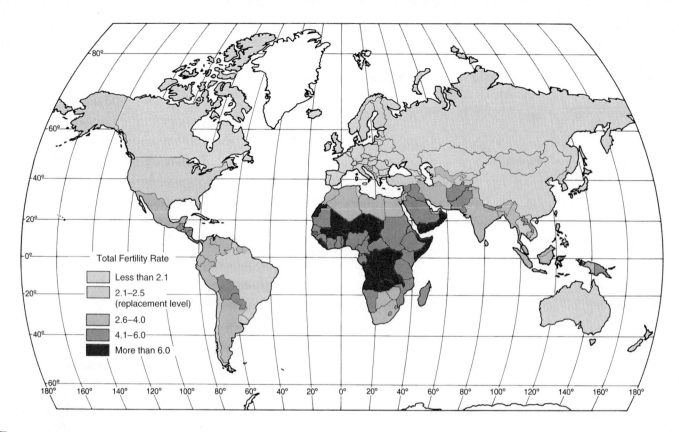

Figure 4.5 **Total fertility rate (TFR)** indicates the average number of children that would be born to each woman if, during her childbearing years, she bore children at the same rate as women of those ages actually did in a given year. Since the TFR is age-adjusted, two countries with identical birth rates may have quite different fertility rates and therefore different prospects for growth. Depending on mortality conditions, a TFR of 2.1 to 2.5 children per woman is considered the "replacement level," at which a population will eventually stop growing.

Source: Data from Population Reference Bureau.

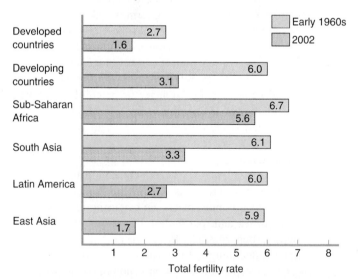

Figure 4.6 **Differential fertility declines.** Fertility has declined most rapidly in Latin America and Asia and much more slowly in sub-Saharan Africa. Developed countries as a group now have below-replacement-level fertility. Europe was far below with a 2002 TFR of 1.4; the United States, however, showed births at just the replacement point of 2.1 at the start of the century.

Sources: Population Reference Bureau and United Nations Population Fund.

of individual states ranged from a low of 1.5 in Cuba to a high of 4.7 in Haiti. The United States 2002 national average fertility rate of 2.0 did not reveal that the TFR for Hispanics was about 3.0, about 2.2 for African Americans, or only 1.9 for Asians and Pacific Islanders.

Death Rates

The **crude death rate (CDR),** also called the **mortality rate,** is calculated in the same way as the crude birth rate: the annual number of events per 1000 population. In the past, a valid generalization was that the death rate, like the birth rate, varied with national levels of development. Characteristically, highest rates (over 20 per 1000) were found in the less developed countries of Africa, Asia, and Latin America; lowest rates (less than 10) were associated with developed states of Europe and Anglo America. That correlation became decreasingly valid as dramatic reductions in death rates occurred in developing countries in the years following World War II. Infant mortality rates and life expectancies improved as antibiotics, vaccinations, and pesticides to treat diseases and control disease carriers were made available in almost all parts of the world and as increased attention was paid to funding improvements in urban and rural sanitary facilities and safe water supplies.

Distinctions between more developed and less developed countries in mortality (Figure 4.7), indeed, have been so reduced

A Population Implosion?

For much of the last half of the 20th century, demographers and economists focused on a "population explosion" and its implied threat of a world with too many people and too few resources of food and minerals to sustain them. By the end of the century, those fears for some observers were being replaced by a new prediction of a world with too few rather than too many people.

That possibility was suggested by two related trends. The first became apparent by 1970 when it was noted that the total fertility rates (TFRs) of 19 countries, almost all of them in Europe, had fallen below the **replacement level**—the level of fertility at which populations replace themselves—of 2.1. Simultaneously, Europe's population pyramid began to become noticeably distorted, with a smaller proportion of young and a growing share of middle-aged and retirement-age inhabitants. The decrease in native working-age cohorts had already, by 1970, encouraged the influx of non-European "guest workers" whose labor was needed to maintain economic growth and to sustain the generous security provisions guaranteed to what was becoming the oldest population of any continent.

Many countries of Western and Eastern Europe sought to reverse their birth rate declines by adopting pronatalist policies. The communist states of the East rewarded pregnancies and births with generous family allowances, free medical and hospital care, extended maternity leaves, and child care. France, Italy, the Scandinavian countries, and others gave similar bonuses or awards for first, second, and later births. Despite those inducements, however, reproduction rates continued to fall. By 2002, 42 of 43 Eu-ropean countries and territories had fertility rates below replacement levels, and the populations of Spain and Italy, for example, are projected to shrink by a quarter between 2000 and 2050. "In demographic terms," France's prime minister remarked, "Europe is vanishing."

Europe's experience soon was echoed in other societies of advanced economic development on all continents. By 1995, Canada, Australia, New Zealand, Japan, Taiwan, South Korea, Singapore, and other older and newly industrializing countries (NICs) registered fertility rates below the replacement level. As they have for Europe, simple projections foretold their aging and declining population. Japan's numbers, for example, will begin to decline in 2006 when its population will be older than Europe's; Taiwan forecasts negative growth by 2035.

The second trend indicating to many that world population numbers should stabilize and even decline during the lifetimes of today's college cohort is a simple extension of the first: TFRs are being reduced to or below the replacement levels in countries at all stages of economic development in all parts of the world. While only 18% of total world population in 1975 lived in countries with a fertility rate below replacement level, nearly 45% did so by the end of the century. By 2015, demographers estimate, half the world's countries and about two-thirds of its population will show TFRs below 2.1 children per woman. Exceptions to the trend are and still will be found in Africa, especially sub-Saharan Africa, and in some areas of South, Central, and West Asia; but even in those regions, fertility rates have been decreasing in recent years. "Powerful globaliz-ing forces [are] at work pushing toward fertility reduction everywhere," was an observation of the French National Institute of Demographic Studies.

That conclusion is plausibly supported by assumptions of the United Nations' 2002 forecast of a decline of long-term fertility rates of most less developed states to an average of 1.9. The same UN assessment envisions that those countries will reach those below-replacement fertilities before 2050. Should these assumptions prove valid, global depopulation could commence by or before midcentury. Between 2040 and 2050, one projection indicates, world population would fall by about 85 million (roughly the amount of its annual growth during much of the 1990s) and shrink further by about 25% with each successive generation.

If the UN low-rate scenario is realized in whole or in part, a much different worldwide demographic and economic future is promised than that prophesied so recently by "population explosion" forecasts. Declining rather than increasing pressure on world food and mineral resources would be in our future along with shrinking rather than expanding world, regional, and national economies. Even the achievement of **zero population growth (ZPG),** a condition for individual countries when births plus immigration equal deaths plus emigration, has social and economic consequences not always perceived by its advocates. These inevitably include an increasing proportion of older citizens, fewer young people, a rise in the median age of the population, and a growing old-age dependency ratio with ever-increasing pension and social services costs borne by a shrinking labor force.

that by 1994 death rates for less developed countries as a group actually dropped below those for the more developed states and have remained lower since. Notably, that reduction did not extend to maternal mortality rates (see "The Risks of Motherhood"). Like crude birth rates, death rates are meaningful for comparative purposes only when we study identically structured populations. Countries with a high proportion of elderly people, such as Denmark and Sweden, would be expected to have higher death rates than those with a high proportion of young people, such as Iceland, assuming equality in other national conditions affecting health and longevity. The pronounced youthfulness of populations in developing countries, as much as improvements in sanitary and health conditions, is an important factor in the recently reduced mortality rates of those areas.

To overcome that lack of comparability, death rates can be calculated for specific age groups. The *infant mortality rate,* for example, is the ratio of deaths of infants aged 1 year or under per 1000 live births:

$$\frac{\text{deaths age 1 year or less}}{\text{1000 live births}}$$

Infant mortality rates are significant because it is at these ages that the greatest declines in mortality have occurred, largely as a result of the increased availability of health services. The drop in

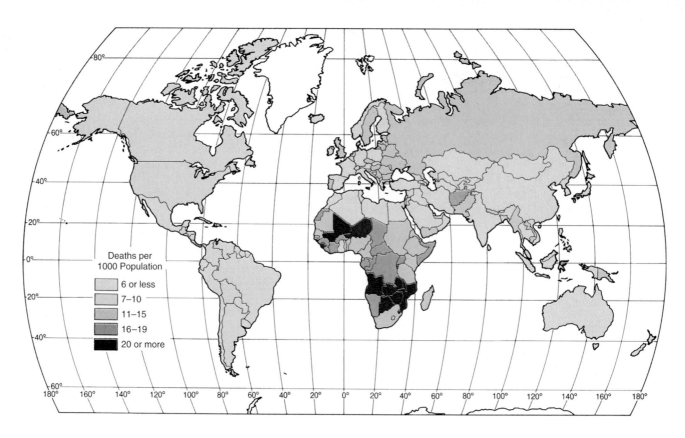

Figure 4.7 **Crude death rates** show less worldwide variability than do the birth rates displayed in Figure 4.3, the result of widespread availability of at least minimal health protection measures and a generally youthful population in the developing countries, where death rates are frequently lower than in "old age" Europe.

Source: Data from Population Reference Bureau.

infant mortality accounts for a large part of the decline in the general death rate in the last few decades, for mortality during the first year of life is usually greater than in any other year.

Two centuries ago, it was not uncommon for 200–300 infants per 1000 to die in their first year. Even today, despite significant declines in those rates over the last 60 years in many countries (Figure 4.8), striking world regional and national variations remain. For all of Africa, infant mortality rates are near 90 per 1000, and individual African states (for example, Liberia, Mozambique and Sierra Leone) showed rates above 130 early in this century. Nor are rates uniform within single countries. The former Soviet Union reported a national infant mortality rate of 23 (1991), but it registered above 110 in parts of its Central Asian region. In contrast, infant mortality rates in Anglo America and Western and Northern Europe are more uniformly in the 4–7 range.

Modern medicine and sanitation have increased life expectancy and altered age-old relationships between birth and death rates. In the early 1950s, only five countries, all in northern Europe, had life expectancies at birth of over 70 years. In the first years·of the 21st century, some 60 countries outside of Europe and North America—though none in sub-Saharan Africa—were on that list. The availability and employment of modern methods of health and sanitation have varied regionally, and the least developed countries have least benefited from them. In such underdeveloped and impoverished areas as much of sub-Saharan

Africa, the chief causes of death other than HIV/AIDS are those no longer of immediate concern in more developed lands: diseases such as malaria, intestinal infections, typhoid, cholera, and especially among infants and children, malnutrition and dehydration from diarrhea.

HIV/AIDS is the tragic and, among developing regions particularly, widespread exception to observed global improvements in life expectancies and reductions in adult death rates and infant and childhood mortalities. AIDS has become the fourth most common cause of death worldwide and is forecast to surpass the Black Death of the 14th century as history's worst-ever epidemic. The World Health Organization estimates 40 million people to be HIV positive early in the 21st century. Some 95% of those infected live in developing countries, and 70% reside in sub-Saharan Africa. In that hardest-hit region, as much as one-fourth of the adult population in some countries is HIV positive, and average life expectancy has been cut drastically. In South Africa, the life expectancy of a baby born in the early 21st century should be 66 years; AIDS has cut that down to 47. In Botswana, it is 36 years instead of 70; in Zimbabwe the decline has been to 43 years from 69. Overall, sub-Saharan life expectancies have been cut by 15 years, and total population by 2015 is now projected to be 60 million less than it would have been in the absence of the disease. Economically, AIDS will cut an estimated 8% off national incomes in the worst-hit sub-Saharan countries by 2010.

The Risks of Motherhood

The worldwide leveling of crude death rates does not apply to pregnancy-related deaths. In fact, the maternal mortality ratio—maternal deaths per 100,000 live births—is the single greatest health disparity between developed and developing countries. According to the World Health Organization, approximately 500,000 women die each year from causes related to pregnancy and childbirth; 99% of them live in less developed states where, as a group, the maternal mortality ratio is some 40 times greater than in the more developed countries. Complications of pregnancy, childbirth, and abortions are the leading slayers of women of reproductive age throughout the developing world, though the incidence of maternal mortality is by no means uniform, as the charts indicate. In Africa, the risk is around 1 death in 16 pregnancies compared with 1 in 110 in Asia and 1 in 2000 in Europe. Country-level differences are even more striking: in Ethiopia, for example, 1 out of every 9 women dies from pregnancy-related complications compared to 1 in 8700 in Switzerland.

Excluding China, less developed countries as a group in the late 1990s had a maternal mortality ratio of 580, and 10% of all deaths were due to perinatal and maternal causes. Although 42% of all maternal deaths occurred in Asia (which accounts for about 60% of the world's births), sub-Saharan African women, burdened with 53% of world maternal mortality, were at greatest statistical risk. There, maternal death ratios in the mid-1990s reached above 1500 in Burundi, Chad, and Somalia and to more than 2000 in Sierra Leone; 1 in 13 women in sub-Saharan Africa dies of maternal causes. In contrast, the maternal mortality ratio in developed countries as a group (including Russia and eastern Europe) is 21, and in some—Ireland, Switzerland, and Sweden, for example—it is as low as 6 to 8 (it was 6 in Canada and 8 in the United States in the late 1990s).

The vast majority of maternal deaths in the developing world are preventable. Most result from causes rooted in the social, cultural, and economic barriers confronting females in their home environment throughout their lifetimes: malnutrition, anemia, lack of

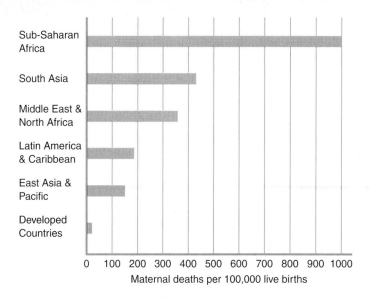

(a)

Regional Ratios of Maternal Deaths

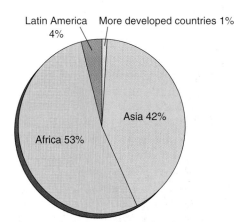

(b)

Regional Shares of Maternal Deaths

Sources: Graph data from WHO.

access to timely basic maternal health care, physical immaturity due to stunted growth, and unavailability of adequate prenatal care or trained medical assistance at birth. Part of the problem is that women are considered expendable in societies where their status is low, although the correlation between women's status and maternal mortality is not exact. In those cultures, little attention is given to women's health or their nutrition,

and pregnancy, although a major cause of death, is simply considered a normal condition warranting no special consideration or management. To alter that perception and increase awareness of the affordable measures available to reduce maternal mortality, 1998 was designated "The Year of Safe Motherhood" by a United Nations interagency group.

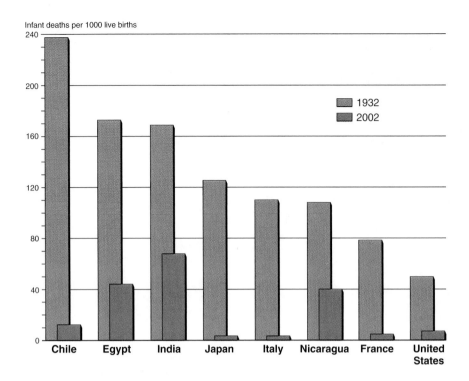

Infant deaths per 1000 live births

Figure 4.8 **Infant mortality rates for selected countries.** Dramatic declines in the rate have occurred in all countries, a result of international programs of health care delivery aimed at infants and children in developing states. Nevertheless, the decreases have been proportionately greatest in the urbanized, industrialized countries, where sanitation, safe water, and quality health care are more widely available.

Sources: Data from U.S. Bureau of the Census and Population Reference Bureau.

Nonetheless, because of their high fertility rates, populations in all sub-Saharan countries except South Africa are still expected to grow significantly between 2000 and 2050, adding nearly 1 billion to the continent's total. Indeed, despite high mortality rates due to HIV/AIDS, the population of the world's 48 least developed countries as a group will, according to UN projections, almost triple between 2000 and 2050, the consequence of their high fertility levels. However, warnings of the rapid spread of the AIDS epidemic in Russia, Ukraine, and South and East Asia—particularly China and India—raise new global demographic concerns even as more hopeful reports of declining infection and mortality rates in some African and Southeast Asian countries are appearing.

Population Pyramids

Another means of comparing populations is through the **population pyramid,** a graphic device that represents a population's age and sex composition. The term *pyramid* describes the diagram's shape for many countries in the 1800s, when the display was created: a broad base of younger age groups and a progressive narrowing toward the apex as older populations were thinned by death. Now many different shapes are formed, each reflecting a different population history (Figure 4.9), and some suggest "population profile" is a more appropriate label. By grouping several generations of people, the pyramids or profiles highlight the impact of "baby booms," population-reducing wars, birth rate reductions, and external migrations.

A rapidly growing country such as Uganda has most people in the lowest age cohorts; the percentage in older age groups declines successively, yielding a pyramid with markedly sloping sides. Typically, female life expectancy is reduced in older cohorts of less developed countries, so that for Uganda, the proportion of females in older age groups is lower than in, for example, Sweden. Female life expectancy and mortality rates may also be affected by cultural rather than economic developmental causes (see "100 Million Women Are Missing"). In Sweden, a wealthy country with a very slow rate of growth, the population is nearly equally divided among the age groups, giving a "pyramid" with almost vertical sides. Among older cohorts, as Austria shows, there may be an imbalance between men and women because of the greater life expectancy of the latter. The impacts of war, as Russia's pyramid vividly shows, are evident in that country's depleted age cohorts and male-female disparities. The sharp contrasts between the composite pyramids of sub-Saharan Africa and Western Europe summarize the differing population concerns of the developing and developed regions of the world; the projection for Botswana suggests the degree to which accepted pyramid shapes can quickly change (Figure 4.10).

The population pyramid provides a quickly visualized demographic picture of immediate practical and predictive value. For example, the percentage of a country's population in each age group strongly influences demand for goods and services within that national economy. A country with a high proportion of young has a high demand for educational facilities and certain types of health delivery services. In addition, of course, a large

Death rates have plummeted, and the benefits of modern medicines, antibiotics, and sanitary practices have enhanced both the quality and expectancy of life in the developed and much of the developing world. Far from being won, however, the struggle against infectious and parasitic diseases is growing in intensity and is, perhaps, unwinnable. More than a half century after the discovery of antibiotics, the diseases they were to eradicate are on the rise, and both old and new disease-causing microorganisms are emerging and spreading all over the world. Infectious and parasitic diseases kill between 17 and 20 million people each year; they officially account for one-quarter to one-third of global mortality and, because of poor diagnosis, certainly are responsible for far more. And their global incidence is rising.

The five leading infectious killers are acute respiratory infections such as pneumonia, diarrheal diseases, tuberculosis, malaria, and measles. In addition, AIDS was killing 3 million or more persons yearly early in this century, far more than measles and as many as malaria. The incidence of infection, of course, is far greater than the occurrence of deaths. Nearly 30% of the world's people, for example, are infected with the bacterium that causes tuberculosis, but only 2 to 3 million are killed by the disease each year. More than 500 million people are infected with such tropical diseases as malaria, sleeping sickness, schistosomiasis, and river blindness, with perhaps 3 million annual deaths. Newer pathogens are constantly appearing, such as those causing Lassa fever, Rift Valley fever, Ebola, Hanta, West Nile virus, and hepatitis C, incapacitating and endangering far more than they kill. In fact, at least 30 previously unknown infectious diseases have appeared since the mid-1970s.

The spread and virulence of infectious diseases are linked to the dramatic changes so rapidly occurring in the earth's physical and social environments. Climate warming permits temperature-restricted pathogens to invade new areas and claim new victims. Deforestation, water contamination, wetland drainage, and other human-induced alterations to the physical environment disturb ecosystems and simultaneously disrupt the natural system of controls that keep infectious diseases in check. Rapid population growth and explosive urbanization, increasing global tourism, population-dislocating wars and migrations, and expanding world trade all increase interpersonal disease-transmitting contacts and the mobility and range of disease-causing microbes, including those brought from previously isolated areas by newly opened road systems and air routes. Add in poorly planned or executed public health programs, inadequate investment in sanitary infrastructures, and inefficient distribution of medical personnel and facilities, and the causative role of humans in many of the current disease epidemics is clearly visible.

In response, a worldwide Program for Monitoring Emerging Diseases (ProMED) was established in 1993 and developed a global on-line infectious disease network linking health workers and scientists in more than 100 countries to battle what has been called a growing "epidemic of epidemics." The most effective weapons in that battle are already known. They include improved health education; disease prevention and surveillance; research on disease vectors and incidence areas (including GIS and other mapping of habitats conducive to specific diseases); careful monitoring of drug therapy; mosquito control programs; provision of clean water supplies; and distribution of such simple and cheap remedies and preventatives as childhood immunizations, oral rehydration therapy, and vitamin A supplementation. All, however, require expanded investment and attention to those spreading infectious diseases—many with newly developed antibiotic-resistant strains—so recently thought to be no longer of concern.

quickly and dramatically increased life expectancies in developing countries. Such imported technologies and treatments accomplished in a few years what it took Europe 50 to 100 years to experience. Sri Lanka, for example, sprayed extensively with DDT to combat malaria; life expectancy jumped from 44 years in 1946 to 60 only 8 years later. With similar public health programs, India also experienced a steady reduction in its death rate after 1947. Simultaneously, with international sponsorship, food aid cut the death toll of developing states during drought and other disasters. The dramatic decline in mortality that had emerged only gradually throughout the European world occurred with startling speed in developing countries after 1950.

Corresponding reductions in birth rates did not immediately follow, and world population totals soared: from 2.5 billion in 1950 to 3 billion by 1960 and 5 billion by the middle 1980s. Alarms about the "population explosion" and its predicted devastating impact on global food and mineral resources were frequent and strident. In demographic terms, the world was viewed by many as permanently divided between developed regions that had made the demographic transition to stable population numbers and the underdeveloped, endlessly expanding ones that had not.

Birth rate levels, of course, unlike life expectancy improvements, depend less on supplied technology and assistance than they do on social acceptance of the idea of fewer children and smaller families (Figure 4.20). That acceptance began to grow broadly but unevenly worldwide even as regional and world population growth seemed uncontrollable. In 1984, only 18% of world population lived in countries with fertility rates at or below replacement levels (that is, countries that had achieved the demographic transition). By 2000, however, 44% lived in such countries, and early in the 21st century it is increasingly difficult to distinguish between developed and developing societies on the basis of their fertility rates. Those rates in many separate Indian states (Kerala and Tamil Nadu, for example) and in such countries as Sri Lanka, Thailand, South Korea, and China are below those of the United States and some European countries. Significant decreases to near the replacement level have also occurred in the space of a single generation in many other Asian and Latin American states with high recent rates of economic growth. Increasingly, it appears, low fertility is becoming a feature of both rich and poor, developed and developing states.

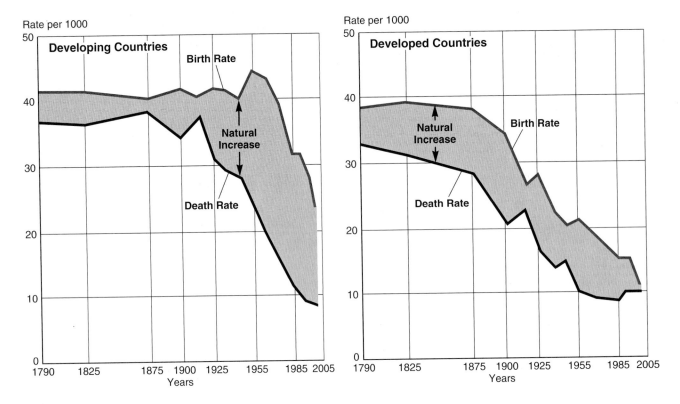

Figure 4.20 **World birth and death rates to 2003.** The "population explosion" after World War II (1939–1945) reflected the effects of drastically reduced death rates in developing countries without simultaneous and compensating reductions in births. By the end of the 20th century, however, three interrelated trends had appeared in many developing world countries: (1) fertility had overall dropped further and faster than had been earlier predicted; (2) contraceptive acceptance and use had increased markedly; and (3) age at marriage was rising. In consequence, the demographic transition had been compressed from a century to a generation in some developing states. In others, fertility decline began to slacken in the mid-1970s but continued to reflect the average number of children—four or more—still desired in many societies.

Source: Revised and redrawn from Elaine M. Murphy, World Population: Toward the Next Century, *revised ed. (Washington, D.C.: Population Reference Bureau, 1989).*

Despite this general substantial convergence in fertility, there still remains a significant minority share of the developing world with birth rates averaging 1.5 to 2 times or more above the replacement level. Indeed, early in the 21st century almost 1.4 billion persons live in countries or regions where total fertility is still 3.5 or greater (a level not considered high, of course, in the early 1950s when only a quarter of the world's population had a TFR below that mark). For the most part, these current high fertility countries and areas are in sub-Saharan Africa and the northern parts of the Indian subcontinent. Although accounting for less than a quarter of world population, high TFR regions collectively, United Nations demographers predict, will provide the majority of world population growth to at least 2050.

The established patterns of both high and low fertility regions tend to be self-reinforcing. Low growth permits the expansion of personal income and the accumulation of capital that enhance the quality and security of life and make large families less attractive or essential. In contrast, in high birth rate regions, population growth consumes in social services and assistance the investment capital that might promote economic expansion. Increasing populations place ever greater demands on limited soil, forest, water, grassland, and cropland resources. As the environmental base deteriorates, productivity declines and population-supporting capacities are so diminished as to make difficult or impossible the economic progress on which the demographic transition depends, an apparent equation of increasing international concern (see "The Cairo Plan").

The Demographic Equation

Births and deaths among a region's population—natural increases or decreases—tell only part of the story of population change. Migration involves the long-distance movement of people from one residential location to another. When that relocation occurs across political boundaries, it affects the population structure of both the origin and destination jurisdictions. The **demographic equation** summarizes the contribution made to regional population change over time by the combination of *natural change* (difference between births and deaths) and *net migration* (difference between in-migration and out-migration).[1] On a global scale, of course, all population change is accounted for by natural change. The impact of migration on the demographic equation increases as the population size of the areal unit studied decreases.

---

[1]See the Glossary definition for the calculation of the equation.

Geography and Public Policy

The Cairo Plan

After a sometimes rancorous 9-day meeting in Cairo in September, 1994, the United Nations International Conference on Population and Development endorsed a strategy for stabilizing the world's population at 7.27 billion by no later than 2015. The 20-year program of action accepted by over 150 signatory countries sought to avoid the environmental consequences of excessive population growth. Its proposals were therefore linked to discussions and decisions of the UN Conference on Environment and Development held in Rio de Janeiro in June, 1992.

The Cairo plan abandoned several decades of top-down governmental programs that promoted "population control" (a phrase avoided by the conference) based on targets and quotas and, instead, embraced for the first time policies giving women greater control over their lives, greater economic equality and opportunity, and a greater voice in reproduction decisions. It recognized that limiting population growth depends on programs that lead women to want fewer children and make them partners in economic development. In that recognition, the Conference accepted the documented link between increased educational access and economic opportunity for women and falling birth rates and smaller families. Earlier population conferences—1974 in Bucharest and 1984 in Mexico City—did not fully address these issues of equality, opportunity, education, and political rights; their adopted goals failed to achieve hoped-for changes in births in large part because women in many traditional societies had no power to enforce contraception and feared their other alternative, sterilization.

The earlier conferences carefully avoided or specifically excluded abortion as an acceptable family planning method. It was the more open discussion of abortion in Cairo that elicited much of the spirited debate that registered religious objections by the Vatican and many Muslim and Latin American states to the inclusion of legal abortion as part of health care, and to language suggesting approval of sexual relations outside of mar-

riage. Although the final text of the conference declaration did not promote any universal right to abortion and excluded it as a means of family planning, some delegations still registered reservations to its wording on both sex and abortion. At conference close, however, the Vatican endorsed the declaration's underlying principles, including the family as "the basic unit of society," the need to stimulate economic growth, and to promote "gender equality, equity, and the empowerment of women."

A special United Nations "Cairo + 5" session in 1999 recommended some adjustments in the earlier agreements. It urged emphasis on measures ensuring safe and accessible abortion in countries where it is legal, called for school children at all levels to be instructed in sexual and reproductive health issues, and told governments to provide special family planning and health services for sexually active adolescents, with particular stress on reducing their vulnerability to AIDS.

By early in the 21st century, positive results of Cairo plan proposals were being seen in declining fertility rates in many of the world's most populous developing countries. Some demographers and many women's health organizations pointedly claim that those declines have little to do with government planning policies. Rather, they assert, current lower and falling fertility rates are the expected result of women assuming greater control over their economic and reproductive lives. The director of the UN population division noted: "A woman in a village making a decision to have one or two or at most three children is a small decision in itself. But . . . compounded by millions and millions . . . of women in India and Brazil and Egypt, it has global consequences."

That women are making those decisions, population specialists have observed, reflects important cultural factors emerging since Cairo. Satellite television brings contraceptive information to even remote villages and shows programs of small, apparently happy families that viewers think of emulating. Increasing urbanization reduces some tradi-

tional family controls on women and makes contraceptives easier to find, and declining infant mortality makes mothers more confident their babies will survive. Perhaps most important, population experts assert, is the dramatic increase in most developing states in female school attendance and corresponding reductions in the illiteracy rates of girls and young women who will themselves soon be making fertility decisions.

Questions to Consider

1. Do you think it is appropriate or useful for international bodies to promote policies affecting such purely personal or national concerns as reproduction and family planning? Why or why not?

2. Do you think that current international concerns over population growth, development, and the environment are sufficiently valid and pressing to risk the loss of long-enduring cultural norms and religious practices in many of the world's traditional societies? Why or why not?

3. The Cairo plan called for sizeable monetary pledges from developed countries to support enhanced population planning in the developing world. For the most part, those pledges have not been honored. Do you think the financial obligations assigned to donor countries are justified in light of the many other international needs and domestic concerns faced by their governments? Why or why not?

4. Many environmentalists see the world as a finite system unable to support ever-increasing populations; to exceed its limits would cause frightful environmental damage and global misery. Many economists counter that free markets will keep supplies of needed commodities in line with growing demand and that science will, as necessary, supply technological fixes in the form of substitutes or expansion of production. In light of such diametrically opposed views of population growth consequences, is it appropriate or wise to base international programs solely on one of them? Why or why not?

Population Relocation

In the past, emigration proved an important device for relieving the pressures of rapid population growth in at least some European countries (Figure 4.21). For example, in one 90-year span, 45% of the natural increase in the population of the British Isles emigrated, and between 1846 and 1935 some 60 million Europeans of all nationalities left that continent. Despite recent massive movements of economic and political refugees across Asian, African, and Latin American boundaries, emigration today provides no comparable relief valve for developing countries. Total population numbers are too great to be much affected by migrations of even millions of people. In only a few countries—Afghanistan, Cuba, El Salvador, and Haiti, for example—have as many as 10% of the population emigrated in recent decades.

Immigration Impacts

Where cross-border movements are massive enough, migration may have a pronounced impact on the demographic equation and result in significant changes in the population structures of both the origin and destination regions. Past European and African migrations, for example, not only altered but substantially created the population structures of new, sparsely inhabited lands of coloniza-tion in the Western Hemisphere and Australasia. In some decades of the late 18th and early 19th centuries, 30% to more than 40% of population increase in the United States was accounted for by immigration. Similarly, eastward-moving Slavs colonized underpopulated Siberia and overwhelmed native peoples.

Migrants are rarely a representative cross section of the population group they leave, and they add an unbalanced age and sex component to the group they join. A recurrent research observation is that emigrant groups are heavily skewed in favor of young singles. Whether males or females dominate the outflow varies with circumstances. Although males traditionally far exceeded females in international flow, in recent years females have accounted for between 40% and 60% of all transborder migrants.

At the least, then, the receiving country will have its population structure altered by an outside increase in its younger age and, probably, unmarried cohorts. The results are both immediate in a modified population pyramid, and potential in future impact on reproduction rates and excess of births over deaths. The origin area will have lost a portion of its young, active members of childbearing years. It perhaps will have suffered distortion in its young adult sex ratios, and it certainly will have recorded a statistical aging of its population. The destination society will likely experience increases in births associated with the youthful newcomers and, in general, have its average age reduced.

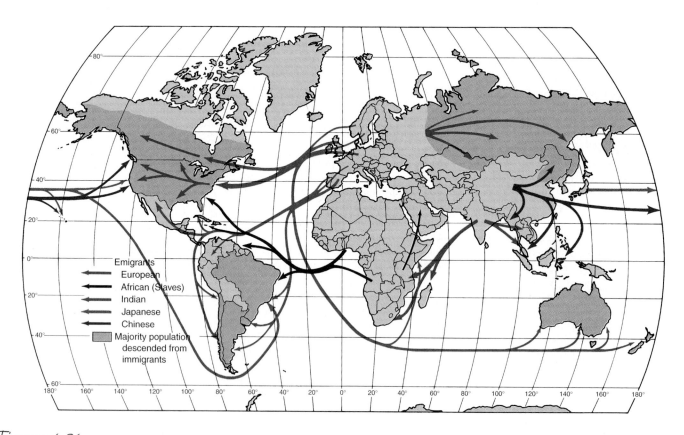

Figure 4.21 **Principal migrations of recent centuries.** The arrows suggest the major free and forced international population movements since about 1700. The shaded areas on the map are regions whose present population is more than 50% descended from the immigrants of recent centuries.

Source: Shaded zones after Daniel Noin, Géographie de la Population *(Paris: Masson, 1979), p. 85.*

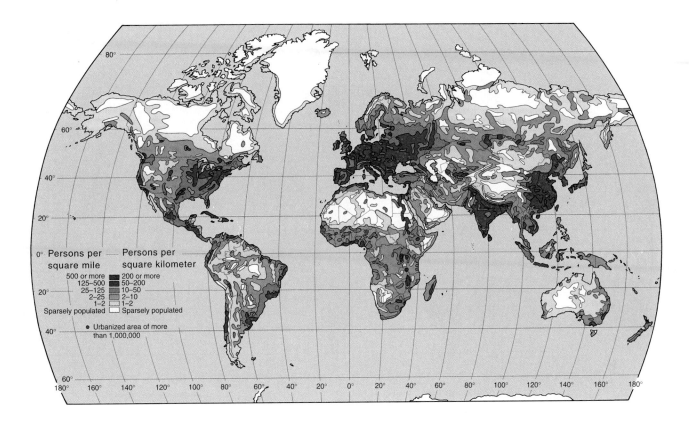

Figure 4.22 **World population density.**

World Population Distribution

The millions and billions of people of our discussion are not uniformly distributed over the earth. The most striking feature of the world population distribution map (Figure 4.22) is the very unevenness of the pattern. Some land areas are nearly uninhabited, others are sparsely settled, and still others contain dense agglomerations of people. A little more than half of the world's people are found—unevenly concentrated, to be sure—in rural areas. Nearly half are urbanites, however, and a constantly growing proportion are residents of very large cities of 1 million or more.

Earth regions of apparently very similar physical makeup show quite different population numbers and densities, perhaps the result of differently timed settlement or of settlement by different cultural groups. Northern and Western Europe, for example, inhabited thousands of years before North America, contain as many people as the United States on 70% less land; the present heterogeneous population of the Western Hemisphere is vastly more dense than was that of earlier Native Americans.

We can draw certain generalizing conclusions from the uneven, but far from irrational, distribution of population shown in Figure 4.22. First, almost 90% of all people live north of the equator and two-thirds of the total dwell in the midlatitudes between 20° and 60° North (Figure 4.23). Second, a large majority of the world's inhabitants occupy only a small part of its land surface. More than half the people live on about 5% of the land, two-thirds on 10%, and almost nine-tenths on less than 20%.

Third, people congregate in lowland areas; their numbers decrease sharply with increases in elevation. Temperature, length of growing season, slope and erosion problems, even oxygen reductions at very high altitudes, all appear to limit the habitability of higher elevations. One estimate is that between 50% and 60% of all people live below 200 meters (650 ft), a zone containing less than 30% of total land area. Nearly 80% reside below 500 meters (1650 ft).

Fourth, although low-lying areas are preferred settlement locations, not all such areas are equally favored. Continental margins have attracted the densest settlement. About two-thirds of world population is concentrated within 500 kilometers (300 mi) of the ocean, much of it on alluvial lowlands and river valleys. Latitude, aridity, and elevation, however, limit the attractiveness of many seafront locations. Low temperatures and infertile soils of the extensive Arctic coastal lowlands of the Northern Hemisphere have restricted settlement there. Mountainous or desert coasts are sparsely occupied at any latitude, and some tropical lowlands and river valleys that are marshy, forested, and disease-infested are also unevenly settled.

Within the sections of the world generally conducive to settlement, four areas contain great clusters of population: East Asia, South Asia, Europe, and northeastern United States/southeastern Canada. The *East Asia* zone, which includes Japan, China, Taiwan, and South Korea, is the largest cluster in both area and numbers. The four countries forming it contain nearly 25% of all people on earth; China alone accounts for one in five

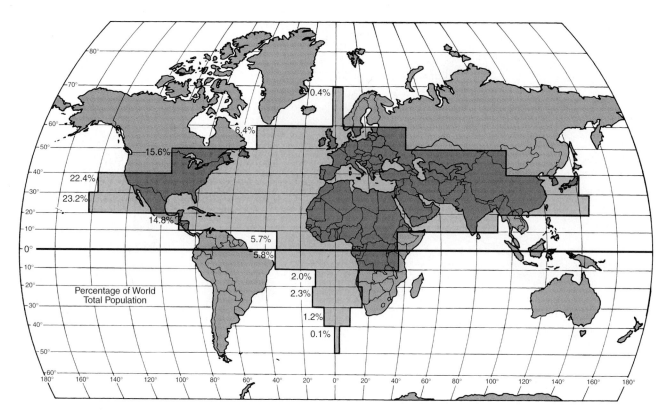

Figure 4.23 **The population dominance of the Northern Hemisphere** is strikingly evident from this bar chart. Only one out of nine people lives south of the equator—not because the Southern Hemisphere is underpopulated, but because it is mainly water.

of the world's inhabitants. The *South Asia* cluster is composed primarily of countries associated with the Indian subcontinent—Bangladesh, India, Pakistan, and the island state of Sri Lanka—though some might add to it the Southeast Asian countries of Cambodia, Myanmar, and Thailand. The four core countries alone account for another one-fifth, 21%, of the world's inhabitants. The South and the East Asian concentrations are thus home to nearly one-half the world's people.

Europe—southern, western, and eastern through Ukraine and much of European Russia—is the third extensive world population concentration, with another 12% of its inhabitants. Much smaller in extent and total numbers is the cluster in *northeastern United States/southeastern Canada.* Other smaller but pronounced concentrations are found around the globe: on the island of Java in Indonesia, along the Nile River in Egypt, and in discontinuous pockets in Africa and Latin America.

The term **ecumene** is applied to permanently inhabited areas of the earth's surface. The ancient Greeks used the word, derived from their verb "to inhabit," to describe their known world between what they believed to be the unpopulated, searing southern equatorial lands and the permanently frozen northern polar reaches of the earth. Clearly, natural conditions are less restrictive than Greek geographers believed. Both ancient and modern technologies have rendered habitable areas that natural conditions make forbidding. Irrigation, terracing, diking, and draining are among the methods devised to extend the ecumene locally (Figure 4.24).

At the world scale, the ancient observation of habitability appears remarkably astute. The **nonecumene,** or *anecumene,* the uninhabited or very sparsely occupied zone, does include the permanent ice caps of the Far North and Antarctica and large segments of the tundra and coniferous forest of northern Asia and North America. But the nonecumene is not continuous, as the ancients supposed. It is discontinuously encountered in all portions of the globe and includes parts of the tropical rain forests of equatorial zones, midlatitude deserts of both the Northern and Southern Hemispheres, and high mountain areas.

Even parts of these unoccupied or sparsely occupied districts have localized dense settlement nodes or zones based on irrigation agriculture, mining and industrial activities, and the like. Perhaps the most striking case of settlement in an environment elsewhere considered part of the nonecumene world is that of the dense population in the Andes Mountains of South America and the plateau of Mexico. Here Native Americans found temperate conditions away from the dry coast regions and the hot, wet Amazon basin. The fertile high basins have served a large population for more than a thousand years.

Even with these locally important exceptions, the nonecumene portion of the earth is extensive. Some 35% to 40% of all the world's land surface is inhospitable and without significant settlement. This is, admittedly, a smaller proportion of the earth than would have qualified as uninhabitable in ancient times or even during the 19th century. Since the end of the Ice Age some

𝓕𝓲𝓰𝓾𝓻𝓮 4.24 Terracing of hillsides is one device to extend a naturally limited productive area. The technique is effectively used here at the Malegcong rice terraces on densely settled Luzon Island of the Philippines.

11,000 to 12,000 years ago, humans have steadily expanded their areas of settlement.

Population Density

Margins of habitation could only be extended, of course, as humans learned to support themselves from the resources of new settlement areas. The numbers that could be sustained in old or new habitation zones were and are related to the resource potential of those areas and the cultural levels and technologies possessed by the occupying populations. The term **population density** expresses the relationship between number of inhabitants and the area they occupy.

Density figures are useful, if sometimes misleading, representations of regional variations of human distribution. The **crude density,** or **arithmetic density,** of population is the most common and least satisfying expression of that variation. It is the calculation of the number of people per unit area of land, usually within the boundaries of a political entity. It is an easily reckoned figure. All that is required is information on total population and total area, both commonly available for national or other political units. The figure can, however, be misleading and may obscure more of reality than it reveals. The calculation is an average, and

a country may contain extensive regions that are only sparsely populated or largely undevelopable (Figure 4.25) along with intensively settled and developed districts. A national average density figure reveals nothing about either class of territory. In general, the larger the political unit for which crude or arithmetic population density is calculated, the less useful is the figure.

Various modifications may be made to refine density as a meaningful abstraction of distribution. Its descriptive precision is improved if the area in question can be subdivided into comparable regions or units. Thus, it is more revealing to know that in 2000, New Jersey had a density of 438 and Wyoming of 3.5 persons per square kilometer (1134 and 9 per sq mi) of land area than to know only that the figure for the conterminous United States (48 states) was 36 per square kilometer (94 per sq mi). If Hawaii and large, sparsely populated Alaska are added, the U.S. density figure drops to 31 per square kilometer (80 per sq mi). The calculation may also be modified to provide density distinctions between classes of population—rural versus urban, for example. Rural densities in the United States rarely exceed 115 per square kilometer (300 per sq mi), while portions of major cities can have thousands of people in equivalent space.

Another revealing refinement of crude density relates population not simply to total national territory but to that area of a country that is or may be cultivated, that is, to *arable* land. When

Figure 4.25 Tundra vegetation and landscape, Ruby Range, Northwest Territories, Canada. Extensive areas of northern North America and Eurasia are part of the one-third or more of the world's land area considered as *nonecumene,* sparsely populated portions of total national territory that affect calculations of arithmetic density.

Table 4.4

Comparative Densities for Selected Countries

Country	Crude Density		Physiological Density[a]		Agricultural Density[b]	
	sq mi	km²	sq mi	km²	sq mi	km²
Argentina	35	14	390	151	39	15
Australia	6	2	115	44	13	5
Bangladesh	2401	929	4333	1673	3118	1204
Canada	8	3	179	69	39	15
China	344	133	2659	1026	1785	689
Egypt	181	70	6480	2502	3100	1197
India	814	314	1657	640	1134	438
Iran	107	41	1015	392	376	145
Japan	872	337	6773	2615	1551	599
Nigeria	355	137	1161	448	642	248
United Kingdom	635	245	2478	957	259	100
United States	80	31	417	161	93	36

[a]Total population divided by area of arable land.
[b]Rural population divided by area of arable land.

Sources: UN Food and Agriculture Organization (FAO), *Production Yearbook;* World Bank, *World Development Indicators;* and Population Reference Bureau, *World Population Data Sheet.*

total population is divided by arable land area alone, the resulting figure is the **physiological density,** which is, in a sense, an expression of population pressure exerted on agricultural land. Table 4.4 makes evident that countries differ in physiological density and that the contrasts between crude and physiological densities of countries point up actual settlement pressures that are not revealed by arithmetic densities alone. The calculation of physiological density, however, depends on uncertain definitions of arable and cultivated land, assumes that all arable land is equally productive and comparably used, and includes only one part of a country's resource base.

Agricultural density is still another useful variant. It simply excludes city populations from the physiological density calculation and reports the number of rural residents per unit of agriculturally productive land. It is, therefore, an estimate of the pressure of people on the rural areas of a country.

Overpopulation

It is an easy and common step from concepts of population density to assumptions about overpopulation or overcrowding. It is wise to remember that **overpopulation** is a value judgment reflecting an observation or conviction that an environment or territory is unable to support its present population. (A related but opposite concept of *underpopulation* refers to the circumstance of too few people to sufficiently develop the resources of a country or region to improve the level of living of its inhabitants.)

Overpopulation is not the necessary and inevitable consequence of high density of population. Tiny Monaco, a principality in southern Europe about half the size of New York's Central Park, has a crude density of some 17,500 people per square kilometer (45,000 per sq mi). Mongolia, a sizable state of 1,565,000 square kilometers (604,000 sq mi) between China and Siberian Russia, has 1.6 persons per square kilometer (4.1 per sq mi); Iran, only slightly larger, has 41 per square kilometer (107 per sq mi). Macao, a former island possession of Portugal off the coast of China, has some 22,000 persons per square kilometer (57,000 per sq mi); the Falkland Islands off the Atlantic coast of Argentina count at most 1 person for every 5 square kilometers (2 sq mi) of territory. No conclusions about conditions of life, levels of income, adequacy of food, or prospects for prosperity can be drawn from these density comparisons.

Overcrowding is a reflection not of numbers per unit area but of the **carrying capacity** of land—the number of people an area can support on a sustained basis given the prevailing technology. A region devoted to efficient, energy-intensive commercial agriculture that makes heavy use of irrigation, fertilizers, and biocides can support more people at a higher level of living than one engaged in the slash-and-burn agriculture described in Chapter 8. An industrial society that takes advantage of resources such as coal and iron ore and has access to imported food will not feel population pressure at the same density levels as a country with rudimentary technology.

Since carrying capacity is related to the level of economic development, maps such as Figure 4.22, displaying present patterns of population distribution and density, do not suggest a correlation with conditions of life. Many industrialized, urbanized countries have lower densities and higher levels of living than do less developed ones. Densities in the United States, where there is a great deal of unused and unsettled land, are considerably lower than those in Bangladesh, where essentially all land is arable and which, with nearly 930 people per square kilometer (over 2400 per sq mi), is the most densely populated nonisland state in the world. At the same time, many African countries have low population densities and low levels of living, whereas Japan combines both high densities and wealth.

Overpopulation can be equated with levels of living or conditions of life that reflect a continuing imbalance between numbers of people and carrying capacity of the land. One measure of that imbalance might be the unavailability of food supplies sufficient in caloric content to meet individual daily energy requirements or so balanced as to satisfy normal nutritional needs. Unfortunately, dietary insufficiencies—with long-term adverse implications for life expectancy, physical vigor, and mental development—are most likely to be encountered in the developing countries, where much of the population is in the younger age cohorts (Figure 4.11).

If those developing countries simultaneously have rapidly increasing population numbers dependent on domestically produced foodstuffs, the prospects must be for continuing undernourishment and overpopulation. Much of sub-Saharan Africa finds itself in this circumstance. Its per capita food production decreased during the 1990s, with continuing decline predicted over the following quarter century as the population-food gap widens (Figure 4.26). The countries of North Africa are similarly strained. Egypt already must import well over half the food it consumes. Africa is not alone. The international Food and Agriculture Organization (FAO) estimates that in 2000, at least 65 separate countries with over 30% of the population of the developing world were unable to adequately feed their inhabitants from their own national territories at the low level of agricultural technology and inputs employed. Even rapidly industrializing

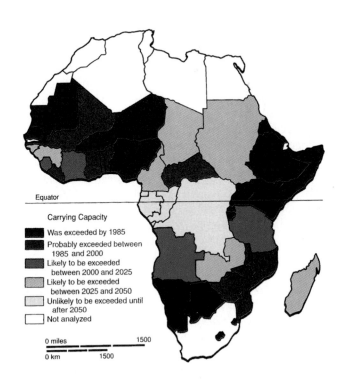

Equator

Carrying Capacity
- ■ Was exceeded by 1985
- ■ Probably exceeded between 1985 and 2000
- ■ Likely to be exceeded between 2000 and 2025
- ▨ Likely to be exceeded between 2025 and 2050
- □ Unlikely to be exceeded until after 2050
- □ Not analyzed

0 miles · · · 1500
0 km · · · 1500

Figure 4.26 **Carrying capacity and potentials in sub-Saharan Africa.** The map assumes that (1) all cultivated land is used for growing food; (2) food imports are insignificant; (3) agriculture is conducted by low technology methods.

Sources: World Bank; United Nations Development Programme. Food and Agriculture Organization (FAO); and Bread for the World Institute.

China, an exporter of grain until 1994, now in most years is a net grain importer.

In the contemporary world, insufficiency of domestic agricultural production to meet national caloric requirements cannot be considered a measure of overcrowding or poverty. Only a few countries are agriculturally self-sufficient. Japan, a leader among the advanced states, is the world's biggest food importer and supplies from its own production only 40% of the calories its population consumes. Its physiological density is high, as Table 4.4 indicates, but it obviously does not rely on an arable land resource for its present development. Largely lacking in either agricultural or industrial resources, it nonetheless ranks well on all indicators of national well-being and prosperity. For countries such as Japan, a sudden cessation of the international trade that permits the exchange of industrial products for imported food and raw materials would be disastrous. Domestic food production could not maintain the dietary levels now enjoyed by their populations and they, more starkly than many underdeveloped countries, would be "overpopulated."

Urbanization

Pressures on the land resource of countries are increased not just by their growing populations but by the reduction of arable land caused by such growth. More and more of world population increase must be accommodated not in rural areas, but in cities that hold the promise of jobs and access to health, welfare, and other public services. As a result, the *urbanization* (transformation from rural to urban status according to individual state's definition of "urban") of population in developing countries is increasing dramatically. Since the 1950s, cities have grown faster than rural areas in nearly all developing states. Indeed, because of the now rapid flow of migrants from countrysides to cities, population growth in the rural areas of the developing world has essentially stopped. Although Latin America, for example, has experienced substantial overall population increase, the size of its rural population is actually declining.

On UN projections, some 97% of all world population increase between 2000 and 2030 will be in urban areas and almost entirely within the developing regions and countries, continuing a pattern established by 1950 (Figure 4.27). In those areas collectively, cities are growing by 3% a year, and the poorest regions are experiencing the fastest growth. By 2020, the UN anticipates, a majority of the population of less developed countries will live in urban areas. In East, West, and Central Africa, for example, cities are expanding by 5% a year, a pace that can double their population every 14 years. Global urban population, just 750 million in 1950, grew to nearly 2.75 billion by century's end and is projected to rise to 5.1 billion by 2030. The uneven results of past urbanization are summarized in Figure 4.28.

The sheer growth of those cities in people and territory has increased pressures on arable land and adjusted upward both arithmetic and physiological densities. Urbanization consumes millions of hectares of cropland each year. In Egypt, for example, urban expansion and new development between 1965 and 1985 took out of production as much fertile soil as the massive Aswan dam on the Nile River made newly available through irrigation

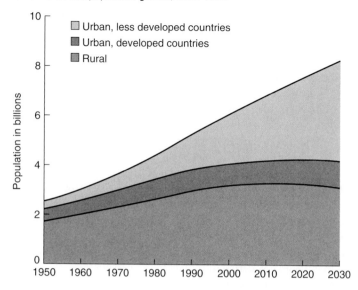

Urban and rural population growth, 1950–2030

Figure 4.27 **Past and projected urban and rural population growth.** According to UN projections, some 65% of the world's total population may be urbanized by 2030.

Redrawn from Population Bulletin *vol. 53, No. 1, Figure 3, p. 12 (Population Reference Bureau, 1998).*

with the water it impounds. And during much of the 1990s, China lost close to 1 million hectares of farmland each year to urbanization, road construction, and industrialization; the pace of such loss continued into the new century. By themselves, some of these developing world cities, often surrounded by concentrations of people living in uncontrolled settlements, slums, and shantytowns (Figure 11.42), are among the most densely populated areas in the world. They face massive problems in trying to provide housing, jobs, education, and adequate health and social services for their residents. These and other matters of urban geography are the topics of Chapter 11.

Population Data and Projections

Population geographers, demographers, planners, governmental officials, and a host of others rely on detailed population data to make their assessments of present national and world population patterns and to estimate future conditions. Birth rates and death rates, rates of fertility and of natural increase, age and sex composition of the population, and other items are all necessary ingredients for their work.

Population Data

The data that students of population employ come primarily from the United Nations Statistical Office, the World Bank, the Population Reference Bureau, and ultimately, from national censuses and sample surveys. Unfortunately, the data as reported may on occasion be more misleading than informative. For much of the

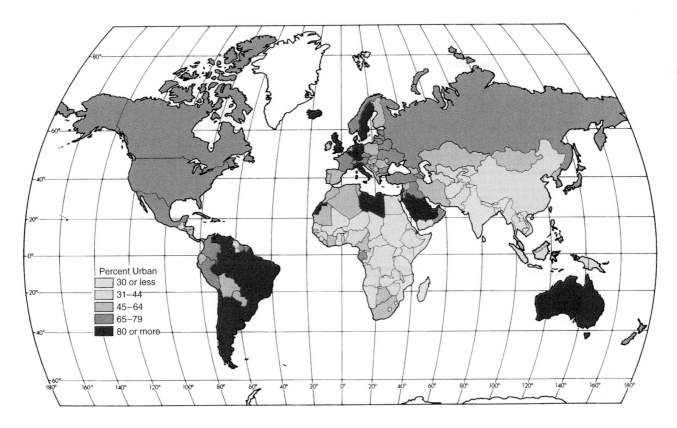

Figure 4.28 **Percentage of national population that is classified as urban.** Urbanization has been particularly rapid in the developing continents. In 1950, only 17% of Asians and 15% of Africans were urban. By 2002, one-third of Africans and nearly 40% of Asians were city dwellers, and collectively, the less developed areas contained almost 70% of the world's city population.

Source: Data from Population Reference Bureau.

developing world, a national census is a massive undertaking. Isolation and poor transportation, insufficiency of funds and trained census personnel, high rates of illiteracy limiting the type of questions that can be asked, and populations suspicious of all things governmental serve to restrict the frequency, coverage, and accuracy of population reports.

However derived, detailed data are published by the major reporting agencies for all national units even when those figures are poorly based on fact or are essentially fictitious. For years, data on the total population, birth and death rates, and other vital statistics for Somalia were regularly reported and annually revised. The fact was, however, that Somalia had never had a census and had no system whatsoever for recording births. Seemingly precise data were regularly reported as well for Ethiopia. When that country had its first-ever census in 1985, at least one data source had to drop its estimate of the country's birth rate by 15% and increase its figure for Ethiopia's total population by more than 20%. And a disputed 1992 census of Nigeria officially reported a population of 88.5 million, still the largest in Africa but far below the generally accepted and widely cited estimates of between 105 and 115 million Nigerians.

Fortunately, census coverage on a world basis is improving. Almost every country has now had at least one census of its population, and most have been subjected to periodic sample surveys (Figure 4.29). However, only about 10% of the developing

Figure 4.29 By the early 21st century, most countries of the developed and developing worlds had conducted a relatively recent census, although some were of doubtful completeness or accuracy. The photo shows an enumerator interviewing a Quito resident during the well-planned Ecuador census of 2001.

world's population live in countries with anything approaching complete systems for registering births and deaths. Estimates are that 40% or less of live births in Indonesia, Pakistan, India, or the Philippines are officially recorded; sub-Saharan Africa has the highest percentage of unregistered births (71%), according to UNICEF. Apparently, deaths are even less completely reported than births throughout Asia. And whatever the deficiencies of Asian states, African statistics are still less complete and reliable. It is, of course, on just these basic birth and death data that projections about population growth and composition are founded.

Population Projections

For all their inadequacies and imprecisions, current data reported for country units form the basis of **population projections,** estimates of future population size, age, and sex composition based on current data. Projections are not forecasts, and demographers are not the social science equivalent of meteorologists. Weather forecasters work with a myriad of accurate observations applied against a known, tested model of the atmosphere. The demographer, in contrast, works with sparse, imprecise, out-of-date, and missing data applied to human actions that will be unpredictably responsive to stimuli not yet evident.

Population projections, therefore, are based on assumptions for the future applied to current data that are, themselves, frequently suspect. Since projections are not predictions, they can never be wrong. They are simply the inevitable result of calculations about fertility, mortality, and migration rates applied to each age cohort of a population now living, and the making of birth rate, survival, and migration assumptions about cohorts yet unborn. Of course, the perfectly valid *projections* of future population size and structure resulting from those calculations may be dead wrong as *predictions*.

Since those projections are invariably treated as scientific expectations by a public that ignores their underlying qualifying assumptions, agencies such as the UN that estimate the population of, say, Africa in the year 2025, do so by not one but by three or more projections: high, medium, and low, for example (see "World Population Projections"). For areas as large as Africa, a medium projection is assumed to benefit from compensating errors and statistically predictable behaviors of very large populations. For individual African countries and smaller populations, the medium projection may be much less satisfying. The usual tendency in projections is to assume that something like current conditions will be applicable in the future. Obviously, the more distant the future, the less likely is that assumption to remain true. The resulting observation should be that the further into the future the population structure of small areas is projected, the greater is the implicit and inevitable error (see Figure 4.13).

World Population Projections

While the need for population projections is obvious, demographers face difficult decisions regarding the assumptions they use in preparing them. Assumptions must be made about the future course of birth and death rates and, in some cases, about migration.

Demographers must consider many factors when projecting a country's population. What is the present level of the birth rate, of literacy, and of education? Does the government have a policy to influence population growth? What is the status of women? What might be the impact of, for example, HIV/AIDS on life expectancies?

Along with these questions must be weighed the likelihood of socioeconomic change, for it is generally assumed that as a country "develops," a preference for smaller families will cause fertility to fall to the replacement level of about two children per woman. But when can one expect this to happen in less developed countries? And for the majority of more developed countries with fertility currently below replacement level, can one assume that fertility will rise to avert eventual disappearance of the population and, if so, when?

Predicting the pace of fertility decline is most important, as illustrated by one earlier set of United Nations long-range projections for Africa. As with many projections, these were issued in a "series" to show the effects of different assumptions. The "low" projection for Africa assumed that replacement level fertility would be reached in 2030, which would put the continent's population at 1.4 billion in 2100. If attainment of replacement-level fertility were delayed to 2065, the population would reach 4.4 billion in 2100. That difference of 3 billion should serve as a warning that using population projections requires caution and consideration of *all* the possibilities.

Unfortunately, demographers usually cast their projections in an environmental vacuum, ignoring the realities of soils, vegetation, water supplies, and climate that ultimately determine feasible or possible levels of population support. Inevitably, different analysts present different assessments of the absolute carrying capacity of the earth. At an unrealistically low level, the World Hunger Project calculated that the world's ecosystem could, with present agricultural technologies and with equal distribution of food supplies, support on a sustained basis no more than 5.5 billion people, a number already far exceeded. Many agricultural economists, in contrast—citing present trends and prospective increases in crop yields, fertilizer efficiencies, and intensification of production methods—are confident that the earth can readily feed 10 billion or more on a sustained basis. Nearly all observers, however, agree that physical environmental realities make unrealistic purely demographically based projections of a world population three or four times its present size.

Population Controls

All population projections include an assumption that at some point in time population growth will cease and plateau at the replacement level. Without that assumption, future numbers become unthinkably large. For the world at unchecked present growth rates, there would be 1 trillion people three centuries from now, 4 trillion four centuries in the future, and so on. Although there is reasonable debate about whether the world is now overpopulated and about what either its optimum or maximum sustainable population should be, totals in the trillions are beyond any reasonable expectation.

Population pressures do not come from the amount of space humans occupy. It has been calculated, for example, that the entire human race could easily be accommodated within the boundaries of the state of Delaware. The problems stem from the food, energy, and other resources necessary to support the population and from the impact on the environment of the increasing demands and the technologies required to meet them. Rates of growth currently prevailing in many countries make it nearly impossible for them to achieve the kind of social and economic development they would like.

Clearly, at some point population will have to stop increasing as fast as it has been. That is, either the self-induced limitations on expansion implicit in the demographic transition will be adopted or an equilibrium between population and resources will be established in more dramatic fashion. Recognition of this eventuality is not new. "[P]estilence, and famine, and wars, and earthquakes have to be regarded as a remedy for nations, as the means of pruning the luxuriance of the human race," was the opinion of the theologian Tertullian during the 2nd century A.D.

Thomas Robert **Malthus** (1766–1834), an English economist and demographer, put the problem succinctly in a treatise published in 1798: All biological populations have a potential for increase that exceeds the actual rate of increase, and the resources for the support of increase are limited. In later publications, Malthus amplified his thesis by noting the following:

1. Population is inevitably limited by the means of subsistence.

2. Populations invariably increase with increase in the means of subsistence unless prevented by powerful checks.

3. The checks that inhibit the reproductive capacity of populations and keep it in balance with means of subsistence are either "private" (moral restraint, celibacy, and chastity) or "destructive" (war, poverty, pestilence, and famine).

The deadly consequences of Malthus's dictum that unchecked population increases geometrically while food production can increase only arithmetically[2] have been reported throughout human history, as they are today. Starvation, the ultimate expression of resource depletion, is no stranger to the past or present. By conservative estimate, some 100 people worldwide

will starve to death during the 2 minutes it takes you to read this page; half will be children under 5. They will, of course, be more than replaced numerically by new births during the same 2 minutes. Losses are nearly always recouped. All battlefield casualties, perhaps 70 million, in all of humankind's wars over the past 300 years equal less than a 1-year replacement period at present rates of natural increase.

Yet, inevitably—following the logic of Malthus, the apparent evidence of history, and our observations of animal populations—equilibrium must be achieved between numbers and support resources. When overpopulation of any species occurs, a population dieback is inevitable. The madly ascending leg of the J-curve is bent to the horizontal, and the J-curve is converted to an S-curve. It has happened before in human history, as Figure 4.30 summarizes. The top of the **S-curve** represents a population size consistent with and supportable by the exploitable resource base. When the population is equivalent to the carrying capacity of the occupied area, it is said to have reached a **homeostatic plateau.**

In animals, overcrowding and environmental stress apparently release an automatic physiological suppressant of fertility. Although famine and chronic malnutrition may reduce fertility in humans, population limitation usually must be either forced or self-imposed. The demographic transition to low birth rates matching reduced death rates is cited as evidence that Malthus's first assumption was wrong: Human populations do not inevitably grow geometrically. Fertility behavior, it was observed, is conditioned by social determinants, not solely by biological or resource imperatives.

Although Malthus's ideas were discarded as deficient by the end of the 19th century in light of the European population experience, the concerns he expressed were revived during the 1950s. Observations of population growth in underdeveloped countries and the strain that growth placed on their resources inspired the viewpoint that improvements in living standards could be achieved only by raising investment per worker. Rapid population growth was seen as a serious diversion of scarce resources away from capital investment and into unending social welfare

---

[2]"Within a hundred years or so, the population can increase from fivefold to twentyfold, while the means of subsistence . . . can increase only from three to five times," was the observation of Hung Liangchi of China, a spatially distant early 19th-century contemporary of Malthus.

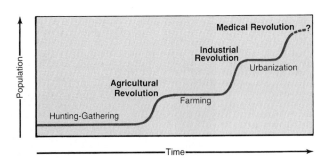

Figure 4.30 The steadily higher *homeostatic plateaus* (states of equilibrium) achieved by humans are evidence of their ability to increase the carrying capacity of the land through technological advance. Each new plateau represents the conversion of the J-curve into an S-curve.

programs. In order to lift living standards, the existing national efforts to lower mortality rates had to be balanced by governmental programs to reduce birth rates. **Neo-Malthusianism,** as this viewpoint became known, has been the underpinning of national and international programs of population limitation primarily through birth control and family planning (Figure 4.31).

Neo-Malthusianism has had a mixed reception. Asian countries, led by China and India, have in general—though with differing successes—adopted family planning programs and policies. In some instances, success has been declared complete. Singapore established its Population and Family Planning Board in 1965, when its fertility rate was 4.9 lifetime births per woman. By 1986, that rate had declined to 1.7, well below the 2.1 replacement level for developed countries, and the board was abolished as no longer necessary. Caribbean and South American countries, even the poorest and most agrarian, have also experienced declining fertility rates, though often these reductions have been achieved despite pronatalist views of governments influenced by the Roman Catholic Church.

Africa and the Middle East have generally been less responsive to the neo-Malthusian arguments because of ingrained cultural convictions among people, if not in all governmental circles, that large families—six or seven children—are desirable. Although total fertility rates have begun to decline in most sub-Saharan African states, they still remain nearly everywhere far above replacement levels. Islamic fundamentalism opposed to birth restrictions also is a cultural factor in the Near East and North Africa. However, the Muslim theocracy of Iran has endorsed a range of contraceptive procedures and developed one of the world's more aggressive family planning programs.

Other barriers to fertility control exist. When first proposed by Western states, neo-Malthusian arguments that family planning was necessary for development were rejected by many less developed countries. Reflecting both nationalistic and Marxist concepts, they maintained that remnant colonial-era social, economic, and class structures rather than population increase hindered development. Some government leaders think there is a correlation between population size and power and pursue pronatalist policies, as did Mao's China during the 1950s and early 1960s. And a number of American economists called *cornucopians* expressed the view, beginning in the 1980s, that population growth is a stimulus, not a deterrent, to development and that human minds and skills are the world's ultimate resource base. Since the time of Malthus, they observe, world population has grown from 900 million to over 6 billion without the predicted dire consequences—proof that Malthus failed to recognize the importance of technology in raising the carrying capacity of the earth. Still higher population numbers, they suggest, are sustainable, perhaps even with improved standards of living for all.

A third view, modifying cornucopian optimism, admits that products of human ingenuity such as the Green Revolution (see p. 282) increases in food production have managed to keep pace with rapid population growth since 1970. But its advocates argue that scientific and technical ingenuity to enhance food production do not automatically appear; both complacency and inadequate

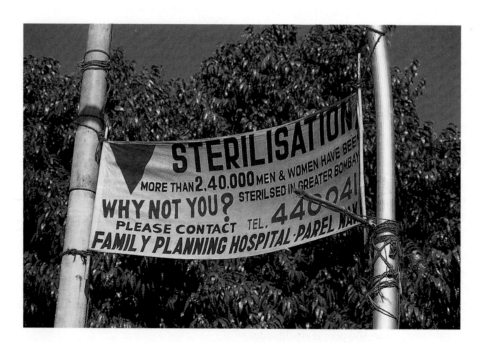

Figure 4.31 A Bombay, India, sign promoting the government's continuing program to reduce the country's high fertility rate. Sterilization is the world's most popular form of birth prevention, and in India, Brazil, and China, a reported one-third or more of all married women have been sterilized. The comparable worldwide married male sterilization rate, in contrast, is 4%.

research support have hindered continuing progress in recent years. And even if further advances are made, they observe, not all countries or regions have the social and political will or capacity to take advantage of them. Those that do not, third-view advocates warn, will fail to keep pace with the needs of their populace and will sink into varying degrees of poverty and environmental decay, creating national and regional—though not necessarily global—crises.

Population Prospects

Regardless of population philosophies, theories, or cultural norms, the fact remains that many or most developing countries are showing significantly declining population growth rates. Global fertility and birth rates are falling to an extent not anticipated by pessimistic Malthusians and at a pace that suggests a peaking of world population numbers sooner—and at lower totals—than previously projected (see "A Population Implosion?" p. 109). In all world regions, steady and continuous fertility declines have been recorded over the past years, reducing fertility from global 5.0-children-per-woman levels in the early 1950s to less than 3 per woman early in the present century.

Momentum

Reducing fertility levels even to the replacement level of about 2.1 births per woman does not mean an immediate end to population growth. Because of the age composition of many societies, numbers of births will continue to grow even as fertility rates per woman decline. The reason is to be found in **demographic** (or **population**) **momentum,** and the key to that is the age structure of a country's population.

When a high proportion of the population is young, the product of past high fertility rates, larger and larger numbers enter the childbearing age each year; that is the case for major parts of the world early in the 21st century. The populations of developing countries are far younger than those of the established industrially developed regions (see Figure 4.11), with about one-third (in Asia and Latin America) to well over 40% (in Africa) below the age of 15. The consequences of the fertility of these young people are yet to be realized. A population with a greater number of young people tends to grow rapidly regardless of the level of childbearing. The results will continue to be felt until the now-youthful groups mature and work their way through the population pyramid.

Inevitably, while this is happening, even the most stringent national policies limiting growth cannot stop it entirely. A country with a large present population base will experience large numerical increases despite declining birth rates. Indeed, the higher fertility was to begin with and the sharper its drop to low levels, the greater will be the role of momentum even after rates drop below replacement. A simple comparison of South Korea and the United Kingdom may serve to demonstrate the point. The two

countries had (in 2002) the same level of fertility, with women averaging about 1.6 children each. Between that year and 2025, the larger population of the U.K. (without considering immigration or the births associated with newcomers) was projected to decline by 2 million persons while the smaller, more youthful South Korea was expected to continue growing, adding 2 million people.

Aging

Eventually, of course, young populations grow older, and even the youthful developing countries are beginning to face the consequences of that reality. The problems of a rapidly aging population that already confront the industrialized economies are now being realized in the developing world as well. Globally, there will be more than 1 billion persons 60 years of age and older by 2025 and nearly 2 billion by 2050, when the world will contain more people aged 60 and above than children under the age of 15. That momentous reversal in relative proportions of young and old already occurred in 1998 in the more developed regions. The progression towards older populations is considered irreversible, the result of the now-global demographic transition from high to low levels of fertility and mortality. The youthful majorities of the past are unlikely to occur again, for globally, the population of older persons early in the century was growing by 2% per year—much faster than the population as a whole—and between 2025–2030, the 60+ growth rate will reach 2.8% per year.

Three-fourths of the mid-century elderly folk will live in the less developed world, for the growth rate of older people is three times as high in developing countries as in the developed ones. In the former, older persons are projected to make up 20% of the population by 2050 in contrast to the 8% over age 60 in the developing world in 2000. Since the pace of aging is much faster in the developing countries, they will have less time than the developed world did to adjust to the consequences of that aging. And those consequences will be experienced at lower levels of personal and national income and economic strength.

In both rich and poor states, the working-age populations will face increasing burdens and obligations. The potential support ratio, or PSR (the number of persons aged 15–64 years per one citizen aged 65 or older), has steadily fallen. Between 1950 and 2000, it dropped from 12 to 9 workers for each older person; by mid-century, the PSR is projected to drop to 4. The implications for social security schemes and social support obligations are obvious and made more serious because the older population itself is aging. By the middle of the century, one-fifth of older persons will be 80 years or older and on average require more support expenditures for health and long-term care than do younger seniors. The consequences of population aging appear most intractable for the world's poorest developing states that generally lack health, income, housing, and social service support systems adequate to the needs of their older citizens. To the social and economic implications of their present population momentum, therefore, developing countries must add the aging consequences of past patterns and rates of growth (Figure 4.32).

Figure 4.32 These senior residents of a Moroccan nursing home are part of the rapidly aging population of many developing countries. Worldwide, the over-60 cohort will number some 22% of total population by 2050 and be larger than the number of children less than 15 years of age. But by 2020, a third of Singapore citizens will be 55 or older, and China will have as large a share of its population over 60—about one in four—as will Europe. Already, the numbers of old people in the world's poorer countries are beginning to dwarf those in the rich world. At the start of the 21st century, there were nearly twice as many persons over 60 in developing countries as in the advanced ones, but most are without the old-age assistance and welfare programs developed countries have put in place.

Summary

Birth, death, fertility, and growth rates are important in understanding the numbers, composition, distribution, and spatial trends of population. Recent "explosive" increases in human numbers and the prospects of continuing population expansion may be traced to sharp reductions in death rates, increases in longevity, and the impact of demographic momentum on a youthful population largely concentrated in the developing world. Control of population numbers historically was accomplished through a demographic transition first experienced in European societies that adjusted their fertility rates downward as death rates fell and life expectancies increased. The introduction of advanced technologies of preventive and curative medicine, pesticides, and famine relief have reduced mortality rates in developing countries without, until recently, always a compensating reduction in birth rates. Recent fertility declines in many developing regions suggest the demographic transition is no longer limited to the advanced industrial countries and promise world population stability earlier and at lower numbers than envisioned just a few years ago.

Even with the advent of more widespread fertility declines, the 6 billion human beings present at the end of the 20th century will still likely grow to near 9 billion by the middle of the 21st. That growth will largely reflect increases unavoidable because of the size and youth of populations in developing countries. Eventually, a new balance between population numbers and carrying capacity of the world will be reached, as it has always been following past periods of rapid population increase.

People are unevenly distributed over the earth. The ecumene, or permanently inhabited portion of the globe, is discontinuous and marked by pronounced differences in population concentrations and numbers. East Asia, South Asia, Europe, and northeastern United States/southeastern Canada represent the world's greatest population clusters, though smaller areas of great density are found in other regions and continents. Since growth rates are highest and population doubling times generally shorter in world regions outside these four present main concentrations, new patterns of population localization and dominance are taking form.

A respected geographer once commented that "population is the point of reference from which all other elements [of geography] are observed." Certainly, population geography is the essential starting point of the human component of the human–environment concerns of geography. But human populations are not merely collections of numerical units; nor are they to be understood solely through statistical analysis. Societies are distinguished not just by the abstract data of their numbers, rates, and trends, but by experiences, beliefs, understandings, and aspirations that collectively constitute that human spatial and behavioral variable called *culture*. It is to that fundamental human diversity that we next turn our attention.

Key Words

For Review

1. How do the *crude birth rate* and the *fertility rate* differ? Which measure is the more accurate statement of the amount of reproduction occurring in a population?

2. How is the *crude death rate* calculated? What factors account for the worldwide decline in death rates since 1945?

3. How is a *population pyramid* constructed? What shape of "pyramid" reflects the structure of a rapidly growing country? Of a population with a slow rate of growth? What can we tell about future population numbers from those shapes?

4. What variations do we discern in the spatial pattern of the *rate of natural increase* and, consequently, of population growth? What rate of natural increase would double population in 35 years?

5. How are population numbers projected from present conditions? Are projections the same as predictions? If not, in what ways do they differ?

6. Describe the stages in the *demographic transition*. Where has the final stage of the transition been achieved? Why do some analysts doubt the applicability of the demographic transition to all parts of the world?

7. Contrast *crude population density, physiological density,* and *agricultural density.* For what differing purposes

might each be useful? How is *carrying capacity* related to the concept of density?

8. What was Malthus's underlying assumption concerning the relationship between population growth and food supply? In what ways do the arguments of *neo-Malthusians* differ from the original doctrine? What governmental policies are implicit in *neo-Malthusianism?*

9. Why is *demographic momentum* a matter of interest in population projections? In which world areas are the implications of demographic momentum most serious in calculating population growth, stability, or decline?

Focus Follow-up

1. **What are some basic terms and measures used by population geographers?** pp. 102–118.

A *cohort* is a population group, usually an age group, treated as a unit. *Rates* record the frequency of occurrence of an event over a given unit of time. Rates are used to trace a wide range of population features and

trends: births, deaths, fertility, infant or maternal mortality, natural increase, and others. Those rates tell us both the present circumstances and likely prospects for national, country group, or world population structures. Population pyramids give visual evidence of the current age and sex cohort structure of countries or country groupings.

2. **What are meant and measured by the demographic transition model and the demographic equation?** pp. 118–124.

The *demographic transition* model traces the presumed relationship between population growth and economic development. In Western countries, the transition model historically displayed four stages:

(a) high birth and death rates; (b) high birth and declining death rates; (c) declining births and reduced growth rates; and (d) low birth and death rates. A fifth stage of population decline is observed for some aging societies. The transition model has been observed to be not fully applicable to all developing states. The *demographic equation* attempts to incorporate cross-border population migration into projections of national population trends.

3. **What descriptive generalizations can be made about world population distributions and densities?** pp. 125–130.

World population is primarily concentrated north of the equator, in lower (below 200 meters) elevations, along continental margins. Major world population clusters include *East Asia* with 25% of the total, *South Asia* with over 20%, *Europe* and *northeastern United States/ southeastern Canada* with significant but lesser shares of world population. Other smaller but pronounced concentrations are found discontinuously on all continents. Within the permanently inhabited areas—the "ecumene"—population densities vary greatly. Highest densities are found in cities; almost one-half of the world's people are urban residents now and the vast majority of world population growth over the first quarter of the 21st century will occur in cities of the developing world.

4. **What are population projections, and how are they affected by various controls on population growth?** pp. 130–135.

Population projections are merely calculations of the future size, age, and sex composition of regional, national, or world populations; they are based on current data and manipulated by varying assumptions about the future. As simple calculations, projections cannot be wrong. They may, however, totally misrepresent what actually will occur because of faulty current data or erroneous assumptions used in their calculation. They may also be invalid because of unanticipated self-imposed or external brakes on population growth, such as changing family size desires or limits on areal carrying capacity that slow or halt current growth trends. Even with such growth limitations, however, population prospects are always influenced greatly by *demographic momentum*, the inevitable growth in numbers promised by the high proportion of younger cohorts yet to enter childbearing years in the developing world, and by the consequences of global population aging.

 Selected References _____

Ashford, Lori S. "New Perspectives on Population: Lessons from Cairo." *Population Bulletin* 50, no. 1. Washington, D.C.: Population Reference Bureau, 1995.

Bongaarts, John. "Population Pressure and the Food Supply System in the Developing World." *Population and Development Review* 22, no. 3 (1996): 483–503.

Brea, Jorge A. "Population Dynamics in Latin America." *Population Bulletin* 58, no. 1. Washington, D.C.: Population Reference Bureau, 2003.

Brown, Lester R., Gary Gardner, and Brian Halweil. *Beyond Malthus: Nineteen Dimensions of the Population Challenge.* New York: W. W. Norton, 1999.

Bulatao, Rodolfo A., and John B. Casterline, eds. *Global Fertility Transition.* Supplement to *Population and Development Review*, vol. 27, 2001.

Caldwell, John C., I. O. Orbulove, and Pat Caldwell. "Fertility Decline in Africa: A New Type of Transition?" *Population and Development Review* 19, no. 2 (1992): 211–242.

Cohen, Joel E. *How Many People Can the Earth Support?* New York: W. W. Norton, 1995.

Daugherty, Helen Ginn, and Kenneth C. W. Kammeyer. *An Introduction to Population.* 2d ed. New York: Guilford Publications, 1995.

Gelbard, Alene, Carl Haub, and Mary M. Kent. "World Population Beyond Six Billion." *Population Bulletin* 54, no. 1. Washington, D.C.: Population Reference Bureau, 1999.

Haub, Carl. "Understanding Population Projections." *Population Bulletin* 42, no. 4. Washington, D.C.: Population Reference Bureau, 1987.

Haupt, Arthur, and Thomas Kane. *Population Handbook.* 4th ed. Washington, D.C.: Population Reference Bureau, 1997.

Himes, Christine L. "Elderly Americans." *Population Bulletin*, 56, no. 4. Washington, D.C.: Population Reference Bureau, 2001.

Hornby, William F., and Melvyn Jones. *An Introduction to Population Geography.* 2d. ed. Cambridge, England: Cambridge University Press, 1993.

King, Russell, ed. *Mass Migrations in Europe: The Legacy and the Future.* London and New York: Belhaven Press and John Wiley & Sons, 1995.

Klasen, Stephan, and Claudia Wink. "A Turning Point in Gender Bias in Mortality? An Update on the Number of Missing Women." *Population and Development Review* 28, no. 2 (June 2002): 285–312.

Lampley, Peter, et al. "Facing the HIV/AIDS Pandemic." *Population Bulletin* 57, no. 3. Washington, D.C.: Population Reference Bureau, 2002.

Lee, James, and Wang Fend. *One Quarter of Humanity: Malthusian Mythology and Chinese Reality 1700–2000.* Cambridge, Mass.: Harvard University Press, 1999.

Martin, Philip, and Jonas Widgren. "International Migration: Facing the Challenge." *Population Bulletin* 57, no. 1. Washington, D.C.: Population Reference Bureau, 2002.

McFalls, Joseph A., Jr. "Population: A Lively Introduction." 3d. ed. *Population Bulletin* 53, no. 3. Washington, D.C.: Population Reference Bureau, 1998.

Newbold, K. Bruce. *Six Billion Plus: Population Issues in the Twenty-First Century.* Lanham, Md.: Rowman and Littlefield, 2002.

Nierenberg, Danielle. *Correcting Gender Myopia: Gender Polity, Women's Welfare, and the Environment.* Worldwatch Paper 161. Washington, D.C.: Worldwatch Institute, 2002.

Olshansky, S. Jay, Bruce Carnes, Richard G. Rogers, and Len Smith. "Infectious Diseases—New and Ancient Threats to World Health." *Population Bulletin* 52, no. 2. Washington, D.C.: Population Reference Bureau, 1997.

Omran, Abdel R., and Farzaneh Roudi. "The Middle East Population Puzzle." *Population Bulletin* 48, no. 1. Washington, D.C.: Population Reference Bureau, 1993.

O'Neill, Brian, and Deborah Balk. "World Population Futures." *Population Bulletin* 56, no. 3. Washington, D.C.: Population Reference Bureau, 2001.

Peters, Gary L., and Robert P. Larkin. *Population Geography: Problems, Concepts, and Prospects.* 7th ed. Dubuque, Ia.: Kendall/Hunt Publishing Company, 2002.

"Population." *National Geographic* (October 1998).

Riley, Nancy E. "Gender, Power, and Population Change." *Population Bulletin* 52, no. 1. Washington, D.C.: Population Reference Bureau, 1997.

Robey, Bryant, Shea O. Rutstein, and Leo Morris. "The Fertility Decline in Developing Countries." *Scientific American* 269 (December 1993): 30–37.

Simon, Julian. *The Ultimate Resource 2.* Princeton, N.J.: Princeton University Press, 1996.

Smil, Vaclav. *Feeding the World: A Challenge for the Twenty-First Century.* Cambridge, Mass.: MIT Press, 2000.

United Nations. *Population and Women.* New York: United Nations, 1996.

United Nations. Department of Economic and Social Affairs. *The World's Women 2000: Trends and Statistics.* New York: United Nations, 2000.

United Nations Population Fund. *The State of World Population.* New York: United Nations, annual.

Visaria, Leela, and Pravin Visaria. "India's Population in Transition." *Population Bulletin* 50, no. 3. Washington, D.C.: Population Reference Bureau, 1995.

World Health Organization. *The World Health Report.* Geneva, Switzerland: WHO, annual.

Xizhe Peng, and Guo Zhigang, eds. *The Changing Population of China.* Oxford, England: Blackwell, 2000.

Websites: The World Wide Web has a tremendous number and variety of sites pertaining to geography. Websites relevant to the subject matter of this chapter appear in the "Web Links" section of the On-line Learning Center associated with this book. Access it at **www.mhhe.com/fellmann8e**

PART Two

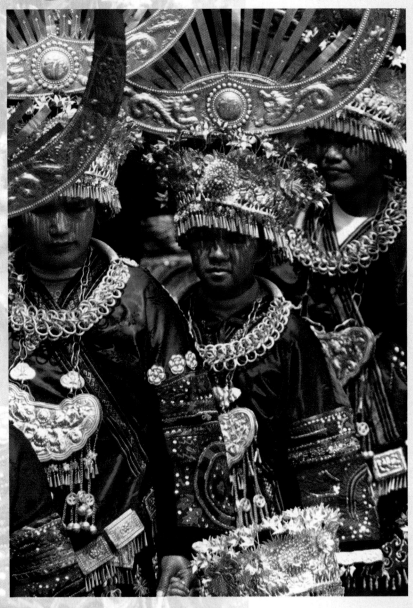

These Hmong dancers of Guizhou Province represent just one of the more than 50 sizable
minority ethnic groups of mainland China.

Patterns of Diversity
and Unity

The concerns of Part One of our study were the cultural processes and spatial interactions of an unevenly distributed world population. The understandings we sought were those that stressed the common characteristics, collective processes, behavioral constants, and unifying traits that form the background to human occupation of the earth and the development of distinctive cultural landscapes upon it.

Our attention now turns to the distinguishing features of the culture groups creating those landscapes. Our concern now is not with common features but with cultural differences and with the spatial cultural mosaic that those differences create. The topics of interest are the principal expressions of unity and diversity among and between different social groups. We ask: In what pronounced ways are populations distinctive? How, if at all, are those elements of distinctiveness interrelated and part of the composite cultures of recognizably different social groups? What notable world and regional spatial patterns of cultural differentiation can we recognize? How do cultural traits and composites change over time and through contact with other, differently constituted groups? And, finally, in what ways and to what extent are cultural contrasts evident in the landscapes built or modified by different social groups?

Although human populations are distinguished one from another in innumerable detailed ways, major points of contrast are relatively few in number and commonly recognized as characteristic traits of distinctive social groups. Languages spoken, religions practiced or espoused, or the composite distinguishing features of ethnic or folk cultural communities are among those major elements of contrast; they will claim our attention in the next three chapters.

Language and religion are prominent threads in the tapestry of culture, serving both to identify and classify individuals within complex societies and to distinguish populations and regions of different tongues and faiths. *Language* is the means of transmission of culture and the medium through which its beliefs and standards are expressed. *Religion* has had a pervasive impact on different culture groups, coloring their perceptions of themselves and their environments and of other groups of different faiths with whom they come in contact. As fundamental components and spatial expressions of culture, language and religion command our attention in Chapter 5.

They are also contributors to the complex of cultural characteristics that distinguish *ethnic groups,* populations set off from other groups by feelings of distinctiveness commonly fostered by some combination of religion, language, race, custom, or nationality. Ethnic groups—either as local indigenous minorities within differently structured majority populations, or as self-conscious immigrant groups in pluralistic societies—represent another form of cultural differentiation of sufficient worldwide importance to require our attention in Chapter 6.

In Chapter 7, we pursue two separate but related themes of cultural geography. The first is that a distinctive and pervasive element of cultural diversity is rooted in *folk culture*—the material and nonmaterial aspects of daily life preserved and transmitted by groups insulated from outside influences through spatial isolation or cultural barriers. The second theme is that the diversity formerly evident in many culturally complex societies is being eroded and leveled by the unification implicit in the spread of *popular cultures* that, at the same time, provide a broadening of the opportunities and choices available to individuals.

Our concerns in Chapters 5–7, then, are the learned behaviors, attitudes, and beliefs that have significant spatial expression and serve in the intricate mosaic of culture as fundamental identifying traits of distinctive social groups.

C H A P T E R

Five

Language and Religion:
Mosaics of Culture

Focus Preview

Language

1. The classification, spread, and distribution of the world's languages; the nature of language change, pp. 144–154.

2. Language standards and variants, from dialects to official tongues, pp. 154–161.

3. Language as cultural identity and landscape relic, pp. 161–165.

Religion

4. The cultural significance and role of religion, pp. 165–166.

5. How world religions are classified and distributed, pp. 166–169.

6. The origins, nature, and diffusions of principal world religions, pp. 169–183.

Opposite: An idol of the Hindu goddess Durga is carried near the Hooghly River in Kolkata (Calcutta), India.

*W*hen God saw [humans become arrogant], he thought of something to bring confusion to their heads: he gave the people a very heavy sleep. They slept for a very, very long time. They slept for so long that they forgot the language they had used to speak. When they eventually woke up from their sleep, each man went his own way, speaking his own tongue. None of them could understand the language of the other any more. That is how people dispersed all over the world. Each man would walk his way and speak his own language and another would go his way and speak in his own language. . . .

God has forbidden me to speak Arabic. I asked God, "Why don't I speak Arabic?" and He said, "If you speak Arabic, you will turn into a bad man." I said, "There is something good in Arabic!" And He said, "No, there is nothing good in it! . . ."

Here, I slaughter a bull and I call [the Muslim] to share my meat. I say, "Let us share our meat." But he refuses the meat I slaughter because he says it is not slaughtered in a Muslim way. If he cannot accept the way I slaughter my meat, how can we be relatives? Why does he despise our food? So, let us eat our meat alone. . . . Why, they insult us, they combine contempt for our black skin with pride in their religion. As for us, we have our own ancestors and our own spirits; the spirits of the Rek, the spirits of the Twic, we have not combined our spirits with their spirits. The spirit of the black man is different. Our spirit has not combined with theirs.[1]

Language and religion are basic components of cultures, the learned ways of life of different human communities. They help identify who and what we are and clearly place us within larger communities of persons with similar characteristics. At the same time, as the words of Chief Makuei suggest, they separate and divide peoples of different tongues and faiths. In the terminology introduced in Chapter 2, language and religion are *mentifacts,* components of the *ideological subsystem* of culture that help shape the belief system of a society and transmit it to succeeding generations. Both within and between cultures, language and religion are fundamental strands in the complex web of culture, serving to shape and to distinguish people and groups.

They are ever-changing strands, for languages and religions in their present-day structure and spatial patterns are simply the temporary latest phase in a continuing progression of culture change. Languages evolve in place, responding to the dynamics of human thought, experience, and expression and to the exchanges and borrowings ever more common in a closely integrated world. They disperse in space, carried by streams of migrants, colonizers, and conquerors. They may be rigorously defended and preserved as essential elements of cultural identity, or abandoned in the search for acceptance into a new society. To trace their diffusions, adoptions, and disappearances is to understand part of the evolving course of historical cultural geography. Religions, too, are dynamic, sweeping across national, linguistic, and cultural boundaries by conversion, conviction, and conquest. Their broad spatial patterns—distinctive culture regions in their own right—are also fundamental in defining the culture realms outlined in Figure 2.4, while at a different scale religious differences may contribute to the cultural diversity and richness within the countries of the world (Figure 5.1).

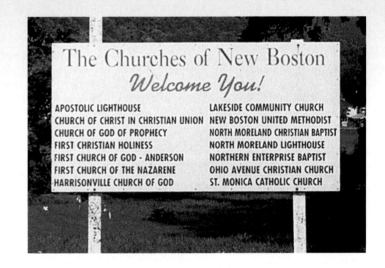

Figure 5.1 Advertised evidence of religious diversity in the United States. The sign details only a few Christian congregations. In reality, the United States has become the most religiously diverse country in the world, with essentially all of the world's faiths represented within its borders.

---

[1]The words of Chief Makuei Bilkuei of the Dinka, a Nilotic people of the southern Sudan. His comments are directed at the attempts to unite into a single people the Arabic Muslims of the north of the Republic of the Sudan with his and other black, Luo-speaking animist and Christian people of the country's southern areas. Recorded by Francis Mading Deng, *Africans of Two Worlds: The Dinka in Afro-Arab Sudan.* Copyright © 1978 Yale University Press, New Haven, CT. Reprinted by permission of the author.

The Geography of Language

Forever changing and evolving, language in spoken or written form makes possible the cooperative efforts, the group understandings, and shared behavior patterns that distinguish culture groups. Language is the most important medium by which culture is transmitted. It is what enables parents to teach their children what the world they live in is like and what they must do to become functioning members of society. Some argue that the language of a society structures the perceptions of its speakers. By the words that it contains and the concepts that it can formulate, language is said to determine the attitudes, the understandings, and the responses of the society to which it belongs.

If that conclusion be true, one aspect of cultural heterogeneity may be easily understood. The more than 6 billion people on earth speak many thousands of different languages. Knowing that more than 1500 languages and language variants are spoken in sub-Saharan Africa gives us a clearer appreciation of the political and social divisions in that continent. Europe alone has some 225 languages and dialects (Figure 5.2). Language is a hallmark of cultural diversity, an often fiercely defended symbol of cultural identity helping to distinguish the world's diverse social groups.

Distribution of Living Languages, 2002

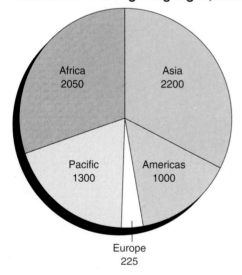

Figure 5.2 **World distribution of living languages, 2002.** Of the perhaps 6800 languages still spoken today, one-third are found in Asia, 30% in Africa, 19% in the Pacific area, 15% in the Americas, and 3% in Europe. Linguists' estimates of the number of languages ever spoken on earth range from 31,000 to as many as 300,000 or more. Assuming the lower estimate or even one considerably smaller, dead languages far outnumber the living. One or two additional tongues, most spoken in the forests of Papua New Guinea or in Indonesia, are lost each week.

Source: Estimates based on Ethnologue *and other sources.*

Classification of Languages

On a clear, dark night, the unaided eye can distinguish between 4000 and 6000 stars, a number comparable to some estimates of the probable total number of the world's languages. In reality, no precise figure is possible, for even today in Africa, Latin America, New Guinea, and elsewhere, linguists are still in the process of identifying and classifying the tongues spoken by isolated peoples. Even when they are well known, languages cannot always be easily or unmistakably recognized as distinctly separate entities.

In the broadest sense, language is any systematic method of communicating ideas, attitudes, or intent through the use of mutually understood signs, sounds, or gestures. For our geographic purposes, we may define **language** as an organized system of spoken words by which people communicate with each other with mutual comprehension. But such a definition fails to recognize the gradations among and between languages or to grasp the varying degrees of mutual comprehension between two or more of them. The language commonly called "Chinese," for example, is more properly seen as a group of distinct but related languages—Mandarin, Cantonese, Hakka, and others—that are as different from each other as are such comparably related European languages as Spanish, Italian, French, and Romanian. "Chinese" has uniformity only in the fact that all of the varied Chinese languages are written alike. No matter how it is pronounced, the same symbol for 'house' or for 'rice,' for example, is recognized by all literate speakers of any Chinese language variant (Figure 5.3). Again, the language known as "Arabic" represents a number of related but distinct tongues, so Arabic spoken in Morocco differs from Palestinian Arabic roughly as Portuguese differs from Italian.

Languages differ greatly in their relative importance, if "importance" can be taken to mean the number of people using them. More than half of the world's inhabitants are native speakers of just eight of its thousands of tongues, and at least half regularly use or have competence in just four of them. That restricted language dominance reflects the reality that the world's linguistic

HOUSE   RICE   TREE

Figure 5.3 All literate Chinese, no matter which of the many languages of China they speak, recognize the same ideographs for house, rice, and tree.

diversity is rapidly shrinking. Of the at most 7000 tongues still remaining, between 20% and 50% are no longer being learned by children and are effectively dead. One estimate anticipates that no more than 600 of the world's current living languages will still be in existence in A.D. 2100. Table 5.1 lists those languages currently spoken as a native or second tongue by 40 million or more people, a list that includes nearly 90% of the world's population. At the other end of the scale are a number of rapidly declining languages whose speakers number in the hundreds or, at most, the few thousands.

The diversity of languages is simplified when we recognize among them related *families*. A **language family** is a group of languages descended from a single, earlier tongue. By varying estimates, from at least 30 to perhaps 100 such families of languages are found worldwide. The families, in turn, may be subdivided into subfamilies, branches, or groups of more closely related tongues. Some 2000 years ago, Latin was the common language spoken throughout the Roman Empire. The fall of the empire in the 5th century A.D. broke the unity of Europe, and regional variants of Latin began to develop in isolation. In the course of the next several centuries, these Latin derivatives, changing and developing as all languages do, emerged as the individual *Romance* languages—Italian, Spanish, French, Portuguese, and Romanian—of modern Europe and of the world colonized by their speakers. Catalan, Sardinian, Provençal, and a few other spatially restricted tongues are also part of the Romance language group.

Family relationship between languages can be recognized through similarities in their vocabulary and grammar. By tracing regularities of sound changes in different languages back through time, linguists are able to reconstruct earlier forms of words and, eventually, determine a word's original form before it underwent alteration and divergence. Such a reconstructed earlier form is said to belong to a **protolanguage.** In the case of the Romance languages, of course, the well-known ancestral tongue was Latin, which needs no such reconstruction. Its root relationship to the Romance languages is suggested by modern variants of *panis*, the Latin word for "bread": *pane* (Italian), *pain* (French), *pan* (Spanish), *pão* (Portuguese), *pîine* (Romanian). In other language families similar word relationships are less confidently traced to their protolanguage roots. For example, the *Germanic* languages, including English, German, Dutch, and the Scandinavian tongues, are related descendants of a less well-known proto-Germanic language spoken by peoples who lived in southern Scandinavia and along the North Sea and Baltic coasts from the Netherlands to western Poland. The classification of languages by origin and historical relationship is called a *genetic classification*.

Further tracing of language roots tells us that the Romance and the Germanic languages are individual branches of an even more extensive family of related languages derived from *proto-Indo-European,* or simply *Indo-European.* Of the principal recognized language clusters of the world, the Indo-European family is the largest, embracing most of the languages of Europe and a large part of Asia, and the introduced—not the native—languages of the Americas (Figure 5.4). All told, languages in the Indo-European family are spoken by about half the world's peoples.

Table 5.1

Languages Spoken by More than 40 Million People, 2003

Language	Millions of Speakers (native plus nonnative)
English	1500
Mandarin (China)	820
Hindi[a] (India, Pakistan)	430
Spanish	359
Bengali (Bangladesh, India)	210
Portuguese	180
Malay-Indonesian	176
Russian	167
German	127
Japanese	126
French	118
Wu (China)	90
Javanese	78
Korean (Korea, China, Japan)	78
Punjabi (India, Pakistan)	72
Yue (Cantonese) (China)	72
Telugu (India)	70
Marathi (India)	69
Vietnamese	69
Tamil (India, Sri Lanka)	67
Urdu[a] (Pakistan, India)	62
Italian	62
Turkish	62
Swahili (East Africa)	50
Min (China)	48
Ukrainian	47
Egyptian Arabic[b]	46
Gujarati (India, Pakistan)	46
Jinyu (China)	45
Polish	44

[a]Hindi and Urdu are basically the same language: Hindustani. Written in the Devangari script, it is called *Hindi*, the official language of India; in the Arabic script it is called *Urdu*, the official language of Pakistan.

[b]Collectively, the many often mutually unintelligible versions of colloquial Arabic are used by at least 200 million native speakers. Classical or literary Arabic, the language of the Koran, is uniform and standardized but is restricted to formal usage as a spoken tongue.

Source: Based on data from *Ethnologue* and others.

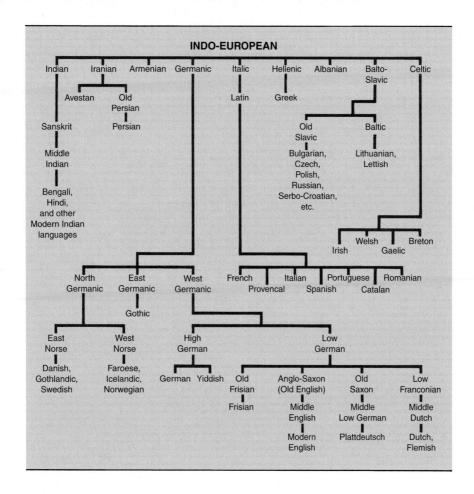

Figure 5.4 **The Indo-European linguistic family tree.** Euskara (Basque), Estonian, Finnish, and Hungarian are the only European languages *not* in the Indo-European family. (See also Figure 5.8.)

By recognizing similar words in most Indo-European tongues, linguists deduce that the Indo-European people—originally hunters and fishers but later becoming pastoralists and learning to grow crops—developed somewhere in eastern Europe or the Ukrainian steppes about 5000 years ago (though some conclude that central Turkey was the more likely site of origin). About 2500 B.C., their society apparently fragmented; they left the homeland, carrying segments of the parent culture in different directions. Some migrated into Greece, others settled in Italy, still others crossed central and western Europe, ultimately reaching the British Isles. Another group headed into the Russian forest lands, and still another branch crossed Iran and Afghanistan, eventually to reach India. Wherever this remarkable people settled, they appear to have dominated local populations and imposed their language on them. For example, the word for sheep is *avis* in Lithuanian, *ovis* in Latin, *avis* in Sanskrit (the language of ancient India), and *hawi* in the tongue used in Homer's Troy. Modern English retains its version in "ewe." All, linguists infer, derive from an ancestral word *owis* in Indo-European. Similar relationships and histories can be traced for other protolanguages.

World Pattern of Languages

The present world distribution of major language families (Figure 5.5) records not only the migrations and conquests of our linguistic ancestors but also the continuing dynamic pattern of human movements, settlements, and colonizations of more recent centuries. Indo-European languages have been carried far beyond their Eurasian homelands from the 16th century onwards by western European colonizers in the Americas, Africa, Asia, and Australasia. In the process of linguistic imposition and adoption, innumerable indigenous languages and language groups in areas of colonization have been modified or totally lost. Most of the estimated 1000 to 2000 *Amerindian* tongues of the Western Hemisphere disappeared in the face of European conquest and settlement (Figure 5.6).

The Slavic expansion eastward across Siberia beginning in the 16th century obliterated most of the *Paleo-Asiatic* languages there. Similar loss occurred in Eskimo and Aleut language areas. Large linguistically distinctive areas comprise the northern reaches of both Asia and America (see Figure 5.5). Their sparse populations are losing the mapped languages as the indigenous

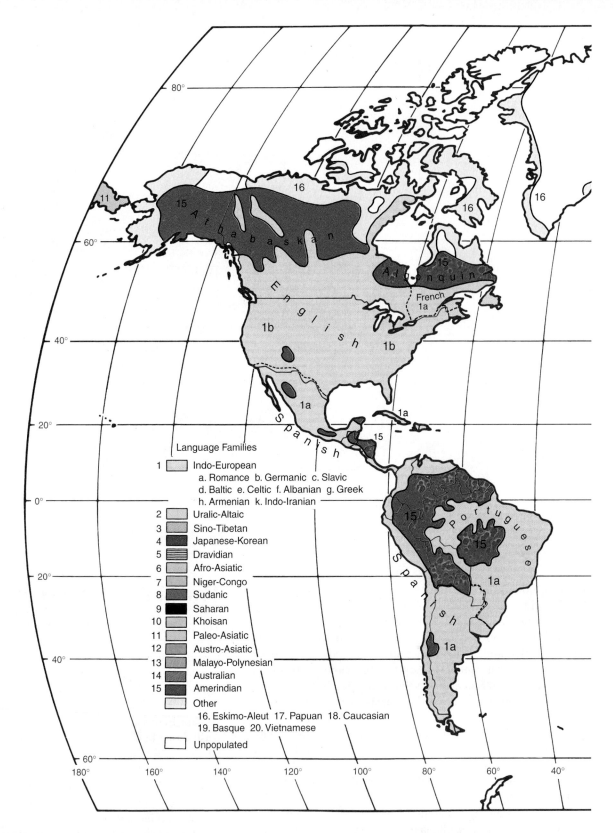

Figure 5.5 **World language families.** Language families are groups of individual tongues that had a common but remote ancestor. By suggesting that the area assigned to a language or language family uses that tongue exclusively, the map pattern conceals important linguistic detail. Many countries and regions have local languages spoken in territories too small to be recorded at this scale. The map also fails to report that the population in many regions is fluent in more than one language or that a second language serves as the necessary vehicle of commerce, education, or government. Nor is important information given about the number of speakers of different languages; the fact that there are more speakers of English in India or Africa than in Australia is not even hinted at by a map at this scale.

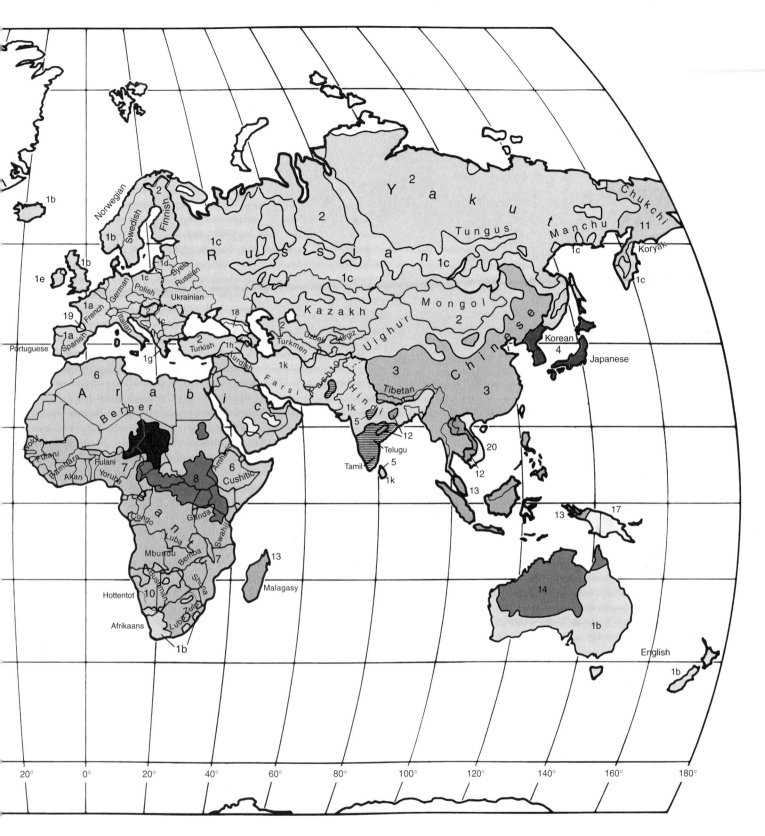

1b

Norwegian
Swedish
Finnish
2

1b

Y²a k u t

Tungus

Chukchi

Manchu

11

1c

Koryak

1c

1b
1e
1c
German
French
Italian
19
1a
Portuguese
1a
Spanish
1g
1d
Byelo
Russian
Polish
Ukrainian
2
1c

R u s s i a n

1c

1c

1c

2

Kazakh

18
Turkish
1h
Kurdish
2
Turkmen
Uzbek Kirgiz
Farsi
Pashtu
1k

Mongol
2

Uighur

3

C h i n e s e

3

Korean
4

Japanese

Arabic
Berber
6

Hindi
Tibetan
1k
5
Telugu
12
Tamil
5
1k

20

12

13

13

17

6
Amharic
Cushitic

Wolof
Fulani
Bambara
Fulani
7
Akan
Yoruba

B a n t u
8
Ganda
Congo
Swahili
Luba
7
Mbundu
Bemba
Bushman
Shona
Hottentot
10
Luba Zulu
Afrikaans
1b

13
Malagasy

14

1b

English

1b

20°   0°   20°   40°   60°   80°   100°   120°   140°   160°   180°

Figure 5.5 (Continued)

Language and Religion: Mosaics of Culture   **149**

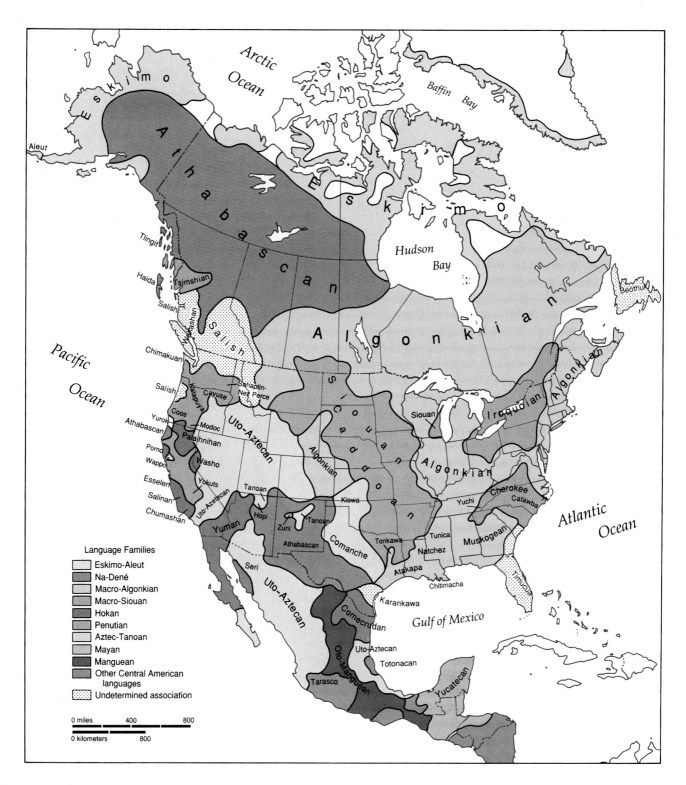

Figure 5.6 **Amerindian language families of North America.** As many as 300 different North American and more than 70 Meso-American tongues were spoken at the time of first European contact. The map summarizes the traditional view that these were grouped into 9 or 10 language families in North America, as many as 5 in Meso-America, and another 10 or so in South America. More recent research, however, suggests close genetic relationships between Native American tongues, clustering them into just 3 families: Eskimo-Aleut in the extreme north and Greenland; Na-Dené in Canada and the U.S. Southwest, and Amerind elsewhere in the hemisphere. Because each family has closer affinities with Asian language groups than with one another, it is suggested that each corresponds to a separate wave of Asian migration to the Americas: the first giving rise to the Amerind family, the second to the Na-Dené, and the last to the Eskimo-Aleut. Many Amerindian tongues have become extinct; others are still known only to very small groups of mostly elderly speakers.

Data from various sources, including C. F. and F. M. Voegelin, Map of North American Indian Languages *(Seattle: University of Washington Press, 1986).*

people adopt the tongues of the majority cultures of which they have been forcibly made a part. In the Southern Hemisphere, the several hundred original *Australian* languages also loom large spatially on the map but have at most 50,000 speakers, exclusively Australian aborigines. Numerically and effectively, English dominates that continent.

Examples of linguistic conquest by non-Europeans also abound. In Southeast Asia, formerly extensive areas identified with different members of the *Austro-Asiatic* language family have been reduced through conquest and absorption by *Sino-Tibetan* (Chinese, Thai, Burmese, and Lao, principally) expansion. Arabic—originally a minor *Afro-Asiatic* language of the Arabian Peninsula—was dispersed by the explosive spread of Islam through much of North Africa and southwestern Asia, where it largely replaced a host of other locally variant tongues and became the official or the dominant language of more than 20 countries and over 250 million people. The more than 300 Bantu languages found south of the "Bantu line" in sub-Saharan Africa are variants of a proto-Bantu carried by an expanding, culturally advanced population that displaced more primitive predecessors (Figure 5.7).

Language Spread

Language spread as a geographical event represents the increase or relocation through time in the area over which a language is spoken. The Bantu of Africa or the English-speaking settlers of North America displaced preexisting populations and replaced as well the languages previously spoken in the areas of penetration. Therefore, we find one explanation of the spread of language families to new areas of occurrence in massive population relocations such as those accompanying the colonization of the Americas or of Australia. That is, languages may spread because their speakers occupy new territories.

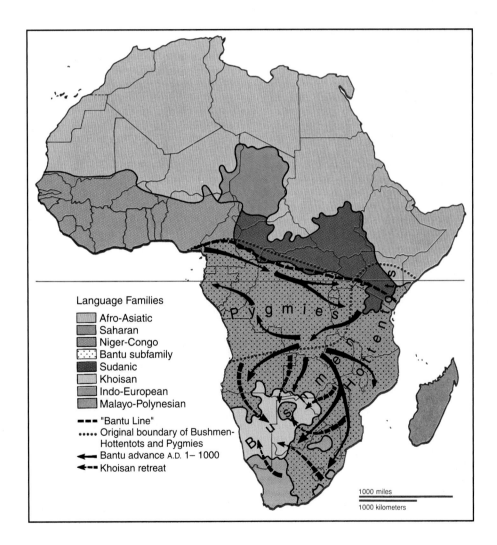

Figure 5.7 **Bantu advance, Khoisan retreat in Africa.** Linguistic evidence suggests that proto-*Bantu* speakers originated in the region of the Cameroon-Nigeria border, spread eastward across the southern Sudan, then turned southward to Central Africa. From there they dispersed slowly eastward, westward, and against slight resistance, southward. The earlier *Khoisan*-speaking occupants of sub-Saharan Africa were no match against the advancing metal-using Bantu agriculturalists. Pygmies, adopting a Bantu tongue, retreated deep into the forests; Bushmen and Hottentots retained their distinctive Khoisan "click" language but were forced out of forests and grasslands into the dry steppes and deserts of the southwest.

Latin, however, replaced earlier Celtic languages in western Europe not by force of numbers—Roman legionnaires, administrators, and settlers never represented a majority population—but by the gradual abandonment of their former languages by native populations brought under the influence and control of the Roman Empire and, later, of the Western Christian church. Adoption rather than eviction of language was the rule followed in perhaps the majority of historical and contemporary instances of language spread. Knowledge and use of the language of a dominating culture may be seen as a necessity when that language is the medium of commerce, law, civilization, and personal prestige. It was on that basis, not through numerical superiority, that Indo-European tongues were dispersed throughout Europe and to distant India, Iran, and Armenia. Likewise, Arabic became widespread in western Asia and North Africa not through massive population relocations but through conquest, religious conversion, and superiority of culture. That is, languages may spread because they acquire new speakers.

Either form of language spread—dispersion of speakers or acquisition of speakers—represents one of the *spatial diffusion* processes introduced in Chapter 2. Massive population relocation in which culture is transported to and made dominant in a new territory is a specialized example of *relocation diffusion*. When the advantages of a new language are discerned and it is adopted by native speakers of another tongue, a form of *expansion diffusion* has occurred along with partial or total *acculturation* of the adopting population. Usually, those who are in or aspire to positions of importance are the first to adopt the new language of control and prestige. Later, through schooling, daily contact, and business or social necessity, other, lower social strata of society may gradually be absorbed into the expanding pool of language adopters.

Such *hierarchical diffusion* of an official or prestigious language has occurred in many societies. In India during the 19th century, the English established an administrative and judicial system that put a very high premium on their language as the sole medium of education, administration, trade, and commerce. Proficiency in it was the hallmark of the cultured and educated person (as knowledge of Sanskrit and Persian had been in earlier periods under other conquerors of India). English, French, Dutch, Portuguese, and other languages introduced during the acquisition of empire retain a position of prestige and even status as the official language in multilingual societies, even after independence has been achieved by former colonial territories. In Uganda and other former British possessions in Africa, a stranger may be addressed in English by one who wishes to display his or her education and social status, though standard Swahili, a second language for many different culture groups, may be chosen if certainty of communication is more important than pride.

As a diffusion process, language spread may be impeded by barriers or promoted by their absence. Cultural barriers may retard or prevent language adoption. Speakers of Greek resisted centuries of Turkish rule of their homeland, and the language remained a focus of cultural identity under foreign domination. Breton, Catalan, Gaelic, and other localized languages of Europe remain symbols of ethnic separateness from surrounding dominant national cultures and controls.

Physical barriers to language spread have also left their mark (see Figure 5.5). Migrants or invaders follow paths of least topographic resistance and disperse most widely where access is easiest. Once past the barrier of the Pamirs and the Hindu Kush mountains, Indo-European tongues spread rapidly through the Indus and Ganges river lowlands of the Indian subcontinent but made no headway in the mountainous northern and eastern border zones. The Pyrenees Mountains serve as a linguistic barrier separating France and Spain. They also house the Basques who speak the only language—*Euskara* in their tongue—in southwestern Europe that survives from pre-Indo-European times (Figure 5.8). Similarly, the Caucasus Mountains between the Black and Caspian seas separate the Slavic speakers to the north and the areas of *Ural-Altaic* languages to the south. At the same time, in their rugged topography they contain an extraordinary mixture of languages, many unique to single valleys or villages, lumped together spatially if not genetically into a separate *Caucasian* language family.

Language Change

Migration, segregation, and isolation give rise to separate, mutually unintelligible languages because the society speaking the parent protolanguage no longer remains unitary. Comparable changes occur normally and naturally within a single language in word meaning, pronunciation, vocabulary, and *syntax* (the way words are put together in phrases and sentences). Because they are gradual, minor, and made part of group use and understanding, such changes tend to go unremarked. Yet, cumulatively, they

Figure 5.8 In their mountainous homeland, the Basques have maintained a linguistic uniqueness despite more than 2000 years of encirclement by dominant lowland speakers of Latin or Romance languages. This sign of friendly farewell gives its message in both Spanish and the Basque language, Euskara.

can result in language change so great that in the course of centuries an essentially new language has been created. The English of 17th-century Shakespearean writings or the King James Bible (1611) sounds stilted to our ears. Few of us can easily read Chaucer's 14th-century *Canterbury Tales,* and 8th-century *Beowulf* is practically unintelligible.

Change may be gradual and cumulative, with each generation deviating in small degree from the speech patterns and vocabulary of its parents, or it may be massive and abrupt. English gained about 10,000 new words from the Norman conquerors of the 11th century. In some 70 years (1558–1625) of literary and linguistic creativity during the reigns of Elizabeth I and James I, an estimated 12,000 words—based on borrowings from Latin, Greek, and other languages—were introduced.

Discovery and colonization of new lands and continents in the 16th and 17th centuries greatly and necessarily expanded English as new foods, vegetation, animals, and artifacts were encountered and adopted along with their existing aboriginal American, Australian, or African names. The Indian languages of the Americas alone brought more than 200 relatively common daily words to English, 80 or more from the North American native tongues and the rest from Caribbean, Central, and South American. More than two thousand more specialized or localized words were also added. *Moose, raccoon, skunk, maize, squash, succotash, igloo, toboggan, hurricane, blizzard, hickory, pecan,* and a host of other names were taken directly into English; others were adopted second hand from Spanish variants of South American native words: *cigar, potato, chocolate, tomato, tobacco, hammock.* More recently, and within a short span of years, new

scientific and technological developments have enriched and expanded the vocabularies not only of English but of all languages spoken by modern societies by adding many words of Greek and Latin derivation.

The Story of English

English itself is a product of change, an offspring of proto-Germanic (see Figure 5.4) descending through the dialects brought to England in the 5th and 6th centuries by conquering Danish and North German Frisians, Jutes, Angles, and Saxons. Earlier Celtic-speaking inhabitants found refuge in the north and west of Britain and in the rugged uplands of what are now Scotland and Wales. Each of the transplanted tongues established its own area of dominance, but the West Saxon dialect of southern England emerged in the 9th and 10th centuries as Standard Old English (Figure 5.9) on the strength of its literary richness.

It lost its supremacy after the Norman Conquest of 1066, as the center of learning and culture shifted northeastward from Winchester to London, and French rather than English became the language of the nobility and the government. When the tie between France and England was severed after the loss of Normandy (1204), French fell into disfavor and English again became the dominant tongue, although now as the French-enriched Middle English used by Geoffrey Chaucer and mandated as the official language of the law courts by the Statute of Pleading (1362). During the 15th and 16th centuries, English as spoken in London emerged as the basic form of Early Modern English.

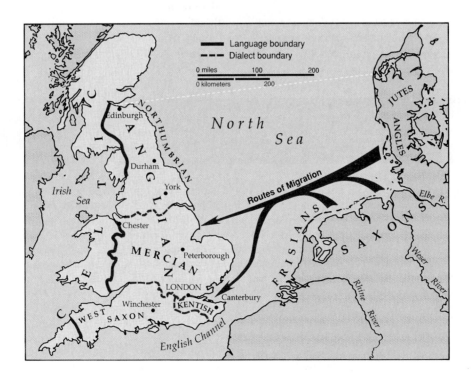

Figure 5.9 **Old English dialect regions.** In structure and vocabulary, Old English brought by the Frisians, Angles, Saxons, and Jutes was purely Germanic, with many similarities to modern German. It owed practically nothing to the Celtic it displaced, though it had borrowings from Latin. Much of Old English vocabulary was lost after the Norman conquest. English today has twice as many words derived from Latin and French as from the Germanic.

During the 18th century, attempts to standardize and codify the rules of English were unsuccessful. But the *Dictionary* of Samuel Johnson (published 1755)—based on cultured language of contemporary London and the examples of major authors—helped establish norms of proper form and usage. A worldwide diffusion of the language resulted as English colonists carried it as settlers to the Western Hemisphere and Australasia; through merchants, conquest, or territorial claim, it established footholds in Africa and Asia. In that spatial diffusion, English was further enriched by its contacts with other languages. By becoming the accepted language of commerce and science, it contributed, in turn, to the common vocabularies of other tongues (see "Language Exchange").

Within some 400 years, English has developed from a localized language of 7 million islanders off the European coast to a truly international language with some 375 million native speakers, perhaps the same number who use it as a second language, and another 750 million who have reasonable competence in English as a foreign language. With roughly 1.5 billion speakers worldwide, English also serves as an official language of some 60 countries (Figure 5.10), far exceeding in that role French (32), Arabic (25), or Spanish (21), the other leading current international languages. At the end of the 20th century, over 78% of Internet Web pages used English (Japanese was second with 2.5%). No other language in history has assumed so important a role on the world scene.

Standard and Variant Languages

People who speak a common language such as English are members of a **speech community,** but membership does not necessarily imply linguistic uniformity. A speech community usually possesses both a **standard language**—comprising the accepted community norms of syntax, vocabulary, and pronunciation—and a number of more or less distinctive *dialects* reflecting the ordinary speech of areal, social, professional, or other subdivisions of the general population.

Standard Language

A dialect may become the standard language through identity with the speech of the most prestigious, highest-ranking, and most powerful members of the larger speech community. A rich literary tradition may help establish its primacy, and its adoption as the accepted written and spoken norm in administration, economic life, and education will solidify its position, minimizing linguistic variation and working toward the elimination of deviant, nonstandard forms. The dialect that emerges as the basis of a country's standard language is often the one identified with its capital or center of power at the time of national development. Standard French is based on the dialect of the Paris region, a variant that assumed dominance in the latter half of the 12th century and was made the only official language in 1539. Castilian Spanish became the standard after 1492 with the Castile-led reconquest of Spain from the Moors and the export of the dialect to the Americas during the 16th century. Its present form, however, is a modified version associated not with Castile but with Madrid, the modern capital of Spain. Standard Russian is identified with the speech patterns of the former capital, St. Petersburg, and Moscow, the current capital. Modern Standard Chinese is based on the Mandarin dialect of Beijing. In England, *Received Pronunciation*—"Oxford English," the speech of educated people of London and southeastern England and used by the British Broadcasting System—was until recently the accepted standard though it is now being modified or replaced by a generalized southern accent called "Estuary English."

Language Exchange

English has a happily eclectic vocabulary. Its foundations are Anglo-Saxon (*was, that, eat, cow*) reinforced by Norse (*sky, get, bath, husband, skill*); its superstructure is Norman-French (*soldier, Parliament, prayer, beef*). The Norman aristocracy used their words for the food, but the Saxon serfs kept theirs for the animals. The language's decor comes from Renaissance and Enlightenment Europe: 16th-century France yielded *etiquette, naive, reprimand* and *police.*

Italy provided *umbrella, duet, bandit* and *dilettante;* Holland gave *cruise, yacht, trigger, landscape,* and *decoy.* Its elaborations come from Latin and Greek: *misanthrope, meditate,* and *parenthesis* all first appeared during the 1560s. In the 20th century, English adopted *penicillin* from Latin, *polystyrene* from Greek, and *sociology* and *television* from both. And English's ornaments come from all round the world: *slogan* and *spree* from Gaelic, *hammock* and *hurricane* from Caribbean languages, *caviar* and *kiosk* from Turkish, *dinghy* and *dungarees* from Hindi, *caravan* and *candy* from Persian, *mattress* and *masquerade* from Arabic.

Redressing the balance of trade, English is sharply stepping up its linguistic exports. Not just the necessary *imotokali* (motor car) and *izingilazi* (glasses) to Zulu; or *motokaa* and *shillingi* (shilling) to Swahili; but also *der Bestseller, der Kommunikations Manager, das Teeshirt* and *der Babysitter* to German; and, to Italian, *la pop art, il popcorn* and *la spray.* In some Spanish-speaking countries you might wear *un sueter* to *el beisbol,* or witness *un nocaut* at *el boxeo.* And in Russia, *biznesmen* prepare a *press rilis* on the *lep-top kompyuter* and print it by *lazerny printer.* Indeed, a sort of global English word list can be drawn up: *airport, passport, hotel, telephone; bar, soda, cigarette; sport, golf, tennis; stop, OK,* and increasingly, *weekend, jeans, know-how, sex-appeal,* and *no problem.*

Excerpted by permission from *The Economist,* London, December 20, 1986, p. 131.

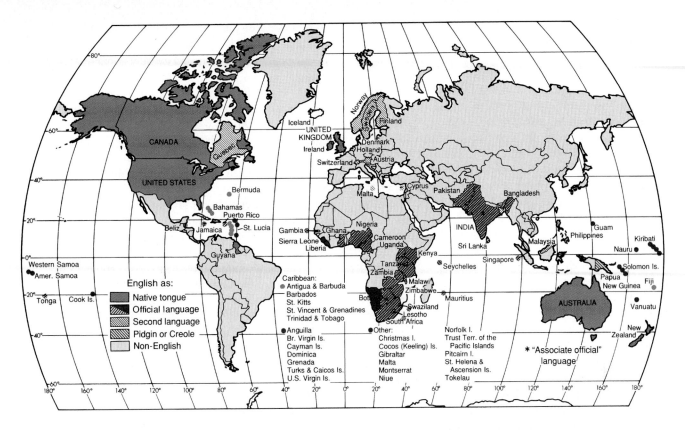

Figure 5.10 **International English.** In worldwide diffusion and acceptance, English has no past or present rivals. Along with French, it is one of the two working languages of the United Nations; some two-thirds of all scientific papers are published in it, making it the first language of scientific discourse, and the accepted language of international air traffic control. English is the sole or joint official language of more nations and territories, some too small to be shown here, than any other tongue. It also serves as the effective unofficial language of administration in other multilingual countries with different formal official languages. "English as a second language" is indicated for countries with near-universal or mandatory English instruction in public schools. The full extent of English penetration of Continental Europe, where over 80% of secondary school students (and 92% of those of European Union states) study it as a second language, is not evident on this map.

Other forces than the political may affect language standardization. In its spoken form, Standard German is based on norms established and accepted in the theater, the universities, public speeches, and radio and television. The Classical or Literary Arabic of the Koran became the established norm from the Indian to the Atlantic Ocean. Standard Italian was derived from the Florentine dialect of the 13th and 14th centuries, which became widespread as the language of literature and economy.

In many societies, the official or unofficial standard language is not the dialect of home or daily life, and populations in effect have two languages. One is their regional dialect they employ with friends, at home, and in local community contacts; the other is the standard language used in more formal situations. In some cases, the contrast is great; regional variants of Arabic may be mutually unintelligible. Most Italians encounter Standard Italian for the first time in primary school. In India, the several totally distinct official regional languages are used in writing and taught in school but have no direct relationship to local speech; citizens must be bilingual to communicate with government officials who know only the regional language but not the local dialect.

Dialects

Just as no two individuals talk exactly the same, all but the smallest and most closely knit speech communities display recognizable speech variants called **dialects.** Vocabulary, pronunciation, rhythm, and the speed at which the language is spoken may set groups of speakers apart from one another and, to a trained observer, clearly mark the origin of the speaker. In George Bernard Shaw's play *Pygmalion,* on which the musical *My Fair Lady* was based, Henry Higgins—a professor of phonetics—is able to identify the London neighborhood of origin of a flower girl by listening to her vocabulary and accent. In many instances, such variants are totally acceptable modifications of the standard language; in others, they mark the speaker as a social, cultural, or regional "outsider" or "inferior." Professor Higgins makes a lady out of the uneducated flower girl simply by teaching her upper-class pronunciation.

Shaw's play tells us dialects may coexist in space. Cockney and cultured English share the streets of London; black English and Standard American are heard in the same school yards throughout the United States. In many societies, **social dialects** denote social class and educational level. Speakers of higher socioeconomic status or educational achievement are most likely to follow the norms of their standard language; less-educated or lower-status persons or groups consciously distinguishing themselves from the mainstream culture are more likely to use the **vernacular**—nonstandard language or dialect native to the locale or adopted by the social group. In some instances, however, as in Germany and German-Switzerland, local dialects are preserved and prized as badges of regional identity.

Different dialects may be part of the speech patterns of the same person. Professionals discussing, for example, medical, legal, financial, or scientific matters with their peers employ vocabularies and formal modes of address and sentence structure that are quickly changed to informal colloquial speech when the conversation shifts to sports, vacations, or personal anecdotes. Even gender may be the basis for linguistic differences added to other determinants of social dialects, with female speakers of many unrelated languages generally acknowledged to use forms of speech considered to be "better" or "more correct" than males of the same social class.

More commonly, we think of dialects in spatial terms. Speech is a geographic variable; each locale is apt to have its own, perhaps slight, language differences from neighboring places. Such differences in pronunciation, vocabulary, word meanings, and other language characteristics tend to accumulate with distance from a given starting point. When they are mapped, they help define the **linguistic geography**—the study of the character and spatial pattern of dialects and languages—of a generalized speech community.

Every dialect feature has a territorial extent. The outer limit of its occurrence is a boundary line called an **isogloss** (the term *isophone* is used if the areal variant is marked by difference in sound rather than word choice), as shown in Figure 5.11. Each isogloss is a distinct entity, but taken together isoglosses give clear map evidence of dialect regions that in their turn may reflect topographic barriers and corridors, long-established political borders, or past migration flows and diffusions of word choice and pronunciation.

Geographic or **regional dialects** may be recognized at different scales. On the world scene, for example, British, American, Indian, and Australian English are all acknowledged distinctive dialects of the same evolving language (see "World Englishes"). Regionally, in Britain alone, one can recognize Southern British English, Northern British English, and Scottish English, each containing several more localized variants. Italy contains the Gallo-Italian and Venetan dialect groups of the north, the Tuscan dialects of the center, and a collection of southern Italian dialects. Japanese has three recognized dialect groups.

Indeed, all long-established speech communities show their own structure of geographic dialects whose number and diversity tend to increase in areas longest settled and most fragmented and isolated. For example, the local speech of Newfoundland—isolated off the Atlantic coast of mainland Canada—retains much of the 17th-century flavor of the four West Counties of England from which the overwhelming majority of its settlers came. Yet the isolation and lack of cultural mixing of the islanders have not led to a general Newfoundland "dialect"; settlement was coastal and in the form of isolated villages in each of the many bays and indentations. There developed from that isolation and the passage of time nearly as many dialects as there are bay settlements, with each dialect separately differing from Standard English in accent, vocabulary, sounds, and syntax. Isolation has led to comparable linguistic variation among the 47,000 inhabitants of the 18 Faeroe Islands between Iceland and Scotland; their Faeroese tongue has 10 dialects.

Dialects in America

Mainland North America had a more diversified colonization than did Newfoundland, and its more mobile settlers mixed and carried linguistic influences away from the coast into the continental interior. Nonetheless, as early as the 18th century, three distinctive dialect regions had emerged along the Atlantic coast of the United States (Figure 5.12) and are evident in the linguistic geography of North America to the present day.

With the extension of settlement after the Revolutionary War, each of the dialect regions expanded inland. Speakers of the Northern dialect moved along the Erie Canal and the Great Lakes. Midland speakers from Pennsylvania traveled down the Ohio River, and the related Upland Southern dialect moved through the mountain gaps into Kentucky and Tennessee. The Coastal Southern dialect was less mobile, held to the east by plantation prosperity and the long resistance to displacement exerted by the Cherokees and the other Civilized Tribes (Figure 5.13).

Once across the Appalachian barrier, the diffusion paths of the Northern dialect were fragmented and blocked by the time they reached the Upper Mississippi. Upland Southern speakers spread out rapidly: northward into the old Northwest Territory, west into Arkansas and Missouri, and south into the Gulf Coast states. But the Civil War and its aftermath halted further major

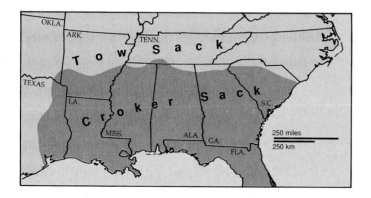

Figure 5.11 **Dialect boundaries.** Descriptive words or terms for common items are frequently employed indicators of dialect difference. The limit of their areas of use is marked by an **isogloss,** such as that shown here for now-obsolete terms describing a coarse sack. Usually such boundary lines appear in clusters or *bundles;* together, they help define the frontier of the dialect under study.

Source: Adapted from Gordon R. Wood, Vocabulary Change: A Study of Variation in Regional Words in Eight of the Southern States *(Carbondale: Southern Illinois University Press, 1971), Map 81, p. 357. Used by permission of the publisher.*

World Englishes

Non-native speakers of English far outnumber those for whom English is the first language. Most of the more than 1 billion people who speak and understand at least some English as a second language live in Asia; they are appropriating the language and remaking it in regionally distinctive fashions to suit their own cultures, linguistic backgrounds, and needs.

It is inevitable that widely spoken languages separated by distance, isolation, and cultural differences will fragment into dialects that, in turn, evolve into new languages. Latin splintered into French, Spanish, Italian, and other Romance languages; the many national variants of spoken Arabic are effectively different tongues. English is similarly experiencing that sort of regional differentiation, shaped by the variant needs and inputs of its far-flung community of speakers, and following the same path to mutual unintelligibility. Although Standard English may be one of or the sole official language of their countries of birth, millions of people around the world claiming proficiency in English or English as their national language cannot understand each other. Even teachers of English from India, Malaya, Nigeria, or the Philippines, for example, may not be able to communicate in their supposedly common tongue—and find cockney English of London utterly alien.

The splintering of spoken English is a fact of linguistic life and its offspring—called "World Englishes" by linguists—defy frequent attempts by different governments to remove localisms and encourage adherence to international standards. Singlish (Singapore English) and Taglish (a mixture of English and Tagalog, the dominant language of the Philippines) are commonly cited examples of the multiplying World Englishes, but equally distinctive regional variants have emerged in India, Malaysia, Hong Kong, Nigeria, the Caribbean, and elsewhere. One linguist suggests that beyond an "inner circle" of states where English is the first and native language—for example, Canada, Australia, United States—lies an "outer circle" where English is a second language (Bangladesh, Ghana, India, Kenya, Pakistan, Zambia, and many others) and where the regionally distinctive World Englishes are most obviously developing. Even farther out is an "expanding circle" of such states as China, Egypt, Korea, Nepal, Saudi Arabia, and others where English is a foreign language and distinctive local variants in common usage have not yet developed.

Each of the emerging varieties of English is, of course, "correct," for each represents a coherent and consistent vehicle for communication with mutual comprehension between its speakers. Each also represents a growing national cultural confidence and pride in the particular characteristics of the local varieties of English, and each regional variant is strengthened by local teachers who do not themselves have a good command of the standard language. Conceivably, these factors may mean that English will fragment into scores or hundreds of mutually unintelligible tongues. But equally conceivably, the worldwide influence of globalized business contacts, the Internet, worldwide American radio and television broadcasts, near-mandatory use of English in scientific publication, and the like will mean a future English more homogeneous and, perhaps, more influenced and standardized by American usage.

Most likely, observers of World Englishes suggest, both divergence and convergence will take place. While use of English as the major language of communication worldwide is a fact in international politics, business, education, and the media, increasingly, speakers of English learn two "dialects"—one of their own community and culture and one in the international context. While the constant modern world electronic and literary interaction between the variant regional Englishes make it likely that the common language will remain universally intelligible, it also seems probable that mutually incomprehensible forms of English will become entrenched as the language is taught, learned, and used in world areas far removed from contact with first-language speakers and with vibrant local economies and cultures independent of the Standard English community. "Our only revenge," said a French official, deploring the declining role of French within the European Union, "is that the English language is being killed by all these foreigners speaking it so badly."

westward movements of the southern dialects. The Midland dialect, apparently so restricted along the eastern seaboard, became, almost by default, the basic form for much of the interior and West of the United States. It was altered and enriched there by contact with the Northern and Southern dialects, by additions from Native American languages, by contact with Spanish culture in the Southwest, and by contributions from the great non-English immigrations of the late 19th and early 20th centuries. Naturally, dialect subregions are found in the West, but their boundary lines—so clear in the eastern interior—become less distinct from the Plains States to the Pacific.

The immigrant contributions of the last centuries are still continuing and growing. In areas with strong infusions of recently arrived Hispanic, Asian, and other immigrant groups, language mixing tends to accelerate language change as more and different non-English words enter the general vocabulary of all Americans. In many cases, those infusions create or perpetuate pockets of linguistically unassimilated peoples whose urban neighborhoods in shops, signage, and common speech bear little resemblance to the majority Anglo communities of the larger metropolitan area. Even as immigrant groups learn and adopt English, there is an inevitable retention of familiar words and phrases and, for many, the unstructured intermixture of old and new tongues into such hybrids as "espanglish."

Local dialects and accents do not display predictable patterns of consistency or change. In ethnically and regionally complex United States, for example, mixed conclusions concerning local speech patterns have been drawn by researchers examining the linguistic results of an increasingly transient population, immigration from other countries and cultures, and the pervasive and presumed leveling effects of the mass media. The distinct evidence of increasing contrasts between the speech patterns and accents of

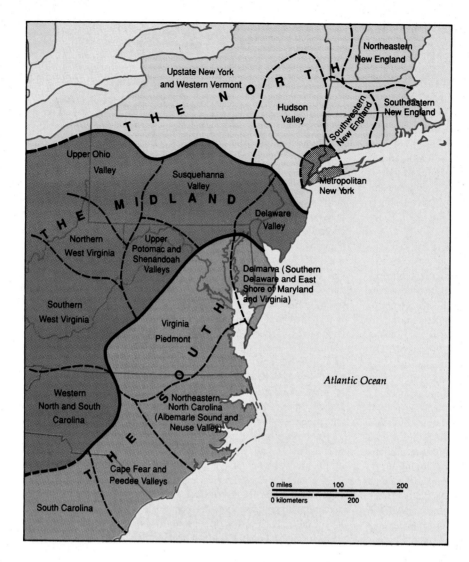

Figure 5.12 **Dialect areas of the eastern United States.** The Northern dialect and its subdivisions are found in New England and adjacent Canada (the international boundary has little effect on dialect borders in Anglo America), extending southward to a secondary dialect area centered on New York City. Midland speech is found along the Atlantic Coast only from central New Jersey southward to central Delaware, but spreads much more extensively across the interior of the United States and Canada. The Southern dialect dominates the East Coast from Chesapeake Bay south.

Source: Redrawn by permission from Hans Kurath, A Word Geography of the Eastern United States *(Ann Arbor: University of Michigan Press, 1949).*

Chicago, New York, Birmingham, St. Louis, and other cities is countered by reports of decreasing local dialect pronunciations in such centers as Dallas and Atlanta that have experienced major influxes of Northerners. And other studies find that some regional accents are fading in small towns and rural areas, presumably because mass media standardization is more influential than local dialect reinforcement as areal populations decline.

Pidgins and Creoles

Language is rarely a total barrier in communication between peoples, even those whose native tongues are mutually incomprehensible. Bilingualism or multilingualism may permit skilled linguists to communicate in a jointly understood third language, but long-term contact between less able populations may require the creation of new language—a pidgin—learned by both parties.

A **pidgin** is an amalgamation of languages, usually a simplified form of one, such as English or French, with borrowings from another, perhaps non-European local language. In its original form, a pidgin is not the mother tongue of any of its speakers; it is a second language for everyone who uses it, a language generally restricted to such specific functions as commerce, administration, or work supervision. For example, such is the variety of languages spoken among the some 270 ethnic groups of the Democratic Republic of the Congo that a special tongue called Lingala, a hybrid of Congolese dialects and French, was created to permit, among other things, issuance of orders to army recruits drawn from all parts of the country.

Pidgins are initially characterized by a highly simplified grammatical structure and a sharply reduced vocabulary, adequate to express basic ideas but not complex concepts. If a pidgin becomes the first language of a group of speakers—who may

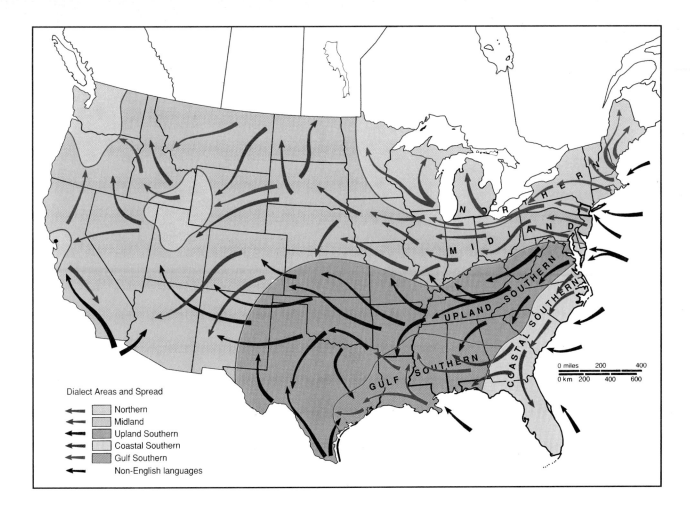

Figure 5.12 **Speech regions and dialect diffusion in the United States.** This generalized map is most accurate for the eastern seaboard and the easternmost diffusion pathways where most detailed linguistic study has been concentrated. West of the Mississippi River, the Midland dialect becomes dominant, though altered through modifications reflecting intermingling of peoples and speech patterns. Northern speech characteristics are still clearly evident in the San Francisco Bay area, brought there in the middle of the 19th century by migrants coming by sea around Cape Horn. Northerners were also prominent among the travelers of the Oregon Trail.

Sources: Based on Raven I. McDavid, Jr. "The Dialects of American English," in W. Nelson Francis, The Structure of American English *(New York: Ronald Press, 1958); "Regional Dialects in the United States,"* Webster's New World Dictionary, *2nd College Edition (New York: Simon and Schuster, 1980); and Gordon R. Wood,* Vocabulary Change *(Carbondale: Southern Illinois University Press, 1971), Map 83, p. 358.*

have lost their former native tongue through disuse—a **creole** has evolved. In their development, creoles invariably acquire a more complex grammatical structure and enhanced vocabulary.

Creole languages have proved useful integrative tools in linguistically diverse areas; several have become symbols of nationhood. Swahili, a pidgin formed from a number of Bantu dialects with major vocabulary additions from Arabic, originated in the coastal areas of East Africa and spread by trade during the period of English and German colonial rules. When Kenya and Tanzania gained independence, they made Swahili the national language of administration and education. Other examples of creolization are Afrikaans (a pidginized form of 17th-century Dutch used in the Republic of South Africa); Haitian Creole (the language of Haiti, derived from the pidginized French used in the slave trade); and Bazaar Malay (a pidginized form of the Malay language, a version of which is the official national language of Indonesia).

Lingua Franca

A **lingua franca** is an established language used habitually for communication by people whose native tongues are mutually incomprehensible. For them it is a *second language,* one learned in addition to the native tongue. Lingua franca, literally "Frankish tongue," was named from the dialect of France adopted as their common tongue by the Crusaders assaulting the Muslims of the Holy Land. Later, it endured as a language of trade and travel in the eastern Mediterranean, useful as a single tongue shared in a linguistically diverse region.

Between 300 B.C. and A.D. 500, the Mediterranean world was unified by Common Greek. Later, Latin became a lingua franca, the language of empire and, until replaced by the vernacular European tongues, of the Church, government, scholarship, and the law. Outside the European sphere, Aramaic served the role from the 5th century B.C. to the 4th century A.D. in the Near East and Egypt; Arabic followed Muslim conquest as the unifying language of that international religion after the 7th century. Mandarin Chinese and Hindi in India both formerly and today have a lingua franca role in their linguistically diverse countries. The immense linguistic diversity of Africa has made regional lingua francas there necessary and inevitable (Figure 5.14), and in a polyglot world, English increasingly serves everywhere as the global lingua franca.

Official Languages

Governments may designate a single tongue as a country's **official language,** the required language of instruction in the schools and universities, government business, the courts, and other official and semiofficial public and private activities. In societies in which two or more languages are in common use **(multilingualism),** such an official language may serve as the approved national lingua franca, guaranteeing communication among all citizens of differing native tongues. In many immigrant societies, such as the United States, only one of the many spoken languages may have implicit or official government sanction (see "An Official U.S. Language?").

Nearly every country in linguistically complex sub-Saharan Africa has selected a European language—usually that of their former colonial governors—as an official language (Figure 5.15), only rarely designating a native language or creole as an alternate official tongue. Indeed, less than 10% of the population of sub-Saharan Africa live in countries with any indigenous African tongue given official status. Nigeria has some 350 clearly different languages and is dominated by three of them: Hausa, Yoruba, and Ibo. For no Nigerian is English a native tongue, yet throughout the country English is the sole language of instruction and the sole official language. Effectively, all Nigerians must learn a foreign language before they can enter the mainstream of national life. Most Pacific Ocean countries, including the Philippines

Figure 5.14 **Lingua francas of Africa.** The importance and extent of competing lingua francas in sub-Saharan Africa change over time, reflecting the spread of populations and the relative economic or political stature of speakers of different languages. In many areas, an individual may employ different lingua francas, depending on activity: dealing with officials, trading in the marketplace, conversing with strangers. Among the elite in all areas, the preferred lingua franca is apt to be a European language. Throughout northern Africa, Arabic is the usual lingua franca for all purposes.

Source: Adapted from Bernd Heine, Status and Use of African Lingua Francas *(Munich, Germany: Weltforum Verlag; and New York: Humanities Press, 1970).*

$\mathcal{F}igure\ 5.15$ **Europe in Africa through official languages.** Both the linguistic complexity of sub-Saharan Africa and the colonial histories of its present political units are implicit in the designation of a European language as the sole or joint "official" language of the different countries.

(with between 80 and 110 Malayo-Polynesian languages) and Papua New Guinea (with over 850 distinct Papuan tongues), have a European language as at least one of their official tongues.

Increasingly, the "purity" of official European languages has been threatened by the popular and widespread inclusion of English words and phrases in everyday speech, press, and television. So common has such adoption become, in fact, that some nearly new language variants are now recognized: *franglais* in France and *Denglish* in Germany are the best-known examples. Both have spurred resistance movements from officially sanctioned language monitors of, respectively, the French Academy and the Institute for the German Language. Poland, Spain, and Latvia are among other European states seeking to preserve the purity of their official languages from contamination by English or other foreign borrowings. Japan's Council on the Japanese Language is doing the same.

In some countries, multilingualism has official recognition through designation of more than a single state language. Canada and Finland, for example, have two official languages (*bilingualism*), reflecting rough equality in numbers or influence of separate linguistic populations comprising a single country. In a few multilingual countries, more than two official languages have been designated. Bolivia and Belgium have three official tongues and Singapore has four. South Africa's constitution designates 11 official languages, and India gives official status to 18 languages at the regional, though not at the national, level.

Multilingualism may reflect significant cultural and spatial divisions within a country. In Canada, the Official Languages Act of 1969 accorded French and English equal status as official languages of the Parliament and of government throughout the nation. French-speakers are concentrated in the Province of Quebec, however, and constitute a culturally distinct population sharply divergent from the English-speaking majority of other parts of Canada (Figure 5.16). Within sections of Canada, even greater linguistic diversity is recognized; the legislature of the Northwest Territories, for example, has eight official languages—six native, plus English and French.

Few countries remain purely *monolingual,* with only a single language of communication for all purposes among all citizens, though most are officially so. Past and recent movements of peoples as colonists, refugees, or migrants have assured that most of the world's countries contain linguistically mixed populations.

Maintenance of native languages among such populations is not assured, of course. Where numbers are small or pressures for integration into an economically and socially dominant culture are strong, immigrant and aboriginal (native) linguistic minorities tend to adopt the majority or official language for all purposes. On the other hand, isolation and relatively large numbers of speakers may serve to preserve native tongues. In Canada, for example, aboriginal languages with large populations of speakers—Cree, Ojibwe, and Inuktitut—are well maintained in their areas of concentration (respectively, northern Quebec, the northern prairies, and Nunavut). In contrast, much smaller language groups in southern and coastal British Columbia have a much lower ratio of retention among native speakers.

Language, Territoriality, and Identity

The designation of more than one official language does not always satisfy the ambitions of linguistically distinct groups for recognition and autonomy. Language is an inseparable part of group identity and a defining characteristic of ethnic and cultural distinction. The view that cultural heritage is rooted in language is well-established and found throughout the world, as is the feeling that losing linguistic identity is the worst and final evidence of discrimination and subjugation. Language has often been the focus of separatist movements, especially of spatially distinct linguistic groups outside the economic heartlands of the strongly centralized countries to which they are attached.

In Europe, highly centralized France, Spain, Britain—and Yugoslavia and the Soviet Union before their dismemberment—experienced such language "revolts" and acknowledged, sometimes belatedly, the local concerns they express. Until 1970, when the ban on teaching regional tongues was dropped, the spoken regional languages and dialects of France were ignored and denied recognition by the state. Since the late 1970s, Spain not only has relaxed its earlier total rejection of Basque and Catalan as regional languages and given state support to instruction in them, but also has recognized Catalan as a co-official language in

Geography and Public Policy

An Official U.S. Language?

Within recent years in Lowell, Massachusetts, public school courses have been offered in Spanish, Khmer, Lao, Portuguese, and Vietnamese, and all messages from schools to parents have been translated into five languages. Polyglot New York City has given bilingual programs in Spanish, Chinese, Haitian Creole, Russian, Korean, Vietnamese, French, Greek, Arabic, and Bengali. In most states, it is possible to get a high-school-equivalency diploma without knowing English because tests are offered in French and Spanish. In at least 39 states, driving tests have been available in foreign languages; California has provided 39 varieties, New York 23, and Michigan 20, including Arabic and Finnish. And as required by the 1965 federal Voting Rights Act, multilingual ballots are provided in some 300 electoral jurisdictions in 30 states.

These, and innumerable other evidences of governmentally sanctioned linguistic diversity, may come as a surprise to those many Americans who assume that English is the of-ficial language of the United States. It isn't; nowhere does the Constitution provide for an official language, and no federal law specifies one. The country was built by a great diversity of cultural and linguistic immigrants who nonetheless shared an eagerness to enter mainstream American life. At the start of the 21st century, a reported 18% of all U.S. residents speak a language other than English in the home. In California public schools, 1 out of 3 students uses a non-English tongue within the family. In Washington, D.C. schools, students speak 127 languages and dialects, a linguistic diversity duplicated in other major city school systems.

Nationwide bilingual teaching began as an offshoot of the civil rights movement in the 1960s, was encouraged by a Supreme Court opinion authored by Justice William O. Douglas, and has been actively promoted by the U.S. Department of Education under the Bilingual Education Act of 1974 as an obligation of local school boards. Its purpose has been to teach subject matter to minority-language children in the language in which they think while introducing them to English, with the hope of achieving English proficiency in 2 or 3 years. Disappointment with the results achieved led to a successful 1998 California anti-bilingual education initiative, Proposition 227, to abolish the program. Similar rejection elsewhere has followed California's lead.

Opponents of the implications of governmentally encouraged multilingual education, bilingual ballots, and ethnic separatism argue that a common language is the unifying glue of the United States and all countries; without that glue, they fear, the process of "Americanization" and *acculturation*—the adoption by immigrants of the values, attitudes, ways of behavior, and speech of the receiving society—will be undermined. Convinced that early immersion and quick proficiency in English is the only sure way for minority newcomers to gain necessary access to jobs, higher education, and full integration into the economic and social life of the country, proponents of "English only" use in public education, voting, and state and local governmental agencies, successfully passed Official English laws or constitutional amendments in 27 states from the late 1980s to 2002.

Although the amendments were supported by sizeable majorities of the voting population, resistance to them—and to their political and cultural implications—has been in

its home region in northeastern Spain. In Britain, parliamentary debates concerning greater regional autonomy in the United Kingdom have resulted in bilingual road and informational signs in Wales, a publicly supported Welsh-language television channel, and compulsory teaching of Welsh in all schools in Wales.

In fact, throughout Europe beginning in the 1980s, nonofficial native regional languages have increasingly not only been tolerated but encouraged—in Western Europe, particularly, as a buffer against the loss of regional institutions and traditions threatened by a multinational "superstate" under the European Union. The Council of Europe, a 41-nation organization promoting democracy and human rights, has adopted a charter pledging encouragement of the use of indigenous languages in schools, the media, and public life. That pledge recognizes the enduring reality that of some 500 million people in Eastern and Western Europe (not including immigrants and excluding the former USSR), more than 50 million speak a local language that is not the official tongue of their country. The language charter acknowledges that cultural diversity is part of Europe's wealth and heritage and that its retention strengthens, not weakens, the separate states of the continent and the larger European culture realm as a whole.

Many other world regions, less permissive than Europe is becoming, have continuing linguistically based conflict. Language has long been a divisive issue in South Asia, for example, leading to wars in Pakistan and Sri Lanka and periodic demands for secession from India by southern states such as Tamil Nadu, where the Dravidian Tamil language is defended as an ancient tongue as worthy of respect as the Indo-European official language, Hindi. In Russia and several other successor states of the former USSR (which housed some 200 languages and dialects), linguistic diversity forms part of the justification for local separatist movements, as it did in the division of Czechoslovakia into Czech- and Slovak-speaking successor states and in the violent dismemberment of former Yugoslavia.

Language on the Landscape: Toponymy

Toponyms—place names—are language on the land, the record of past inhabitants whose namings endure, perhaps corrupted and disguised, as reminders of their existence and their passing. **Toponymy** is the study of place names, a special interest of linguistic geography. It is also a revealing tool of historical cultural geography, for place names become a part of the cultural landscape that remains long after the name givers have passed from the scene.

(Continued)

every instance strong and persistent. Ethnic groups, particularly Hispanics, who are the largest of the linguistic groups affected, charged that they were evidence of blatant Anglo-centric racism, discriminatory and repressive in all regards. Some educators argued persuasively that all evidence proved that while immigrant children eventually acquire English proficiency in any event, they do so with less harm to their self-esteem and subject matter acquisition when initially taught in their own language. Business people with strong minority labor and customer ties and political leaders—often themselves members of ethnic communities or with sizable minority constituencies—argued against "discriminatory" language restrictions.

And historians noted that it had all been unsuccessfully tried before. The anti-Chinese Workingmen's Party in 1870s California led the fight for English-only laws in that state. The influx of immigrants from central and southeastern Europe at the turn of the century led Congress to make oral English a requirement for naturalization, and anti-German sentiment during and after World War I led some states to ban any use of German. The Supreme Court struck down those laws in 1923, ruling that the "protection of the Constitution extends to all, to those who speak other languages as well as to those born with English on their tongue." Following suit, some of the recent state language amendments have also been voided by state or federal courts. In ruling its state's English-only law unconstitutional, Arizona's Supreme Court in 1998 noted it "chills First Amendment rights."

To counter those judicial restraints and the possibility of an eventual multilingual, multicultural United States in which English and, likely, Spanish would have co-equal status and recognition, U.S. English—an organization dedicated to the belief that "English is, and ever must remain, the only official language of the people of the United States"—actively supports the proposed U.S. Constitutional amendment first introduced in Congress by former Senator S. I. Hayakawa in 1981, and resubmitted by him and others in subsequent years. The proposed amendment would simply establish English as the official national language but would impose no duty on people to learn English and would not infringe on any right to use other languages. Whether or not these modern attempts to designate an official U.S. language eventually succeed, they represent a divisive subject of public debate affecting all sectors of American society.

Questions to Consider

1. Do you think multiple languages and ethnic separatism represent a threat to U.S. cultural unity that can be avoided only by viewing English as a necessary unifying force? Or do you think making English the official language might divide its citizens and damage its legacy of tolerance and diversity? Why or why not?

2. Do you feel that immigrant children would learn English faster if bilingual classes were reduced and immersion in English was more complete? Or do you think that a slower pace of English acquisition is acceptable if subject matter comprehension and cultural self-esteem are enhanced? Why or why not?

3. Do you think Official English laws serve to inflame prejudice against immigrants or to provide all newcomers with a common standard of admission to the country's political and cultural mainstream?

In England, for example, place names ending in *chester* (as in Winchester and Manchester) evolved from the Latin *castra,* meaning "camp." Common Anglo-Saxon suffixes for tribal and family settlements were *ing* (people or family) and *ham* (hamlet or, perhaps, meadow) as in Birmingham or Gillingham. Norse and Danish settlers contributed place names ending in *thwaite* ("meadow") and others denoting such landscape features as *fell* (an uncultivated hill) and *beck* (a small brook). The Celts, present in Europe for more than 1000 years before Roman times, left their tribal names in corrupted form on territories and settlements taken over by their successors. The Arabs, sweeping out from Arabia across North Africa and into Iberia, left their imprint in place names to mark their conquest and control. *Cairo* means "victorious," *Sudan* is "the land of the blacks," and *Sahara* is "wasteland" or "wilderness." In Spain, a corrupted version of the Arabic *wadi,* "watercourse," is found in *Guadalajara* and *Guadalquivir.*

In the New World, not one people but many placed names on landscape features and new settlements. In doing so they remembered their homes and homelands, honored their monarchs and heroes, borrowed and mispronounced from rivals, followed fads, recalled the Bible, and adopted and distorted Amerindian names.

Homelands were recalled in New England, New France, or New Holland; settlers' hometown memories brought Boston, New Bern, New Rochelle, and Cardiff from England, Switzerland, France, and Wales. Monarchs were remembered in Virginia for the Virgin Queen Elizabeth, Carolina for one English king, Georgia for another, and Louisiana for a king of France. Washington, D.C.; Jackson, Mississippi and Michigan; Austin, Texas; and Lincoln, Illinois memorialized heroes and leaders. Names given by the Dutch in New York were often distorted by the English; Breukelyn, Vlissingen, and Haarlem became Brooklyn, Flushing, and Harlem. French names underwent similar twisting or translation, and Spanish names were adopted, altered, or, later, put into such bilingual combinations as Hermosa Beach. Amerindian tribal names—the Yenrish, Maha, Kansa—were modified, first by French and later by English speakers—to Erie, Omaha, and Kansas. A faddish "Classical Revival" after the Revolution gave us Troy, Athens, Rome, Sparta, and other ancient town names and later spread them across the country (see Figure 7.31). Bethlehem, Ephrata, Nazareth, and Salem came from the Bible. Names adopted were transported as settlement moved westward across the United States (Figure 5.17).

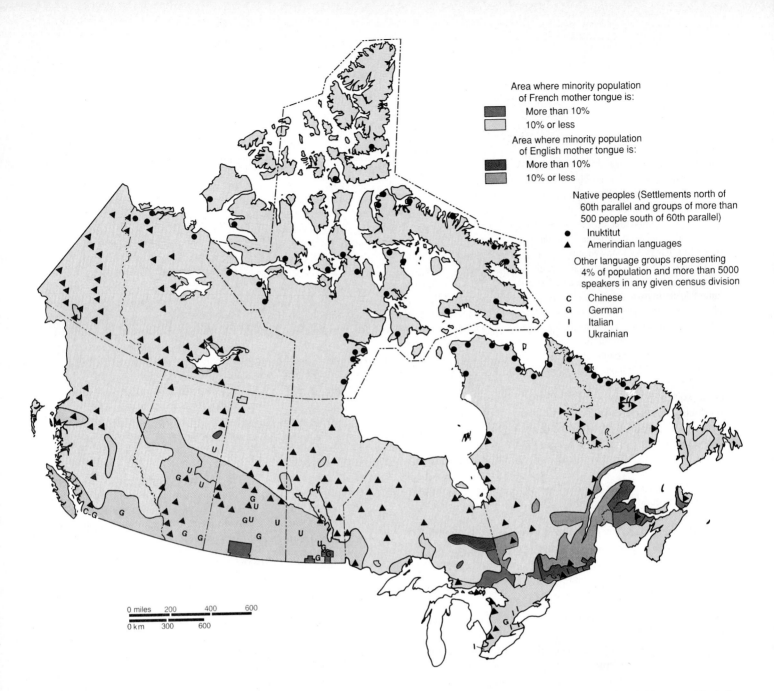

Legend:

Area where minority population of French mother tongue is:
- More than 10%
- 10% or less

Area where minority population of English mother tongue is:
- More than 10%
- 10% or less

Native peoples (Settlements north of 60th parallel and groups of more than 500 people south of 60th parallel)
- ● Inuktitut
- ▲ Amerindian languages

Other language groups representing 4% of population and more than 5000 speakers in any given census division
- C Chinese
- G German
- I Italian
- U Ukrainian

0 miles 200 400 600
0 km 300 600

Figure 5.16 **Bilingualism and diversity in Canada.** The map shows areas of Canada that have a minimum of 5000 inhabitants and include a minority population identified with an official language.

Source: Commissioner of Official Languages, Government of Canada.

Place names, whatever their language of origin, frequently consist of two parts: *generic* (classifying) and *specific* (modifying or particular). *Big River* in English is found as *Rio Grande* in Spanish, *Mississippi* in Algonquin, and *Ta Ho* in Chinese. The *order* of generic and specific, however, may alter between languages and give a clue to the group originally bestowing the place name. In English, the specific usually comes first: *Hudson River, Bunker Hill, Long Island.* When, in the United States, we find *River Rouge* or *Isle Royale,* we also find evidence of French settlement—the French reverse the naming order. Some generic names can be used to trace the migration paths across the United

States of the three Eastern dialect groups (see Figure 5.12). Northern dialect settlers tended to carry with them their habit of naming a community and calling its later neighbors by the same name modified by direction—Lansing and East Lansing, for example. *Brook* is found in the New England settlement area, *run* is from the Midland dialect, *bayou* and *branch* are from the Southern area.

European colonists and their descendants gave place names to a physical landscape already adequately named by indigenous peoples. Those names were sometimes adopted, but often shortened, altered, or—certainly—mispronounced. The vast territory that local Amerindians called "Mesconsing," meaning "the long

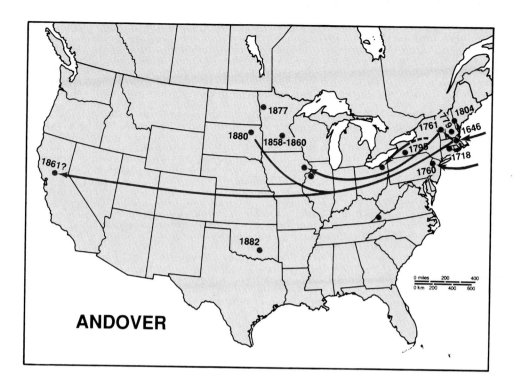

Figure 5.17 **Migrant Andover.** Place names in a new land tend to be transportable, carried to new locales by migrating town founders. They are a reminder of the cultural origins and diffusion paths of settlers. Andover, a town name from England, was brought to New England in 1646 and later carried westward.

Source: With kind permission of the American Name Society.

river," was recorded by Lewis and Clark as "Quisconsing," later to be further distorted into "Wisconsin." *Milwaukee* and *Winnipeg, Potomac* and *Niagara, Adirondack, Chesapeake, Shenandoah,* and *Yukon;* the names of 27 of the 50 United States; and the present identity of thousands of North American places and features, large and small, had their origin in Native American languages.

In the Northwest Territories of Canada, Indian and Inuit (Eskimo) place names are returning. The town of Frobisher Bay has reverted to its Eskimo name *Iqaluit* ("place of the fish"); Resolute Bay becomes *Kaujuitok* ("place where the sun never rises") in Inuktitut, the lingua franca of the Canadian Eskimos; the Jean Marie River returns to *Tthedzehk'edeli* ("river that flows over clay"), its earlier Slavey name. These and other official name changes reflect the decision of the territory's Executive Council

that community preference will be the standard for all place names, no matter how entrenched European versions might be.

It was a decision that recognized the importance of language as a powerful unifying thread in the culture complex of peoples. In India, for example, the changing of various long-accepted municipal place names—*Mumbai* instead of Bombay, *Chennai* but not Madras, or *Thiruvananthapuram* replacing Trivandrum—demonstrates both post-colonial pride and growing Hindu nationalism. Language may serve as a fundamental evidence of ethnicity and be the fiercely defended symbol of the history and individuality of a distinctive social group. Spanish Americans demand the right of instruction in their own language, and Basques wage civil war to achieve a linguistically based separatism. Indian states were adjusted to coincide with language boundaries, and the Polish National Catholic Church was created in America, not Poland, to preserve Polish language and culture in an alien environment.

 *Patterns of Religion*

Religion, like language, is a symbol of group identity and a cultural rallying point. Religious enmity forced the partition of the Indian subcontinent between Muslims and Hindus after the departure of the British in 1947. French Catholics and French Huguenots (Protestants) freely slaughtered each other in the name

of religion in the 16th century. English Roman Catholics were hounded from their country after the establishment of the Anglican Church. Religion has continued to be a root cause of many local and regional conflicts throughout the world during the 20th and into the 21st century, as Chief Makuei's words opening this chapter suggest, including confrontations among Catholic and Protestant Christian groups in Northern Ireland; Muslim sects in Lebanon, Iran, Iraq, and Algeria; Muslims and Jews in Palestine; Christians and Muslims in the Philippines and Lebanon; and Buddhists and Hindus in Sri Lanka. More peacefully, in the name of their beliefs, American Amish, Hutterite, Shaker, and other religious communities have isolated themselves from the secular world and pursued their own ways of life.

Religion and Culture

Unlike language, which is an attribute of all people, religion varies in its cultural role—dominating among some societies, unimportant or denied totally in others. All societies have *value systems*—common beliefs, understandings, expectations, and controls—that unite their members and set them off from other, different culture groups. Such a value system is termed a **religion** when it involves systems of formal or informal worship and faith in the sacred and divine. Religion may intimately affect all facets of a culture. Religious belief is by definition an element of the ideological subsystem; formalized and organized religion is an institutional expression of the sociological subsystem. And religious beliefs strongly influence attitudes toward the tools and rewards of the technological subsystem.

Nonreligious value systems can exist—humanism or Marxism, for example—that are just as binding on the societies that espouse them as are more traditional religious beliefs. Even societies that largely reject religion—that are officially atheistic or secular—are strongly influenced by traditional values and customs set by predecessor religions, in days of work and rest, for example, or in legal principles.

Since religions are formalized views about the relation of the individual to this world and to the hereafter, each carries a distinct conception of the meaning and value of this life, and most contain strictures about what must be done to achieve salvation. These beliefs become interwoven with the traditions of a culture. For Muslims, the observance of the *sharia* (law) is a necessary part of *Islam,* submission to Allah (see Figure 5.27). In classical Judaism, the keeping of the *Torah,* the Law of Moses, involved ritual and moral rules of holy living. For Hindus, the *dharma,* or teaching, includes the complex laws enunciated in the ancient book of Manu. Ethics of conduct and humane relations rather than religious rituals are central to the Confucian tradition of China, while the Sikh *khalsa,* or holy community, is defined by various rules of observance, such as prohibiting the cutting of one's hair.

Economic patterns may be intertwined with past or present religious beliefs. Traditional restrictions on food and drink may affect the kinds of animals that are raised or avoided, the crops that are grown, and the importance of those crops in the daily diet. Occupational assignment in the Hindu caste system is in part religiously supported. In many countries, there is a state religion—that is, religious and political structures are intertwined. Buddhism, for example, has been the state religion in Myanmar, Laos, and Thailand. By their official names, the Islamic Republic of Pakistan and the Islamic Republic of Iran proclaim their identity of religion and government. Despite Indonesia's overwhelming Muslim majority, that country sought and formerly found domestic harmony by recognizing five official religions and a state ideology—*pancasila*—whose first tenet is belief in one god.

The landscape imprint of religions may be both obvious and subtle. The structures of religious worship—temples, churches, mosques, stupas, or cathedrals—landscape symbols such as shrines or statues, and such associated land uses as monasteries may give an immediately evident and regionally distinctive cultural character to an area. "Landscapes of death" may also be visible regional variables, for different religions and cultures dispose of their dead in different manners. Cemeteries are significant and reserved land uses among Christians, Jews, and Muslims who typically bury their deceased with headstones or other markers and monuments to mark graves. Egyptian pyramids or elaborate mausoleums like the Taj Mahal are more grandiose structures of entombment and remembrance. On the other hand, Hindus and Buddhists have traditionally cremated their dead and scattered their ashes, leaving no landscape evidence or imprint.

Some religions may make a subtle cultural stamp on the landscape through recognition of sacred places and spaces not otherwise built or marked. Grottos, lakes, single trees or groves, such rivers as the Ganges or Jordan, or special mountains or hills, such as Mount Ararat or Mount Fuji, are examples that are unique to specific religions and express the reciprocal influences of religion and environment.

Classification of Religion

Religions are cultural innovations. They may be unique to a single culture group, closely related to the faiths professed in nearby areas, or derived from or identical to belief systems spatially far removed. Although interconnections and derivations among religions can frequently be discerned—as Christianity and Islam can trace descent from Judaism—family groupings are not as useful to us in classifying religions as they were in studying languages. A distinction between **monotheism,** belief in a single deity, and **polytheism,** belief in many gods, is frequent, but not particularly spatially relevant. Simple territorial categories have been offered recognizing origin areas of religions: Western versus Eastern, for example, or African, Far Eastern, or Indian. With proper detail such distinctions may inform us where particular religions had their roots but do not reveal their courses of development, paths of diffusion, or current distributions.

Our geographic interest in the classification of religions is different from that of, say, theologians or historians. We are not so concerned with the beliefs themselves or with their birthplaces (though both help us understand their cultural implications and areal arrangements). We are more interested in religions' patterns and processes of diffusion once they have developed, with the spatial distributions they have achieved, and with the impact of the practices and beliefs of different religious systems on the landscape. To satisfy at least some of those interests, geographers have found it useful to categorize religions as *universalizing, ethnic,* or *tribal (traditional).*

Christianity, Islam, and Buddhism are the major world **universalizing religions,** faiths that claim applicability to all humans and that seek to transmit their beliefs through missionary work and conversion. Membership in universalizing religions is open to anyone who chooses to make some sort of symbolic commitment, such as baptism in Christianity. No one is excluded because of nationality, ethnicity, or previous religious belief.

Ethnic religions have strong territorial and cultural group identification. One usually becomes a member of an ethnic religion by birth or by adoption of a complex lifestyle and cultural identity, not by simple declaration of faith. These religions do not usually proselytize, and their members form distinctive closed communities identified with a particular ethnic group or political unit. An ethnic religion—for example, Judaism, Indian Hinduism, or Japanese Shinto—is an integral element of a specific culture; to be part of the religion is to be immersed in the totality of the culture.

Tribal, or **traditional religions,** are special forms of ethnic religions distinguished by their small size, their unique identity with localized culture groups not yet fully absorbed into modern society, and their close ties to nature. **Animism** is the name given to their belief that life exists in all objects, from rocks and trees to lakes and mountains, or that such inanimate objects are the abode of the dead, of spirits, and of gods. **Shamanism** is a form of tribal religion that involves community acceptance of a *shaman,* a religious leader, healer, and worker of magic who, through special powers, can intercede with and interpret the spirit world.

Patterns and Flows

The nature of the different classes of religions is reflected in their distributions over the world (Figure 5.18) and in their number of adherents. Universalizing religions tend to be expansionary, carrying their message to new peoples and areas. Ethnic religions, unless their adherents are dispersed, tend to be regionally confined or to expand only slowly and over long periods. Tribal religions tend to contract spatially as their adherents are incorporated increasingly into modern society and converted by proselytizing faiths.

As we expect in human geography, the map records only the latest stage of a constantly changing cultural reality. While established religious institutions tend to be conservative and resistant to change, religion as a culture trait is dynamic. Personal and collective beliefs may alter in response to developing individual and societal needs and challenges. Religions may be imposed by conquest, adopted by conversion, or be defended and preserved in the face of surrounding hostility or indifference.

The World Pattern

Figure 5.18 (at this scale) cannot present a full picture of current religious affiliation or regionalization. Few societies are homogeneous, and most modern ones contain a variety of different faiths or, at least, variants of the dominant professed religion. Some of

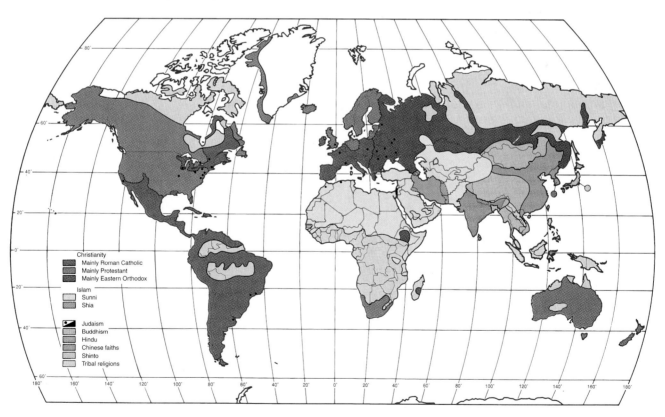

Figure 5.18 **Principal world religions.** The assignment of individual countries to a single religion category conceals a growing intermixture of faiths in European and other western countries that have experienced recent major immigration flows. In some instances, those influxes are altering the effective, if not the numerical, religious balance. In nominally Christian, Catholic France, for example, low church-going rates suggest that now more Muslims than practicing Catholics reside there and, considering birth rate differentials, that someday Islam may be the country's predominant religion as measured by the number of practicing adherents. Secularism—rejection of religious belief—is common in many countries but is not locationally indicated on this map.

those variants in many religions are intolerant or antagonistic toward other faiths or toward those sects and members of their own faith deemed insufficiently committed or orthodox (see "Militant Fundamentalism").

Frequently, members of a particular religion show areal concentration within a country. Thus, in urban Northern Ireland, Protestants and Catholics reside in separate areas whose boundaries are clearly understood and respected. The "Green Line" in Beirut, Lebanon, marked a guarded border between the Christian East and the Muslim West sides of the city, while within the country as a whole regional concentrations of adherents of different faiths and sects are clearly recognized (Figure 5.19). Religious diversity within countries may reflect the degree of toleration a majority culture affords minority religions. In dominantly (90%) Muslim Indonesia, Christian Bataks, Hindu Balinese, and Muslim Javanese for many years lived in peaceful coexistence. By contrast, the fundamentalist Islamic regime in Iran has persecuted and executed those of the Baha'i faith.

Data on religious affiliation are not precise. Most nations do not have religious censuses, and different religious groups differently and inconsistently report their membership. When communism was supreme in the former Soviet Union and Eastern European countries, official atheism dissuaded many from openly professing or practicing any religion; in nominally Christian Europe and North America, many who claim to be believers are not active church members and others renounce religion altogether.

More than half of the world's population probably adheres to one of the major universalizing religions: Christianity, Islam, or Buddhism. Of these three, Figure 5.18 indicates, Christianity and Islam are most widespread; Buddhism is largely an Asian religion. Hinduism, the largest ethnic faith, is essentially confined to the Indian subcontinent, showing the spatial restriction characteristic of most ethnic and traditional religions even when found outside of their homeland area. Small Hindu emigrant communities in Africa, southeast Asia, England, or the United States, for example, tend to remain isolated even in densely crowded urban

Militant Fundamentalism

The term *fundamentalism* entered the social science vocabulary in the late 20th century to describe any religious orthodoxy that is revivalist and ultraconservative in nature. Originally it designated an American Christian movement named after a set of volumes—*The Fundamentals: A Testimony of the Truth*—published between 1910 and 1915 and embracing both absolute religious orthodoxy and a commitment to inject its beliefs into the political and social arenas. More recently, fundamentalism has become a generic description for all religious movements that seek to regain and publicly institutionalize traditional social and cultural values that are usually rooted in the teachings of a sacred text or written dogma.

Springing from rejection of the secularist tendencies of modernity, fundamentalism is now found in every dominant religion wherever a western-style society has developed, including Islam, Hinduism, Judaism, Sikhism, Buddhism, Confucianism, and Zoroastrianism. Fundamentalism is therefore a reaction to the modern world; it represents an effort to draw upon a "golden age" religious tradition in order to cope with and counteract an already changing society that is denounced as trying to erase the true faith and traditional religious values. The near-universality of fundamentalist movements is seen by some as another expression of a

widespead rebellion against the presumed evils of secular globalization.

Fundamentalists always place a high priority on doctrinal conformity and its necessity to achieve salvation. Further, they are convinced of the unarguable correctness of their beliefs and the necessity of the unquestioned acceptance of those beliefs by the general society. Since the truth is knowable and indisputable, fundamentalists hold, there is no need to discuss it or argue it in open forum. To some observers, therefore, fundamentalism is by its nature undemocratic and states controlled by fundamentalist regimes combining politics and religion, of necessity, stifle debate and punish dissent. In the modern world, that rigidity seems most apparent in Islam where, it is claimed, "all Muslims believe in the absolute inerrancy of the Quran . . ." (*The Islamic Herald,* April, 1995) and several countries—for example, the Islamic Republic of Iran and the Islamic Republic of Pakistan—proclaim by official name their administrative commitment to religious control.

In most of the modern world, however, such commitment is not overt or official, and fundamentalists often believe that they and their religious convictions are under mortal threat. They view modern secular society—with its assumption of equality of competing voices and values—as trying to eradicate the

true faith and religious verities. Initially, therefore, every fundamentalist movement begins as an intrareligious struggle directed against its own co-religionists and countrymen in response to a felt assault by the liberal or secular society they inhabit. At first, the group may blame its own weakness and irresolution for the oppression they feel and the general social decay they perceive. To restore society to its idealized standards, the aroused group may exhort its followers to ardent prayer, ascetic practices, and physical or military training.

If it is unable peacefully to impose its beliefs on others, the fundamentalist group—seeing itself as the savior of society—may justify other more extreme actions against perceived oppressors. Initial protests and nonviolent actions may escalate to attacks on corrupt public figures thwarting their vision and to outright domestic guerrilla warfare. When an external culture or power—commonly a demonized United States—is seen as the unquestioned source of the pollution and exploitation frustrating their social vision, some fundamentalists have been able to justify any extreme action and personal sacrifice for their cause. In their struggle it appears an easy progression from domestic dispute and disruption to international terrorism.

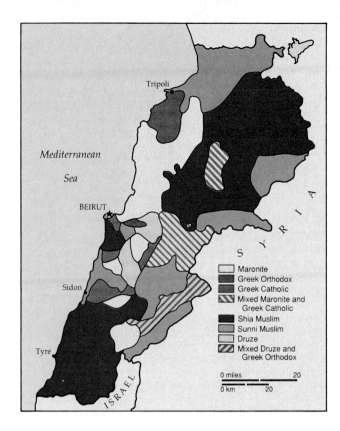

Figure 5.19 **Religious regions of Lebanon.** Religious territoriality and rivalry contributed to a prolonged period of conflict and animosity in this troubled country.

Map legend:
- Maronite
- Greek Orthodox
- Greek Catholic
- Mixed Maronite and Greek Catholic
- Shia Muslim
- Sunni Muslim
- Druze
- Mixed Druze and Greek Orthodox

areas. Although it is not localized, Judaism is also included among the ethnic religions because of its identification with a particular people and cultural tradition.

Extensive areas of the world are peopled by those who practice tribal or traditional religions, often in concert with the universalizing religions to which they have been outwardly converted. Tribal religions are found principally among peoples who have not yet been fully absorbed into modern cultures and economies or who are on the margins of more populous and advanced societies. Although the areas assigned to tribal religions in Figure 5.18 are large, the number of adherents is small and declining.

One cannot assume that all people within a mapped religious region are adherents of the designated faith or that membership in a religious community means active participation in its belief system. **Secularism,** an indifference to or rejection of religion and religious belief, is an increasing part of many modern societies, particularly of the industrialized nations and those now or formerly under communist regimes. In England, for example, the state Church of England claims 20% of the British population as communicants, but only 2% of the adult population attends its Sunday services. Two-thirds of the French describe themselves as Catholic, and less than 5% regularly go to church. Even in devoutly Roman Catholic South American states, low church attendance attests to the rise of at least informal secularism. In Colombia, only 18% of people attend Sunday services; in Chile,

the figure is 12%, in Mexico 11%, and Bolivia 5%. Most early 21st century estimates put the world number of the nonreligious at between 800 million and 1 billion. Official governmental policies of religious tolerance or constitutionally mandated neutrality (as in the cases of the United States or India, for example) are, of course, distinct from purely personal and individual elections of secular or nonreligious beliefs.

The Principal Religions

Each of the major religions has its own unique mix of cultural values and expressions, each has had its own pattern of innovation and spatial diffusion (Figure 5.20), and each has had its own impact on the cultural landscape. Together they contribute importantly to the worldwide pattern of human diversity.

Judaism

We begin our review of world faiths with **Judaism,** whose belief in a single God laid the foundation for both Christianity and Islam. Unlike its universalizing offspring, Judaism is closely identified with a single ethnic group and with a complex and restrictive set of beliefs and laws. It emerged some 3000 to 4000 years ago in the Near East, one of the ancient culture hearth regions (see Figure 2.15). Early Near Eastern civilizations, including those of Sumeria, Babylonia, and Assyria, developed writing, codified laws, and formalized polytheistic religions featuring rituals of sacrifice and celebrations of the cycle of seasons.

Judaism was different. The Israelites' conviction that they were a chosen people, bound with God through a covenant of mutual loyalty and guided by complex formal rules of behavior, set them apart from other peoples of the Near East. Theirs became a distinctively *ethnic* religion, the determining factors of which are descent from Israel (the patriarch Jacob), the Torah (law and scripture), and the traditions of the culture and the faith. Early military success gave the Jews a sense of territorial and political identity to supplement their religious self-awareness. Later conquest by nonbelievers led to their dispersion (*diaspora*) to much of the Mediterranean world and farther east into Asia by A.D. 500 (Figure 5.21).

Alternately tolerated and persecuted in Christian Europe, occasionally expelled from countries, and usually, as outsiders of different faith and custom, isolated in special residential quarters (ghettos), Jews retained their faith and their sense of community even though two separate branches of Judaism developed in Europe during the Middle Ages. The Sephardim were originally based in the Iberian Peninsula and expelled from there in the late 15th century; with ties to North African and Babylonian Jews, they retained their native Judeo-Spanish language (Ladino) and culture. Between the 13th and 16th centuries, the Ashkenazim, seeking refuge from intolerable persecution in western and central Europe, settled in Poland, Lithuania, and Russia (Figure 5.21). It was from eastern Europe that many of the Jewish immigrants to the United States came during the later 19th and early 20th centuries, though

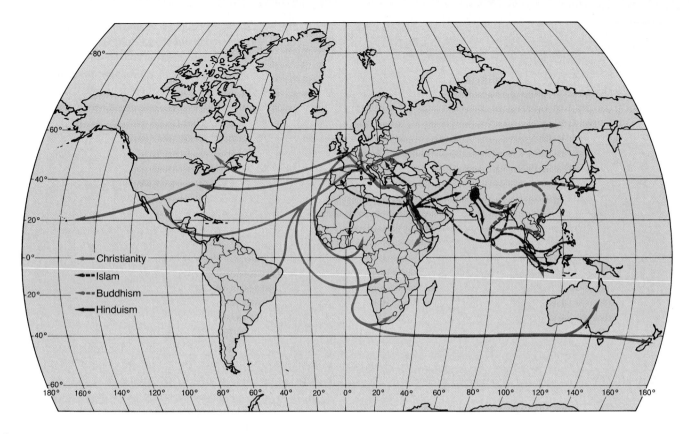

Figure 5.20 **Innovation areas and diffusion routes of major world religions.** The monotheistic (single deity) faiths of Judaism, Christianity, and Islam arose in southwestern Asia, the first two in Palestine in the eastern Mediterranean region and the last in western Arabia near the Red Sea. Hinduism and Buddhism originated within a confined hearth region in the northern part of the Indian subcontinent. Their rates, extent, and directions of diffusions are suggested here and detailed on later maps.

German-speaking areas of central Europe were also important source regions.

The Ashkenazim constitute perhaps 80% of all Jews in the world and differ from the Sephardim in cultural traditions (for example, their widespread use of Yiddish until the 20th century) and liturgy. Both groups are present in roughly equal numbers in Israel. The mass destruction of Jews in Europe before and during World War II—the Holocaust—drastically reduced their representation among that continent's total population.

The establishment of the state of Israel in 1948 was a fulfillment of the goal of *Zionism,* the belief in the need to create an autonomous Jewish state in Palestine. It demonstrated a determination that Jews not lose their identity by absorption into alien cultures and societies. The new state represented a reversal of the preceding 2000-year history of dispersal and relocation diffusion. Israel became largely a country of immigrants, an ancient homeland again identified with a distinctive people and an ethnic religion.

Judaism's imprint on the cultural landscape has been subtle and unobtrusive. The Jewish community reserves space for the practice of communal burial; the spread of the cultivated citron in the Mediterranean area during Roman times has been traced to Jewish ritual needs; and the religious use of grape wine assured the cultivation of the vine in their areas of settlement. The synagogue as place of worship has tended to be less elaborate than its

Christian counterpart. The essential for religious service is a community of at least 10 adult males, not a specific structure.

Christianity

Christianity had its origin in the life and teachings of Jesus, a Jewish preacher of the 1st century of the modern era, whom his followers believed was the messiah promised by God. The new covenant he preached was not a rejection of traditional Judaism but a promise of salvation to all humankind rather than to just a chosen people.

Christianity's mission was conversion. As a universal religion of salvation and hope, it spread quickly among the underclasses of both the eastern and western parts of the Roman Empire, carried to major cities and ports along the excellent system of Roman roads and sea lanes (Figure 5.22). *Expansion diffusion* followed the establishment of missions and colonies of converts in locations distant from the hearth region. Important among them were the urban areas that became administrative seats of the new religion. For the Western Church, Rome was the principal center for dispersal, through *hierarchical diffusion,* to provincial capitals and smaller Roman settlements of Europe. From those nodes and from monasteries established in pagan rural areas, *contagious diffusion* disseminated Christianity throughout the continent. The acceptance of Christianity as the

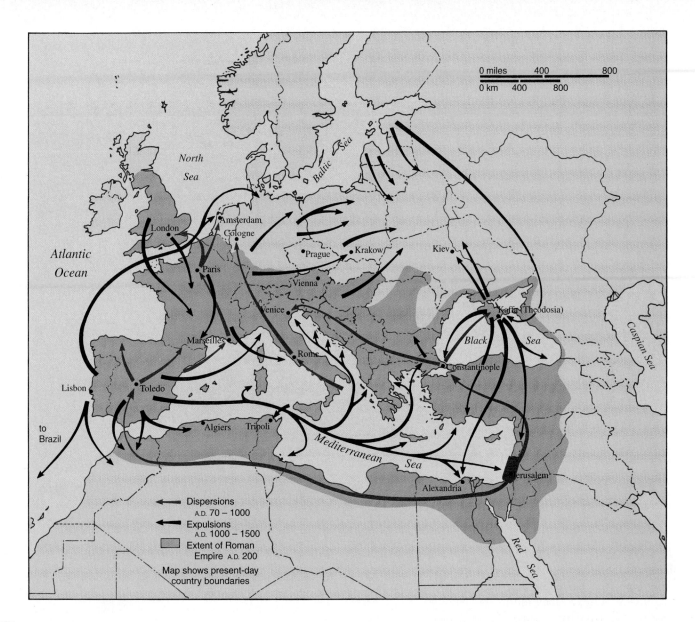

Figure 5.21 **Jewish dispersions, A.D. 70–1500.** A revolt against Roman rule in A.D. 66 was followed by the destruction of the Jewish Temple 4 years later and an imperial decision to Romanize the city of Jerusalem. Judaism spread from the hearth region by *relocation diffusion,* carried by its adherents dispersing from their homeland to Europe, Africa, and eventually in great numbers to the Western Hemisphere. Although Jews established themselves and their religion in new lands, they did not lose their sense of cultural identity and did not seek to attract converts to their faith.

state religion of the empire by the Emperor Constantine in A.D. 313 was also an expression of hierarchical diffusion of great importance in establishing the faith throughout the full extent of the Roman world. Finally, and much later, *relocation diffusion* brought the faith to the New World with European settlers (see Figure 5.18).

The dissolution of the Roman Empire into a western and an eastern half after the fall of Rome also divided Christianity. The Western Church, based in Rome, was one of the very few stabilizing and civilizing forces uniting western Europe during the Dark Ages. Its bishops became the civil as well as ecclesiastical authorities over vast areas devoid of other effective government. Parish churches were the focus of rural and urban life, and the cathedrals replaced Roman monuments and temples as the sym-

bols of the social order (Figure 5.23). Everywhere, the Roman Catholic Church and its ecclesiastical hierarchy were dominant.

Secular imperial control endured in the eastern empire, whose capital was Constantinople. Thriving under its protection, the Eastern Church expanded into the Balkans, eastern Europe, Russia, and the Near East. The fall of the eastern empire to the Turks in the 15th century opened eastern Europe temporarily to Islam, though the Eastern Orthodox Church (the direct descendant of the Byzantine state church) remains, in its various ethnic branches, a major component of Christianity.

The Protestant Reformation of the 15th and 16th centuries split the church in the west, leaving Roman Catholicism supreme in southern Europe but installing a variety of Protestant denominations and national churches in western and northern Europe.

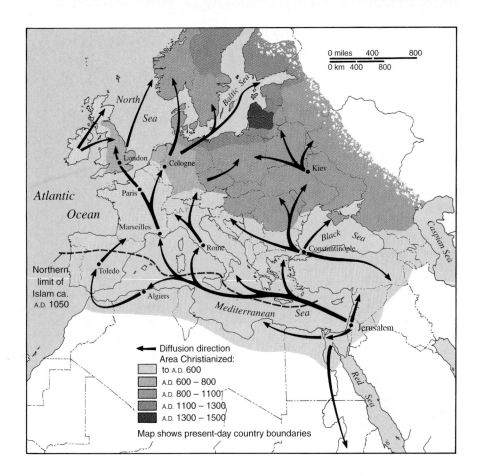

Figure 5.22 **Diffusion paths of Christianity, A.D. 100–1500.** Routes and dates are for Christianity as a composite faith. No distinction is made between the Western Church and the various subdivisions of the Eastern Orthodox denominations.

The split was reflected in the subsequent worldwide dispersion of Christianity. Catholic Spain and Portugal colonized Latin America, taking both their languages and the Roman church to that area (see Figure 5.20), as they did to colonial outposts in the Philippines, India, and Africa. Catholic France colonized Quebec in North America. Protestants, many of them fleeing Catholic or repressive Protestant state churches, were primary early settlers of Anglo America, Australia, New Zealand, Oceania, and South Africa.

In Africa and Asia, both Protestant and Catholic missions attempted to convert nonbelievers. Both achieved success in sub-Saharan Africa, though traditional religions are shown on Figure 5.18 as dominant through much of that area. Neither was particularly successful in China, Japan, or India, where strong ethnic religious cultural systems were barriers largely impermeable to the diffusion of the Christian faith. Although accounting for nearly one-third of the world's population and territorially the most extensive belief system, Christianity is no longer numerically important in or near its original hearth. Nor is it any longer dominated by Northern Hemisphere adherents. In 1900, 80% of all Christians lived in Europe and North America; in 2000, two-thirds of an estimated 2.1 billion total lived elsewhere—in South America, Africa, and Asia.

Regions and Landscapes of Christianity

All of the principal world religions have experienced theological, doctrinal, or political divisions; frequently these have spatial expression. In Christianity, the early split between the Western and Eastern Churches was initially unrelated to dogma but nonetheless resulted in a territorial separation still evident on the world map. The later subdivision of the Western Church into Roman Catholic and Protestant branches gave a more intricate spatial patterning in western Europe that can only be generally suggested at the scale of Figure 5.18. Still more intermixed are the areal segregations and concentrations that have resulted from the denominational subdivisions of Protestantism.

In Anglo America, the beliefs and practices of various immigrant groups and the innovations of domestic congregations have created a particularly varied spatial patterning (Figure 5.24), though intermingling rather than rigid territorial division is characteristic of the North American, particularly United States, scene (see Figure 5.1). While 85% of Canadian Christians belong to one of three denominations (Roman Catholic, Anglican, or United Church of Canada), it takes at least 20 denominations to account for 85% of religious adherents in America. Nevertheless, for the United States, one observer has suggested a pattern of

Figure 5.23 The building of Notre Dame Cathedral of Paris, France, begun in 1163, took more than 100 years to complete. Perhaps the best known of the French Gothic churches, it was part of the great period of cathedral construction in Western Europe during the late 12th and the 13th centuries. Between 1170 and 1270, some 80 cathedrals were constructed in France alone. The cathedrals were located in the center of major cities; their plazas were the sites of markets, public meetings, morality plays, and religious ceremonies. They were the focus of public and private life and the symbol not only of the faith but of the pride and prosperity of the towns and regions that erected them.

"religious regions" of the country (Figure 5.25) that, he believes, reflects a larger cultural regionalization of the United States.

Strongly French-, Irish-, and Portuguese-Catholic New England, the Hispanic-Catholic Southwest, and the French-Catholic vicinity of New Orleans (evident on Figure 5.25) are commonly recognized regional subdivisions of the United States. Each has a cultural identity that includes, but is not limited to, its dominant religion. The western area of Mormon (more properly, Church of Jesus Christ of Latter-day Saints, or LDS) cultural and religious dominance is prominent and purely American. The Baptist presence in the South and that of the Lutherans in the Upper Midwest (see Figure 5.24a) help determine the boundaries of other distinctive composite regions. The zone of cultural mixing across the center of the country from the Middle Atlantic states to the western LDS region—so evident in the linguistic geography of the United States (see Figure 5.13)—is again apparent on the map of religious affiliation. No single church or denomination dominates, a characteristic as well of the Far Western zone.

Indeed, in no large section of the United States is there a denominational dominance to equal the overwhelming (over 88%) Roman Catholic presence in Quebec suggested, on Figure 5.24b, by the absence of any "second rank" religious affiliation. The

"leading" position of the United Church of Canada in the Canadian West or of the Anglican Church in the Atlantic region of Newfoundland is much less commanding. Much of interior Canada shows a degree of cultural mixing and religious diversity only hinted at by Figure 5.24b, where only the largest church memberships are noted.

The mark of Christianity on the cultural landscape has been prominent and enduring. In pre-Reformation Catholic Europe, the parish church formed the center of life for small neighborhoods of every town, and the village church was the centerpiece of every rural community. In York, England, with a population of 11,000 in the 14th century, there were 45 parish churches, one for each 250 inhabitants. In addition, the central cathedral served simultaneously as a glorification of God, a symbol of piety, and the focus of religious and secular life. The Spanish Laws of the Indies (1573) perpetuated that landscape dominance in the New World, decreeing that all Spanish American settlements should have a church or cathedral on a central plaza (Figure 5.26a).

While in Europe and Latin America a single dominant central church was the rule, North American Protestantism placed less importance on the church edifice as a monument and urban symbol. The structures of the principal denominations of colonial

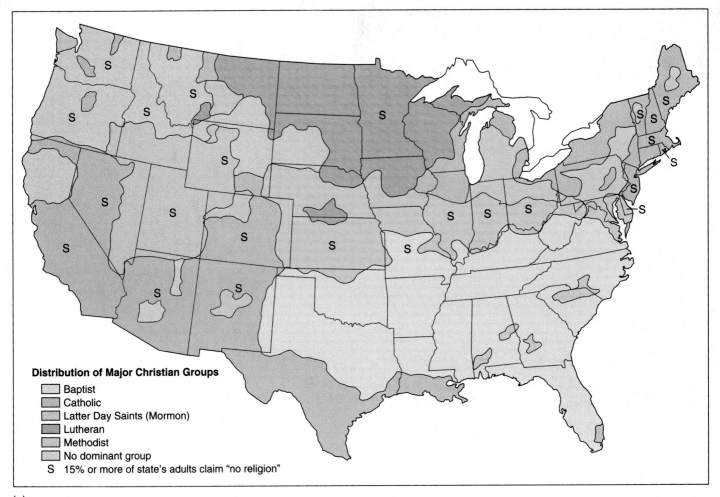

Distribution of Major Christian Groups

- Baptist
- Catholic
- Latter Day Saints (Mormon)
- Lutheran
- Methodist
- No dominant group
- S — 15% or more of state's adults claim "no religion"

(a)

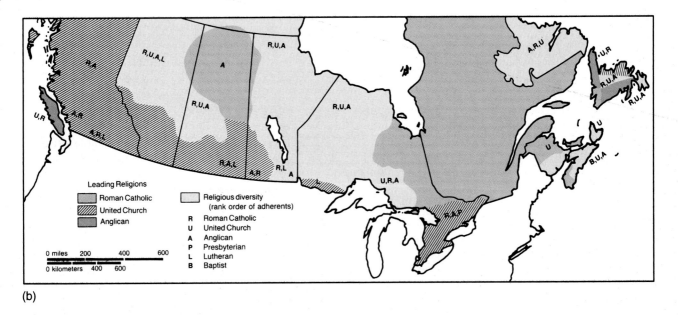

Leading Religions

- Roman Catholic
- United Church
- Anglican
- Religious diversity (rank order of adherents)

R — Roman Catholic
U — United Church
A — Anglican
P — Presbyterian
L — Lutheran
B — Baptist

0 miles 200 400 600
0 kilometers 400 600

(b)

Figure 5.24 (*a*) **Religious affiliation in the conterminous United States.** The greatly generalized areas of religious dominance shown conceal the reality of immense diversity of church affiliations throughout the United States. "Major" simply means that the indicated category had a higher percentage response than any other affiliation; in practically no case was that as much as 50%. A sizable number of Americans claim to have "no religion." Secularism (marked by S on the map) is particularly prominent in the Western states, in the industrial Midwest, and in the Northeast. (*b*) **Religious affiliation in Canada.** The richness of Canadian religious diversity is obscured by the numerical dominance of a small number of leading Christian denominations.

Sources: (a) Redrawn with permission from "Christian Denominations in the Conterminous United States," in Historical Atlas of the Religions of the World, *ed. Isma'il R. al-Faruqi and David E. Sopher (New York: Macmillan, 1974); (b) Based on Statistics Canada,* Population: Religion *(Ottawa, 1984): and* The National Atlas of Canada.

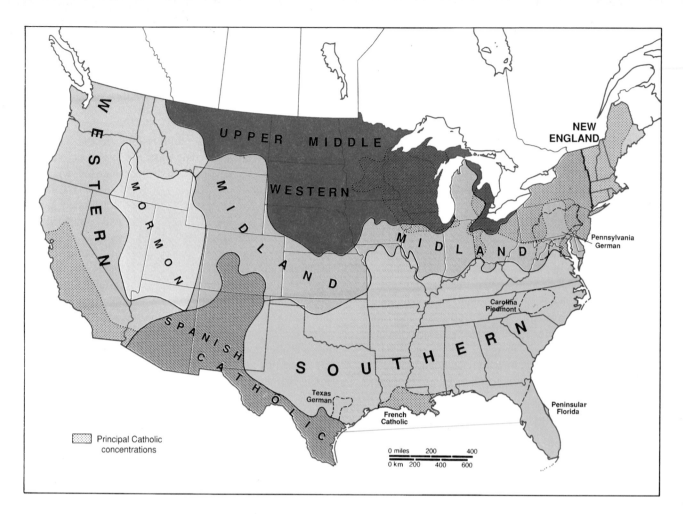

Figure 5.25 **Major religious regions of the United States.**

Source: Redrawn with permission from Annals of the Association of American Geographers, *Wilbur Zelinsky, Vol. 51, Association of American Geographers, 1961.*

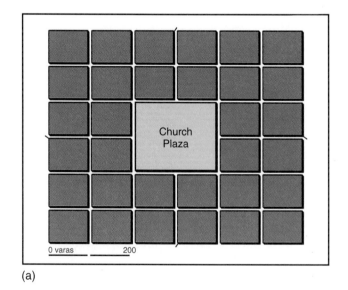

(a)

(b)

Figure 5.26 In Christian societies, the church assumes a prominent central position in the cultural landscape. (*a*) By royal decree, Spanish planned settlements in the New World were to focus on cathedral and plaza centered within a gridiron street system. On average, 1 vara equals about 84 centimeters 33 inches). (*b*) Individually less imposing than the central cathedral of Catholic areas, the several Protestant churches common in small and large Anglo American towns collectively constitute an important land use, frequently seeking or claiming space in the center of the community. The distinctive New England spired church became a model for Protestant edifices elsewhere in the United States and a symbol of religion in national life.

New England were, as a rule, clustered at the village center (Figure 5.26b), and that centrality remained a characteristic of small-town America to the present. In earlier periods, too, they were often adjoined by a cemetery, for Christians—in common with Muslims and Jews—practice burial in areas reserved for the dead. In Christian countries in particular, the cemetery—whether connected to the church, separate from it, or unrelated to a specific denomination—has traditionally been a significant land use within urban areas. Frequently, the separate cemetery, originally on the outskirts of the community, becomes with urban expansion a more central land use and often one that distorts or blocks the growth of the city.

One striking aspect of the U.S. religious landscape is the great number of different churches and church-related structures in a wide diversity of architectural styles that are found throughout urban areas. In addition to the formal structures of the established denominations, innumerable storefront churches—mainly associated with poorer neighborhoods, changing ethnic urban communities (often composed of Muslim, Hindu, Buddhist, and other immigrant groups) and splinter Protestant sects—have become part of the American scene. Nearly uniquely in the United States, the "megachurch" with thousands of congregants and the near-mandatory parking lot for all churches regardless of size are promi-nent in the land use structure of the urban and suburban scene. Also distinctly, though not exclusively, American is the proliferation of religious and denominational signage (see Figure 5.1) on city buildings, storefronts, or highway billboards.

Islam

Islam—the word means "submission" (to the will of God)—springs from the same Judaic roots as Christianity and embodies many of the same beliefs: There is only one God, who may be revealed to humans through prophets; Adam was the first human; Abraham was one of his descendants. Mohammed is revered as the prophet of *Allah* (God), succeeding and completing the work of earlier prophets of Judaism and Christianity, including Moses, David, and Jesus. The Koran, the word of Allah revealed to Mohammed, contains not only rules of worship and details of doctrine but also instructions on the conduct of human affairs. For fundamentalists, it thus becomes the unquestioned guide to matters both religious and secular. Observance of the "five pillars" (Figure 5.27) and surrender to the will of Allah unites the faithful into a brotherhood that has no concern with race, color, or caste.

That law of brotherhood served to unify an Arab world sorely divided by tribes, social ranks, and multiple local deities.

Figure 5.27 Worshipers gathered during *hajj,* the annual pilgrimage to Mecca. The black structure is the Ka'ba, the symbol of God's oneness and of the unity of God and humans. Many rules concerning daily life are given in the Koran, the holy book of the Muslims. All Muslims are expected to observe the five pillars of the faith: (1) repeated saying of the basic creed; (2) prayers five times daily, facing Mecca; (3) a month of daytime fasting (Ramadan); (4) almsgiving; and, (5) if possible, a pilgrimage to Mecca.

Mohammed was a resident of Mecca but fled in A.D. 622 to Medina, where the Prophet proclaimed a constitution and announced the universal mission of the Islamic community. That flight—*Hegira*—marks the starting point of the Islamic (lunar) calendar. By the time of Mohammed's death in 11 A.H. (anno—the year of—Hegira, or A.D. 632), all of Arabia had joined Islam. The new religion swept quickly by *expansion diffusion* outward from that source region over most of Central Asia and, at the expense of Hinduism, into northern India (Figure 5.28).

The advance westward was particularly rapid and inclusive in North Africa. In western Europe, 700 years of Muslim rule in much of Spain were ended by Christian reconquest in 1492. In eastern Europe, conversions made under an expansionary Ottoman Empire are reflected in Muslim components in Bosnia and Kosovo regions of former Yugoslavia, in Bulgaria, and in the 70% Muslim population of Albania. Later, by *relocation diffusion*, Islam was dispersed into Indonesia, southern Africa, and the Western Hemisphere. Muslims now form the majority population in 39 countries.

Asia has the largest absolute number and Africa the highest proportion of Muslims among its population—more than 42%. Islam, with an estimated 1.25 billion adherents worldwide, is the fastest-growing major religion at the present time and a prominent element in recent and current political affairs. Sectarian hatreds fueled the 1980–1988 war between Iran and Iraq; Afghan *mujahedeen*—"holy warriors"—found inspiration in their faith to resist Soviet occupation of their country, and Chechens drew strength from Islam in resisting the Russian assaults on their Caucasian homeland during the 1990s and after. Islamic fundamentalism led to the 1979 overthrow of Iran's shah. Muslim separatism is a recurring theme in Philippine affairs, and militant groups seek establishment of religiously rather than secularly based governments in several Muslim states. Extremist Muslim militants carried out the September 11, 2001, World Trade Center attack and other acts of terrorism.

Islam initially united a series of separate tribes and groups, but disagreements over the succession of leadership after the Prophet led to a division between two groups, Sunnis and Shi'ites. Sunnis, the majority (80% to 85% of Muslims) recognize the first four *caliphs* (originally, "successor" and later the title of the religious and civil head of the Muslim state) as Mohammed's rightful successors. The Shi'ites reject the legitimacy of the first three and believe that Muslim leadership rightly belonged to the fourth caliph, the Prophet's son-in-law, Ali, and his descendants. At the start of the 21st century, Sunnis constitute the majority of Muslims in all countries except Iran, Iraq, Bahrain, and perhaps Yemen.

The mosque—place of worship, community club house, meeting hall, and school—is the focal point of Islamic communal life and the primary imprint of the religion on the cultural landscape. Its principal purpose is to accommodate the Friday communal service mandatory for all male Muslims. It is the congregation rather than the structure that is important. Small or poor communities are as well served by a bare whitewashed room

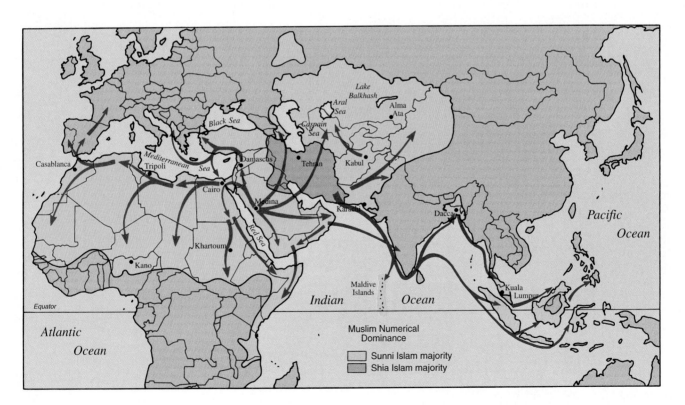

Figure 5.28 **Spread and extent of Islam.** Islam predominates in over 35 countries along a band across northern Africa to Central Asia, northwestern China, and the northern part of the Indian subcontinent. Still farther east, Indonesia has the largest Muslim population of any country. Islam's greatest development is in Asia, where it is second only to Hinduism, and in Africa, where some observers suggest it may be the leading faith. Current Islamic expansion is particularly rapid in the Southern Hemisphere.

as are larger cities by architecturally splendid mosques with domes and minarets. The earliest mosques were modeled on or converted from Christian churches. With time, however, Muslim architects united Roman, Byzantine, and Indian design elements to produce the distinctive mosque architecture found throughout the world of Islam. With its perfectly proportioned, frequently gilded or tiled domes, its graceful, soaring towers and minarets (from which the faithful are called to prayer), and its delicately wrought parapets and cupolas, the carefully tended mosque is frequently the most elaborate and imposing structure of the town (Figure 5.29).

Hinduism

Hinduism is the world's oldest major religion. Though it has no datable founding event or initial prophet, some evidence traces its origin back 4000 or more years. Hinduism is not just a religion but an intricate web of religious, philosophical, social, economic, and artistic elements comprising a distinctive Indian civilization. Its estimated 850 million to 1 billion adherents are largely confined to India, where it claims 80% of the population.

Hinduism derives its name from its cradle area in the valley of the Indus River. From that district of present-day Pakistan, it spread by *contagious diffusion* eastward down the Ganges River and southward throughout the subcontinent and adjacent regions by amalgamating, absorbing, and eventually supplanting earlier native religions and customs (see Figure 5.20). Its practice eventually spread throughout southeastern Asia, into Indonesia, Malaysia, Cambodia, Thailand, Laos, and Vietnam, as well as into neighboring Myanmar (Burma) and Sri Lanka. The largest Hindu temple complex is in Cambodia, not India, and Bali remains a Hindu pocket in dominantly Islamic Indonesia. Hinduism's more recent growing presence in western Europe and North America reflects a *relocation diffusion* of its adherents.

No common creed, single doctrine, or central ecclesiastical organization defines the Hindu. A Hindu is one born into a caste, a member of a complex social and economic—as well as religious—community. Hinduism accepts and incorporates all forms of belief; adherents may believe in one god or many or none. It emphasizes the divinity of the soul and is based on the concepts of reincarnation and passage from one state of existence to another in an unending cycle of birth and death in which all

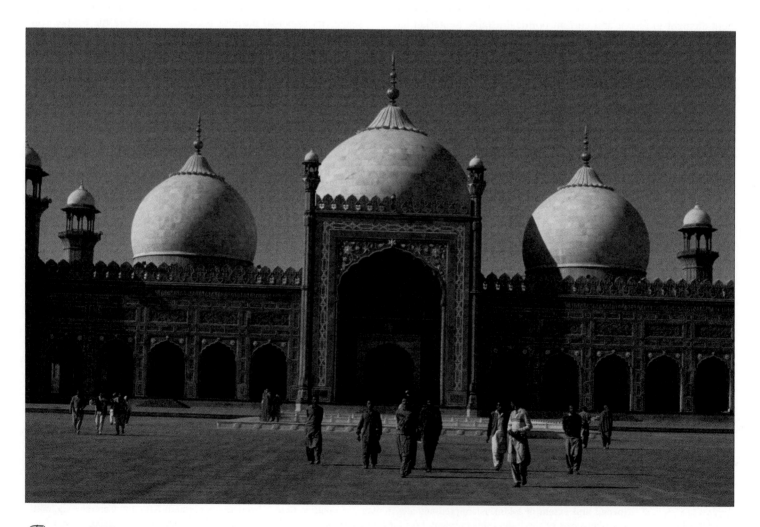

Figure 5.29 The common architectural features of the mosque make it an unmistakable landscape evidence of the presence of Islam in any local culture. The Badashi Mosque in Lahore, Pakistan, would not be out of place architecturally in Muslim Malaysia or Indonesia.

living things are caught. One's position in this life is determined by one's *karma,* or deeds and conduct in previous lives. Upon that conduct depends the condition and the being—plant, animal, or human—into which a soul, after a stay in heaven or hell, is reborn. All creatures are ranked, with humans at the top of the ladder. But humans themselves are ranked, and the social caste into which an individual is born is an indication of that person's spiritual status. The goal of existence is to move up the hierarchy, eventually to be liberated from the cycle of rebirth and redeath and to achieve salvation and eternal peace through union with the *Brahman,* the universal soul.

The **caste** (meaning "birth") structure of society is an expression of the eternal transmigration of souls. For the Hindu, the primary aim of this life is to conform to prescribed social and ritual duties and to the rules of conduct for the assigned caste and profession. Those requirements comprise that individual's *dharma*—law and duties. To violate them upsets the balance of society and nature and yields undesirable consequences. To observe them improves the chance of promotion at the next rebirth. Traditionally, each craft or profession is the property of a particular caste: brahmins (scholar-priests), kshatriyas (warrior-landowners), vaishyas (businessmen, farmers, herdsmen), sudras (servants and laborers). Harijans, *untouchables* for whom the most menial and distasteful tasks were reserved, and backwoods tribes—together accounting for around one-fifth of India's population—stand outside the caste system. The castes are subdivided into thousands of *jati* groups defined by geography and occupation. Caste rules define who you can mingle with, where you can live, what you may wear, eat, and drink, and how you can earn your livelihood.

The practice of Hinduism is rich with rites and ceremonies, festivals and feasts, processions and ritual gatherings of literally millions of celebrants. It involves careful observance of food and marriage rules and the performance of duties within the framework of the caste system. Pilgrimages to holy rivers and sacred places are thought to secure deliverance from sin or pollution and to preserve religious worth (Figure 5.30). In what is perhaps the largest periodic gathering of humans in the world, millions of Hindus of all castes, classes, and sects gather about once in 12 years for ritual washing away of sins in the Ganges River near Allahabad. Worship in the temples and shrines that are found in every village and the leaving of offerings to secure merit from the gods are required. The doctrine of *ahimsa*—also fundamental in Buddhism—instructs Hindus to refrain from harming any living being.

Temples and shrines are everywhere; their construction brings merit to their owners—the villages or individuals who paid for them. Temples must be erected on a site that is beautiful and auspicious, in the neighborhood of water since the gods will not come to other locations. Within them, innumerable icons of gods in various forms are enshrined, the objects of veneration, gifts, and daily care. All temples have a circular spire as a

Figure 5.30 Pilgrims at dawn worship in the Ganges River at Varanasi (Banares), India, one of the seven most sacred Hindu cities and the reputed earthly capital of Siva, Hindu god of destruction and regeneration. Hindus believe that to die in Varanasi means release from the cycle of rebirth and permits entrance into heaven.

Figure 5.31 The Hindu temple complex at Khajraho in central India. The creation of temples and the images they house has been a principal outlet of Indian artistry for more than 3000 years. At the village level, the structure may be simple, containing only the windowless central cell housing the divine image, a surmounting spire, and the temple porch or stoop to protect the doorway of the cell. The great temples, of immense size, are ornate extensions of the same basic design.

reminder that the sky is the real dwelling place of the god who temporarily resides within the temple (Figure 5.31). The temples, shrines, daily rituals and worship, numerous specially garbed or marked holy men and ascetics, and the ever-present sacred animals mark the cultural landscape of Hindu societies—a landscape infused with religious symbols and sights that are part of a total cultural experience.

Buddhism

Numerous reform movements have derived from Hinduism over the centuries, some of which have endured to the present day as major religions on a regional or world scale. *Jainism,* begun in the 6th century B.C. as a revolt against the authority of the early Hindu doctrines, rejects caste distinctions and modifies concepts of karma and transmigration of souls; it counts perhaps 4 million adherents. *Sikhism* developed in the Punjab area of northwestern India in the late 15th century A.D., rejecting the formalism of both Hinduism and Islam and proclaiming a gospel of universal toleration. The great majority of some 20 million Sikhs still live in India, mostly in the Punjab, though others have settled in Malaysia, Singapore, East Africa, the United Kingdom, and North America.

The largest and most influential of the dissident movements has been **Buddhism,** a universalizing faith founded in the 6th

century B.C. in northern India by Siddhartha Gautama, the Buddha (*Enlightened One*). The Buddha's teachings were more a moral philosophy that offered an explanation for evil and human suffering than a formal religion. He viewed the road to enlightenment and salvation to lie in understanding the "four noble truths": existence involves suffering; suffering is the result of desire; pain ceases when desire is destroyed; the destruction of desire comes through knowledge of correct behavior and correct thoughts. In Buddhism, which retains the Hindu concept of *karma,* the ultimate objectives of existence are the achievement of *nirvana,* a condition of perfect enlightenment, and cessation of successive rebirths. The Buddha instructed his followers to carry his message as missionaries of a doctrine open to all castes, for no distinction among people was recognized. In that message, all could aspire to ultimate enlightenment, a promise of salvation that raised the Buddha in popular imagination from teacher to savior and Buddhism from philosophy to universalizing religion.

Contact or *contagious diffusion* spread the belief system throughout India, where it was made the state religion in the 3rd century B.C. It was carried elsewhere into Asia by missionaries, monks, and merchants. While expanding abroad, Buddhism began to decline at home as early as the 4th century A.D., slowly but irreversibly reabsorbed into a revived Hinduism. By the 8th century, its dominance in northern India was broken by conversions to

Islam; by the 15th century, it had essentially disappeared from all of the subcontinent.

Present-day spatial patterns of Buddhist adherence reflect the schools of thought, or *vehicles,* that were dominant during different periods of dispersion of the basic belief system (Figure 5.32). Earliest, most conservative, and closest to the origins of Buddhism was *Theravada* (Vehicle of the Elders) Buddhism, which was implanted in Sri Lanka and Southeast Asia beginning in the 3rd century B.C. Its emphasis is on personal salvation through the four noble truths; it mandates a portion of life to be spent as monk or nun.

Mahayana (Greater Vehicle) was the dominant tradition when Buddhism was accepted into East Asia—China, Korea, and Japan—in the 4th century A.D. and later. Itself subdivided and diversified, Mahayana Buddhism considers the Buddha divine and, along with other deities, a savior for all who are truly devout. It emphasizes meditation (contemplative Zen Buddhism is a variant form), does not require service in monasteries, and tends to be more polytheistic and ritualistic than does Theravada Buddhism.

Vajrayana (the Diamond Vehicle) was dominant when the conversion of Tibet and neighboring northern areas began, first in the 7th century and again during the 10th and 11th centuries as a revived Lamaist tradition. That tradition originally stressed self-discipline and conversion through meditation and the study of philosophy, but it later became more formally monastic and ritualistic, elevating the Dalai Lama as the reincarnated Buddha, who became both spiritual and temporal ruler. Before Chinese conquest and the flight of the Dalai Lama in 1959, as many as one out of four or five Tibetan males was a monk whose celibacy helped keep population numbers stable. Tibetan Buddhism was further dispersed, beginning in the 14th century, to Mongolia, northern China, and parts of southern Russia.

In all of its many variants, Buddhism imprints its presence vividly on the cultural landscape. Buddha images in stylized human form began to appear in the 1st century A.D. and are common in painting and sculpture throughout the Buddhist world. Equally widespread are the three main types of buildings and monuments: the *stupa,* a commemorative shrine; the temple or pagoda enshrining an image or relic of the Buddha; and the monastery, some of them the size of small cities (Figure 5.33). Common, too, is the *bodhi* (or *bo*) tree, a fig tree of great size and longevity. Buddha is said to have received enlightenment seated under one of them at Bodh Gaya, India, and specimens have been planted and tended as an act of reverence and symbol of the faith throughout Buddhist Asia.

Buddhism has suffered greatly in Asian lands that came under communist control: Inner and Outer Mongolia, Tibet, North Korea, China, and parts of Southeast Asia. Communist

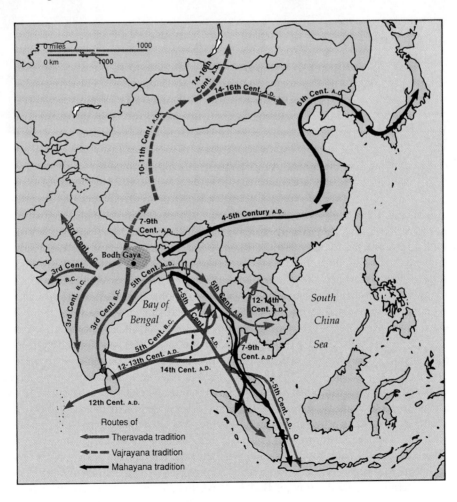

Figure 5.32 **Diffusion paths, times, and "vehicles" of Buddhism.**

Figure 5.33 The golden stupas of the Swedagon pagoda, Yangon, Myanmar (Rangoon, Burma).

governments abolished the traditional rights and privileges of the monasteries. In those states, monks were no longer prominent in numbers or presence; Buddhist religious buildings were taken over by governments and converted into museums or other secular uses, abandoned, or destroyed. In consequence, the number of adherents of Buddhism can now be only roughly and uncertainly estimated, with world totals commonly assumed to lie between 225 million and 500 million.

East Asian Ethnic Religions

When Buddhism reached China from the south some 1500 to 2000 years ago and was carried to Japan from Korea in the 7th century, it encountered and later amalgamated with already well established ethical belief systems. The Far Eastern ethnic religions are **syncretisms,** combinations of different forms of belief and practice. In China, the union was with Confucianism and Taoism, themselves becoming intermingled by the time of Buddhism's arrival. In Japan, it was with Shinto, a polytheistic animism and shamanism.

Chinese belief systems address not so much the hereafter as the achievement of the best possible way of life in the present existence. They are more ethical or philosophical than religious in the pure sense. Confucius (K'ung Fu-tzu), a compiler of traditional wisdom who lived about the same time as Gautama Buddha, emphasized the importance of proper conduct—between ruler and subjects and among family members. The family was extolled as the nucleus of the state, and filial piety was the loftiest of virtues. There are no churches or clergy in **Confucianism,** though its founder believed in a Heaven seen in naturalistic terms, and the Chinese custom of ancestor worship as a mark of gratitude and respect was encouraged. After his death, the custom was expanded to include worship of Confucius himself in temples erected for that purpose. That worship became the official state religion in the 2nd century B.C., and for some 2000 years—until the start of the 20th century A.D.—Confucianism, with its emphasis on ethics and morality rooted in Chinese traditional wisdom, formed the basis of the belief system of China.

It was joined by, or blended with, **Taoism,** an ideology that according to legend was first taught by Lao-tsu in the 6th century B.C.

Its central theme is *Tao*, the Way, a philosophy teaching that eternal happiness lies in total identification with nature and deploring passion, unnecessary invention, unneeded knowledge, and government interference in the simple life of individuals. Beginning in the 1st century A.D. this philosophical naturalism was coupled with a religious Taoism involving deities, spirits, magic, temples, and priests. Buddhism, stripped by Chinese pragmatism of much of its Indian otherworldliness and defining a *nirvana* achievable in this life, was easily accepted as a companion to these traditional Chinese belief systems. Along with Confucianism and Taoism, Buddhism became one of the honored Three Teachings, and to the average person, there was no distinction in meaning or importance between a Confucian temple, Taoist shrine, or Buddhist stupa.

Buddhism also joined and influenced Japanese Shinto, the traditional religion of Japan that developed out of nature and an- cestor worship. **Shinto**—The Way of the Gods—is basically a structure of customs and rituals rather than an ethical or moral system. It observes a complex set of deities, including deified emperors, family spirits, and the divinities residing in rivers, trees, certain animals, mountains, and, particularly, the sun and moon. Buddhism, at first resisted, was later intertwined with traditional Shinto. Buddhist deities were seen as Japanese gods in a different form, and Buddhist priests formerly but no longer assumed control of most Shinto shrines. More recently, Shinto divested itself of many Buddhist influences and became, under the reign of the Emperor Meiji (1868–1912), the official state religion, emphasizing loyalty to the emperor. The centers of worship are the numerous shrines and temples in which the gods are believed to dwell and which are approached through ceremonial *torii*, or gateway arches (Figure 5.34).

Figure 5.34 A Shinto shrine, Nikko Park, Honshu Island, Japan.

Summary

Language and religion are basic threads in the web of culture. They serve to identify and categorize individuals within a single society and to separate peoples and nations of different tongues and faiths. By their pronunciation and choice of words, we quickly recognize districts of origin and educational levels of speakers of our own language and easily identify those who originally had different native tongues. In some societies, religion may serve as a similar identifier of individuals and groups who observe distinctive modes or rhythms of life dictated by their separate faiths. Both language and religion are mentifacts, parts of the ideological subsystem of culture; both are transmitters of culture as well as its identifiers. Both have distinctive spatial patterns—reflecting past and present processes of spatial interaction and diffusion—that are basic to the recognition of world culture realms.

Languages may be grouped genetically—by origin and historical development—but the world distribution of language families depends as much on the movement of peoples and histories of conquest and colonization as it does on patterns of linguistic evolution. Linguistic geography studies spatial variations in languages, variations that may be minimized by encouragement of standard and official languages or overcome by pidgins, creoles, and lingua francas. Toponymy, the study of place names, helps document that history of movement.

Religion is a less pronounced identifier or conveyer of culture than is language. While language characterizes all peoples, religion varies in its impact and influence on culture groups. Some societies are dominated in all aspects by their controlling religious belief: Hindu India, for example, or Islamic Iran. Where religious beliefs are strongly held, they can unite a society of adherents and divide nations and peoples holding divergent faiths. Although religions do not lend themselves to easy classification,

their patterns of distribution are as distinct and revealing as are those of languages. They, too, reflect past and present patterns of migration, conquest, and diffusion, part of the larger picture of dynamic cultural geography.

While each is a separate and distinct thread of culture, language and religion are not totally unrelated. Religion can influence the spread of languages to new peoples and areas, as Arabic, the language of the Koran, was spread by conquering armies of Muslims. Religion may conserve as well as disperse language. Yiddish remains the language of religion in Hasidic Jewish communities; church services in German or Swedish, and school instruction in them, characterize some Lutheran congregations in Anglo America. Until the 1960s, Latin was the language of liturgy in the Roman Catholic Church and Sanskrit remains the language of the Vedas, sacred in Hinduism. Sacred texts may demand the introduction of an alphabet to nonliterate societies: the Roman alphabet follows Christian missionaries, Arabic script accompanies Islam. The Cyrillic alphabet of eastern Europe was developed by missionaries. The tie between language and religion is not inevitable. The French imposed their language but not their religion on Algeria; Spanish Catholicism but not the Spanish language became dominant in the Philippines.

Language and religion are important and evident components of spatial cultural variation. They are, however, only part of the total complex of cultural identities that set off different social groups. Prominent among those identities is that of *ethnicity,* a conviction of members of a social group that they have distinctive characteristics in common that significantly distinguish and isolate them from the larger population among which they reside. Our attention next turns in Chapter 6 to the concept and patterns of ethnicity, a distinctive piece in the mosaic of human culture.

Key Words

animism 167	language 145	Shinto 183
Buddhism 180	language family 146	social dialect 156
caste 179	lingua franca 159	speech community 154
Christianity 170	linguistic geography 156	standard language 154
Confucianism 182	monotheism 166	syncretism 182
creole 159	multilingualism 160	Taoism 182
dialect 155	official language 160	toponym 162
ethnic religion 167	pidgin 158	toponymy 162
geographic (regional) dialect 156	polytheism 166	tribal (traditional) religion 167
Hinduism 178	protolanguage 146	universalizing religion 167
Islam 176	religion 166	vernacular 156
isogloss 156	secularism 169	
Judaism 169	shamanism 167	

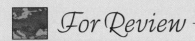 *For Review*

1. Why might one consider language the dominant differentiating element of culture separating societies?

2. In what way can religion affect other cultural traits of a society? In what cultures or societies does religion appear to be a growing influence? What might be the broader social or economic consequences of that growth?

3. In what way does the concept of *protolanguage* help us in linguistic classification? What is meant by *language family*? Is *genetic* classification of language an unfailing guide to spatial patterns of languages? Why or why not?

4. What spatial diffusion processes may be seen in the prehistoric and historic spread of languages? What have been the consequences of language spread on world linguistic diversity?

5. In what ways do *isoglosses* and the study of *linguistic geography* help us understand other human geographic patterns?

6. Cite examples that indicate the significance of religion as a cultural dominant in the internal and foreign relations of nations.

7. How does the classification of religions as *universalizing, ethnic,* or *tribal* help us to understand their patterns of distribution and spatial diffusion?

8. What connection, if any, do you see between language, religion, and intergroup rivalry and violence in the contemporary world?

 Focus Follow-up

Language

1. **How are the world's languages classified and distributed?** pp. 144–154.

   The some 6000 languages spoken today may be grouped within a limited number of language families that trace their origins to common protolanguages. The present distribution of tongues reflects the current stage of continuing past and recent dispersion of their speakers and their adoption by new users. Languages change through isolation, migration, and the passage of time.

2. **What are standard languages and what kinds of variants from them can be observed?** pp. 154–161.

   All speakers of a given language are members of its speech community, but not all use the language uniformly. The standard language is that form of speech that has received official sanction or acceptance as the "proper" form of grammar and pronunciation. Dialects, regional and social, represent nonstandard or vernacular variants of the common tongue. A pidgin is a created, composite, simple language designed to promote exchange between speakers of different tongues. When evolved into a complex native language of a people, the pidgin has become a creole. Governments may designate one or more official state languages (including, perhaps, a creole such as Swahili).

3. **How does language serve as a cultural identifier and landscape artifact?** pp. 161–165.

   Language is a mentifact, a part of the ideological subsystem of culture. It is, therefore, inseparable from group identity and self-awareness. Language may also be divisive, creating rifts within multilingual societies when linguistic minorities seek recognition or separatism. Toponyms (place names) record the order past and present occupants have tried to place on areas they inhabit or transit. Toponymy in tracing that record becomes a valuable tool of historical cultural geography.

Religion

4. **What is the cultural role of religion?** pp. 165–166.

   Like language, religion is a basic identifying component of culture, a mentifact that serves as a cultural rallying point. Frequently, religious beliefs and adherence divide and alienate different groups within and among societies. Past and present belief systems of a culture may influence its legal norms, dietary customs, economic patterns, and landscape imprints.

5. **How are religions classified and distributed?** pp. 166–169.

   As variable cultural innovations, religions do not lend themselves to easy clustering or classification. Distinctions among universalizing, ethnic, and traditional religions have some geographic significance, but geographers are more interested in religions' spatial patterns and diffusion processes and landscape impacts than in their theologies. Those patterns reflect their origin areas, the migrations and conquests achieved by their past adherents, and the converts they have attracted in home and distant areas.

6. **What are the principal world religions and how are they distinguished in patterns of innovation, diffusion, and landscape imprint?** pp. 169–183.

   The text briefly traces those differing origins, spreads, and cultural landscape impacts of Judaism, Christianity, Islam, Hinduism, Buddhism, and certain East Asian ethnic religions.

Selected References

Beinart, Haim. *Atlas of Medieval Jewish History*. New York: Simon & Schuster, 1992.

Bryson, Bill. *The Mother Tongue: English and How It Got That Way*. New York: Morrow, 1990.

Cartwright, Don. "Expansion of French Language Rights in Ontario, 1968–1993." *The Canadian Geographer/Le Géographe canadien* 40, no. 3 (1996): 238–257.

Carver, Craig. *American Regional Dialects: A Word Geography*. Ann Arbor: University of Michigan Press, 1987.

Chadwick, Henry, and G. R. Evans, eds. *Atlas of the Christian Church*. New York: Facts on File Publications, 1987.

Comrie, Bernard, ed. *The World's Major Languages*. New York: Oxford University Press, 1990.

Cooper, Robert L., ed. *Language Spread: Studies in Diffusion and Social Change*. Bloomington: Indiana University Press, 1982.

Crystal, David, ed. *The Cambridge Encyclopedia of the English Language*. Cambridge, England: Cambridge University Press, 1995.

Crystal, David. *The Cambridge Encyclopedia of Language*. 2d ed. Cambridge, England: Cambridge University Press, 1997.

Crystal, David. "Vanishing Languages." *Civilization*. February/March, 1997: 40–45.

Crystal, David. *Language Death*. Cambridge, England: Cambridge University Press, 2000.

Dutt, Ashok K., and Satish Davgun. "Patterns of Religious Diversity." In *India: Cultural Patterns and Processes*, edited by Allen G. Noble and Ashok K. Dutt, pp. 221–246. Boulder, Colo.: Westview Press, 1982.

Edwards, Viv. *Language in a Black Community*. San Diego, Calif.: College-Hill Press, 1986.

Encyclopedia of World Religions. Wendy Doniger, consulting editor. Springfield, Mass.: Merriam-Webster, Inc., 1999.

al-Faruqi, Isma'il R., and Lois L. al-Faruqi. *The Cultural Atlas of Islam*. New York: Macmillan, 1986.

al-Faruqi, Isma'il R., and David E. Sopher, eds. *Historical Atlas of the Religions of the World*. New York: Macmillan, 1974.

"Focus: Geography and Names." *The Professional Geographer* 49, no. 4 (1997): 465–500.

Freeman-Grenville, G. S. P., and Stuart Munro-Hay. *Historical Atlas of Islam*. New York: Continuum, 2002.

Gamkrelidze, Thomas V., and V. V. Ivanov. "The Early History of Indo-European Languages." *Scientific American* 262 (March, 1990): 110–116.

Gaustad, Edwin Scott, and Philip L. Barlow, with Richard W. Dishno. *The New Historical Atlas of Religion in America*. New York: Oxford University Press, 2001.

Hall, Robert A., Jr. *Pidgin and Creole Languages*. Ithaca, N.Y.: Cornell University Press, 1966.

Halvorson, Peter L., and William M. Newman. *Atlas of Religious Change in America, 1952–1990*. Atlanta, Ga.: Glenmary Research Center, 1994.

Hill, Samuel S. "Religion and Region in America." *Annals, American Academy of Political and Social Science* 480 (July, 1985): 132–141.

Jones, Dale E., et al. *Religious Congregations and Membership in the United States 2000*. Atlanta, Ga.: Glenmary Research Center, 2002.

Journal of Cultural Geography (Popular Culture Association and The American Cultural Association) 7, no. 1 (Fall/Winter, 1986). Special issue devoted to geography and religion.

Kaplan, David H. "Population and Politics in a Plural Society: The Changing Geography of Canada's Linguistic Groups." *Annals of the Association of American Geographers* 84, no. 1 (March, 1994): 46–67.

Katzner, Kenneth. *The Languages of the World*. Rev. ed. London: Routledge & Kegan Paul, 1986.

Key, Mary Ritchie. *Male/Female Language*. Metuchen, N.J.: The Scarecrow Press, 1975.

King, Noel Q. *African Cosmos: An Introduction to Religion in Africa*. Belmont, Calif.: Wadsworth, 1986.

Lind, Ivan. "Geography and Place Names." In *Readings in Cultural Geography*, edited by Philip L. Wagner and Marvin Mikesell, pp. 118–128. Chicago: University of Chicago Press, 1962.

McCrum, Robert, William Cran, and Robert MacNeil. *The Story of English*. New York: Elizabeth Sifton Books/Viking, 1986.

Moseley, Christopher, and R. E. Asher, eds. *Atlas of the World's Languages*. London, England, and New York: Routledge, 1994.

"Native American Geographic Names." Special issue of *Names*, vol. 44, no. 4 (Dec., 1996).

Newman, William M., and Peter Halvorson. *Atlas of American Religion: The Denominational Era, 1776–1990*. Walnut Creek, Calif.: Alta Mira Press, 2000.

Numrich, Paul D. "Recent Immigrant Religions in a Restructuring Metropolis: New Religious Landscapes in Chicago." *Journal of Cultural Geography* 17, no. 1 (Fall/Winter, 1997): 55–76.

Park, Chris. *Sacred Worlds: An Introduction to Geography and Religion*. New York: Routledge, 1994.

Rayburn, Alan. *Naming Canada: Stories about Place Names from Canadian Geographic*. Toronto: University of Toronto Press, 1994.

Renfrew, Colin. "World Linguistic Diversity." *Scientific American* 270, no. 1 (January, 1994): 116–123.

Scott, Jamie, and Paul Simpson-Housley, eds. *Sacred Places and Profane Spaces: Essays in the Geographics of Judaism, Christianity, and Islam*. New York: Greenwood Press, 1991.

Simoons, Frederick J. *Eat Not This Flesh: Food Avoidances from Prehistory to the Present.* 2d ed. Madison: University of Wisconsin Press, 1994.

Sloane, David Charles. *The Last Great Necessity: Cemeteries in American History.* Baltimore, Md.: Johns Hopkins University Press, 1991.

Sopher, David E. *The Geography of Religions.* Englewood Cliffs, N.J.: Prentice Hall, 1967.

Stewart, George R. *Names on the Globe.* New York: Oxford University Press, 1975.

Stewart, George R. *Names on the Land.* 4th ed. San Francisco: Lexikos, 1982.

Trudgill, Peter. *On Dialect: Social and Geographical Perspectives.* Oxford, England: Basil Blackwell, 1983.

Weatherford, Jack. *Native Roots: How the Indians Enriched America.* New York: Ballantine Books, 1991.

Williamson, Juanita V., and Virginia M. Burke, eds. *A Various Language: Perspectives on American Dialects.* New York: Holt, Rinehart and Winston, 1971.

Wurm, Stephen A., ed., Ian Heyward, cart. *Atlas of the World's Languages in Danger of Disappearing.* 2d ed. Paris: UNESCO, 2001.

Zelinsky, Wilbur. "Some Problems in the Distribution of Generic Terms in the Place-Names of the Northeastern United States." *Annals of the Association of American Geographers* 45 (1955): 319–349.

Zelinsky, Wilbur. "The Uniqueness of the American Religious Landscape." *Geographical Review* 91, no. 3 (July, 2001): 565–585.

Websites: The World Wide Web has a tremendous number and variety of sites pertaining to geography. Websites relevant to the subject matter of this chapter appear in the "Web Links" section of the On-line Learning Center associated with this book. Access it at **www.mhhe.com/fellmann8e**

Six

Ethnic Geography:
Threads of Diversity

Focus Preview

1. Ethnicity, ethnic diversity, and the changing immigration streams to multiethnic Anglo America, pp. 190–196.

2. Acculturation and the persistence of ethnic clusters and identities in Anglo America and elsewhere, pp. 196–208.

3. Anglo American and world urban ethnic diversity and patterns of segregation, pp. 209–216.

4. The landscape impacts and residues of ethnic diversity, pp. 216–223.

Opposite: Performing the Blessing Dance in Long Beach, California, during the Cambodian New Year celebration.

e must not forget that these men and women who file through the narrow gates at Ellis Island, hopeful, confused, with bundles of misconceptions as heavy as the great sacks upon their backs—we must not forget that these simple, rough-handed people are the ancestors of our descendants, the fathers and mothers of our children.

So it has been from the beginning. For a century a swelling human stream has poured across the ocean, fleeing from poverty in Europe to a chance in America. Englishman, Welshman, Scotchman, Irishman; German, Swede, Norwegian, Dane; Jew, Italian, Bohemian, Serb; Syrian, Hungarian, Pole, Greek—one race after another has knocked at our doors, been given admittance, has married us and begot our children. We could not have told by looking at them whether they were to be good or bad progenitors, for racially the cabin is not above the steerage, and dirt, like poverty and ignorance, is but skin-deep. A few hours, and the stain of travel has left the immigrant's cheek; a few years, and he loses the odor of alien soils; a generation or two, and these outlanders are irrevocably our race, our nation, our stock.[1]

The United States is a cultural composite—as increasingly are most of the countries of the world. North America's peoples include aborigine and immigrant, native born and new arrival. Had this chapter's introductory passage been written in the 21st century rather than early in the 20th, the list of foreign origins would have been lengthened to include many Latin American, African, and Asian countries as well as the European sources formerly most common.

The majority of the world's societies, even those outwardly seemingly most homogeneous, house distinctive **ethnic groups,** populations that feel themselves bound together by a common origin and set off from other groups by ties of culture, race, religion, language, or nationality. Ethnic diversity is a near-universal part of human geographic patterns; the current some 200 or so independent countries are home to at least 5000 ethnic groups. European states house increasing numbers of African and Asian immigrants and guest workers from outside their borders and have effectively become multiethnic societies. Refugees and jobseekers are found in alien lands throughout both hemispheres (Figure 6.1). Cross-border movements and resettlements in Southeast Asia and Africa are well-reported current events. European colonialism created pluralistic societies in tropical lands through introduction of both ruling elites and, frequently, nonindigenous laboring groups. Polyethnic Russia, Afghanistan, China, India, and most African countries have native—rather than immigrant—populations more characterized by racial and cultural diversity than by uniformity. Tricultural Belgium has a nearly split personality in matters political and social. The idea of an ethnically pure nation-state is no longer realistic.

Like linguistic and religious differences within societies, such population interminglings are masked by the "culture realms" shown in Figure 2.4 but are, at a larger scale, important threads in the cultural-geographic web of our complex world. The multiple movements, diffusions, migrations, and mixings of peoples of different origins making up that world are the subject of **ethnic geography.** Its concerns are those of spatial distributions and interactions of ethnic groups however defined, and of the cultural characteristics and influences underlying them.

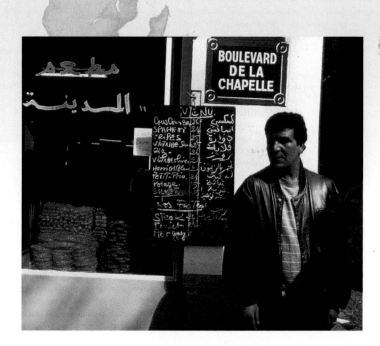

Figure 6.1 "Guest workers"—frequently called by their German name, *Gastarbeiter*—have substantially altered the ethnic mix in formerly unicultural cities of Western Europe. The restaurant shown here is in an Algerian neighborhood of Paris, France. On average, foreigners comprise nearly 10% of Western Europe's labor force. They form the majority of the work force in many Middle Eastern countries; between 60% and 90% of the workers of the Persian Gulf countries of Bahrain, Kuwait, Oman, Qatar, Saudi Arabia, and the United Arab Emirates are foreigners.

Culture, we saw in Chapter 2, is the composite of traits making up the way of life of a human group—collective beliefs, symbols, values, forms of behavior, and complexes of such nonmaterial and material traits as social customs, language, religion, food habits, tools, structures, and the like. Culture is

---

[1]From Walter E. Weyl, "The New Americans," *Harper's Magazine* 129 (1914): 615. Copyright © 1914 Harper's Magazine Foundation, New York, N.Y.

learned; it characterizes the group and distinguishes it from all other groups that have collectively created and transmitted to its children still other "ways of life." *Ethnicitiy*, in contrast, is simply the summary term of identification assigned to a large group of people recognized as sharing the traits of a distinctive common culture. It is always based on a firm understanding by members of a group that they are in some fundamental ways different from others who do not share their distinguishing characteristics or cultural heritage.

Ethnicity is, at root, a spatial concept. Ethnic groups are associated with clearly recognized territories—either larger homeland districts or smaller rural or urban enclaves—in which they are primary or exclusive occupants and upon which they have placed distinctive cultural marks. Since territory and ethnicity are inseparable concepts, ethnicity becomes an important concern in the cultural patterning of space and clearly an item of human geographic interest. Further, since ethnicity is often identified with language or religious practices setting a minority group off from a surrounding majority culture, consideration of ethnicity flows logically from the discussions of language and religion in Chapter 5.

Our examination of ethnic patterns will concentrate on Anglo America (the United States and Canada), an area originally occupied by a multitude of territorially, culturally, and linguistically distinctive Native American people who were overwhelmed and displaced by immigrants—and their descendants—representing a wide spectrum of the Old World's ethnic groups. While Anglo America lacks the homelands that gave territorial identity to immigrant ethnics in their countries of origin, it has provided a case study of how distinctive culture groups partition space and place their claims and imprints on it. It shows, as well, the durability of the idea of ethnic distinction even under conditions and national myths that emphasize intermixing and homogenization of population as the accepted norm. Examples drawn from other countries and environments will serve to highlight ways in which American-based generalizations may be applied more broadly or in which the North American experience reflects a larger world scene.

Ethnic Diversity and Separatism

Each year on a weekend in May, New York City celebrated its cultural diversity and vitality by closing off to all but pedestrian traffic a 1-mile stretch of street to conduct the Ninth Avenue International Festival. Along the reserved route from 37th to 57th streets, a million or more New Yorkers came together to sample the foods, view the crafts, and hear the music of the great number of the world's cultures represented among the citizens of the city. As a resident of the largest U.S. metropolis, each of the merchants and artists contributing one of the several hundred separate storefront, stall, or card-table displays of the festival became a member of the Anglo American culture realm. Each, however, preserved a distinctive small-group identity within that larger collective "realm" (Figure 6.2).

The threads of diversity exhibited in the festival are expressions of **ethnicity,** a term derived from the Greek word *ethnos,* meaning a "people" or "nation." Intuitively we recognize that the

Figure 6.2 The annual Ninth Avenue International Fair in New York City became one of the largest of its kind. Similar festivals celebrating America's ethnic diversity are found in cities and small towns across the country.

literal translation is incomplete. Ethnic groups are composed of individuals who share some prominent traits or characteristics, some evident physical or social identifications setting them apart both from the majority population and from other distinctive minorities among whom they may live.

No single trait denotes ethnicity. Group recognition may be based on language, religion, national origin, unique customs, or—improperly—an ill-defined concept of "race" (see "The Matter of Race"). Whatever may establish the identity of a group, the common unifying bonds of ethnicity are a shared ancestry and cultural heritage, the retention of a set of distinctive traditions, and the maintenance of in-group interactions and relationships. The principal racial and ethnic groups of the United States are identified in Tables 6.1 and 6.2 and of Canada in Table 6.4.

Ethnocentrism is the term describing a tendency to evaluate other cultures against the standards of one's own. It implies the feeling that one's own ethnic group is superior. Ethnocentrism can divide multiethnic societies by establishing rivalries and provoking social and spatial discord and isolation. It can, as well, be a sustaining and identifying emotion, giving familiar values and support to the individual in the face of the complexities of life. The ethnic group maintains familiar cultural institutions and shares traditional food and music. More often than not, it provides the friends, spouses, business opportunities, and political identification of ethnic group members.

Territorial isolation is a strong and supporting trait of ethnic separatism and assists individual groups to retain their identification. In Europe, Asia, and Africa, ethnicity and territorial identity

Human populations may be differentiated from one another on any number of bases: gender, nationality, stage of economic development, and so on. One common form of differentiation is based on recognizable inherent physical characteristics, or *race.*

A **race** is usually understood to be a population subset whose members have in common some hereditary biological characteristics that set them apart physically from other human groups. The spread of human beings over the earth and their occupation of different environments was accompanied by the development of physical variations in skin pigmentation, hair and eye color, hair texture, facial characteristics, blood composition, and other traits largely related to soft tissue. Some subtle skeletal differences among peoples also exist.

Such differences formed the basis for the segregation, by some anthropologists, of humanity into different racial groups, although recent research indicates that the greatest genetic variation between any racial groups ever identified is far less than the variation within any single population. Caucasoid, Negroid, Mongoloid, Amerindian, Australoid and other races have been recognized in a process of arbitrary classification invention, modification, and refinement that began at least two centuries ago. Racial differentiation as commonly understood—based largely on surface appearance—is old and can reasonably be dated at least to the Paleolithic (100,000 to about 11,000 years ago) spread and isolation of population groups.

Although racial classifications vary by author, most are derived from recognized geographical variations of populations. Thus,

Mongoloids are associated with northern and eastern Asia; Australoids are the aboriginal people of Australasia; Amerindians developed in the Americas, and so on. If all of humankind belongs to a single species that can freely interbreed and produce fertile offspring, how did this areal differentiation by race occur? Why is it that despite millennia of mixing and migration, people with distinct combinations of physical traits appear to be clustered in particular areas of the world?

Two causative forces appear to be most important. First, through evolutionary **natural selection** or **adaptation,** characteristics are transmitted that enable people to adapt to particular environment conditions, such as climate. Studies have suggested some plausible relationship between, for example, solar radiation and skin color, and between temperature and body size. In tropical climates, for example, it persumably is advantageous to be short since that means a greater bodily surface area for sweat to evaporate. In frigid Arctic regions, it is suggested, Inuits and other native populations developed round heads and bodies with increased bodily volume and decreased evaporative surface area.

The second force, **genetic drift,** refers to a heritable trait (such as flatness of face) that appears by chance in one group and becomes accentuated through inbreeding. If two populations are too separated spatially for much interaction to occur (*isolation*), a trait may develop in one but not in the other. Unlike natural selection, genetic drift differentiates populations in nonadaptive ways.

Natural selection and genetic drift promote racial differentiation. Countering them

is **gene flow** via interbreeding (also called *admixture*), which acts to homogenize neighboring populations. Genetically, it has been observed, there is no such thing as a "pure" race since people breed freely outside their local group. Opportunities for interbreeding, always part of the spread and intermingling of human populations, have accelerated with the growing mobility and migrations of people in the past few centuries. While we may have an urge to group humans "racially," we cannot use biology to justify it, and anthropologists have largely abandoned—and geneticists dismissed—the idea of race as a scientific concept.

Nor does race have meaningful application to any human characteristics that are culturally acquired. That is, race is *not* equivalent to ethnicity or nationality and has no bearing on differences in religion or language. There is no "Irish" or "Hispanic" race, for example. Such groupings are based on culture, not genes. Culture summarizes the way of life of a group of people, and members of the group may adopt it irrespective of their individual genetic heritage, or race. Nevertheless, despite the fact that the older view of race as a biological category has been thoroughly discredited, race and ethnicity remain as defining and divisive realities in American society. Both are deeply rooted in individual and group consciousness, and both are strongly ingrained in the country's social and institutional life. It has been observed that while biological notions of race have little meaning, the society itself is extremely "racialized."

are inseparable. Ethnic minorities are first and foremost associated with *homelands.* This is true of the Welsh, Bretons, and Basques of Western Europe; the Slovenes, Croatians, or Bosnians of Eastern Europe; the non-Slavic "nationalities" of Russia; and the immense number of ethnic communities of South and Southeast Asia. These minorities have specific spatial identity even though they may not have political independence.

Where ethnic groups are intermixed and territorial boundaries imprecise—former Yugoslavia is an example—or where a single state contains disparate, rival populations—the case of many African and Asian countries—conflict among groups can be serious if peaceful relations or central governmental control break down. "Ethnic cleansing," a polite term with grisly implications, has become a past or present justification and objective

for civil conflict in parts of the former Soviet Union and Eastern Europe and in several African and southeast Asian countries. The Holocaust slaughter of millions of Jews before and during World War II in Western and Eastern Europe was an extreme case of ethnic extermination, but comparable murderous assaults on racial or cultural target populations by conquering or controlling groups are as old as human history. Such "cleansing" involves, through mass genocide, the violent elimination of a target ethnic group from a particular geographic or political area to achieve racial or cultural homogeneity and expanded settlement area by the perpetrating state or ethnic group. Its outcome is not only an alteration of the ethnic composition of regions and states, but of the ethnic mix in, usually, adjacent areas and countries to which assaulted and displaced populations have fled as refugees.

Table 6.1
U.S. Resident Population by Race and Hispanic Origin, 2000

Race	Number (millions)	Percent of U.S. Population
One race	274.6	97.6
White	211.5	75.1
Black or African American	34.7	12.3
Asian	10.2	3.6
American Indian and Alaska Native	2.5	0.9
Native Hawaiian and Other Pacific Islander	0.4	0.1
Some other race	15.4	5.5
Two or more races	6.8	2.4
Total Population	**281.4**	**100.0**
Hispanic or Latino[a]	35.3	12.5

Note: Race as reported reflects the self-identification of respondents.
[a]Persons of Hispanic origin may be of any race.
Source: U.S. Bureau of the Census.

Table 6.2
Leading U.S. Ancestries Reported, Census 2000

Ancestry	Number (000)	Percent of Total Population
German	46,489	16.5
Irish	33,067	11.7
English	28,265	10.0
Mexican	20,641	7.3
Italian	15,943	5.7
French	10,012	3.6
Polish	9054	3.2
Scottish	5423	1.9
Scotch-Irish	5226	1.9
Swedish	4339	1.5

Note: More than 20 million persons indicated "United States" or "American" as their ancestry. The tabulation is based on self-identification of respondents, not on objective criteria. Many persons reported multiple ancestries and were tabulated by the Census Bureau under each claim.
Source: Census 2000 Supplementary Survey, "Selected Social Characteristics," and Census Summary File 1.

Few identifiable homelands exist within the North American cultural mix. However, the "Chinatowns" and "Little Italys" as created enclaves within North American cities have provided both the spatial refuge and the support systems essential to new arrivals in an alien culture realm. Asian and West Indian immigrants in London and other English cities and foreign *guest workers*—originally migrant and temporary laborers, usually male—that reside in Continental European communities assume similar spatial separation. While serving a support function, this segregation is as much the consequence of the housing market and of public and private restriction as it is simply of self-selection. In Southeast Asia, Chinese communities remain aloof from the majority culture not as a transitional phase to incorporation with it but as a permanent chosen isolation.

By retaining what is familiar of the old in a new land, ethnic enclaves have reduced cultural shock and have paved the way for the gradual process of adaptation that prepares both individuals and groups to operate effectively in the new, larger **host society.** The traditional ideal of the United States "melting pot," in which ethnic identity and division would be lost and full amalgamation of all minorities into a blended, composite majority culture would occur, was the expectation voiced in the chapter-opening quotation. For many even long-resident ethnic groups, however, that ideal has not become a reality.

Recent decades have seen a resurgence of cultural pluralism and an increasing demand for ethnic autonomy not only in North America but also in multiethnic societies around the world (see "Nations of Immigrants," p. 194). At least, recognition is sought for ethnicity as a justifiable basis for special treatment in the allo-

cation of political power, the structure of the educational system, the toleration or encouragement of minority linguistic rights, and other evidences of group self-awareness and promotion. In some multiethnic societies, second- and third-generation descendants of immigrants, now seeking "roots" and identity, embrace the ethnicity that their forebears sought to deny.

Immigration Streams

The ethnic diversity found on the Anglo American scene today is the product of continuous flows of immigrants—some 70 million of them by the start of the 21st century—representing, at different periods, movements to this continent of members of nearly all of the cultures and races of the world (Figure 6.3). For the United States, that movement took the form of three distinct immigrant waves, all of which, of course, followed much earlier Amerindian arrivals.

The first wave, lasting from pioneer settlement to about 1870, was made up of two different groups. One comprised white arrivals from western and northern Europe, with Britain and Germany best represented. Together with the Scots and Scotch-Irish, they established a majority society controlled by Protestant Anglo-Saxons and allied groups. The Europeans dominated numerically the second group of first-wave immigrants, Africans brought involuntarily to the New World, who made up nearly 20% of U.S. population in 1790. The mass immigration that occurred beginning after the middle of the 19th century

Geography and Public Policy

Nations of Immigrants

Americans, steeped in the country's "melting pot" myth and heritage, are inclined to forget that many other countries are also "nations of immigrants" and that their numbers are dramatically increasing. In the United States, Canada, Australia, and New Zealand, early European colonists and, later, immigrants from other continents overwhelmed indigenous populations. In each, immigration has continued, contributing not only to national ethnic mixes but maintaining or enlarging the proportion of the population that is foreign born. In Australia, as one example, that proportion now equals 25%; for Canada it is some 18%.

In Latin America, foreign population domination of native peoples was and is less complete and uniform than in Anglo America. While in nearly all South and Central American states, European and other nonnative ethnic groups dominate the social and economic hierarchy, in many they constitute only a minority of the total population. In Paraguay, for example, the vast majority of inhabitants are native Paraguayans who pride themselves on their Native American descent, and Amerindians comprise nearly half the population of Peru, Bolivia, and Ecuador. But European ethnics make up over 90% of the population of Argentina, Uruguay, Costa Rica, and southern Chile, and about 50% of the inhabitants of Brazil.

The original homelands of those immigrant groups are themselves increasingly becoming multiethnic, and several European countries are now home to as many or more of the foreign-born proportionately than is the United States. Some 20% of Switzerland's population, 13% of France's, 10% of Sweden's, and over 9% of Germany's are of foreign birth, compared with America's 11%. Many came as immigrants and refugees fleeing unrest or poverty in post-communist Eastern Europe. Many are "guest workers" and their families who were earlier recruited in Turkey and North Africa; or they are immigrants from former colonial or overseas territories in Asia, Africa, and the Caribbean. More than 7% of Germany's inhabitants come from outside the European Union, as do over 3% of Holland's and Belgium's.

The trend of ethnic mixing is certain to continue and accelerate. Cross-border movements of migrants and refugees in Africa, Asia, the Americas, as well as in Europe are continuing common occurrences, reflecting growing incidences of ethnic strife, civil wars, famines, and economic hardships. But of even greater long-term influence are the growing disparities in population numbers and economic wealth between the older developed states and the developing world. The population of the world's poorer countries is growing twice as fast as Europe's of the late 19th century, when that continent fed the massive immigration streams across the Atlantic. The current rich world, whose population is projected to stabilize well below 1.5 billion, will increasingly be a magnet for those from poorer countries where numbers will rise from some 4 billion to more than 6.5 billion by A.D. 2025 and to nearly 8 billion in a half-century. The economic and population pressures building in the developing world insure greater international and intercontinental migration and a rapid expansion in the numbers of "nations of immigrants."

Many of those developed host countries are beginning to resist that flow. Although the Universal Declaration of Human Rights declares individuals are to be free to move within or to leave their own countries, no right of admittance to any other country is conceded. Political asylum is often—but not necessarily—granted; refugees or migrants seeking economic opportunity or fleeing civil strife or starvation have no claims for acceptance. Increasingly, they are being turned away. The Interior Minister of France advocates "zero immigration"; Germany's government closed its doors in 1993 by increasing border controls and changing its constitutional right to asylum; Britain in 1994 tightened immigration rules even for foreign students and casual workers. And all European Union countries—which have no common EU policies on illegal immigration—have measures for turning back refugees who come via another EU country. In 1995, the EU's members materially narrowed the definition of who may qualify for asylum. Additional individual and collective restrictions have been enforced during the later 1990s and into the 21st century.

Nor is Europe alone. Hong Kong ejects Vietnamese refugees; Congo orders Rwandans to return to their own country; India tries to stem the influx of Bangladeshis; the United States rejects "economic refugees" from Haiti. Algerians are increasingly resented in France as their numbers and cultural presence increase. Turks feel the enmity of a small but violent group of Germans, and East Indians and Africans find growing resistance among the Dutch. In many countries, policies of exclusion or restriction appear motivated by unacceptable influxes of specific racial, ethnic, or national groups.

Questions to Consider

1. Do you think all people everywhere should have a universal right of admittance to a country of choice equivalent to their declared right to depart their homelands? Why or why not?

2. Do you think it appropriate that destination states make a distinction between political and economic refugees? Why or why not?

3. Do you think it legitimate for countries to establish immigration quotas based on national origin or to classify certain potential immigrants as unacceptable or undesirable on the grounds that their national, racial, or religious origins are incompatible with the culture of the prospective host country? Why or why not?

Figure 6.3 Although it was not opened until 1892, New York harbor's Ellis Island—the country's first federal immigration facility—quickly became the symbol of all the migrant streams to the United States. By the time it was closed in late 1954, it had processed 17 million immigrants. Today their descendants number over 100 million Americans. A major renovation project was launched in 1984 to restore Ellis Island as a national monument.

began to reduce both the northwest European dominance of American society and the percentage of blacks within the growing total population.

That second immigrant wave, from 1870 to 1921, was heavily weighted in favor of eastern and southern Europeans, who comprised more than 50% of new arrivals by the end of the 19th century. The second period ended with congressional adoption of a quota system regulating both the numbers of individuals who would be accepted and the countries from which they could come. That system, plus a world depression and World War II (1939–1945), greatly slowed immigration until a third-wave migration was launched during the 1960s. At that time the old national quota system of immigrant regulation was replaced by one more liberal in its admission of Latin Americans. Along with more recent Asian arrivals, they became the largest segment of new arrivals. The changing source areas of the newcomers are traced in Table 6.3 and Figure 6.4.

Canada experienced three quite different immigration streams. Until 1760, most settlers came from France. After that date, the pattern abruptly altered as a flood of United Kingdom (English, Irish, and Scottish) immigrants arrived. Many came by way of the United States, fleeing, as Loyalists, to Canada during and after the American Revolutionary War. Others came directly from overseas. Another pronounced shift in arrival pattern occurred during the 20th century as the bulk of new immigrants began originating in Continental Europe and, more recently, in other continents. By 2001, 18% of all Canadians had been born outside of the country, and immigration accounted for more than one-half of Canada's population growth between 1996 and 2001. The ten leading ethnicities in 2001 (of more than 200 different ethnic origins reported) are listed in Table 6.4.

Table 6.3

Immigrants to the United States: Major Flows by Origin

Ethnic Groups	Time Period	Numbers in Millions (approximate)
Blacks	1650s–1800	1
Irish	1840s and 1850s	1.75
Germans	1840s–1880s	4
Scandinavians	1870s–1900s	1.5
Poles	1880s–1920s	1.25
East European Jews	1880s–1920s	2.5
Austro-Hungarians	1880s–1920s	4
Italians	1880s–1920s	4.75
Mexicans	1950s–2000	10.5
Cubans	1960s–2000	1.3
Asians	1960s–2000	7.5

The United States' cultural diversity has increased as its immigration source regions have changed from the traditional European areas to Latin America and Asia. The dominant European ethnic groups had completed their major periods of arrival in the United States by the 1920s, and immigration essentially halted until after World War II. Except for a spurt of legal and illegal immigration from Eastern Europe and Russia after 1990, the modest postwar revival of inflow from Europe went largely

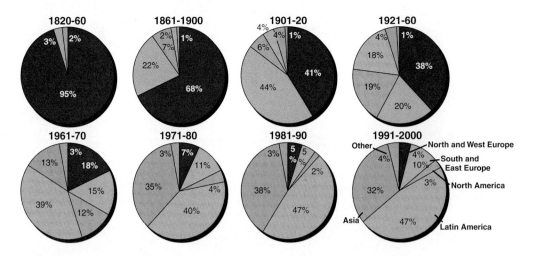

Figure 6.4 **Legal immigrants admitted to the United States by region of origin, 1820–2000.** The diagrams clearly reflect the dramatic change in geographic origins of immigrants. After 1965, immigration restrictions based on national origin were shifted to priorities based on family reunification and needed skills and professions. Those priorities underwent Congressional reconsideration in 1995 and 1996. What is not shown is the dramatic increase in the *total* numbers of entrants to United States in the 1980s and 1990s, years that witnessed the highest legal and illegal immigrant and refugee volumes in the nation's history.

Data from Leon F. Bouvier and Robert W. Gardner, "Immigration to the United States: The Unfinished Story," Population Bulletin 41, no. 4 (Washington, D.C.: Population Reference Bureau, 1986); and Immigration and Naturalization Service.

Table 6.4

Canadian Population Ranked by Claimed Ethnic Origin, 2001

Rank	Ethnic Group
1	Canadian
2	English
3	French
4	Scottish
5	Irish
6	German
7	Italian
8	Chinese
9	Ukrainian
10	North American Indian

Source: Statistics Canada.

unnoticed as the new entrants affiliated with already assimilated groups of the same cultural background.

More recent expanded immigration from new source regions has increased the number of visible and vocal ethnic communities and the regions of the country housing significant minority populations. Simultaneously, the proportion of foreign-born residents has increased in the U.S. population mix. In 1920, at the end of the period of the most active European immigration, more than 13% of the American population had been born in another country. That percentage declined each decade until a low of 4.8%

foreign born was reported in 1970. So great was the inflow of aliens after 1970, however, that by 2002 an all-time high of some 32.5 million people—11.5% of the population—had been born abroad, and over 30% of total population growth of the country between 1990 and 2000 was accounted for by legal and illegal immigration. Illegal immigration alone appears to have totaled more than 6 million during the 1990s, with more than 700,000 undocumented arrivals entering per year after 1995.

Individual cities and counties showed very high concentrations of the foreign born at the end of the century. New York City, for example, received one million immigrants in the 1990s, and by 2000, 40% of its population had been born abroad. Similar proportionate immigration flows and foreign-born ratios were recorded for Dade County (Miami) Florida, the Silicon Valley, California, counties of San Mateo and Santa Clara, and others. Monterey Park, California, has a population that is 60% Asian, the vast majority recent Chinese immigrants.

As had been the case during the 19th century, growing influxes from new immigrant source regions prompted movements to halt the flow and to preserve the ethnic status quo (see "Backlash," Chapter 3, p. 90).

Acculturation and Assimilation

In the United States, at least, the sheer volume of multiple immigration streams makes the concept of "minority" suspect when no single "majority" ethnic group exists (see Table 6.2). Indeed, high rates of immigration and declining birth rates among white Americans have placed the country on the verge of becoming a state with no racial—as well as no ethnic—majority. No later than 2050, current trends promise, America will be truly multiracial, with no group constituting more than 50% of the total population. Even now, American society is a composite of unity and diversity with immigrants being both shaped by and shaping the larger community they joined.

Amalgamation theory is the formal term for the traditional "melting pot" concept of the merging of many immigrant cultural heritages into a composite American mainstream. Popular and accepted in the late 19th and early 20th centuries, amalgamation theory has more recently been rejected by many as unrealistic in light of current widespread social and cultural tensions. Recent experience in western European countries and Anglo America—destination areas of multiple immigration flows—indicates that strongly retained and defended cultural identities are increasingly the rule and that a militant multiculturalism rather than voluntary amalgamation is a more realistic description of current conditions in immigrant destination countries. Such cultural separatism is buttressed by the current ease—through radio, telephone, Internet, television, and rapid transportation—of communication and identification with the homeland societies of immigrants who no longer are essentially divorced from their past to make new lives in an alien land. The old "melting pot" concept of America has largely dissolved, replaced with a greater emphasis on preserving the diverse cultural heritages of the country's many ethnic components.

Nonetheless, as we shall see, all immigrant groups after the first found a controlling culture in place, with accustomed patterns of behavior and response and a dominating language of the workplace and government. The customs and practices familiar and expected among those already in place had to be learned by newcomers if they were to be accepted. The process of **acculturation** is that of the adoption by the immigrants of the values, attitudes, ways of behavior, and speech of the receiving society. In the process, the ethnic group loses its separate cultural identity as it accepts over time the culture of the larger host community. Although acculturation most usually involves a minority group adopting the patterns of the dominant population, the process can be reciprocal. That is, the dominant group may also adopt at least some patterns and practices typical of the minority group.

Acculturation is a slow process for many immigrant individuals and groups, and the parent tongue may of choice or necessity be retained as an ethnically identifying feature even after fashions of dress, food, and customary behavior have been substantially altered in the new environment. In 2000, some 18% of U.S. census respondents reported speaking a language other than English in the home; for 60% of them, that language was Spanish. In the light of recent immigration trends, we can assume that the number of people speaking a foreign language at home is increasing. The retention of the native tongue is encouraged rather than hindered by American civil rights regulations that give to new immigrants the right to bilingual education and (in some cases) special assistance in voting in their own language (see "An Official U.S. Language?", p. 162).

The language barrier that has made it difficult for foreign-born groups, past and present, to gain quick entrance to the labor force has encouraged their high rate of initiation of or entry into small businesses. The consequence has been a continuing stimulus to the American economy and, through the creation of family-held neighborhood enterprises, the maintenance of the ethnic

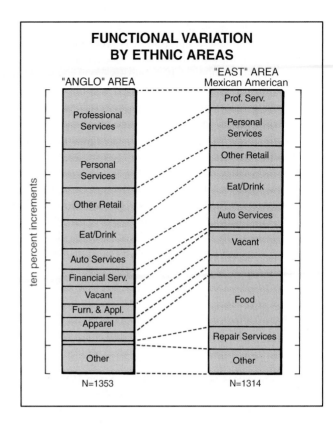

Figure 6.5 Variations in business establishments in Anglo and Mexican American neighborhoods of Los Angeles in the late 1960s. Although the total populations of the two areas were comparable, the Mexican American community had over three times more food stores because of the dominance of corner grocery stores over supermarkets. Bakeries (*tortillerías*) were a frequent expression of ethnic dietary habits. Neighborhood businesses conducted in Spanish and related to the needs of the community were the rule. Anglo neighborhoods, because of greater affluence, had larger numbers of professional services (doctors, lawyers) available. Comparable current "business directory" contrasts among the many and varied Asian, Latin American, and African immigrant communities themselves and between them and the established older majority integrated neighborhoods mark the American city scene today.

Source: Redrawn by permission from Annals of the Association of American Geographers, *Keith D. Harries, Vol. 61, p.739, Association of American Geographers, 1971.*

character of immigrant communities (Figure 6.5). The result has also been the gradual integration of the new arrivals into the economic and cultural mainstream of American society.

When that integration is complete, **assimilation** has occurred. Full assimilation may be seen as a two-part process. **Behavioral** (or **cultural**) **assimilation** is the rough equivalent of acculturation; it implies integration into a common cultural life through shared experience, language, intermarriage, and sense of history. **Structural assimilation** refers to the fusion of immigrant ethnics with the groups, social systems, and occupations of the host society and the adoption of common attitudes and values. The extent of structural assimilation is frequently measured by the degree of residential segregation that sets off the minority group from the larger general community. Employment segregation and intermarriage rates are also indicative. For most of the "old" (pre-1921 European) immigrants and their descendants, both forms of assimilation are complete. For most of the "new" (post-1960s) immigrants, acculturation is proceeding or has already occurred, but for many of them and for racial minorities as well, structural assimilation has been elusive.

Assimilation does not necessarily mean that ethnic consciousness or awareness of racial and cultural differences is lost. *Competition theory,* in fact, suggests that as ethnic minorities begin to achieve success and enter into mainstream social and economic life, awareness of ethnic differences may be heightened. Frequently, ethnic identity may be most clearly experienced and expressed by those who can most successfully assimilate but who choose to promote group awareness and ethnic mobilization movements. That promotion, the theory holds, is a reflection of pressures of American urban life and the realities of increased competition. Those pressures transform formerly isolated groups into recognized, self-assertive ethnic minorities pursuing goals and interests dependent on their position within the larger society.

While in the United States it is usually assumed that acculturation and assimilation are self-evidently advantageous, Canada established multiculturalism in the 1970s as the national policy designed to reduce tensions between ethnic and language groups and to recognize that each thriving culture is an important part of the country's priceless personal resources. Since 1988, multiculturalism has been formalized by act of the Canadian parliament and supervised by a separate government ministry. An example of its practical application can be seen in the way Toronto, Canada's largest and the world's most multicultural city with 44% of its residents (2001) foreign-born, routinely sends out property tax notices in six languages—English, French, Chinese, Italian, Greek, and Portuguese. Nevertheless, Canada—which takes in more immigrants per capita than any other industrialized country—began in 1995 to reduce the number of newcomers it was prepared to admit.

Both Canada and the United States seek to incorporate their varied immigrant minorities into composite national societies. In other countries, quite different attitudes and circumstances may prevail when indigenous—not immigrant—minorities feel their cultures and territories threatened. The Sinhalese comprise 75% of Sri Lanka's population, but the minority Tamils have waged years of guerrilla warfare to defend what they see as majority threats to their culture, rights, and property. In India, Kashmiri nationalists fight to separate their largely Muslim valley from the Hindu majority society. Expanding ethnic minorities made up nearly 8.5% of China's 2000 population total. Some, including Tibetans, Mongols, and Uighurs, face assimilation largely because of massive migrations of ethnic Chinese into their traditional homelands. And in many multiethnic African countries, single-party governments seek to impose a sense of national unity on populations whose primary and nearly unshakable loyalties are rooted in their tribes and regions and not the state that is composed of many tribes. Across the world, conflicts between ethnic groups within states have proliferated in recent years. Armenia, Azerbaijan, Burma, Burundi, Ethiopia, Indonesia, Iraq, Russia, Rwanda, and the former Yugoslavia are others in a long list of countries where ethnic tensions have erupted into civil conflict.

Basques and Catalans of Spain and Corsicans, Bretons, and Normans of France have only recently seen their respective cen-tral governments relax strict prohibitions on teaching or using the languages that identified those ethnic groups. On the other hand, in Bulgaria, ethnic Turks, who unofficially comprise 10% of the total population, officially ceased to exist in 1984 when the government obliged Turkish speakers and Muslims to replace their Turkish and Islamic names with Bulgarian and Christian ones. The government also banned their language and strictly limited practice of their religion. The intent was to impose an assimilation not sought by the minority.

Elsewhere, ethnic minorities—including immigrant minorities—have grown into majority groups, posing the question of who will assimilate whom. Ethnic Fijians sought to resolve that issue by staging a coup to retain political power when the majority immigrant ethnic Indians came to power by election in 1987 and another in 2000 after the election of an ethnic-Indian prime minister. As these and innumerable other examples from all continents demonstrate, Anglo American experiences and expectations have limited application to other societies differently constituted and motivated.

Areal Expressions of Ethnicity

Throughout much of the world, the close association of territoriality and ethnicity is well recognized, accepted, and often politically disruptive. Indigenous ethnic groups have developed over time in specific locations and, through ties of kinship, language, culture, religion, and shared history, have established themselves in their own and others' eyes as distinctive peoples with defined homeland areas. The boundaries of most countries of the world encompass a number of racial or ethnic minorities, whose demands for special territorial recognition have increased rather than diminished with advances in economic development, education, and self-awareness (Figure 6.6).

The dissolution of the Soviet Union in 1991, for example, not only set free the 14 ethnically based union republics that formerly had been dominated by Russia and Russians, but also opened the way for many smaller ethnic groups to seek recognition and greater local control from the majority populations, including Russians, within whose territory their homelands lay. In Asia, the Indian subcontinent was subdivided to create separate countries with primarily religious-territorial affiliations, and the country of India itself has adjusted the boundaries of its constituent states to accommodate linguistic-ethnic realities. Other continents and countries show a similar acceptance of the importance of ethnic territoriality in their administrative structure. In some cases, as in the dismemberment of the Austro-Hungarian Empire after World War I, the recognition of ethnically defined homelands was the basis of new country formations (see "The Rising Tide of Nationalism").

With the exceptions of some—largely Canadian—Native American groups and of French Canadians, there is not the coincidence in Anglo America between territorial claim and ethnic-racial distinctiveness so characteristic elsewhere in the world (Figure 6.7). The general absence of such claims is the result of the immigrant nature of American society. Even the Native

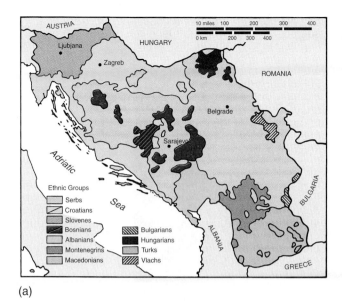

(a)

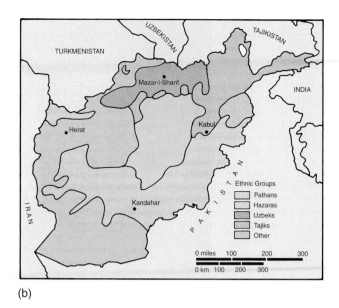

(b)

Figure 6.6 (*a*) **Ethnicity in former Yugoslavia.** Yugoslavia was formed after World War I (1914–1918) from a patchwork of Balkan states and territories, including the former kingdoms of Serbia and Montenegro, Bosnia-Herzegovina, Croatia-Slavonia, and Dalmatia. The authoritarian central government created in 1945 began to disintegrate in 1991 as non-Serb minorities voted for regional independence. In response, Serb guerillas backed by the Serb-dominated Yugoslav military engaged in a policy of territorial seizure and "ethnic cleansing" to secure areas claimed as traditional Serb "homelands." Religious differences between Eastern Orthodox, Roman Catholic, and Muslim adherents compound the conflicts rooted in nationality. (*b*) **Afghanistan** houses Pathan, Tajik, Uzbek, and Hazara ethnic groups speaking Pashto, Dari Persian, Uzbek, and several minor languages, and split between majority Sunni and minority Shia Moslem believers. Following Soviet military withdrawal in 1989, conflict between various Afghan groups hindered the establishment of a unified state and government.

The Rising Tide of Nationalism

The end of the 20th and start of the 21st centuries are witness to spreading ethnic self-assertion and demands for national independence and cultural purification of homeland territories. To some, these demands and the conflicts they frequently engender are the expected consequences of the decline of strong central governments and imperial controls. It has happened before. The collapse of the Roman and the Holy Roman Empires were followed by the emergence of the nation-states of medieval and Renaissance Europe. The fall of Germany and the Austro-Hungarian Empire after World War I saw the creation of new ethnically based countries in Eastern Europe. The brief decline of post-czarist Russia permitted freedom for Finland, and for 20 years for Estonia, Latvia, and Lithuania. The disintegration of British, French, and Dutch colonial control after World War II resulted in new state formation in Africa, South and East Asia, and Oceania.

Few empires have collapsed as rapidly and completely as did that of the Soviet Union and its Eastern European satellites in the late 1980s and early 1990s. In the subsequent loss of strong central authority, the ethnic nationalisms that communist governments had for so long tried to suppress asserted themselves in independence movements. At one scale, the Commonwealth of Independent States and the republics of Estonia, Latvia, Lithuania, and Georgia emerged from the former Soviet Union. At a lesser territorial scale, ethnic animosities and assertions led to bloodshed in the Caucasian republics of the former USSR, in former Yugoslavia (see Figure 6.6a), in Moldova, and elsewhere, while Czechs and Slovaks agreed to peacefully go their separate ways at the start of 1993.

Democracies, too, at least before legal protections for minorities are firmly in place, risk disintegration or division along ethnic, tribal, or religious lines. African states with their multiple ethnic loyalties (see Figure 12.5) have frequently used those divisions to justify restricting political freedoms and continuing one-party rule. However, past and present ethnically inspired civil wars and regional revolts in Somalia, Ethiopia, Nigeria, Uganda, Liberia, Angola, Rwanda, Burundi, and elsewhere show the fragility of the political structure on that continent.

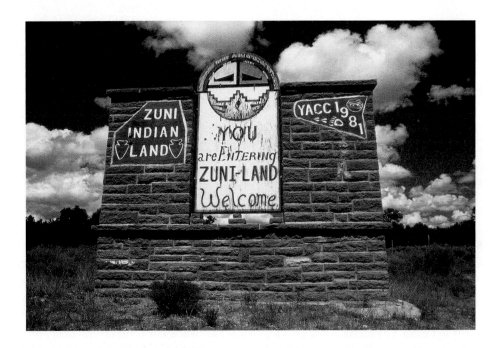

Figure 6.7 Although all of North America was once theirs alone, Native Americans have become now part of a larger cultural mix. In the United States, their areas of domination have been reduced to reservations found largely in the western half of the country and to the ethnic provinces shown in Figure 6.11. These are often areas to which Amerindian groups were relocated, not necessarily the territories occupied by their ancestors at the time of European colonization.

Amerindians were never a single ethnic or cultural group and cannot be compared to a European national immigrant group in homogeneity. Arriving over many thousands of years, from many different origin points, with different languages, physical characteristics, customs, and skills, they are in no way comparable to a culturally uniform Irish or Slovak ethnic group arriving during the 19th century or Salvadorans or Koreans during the 20th. Unlike most other minorities in the American amalgam, Amerindians have generally rejected the goal of full and complete assimilation into the national mainstream culture.

The reported Native American population increased sharply between 1960 and 2000, in part due to high fertility rates but also, importantly, to changing census racial classifications. Perhaps because of expanded ethnic awareness and pride, many more citizens now claim a Native American identity; in the 2000 census, about 2 million persons opted for the single Native American category, and another 2 million checked that and something else.

American "homeland" reservations in the United States are dispersed, noncontiguous, and in large part artificial impositions.[2] The spatial pattern of ethnicity that has developed is therefore more intricate and shifting than in many other pluralistic societies. It is not based on absolute ethnic dominance but on interplay between a majority culture and, usually, several competing minority groups. It shows the enduring consequences of early settlement and the changing structure of a fluid, responsive, freely mobile North American society.

Charter Cultures

Although, with the Canadian French and Native American exceptions noted, no single ethnic minority homeland area exists in present-day Anglo America, a number of separate social and ethnic groups are of sufficient size and regional concentration to have put their impress on particular areas. Part of that imprint results from what the geographer Wilbur Zelinsky termed the "doctrine of **first effective settlement.**" That principle holds that

Whenever an empty territory undergoes settlement, or an earlier population is dislodged by invaders, the specific characteristics of the first group able to effect a viable, self-perpetuating society are of crucial significance for the later social and cultural geography of the area, no matter how tiny the initial band of settlers may have been.[3]

On the North American stage, the English and their affiliates, although few in number, were the first effective entrants in the eastern United States and shared with the French that role in eastern Canada. Although the French were ousted from parts of Seaboard Canada, they retained their cultural and territorial dominance in Quebec Province, where today their political power and ethnocentricity foster among some the determination to achieve separate nationhood. In the United States, British immigrants (English, Welsh, Scottish, and Scotch-Irish) constituted the main portion of the new settlers in eastern Colonial America and retained their significance in the immigrant stream until after 1870.

The English, particularly, became the **charter group,** the dominant first arrivals establishing the cultural norms and

---

[2]In Canada, a basic tenet of Aboriginal policy since 1993 has been the recognition of the inherent right of self-government under Section 35 of the Canadian Constitution. The new territory of Nunavut, the central and eastern portion of the earlier Northwest Territories, is based largely on Inuit land claims and came into existence as a self-governing district in 1999.

[3]*The Cultural Geography of the United States.* Rev. ed. (Englewood Cliffs, N.J.: Prentice-Hall, 1992), p. 13.

standards against which other immigrant groups were measured. It is understandable, then, in the light of Zelinsky's "doctrine," that English became the national language; English common law became the foundation of the American legal system; British philosophers influenced the considerations and debates leading to the American Constitution; English place names predominate in much of the country; and the influence of English literature and music remains strong. By their early arrival and initial dominance, the British established the majority culture of the Anglo American realm; their enduring ethnic impact is felt even today.

Somewhat comparable to the British domination in the East is the Hispanic influence in the Southwest. Mexican and Spanish explorers established settlements in New Mexico a generation before the Pilgrims arrived at Plymouth Rock. Spanish-speaking El Paso and Santa Fe were prospering before Jamestown, Virginia, was founded in 1607. Although subsequently incorporated into an expanding "Anglo"-controlled cultural realm and dominated by it, the early established Hispanic culture, reinforced by continuing immigration, has proved enduringly effective. From Texas to California, Spanish-derived social, economic, legal, and cultural institutions and traditions remain an integral part of contemporary life—from language, art, folklore, and names on the land through Spanish water law to land ownership patterns reflecting Spanish tenure systems.

Ethnic Clusters

Because the British already occupied much of the agricultural land of the East, other, later immigrant streams from Europe were forced to "leapfrog" those areas and seek settlement opportunities in still-available productive lands of the interior and western United States and Canada. The Scandinavians of the North Central states, the Germans in the Appalachian uplands, the upper Middle West, and Texas, various Slavic groups farther west on the Plains, and Italians and Armenians in California are examples of later arrivals occupying, ethnically influencing, and becoming identified with different sections of the United States even as they remained part of a larger cultural realm dominated by British roots. Such areas of ethnic concentration are known as **ethnic islands,** the dispersed and rural counterparts of urban ethnic neighborhoods (Figure 6.8).

Characterized usually by a strong sense of community, ethnic islands frequently placed their distinctive imprint on the rural

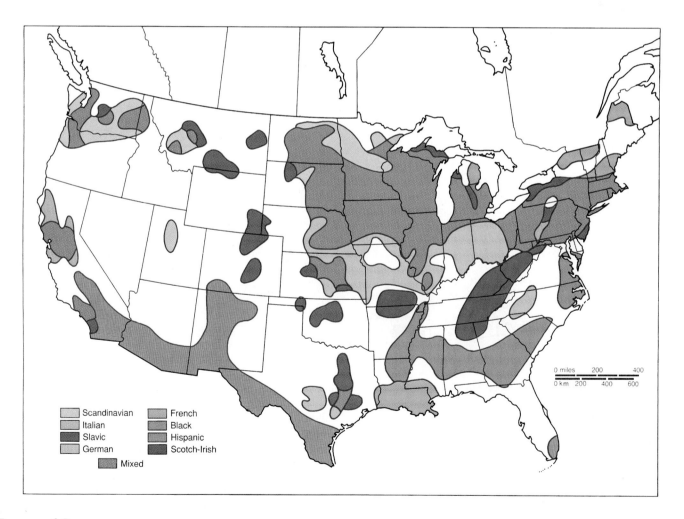

Figure 6.8 **Ethnic islands in the United States.**

Source: Redrawn Settlement Patterns in Missouri *by Russel L. Gerlach, by permission of University of Missouri Press © 1986 by Curators of the University of Missouri.*

landscape by retaining home-country barn and house styles and farmstead layouts, while their inhabitants may have retained their own language, manner of dress, and customs. With the passing of the generations, rural ethnic identity has tended to diminish, and 20th-century adaptations and dispersions have occurred. When long-enduring through spatial isolation or group determination, ethnic islands have tended to be considered landscape expressions of folk culture rather than purely ethnic culture; we shall return to them in that context in Chapter 7.

Similar concentrations of immigrant arrivals are found in Canada. Descendants of French and British immigrants dominate its ethnic structure, both occupying primary areas too large to be considered ethnic islands. British origins are most common in all the provinces except Quebec, where 75% of the population is of French descent and over 80% of French Canadians make their home. French descendants are the second-largest ethnic group in Atlantic Canada and Ontario but fall to fifth or sixth position among minorities in the western provinces. Chinese have concentrated in British Columbia, Italians in Ontario and Quebec, and Ukrainians are the third-largest minority in the Prairie Provinces. The ethnic diversity of that central portion of Canada is suggested by Figure 6.9.

European immigrants arriving by the middle of the 19th century frequently took up tracts of rural land as groups rather than as individuals, assuring the creation of at least small ethnic islands. German and Ukrainian Mennonites in Manitoba and Saskatchewan, for example; Doukhobors in Saskatchewan; Mennonites in Alberta; Hutterites in South Dakota, Manitoba, Saskatchewan, and Alberta; the Pennsylvania Dutch (whose name is a corruption of *Deutsch*, or "German," their true nationality); Frisians in Illinois; and other ethnic groups settled as collectives. They sometimes acted on the advice and the land descriptions reported by advance agents sent out by the group. In most cases, sizable extents of rural territory received the imprint of a group of immigrants acting in concert.

Such **cluster migration** was not unique to foreign colonies. In a similar fashion, a culturally distinctive American group, the Latter-day Saints (Mormons), placed their enduring mark as the first and dominant settlers on a large portion of the West, focusing on Utah and adjacent districts (Figure 6.10). In general, however, later in the century and in the less arable sections of the western United States, the disappearance of land available for homesteading and the changing nature of immigrant flows reduced the incidence of cluster settlement. Impoverished individuals rather than financially

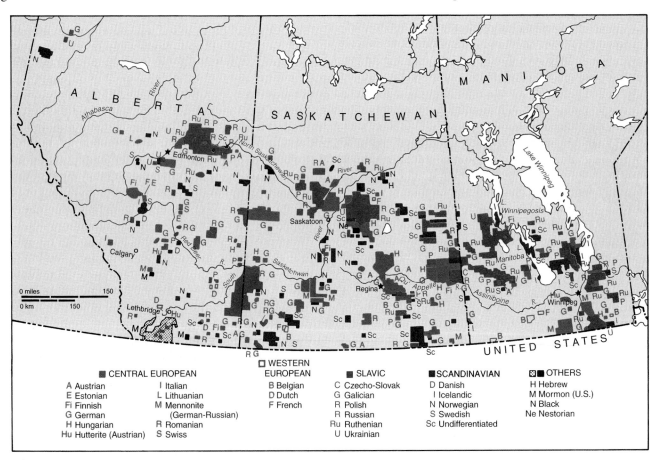

Figure 6.9 **Ethnic diversity in the Prairie Provinces of Canada.** In 1991, 69% of all Canadians claimed some French or English ancestry. For the Prairie Provinces with their much greater ethnic mixture, only 15% declared any English or French descent. Immigrants comprise a larger share of Canadian population than they do of the U.S. population. Early in the 20th century, most newcomers located in rural western Canada and by 1921 about half the population of the Prairie Provinces was foreign born. Later immigrants concentrated in the major metropolitan centers. In 2001, some 40% of Toronto's population was foreign born and 35% of Vancouver's. In the period 1981 to 1991, 48% of Canada's immigrants were from Asia and only 25% from Europe, the traditionally dominant source region. From 1991 to 2001, the disparity increased: to 58% and 20%.

Source: D. G. G. Kerr, A Historical Atlas of Canada, *2nd edition, 1966. Thomas Nelson & Sons Ltd., 1966.*

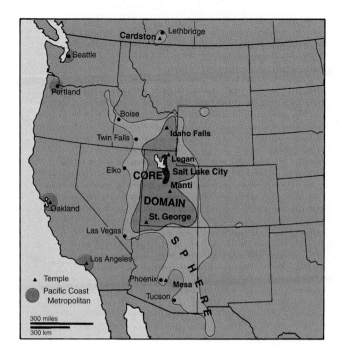

Figure 6.10 **The Mormon culture region** as defined by D. W. Meinig. To express the observed spatial gradations in Mormon cultural dominance and to approximate its sequential spread, Professor Meinig defined the Salt Lake City *core* region of Mormon culture as "a centralized zone of concentration . . . and homogeneity." The broader concept of *domain* identifies "areas in which the . . . culture is dominant" but less intensive than in the core. The *sphere* of any culture, Meinig suggests, is the zone of outer influence, where only parts of the culture are represented or where the culture's adherents are a minority of the total population.

Source: Redrawn with permission from Annals of the Association of American Geographers, *D. W. Meinig, Vol. 55, p. 214, Association of American Geographers, 1965.*

solid communities sought American refuge and found it in urban locations and employment.

While cluster migration created some ethnic concentrations of Anglo America, others evolved from the cumulative effect of **chain migration**—the assemblage in one area of the relatives, friends, or unconnected compatriots of the first arrivals, attracted both by favorable reports and by familiar presences in specific locales of the New World (see also p. 88). Although such chain migration might not affect sizable districts, it could and did place a distinctive imprint on restricted rural ethnic islands and, particularly, urban areas. "Chinatown," "Little Sicily," and other urban enclaves, the concentration of Arab Americans in Dearborn, Michigan, and the Italian and Armenian farm communities of California's Central Valley, are examples of chain migrations and congregate settlement.

Black Dispersions

Some entire regions of North America—vastly larger than the distinctive ethnic islands—have become associated with larger ethnic or racial aggregations numbering in the thousands or millions. Such **ethnic provinces** include French Canadians in Quebec; African Americans in the United States Southeast; Native

Americans in Oklahoma, the Southwest, the Northern Plains and Prairie Provinces; and Hispanics in the southern border states of the United States West (Figure 6.11). The identification of distinctive communities with extensive regional units persists, even though ethnicity and race have not been fully reliable bases for regionalization in North America. Cultural, ethnic, and racial mixing has been too complete to permit United States counterparts of Old World ethnic homelands to develop, even in the instance of the now-inappropriate association of African Americans with southern states.

African Americans, involuntary immigrants to the continent, were nearly exclusively confined to rural areas of the South and Southeast prior to the Civil War (Figure 6.12). Even after emancipation, most remained on the land in the South. During the 20th century, however, established patterns of southern rural residence and farm employment underwent profound changes, although southern regionalization of blacks is still evident (Figure 6.13). The decline of subsistence farming and share-cropping, the

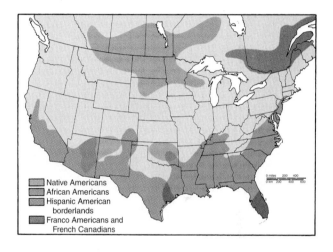

Figure 6.11 **Four North American ethnic groups and their provinces.** Note how this generalized map differs from the more detailed picture of ethnic distributions shown in Figure 6.8.

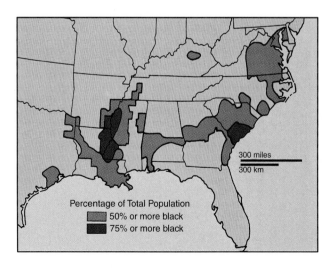

Figure 6.12 **African American concentrations, 1850.**

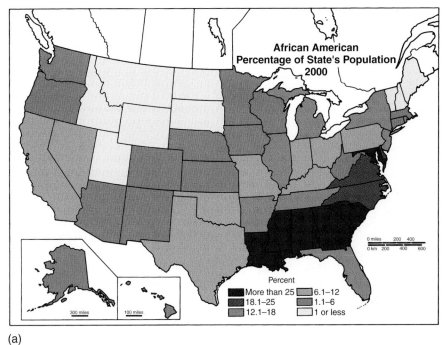

(a)

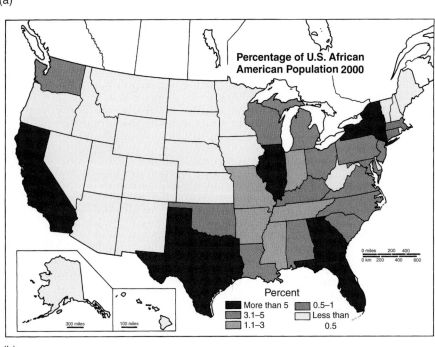

(b)

Figure 6.13 **Evidences of African American concentration, 2000.** (*a*) The top map indicates the importance of African Americans in the total population of the individual states. They are particularly significant in the largely rural, relatively low-population states of the Southeast in a pattern reminiscent of their distribution in 1850 (Figure 6.12). (*b*) The bottom map makes clear the African American response to employment opportunities now more widely available than 150 years ago in the urbanized, industrialized states of the Northeast, Midwest, and California. However, the South was still the home of almost 55% of African Americans in 2000, reflecting both tradition and a pronounced return migration in the later 20th century.

Source: Data from U.S. Bureau of the Census.

mechanization of southern agriculture, the demand for factory labor in northern cities starting with World War I (1914–1918), and the general urbanization of the American economy all affected traditional patterns of black residence and livelihood.

Between 1940 and 1970, more than 5 million black Americans left their traditional southeastern concentration—the largest internal ethnic migration ever experienced in America. A modest return migration of, particularly, middle-class African Americans that began in the 1970s picked up speed during the 1980s and became a major flow in the 1990s. The first-of-century pace of that return flow suggests a net inflow to the South of some 3 million African Americans between 1975 and 2010—more than half of the post-1940 out-migration.

The growing African American population (over 12% of all Americans in 2000) has become more urbanized than the general population; 86% were residents of metropolitan areas in 1999, compared to 75% for all Americans combined. Although recent national economic trends, including industrial growth in the Sunbelt, have encouraged a reverse migration, almost half of African Americans in 2000 resided outside the South (Figure 6.13).

Black Americans, like Asian Americans and Hispanics, have had thrust on them an assumed common ethnicity that does not, in fact, exist. Because of prominent physical or linguistic characteristics, quite dissimilar ethnic groups have been categorized by the white, English-speaking majority in ways totally at odds with the realities of their separate national origins or cultural inheritances. Although the U.S. Census Bureau makes some attempt to subdivide Asian ethnic groups—Chinese, Filipino, and Korean, for example—these are distinctions not necessarily recognized by members of the white majority. But even the Census Bureau, in its summary statistics, has treated "Black" and "Hispanic Origin" as catchall classifications that suggest ethnic uniformities where none necessarily exist.

In the case of African Americans, such clustering is of decreasing relevance. Their spatial mobility was encouraged by the industrial urban labor requirements first apparent during World War I and continuing through the Vietnam era of the 1960s. Government intervention, mandating and promoting racial equality, further deracialized the economic sector. As a result, the black community has become subdivided along socioeconomic rather than primarily regional lines. No common native culture united the slaves brought to America; few of their transported traits or traditions could endure the generations of servitude. By long residence and separate experiences, African Americans have become as differentiated as comparably placed ethnics of any other heritage. In fact, ethnic diversity among them is increasing. Census estimates are that by 2010 as many as 10% of Americans of African descent will be recent immigrants from various regions of Africa or the Caribbean.

Hispanic Concentrations

Similarly, the members of the multiracial, multinational, and multicultural composite population lumped by the Census Bureau into the single category of "Hispanic or Latino" are not a homogeneous group either. Indeed, it was the Census Bureau, not ethnicity or culture, that created the concept and distinct statistical and social category of "Hispanics." Prior to 1980 no such composite group existed; since then it has been statistically isolated from other ethnic or racial groupings with which its arbitrarily assigned members might otherwise identify. Although the Census assures that Hispanics "can be members of any race," it removes them officially from any other association. For example, although in 2000, 4 million California Hispanics—40% of their total number—specifically marked "white" as their race on the Census form, denying them that identity changed whites statistically at least from a 59.5% majority to a minority of the state's residents.

Hispanic Americans represent as much diversity within the assumed uniform group as they do between that group and the rest of the population. By commonly used racial categories, they may also be white, black, or Native American; nearly 60% of Hispanic Americans, in fact, report themselves to be white. Individually, they are highly diversified by country and culture of origin. Collectively, they also constitute the most rapidly growing minority component of U.S. residents—increasing nearly 57% (to over 35.3 million) between 1990 and 2000 to surpass African Americans as the largest minority, as Table 6.5 indicates. Indeed, by 1990 Hispanics had already outnumbered blacks in four of the country's ten largest cities, and by 2000 they exceeded African Americans in seven of the top ten. By mid-2002 the Hispanic population grew to 37 million—nearly 13% of U.S. population.

Mexican Americans account for nearly 60% of all Hispanic Americans (Table 6.6). About 8.8 million of them in 2000 had been born in Mexico, representing just under 30% of the total U.S. foreign-born population. They are overwhelmingly located in the five southwestern states that constitute the ethnic province called the Hispanic American borderland (see Figure 6.11). Beginning in the 1940s, the Mexican populations in the United States became increasingly urbanized and dispersed, losing their earlier primary identification as agricultural *braceros* (seasonal laborers) and as residents of the rural areas of Texas, New Mexico, and Arizona. California rapidly increased its Mexican American populations

Table 6.5

Actual and Projected United States Population Mix: 2000, 2025, and 2050

Population Group (One race options)	Percent of Total		
	2000	2025	2050
Non-Hispanic White	69.1	62.0	52.8
Hispanic or Latino	12.5	18.2	24.5
Black or African American	12.1	12.9	13.2
Asian/Pacific Islander	3.7	6.2	8.9
Native American	0.9	0.8	0.8

Note: Black, Asian, and Native American categories exclude Hispanics, who may be of any race.

Source: U.S. Bureau of the Census, Population Projection Program. Totals do not round to 100%.

Table 6.6
Composition of U.S. Hispanic Population, 2000

Hispanic Subgroup	Number (millions)	Percent
Mexican	20.6	58.5
Puerto Rican	3.4	9.6
Cuban	1.2	3.5
Dominican	0.8	2.2
Central American	1.7	4.8
South American	1.4	3.8
Spanish	0.1	0.3
Other Hispanic origin[a]	6.1	17.3
Total Hispanic or Latino	**35.3**	**100**

[a]"Other Hispanics" includes those with origins in Spain or who identify themselves as "Hispanic," "Latino," "Spanish American," etc.

Source: U.S. Bureau of Census.

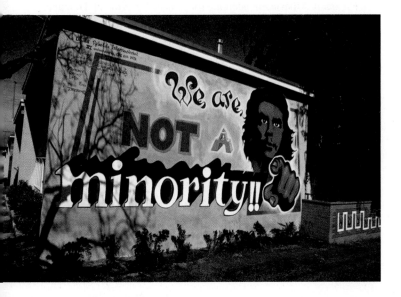

Figure 6.14 A proudly assertive street mural in the Boyle Heights, Los Angeles, *barrio*. Half of Los Angeles' population in the early 21st century was Hispanic and overwhelmingly Mexican American. Their impact on the urban landscape—in choice of house colors, advertising signs, street vendors, and colorful wall paintings—is distinctive and pervasive.

(Figure 6.14), as did the Midwest, particularly the chain of industrial cities from southeastern Wisconsin through metropolitan Chicago to Detroit. Wherever they settled in the United States, Mexican immigrants in 2000 represented a loss to their home country of 12% of its total labor force and 30% of its holders of doctorate degrees.

Mexican Americans, representing a distinctive set of cultural characteristics, have been dispersing widely across the United States, though increases in the Midwestern states have been particularly noticeable. In similar fashion, immigrants from equally distinctive South, Central, and Caribbean American countries have been spreading out from their respective initial geographic concentrations. Puerto Ricans, already citizens, first localized in New York City, now the largest Puerto Rican city anywhere in numerical terms. Since 1940, however, when 88% of mainland Puerto Ricans were New Yorkers, there has been an outward dispersal primarily to other major metropolitan areas of the northeastern part of the country. The old industrial cities of New Jersey (Jersey City, Newark, Paterson, Passaic, and Hoboken); Bridgeport and Stamford, Connecticut; the Massachusetts cities of Lowell, Lawrence, and Brockton; and Chicago and other central cities and industrial satellites of the Midwest have received the outflow. By the end of the 1990s, New York City retained only about one-quarter of the mainland Puerto Ricans.

Miami and Dade County, Florida, play the same magnet role for Cubans as New York City earlier did for Puerto Ricans. The first large-scale movement of Cuban refugees from the Castro revolution occurred between 1959 and 1962. There followed a mixed period lasting until 1980 when emigration was alternately permitted and prohibited by the Cuban government. Suddenly and unexpectedly, in April, 1980, a torrent of Cuban migration was released through the small port of Mariel. Although their flow was stopped after only 5 months, some 125,000 *Marielitos* fled from Cuba to the United States. A 1994 accord between the United States and Cuba allows for a steady migration of at least 20,000 Cubans each year, assuring strong Cuban presence in Florida, where 67% of Cuban Americans reside among a growing number of other, largely Central American immigrants, particularly in Miami's "Little Havana" community.

Early in the period of post-1959 Cuban influx, the federal government attempted a resettlement program to scatter the new arrivals around the United States. Some remnants of that program are still to be found in concentrations of Cubans in New York City, northern New Jersey, Chicago, and Los Angeles. The majority of early and late arrivals from Cuba, however, have settled in the Miami area.

Immigrants from the Dominican Republic, many of them undocumented and difficult to trace, appear to be concentrating in the New York City area. Within that same city, Central and South Americans have congregated in the borough of Queens, with the South American contingent, particularly Colombians, settling in the Jackson Heights section. Elsewhere, Central American Hispanics also tend to cluster. Los Angeles is estimated to hold some 40% of Central American immigrants; other concentrations include San Francisco, New York City, and Washington, D.C. Each concentration differs in its country of origin. Most Nicaraguans are found in the Miami area, most Hondurans in New Orleans. As noted, migrants from the Dominican Republic seek refuge in New York City; Salvadoran and Guatemalan migrants have dispersed themselves more widely.

New arrivals tend to follow the paths of earlier countrymen. Chain migration and the security and support of an ethnically distinctive halfway community are as important for recent immigrants as for their predecessors of earlier times and different cultures. As the residential concentrations of the different Central American subgroups suggest, Hispanics as a whole are more urbanized than are non-Hispanic populations of the United States. In 2000, over 91% of Hispanic households were in metropolitan areas compared to 78% for non-Hispanic whites and 86% for blacks. Particularly the urbanized Hispanic population, it has been observed, appears confronted by two dominating but opposite trends. One is a drive toward conventional assimilation within American society. The other is consignment to a pattern of poverty, isolation, and, perhaps, cultural alienation from mainstream American life. Because of their numbers, which trend Hispanics follow will have significant consequences for American society as a whole.

Asian Contrasts

Since 1965 and the abolition of earlier immigration limits, the Asian American population has grown from 1.5 million to nearly 12 million (including mixed race options) in 2000; it is projected to grow to 20 million by 2020. Once largely U.S.-born and predominantly of Japanese and Chinese heritage, the Asian American population is now largely foreign-born and, through multiple national origins, is increasingly heterogeneous. Major sending home countries include Korea, the Philippines, Vietnam, India, Thailand, and Pakistan, in addition to continuing arrivals from China. Though second to Hispanics in numbers of new arrivals, Asians still comprised nearly one-third of the legal immigrant flow to the United States between 1990 and 2000.

Their inflow was encouraged, first, by changes in immigration law that abolished the older national origins system and favored family reunification as an admission criterion. Educated Asians, taking advantage of professional preference categories in the immigration laws to move to the United States (or remain here on adjusted student visas), could become citizens after 5 years and send for immediate family and other relatives without restriction. They, in turn, after 5 years, could bring in other relatives. Chain migration was an important agency. As a special case, the large number of Filipino Americans is related to U.S. control of the Philippines between 1899 and 1946. In the early part of the last century, Filipino workers were brought to Hawaii to work on sugar plantations, to California to labor on farms, or to Alaska to work in fish canneries. During World War II, Filipinos who served under the U.S. military were granted citizenship; immigration continues to be common today, especially for Filipino professionals.

Second, the flood of Southeast Asian refugees admitted during 1975–1980 under the Refugee Resettlement Program after the Vietnam War swelled the Asian numbers in the United States by over 400,000, with 2.4 million more Asian immigrants admitted between 1980 and 1990. At the start of the 21st century, nearly 28% of the U.S. foreign-born population were from Asia. Canada shows a similar increase in the immigrant flow from that continent. Although the annual share of immigrants coming from Asia to Canada never exceeded 5% during the 1950s, between 1991 and 2001, 58% of new arrivals were of Asian birth.

Asia is a vast continent; successive periods of immigration have seen arrivals from many different parts of it, representing totally different ethnic groups and cultures. The major Asian American populations are detailed in Table 6.7, but even these groups are not homogeneous and cannot suggest the great diversity of other ethnic groups—Bangladeshi, Burmese, Nepalese, Sri Lankan, Mien, Indonesians of great variety, and many more—who have joined the American realm. Although settled in all sections of the country and, like Hispanic Americans, differently localized by ethnic group, Asian Americans as a whole are relatively concentrated in residence—far more so than the rest of the population.

In 2000, about half of them resided in the West (and over 35% in California alone), where only 22% of all Americans lived; 36% of the whole population lived in the South, but only 19% of Asian Americans were found there. Japanese and Filipinos are particularly concentrated in the western states, where more than half of the Chinese Americans are also found. Only some 20% of all Asian Americans lived in the Northeast, but about one-third of the country's Asian Indians were localized there. Certain groups clearly indicate the tendency of Asian Americans to cluster: almost one-half of Filipinos are found in California, for example, as are 40% of the country's Vietnamese.

At a different scale, 18% of America's Koreans and 16% of its Filipinos found residence in the Los Angeles–Long Beach metropolitan area alone, and the largest Vietnamese community outside of Vietnam itself is in Orange County, south of Los Angeles. In whatever part of the country they settled, Asian Americans (and Pacific Islanders) were drawn to metropolitan areas,

Table 6.7

U.S. Leading Asian Populations by Ethnicity,[a] 2000

Ethnicity	Number (000)	Percent of Asian American Total
Chinese	2734.8	23.0
Filipino	2364.8	19.9
Asian Indian	1899.6	16.0
Korean	1228.4	10.3
Vietnamese	1223.7	10.3
Japanese	1148.9	9.6
Cambodian	206.1	1.7
Pakistani	204.3	1.7
Laotian	198.2	1.6
Hmong	186.3	1.6
Thai	150.3	1.3
Taiwanese	144.8	1.2
Other Asian	208.6	1.8
Total	**11,898.8**	**100**

[a]Ethnicity as reported by respondents, including claimed combination ethnicities.
Source: U. S. Bureau of the Census.

where 96% of them lived at the start of the century—than half in suburban districts. Although their metropolitan affinities have remained constant, the trend over time has been for greater dispersal around the country. In 1990, the eight metropolitan areas with the largest concentrations had 47% of the Asian American population; by 2000, they held only 41%.

Immigrant Gateways and Clusters

Although new immigrants may ultimately seek residence in all parts of the United States, over the short term, immigrant concentrations rather than dispersals are the rule. Initially, most immigrants tend to settle near their points of entry (that is, nearest their country of origin) or in established immigrant communities. Family ties and job availability may replace or reinforce those primary draws. Five states—California, Texas, New York, Illinois, and New Jersey have experienced the largest increases in their foreign-born populations. Together, they housed (2000) almost 70% of America's total immigrant numbers, but only 36% of its native-born residents. Their attraction of newcomers, however, decreased over the last years of the 20th century: while they sustained 87% of the country's foreign-born gains in the 1980s, their share decreased to 60% of newcomer increases in the 1990s.

Certain metropolitan areas similarly experienced disproportionate immigrant gains during the 1990s. New York and Los Angeles received the most foreign-born, followed by San Francisco, Chicago, Miami, Dallas, Houston, and Washington. As a group, these eight metropolitan areas accounted for half of the country's foreign-born growth in the last decade of the 20th century and housed 57% of America's foreign-born population. These magnet cities contain established immigrant networks that offer social and economic support to new arrivals drawn to them by chain migration flows. Those attractions are not permanent and census evidence suggests that immigrant diffusion is occurring in areas where the native-born labor supply does not satisfy market needs for both low-skilled and technically trained workers and as the socioeconomic condition of immigrants improves and their residential choices are less influenced by ethnic community considerations.

French Uniformity

The stamp of the French charter group on the ethnic province of French Canada is overwhelming. Quebec Province—with ethnic extensions into New Brunswick and northernmost Maine—is the only extensive region of North America (except northern Canadian Native American homelands) where regional delimitation on purely ethnic lines is possible or appropriate. In language, religion, legal principles, system of land tenure, the arts, cuisine, philosophies of life, and landscapes of rural and urban occupance, Quebec stands apart from the rest of Canada (Figure 6.15). Its distinctiveness and self-assertion have won it special consideration and treatment within the political structure of the country.

Although the *Canadiens* of Quebec were the charter group of eastern Canada and for some 200 years the controlling population, they numbered only some 65,000 when the Treaty of Paris ended the North American wars between the British and the

French in 1763. That treaty, however, gave them control over three primary aspects of their culture and lives: language, religion, and land tenure. From these, they created their own distinctive and enduring ethnic province of some 1.5 million square kilometers (600,000 sq mi) and 7.4 million people, more than 80% of whom have French as their native tongue (see Figure 5.16) and adhere to the Roman Catholic faith. Quebec City is the cultural heart of French Canada, though the bilingual Montreal metropolitan area with a population of 3.5 million is the largest center of Quebec Province. The sense of cultural identity prevalent throughout French Canada imparted a spirit of nationalism not similarly expressed in other ethnic provinces of North America. Laws and guarantees recognizing and strengthening the position of French language and culture within the province assure the preservation of this distinctive North American cultural region, even if the movement for full political separation from the rest of Canada is not successful.

(a)

(b)

Figure 6.15 (a) The hotel Château Frontenac stands high above the lower older portion of Quebec City, where many streets show the architecture of French cities of the 18th century carried over to the urban heart of modern French Canada. (b) Rural Richelieu Valley in the Eastern Townships of Quebec Province.

Urban Ethnic Diversity and Segregation

"Little Havanas" and "Little Koreas" have joined the "China-towns," "Little Italys," and "Germantowns" of earlier eras as part of the American urban scene. The traditional practice of selective concentration of ethnics in their own frequently well-defined sub-communities is evidence of a much more inclusive, sharply defined social geography of urban America, in which ethnic neighborhoods have been a pronounced, enduring feature.

Protestant Anglo Americans created from colonial times the dominating host culture—the charter group—of urban North America. To that culture the mass migrations of the 19th and early 20th centuries brought individuals and groups representative of different religious and ethnic backgrounds, including Irish Catholics, eastern European Jews, and members of every nationality, ethnic stock, and distinctive culture of central, eastern, and southern Europe. To them were added, both simultaneously and subsequently, newcomers from Asia and Latin America and such urbanizing rural Americans as Appalachian whites and Southern blacks.

Each newcomer element sought both accommodation within the urban matrix established by the charter group and acceptable relationships with other in-migrant ethnic groups. That accommodation has characteristically been achieved by the establishment of the ethnic community or neighborhood—an area within the city where a particular culture group aggregates, which it dominates, and which may serve as the core area from which diffusion or absorption into the host society can occur. The rapidly urbanizing, industrializing society of 19th-century America became a mosaic of such ethnic enclaves. Their maintenance as distinctive social and spatial entities depended on the degree to which the assimilation of their population occurred. Figure 6.16 shows the more recent ethnic concentrations that developed by the start of the 21st century in one major American city. The increasing subdivision of the immigrant stream and the consequent reduction in the size of identified enclaves make comparable maps of older U.S. cities such as New York and Chicago nearly unintelligibly complex.

Immigrant neighborhoods are a measure of the **social distance** that separates the minority from the charter group. The greater the perceived differences between the two groups, the greater the social distance and the less likely is the charter group to easily accept or assimilate the newcomer. Consequently, the ethnic community will endure longer as a place both of immigrant refuge and of enforced segregation.

Segregation is a shorthand expression for the extent to which members of an ethnic group are not uniformly distributed in relation to the rest of the population. A commonly employed measure quantifying the degree to which a distinctive group is segregated is the segregation index or *index of residential dissimilarity*. It indicates the percentage difference between the distribution of two component groups of a population, with a theoretical range of values from 0 (no segregation) to 100 (complete segregation). For example, according to the 2000 Census, the index of dissimilarity in the New York City metropolitan area was a very high 81.8, meaning that nearly 82% of all blacks (or whites) would have to move to different census tracts before the two groups are equally distributed across the set of tracts. Evidence from cities throughout the world makes clear that most ethnic minorities tend to be sharply segregated from the charter group and that segregation on racial or ethnic lines is usually greater than would be anticipated from the socioeconomic levels of the groups involved. Further, the degree of segregation varies among cities in the same country and among different ethnic mixes within each city.

Among major U.S. metropolitan areas in 2000, for example, Chicago, Illinois, had a black–white segregation index of 81, while for Raleigh-Durham, North Carolina, it was 46. Within the Detroit metropolitan area, on the other hand, the black–white index of residential dissimilarity was 85, but the Hispanic–white and Asian–white indexes were much lower, each at 46. In the country as a whole in 2000, the typical white neighborhood was nearly 83% white, and the typical African American lived in a neighborhood that was 54% black. On average, Hispanics resided in areas 42% Hispanic and Asians in communities that were only 19% Asian. Collectively, blacks, Hispanics, and Asians lived in more integrated neighborhoods than did whites.

Each world region and each country, of course, has its own patterns of national and urban immigration and immigrant residential patterns. Even when those population movements involve distinctive and contrasting ethnic groups, American models of spatial differentiation may not be applicable.

Foreign migrants to West European cities, for example, frequently do not have the same expectations of permanent residence and eventual amalgamation into the host society as their American counterparts. Many came under labor contracts with no initial legal assurance of permanent residence. Although many now have been joined by their families, they often find citizenship difficult to acquire; in Germany, even German-born children of "guest workers" are considered aliens. Their residential choices are consequently influenced by difficulties or disinterest in integration or amalgamation, a high degree of migrant self-identity, restriction to housing units or districts specially provided for them, and the locational pull of chain migration. Culture and religion are important in that regard as even small ethnically homogeneous groups, confined perhaps to part of a city block or to a single apartment building, help to maintain the lifestyle and support systems of home territories.

The Islamic populations from North Africa and Turkey tend to be more tightly grouped and defensive against the surrounding majority culture of western European cities than do African or south and east European Christian migrants. France, with some 5 million Muslim residents, most of them from North Africa, has tended to create bleak, distant outer city ghettoes in which Arab legal and illegal immigrants remain largely isolated from mainstream French life.

Racial and ethnic divisions appear particularly deep and divisive in Britain. A British government report of 2001 claimed that in Britain, whites and ethnic minorities lead separate lives with no social or cultural contact and no sense of belonging to the same nation. Residential segregation in public housing and inner-city areas was compounded by deep social polarization. The 7.1% (2001) of British population that was nonwhite—largely Caribbean and Asian in origin—and the white majority, the

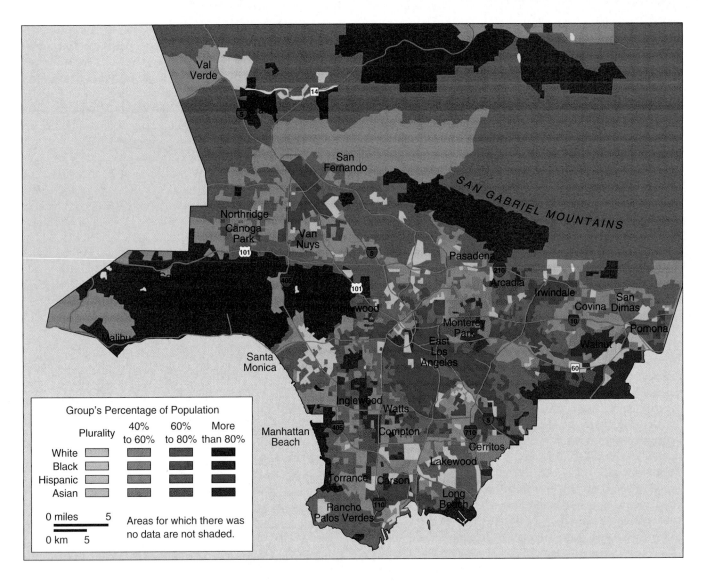

Figure 6.16 **Racial/ethnic patterns in Los Angeles County, 2000,** are greatly generalized on this map, which conceals much of the complex intermingling of different ethnic groups in several sections of Los Angeles city. However, the tendency of people to cluster in distinct neighborhoods by race and ethnicity is clearly evident.

From The New York Times, *March 30, 2001, p. A16. Copyright © 2001 The New York Times Co. Reprinted by permission.*

report concluded, "operate on the basis of a series of parallel lives . . . that often do not seem to touch at any point," assigning blame for the situation on "communities choosing to live in separation rather than integration" (see "The Caribbean Map in London"). The Home Secretary observed on the basis of the report that many "towns and cities lack any sense of civic identity or shared values."

Rapid urbanization in multiethnic India has resulted in cities of extreme social and cultural contrasts. Increasingly, Indian cities feature defined residential colonies segregated by village and caste origins of the immigrants. Chain migration has eased the influx of newcomers to specific new and old city areas; language, custom, religion, and tradition keep them confined. International and domestic migration within ethnically diverse Africa has had a similar residential outcome. In the Ivory Coast, for ex-

ample, the rural-to-urban population shift has created city neighborhoods defined on tribal and village lines. Worldwide in all continental and national urban contexts, the degree of immigrant segregation is at least in part conditioned by the degree of social distance felt between the newcomer population and the other immigrant and host societies among whom residential space is sought.

Constraints on assimilation and the extent of discrimination and segregation are greater for some minorities than for others. In general, the rate of assimilation of an ethnic minority by the host culture depends on two sets of controls: *external,* including attitudes toward the minority held by the charter group and other competing ethnic groups, and *internal* controls of group cohesiveness and defensiveness.

The Caribbean Map in London

Although the movement [to England] from the West Indies has been treated as if it were homogeneous, the island identity, particularly among those from the small islands, has remained strong. . . . [I]t is very evident to anyone working in the field that the process of chain migration produced a clustering of particular island or even village groups in their British destination. . . .

The island identities have manifested themselves on the map of London. The island groups can still be picked out in the clusters of settlements in different parts of the city. There is an archipelago of Windward and Leeward islanders north of the Thames; Dominicans and St. Lucians have their core areas in Paddington and Notting Hill; Grenadians are found in the west in Hammersmith and Ealing; Montserratians are concentrated around Stoke Newington, Hackney and Finsbury Park; Antiguans spill over to the east in Hackney, Waltham Forest and Newham; south of the river is Jamaica.

That is not to say that Jamaicans are found only south of the river or that the only West Indians in Paddington are from St. Lucia. The mixture is much greater than that. The populations overlap and interdigitate: there are no sharp edges. . . . [Nevertheless, north of the river] there is a west-east change with clusters of Grenadians in the west giving way to St Lucians and Dominicans in the inner west, through to Vincentians and Montserratians in the inner north and east and thence to Antiguans in the east.

Source: Ceri Peach, "The Force of West Indian Island Identity in Britain," in *Geography & Ethnic Pluralism*, ed. Colin Clarke, David Ley, and Ceri Peach. (London: George Allen & Unwin, 1984).

External Controls

When the majority culture or rival minorities perceive an ethnic group as threatening, the group tends to be spatially isolated by external "blocking" tactics designed to confine the rejected minority and to resist its "invasion" of already occupied urban neighborhoods. The more tightly knit the threatened group, the more adamant and overt are its resistance tactics. When confrontation measures (including, perhaps, threats and vandalism) fail, the invasion of charter-group territory by the rejected minority proceeds until a critical percentage of newcomer housing occupancy is reached. That level, the **tipping point,** may precipitate a rapid exodus by the former majority population. Invasion, followed by succession, then results in a new spatial pattern of ethnic dominance according to models of urban social geography developed for American cities and examined in Chapter 11, models less applicable to the European scene.

Racial or ethnic discrimination in urban areas generally expresses itself in the relegation of the most recent, most alien, most despised minority to the poorest available housing. That confinement has historically been abetted by the concentration of the newest, least assimilated ethnic minorities at the low end of the occupational structure. Distasteful, menial, low-paying service and factory employment unattractive to the charter group is available to those new arrivals even when other occupational avenues may be closed. The dockworkers, street cleaners, slaughterhouse employees, and sweatshop garment workers of earlier America had and have their counterparts in other regions. In England, successive waves of West Indians and Commonwealth Asians took the posts of low-pay hotel and restaurant service workers, transit workers, refuse collectors, manual laborers, and the like; Turks in German cities and North Africans in France play similar low-status employment roles.

In the United States, there has been a spatial association between the location of such employment opportunities—the inner-city central business district (CBD) and its margins—and the location of the oldest, most dilapidated, and least desirable housing. Proximity to job opportunity and the availability of cheap housing near the CBD, therefore, combined to concentrate the U.S. immigrant slum near the heart of the 19th-century central city. In the second half of the 20th century, the suburbanization of jobs, the rising skill levels required in the automated offices of the CBD, and the effective isolation of inner-city residents by the absence of public transportation or their inability to pay for private transport maintained the association of the least competitive minorities and the least desirable housing area. But now those locations lack the promise of entry-level jobs formerly close at hand.

That U.S. spatial association does not necessarily extend to other cultures and urban environments. In Latin American cities, newest arrivals at the bottom of the economic and employment ladder are most apt to find housing in squatter or slum areas on the outskirts of the urban unit (see Figure 11.43), prestigious housing claims room near the city center. European cities, too, have retained a larger proportion of upper income groups at the urban center than have their American counterparts, with a corresponding impact on the distribution of lower-status, lower-income housing (see Figure 11.38). In French urban agglomerations, at least, the outer fringes frequently have a higher percentage of foreigners than the city itself.

Internal Controls

Although part of the American pattern of urban residential segregation may be explained by the external controls of host-culture resistance and discrimination, the clustering of specific groups into discrete, ethnically homogeneous neighborhoods is best understood as the result of internal controls of group defensiveness and conservatism. The self-elected segregation of ethnic groups can be seen to serve four principal functions—defense, support, preservation, and "attack."

First, it provides *defense,* reducing individual immigrant isolation and exposure by physical association within a limited area. The walled and gated Jewish quarters of medieval European cities have their present-day counterparts in the clearly marked and defined "turfs" of street gang members and the understood exclusive domains of the "black community," "Chinatown," and other ethnic or racial neighborhoods. In British cities, it has been observed that West Indians and Asians fill identical slots in the British economy and reside in the same sorts of areas, but they tend to avoid living in the *same* areas. West Indians avoid Asians; Sikhs isolate themselves from Muslims; Bengalis avoid Punjabis. In London, patterns of residential isolation even extend to West Indians of separate island homelands, as "The Caribbean Map in London" makes clear.

Their own defined ethnic territory provides members of the group with security from the hostility of antagonistic social groups, a factor also underlying the white flight to "garrison" suburbs. That outsiders view at least some closely defined ethnic communities as homogeneous, impenetrable, and hostile is suggested by Figure 6.17, a "safety map" of Manhattan published in the newspaper *l'Aurore* for the guidance of French tourists.

Second, the ethnic neighborhood provides *support* for its residents in a variety of ways. The area serves as a halfway station between the home country and the alien society, to which admittance will eventually be sought. It acts as a place of initiation and indoctrination, providing supportive lay and religious ethnic institutions, familiar businesses, job opportunities where language barriers are minimal, and friendship and kinship ties to ease the transition to a new society.

Third, the ethnic neighborhood may provide a *preservation* function, reflecting the ethnic group's positive intent to preserve and promote such essential elements of its cultural heritage as language and religion. The preservation function represents a disinclination to be totally absorbed into the charter society and a desire to maintain those customs and associations seen to be essential to the conservation of the group. For example, Jewish dietary laws are more easily observed by, or exposure to potential marriage partners within the faith is more certain in, close-knit communities than when individuals are scattered.

Finally, ethnic spatial concentration can serve what has been termed the *attack* function, a peaceful and legitimate search for, particularly, political representation by a concentration of electoral power. Voter registration drives among African and Hispanic Americans represent concerted efforts to achieve the promotion of group interests at all governmental levels.

Shifting Ethnic Concentrations

Ethnic communities once established are not necessarily, or even usually, permanent. For Europeans who came in the 19th and early 20th centuries and for more recent Hispanic and Asian immigrants, high concentrations were and are encountered in neighborhoods of first settlement (see "Colonies of Immigrants"). Second generation neighborhoods usually become far more mixed. The 2000 census reveals that older dominant urban ethnic groups in places called, for example, "Little Italy" rarely exceeded 50% as middle and upper-middle class members of the immigrant group move on. That mobility pattern appears to be repeating among Asian and Latino groups, but only or most clearly where those groups collectively account for a relatively small share of the total metropolitan

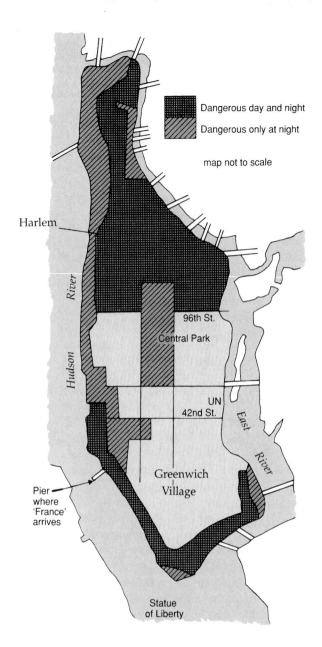

Figure 6.17 **A "safety map" of Manhattan.** According to the editors of the French newspaper *l'Aurore*, all of western and southern Manhattan fringe areas and any place north of 96th street—then (late 1970s) all largely ethnic and racial minority districts—were best avoided, particularly at night.

Source: From l'Aurore *as reproduced in Peter Jackson,* Ethnic Groups and Boundaries, *School of Geography, Oxford University, 1980.*

area population. Black segregation and black communities, in contrast, appear more pronounced and permanent. Continuing racial or ethnic concentrations of 80% or 90% occur almost exclusively in African American neighborhoods where the dominant group remains isolated by segregation and white flight.

Ethnic congregations initially identified with particular central city areas are frequently or usually displaced by different newcomer groups. With recent diversified immigration, older homogeneous ethnic neighborhoods have become highly subdivided and polyethnic. In Los Angeles, for example, the great wave of

Colonies of Immigrants

In the following extract from his 1904 book *Poverty*, Robert Hunter conveys a sense of the ethnic diversity found in American cities:

[In American cities] great colonies, foreign in language, customs, habits, and institutions, are separated from each other and from the distinctly American groups on national or racial lines. By crossing the Bowery one leaves behind him the great Jewish colony made up of Russians, Poles, and Roumanians and passes into Italy; to the northeast lies a little Germany; to the southwest a colony of Syrians; to the west lies an Irish community, a settlement of negroes, a remnant of the old native American stock; to the south lie a Chinese and a Greek colony. On Manhattan alone, either on the extreme west side or the extreme east side, there are other colonies of the Irish, the Jews, and the Italians, and, in addition, there is a large colony of Bohemians. In Chicago there are the same foreign poor. To my own knowledge there are four Italian colonies, two Polish, a Bohemian, an Irish, a Jewish, a German, a negro, a Chinese, a Greek, a Scandinavian, and other colonies. So it is also in Boston and many other cities. In New York alone there are more persons of German descent than persons of native descent, and the German element is larger than in any city of Germany except Berlin. There are nearly twice as many Irish as in Dublin, about as many Jews as in Warsaw, and more Italians than in Naples or Venice. . . .

To live in one of these foreign communities is actually to live on foreign soil. The thoughts, feelings, and traditions which belong to the mental life of the colony are often entirely alien to an American. The newspapers, the literature, the ideals, the passions, the things which agitate the community are unknown to us except in fragments. . . .

While there is a great movement of population from all parts of the old world to all parts of the new, the migration to the United States is the largest and the most conspicuous. Literally speaking, millions of foreigners have established colonies in the very hearts of our urban and industrial communities. . . . In recent years the flow of immigrants to the cities, where they are not needed, instead of to those parts of the country where they are needed, has been steadily increasing. Sixty-nine percent of the present immigration avows itself as determined to settle either in the great cities or in certain communities of the four great industrial states, Massachusetts, New York, Pennsylvania, and Illinois. According to their own statements, nearly 60 percent of the Russian and Polish Jews intend to settle in the largest cities. As a matter of fact, those who actually do settle in cities are even more numerous than this percentage indicates. As the class of immigrants, drawn from eastern and southern Europe, Russia, and Asia, come in increasing numbers to the United States, the tendency to settle in cities likewise increases.

Source: Robert Hunter, *Poverty*. (New York: Macmillan, 1904.)

immigrants from Mexico, Central America, and Asia has begun to push African Americans out of Watts and other well-established black communities, converting them from racially exclusive to multicultural areas. In New York, the Borough of Queens, once the stronghold of European ethnics, has now become home to more than 110 different, mainly non-European nationalities. In Woodside in Queens, Latin Americans and Koreans are prominent among the many replacements of the formerly dominant German and Irish groups. Elsewhere within the city, West Indians now dominate the old Jewish neighborhoods of Flatbush; Poles and Dominicans and other Central Americans have succeeded Germans and Jews in Washington Heights. Manhattan's Chinatown expands into old Little Italy, and a new Little Italy emerges in Bensonhurst.

Further, the new ethnic neighborhoods are intermixed in a way that enclaves of the early 20th century never were. The restaurants, bakeries, groceries, specialty shops, their customers and owners from a score of different countries and even different continents are now found within a two- or three-block radius. In the Kenmore Avenue area of East Los Angeles, for example, a half-square-mile (1.3 km²) area of former Anglo neighborhood now houses over 9000 people representing Hispanics and Asians of widely varied origin along with Pacific Islanders, Amerindians, African Americans, and a scattering of native-born whites.

Students in the neighborhood school come from 43 countries and speak 23 languages, a localized ethnic intermixture unknown in the communities of single ethnicity so characteristic of earlier stages of immigration to the United States, as the excerpt "Colonies of Immigrants" suggests.

The changing ethnic spatial pattern is not yet clear or certain. Increasing ethnic diversity coupled with continuing immigration flow has, in some instances, expanded rather than reduced patterns of urban group segregation. The tendency for separate ethnic groups to cluster for security, economic, and social reasons cannot be effective if a great many relatively small numbers of different ethnic groups find themselves in a single city setting. Intermixture is inevitable when individual groups do not achieve the critical mass necessary to establish a true identifiable separate community. But as continuing immigration and natural increase allow groups to expand in size, they are able to create more distinctive self-selected ethnic clusters and communities. The 2000 census clearly shows the New York region, for example, to be more ethnically diverse and more segregated than was suspected during the 1990s, with multiple clearly recognizable enclaves and districts each with its own distinctive ethnic or racial composition and character. Immigration growth during the preceding decade yielded not only greater ethnic diversity but greater evident segregation as well.

Even when an ethnic community rejects or is denied assimilation into the larger society, it may both relocate and retain its coherence. "Satellite Chinatowns" are examples of migration from city centers outward to the suburbs or to outer boroughs—in Los Angeles' San Gabriel Valley, stretching in a 20-mile swath eastward from Alhambra and Monterey Park to West Covina and Diamond Bar; in San Francisco, from the downtown area along Grant Avenue to the Richmond district 3 miles away. In New York City, the satellite move was from the still-growing Canal Street area in lower Manhattan to Flushing, about 15 miles away (Figure 6.18) and to Elmhurst which, with immigrants from 114 different countries, is the city's most ethnically diverse neighborhood. Other growing, older ethnic communities—needing more space and containing newly affluent and successful members able to compete for better housing elsewhere—have followed a similar pattern of subdivision and relocation. One result has been termed the *ethnoburb,* a fully structured socioeconomic and political suburban community with a significant, though not necessarily exclusive, concentration of a single ethnic group. For some ethnics, assimilation in job and society does not reduce the need for community identity.

Typologies and Spatial Results

When both the charter group and the ethnic group perceive the social distance separating them to be small, the isolation caused by external discriminatory and internal cohesiveness controls is temporary, and developed ethnic residential clusters quickly give way to full assimilation. While they endure, the clusters may be termed **colonies,** serving essentially as points of entry for members of the particular ethnic group. They persist only to the extent that new arrivals perpetuate the need for them. In American cities, many European ethnic colonies began to lose their vitality

and purpose with the reduction of European immigration flows after the 1920s.

When an ethnic cluster does persist because its occupants choose to preserve it, their behavior reflects the internal cohesiveness of the group and its desire to maintain an enduring **ethnic enclave** or neighborhood. When the cluster is perpetuated by external constraints and discriminatory actions, it has come to be termed a **ghetto.** In reality, the colony, the enclave, and the ghetto are spatially similar outcomes of ethnic concentrations whose origins are difficult to document. Figure 6.19 suggests the possible spatial expressions of these three recognized ethnic-cluster models.

Both discrimination and voluntarism determine the changing pattern of ethnic clustering within metropolitan areas. Where forced segregation limits residential choices, ethnic or racial minorities may be confined to the older, low-cost housing areas, typically close to the city center. Growing ethnic groups that maintain voluntary spatial association frequently expand the area of their dominance by growth outward from the core of the city in a radial pattern. That process has long been recognized in Chicago (Figure 6.20) and has, in that and other cities, typically been extended beyond the central city boundaries into at least the inner fringe of the suburbs.

African Americans have, traditionally, found strong resistance to their territorial expansion from the Anglo charter group, though white-black urban relations and patterns of black ghetto formation and expansion have differed in different sections of the country. A revealing typology of African American ghettos is outlined in Figure 6.21. In the South, the white majority, with total control of the housing market, was able to assign residential space to blacks in accordance with white, not black, self-interest. In the *early southern* ghetto of such pre-Civil War cities as Charleston and New Orleans, African Americans were assigned

Figure 6.18 The Flushing, Queens, area of New York City contains one of the developing "satellite Chinatowns." Like those in other cities, it reflects both the pressures exerted by a growing Chinese community on their older urban enclaves and the suburbanization of an affluent younger generation that still seeks community coherence.

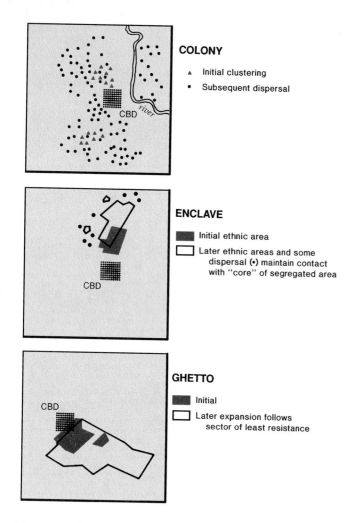

Figure 6.19 **Types of ethnic areas.** Unlike the truly suburban *ethnoburb*, the traditionally recognized ethnic concentrations shown here are all contained within the built-up area of the central city.

Source: David T. Herbert and Colin J. Thomas, Urban Geography, *London: David Fulton Publishers, 1987. Redrawn by permission.*

small dwellings in alleys and back streets within and bounding the white communities where they worked as (slave) house and garden servants. The *classic southern* ghetto for newly free blacks was composed of specially built, low-quality housing on undesirable land—swampy, perhaps, or near industry or railroads—and was sufficiently far from better-quality white housing to maintain full spatial and social segregation.

In the North, on the other hand, African Americans were open competitors with other claimants for space in a generalized housing market. The *early northern* ghetto represented a "toehold" location in high-density, aged, substandard housing on the margin of the central business district. The *classic northern* ghetto is a more recent expansion of that initial enclave to surround the CBD and to penetrate, through invasion and succession, contiguous zones as far as the numbers and the rent-paying ability of the growing African American community will carry. Finally, in new western and southwestern cities not tightly hemmed in by resistant white neighborhoods or suburbs, the black community may display a linear expansion from the CBD to the suburban fringe.

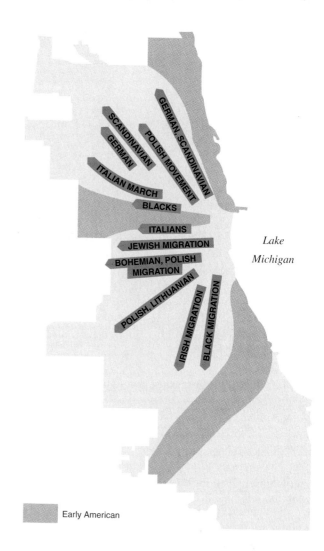

Figure 6.20 **The outward expansion of racial and nationality groups in Chicago.** "Often," Samuel Kincheloe observed in the 1930s, "[minority] groups first settle in a deteriorated area of a city somewhere near its center, then push outward along the main streets." More recently, many—particularly young, innovative, and entrepreneurial—immigrants have avoided traditional first locations in central cities and from their arrival have settled in metropolitan area suburbs and outlying cities where economic opportunity and quality of life is perceived as superior to conditions in the primary inner city.

Source: The American City and its Church *by Samuel Kincheloe. Copyright 1938 by Friendship Press, New York.*

Native-Born Dispersals

Immigration flows to the United States during the last third of the 20th century—unlike those of earlier mass-immigration periods—have begun to affect both the broad regional ethnic makeup of the United States and the internal migration pattern of native-born Americans. The spatial consequence has been dubbed a "demographic balkanization," a pronounced and apparently reinforcing areal segmentation of population by race/ethnicity, economic status, and age across extended metropolitan areas and larger regions of the country.

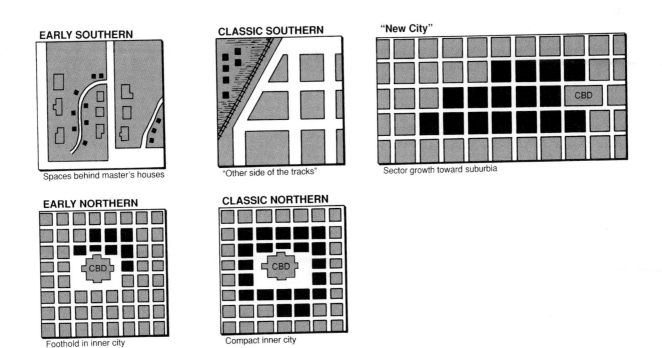

Figure 6.21 **A typology of black ghettos in the United States.**

Source: David T. Herbert and Colin J. Thomas, Urban Geography, *London: David Fulton Publishers, 1987. Redrawn by permission.*

Early 20th-century immigration streams resulted, as we have seen, in temporary ethnic segregation by urban neighborhoods and between central cities and suburbs. Immigration legislation of 1965 dropped the national-origin quotas that had formerly favored European immigrants, replacing that with a more inclusive formula emphasizing family reunification. That change, plus economic and political pressures in many countries of Asia and Latin America, has swelled the influx of poorer, less-skilled Asians and Hispanics. Highly dependent on family members and friends for integration into the informal and formal American job market, the new arrivals are drawn to primary port-of-entry metropolitan areas by chain migration links. In those areas where immigrants account for most of the present and prospective population growth, the trend is toward increasingly multicultural, younger, and poorer residents and dominantly of Hispanic and Asian origins.

The high degree of areal concentration of recent immigrant groups initiated a selective native-born, particularly white, retreat, not only fleeing the cities for the suburbs but leaving entire metropolitan areas and states. California, with nearly one-quarter of its population foreign-born in the mid-1990s, saw a departure of one native-born white or black resident for nearly each foreign-born arrival. Individual urban areas echoed California's state experience. The New York, Chicago, Los Angeles, Houston, and Boston metropolitan areas—5 of the top 11 immigrant destinations—lost 9 native residents for every 10 immigrant arrivals. For whites, top destinations were to cities and states away from coastal and southern border immigrant entry points, from San Francisco to Houston in the West, Boston to Washington plus Miami in the East, and the Chicago district in the interior. African Americans, too, are leaving most of the high-immigration metropolitan areas

with Atlanta, Georgia, the preferred destination. A visible spatial consequence, then, of recent patterns of U.S. immigration and settlement is a decline of the older ideal and reality of immigrant assimilation and of racial and cultural urban mixtures. Instead, the emerging pattern is one of increasing wholesale segregation and isolation by metropolitan areas and segments of the country. Immigrant assimilation may now be more difficult than in the past and social and political divisions more pronounced and enduring.

Cultural Transfer

Immigrant groups arrive at their destinations with already existing sets of production techniques and skills. They bring established ideas of "appropriate" dress, foods, and building styles, and they have religious practices, marriage customs, and other cultural expressions in place and ingrained. That is, immigrants carry to their new homes a full complement of artifacts, sociofacts, and mentifacts. They may modify, abandon, or even pass these on to the host culture, depending on a number of interacting influences: (1) the background of the arriving group; (2) its social distance from the charter group; (3) the disparity between new home and origin-area environmental conditions; (4) the importance given by the migrants to the economic, political, or religious motivations that caused them to relocate; and (5) the kinds of encountered constraints that force personal, social, or technical adjustments on the new arrivals.

Immigrant groups rarely transferred intact all of their culture traits to North America. Invariably, there have been modifications as a result of the necessary adjustment to new circumstances or

physical conditions. In general, if a transplanted cultural element was usable in the new locale, it was retained. Simple inertia suggested there was little reason to abandon the familiar and comfortable when no advantage accrued. If a trait or a cultural complex was essential to group identity and purpose—the religious convictions of the rural Amish, for example, or of urban Hasidic Jews—its retention was certain. But ill-suited habits or techniques would be abandoned if superior American practices were encountered, and totally inappropriate practices would be dropped. German settlers in Texas, for example, found that the vine and the familiar midlatitude fruits did not thrive there. Old-country agricultural traditions were, they discovered, not fully transferable and had to be altered.

Finally, even apparently essential cultural elements may be modified in the face of unalterable opposition from the majority population. Although American in origin, the Latter-day Saints (Mormons) were viewed as outsiders whose practice of polygamy was alien and repugnant. To secure political and social acceptance, church members abandoned that facet of their religious belief. More recently, the some 30,000 Hmong and Mien tribespeople who settled in the Fresno, California, area after fleeing Vietnam, Thailand, and Laos found that their traditional practices of medicinal use of opium, of "capturing" young brides, and of ritual slaughtering of animals brought them into conflict with American law and customs and with the more Americanized members of their own culture group.

Every relocated ethnic group is subject to forces of attraction and rejection. The former tend toward assimilation into the host society; the latter, innate to the group, encourage retention of its self-identity. Acculturation tends to be responsive to economic advantage and to be accelerated if the immigrant group is in many basic traits similar to the host society, if it is relatively well educated, relatively wealthy, and finds political or social advantages in being "Americanized."

Rejection factors internal to the group that aid in the retention of cultural identification include the element of isolation. The immigrant group may seek physical separation in remote areas or raise barriers of a social nature to assure its separation from corrupting influences. Social isolation can be effective even in congested urban environments if it is buttressed by distinctive costume, beliefs, or practices (Figure 6.22). Group segregation may even result in the retention of customs, clothing, or dialects discarded in the original home area.

Rejection factors may also involve **culture rebound,** a belated adoption of group consciousness and reestablishment of identifying traits. These may reflect an attempt to reassert old values and to achieve at least a modicum of social separation. The wearing of dashikis, the adoption of "Afro" hairstyles, the popularity of Ghanian-origin kente cloth, or the celebration of Kwanzaa by American blacks seeking identification with African roots are examples of culture rebound. Ethnic identity is fostered by the nuclear family and ties of kinship, particularly when reinforced by

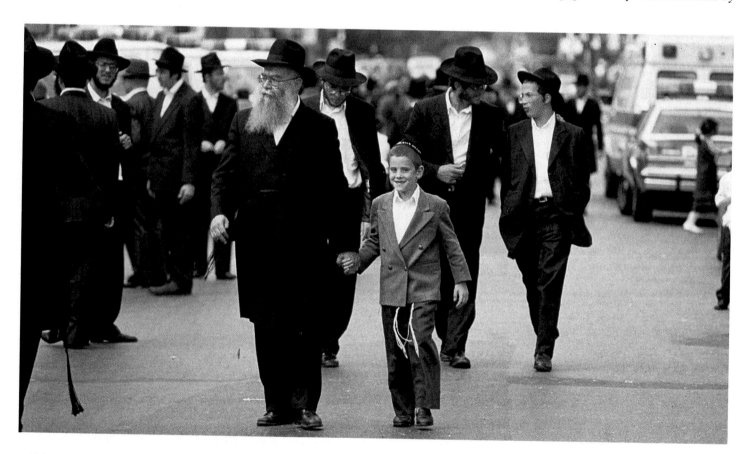

Figure 6.22 Ultra-orthodox Hasidim, segregating themselves by dress and custom, seek social isolation and shun corrupting outside influences even in the midst of New York City's congestion.

Figure 6.23 These young girls, dressed in traditional garb for a Los Angeles Greek Orthodox Church festival, show the close association of ethnicity and religion in the American mosaic.

residential proximity. It is preserved by such group activities as distinctive feasts or celebrations and by marriage customs; by ethnically identified clubs, such as the Turnverein societies of German communities or the Sokol movement of athletic and cultural centers among the Czechs; and by ethnic churches (Figure 6.23).

The Ethnic Landscape

Landscape evidence of ethnicity may be as subtle as the greater number and size of barns in the German-settled areas of the Ozarks or the designs of churches or the names of villages. The evidence may be as striking as the buggies of the Amish communities, the massive Dutch (really, German-origin) barns of southeastern Pennsylvania (Figure 6.24), or the adobe houses of Mexican American settlements in the Southwest. The ethnic landscape, however defined, may be a relic, reflecting old ways no longer pursued. It may contain evidence of artifacts or designs imported, found useful, and retained. In some instances, the physical or customary trappings of ethnicity may remain unique to one or a very few communities. In others, the diffusion of ideas or techniques may have spread introductions to areas beyond their initial impact. The landscapes and landscape evidences explored by cultural geographers are many and complex. The following paragraphs seek merely to suggest the variety of topics pursued in tracing the landscape impacts evident from the cultural diversity of Anglo America.

Figure 6.24 The Pennsylvania Dutch barn, with its origins in southern Germany, has two levels. Livestock occupy the ground level; on the upper level, reached by a gentle ramp, are the threshing floor, haylofts, and grain and equipment storage. A distinctive projecting forebay provides shelter for ground-level stock doors and unmistakably identifies the Pennsylvania Dutch barn. The style, particularly in its primitive log form, was exported from its eastern origins, underwent modification, and became a basic form in the Upland (i.e., off the Coastal Plain) South, Ohio, Indiana, Illinois, and Missouri. An example of a distinctive ethnic imprint on the landscape, the Pennsylvania Dutch barn also became an example of cultural transfer from an immigrant group to the charter group.

Land Survey

The charter group of any area had the option of designing a system for claiming and allotting land appropriate to its needs and traditions. For the most part, the English established land-division policies in the Atlantic Seaboard colonies. In New England, basic land grants were for "towns," relatively compact blocks ideally 6 miles (9.7 km) square. The established central village, with its meeting house and its commons area, was surrounded by larger fields subdivided into strips for allocation among the community members (Figure 6.25). The result was a distinctive pattern of nucleated villages and fragmented farms.

From Pennsylvania southward, the original royal land grants were made to "proprietors," who in turn sold or allotted holdings to settlers. In the southern colonies, the occupants claimed land in amounts approved by the authorities but unspecified in location. The land evaluated as best was claimed first, poor land was passed over, and parcel boundaries were irregular and unsystematic. The *metes-and-bounds* system of property description of the region, based largely on landform or water features or such temporary landscape elements as prominent trees, unusual rocks, or cairns, led to boundary uncertainty and dispute (Figure 6.26). It also resulted in "topographic" road patterns, such as those found in Pennsylvania and other eastern states, where routes are often controlled by the contours of the land rather than the regularity of a geometric survey.

When independence was achieved, the federal government decided that the public domain should be systematically surveyed and subdivided before being opened for settlement. The resulting township and range *rectangular survey* system, adopted in the Land Ordinance of 1785, established survey lines oriented in the cardinal directions and divided the land into townships 6 miles (9.7 km) square which were further subdivided into sections 1 mile (1.6 km) on a side (Figure 6.26). The resultant rectilinear system of land subdivision and ownership was extended to all parts of the United States ever included within the public domain, creating the basic checkerboard pattern of minor civil divisions, the regular pattern of section-line and quarter-line country roads, the block patterns of fields and farms, and the gridiron street systems of American towns and cities.

Elsewhere in North America, the French and the Spanish constituted charter groups and established their own traditions of land description and allotment. The French impress has been particularly enduring. The *long-lot* system was introduced into the St. Lawrence Valley and followed French settlers wherever they established colonies in the New World: the Mississippi Valley, Detroit, Louisiana, and elsewhere. The long lot was a surveyed elongated holding typically about 10 times longer than wide,

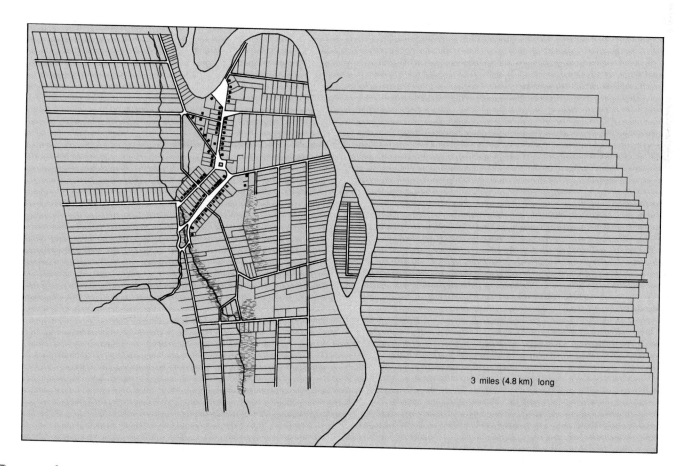

3 miles (4.8 km) long

Figure 6.25 **Wethersfield, Connecticut: 1640–1641.** The home lot and field patterns of 17th-century Wethersfield were typical of villages of rural New England.

Source: Charles M. Andrews, "The River Towns of Connecticut," in Johns Hopkins University Studies in Historical and Political Science, *7th series, VII-VIII-IX (1899), opposite p. 5.*

Figure 6.26 **A contrast in survey systems.** The original metes-and-bounds property survey of a portion of the Virginia Military District of western Ohio is here contrasted with the regularity of surveyor's townships, made up of 36 numbered sections, each one mile (1.6 km) on a side.

Source: Redrawn by permission from Original Survey and Land Subdivision, *Monograph Series No. 4, Norman J. W. Thrower, p. 46, Association of American Geographers, 1966.*

stretching far back from a narrow river frontage (Figure 6.27). The back of the lot was indicated by a roadway roughly parallel to the line of the river, marking the front of a second series (or *range*) of long lots. The system had the advantage of providing each settler with a fair access to fertile land along the floodplain, lower-quality river terrace land, and remote poorer-quality back areas on the valley slopes serving as woodlots. Dwellings were built at the front of the holding, in a loose settlement alignment called a *côte,* where access was easy and the neighbors were close.

Although English Canada adopted a rectangular survey system, the long lot became the legal norm in French Quebec, where it controls land survey even in areas where river access is not significant. In the Rio Grande Valley of New Mexico and Texas, Spanish colonists introduced a similar long-lot system.

Settlement Patterns

The U.S. rural settlement pattern has been dominated by isolated farmsteads dispersed through the open countryside. It is an arrangement conditioned by the block pattern of land survey, by the homesteading tradition of "proving up" claims through residence on them, and by the regular pattern of rural roads. Other survey systems, of course, permitted different culturally rooted settlement choices. The French and Hispanic long lots encouraged the alignment of closely spaced, but separated, farmsteads along river or road frontage (Figure 6.28). The New England village reflected the transplanting of an English tradition. Agricultural villages were found as well in Mormon settlement areas, in the Spanish American Southwest, and as part of the cultural landscapes established by early communistic or religious communities, such as the Oneida Community of New York; the Rappites's Harmony, Indiana; Fountaingrove, California; and other, mostly short-lived "utopias" of the 19th and early 20th centuries.

To encourage the settlement of the prairies, the Mennonites were granted lands in Manitoba not as individuals but as communities. Their established agricultural villages with surrounding communal fields (Figure 6.29) re-created in North America the landscape of their European homelands. (Ethnically German, some Mennonites had colonized in Russia and Ukraine before relocating in North America.)

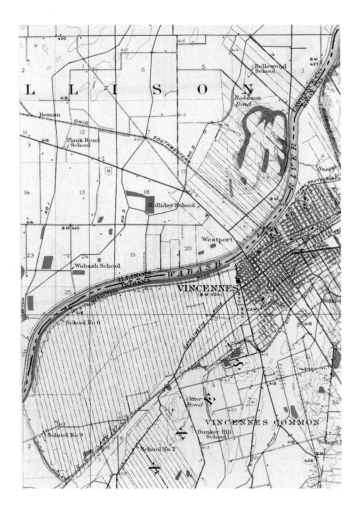

Figure 6.27 **A portion of the Vincennes, Indiana–Illinois topographic quadrangle** (1944) showing evidence of original French long-lot survey. Note the importance of the Wabash River in both long-lot and Vincennes street-system orientations. This U.S. Geological Survey map was originally published at the fractional scale of 1:62,500.

Source: U.S. Geological Survey map.

Ethnic Regionalism

Other world regions display even more pronounced contrasts in the built landscape, reflecting the more entrenched homeland pattern of long-established ethnic regionalism. In areas of intricate mix-tures of peoples—eastern and southeastern Europe, for example—different house types, farmstead layouts, even the use of color can distinguish for the knowledgeable observer the ethnicity of the local population. The one-story "smoking-room" house of the northern Slavs with its covered entrance hall and stables all under one roof marks their areas of settlement even south of the Danube River. Blue-painted one-story, straw-roofed houses indicate Croatian communities. In the Danube Basin, areas of Slovene settlement are distinguished by the Pannonian house of wood and straw-mud. In Spain, the courtyard farmstead marks areas of Moorish influence just as white stucco houses trimmed with dark green or ochre paint on the shutters indicates Basque settlement.

It is impossible to delineate ethnic regions of the United States that correspond to the distinctive landscapes created by sharply contrasting cultural groups in Europe or other world areas. The reason lies in the mobility of Americans, the degree of acculturation and assimilation of immigrants and their offspring, and the significance of other than ethnic considerations in shaping the activities, the associations, and the material possessions of participants in an urbanized, mass communication society. What can be attempted is the delimitation of areas in which particular immigrant-group influences have played a recognizable or determinant role in shaping tangible landscapes and intangible regional "character."

The "melting pot" producing a uniform cultural amalgam, we have seen, has been more American myth than reality. Therefore, there has occurred an inevitable, persistent disparity between the landscapes created by diverse immigrant groups and the national uniformity implicit either in the doctrine of first effective settlement or the concept of amalgamation. That disparity was summarized by Wilbur Zelinsky, as shown in Figure 6.30. The cultural areas are European in origin and can be seen as the expansionary product of three principal colonial culture hearths of the Atlantic Seaboard: the *New England,* the *South,* and the *Midland.* As the figure indicates, the "Middle West" is the product of the union of all three colonial regions. The popularly conceived American "West" probably exists not as a separate unit but as a set of subregions containing cross sections of national population with cultural mixing, but as yet with no achieved cultural uniformity.

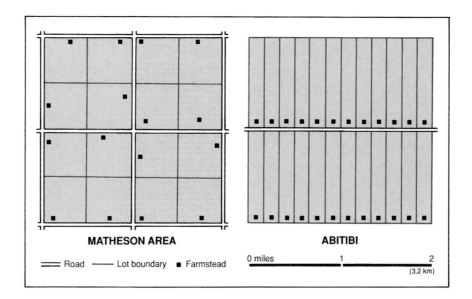

MATHESON AREA   ABITIBI

═══ Road ─── Lot boundary ■ Farmstead

0 miles   1   2
                         (3.2 km)

Figure 6.28 **Land survey in Canada.** Adjacent areas of Canada demonstrate the effects of different survey systems and cultural heritages on rural settlement patterns. The regular plots of Ontario in English Canada (left map) display the isolated farmsteads characteristic of much of rural Anglo America. The long-lot survey of Quebec in French Canada (right map) shows the lot-front alignments of rural dwellings.

Source: Redrawn by permission from Annals of the Association of American Geographers, *George I. McDermott, Vol. 51, p. 263, Association of American Geographers, 1961.*

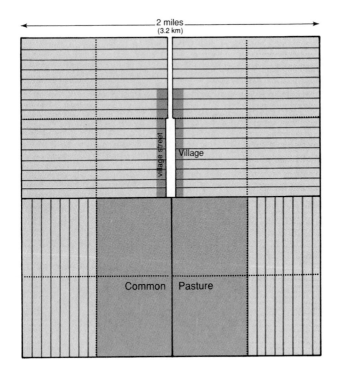

2 miles
(3.2 km)

village street

Village

Common | Pasture

Figure 6.29 **A transplanted ethnic landscape.** The German-speaking Mennonites settled in Manitoba in the 1870s and recreated the agricultural village of their European homeland. Individual farmers were granted strip holdings in the separate fields to be farmed in common with the other villagers. The farmsteads themselves, with elongated rear lots, were aligned along both sides of a single village street in an Old World pattern.

Source: Redrawn from Carl A. Dawson, Group Settlement: Ethnic Communities in Western Canada, *Vol. 7, Canada Frontiers of Settlement (Toronto: Macmillan Company of Canada, 1936), p. 111.*

REGION	APPROXIMATE DATES OF SETTLEMENT AND FORMATION	MAJOR SOURCES OF CULTURE (listed in order of importance)
NEW ENGLAND		
1a. Nuclear New England	1620–1750	England
1b. Northern New England	1750–1830	Nuclear New England, England
THE MIDLAND		
2a. Pennsylvania Region	1682–1850	England and Wales, Rhineland, Ulster, 19th-Century Europe
2b. New York Region or New England Extended	1624–1830	Great Britain, New England, 19th-Century Europe, Netherlands
THE SOUTH		
3a. Early British Colonial South	1607–1750	England, Africa, British West Indies
3b. Lowland or Deep South	1700–1850	Great Britain, Africa, Midland, Early British Colonial South, aborigines
3b-1. French Louisiana	1700–1760	France, Deep South, Africa, French West Indies
3c. Upland South	1700–1850	Midland, Lowland South, Great Britain
3c-1. The Bluegrass	1770–1800	Upland South, Lowland South
3c-2. The Ozarks	1820–1860	Upland South, Lowland South, Lower Middle West
THE MIDDLE WEST		
4a. Upper Middle West	1800–1880	New England Extended, New England, 19th-Century Europe, British Canada
4b. Lower Middle West	1790–1870	Midland, Upland South, New England Extended, 19th-Century Europe
4c. Cutover Area	1850–1900	Upper Middle West, 19th-Century Europe

REGION	APPROXIMATE DATES OF SETTLEMENT AND FORMATION	MAJOR SOURCES OF CULTURE (listed in order of importance)
THE WEST		
5a. Upper Rio Grande Valley	1590–	Mexico, Anglo America, aborigines
5b. Willamette Valley	1830–1900	Northeast U.S.
5c. Mormon Region	1847–1890	Northeast U.S., 19th-Century Europe
5d. Central California	(1775–1848) (Mexico) 1840–	Eastern U.S., 19th Century Europe, Mexico, East Asia
5e. Colorado Piedmont	1860–	Eastern U.S., Mexico
5f. Southern California	(1760–1848) (Mexico) 1880–	Eastern U.S., 19th and 20th-Century Europe, Mormon Region, Mexico, East Asia
5g. Puget Sound	1870–	Eastern U.S., 19th and 20th-Century Europe, East Asia
5h. Inland Empire	1880–	Eastern U.S., 19th and 20th-Century Europe
5i. Central Arizona	1900–	Eastern U.S., Southern California, Mexico
REGIONS OF UNCERTAIN STATUS OR AFFILIATION		
A. Texas	(1690–1836) (Mexico) 1821–	Lowland South, Upland South, Mexico, 19th-Century Central Europe
B. Peninsular Florida	1880–	Northeast U.S., the South, 20th-Century Europe, Antilles
C. Oklahoma	1890–	Upland South, Lowland South, aborigines, Middle West

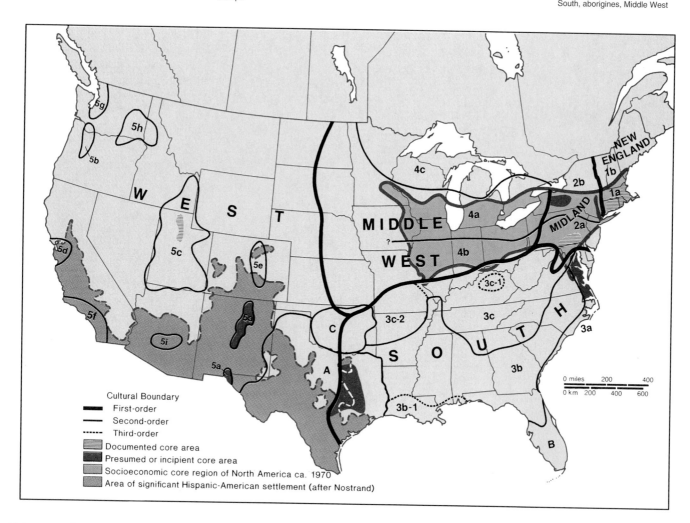

Figure 6.30 **Culture areas of the United States** based on multiple lines of evidence.

From Wilbur Zelinsky, The Cultural Geography of the United States, *© 1992, pp. 118–119. Reprinted by permission of Prentice-Hall, Inc., Upper Saddle River, N.J.*

Ethnic Geography: Threads of Diversity

Summary

Ethnic diversity is a reality in most countries of the world and is increasing in many of them. Immigration, refugee streams, guest workers, and job seekers all contribute to the mixing of peoples and cultures in an area. The mixing is not complete, however. Ethnicity—affiliation in a group sharing common identifying cultural traits—is fostered by territorial separation or isolation. In much of the world, that separation identifies home territories within which the ethnic group is dominant and with which it is identified. In societies of immigrants—Anglo America, for example—such homelands are replaced by ethnic colonies, enclaves, ghettos, and ethnoburbs of self-selected or imposed separation from the larger host society. Cluster migration helped establish such colonies in rural America; chain migration encouraged their development in cities.

The 19th- and early 20th-century American central city displayed pronounced areal segregation as immigrant groups established and clung to protective ethnic neighborhoods while they gradually adjusted to the host culture. A continual population restructuring of urban areas occurred as older groups underwent acculturation, amalgamation, or assimilation, and new groups entered the urban social mix. The durability of ethnic neighborhoods has depended, among other considerations, on the degree of social distance separating the minority group from the host culture and on the significance the immigrant group places on long-term maintenance of their own cultural identity. That is, ethnic communities have been the product of both external and internal forces.

In other world regions, similar spatial separation of immigrant groups by racial, cultural, national, tribal, or village origin within the alien city is common. In Europe, because of the uncertain legal and employment status of many foreign populations and the restricted urban housing market they enter, ethnic enclaves have taken a different form, extent, and level of segregation than has been the case in Anglo America.

Ethnicity is one of the threads of diversity in the spatial cultural fabric. Throughout the world, ethnic groups have imprinted their presence on the landscapes in which they have developed or to which they have transported their culture. In land division, house and farm building style, settlement patterns, and religious structures, the beliefs and practices of distinctive groups are reflected in the cultural landscape. Ethnicity is not, of course, the sole thread in the regional tapestry of societies. Folk culture joins ethnicity as a force creating distinctions between peoples and imparting special character to area. Countering those culturally based sources of separation is the behavioral unification and reduction of territorial distinctiveness that result from the leveling impact of popular culture. It is to these two additional strands in the cultural fabric—folk and popular culture—that we next turn our attention in Chapter 7.

Key Words

acculturation 197
adaptation 192
amalgamation theory 197
assimilation 197
behavioral (cultural) assimilation 197
chain migration 203
charter group 200
cluster migration 202
colony 214
culture rebound 217

ethnic enclave 214
ethnic geography 190
ethnic group 190
ethnic island 201
ethnicity 191
ethnic province 203
ethnocentrism 191
first effective settlement 200
gene flow 192
genetic drift 192

ghetto 214
host society 193
natural selection 192
race 192
segregation 209
social distance 209
structural assimilation 197
tipping point 211

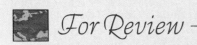 *For Review*

1. How does *ethnocentrism* contribute to preservation of group identity? In what ways might an ethnic group sustain and support new immigrants?

2. How are the concepts of *ethnicity* and *culture* related?

3. What have been some of the principal time patterns of immigration flows into the United States? Into Canada? How are those patterns important to an understanding of present-day social conflicts in either or both countries?

4. How may *segregation* be measured? Does ethnic segregation exist in the cities of world areas outside of North America? If so, does it take different form than in American cities?

5. What forces external to ethnic groups help to create and perpetuate immigrant neighborhoods? What functions beneficial to immigrant groups do ethnic communities provide?

6. What kinds of land surveys were important in the allocation of property in the North American culture realm? With which *charter groups* were the different survey systems associated? How did survey systems affect settlement patterns?

 Focus Follow-up

1. **What are the implications and bases of "ethnicity," and how have historic immigration streams shaped Anglo American multiethnicity?** pp. 190–196.

Ethnicity implies a "people" or "nation," a large group classified according to common religious, linguistic, or other aspects of cultural origin or background, or, often, to racial distinctions. In common with nearly all countries, the United States and Canada are multiethnic. Past and current immigration streams—earlier primarily European, more recently Asian and Latin American—have intricately mixed their populations.

2. **How were the dominant Anglo American culture norms established, and how complete spatially and socially are its ethnic minorities integrated?** pp. 196–208.

The first effective settlers of Anglo America created its English-rooted charter culture to which other, later immigrant groups were expected to conform. Assimilation or acculturation has not been complete, and areal expressions of ethnic differentiation persist in America in the form of ethnic islands, provinces, or regional concentrations. French Canadian, black, Amerindian, Hispanic, Asian American, and other, smaller groups display recognizable areal presences. Among immigrant groups, those concentrations may result from cluster and chain migration.

3. **What patterns of ethnic diversity and segregation exist in the world's urban areas, and how are they created or maintained?** pp. 209–216.

Ethnic communities, clusters, and neighborhoods are found in cities worldwide. They are a measure of the social distance that separates minority from majority or other minority groups. Segregation measures the degree to which culture groups are not uniformly distributed within the total population. Although different world regions show differing patterns, all urban segregation is based on external restrictions of isolation and discrimination or ethnic group internal separatism controls of defense, mutual support, and cultural preservation. Ethnic colonies, enclaves, ghettos, and ethnoburbs are the spatial result.

4. **What have been some of the cultural landscape consequences of ethnic concentrations in Anglo America and elsewhere?** pp. 216–223.

Landscape evidence of ethnicity may be subtle or pronounced. In Anglo America, differing culturally based systems of land survey and allocation—such as metes-and-bounds, rectangular, or long lot—of earlier groups may still leave their landscape impacts. Clustered and dispersed rural settlement customs; house and barn types and styles; distinctive, largely urban, "Chinatowns," "Little Havanas," and other cultural communities; and even choices in dwelling-house colors or urban art are landscape imprints of multiethnicity in modern societies.

 Selected References

Allen, James P., and Eugene Turner. "Spatial Patterns of Immigrant Assimilation." *Professional Geographer* 48, no. 2 (1996): 140–155.

Arreola, Daniel D. "Urban Ethnic Landscape Identity." *Geographical Review* 85, no. 3 (1995): 527–543.

Berry, Kate A., and Martha L. Henderson, eds. *Geographical Identities of Ethnic America*. Reno: University of Nevada Press, 2002.

Brewer, Cynthia, and Trudy Suchan. *Mapping Census 2000: The Geography of U.S. Diversity.* Washington, D.C.: U.S. Bureau of the Census, 2001.

Clark, William A. V. *The California Cauldron: Immigration and the Fortunes of Local Communities.* New York: Guilford, 1998.

Clark, William A. V. *Immigrants and the American Dream: Remaking the Middle Class.* New York: Guilford, 2003.

Conzen, Michael P. "Ethnicity on the Land." In *The Making of the American Landscape,* edited by Michael P. Conzen, pp. 221–248. Boston: Unwin Hyman, 1990.

Frantz, Klaus, and Robert A. Sauder, eds. *Ethnic Persistence and Change in Europe and America: Traces in Landscape and Society.* Innsbruck, Austria: University of Innsbruck, 1996. *Veröffentlichungen der Universität Innsbruck* 213.

Harris, Chauncy. "New European Countries and Their Minorities." *Geographic Review* 83 no. 3 (1993): 301–320.

Harris, Cole. "French Landscapes in North America." In *The Making of the American Landscape,* edited by Michael P. Conzen, pp. 63–79. Boston: Unwin Hyman, 1990.

Hornbeck, David. "Spanish Legacy in the Borderlands." In *The Making of the American Landscape,* edited by Michael P. Conzen, pp. 51–62. Boston: Unwin Hyman, 1990.

"International Migration and Ethnic Segregation: Impacts on Urban Areas." Special issue of *Urban Studies* 35, no. 3 (March 1998).

Kaplan, David H., and Steven R. Halloway. *Segregation in Cities.* Washington, D.C.: Association of American Geographers, 1998.

Lee, Sharon M. "Asian Americans: Diverse and Growing." *Population Bulletin* 53, no. 2. Washington, D.C.: Population Reference Bureau, 1998.

Martin, Philip, and Elizabeth Midgley. "Immigration: Shaping and Reshaping America." *Population Bulletin* 58, no. 2. Washington, D.C.: Population Reference Bureau, 2003.

Martin, Philip, and Elizabeth Midgley. "Immigration to the United States: Journey to an Uncertain Destination." *Population Bulletin* 54, no. 2. Washington, D.C.: Population Reference Bureau, 1999.

McKee, Jesse O., ed. *Ethnicity in Contemporary America: A Geographical Appraisal.* 2d rev. ed. Lanham, Md.: Rowman & Littlefield, 2000.

Noble, Allen G., ed. *To Build in a New Land: Ethnic Landscapes in North America.* Baltimore, Md.: Johns Hopkins University Press, 1992.

Nostrand, Richard L. *The Hispano Homeland.* Norman: University of Oklahoma Press, 1992.

Nostrand, Richard, and Lawrence Estaville, eds. *Homelands: A Geography of Culture and Place Across America.* Baltimore, Md.: Johns Hopkins University Press, 2002.

O'Hare, William P. "America's Minorities: The Demographics of Diversity." *Population Bulletin* 47, no. 4. Washington, D.C.: Population Reference Bureau, 1992.

Pinal, Jorge del, and Audrey Singer. "Generations of Diversity: Latinos in the United States." *Population Bulletin* 52, no. 3. Washington, D.C.: Population Reference Bureau, 1997.

Pollard, Kelvin M., and William P. O'Hare. "America's Racial and Ethnic Minorities." *Population Bulletin* 54, no. 3. Washington, D.C.: Population Reference Bureau, 1999.

Portes, Alejandro, and Ruben G. Rumbaut. *Immigrant America.* 2d ed. Berkeley: University of California Press, 1996.

Price, Edward T. *Dividing the Land: Early American Beginnings of Our Private Property Mosaic.* Geography Research Paper 238. Chicago: University of Chicago Press, 1995.

Pritzker, Barry M. *A Native American Encyclopedia: History, Culture, and Peoples.* New York: Oxford University Press, 2000.

Riche, Martha F. "America's Diversity and Growth: Signposts for the 21st Century." *Population Bulletin* 55, no. 2. Washington, D.C.: Population Reference Bureau, 2000.

Roseman, Curtis C., Gunther Thieme, and Hans Dieter Laux, eds. *EthniCity: Geographic Perspectives on Ethnic Change in Modern Cities.* Lanham, Md.: Rowman & Littlefield, 1995.

Shinagawa, Larry Hajime, and Michael Jang. *Atlas of American Diversity.* Walnut Creek, Calif.: Altamira Press/Sage Publications, 1998.

Shumway, J. Matthew, and Richard H. Jackson. "Native American Population Patterns." *Geographical Review* 85, no. 2 (1995): 185–201.

Waldinger, Roger, ed. *Strangers at the Gates: New Immigrants in Urban America.* Berkeley, Calif.: University of California Press, 2001.

Zelinsky, Wilbur. *The Enigma of Ethnicity: Another American Dilemma.* Iowa City: University of Iowa Press, 2001.

Websites: The World Wide Web has a tremendous number and variety of sites pertaining to geography. Websites relevent to the subject matter of this chapter appear in the "Web Links" section of the On-line Learning Center associated with this book. Access it at **www.mhhe.com/fellmann8e**

Seven

Folk and Popular Culture:
Diversity and Uniformity

Focus Preview

A. Folk Culture

1. Anglo American hearths and folk building traditions, pp. 230–242.

2. Nonmaterial folk culture: foods, music, medicines, and folklore, pp. 242–249.

3. Folk regions and regionalism, pp. 249–251.

B. Popular Culture

4. The nature and patterns of popular culture: inside the mall and out, pp. 251–257.

5. Diffusion and regionalism in popular culture, pp. 257–262.

Opposite: Performing the horn dance at Staffordshire, England. Folk culture and traditions are cherished and preserved in all societies.

n rural and frontier America before 1850, the games peo-ple played were local, largely unorganized individual and team contests. Running, wrestling, weight lifting, shooting, or—if the Native American influence had been strong—shinny (field hockey), kickball, or lacrosse. In the growing cities, rowing, boxing, cricket, fencing, and the like involved the athleti-cally inclined, sometimes as members of sporting clubs and spon-sored teams. Everywhere, horse racing was an avid interest. In the countryside, sports and games relieved the monotony and iso-lation of life and provided an excuse, after the contests, for meet-ing friends, feasting, and dancing. Purely local in participation, games reflected the ethnic heritage of the local community—the games of the homeland—as well as the influence of the American experience. In the towns, they provided the outdoor recreation and exercise otherwise denied to shop-bound clerks and artisans. Without easy transportation, contests at a distance were difficult and rare; without easy communication, sports results were of local interest only.

The railroad and the telegraph changed all that. Teams could travel to more distant points, and scores could be immedi-ately known to supporters at home and rivals in other cities. Baseball clubs were organized during the 1850s throughout the East and the Middle West. The establishment of the National As-sociation of Base Ball Players in 1857 followed shortly after the railroad reached Chicago, and even before the Civil War, New York teams were competing throughout that state. After the war, the expanding rail network turned baseball into a national craze. The National League was organized in 1876; Chicago, Boston, New York, Washington, Kansas City, Detroit, St. Louis, and Philadelphia all had professional teams by the 1880s, and innu-merable local leagues were formed. Horse racing, prizefighting, amateur and professional cycling races, and intercollegiate sports—football, baseball, rowing, and track and field contests—pitted contestants and drew crowds over long distances. Sports and games had been altered from small-group participations to national events. They were no longer purely local, traditional, in-formal expressions of community culture; rather, organized sport had emerged as a unifying, standardized expression of national popular culture (Figure 7.1).

The kaleidoscope of culture presents an endlessly changing de-sign, different for every society, world region, and national unit, and different over time. Ever present in each of its varied pat-terns, however, are two repeated fragments of diversity and one spreading color of uniformity. One distinctive element of diver-sity in many societies derives from *folk* culture—the material and nonmaterial aspects of daily life preserved by smaller groups par-tially or totally isolated from the mainstream currents of the larger society around them. A second source of diversity in com-posite societies, as we saw in Chapter 6, is surely and clearly pro-vided by *ethnic* groups, each with its distinctive characterizing heritage and traditions and each contributing to the national cul-tural mix. Finally, given time, easy communication, and common interests, *popular* culture may provide a unifying and liberating coloration to the kaleidoscopic mix, reducing differences between formerly distinctive groups though perhaps not totally eradicating them. These three elements—folk, ethnic, and popular—of the cultural mosaic are intertwined. We will trace their connections particularly in the Anglo American context, where diversified im-migration provided the ethnic mix, frontier and rural isolation en-couraged folk differentiation, and modern technology produced the leveling of popular culture. Along the way, we will see evi-dences of their separate influences in other societies and other culture realms.

Folk Cultural Diversity and Regionalism

Folk connotes traditional and nonfaddish, the characteristic or product of a homogeneous, cohesive, largely self-sufficient group that is essen-tially isolated from or resistant to outside influences, even of a larger society surrounding it. **Folk culture,** therefore, may be defined as the collective heritage of institutions, customs, skills, dress, and way of life of a small, stable, closely knit, usually rural community. Tradition con-trols folk culture, and resistance to change is strong. The homemade and handmade dominate in tools, food, music, story, and ritual. Build-ings are erected without architect or blueprint, but with plan and pur-pose clearly in mind and by a design common to the local society using locally available building materials. When, as in Anglo America, folk culture may represent a modification of imported ideas and tech-niques, local materials often substitute for a less-available original substance even as the design concepts are left unchanged.

Figure 7.1 Spectator sports emerged as a major element in American popular culture following the Civil War. The Cincinnati Red Stockings of 1869, shown in this print, were the first openly professional baseball team; the National League was established in 1876. Mark Twain, an early fan, wrote: "Baseball is the very symbol, the outward and visible expression of the drive and push and struggle of the raging, tearing, booming nineteenth century." Organized football was introduced as a college sport—also in 1869—when Rutgers played Princeton in the first intercollegiate game.

Folk life is a cultural whole composed of both tangible and intangible elements. **Material culture** is made up of physical, visible things: everything from musical instruments to furniture, tools, and buildings. Collectively, material culture comprises the **built environment,** the landscape created by humans. At a different scale, it also constitutes the contents of household and workshop. **Nonmaterial culture,** in contrast, is the intangible part, the mentifacts and sociofacts expressed in oral tradition, folk song and folk story, and customary behavior. Ways of speech, patterns of worship, outlooks and philosophies are parts of the nonmaterial component passed to following generations by teachings and examples.

Within Anglo America, true folk societies no longer exist; the universalizing impacts of industrialization, urbanization, and mass communication have been too pervasive for their full retention. Generations of intermixing of cultures, of mobility of peo-

ples, and of leveling public education have altered the meaning of *folk* from the identification of a group to the recognition of a style, an article, or an individual preference in design and production. The Old Order Amish, with their rejection of electricity, the internal combustion engine, and other "worldly" accoutrements in favor of buggy, hand tools, and traditional dress are one of the least altered—and few—folk societies of the United States (Figure 7.2).

Canada, on the other hand, with as rich a mixture of cultural origins as the United States, has kept to a much later date clearly recognizable ethnically unique folk and decorative art traditions. One observer has noted that nearly all of the national folk art traditions of Europe can be found in one form or another well preserved and practiced somewhere in Canada. From the earliest arts and crafts of New France to the domestic art forms and folk artifacts of the Scandinavians, Germans, Ukrainians, and others who settled in western Canada in the late 19th and early 20th centuries, folk and ethnic are intertwined through transference of traditions from homelands and their adaptation to the Canadian context.

Folk culture today is more likely to be expressed by individuals than by coherent, isolated groups. The collector of folk songs, the artist employing traditional materials and styles, the artisan producing in wood and metal products identified with particular groups or regions, the quilter working in modern fabrics the designs of earlier generations all are perpetuating folk culture: material culture if it involves "things," nonmaterial if the preserved tradition relates to song, story, recipe, or belief. In this respect, each of us bears the evidence of folk life. Each of us uses proverbs traditional to our family or culture; each is familiar with and can repeat childhood nursery rhymes and fables. We rap wood for luck and likely know how to make a willow whistle, how to plant a garden by phases of the moon, and what is the "right" way to prepare a favorite holiday dish.

When many persons share at least some of the same folk **customs**—repeated, characteristic acts, behavioral patterns, artistic traditions, and conventions regulating social life—and when those customs and artifacts are distinctively identified with any area long inhabited by a particular group, a *folk culture region* may be recognized. As with landscape evidence of ethnicity, folk culture in its material and nonmaterial elements may be seen to vary over time and space and to have hearth regions of origin and paths of diffusion.

Indeed, in many respects, ethnic geography and folk geography are extensions of each other and are logically intertwined. The variously named "Swiss" or "Mennonite" or "Dutch" barn introduced into Pennsylvania by German immigrants has been cited as physical evidence of ethnicity; in some of its many modifications and migrations, it may also be seen as a folk culture artifact of Appalachia. The folk songs of, say, western Virginia can be examined either as nonmaterial folk expressions of the Upland South or as evidence of the ethnic heritage derived from rural English forebears. In the New World the debt of folk culture to ethnic origins is clear and persuasive. With the passage of time, of course, the dominance of origins recedes and new cultural patterns and roots emerge.

(a)

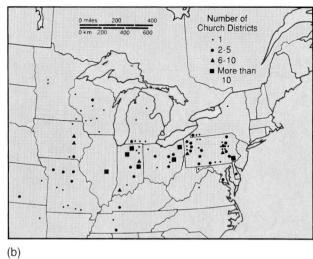

(b)

Figure 7.2 (*a*) Motivated by religious conviction that the "good life" must be reduced to its simplest forms, Old Order Amish communities shun all modern luxuries of the majority secular society around them. Children use horse and buggy, not school bus or automobile, on their daily trip to this rural school in east central Illinois. (*b*) **Distribution of Old Order Amish communities** in the United States.

Source: (b) Redrawn by permission from Annals of the Association of American Geographers, *William K. Crowley, Vol. 68, p. 262, Association of American Geographers, 1978.*

Anglo American Hearths

Anglo America is an amalgam of peoples who came as ethnics and stayed as Americans or Canadians. They brought with them more than tools and household items and articles of dress. Importantly, they brought clear ideas of what tools they needed, how they should fashion their clothes, cook their food, find a spouse, and worship their deity. They knew already the familiar songs to be sung and stories to be told, how a house should look and a barn be raised. They came, in short, with all the mentifacts and sociofacts to shape the artifacts of their way of life in their new home (Figure 7.3). (Mentifacts, sociofacts, and artifacts are discussed in Chapter 2.)

Their trappings of material and nonmaterial culture frequently underwent immediate modification in the New World. Climates and soils were often different from their homelands; new animal and vegetable foodstuffs were found for their larders. Building materials, labor skills, and items of manufacture available at their origins were different or lacking at their destinations. What the newcomers brought in tools and ideas they began to modify as they adapted and adjusted to different American materials, terrains, and potentials. The settlers still retained the essence and the spirit of the old but made it simultaneously new and American.

The first colonists, their descendants, and still later arrivals created not one but many cultural landscapes of America, defined by the structures they built, the settlements they created, and the regionally varied articles they made or customs they followed. The natural landscape of America became settled, and superimposed on the natural landscape as modified by its Amerindian occupants were the regions of cultural traits and characteristics of the European immigrants (see "Vanished American Roots"). In

Figure 7.3 Reconstructed Plimoth Plantation. The first settlers in the New World carried with them fully developed cultural identities. Even their earliest settlements reflected established ideas of house and village form. Later, they were to create a variety of distinctive cultural landscapes reminiscent of their homeland areas, though modified by American environmental conditions and material resources.

their later movements and those of their neighbors and offspring, they left a trail of landscape evidence from first settlement to the distant interior locations where they touched and intermingled.

The early arrivers established footholds along the East Coast. Their settlement areas became cultural hearths, nodes of introduction into the New World—through *relocation diffusion*—of concepts and artifacts brought from the Old. Locales of innovation in a new land rather than areas of new invention, they were—exactly as their ancient counterparts discussed in Chapter 2—source regions from which relocation and *expansion diffusion* carried their cultural identities deeper into the continent (Figure 7.4). Later arrivals, as we have seen in Chapter 6, not only added their own evidence of passage to the landscape but often set up independent secondary hearths in advance of or outside of the main paths of diffusion.

Each of the North American hearths had its own mix of peoples and, therefore, its own landscape distinctiveness. French settlement in the lower St. Lawrence Valley re-created there the long lots and rural house types of northwestern France. Upper Canada was English and Scottish with strong infusions of New England folk housing carried by Loyalists leaving that area

Vanished American Roots

America, like every other world region, had its own primitive, naïve, and indigenous original architecture. But this was the architecture of Indians—the bark houses of the Penobscots, the long houses of the Iroquois, the tipis of the Crows, the mounds of the Mandans, the pueblos of the Zuñi, the hogans of the Navajos, the [plank] dwellings of Puget Sound.

Some of these were even elegant, many contained seeds of promise; but we swept them all aside. Indian words and Indian foods passed into the American culture but nothing important from the Indian architecture, save a belated effort to imitate the form but not the function of the pueblos. (The so-called "Spanish" architecture of the Hispanic borderlands and northern Mexico, however—

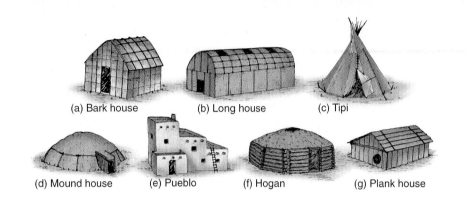

(a) Bark house   (b) Long house   (c) Tipi

(d) Mound house   (e) Pueblo   (f) Hogan   (g) Plank house

adobe-walled with small windows and flat-roofs supported by wooden beams—was of Amerindian, not European, origin.)

Source: From John Burchard and Albert Bush-Brown, *The Architecture of America: A Social and Cultural History,* (Boston: Little, Brown and Company, 1961), p. 57. © 1961, The American Institute of Architects.

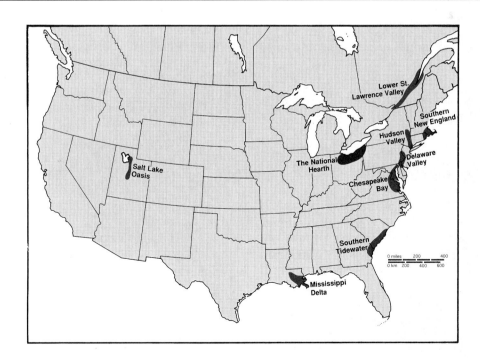

Figure 7.4 **Early Anglo American culture hearths.** The interior "national hearth," suggested by Richard Pillsbury, represents a zone of coalescence in the eastern Midwest, from which composite housing ideas dispersed farther into the interior.

Sources: Based on Allen G. Noble, Wood, Brick, and Stone, *vol. 1 (Amherst: University of Massachusetts Press, 1984); and* Annals of the Association of American Geographers, *Richard Pillsbury, vol. 60, p. 446, Association of American Geographers, 1970.*

during the Revolutionary War. Southern New England bore the imprint of settlers from rural southern England, while the Hudson Valley hearth showed the impress of Dutch, Flemish, English, German, and French Huguenot settlers.

In the Middle Atlantic area, the Delaware River hearth was created by a complex of English, Scotch-Irish, Swedish, and German influences. The Delaware Valley below Philadelphia also received the eastern Finns, or Karelians, who introduced, according to one viewpoint, the distinctive "backwoods" lifestyles, self-sufficient economies, and log-building techniques and house designs of their forested homeland. It was their pioneering "midland" culture that was the catalyst for the rapid advance of the frontier and successful settlement of much of the interior of the continent and, later, of the Pacific Northwest.

Coastal Chesapeake Bay held English settlers, though Germans and Scotch-Irish were added elements away from the major rivers. The large landholdings of the area dispersed settlement and prevented a tightly or clearly defined culture hearth from developing, although distinctive house types that later diffused outward did emerge there. The Southern Tidewater hearth was dominantly English modified by West Indian, Huguenot, and African influences. The French again were part of the Delta hearth, along with Spanish and Haitian elements.

Later in time and deeper in the continental interior, the Salt Lake hearth marks the penetration of the distant West by the Mormons, considered an ethnic group by virtue of their self-identity through religious distinctiveness. Spanish American borderlands, the Upper Midwest Scandinavian colonies, English Canada, and the ethnic clusters of the Prairie Provinces could logically be added to the North American map of distinctive immigrant culture hearths.

The ethnic hearths gradually lost their identification with immigrant groups and became source regions of American architecture and implements, ornaments and toys, cookery and music. The evidence of the homeland was there, but the products became purely indigenous. In the isolated, largely rural American hearth regions, the ethnic culture imported from the Old World was partially transmuted into the folk culture of the New.

Folk Building Traditions

People everywhere house themselves and, if necessary, provide protection for their domesticated animals. Throughout the world, native rural societies established types of housing, means of construction, and use of materials appropriate to their economic and family needs, the materials and technologies available to them, and the environmental conditions they encountered. Because all these preconditions are spatially variable, rural housing and settlement patterns are comparably varied, a diversity increasingly lost as standardization of materials (corrugated metal, poured concrete, cinder block, and the like) and of design replace the local materials and styles developed through millennia by isolated folk societies.

The world is not yet, of course, totally homogenized. The family compound of the Bambara of Mali (Figure 7.5) is obvi-

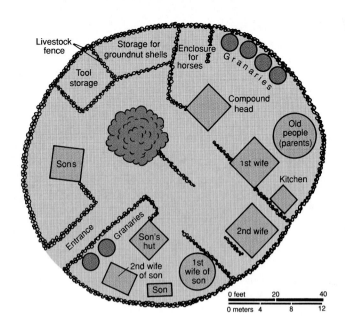

Figure 7.5 The extended family compound of the Bambara of Mali.

Source: Redrawn with permission from Reuben K. Udo, The Human Geography of Tropical Africa *(Ibadan: Heinemann Education Books (Nigeria) Ltd., 1982), p. 50.*

ously and significantly different from the farmstead of a North American rural family. The Mongol or Turkic *yurt* or *ger,* a movable low, rounded shelter of felt, skin, short poles, and rope, is a housing solution adapted to the needs and materials of nomadic herdsmen of the Asian grasslands (Figure 7.6a). A much different solution with different materials is reached by the Maasai, a similar nomadic herding society but of the grasslands of eastern Africa. Their temporary home was traditionally the *manyatta,* an immovable low, rounded hut made of poles, mud, and cow dung that was abandoned as soon as local grazing and water supplies were consumed (Figure 7.6b). As the structures in Figure 7.7 can only slightly suggest, folk housing solutions in design and materials provide a worldwide mosaic of nearly infinite diversity and ingenuity.

Within the Anglo American realm, although architectural diversity does not reach global proportions, the variety of ethnic and regional origins of immigrant streams and the differences in encountered environmental conditions assured architectural contrasts among the several settlement hearths of the New World. The landscapes of structures and settlements creating those contrasts speak to us of their creators' origins, travels, adaptations to new locales, and importations and retentions of the habits and customs of other places. One of the joys of travel in a world region as internally diverse as that of North America is to observe the variations in its cultural landscape, to listen to the many voices that tell of its creation through houses, barns, farmsteads, and village designs.

The folk cultural heritage is now passing; old farm structures are replaced or collapse with disuse as farming systems change. Old houses are removed, remodeled, or abandoned, and the modern, the popular, or the faddish everywhere replaces the evidences of first occupants. A close-knit community may preserve

(a)

(b)

Figure 7.6 (*a*) A Uighur *yurt* in Xinjiang Province, China; (*b*) the Maasai *manyatta.*

(a)

(b)

(c)

(d)

(e)

Figure 7.7 The common characteristics of preindustrial folk housing are an essential uniformity of design within a culture group and region, a lack of differentiation of interior space, a close adaptation to the conditions of the natural environment, and frequently ingenious use of available materials and response to the dictates of climate or terrain. (*a*) Stone house of Nepal; (*b*) Icelandic sod farm house; (*c*) reed dwelling of the Uros people on Lake Titicaca, Peru; (*d*) a Dogon village in Mali, West Africa; (*e*) traditional housing on Nias Island, off the west coast of Sumatra, Indonesia.

(a) and (b) Courtesy of Colin E. Thorn.

Folk and Popular Culture: Diversity and Uniformity   **235**

the past by resisting the present, but except where the efforts of preservationists have been successful in retaining and refurbishing one or a few structures or where outdoor museums and re-creations have been developed, the landscapes—the voices—of the past are gradually lost. Many of those fading voices first took on their North American accents in the culture hearths suggested in Figure 7.4. They are still best heard in the house types associated with them.

The Northern Hearths

Vernacular house styles—those built in traditional form but without formal plans or drawings—were part of the material culture of early colonists that met new conditions in America. In the Northeast, colder, snowier winters posed different environmental challenges than did the milder, frequently wetter climates of northwestern Europe, and American stone and timber were more accessible and suitable construction materials than the clay and

thatch common in the homeland. Yet the new circumstances at first affected not at all, or only slightly, the traditional housing forms.

The Lower St. Lawrence Valley

The St. Lawrence Valley (see Figure 7.4) remains as one of the few areas with structural reminders of a French occupation that spread widely but impermanently over eastern Anglo America. There, in French Canada, beginning in the middle of the 17th century, three major house types were introduced. All were styles still found in western France today.

In the lower valley below Quebec City, *Norman cottages* appear as near-exact replicas of houses of the Normandy region of northern France, with immense hipped roofs steeply pitched to wide or upturned (bell-cast) eaves (Figure 7.8a). The *Quebec cottage,* more widely distributed and more varied in construction materials than the Norman cottage, featured two unequal rooms; a steeply pitched (but gabled) roof with wide, overhanging eaves; and, frequently, an elevated front porch or galley. External walls

(a)

(b)

(c)

(d)

Figure 7.8 Buildings of the Lower St. Lawrence hearth region. (*a*) the Norman cottage; (*b*) the Quebec cottage; (*c*) the Montreal house; (*d*) the Quebec long barn.

Sources: (a), (b), and (c) Courtesy of John A. Jakle. (d) Reprinted from Wood, Brick, and Stone, Volume 2, The North American Settlement Landscape *by Allen G. Noble. Drawing by M. Margaret Geib. Copyright © 1984 by the University of Massachusetts Press. Reprinted by permission.*

were built of mortared and whitewashed stone rubble or, often, were framed and sheathed with sawn weatherboard (Figure 7.8b). The *Montreal house*—so named because of its concentration in the Upper St. Lawrence Lowland—was a larger stone structure more characteristic of the crowded city than of the open countryside (Figure 7.8c).

Along with their house styles, the French brought the characteristic *Quebec long barn,* stretching 50 or more feet wide with several bays and multiple barn functions efficiently contained within a single structure. It was an attractive design for keeping the farmer indoors in bitterly cold Canadian winters, though weather protection was not the primary purpose of the French original (Figure 7.8d). While the St. Lawrence Valley house types were found in other areas of French settlement in North America—Louisiana, the St. Genevieve area of Missouri, northern Maine—the long barn was not carried outside of French Canada.

Southern New England

The rural southern English colonists who settled in New England carried memories of the heavily framed houses of their home counties: sturdy posts and stout horizontal beams held together by simple joinery and sided by overlapping clapboards. The series of New England vernacular houses that emerged in the new settlements all displayed that construction and were further distinguished by steep roofs and massive central chimneys.

Among the primary house types evolved in the New England hearth were (1) the *garrison house,* whose characteristic second-floor overhang was a relict of urban house design in medieval Europe (Figure 7.9a); (2) the *saltbox house,* with an asymmetrical gable roof covering a shed or lean-to addition giving extra rooms on the first floor to the rear (Figure 7.9b); and (3) the *New England large house* of up to ten rooms with lobby entrance, central chimney, and a symmetrical gable roof. Later, the central-chimney design gave way completely to the *Georgian* style with paired

(a)

(b)

(c)

(d)

Figure 7.9 New England house types. (*a*) The garrison house; (*b*) the saltbox house; (*c*) the Georgian-style variant of the New England large house; (*d*) An upright-and-wing house, the wing representing a one-story extension of the basic gable-front house plan.

(b) and (d) Courtesy of John A. Jakle.

chimneys that reinforced the sense of balance (Figure 7.9c). The New England hearth also created the *gable-front* house and its variation, the *upright-and-wing,* or *lazy-T,* house (Figure 7.9d), modified versions of which became landscape staples in both rural and urban areas from western New York into the Middle West.

The Hudson Valley

An area settled by a complex mixture of Dutch, French, Flemish, English, and German settlers, the Hudson Valley showed a comparable mixture of common house forms. However, the first European settlers of consequence were the Dutch. Beginning in 1614, they dominated the Valley until the English took control of the region in the late 17th century, though Dutch-style homes were built into the 18th century and some still are found in the region (Figure 7.10). Dutch homes were characterized by a steep roof, often with flared eaves. Most were side-gabled, one-story stone structures. A common feature was the split "Dutch door" with an upper half allowing air to enter the building while the closed lower half kept children inside and animals out.

The Middle Atlantic Hearths

The Middle Atlantic hearths were ethnically diverse sites of vernacular architecture more influential on North American housing styles than any other early settlement areas. The log cabin, later carried into Appalachia and the trans-Appalachian interior, evolved there. There, too, was introduced what would later be called the *I house*—a two-story structure one room deep with two rooms on each floor that became prominent in the Upper South and the Lower Middle West in the 19th century.

Figure 7.10 This New York Dutch house shares similarities, especially its outswept eaves, with the Quebec cottage of Canada's St. Lawrence Valley. The French Canadian house, with its roots in Normandy, and this Dutch American house of the Colonial Hudson Valley, with its roots in Flanders, issue from vernacular designs of the North European Plain.

Photo by Jon C. Malinowski.

The Delaware Valley

Dutch and Swedish settlers were less successful in colonizing the Delaware Valley hearth area than were the English Quakers and Germans who arrived in the late 17th and early 18th centuries. These latter were joined by Finns, Welsh, and Scotch-Irish, each of whom contributed to the diversity of stone, brick, frame, and log housing of the district. Urban Quakers, arriving with the memory of the Great London Fire of 1666 still fresh in mind, built in red brick small versions of the Georgian-style houses then becoming fashionable in England. Germans, Scandinavians, and, particularly, eastern Finns introduced the first New World log houses in the 17th century in Delaware and New Jersey. It was they, not the English colonists, who gave America that frontier symbol.

The Delaware Valley hearth (sometimes called the Pennsylvania hearth) is particularly noted for two vernacular house designs: the *four-over-four house*—so called in reference to its basic two-story floor plan with four rooms up and four down (Figure 7.11a); and the classic *I house,* almost always a "two-over-two" arrangement (Figure 7.11b). The valley also made a major contribution to American architecture in the form of a barn—or rather a series of related designs of the German bank barn. Unlike the earlier English and Dutch barns, which were essentially crop oriented, the bank barn combined animal shelter with the grain storage and threshing functions. The variously named German, Pennsylvania, Dutch, or Schweizer (Swiss) barn—in its several versions—was carried from its Pennsylvania origins to the continental interior and from the southern Appalachians northward to Ontario (see Figure 6.24). Perhaps no other hearth region had as widespread an influence on American vernacular architecture as did the Pennsylvania hearth. Migrants from there carried their material culture southwestward along the Great Valley of Virginia, as well as due west into the Ohio Valley.

Chesapeake Bay

The area of dominantly English and Scotch-Irish settlement around Chesapeake Bay was rural and nearly devoid of large cities. Its settlers initially introduced wood-framed houses, though brick construction became increasingly common. Both building types featured raised foundations, outside end chimneys, and one-deep floor plans. Kitchens were often detached, and by the 18th century adaptation to the more southerly temperature conditions was reflected in added front porches and front-to-rear ventilation passages. Popular throughout the Middle Atlantic hearth regions, the classic *I house* was also part of the vernacular architecture of the Chesapeake Bay hearth (Figure 7.11b); it was early carried into the Upper South and, after the 1850s, into the interior. Sometimes of brick but overwhelmingly of frame construction, its builders and building materials were brought by the new railroads to Indiana, Illinois, and Iowa (the *I*'s after which the house was named).

The Southern Hearths

Both climate and a new ethnic cultural mix altered the form of vernacular housing in the southern hearths along the Atlantic Coast and in the Gulf and Delta areas. Although local responses

(a)

(b)

Figure 7.11 House types of the Middle Atlantic hearths. (*a*) Four-over-four house; (*b*) the traditional or classic I house, with its two rooms on each floor separated by central hallways, had a varying number of façade openings and, usually, end chimneys located in the standard gable roof, but all symmetrically organized. This brick version, characteristic of the Upper South, has a detached summer kitchen.

(a) Courtesy of John A. Jakle.

to these influences varied, the overall result was housing in a different style for different needs in the South than in the North.

The Southern Tidewater

Along the southeastern Atlantic coastal region of South Carolina and Georgia, European settlers faced problems of heat, humidity, and flooding not encountered farther to the north. The malaria, mosquitoes, and extreme heat plaguing their inland plantations during the summer caused the wealthy to prefer hot-season residence in coastal cities such as Charleston, where sea breezes provided relief. The result was the characteristic *Charleston single house,* a name related to its single row of three or four rooms ranged from front to back and lined on the outside of each floor by a long veranda along one side of the structure (Figure 7.12).

The Mississippi Delta

The French, dominant in the Lower St. Lawrence Valley far to the north, established their second North American culture hearth in New Orleans and along the lower Mississippi during the 18th century (see Figure 7.8). There, French influences from Nova Scotia and the French Caribbean islands—Haiti, specifically— were mixed with Spanish and African cultural contributions. Again, heat and humidity were an environmental problem requiring distinctive housing solutions. The *grenier house* emerged as the standard design for rural Louisiana. Usually of frame construction with cypress siding, the structures were raised on posts or pillars several feet off the ground for cooling and protection against floodwaters, ground rot, and termites.

The *shotgun house* is a simple, inexpensive, and efficient house style identified with the delta area but owing its origin to Africa by way of Port-au-Prince, Haiti, and an introduction into America by free Haitian blacks who settled in the delta before the middle of the 19th century. With its narrow gable front, its considerable length of three or four rooms, and its front-to-back

Figure 7.12 The Charleston single house.

alignment of all room door openings (Figure 7.13), it was quickly and cheaply made of sawed lumber and found favor far beyond the delta area as affordable urban and rural housing.

Interior and Western Hearths

Other immigrant groups, some from the eastern states, others from abroad—and all encountering still different environmental circumstances and building material sources—made their impress on local areas of the interior and North American west. Settlers of many different origins on the Great Plains initially built sod dugouts or sod or rammed earth houses in the absence of native timber stands. Later, after the middle of the century, "balloon frame" construction, utilizing newly available cheap wire nails

(a)

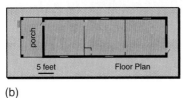

(b)

Figure 7.13 (*a*) Shotgun cottages in Claiborne County, Mississippi; (*b*) one variant of a shotgun cottage floor plan.

(*a*) *Courtesy of John A. Jakle.*

and light lumber milled to standard dimensions, became the norm in the interior where heavy timbers for traditional post and beam construction were not available. The strong, low-cost housing the new techniques and materials made possible owed less to the architectural traditions of eastern America than it did to the simplicity and proportional dimensions imposed by the standardized materials. Midwest vernacular house types developed—including the one-story gabled rectangle, double-wing, and two-story foursquare farmhouses, quickly constructed by local carpenters or the farmers themselves.

The thick-walled *Spanish adobe house,* long and single-storied with a flat or low-pitched earth-covered roof entered Anglo America through the Hispanic borderlands (see Figure 6.11) but in most of its features owed more to indigenous Pueblo Indian design than to Spanish origins. In the Far West, Hispanic and Russian influences were locally felt, although housing concepts imported from the humid East predominated. In the Utah area, Mormon immigrants established the *central-hall house,* related to both the I house and the four-over-four house, as the dominant house type.

A variety of ethnic and architectural influences met and intermingled in the Pacific Northwest. French Canadians produced a closely knit ethnic settlement on the Willamette River at French Prairie (between Salem and Portland, Oregon). Chinese came to the coal mines of Vancouver Island in the 1860s; later, thousands were employed in the construction of the Northern Pacific Railroad. By the 1870s, an architecturally distinctive Chinatown was centered around the foot of Yesler Way and Occidental Avenue in Seattle, and similar enclaves were established in Tacoma, Portland, and other urban centers. But most immigrants to the British

Columbia—Washington—Oregon regions were of Anglo American, not foreign birth, and the vast majority on the U.S. side came from midwestern roots, representing a further westward migration of populations whose forebears (or who themselves) were part of the Middle Atlantic culture hearths. Some—the earliest—carried to the Oregon and Washington forested regions the "midland" American backwoods pioneer culture and log-cabin tradition first encountered in the Delaware Valley hearth; others brought the variety of housing styles already well represented in the continental interior.

Architectural Diffusions

These vernacular architectural origins and movements were summarized by the cultural geographer Fred Kniffen, who thought that house types of the eastern United States and ultimately of much of Anglo America could be traced to three source regions on the Atlantic Coast, each feeding a separate diffusion stream: New England, Middle Atlantic, and Southern Coastal (Figure 7.14). *New England,* he argued, gave rise to a series of evolving house types based on a simple English original, variants of which spread westward with the settlers across New York, Ohio, Indiana, and Illinois and into Wisconsin and Iowa.

The principal contributions of the *Middle Atlantic* source region were the English *I house* and the Finnish-German log building. Its major diffusion directions were southward along the Appalachian Uplands, with offshoots in all directions, and westward across Pennsylvania. Multiple paths of movement from this hearth converged in the Ohio Valley Midwest, creating an interior "national hearth" of several intermingled streams (see Figure 7.4), and from there spread north, south, and west. In this respect, the narrow Middle Atlantic region played for vernacular architecture the same role its Midland dialect did in shaping the linguistic geography of the United States, as discussed on pages 156–157.

The earliest diffusion from the Middle Atlantic hearth was the backwoods frontier culture that carried rough log carpentry to all parts of the forested East and, eventually, westward to the northern Rockies and the Pacific Northwest. The identifying features of that building tradition were the dogtrot and saddlebag house plans and double-crib barn designs. The basic unit of both house and barn was a rectangular "pen" ("crib" if for a barn) of four log walls that characteristically stood in tandem with an added second room that joined the first at the chimney end of the house. The resultant two-room central chimney design was called a *saddlebag house.* Another even more common expansion of the single-pen cabin was the *dogtrot* (Figure 7.15), a simple roofing-over of an open area left separating the two pens facing gable toward gable. Log construction techniques and traditions were carried across the intervening grasslands to the wooded areas of the mountains and the Pacific Coast during the 19th century. The first log buildings of settlements and farmsteads of the Oregon territory, for example, were indistinguishable from their eastern predecessors of the preceding century.

The third source area, in the Lower Chesapeake, spread its remarkably uniform influence southward as the *Southern Coastal*

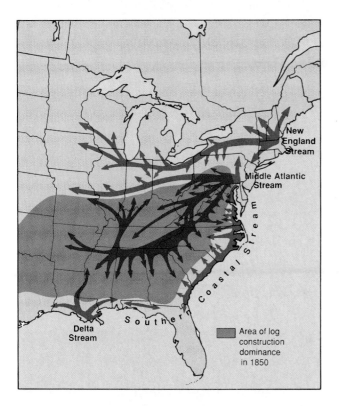

(a)

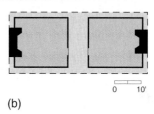

(b)

Figure 7.14 **Architectural source areas and the diffusion of building methods** from the Atlantic Seaboard hearths. The map emphasizes log and frame construction as of 1850. The variation in the width of stream paths suggests the strength of the influence of the various hearths on vernacular housing away from the coast. The Southern Coastal Stream was limited in its influence to the coastal plain. The Delaware Valley hearth not only exerted a strong impact on the Upland South but also became—along with other Middle Atlantic hearths—the dominant vernacular housing source for the lower Middle West and the continental interior. By 1850, and farther to the west, new expansion cores were emerging around Salt Lake City, in coastal California, and in the Willamette Valley area of Oregon—all bearing the imprint of housing designs that first emerged in eastern hearths.

Sources: F. Kniffen, Annals of the Association of American Geographers, Vol. 55:560, 1965; Fred Kniffen and Henry Glassie, "Building in Wood in the Eastern United States" in Geographical Review 56:60, © 1966 The American Geographical Society; and Terry G. Jordan and Matti Kaups, The American Backwoods Frontier, pp. 8–9, © 1989 The Johns Hopkins University Press.

Figure 7.15 The "dogtrot" house.

Folk Fencing in Anglo America

Fencing, a nearly essential adjunct of agricultural land use, has been used as an indicator of the folk cultural traditions of farm populations, as a guide to settlement periods and stages, and as evidence of the resources and environmental conditions the settlers found. The stone fence, for example, was an obvious response to the need to clear glacial fieldstone before cultivation, and over 250,000 miles of stone fences were in existence in the 1870s in New York and New England. Elsewhere, angular or flat stones in sedimentary rock areas—in southern Ohio, Indiana, or parts of Kentucky, for example—made fences easier to build and maintain than did the rounded glacial boulders of New England.

The heavily forested eastern hearth regions provided an abundance of timber and poles for a variety of wooden fences, many based on European models. The "buck" fence, identical to some French folk fences, was reproduced in North America from French Canada southward into the Southern Appalachians (Figure 7.16a). The wattle fence of interlaced poles and branches (Figure 7.16b) was common in medieval Europe, known and briefly used by early settlers of Massachusetts and Virginia, but not found elsewhere in North America.

The angled-rail, "snake," or "worm" fence was a dominant American fence form for much of the 19th century until increasing labor costs and wood scarcity made it uneconomical. Its earlier attraction was its ease of construction—it required no post holes—and its use of abundant farm-produced wood; it was widely found in the South and in the eastern portion of the

Stream, diffusing its impact inland along numerous paths into the Upland South (Figure 7.14). In that area of complex population movements and topographically induced isolations, source area architectural styles were transformed into truly indigenous local folk housing forms. By 1850, diffusions from the eastern architectural hearths had produced a clearly defined folk housing geography in the eastern half of the United States and subsequently, by relocation and expansion diffusion, had influenced vernacular housing throughout Anglo America. The French and Caribbean influences of the *Delta Stream,* in contrast, were much more restricted and localized.

Figure 7.16 Folk fencing of the eastern United States. (*a*) A buck fence; (*b*) a wattle or woven fence; (*c*) the angled-rail, snake, or worm fence; (*d*) a post-and-rail fence.

Middle West (Figure 7.16c). Indeed, wherever the backwoods pioneers temporarily settled, the zigzag log fence enclosed their forest clearings. The design was carried into the Pacific Northwest from Oregon to British Columbia, to be replaced only as new generations of farmers appeared and farm woodlots were reduced.

The post-and-rail fence, a form that consumed less land and fewer rails than did the angled fence, was particularly popular in southern New England and the Delaware Valley areas (Figure 7.16d). After the establishment of an American steel industry during the last half of the 19th century—and in the grasslands of the Great Plains—wire fencing, barbed wire in particular, became the commonly encountered form of stock enclosure or crop protection. Briefly, however, both sod fences and hedge fences (the Osage orange in particular) were popular from the forest margin westward to the mountains.

Nonmaterial Folk Culture

Houses burn, succumb to rot, are remodeled beyond recognition, or are physically replaced. Fences, barns, and outbuildings are similarly transitory features of the landscape, lost or replaced by other structures in other materials for other purposes as farms mechanize, consolidate, no longer rear livestock, are abandoned, or are subdivided for urban expansion. The folk housing and farm buildings that seem so solid a part of the built environment are, in reality, but temporary features of it. Impermanent, too, for the most part, are the tools and products of the folk craft worker. Some items of daily life and decoration may be preserved in museums or as household heirlooms; others may be exchanged among collectors of antiques. But inevitably, material folk culture is lost as the artifacts of even isolated groups increasingly are replaced by products of modern manufacture and standardized design.

In many ways, more permanent records of our folk heritages and differentials are to be found in the intangibles of our lives, in the nonmaterial folk culture that all of us possess and few recognize. Although ways of life change and the purposes of old tools are forgotten, favorite foods and familiar songs endure. Nonmaterial characteristics may more indelibly mark origins and flows of cultures and peoples than physical trappings outmoded, replaced, or left behind.

One important aspect of folk geography is the attention it pays to the spatial association of culture and environment. Folk societies, because of their subsistence, self-reliant economies,[1] and limited technologies, are deemed particularly responsive to physical environmental circumstances (see "Subsistence Household Economies"). Thus, foodstuffs, herbs, and medicinal plants naturally available or able to be grown locally—as well as shelter—have been especially subject to folk geographic study. Less immediately connected to the environment, but important indicators of the backgrounds and memories of a social group, are the stories, fables, and music traditional to it and transmitted within it.

[1]In which people produce most or all of the goods to satisfy their own and their family's needs; little or no exchange occurs outside of the immediate or extended family.

Subsistence Household Economies

Whatever may be folk societies' regional differences based on the varying environments they occupy or the differing cultures they express, all have in common an economy of self-sufficiency based on family or small group cooperative effort. In the hearth regions and along the diffusion paths of settlement in Anglo America, the basic subsistence unit was the individual household. The husband and wife were equal partners in the enterprise, producing their own food, clothing, housing, furniture, and such necessary household goods as candles and soap. Beverly Sanders describes the gender division of labor in colonial households throughout much of the eastern hearth regions and the complex and arduous tasks assigned to women, who of necessity were masters of a wide range of folk crafts, artifacts, and skills. The essentials of women's responsibilities did not change in later years in the settlement of the United States or Canadian prairies or of the Northwest.

> By and large the man was responsible for building the house and furniture, clearing the land and planting crops, [and] slaughtering the larger animals. The woman was responsible for feeding and clothing all household members, manufacturing household necessities, housecleaning, nursing and child care. . . .
>
> Feeding a family involved far more than just cooking. The first care . . . was tending the fire. Throughout the year, the woman had to produce most of the food as well as cook it. She generally planted and tended a kitchen garden in which she grew vegetables that could be stored in cold cellars for the winter, and others that could be dried. Fruits such as apples and berries were dried or preserved. The homemaker also cared for the barnyard animals. In order to have a chicken in the kettle for dinner, [she] had to slaughter, pluck and clean the fowl the same day. When cows and pigs were slaughtered she had the gigantic task of salting the beef to preserve it, and smoking the pig meat into ham and bacon. . . . Homemakers pickled a wide variety of foods. . . .
>
> Like food, clothing had to be made "from scratch." Fur and leather were popular materials for clothing [on the frontier] because they didn't need to be woven into cloth. Making linen cloth from the flax plant was a painstaking process that could take as long as sixteen months—from planting flax to spinning and bleaching thread. In wool making, the fleece from the sheep was cleaned, oiled, and then combed to draw out the fibers to be spun into thread. Women wove the linen and woolen threads into cloth on a hand loom. . . . Clothing took a long time to make because it was generally worked on in odd hours that women could spare from more pressing chores.
>
> . . . [C]andle making had to be done during the busy autumn season before the long dark winter set in. Most were made from tallow, a rendered animal fat which was melted with boiling water in a large heavy kettle. Rows of candle wicks . . . were dipped into the melted tallow, cooled, then dipped again and again. Since candles were always scarce, they were stored away very carefully, and used very sparingly.
>
> The most unpleasant chores of all were the cleaning of houses, clothing and people. Water was scarce and had to be carried into the house from a nearby stream. The cleaning agent—a soft soap—was manufactured by a tedious process that involved the combination of animal grease and lye, a caustic substance derived from ashes. . . . The busy homemaker could not possibly do laundry and housecleaning every day, but rather set aside special days for it once a month, or even once in three months. . . .
>
> Here is one day's work in the year 1775, set down in the diary of . . . a Connecticut girl:
>
> Fix'd gown for Prude,—Mend Mother's Riding-hood,—Spun short thread,—Fix'd two gowns for Welsh's girls,—Carded tow,—Spun linen,—Worked on Cheesebasket,—Hatchle'd flax with Hannah, we did 51 lbs. apiece,—Pleated and ironed,—Read a Sermon of Dodridge's,—Spooled a piece,—Milked the cows,—Spun linen, did 50 knots,—Made a Broom of Guinea wheat straw,—Spun thread to whiten,—Set a Red dye,—Had two Scholars from Mrs. Taylor's,—I carded two pounds of whole wool and felt Nationaly,—Spun harness twine,—Scoured the pewter.

Source: Beverly Sanders, *Women in the Colonial Era and the Early American Republic 1607–1820,* (Newton, MA: WEEA/Education Development Center, 1979) pp. 18–21.

Folk Food and Drink Preferences

Folk and Customary Foods

Cuisine, meaning the selection of foods and the style of cookery, is one of the most evident and enduring of the elements distinguishing cultural groups. Ethnic foods are the mainstay and the attraction of the innumerable fairs and "fests" held throughout the United States and Canada to celebrate the traditions of locally important groups. In the case of ethnic foods, of course, what is celebrated is the retention in a new environment of the food preferences, diets, and recipes that had their origin in a distant homeland. Folk food habits, on the other hand, are products of local circumstances; the dietary inputs are the natural foods derived from hunting, gathering, and fishing or the cultivated foods and domestic animals suited to the environmental conditions locally encountered.

The distinction between folk and ethnic is no clearer in foods than it is in other aspects of regional culture. Three observations may be made on this point. First, most societies have until recent times been intimately and largely concerned with food production on an individual and family basis. The close ties of people to environment—*folk* ties—are therefore particularly evident in food gathering and growing (Figure 7.17).

Second, most areas of the world have been occupied by a complex mix of peoples migrating in search of food and carrying food habits and preferences with them in their migrations. In the Americas, Australia, New Zealand, and a few other regions of recent colonization and diversified settlement, we are aware of these differing *ethnic* origins and the recipes and customs they imply. In other world regions, ethnic and cultural intermixture is less immediately apparent. In Korea, for example, what outsiders see as a distinctive ethnic cookery best known, perhaps, for *kimch'i*—brined, pickled, and spiced vegetables in endless combinations and uniquely Korean—also incorporates Japanese and Chinese foodstuffs and dishes.

Third, food habits are not just matters of sustenance but are intimately connected with the totality of *culture* or *custom.* People eat what is available and also what is, to them, edible. Sheep's brains and eyeballs, boiled insects, animal blood, and pig intestines, which are delicacies in some cultures, may be abominations to others unfamiliar with the culture that offers them as special treats to guests. (For a special case of folk food habit, see "A Taste for Dirt.") Further, in most societies food and eating are considered a social, not just a personal, experience. Among Slavic peoples, to offer a guest bread and salt is a mark of esteem and welcome, and the bountiful and specially prepared meal as the mark of hospitality is common in nearly all cultures.

The interconnections between the folk, the ethnic, and the customary in food habits and preferences are evident in the Anglo American scene of mixed settlement and environmental diversity. Of course, the animals and plants nurtured, the basic recipes followed and flavorings added, and the specialized festive dishes of American folk groups have ethnic origins. Many originated abroad and were carried to and preserved in remote New World areas. Many were derived from the larder of the Amerindians and often varyingly used in different regional contexts. Turkey, squash, pumpkin, and cranberries were among them as was the

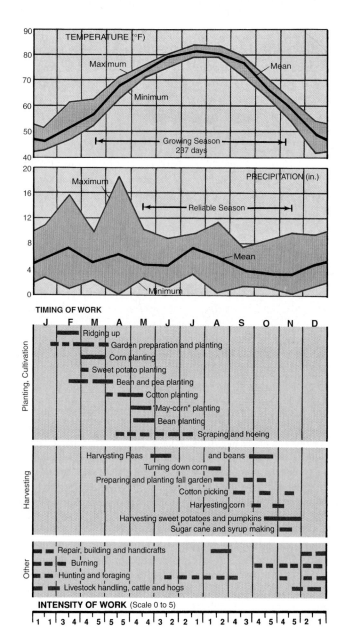

Figure 7.17 **The traditional "annual round" of folk culture farming** in the Upland South area of eastern Louisiana. The system and sequence of farming activities has varied little since the area was first settled around 1800. Frost danger dates and the phases of the moon are important in determining exact planting times. The corn, peas, and sweet potatoes assure the Upland farmer subsistence for family and animals. "The prudent folk farmer provides for subsistence first; then he turns to money crops"—in this case, cotton.

Source: Milton Newton, Jr., "The Annual Round in the Upland South: The Synchronization of Man and Nature through Culture," Pioneer America 3, No. 2 (Akron, Ohio: Pioneer Society of America, 1971), p. 65. Redrawn by permission.

corn (maize) that appeared with time as Southern grits, Southwestern tortillas, and everywhere south of Pennsylvania as the American replacement for wheat in the making of bread. Such classic American dishes as Brunswick stew (a thick stew made with vegetables and two meats, such as squirrel and rabbit or chicken), the clambake, smoked salmon, cornflakes, and beef jerky were originally Indian fare. Gradually, the environmental

A Taste for Dirt

Hundreds of millions of people throughout the world eat dirt, usually fine clays, in a custom—called *geophagy*—so widespread it is usually considered to be within the range of normal human behavior. The practice is particularly common in sub-Saharan Africa, where hundreds of farmer and herder cultures consume dirt and, in some cases, sand. Africans brought as slaves to the United States carried the habit with them, and it is now prevalent among their descendants in the American South. It is also found widely in Asia, the Middle East, and parts of Latin America.

Wherever the custom is practiced, it is most common among pregnant women, though it may be more usual than reported among males as well. Some data indicate that from 30% to 50% of expectant mothers in large areas of Africa and among rural blacks in sections of the American South eat clay, as do hundreds of millions of women elsewhere in the world. Since dozens of animal species also consume clay, it is usually assumed that geophagy can supply minerals otherwise deficient in diets or can counteract nausea or diarrhea. (One of the most popular and widely available commercial remedies for diarrhea is based upon the clay *kaolin*). Heavy, chronic clay consumption can also cause serious ailments, including intestinal blockages, anemia, growth retardation, and zinc deficiency among some practitioners of the habit, however.

As a folk food, clay may be specially selected for flavor (a sour taste is preferred). It is often mixed with bread dough or with vinegar and salt and baked, cooked, or smoked like bacon. It may also be used as a condiment or neutralizer. Nearly all varieties of wild potatoes growing in the Andes Mountains of South America (where the potato is native) contain toxic chemicals, as do some of the species cultivated and regularly eaten by the Indian populations there. To counteract intestinal distress caused by consuming the tubers, Indians either leach out the chemicals or, commonly, eat the potatoes with a dip made of clay and a mustardlike herb. That practice may have made it possible to domesticate this important food crop. Amerindians of the southwestern United States similarly use clay as a condiment with toxic wild tubers and acorns.

Like many other folk customs, geophagy is identified with specific groups rather than universally practiced. Like others, it seems to persist despite changes in the earlier dietary circumstances that may originally have inspired it and despite (in the United States) growing social pressure condemning the habit. Southern relatives may send favored varieties of clay to women who have moved to northern cities. Others, yielding to refined sensibilities, may substitute Argo starch, which has similar properties, for the clay consumption habits of their culture group.

---

influences, isolation, and time spans implicit in the concept of folk culture created culinary distinctions among populations recognized as American rather than ethnic immigrants.

Shelves of cookbooks mark the general distinctions of folk cuisines of the United States. Broad categories of New England, Creole, Southern, Chesapeake, Southwestern, and other regional cookery may be further refined into cookbooks containing Boston, Pennsylvania Dutch, Charleston, New Orleans, Tidewater, and other more localized recipes. Specific American dishes that have achieved fame and wide acceptance developed locally in response to food availability. New England seafood chowders and baked beans; southern pone, johnnycake, hush puppies, and other corn- (maize-) based dishes; the wild rice of the Great Lakes states; Louisiana crayfish (crawfish); southern gumbo; and salmon and shellfish dishes of the Pacific coast are but a few of many examples of folk foods and recipes originally and still characteristic of specific cultural areas but subsequently made part of national food experience. Others, once locally known, effectively disappear as the culture or foodstuff source is lost. The "fern pie" of Oregon's frontier past and "pigeon pie" made with the now-extinct passenger pigeon are among many examples.

Drink

In the United States, drink also represents an amalgam between ethnic imports and folk responses and emphases. A colonial taste for rum was based on West Indian and Tidewater sugarcane and molasses. European rootstock was introduced, with mixed results, to develop vineyards in most seaboard settlements; the native scuppernong grape was tried for wine making in the South. Peach, cherry, apple, and other fruit brandies were distilled for home consumption. Whiskey was a barley-based import accompanying the Scots and the Scotch-Irish to America, particularly to the Appalachians. In the New World, the grain base became native corn (maize), and whiskey making became a deeply rooted folk custom integral to the subsistence economy.

Whiskey also had cash economy significance. Small farmers of isolated areas far from markets converted part of their corn and rye crops into whiskey to produce a concentrated and valuable commodity conveniently transportable by horseback over bad roads (see "Transferability," p. 67). Such farmers viewed a federal excise tax imposed in 1791 on the production of distilled spirits as an intolerable burden not shared by those who could sell their grain directly. The tax led, first, to a short-lived tax revolt, the Whiskey Rebellion of 1794, in western Pennsylvania and, subsequently, to a tradition of moonshining—producing untaxed liquor in unlicensed stills. Figure 7.18 suggests the close association between its isolated Appalachian upland environment and illicit whiskey production in East Tennessee in the 1950s.

Folk Music

North American folk music began as transplants of familiar Old World songs carried by settlers to the New World. Each group of immigrants established an outpost of a European musical community, making the American folk song, in the words of Alan Lomax, "a museum of musical antiques from many lands." But the imported songs became Americanized, hybridization between musical traditions occurred, and American experience added its own songs of frontier life, of farming, courting, and laboring (see "The American Empire of Song"). Eventually, distinctive American styles of folk music and recognizable folk song cultural regions developed (Figure 7.19).

The *Northern* song area—including the Maritime Provinces of Canada, New England, and the Middle Atlantic states—in general featured unaccompanied solo singing in clear, hard tones. Its ballads were close to English originals, and the British connection was continuously renewed by new immigrants, including Scots and Irish. The traditional ballads and current popular songs brought by British immigrants provided the largest part of the Anglo Canadian folk song heritage. On both sides of the border,

the fiddle was featured at dances, and in the States, fife-and-drum bands became common in the early years of the Republic.

The *Southern Backwoods and Appalachian* song area, extending westward to East Texas, involved unaccompanied, high-pitched, and nasal solo singing. The music, based on English tradition and modified by Appalachian "hardscrabble" life, developed in isolation in upland and lowland settlement areas. Marked by moral and emotional conflict with an undercurrent of haunting melancholy, the backwoods style emerged in the modern period as the roots of the distinctive and popular genre of "country" music.

The northern and southern traditions abutted in a transition zone along the Ohio Valley but blended together across the Mississippi to create the *Western* song area. There, factual narrative songs reflected the experiences of cowboy, riverman, sodbuster, and gold seeker. Natural beauty, personal valor, and feminine purity were recurring themes. Many songs appeared as reworked

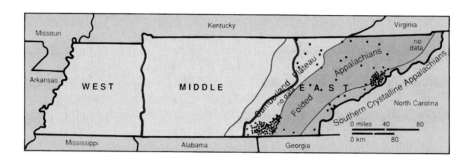

Figure 7.18 In the mid-1950s, official estimates put weekly moonshine production at 24,000 gallons in mountainous eastern Tennessee, at 6000 gallons in partially hilly middle Tennessee, and at 2000 gallons in flat western Tennessee. The map shows the approximate number of stills seized each month at that time in East Tennessee. Each dot indicates one still.

Source: Redrawn by permission from Loyal Durand, "Mountain Moonshining in East Tennessee," Geographical Review 46 (New York: American Geographical Society, 1956), p. 171.

The American Empire of Song

The map sings. The chanteys surge along the rocky Atlantic seaboard, across the Great Lakes and round the moon-curve of the Gulf of Mexico. The paddling songs of the French-Canadians ring out along the Saint Lawrence and west past the Rockies. Beside them, from Newfoundland, Nova Scotia, and New England, the ballads, straight and tall as spruce, march towards the West.

Inland from the Sea Islands, slave melodies sweep across the whole South from the Carolinas to Texas. And out on the shadows of the Smoky and Blue Ridge mountains the old ballads, lonesome love songs, and hoedowns echo through the upland South into the hills of Arkansas and Oklahoma.

There in the Ozarks the Northern and Southern song families swap tunes and make a marriage.

The Texas cowboys roll the little doughies [sic] north to Montana, singing Northern ballads with a Southern accent. New roads and steel rails lace the Southern backwoods to the growl and thunder of Negro chants of labour—the axe songs, the hammer songs, and the railroad songs. These blend with the lonesome hollers of levee-camp mule-skinners to create the blues, and the blues, America's *cante hondo*, uncoils its subtle, sensual melancholy in the ear of all the states, then all the world.

The blues roll down the Mississippi to New Orleans, where the Creoles mix the musical gumbo of jazz—once a dirty word, but now a symbol of musical freedom for the West. The Creoles add Spanish pepper and French sauce and blue notes to the rowdy tantara of their reconstruction-happy brass bands, stir up the hot music of New Orleans and warm the weary heart of humanity. . . . These are the broad outlines of America's folk-song map.

"Introduction" from *Folk Songs of North America* by Alan Lomax. Copyright © 1960 by Alan Lomax. Used by permission of Doubleday, a division of Random House, Inc.

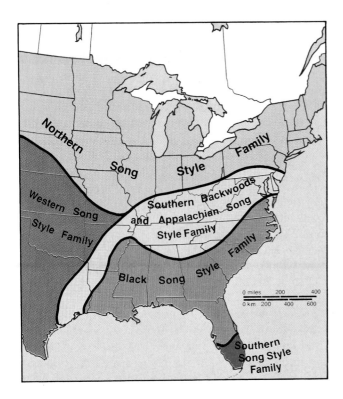

Figure 7.19 **Folk song regions of eastern United States.** Alan Lomax has indirectly outlined folk culture regions of the eastern United States by defining areas associated with different folk song traditions.

Source: Redrawn "Map depicting folk song regions of the Eastern U.S." by Rafael Palacios, from Folk Songs of North America *by Alan Lomax. Copyright © 1960 by Alan Lomax. Used by permission.*

lumberjack ballads of the North or other modifications from the song traditions of the eastern United States.

Imported songs are more prominent among the traditional folk tunes of Canada than they are in the United States; only about one-quarter of Canadian traditional songs were composed in the New World. Most native Canadian songs—like their U.S. counterparts—reflected the daily lives of ordinary folk. In Newfoundland and along the Atlantic coast, those lives were bound up with the sea, and songs of Canadian origin dealt with fishing, sealing, and whaling. Particularly in Ontario, it was the lumber camps that inspired and spread folk music. Anglo Canadian songs show a strong Irish character in pattern and tune and traditionally were sung solo and unaccompanied.

The *Black* folk song tradition, growing out of racial and economic oppression, reflects a union of Anglo American folk song, English country dancing, and West African musical patterns. The African American folk song of the rural South or the northern ghetto was basically choral and instrumental in character; hands and feet were used to establish rhythm. A strong beat, a leader-chorus style, and deep-pitched mellow voices were characteristic.

Lomax dealt with and mapped only English-language folk song styles. To round out the North American scene, mention must also be made of French Canadian river and fur trader songs of the Northeast and the strong Mexican American musical tradition still vital and spreading in the Southwest.

Different folk music traditions metamorphosed and spread in the 20th century as distinctive styles of popular music. Jazz emerged in New Orleans in the later 19th century as a union of minstrel show ragtime and the blues, a type of southern black music based on work songs and spirituals. Urban blues—performed with a harsh vocal delivery accompanied by electric guitars, harmonicas, and piano—was a Chicago creation, brought there largely by artists from Mississippi. Country music spread from its southern white ancestral areas with the development of the radio and the phonograph in the 20th century. It became commercialized, electrified, and amplified but remained at core modified folk music (Figure 7.20). Bluegrass style, a high-pitched derivative of Scottish bagpipe sound and church congregation singing tradition, is performed unamplified, true to its folk origins. Bluegrass identification with commercial singing groups bearing identities derived from place names emphasizes the ties of the people, the performers, and the land in the folk tradition.

As these examples of musical style and tradition show, the ethnic merges into the folk, and the folk blends into the popular—in music and in many other elements of culture. On the other hand, Anglo American religious folk songs have become less popular and more spatially confined in an era of popular culture. The white spirituals, diffusing from their 18th-century New England hearth, covered much of the eastern United States during the 19th century, before contracting—far from their original core region—into the Lowland South during the 20th century (Figure 7.21).

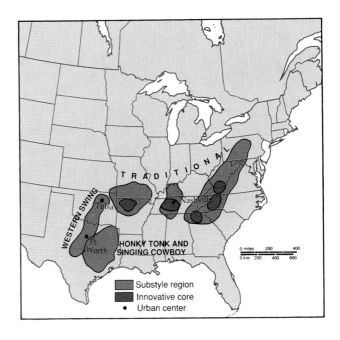

Figure 7.20 **Country music refuge areas.** Traditional "old-timey" country music, little changed from that of the 18th and 19th centuries, was preserved into the 20th in five pockets of the Upland South, according to George Carney, before it turned modern and popular after World War II.

Source: Redrawn by permission from George O. Carney, "Country Music and the South," Journal of Cultural Geography *1 (Fall/Winter 1980): 25.*

Folk and Popular Culture: Diversity and Uniformity   **247**

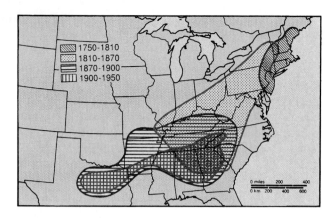

Figure 7.21 **The southward shift of white spirituals.** The southward move and territorial contraction of the white spiritual folk song tradition in America is clearly shown on this map.

Source: From Journal of American Folklore *65:258, October–December, 1952.*

The making of musical instruments is a recurring part of material folk culture traditions. For example, the zither was brought to the United States from northern and central Europe, but as the Pennsylvania Germans carried it southward into the southern Appalachians, it became the American-made three- or four-stringed strummed or plucked dulcimer. The banjo has clear African origins, but by the end of the 19th century it had become a characteristically American folk instrument, versions of which—five-stringed and fretless—were homemade throughout the Southern Uplands. The fiddle was the preeminent Canadian folk instrument and—along with the bagpipe in Scottish communities—was the most common accompaniment for dancing. Both instruments were frequently homemade.

Folk Medicines and Cures

All folk societies—isolated and close to nature—have developed elaborate diversities of medicines, cures, and folk health wisdom based on the plants, barks, leaves, roots, and fruits of their areas of settlement and familiarity. Indeed, botanicals are a fundamental part of modern medicine, most of them initially employed by folk societies: quinine from the bark of the South American cinchona tree, for example, or digitalis, a heart stimulant derived from the dried leaf of the common foxglove.

In Anglo America, folk medicine derived from the common wisdom of both the Old and the New World. The settlers assigned medicinal values to many garden herbs and spices they brought from Europe and planted everywhere in the eastern hearths. Basil, widely used in soups, stews, and salads, also was considered effective heart medicine and a cure for melancholy; thyme tea eased sorrow. Parsley was thought to be generally healthful, and fennel seeds, leaves, and roots properly prepared were deemed appropriate treatment for obesity. Sage was a specific for colds, balm was made into a tea for breaking fevers, marjoram was a cure-all for coughs, bronchitis, dropsy, and yellow jaundice. Lemon balm and rosemary were thought to prevent

baldness, and boiled chervil roots, eaten cold, were healthful for the aged.

Native Americans provided Europeans with information and example of the curative values of a whole new set of plants and practices. Sassafras was a cure-all widely known in all the colonies—good as a purgative, as an ointment for bruises, as a "blood purifier," and as a means of curing fevers. Bearberry, a variety of cranberry, was an astringent and diuretic. Boneset (*Eupatoria*) was an emetic and purgative also used to cure intermittent fevers, arthritis, rheumatism, and gout. Goldenrod was a specific for fevers, pain in the chest, boils, and colds. American hemp was thought beneficial in dropsy, rheumatism, asthma, and as a diuretic. The list runs to the hundreds of medicinal plants from all sections of the country, cures (and practices such as sweat baths and cauterization) that were widely adopted by European newcomers, made part of their folk medicine and, in the case of some 200 different plants, incorporated into modern drug compendia.

Folkloric magic and symbolism were (and are) also important. Diseases of the head are best cured by tops of plants, while roots are specifics for leg problems. Brain fever should be treated by nut meats that resemble the brain. Scarlet fever should respond to wrapping patients in scarlet blankets and doctoring them with red medicine; yellow plants are good for jaundice. Wormroot, the Indians taught, is good for worms, and the red juice in bloodroot prevents bleeding.

Folk medicines, cures, and health wisdom in the United States have been best developed and preserved in the Upland South and Southern Appalachia, along the Texas-Mexican boundary in the Hispanic borderlands, and in the rural West among both white and Indian populations. But primitive peoples everywhere have their known cures and their folk wisdoms in matters medicinal, and all of us in our everyday references may unknowingly include reminders of that knowledge: "The hair of the dog that bit you" was not originally a recommendation for treating a hangover but an accepted remedy for curing the bite of a mad dog.

The Oral Folk Tradition

Folklore is the oral tradition of a group. It refers to ways of talking and interacting and includes proverbs, prayers, common expressions, and particularly, superstitions, beliefs, narrative tales, and legends. It puts into words the basic shared values of a group and informally expresses its ideals and codes of conduct. Folklore serves, as well, to preserve old customs and tales that are the identity of the folk group. The Brothers Grimm recorded German fairy tales early in the 19th century to trace the old mythologies and beliefs of the German people, not for the entertainment of the world's children.

Immigrant groups settling in the Americas brought with them different well-developed folklore traditions, each distinctive not only to the ethnic group itself but even to the part of the home country from which it came. In the New World, the established folklores of home areas became intermixed. The countries of North and South America contain many coexisting and interacting folklore traditions brought by early European colonizers, by transported African slaves, and by later diversified immigrant groups from both Europe and Asia.

The imported folk traditions serve to identify the separate groups in pluralistic societies. In some instances, the retention of folk identity and belief is long term because particular groups—Pennsylvania Dutch, Old Order Amish, and the Hasidim of Brooklyn, for example—isolate themselves from mainstream American culture. Other groups—in Appalachia, the Missouri Ozarks, and the Louisiana bayous—may retain or develop distinctive folklore traditions because isolation was thrust on them by remoteness or terrain.

Where immigrant groups intermixed, however, as in most New World countries, *syncretism*—the merging or fusion of different traditions—is characteristic. Old World beliefs, particularly in magic, begin to recede and are lost to later generations. Proverbs begin to be shared, common short jokes replace long folk tales as both entertainment and devices of instruction or ridicule of deviant behavior, and literacy reduces dependence on the reports and repetitions of knowledgeable elders. **Folkways**—the learned behavior shared by a society that prescribes accepted and common modes of conduct—become those of the country as a whole as acculturation and popularization dictate the ways of life of all.

With the passage of time, too, a new folklore of legend, myth, and hero develops. In the United States, Washington and the cherry tree, Patrick Henry's plea for liberty, the exploits of Jim Bowie or Davy Crockett, the song of John Henry the steel-driving man, or tales of Paul Bunyan become the common property and heritage of all Americans—a new national folklore that transcends regional boundaries or immigrant origins.

Folk Cultural Regions of Eastern United States

A small set of hearths or source regions of folk culture origin and dispersal have been recognized for the eastern United States. They are indicated in Figure 7.22. The similarity of the hearth locations and diffusion routes to the pattern of ethnically based architectural regions and flows shown in Figure 7.14 is unmistakable and a reminder that in the American context, "folk"

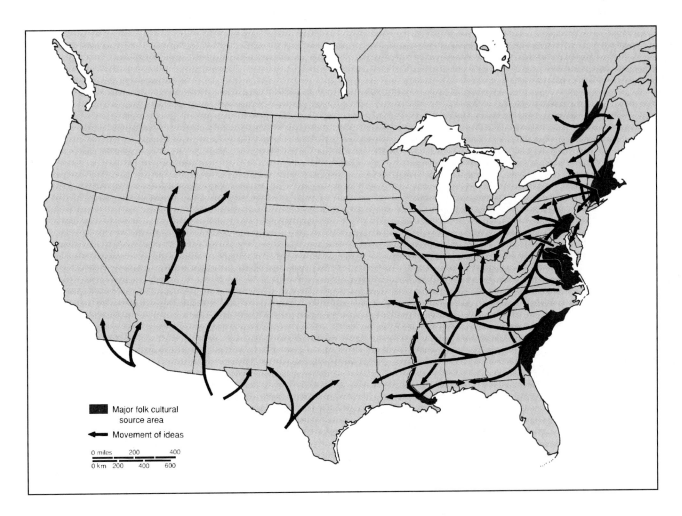

Figure 7.22 **American folk culture hearths and diffusions.**

Sources: Based on Henry Glassie, Pattern in the Material Folk Culture of the Eastern United States, *pp. 37–38, copyright © 1968 by the Trustees of the University of Pennsylvania; and Michael P. Conzen, ed.,* The Making of the American Landscape *(Winchester, Mass.: Unwin Hyman, 1990), 373.*

and "ethnic" are intertwined and interchangeable when traced back to first settlement. Frontier settlers carrying to new, interior locations the artifacts and traditions of those hearth areas created a small set of indistinctly bounded eastern folk cultural regions (Figure 7.23). Although they have become blurred as folk traditions have died, their earlier contributions to American folk diversity remain clear.

From the small *Mid-Atlantic* region, folk cultural items and influences were dispersed into the North, the Upland South, and the Midwest. Southeastern Pennsylvania and the Delaware Valley formed its core and the Pennsylvania Dutch determined much of the Mid-Atlantic region's character. The eastern Finns added their log-building techniques and subsistence lifestyles. Furniture styles, log construction, decorative arts, house and barn types, and distinctive "sweet" cookery were among the purely European imports converted in the Mid-Atlantic hearth to American folk expressions.

The folk culture of the *Lowland South,* by contrast, derived from English originals and African admixtures. French influences in the Louisiana coastal extension and some down-slope migrations from the highland areas add to the amalgam. Dogtrot and I houses became common; English cuisine was adapted to include black-eyed peas, turnip greens, sweet potatoes, small-bird pies, and syrups from sugarcane and sorghum. African origins influenced the widespread use of the banjo in music.

The *Upland South* showed a mixture of influences carried up from the Tidewater and brought south from the Mid-Atlantic folk region along the Appalachian highlands by settlers of German and Scotch-Irish stock. The sheltered isolation of the Upland South and its Ozark outlier encouraged the retention of traditional folk culture long after it had been lost in more accessible and exposed locations. Log houses and farm structures, rail fences, traditional art and music, and home-crafted quilts and furniture make the Upland South region a prime repository of folk artifacts and customs in the United States.

The *North*—dominated by New England, but including New York State, English Canada, Michigan, and Wisconsin—showed a folk culture of decidedly English origin. The saltbox house and Boston baked beans in pots of redware and stoneware are among characteristic elements. The New England-British domination is locally modified by French Canadian and central European influences.

The *Midwest*—a conglomerate of inputs from the Upland South, from the North, and, particularly, from the Mid-Atlantic region—is the least distinctive, most intermixed and Americanized of the cultural regions. Everywhere the interior contains evidences, both rural and urban, of artifacts carried by migrants from the eastern hearths and by newly arriving European immigrants. Folk geography in the Midwest is more the occasional discovery of architectural relics more or less pure in form, though frequently dilapidated, or the recognition of such unusual cultural pockets as those of the Amish, than it is a systematic survey of a defined cultural region.

The Passing of Folk Cultural Regionalism

By the early 20th century, the impacts of immigrant beginnings, settlement diffusions, and ethnic modifications had made themselves felt in a pattern of regionally differentiated rural cultural landscapes. The cities of the eastern and midwestern parts of the country were socially a world apart from the farms. Brash and booming with the economic success of rampant industrialization, the cities were in constant flux. Building and rebuilding, adding and absorbing immigrants and rural in-migrants, increasingly interconnected by passenger and freight railroad and by the national economic unification important since the 1870s, they were far removed in culture, outlook, and way of life from the agricultural areas in which they were physically but not emotionally located.

It was in the countryside that the most pronounced effects of regional cultural differentiation were to be discerned. Although the flow of young people to the city, responding to the push of farm mechanization and the pull of urban jobs and excitement, was altering traditional social orders and rhythms, the automobile, electrical appliances, and the lively mass medium of radio had not as yet obscured the distinction between urban and rural. The family farm, kinship and community ties, the traditions,

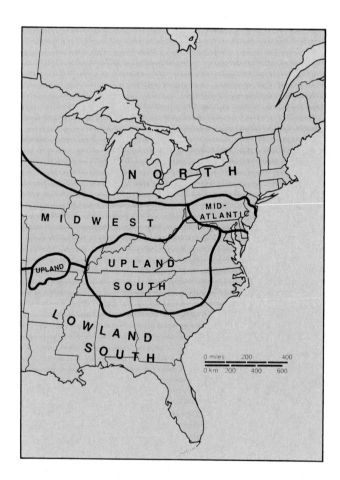

Figure 7.23 **Material folk culture regions of the eastern United States.**

Redrawn with permission from Henry Glassie, Pattern in the Material Folk Culture of the Eastern United States, p. 39, copyright © 1968 by the Trustees of the University of Pennsylvania.

ways of life, and artifacts of small town and rural residence still existed as regionally varied composites. But those ways and artifacts, and the folk cultural regions they defined, were all being eroded and erased with the passage of time and the modernization of all segments of North American life and culture.

Regional character is a transient thing. New peoples, new economic challenges, and new technologies serve as catalysts of rapid change. By World War I and the Roaring Twenties, the automobile, the radio, motion pictures, and a national press began to homogenize America. The slowing of the immigrant stream and second-generation absorption of the common national culture served to blur and obliterate some of the most regionally distinctive cultural identifications. Mechanization, mass production, and mass distribution through mail order and market town diminished self-sufficiency and household crafts. Popular culture began to replace traditional culture in everyday life throughout the United States and Canada.

Patterns of Popular Culture

In 1728, Mary Stith of Virginia wrote to a friend, then in England, "When you come to London pray favour me in your choice of a suit . . . suitably dressed with . . . whatever the fashion requires." In the 1750s, George Washington wrote to his British agent, Thomas Knox, to request "two pair of Work'd Ruffles . . . ; if work'd Ruffles shou'd be out of fashion send such as are not . . . ," noting "whatever Goods you may send me . . . you will let them be fashionable." In the 1760s, he asked another agent, Charles Lawrence, to "send me a Suit of handsome Cloth Cloaths. I have no doubts but you will choose a fashionable coloured Cloth as well as a good one and make it in the best taste. . . ." The American gentry might be distant and isolated, but they did not wish to be unstylish.

The leading American women's magazine of the middle 19th century was *Godey's Lady's Book,* featuring hand-colored pictures of the latest foreign and American fashions in clothing and articles about household furnishings in the newest styles. Its contents influenced ladies of fashion in cities and towns throughout the settled United States. The Montgomery Ward and Sears, Roebuck catalogs appearing in the late 19th century served the same purpose for more ordinary goods, garments, and classes of customers. Popular culture, based on fashions, standards, or fads developed in national centers of influence and prestige, became an important reality over wide areas and across social strata.

By general understanding, popular culture stands in opposition to folk or ethnic culture. The latter two suggest individuality, small group distinctiveness, and above all, tradition. **Popular culture,** in contrast, implies the general mass of people, primarily but not exclusively urban based, constantly adopting, conforming to, and quickly abandoning ever-changing common modes of behavior and fads of material and nonmaterial culture. Popular culture presumably substitutes for and replaces folk and ethnic differences.

For some, the term "popular culture" is not sufficiently precise to define the realities and trends of modern ways of life. They suggest a further distinction should be made between *popular culture* and *mass culture.* The former they view as participatory, involving people in active engagement in the activities and events that were formerly central in American life but now are diminished in relative importance. Amateur athletics; county-fair attendance; barn and ballroom dancing; viewing live theatrical and vaudeville performances; membership in community social, charitable, or political clubs; and the like exemplify the involvement and interaction of true popular culture. That kind of social and personal interaction, it is held, declined markedly after the middle of the 20th century. It was presumably largely replaced by a mass culture comprised of passive and solitary activities, typified by television viewing of game shows and sporting events in which the individual is manipulated and shaped by mass media without the necessity or opportunity to participate or interact with other participants.

Whether or not such a subdivision is useful, the basic characteristic of popular culture as a whole remains certain: it is pervasive, involving the vast majority of a population in similar consumption habits, exposing them to similar recreational choices, and leading them to similar behavioral patterns. Those restrictive similarities are the product of mass production of goods and services and their willing acceptance by the majority of the population as the normal, expected, and desired ingredients of daily life.

Many details of popular culture, we should remember, derive from regional folk cultural traits. For example, universally enjoyed popular and spectator sports such as soccer, football, golf, and tennis originated as local and regional folk games, many of them hundreds of years old. Similarly, musicologists easily trace most recent and current musical styles and fads to earlier folk and ethnic music genres.

In many respects, therefore, the distinctions between folk, ethnic, popular, and mass culture become blurred, and differences between local and universal lose their usual meaning. Presumably, all mass-produced consumer goods should be equally available to all segments of a society. We know, however, that regional tastes and consumption choices differ in, for example, the relative popularity and per capita consumption of different soft drinks (Dr. Pepper more favored in the southern states than elsewhere) or various processed foods (Puerto Rico's high per capita consumption of Cheez Whiz). These mark regionally distinctive choices among widely available similar products. Similar "regionalism" can be found throughout the popular culture realm. The disinterest expressed until the last decade of the 20th century in soccer in the United States and the intense passion aroused by the game nearly everywhere else in the world is a common example, as is the identification of cricket with Britain and the countries of its former empire while other societies are unfamiliar with the game.

Such exceptions do not void the generalization that popular culture becomes the way of life of the mass of the population, reducing, though perhaps not eliminating, regional folk and ethnic differences. It becomes both a leveling and a liberating force, obliterating those locally distinctive lifestyles and material and nonmaterial cultures that develop when groups remain isolated and ethnocentric. Uniformity is substituted for differentiation, and group identity is eroded. At the same time, however, the individual is liberated through exposure to a much broader range of available opportunities—in clothing, foods, tools, recreations, and lifestyles—than ever were available in a cultural environment controlled by the restrictive and limited choices imposed by custom and isolation. Although broad areal uniformity may displace localisms, it is a cultural uniformity vastly richer in content and possibilities than any it replaces.

That uniformity is frequently, though not exclusively, associated with national populations: the American or Canadian way of life distinguished from the English, the Japanese, or others. Even these distinctions are eroding as popular culture in many aspects

of music, movies, sports (soccer, for example), and the like becomes internationalized (Figure 7.24). Popular culture becomes dominant with the wide dissemination of common influences and with the mixing of cultures that force both ethnic and folk communities to become aware of and part of a larger homogeneous society. The result is a material and nonmaterial cultural mix that is not necessarily better or worse than the folk and ethnic cultures lost. It is, however, certainly and obviously different from the traits and distinctions of the past.

National Uniformities and Globalization

Landscapes of popular culture tend to acquire uniformity through the installation of standardized facilities. Within the United States, for example, national motel chains announced by identical signs, advertised by repetitious billboards, and featuring uniform facilities and services may comfort travelers with the familiar but also deny them the interest of regional contrast. Fast-food restaurants—franchised, standardized, and merchandised as identical—carry single logos, building designs, and menus across cultural boundaries and national borders (Figure 7.25). They provide the assurance of the known and the tolerable but insulate the palate from the regionally distinctive. Even food outlets identified with

Figure 7.24 Soccer has become the world's most popular sport, with a television audience approaching 2 billion worldwide for the quadrennial World Cup soccer finals. The game here is being played in Madrid, Spain.

Figure 7.25 Western fast-food chains, classics of standardized popular culture, have gone international—and bilingual—as this KFC outlet in Xian, China, reveals.

Photo by Jon C. Malinowski.

ethnics have lost their cultural character. The pizza has become American, not Italian (Figure 7.26), just as the franchised Mexican American taco and burrito have escaped their regional and ethnic confines and been carried nationwide. Chain gas stations, discount stores, and other enterprises carry on the theme of familiarity of outlet and standardization of product and service wherever one resides or journeys.

Many of these Anglo American elements of popular culture are oriented toward the automobile, the ubiquitous means of local and interregional travel (Figure 7.27). Advertisements' distinctiveness of design assures instant recognition, and their clustering along highways and main streets guarantees that whatever the incidence of regional character still remaining, the public face of town and highway is everywhere the same (Figure 7.28).

Those uniformities are transitory. While folk cultures have ingrained traditions that change only slowly and locally, popular culture tends to change rapidly and uniformly over wide expanses. That is, popular culture diffuses rapidly, even instantaneously, in our age of immediate global communication and sharing of ideas through television, radio, and the Internet. Those same media and means assure the widespread quick replacement of the old fads with new. The globalization of popular culture is commonly recognized. In clothing styles and fashion trade names, near-universal display of American movies and television shows, worldwide acceptance of the cultural norms of urban life and western business conduct and institutions, and the global spread of soft drink signs and golden arches testify to the international standardization of life and the quick adoption of changing tastes and practices.

That standardization, of course, is not complete. National and regional cultural contrasts remain embedded in urban and rural landscapes, and seemingly universal popular icons are always

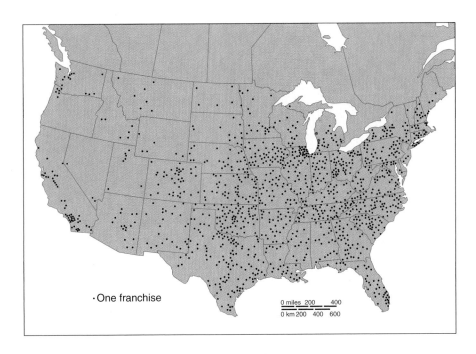

· One franchise

0 miles 200　　400
0 km 200　400　600

Figure 7.26 **The locations of pizza parlors of a single national chain.**

Source: Floyd M. Henderson and J. Russel, unpublished drawing.

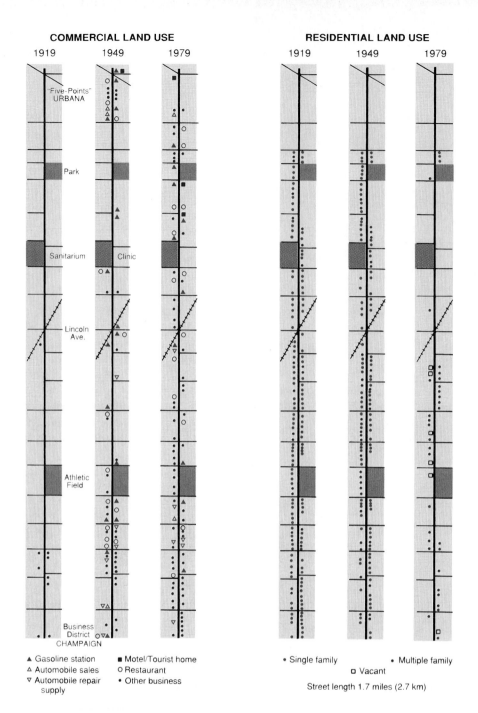

COMMERCIAL LAND USE

1919 1949 1979

RESIDENTIAL LAND USE

1919 1949 1979

▲ Gasoline station   ■ Motel/Tourist home
△ Automobile sales   ○ Restaurant
▽ Automobile repair   • Other business
   supply

• Single family   • Multiple family
   □ Vacant

Street length 1.7 miles (2.7 km)

Figure 7.27 **A street transformation.** A residential street becomes a commercial strip in Champaign-Urbana, Illinois. After 60 years of automobile traffic on this main arterial street, residential land use had been replaced almost entirely by commercial uses, all depending on drive-in customer access or catering to automotive needs themselves. The major change in land use came after 1949.

Source: John A. Jakle and Richard L. Mattson, "The Evolution of a Commercial Strip," Journal of Cultural Geography *1 (Spring/Summer 1981):14, 20. Redrawn by permission.*

differentially adapted and modified for easy acceptance by different national societies. Domino's and Pizza Hut, for example, have a combined total of some 6000 overseas outlets in over 100 countries but do not serve a standard product worldwide. Pizza in India likely will be ordered with spicy chicken sausage or pickled ginger. In Japan, a best seller is pizza topped with potatoes, mayonnaise, and ham or bacon bits. Hong Kong customers prefer their pizza flavoring to be Cajun spices; Thais favor hot spices mixed with lemon grass and lime; while in England preferred toppings

include sweet corn and tuna. The store signs and designs may be universal; the product varies to fit local tastes.

Yet, on the world scene, globalized cultural amalgamation is increasingly evident though not universally welcomed. Imagine this scene: Wearing a Yankees baseball cap, a GAP shirt, Levis, and Reebok shoes, a teenager in Lima, Peru, goes with her friends to see the latest thriller. After the movie, they plan to eat at a nearby McDonald's. Meanwhile, her brother sits at home, listening to the latest American music on his Sony Walkman while

Figure 7.28 For some, the advertisements for commercialized popular culture constitute visual blight.

playing a video game on the family's computer. The activities of both young people are evidence of the globalization of popular culture that is Western and particularly American in origin. U.S. movies, television shows, software, music, food, brand names, and fashions are marketed worldwide. They influence the beliefs, tastes, and aspirations of people in virtually every country, though their effect is most pronounced on the young. They, rather than their elders, want to emulate the stars they see in movies and on MTV and to adopt what they think are up-to-date lifestyles, manners, and modes of dress. They are also the group most apt to use English words and slang in everyday conversation, though the use of English as the worldwide medium of communication in economics, technology, and science is an even broader indication of current cultural merging.

The globalization of popular culture (and the dominance of English in popular and professional use) is resented by many people, rejected by some, and officially opposed or controlled by certain governments. Iran, Singapore, China, and other states, for example, attempt to restrict Western radio and television programming from reaching their people. Governments of many countries—Saudi Arabia, Myanmar, Laos, Yemen, China, and the United Arab Emirates among them—impose Internet surveillance and censorship and demand that U.S.-based search engines filter content to conform with official restrictions and limitations. China, with over 50 million citizens with access to the World Wide Web, began regulating Internet access in the mid-1990s; more recently, even the Western European countries of Germany and France have demanded that American search engines exclude numerous disapproved websites from the German and French versions of their indexes.

In other instances, religious and cultural conservatives may decry what they see as the imposition of Western values, norms, and excesses through such mass culture industries as advertising, the media, professional sports, and the like. Whether or not movies, music, television programming, or clothing fads accurately reflect the essence of Western culture, critics argue that they force on other societies, as normal and unquestioned, alien values of materi-

alism, innovation, self-indulgence, sexuality, and defiance of authority and tradition. More basically, perhaps, globalization of popular culture is seen as a form of dominance made possible by Western control of the means of communication and by self-proclaimed Western technical, educational, and social superiority. What may be accepted or sought by the young and better educated in many societies may simultaneously be strongly resisted by those of the same societies more traditional in outlook and belief.

The Shopping Mall

Within Anglo America and, increasingly, other world regions, the apparent exterior sameness of popular culture has been carried indoors into the design, merchants, and merchandise of the shopping mall. Major regional malls have been created in every part of North America that boasts a metropolitan population large enough to satisfy their carefully calculated purchasing-power requirements. Local and neighborhood malls extend the concept to smaller residential entities. With their mammoth parking lots and easy access from expressways or highways, America's 38,000 large and small malls are part of the automobile culture that helped create them after World War II. Increasingly, however, they stand in standardized separation from the world of movement and of regional contrast. Enclosed, temperature controlled, without windows or other acknowledgment of a world outside, they cater to a full range of homogenized shopping and consuming wants with a repetitive assemblage of brand-name products available in a uniform collection of national chain outlets.

Some assume monumental size, approximating the retail space contained in the central business districts of older medium-sized and large cities (Figure 7.29). For example, the

Figure 7.29 Massive, enclosed, and buffered from its surroundings, the modern metropolitan shopping mall is an external and internal built environment that summarizes the contrasts between popular culture standardization and folk and ethnic cultural individuality. This mall is in the center city area of Philadelphia, Pennsylvania.

West Edmonton Mall in Edmonton, Alberta, was at its completion in 1986 the world's largest shopping mall, containing 836 stores, 110 restaurants, 20 movie theaters, a 360-room hotel, plus such other recreational features as roller coasters, carousels and other rides, a miniature golf course, a water slide, and a hockey rink. A slightly smaller U.S. counterpart opened in Bloomington, Minnesota, in 1992. More recently, expansion of established malls has outpaced development of new ones, and by the end of 1995 both Woodfield Mall near Chicago and the King of Prussia Mall near Philadelphia claimed the "world's largest" title after their renovation and enlargement. Malls are, it has been suggested, an idealized, Disneyland version of the American myth of small-town sanitized intimacy, itself a product of popular culture.

The ubiquity of malls and the uniformity of their goods are clearly reflected in items of clothing. Fashion replaces personal preference, social position, occupation, or tradition as the arbiter of type or design of clothing. Whatever may be dictated nationally—miniskirt, leisure suit, designer jeans, or other fad—is instantly available locally, hurried to market by well-organized chains responding to well-orchestrated customer demand. A few national or international fashion centers dictate what shall be worn, a few designer names dominate the popularly acceptable range of choices. Since popular culture is, above all, commercialized culture, a market success by one producer is instantly copied by others. Thus, even the great number of individual shops within the mall is only an assurance of variations on the same limited range of clothing (or other) themes, not necessarily of diversity of choice. Yet, of course, the very wealth of variations and separate items permits an individuality of choice and selection of image not possible within constrained and controlled folk or ethnic groups.

Even culture in the sense of the arts is standardized within the malls. Chain bookstores offer identical best-sellers and paperbacks; multiscreen cinemas provide viewing choice only among the currently popular films; gift shops have nationally identical selections of figurines and pressed glass; and art stores stock similar lines of prints, photographs, and posters. It has been noted that Americans, at least, spend more of their time within malls than anywhere else except home and work. It is not unlikely that a standardized popular culture is at least in part traceable to the homogenized shopping mall. By the late 1990s, the growing market dominance of a limited number of national chains of "super" stores and discount outlets—the Wal-Marts, K-Marts, and the like—were noticeably eroding the customer volume at shopping malls of all sizes and further reducing the number and standardizing the array of clothing styles and brands and other common items universally available.

Outside the Mall

Nonmaterial tastes and recreations are, in popular culture, subject to the same widespread uniformities as are the goods available within repetitive shopping complexes. Country music, we saw, was culturally associated with the Upland South. It has long since lost that regional exclusivity, and Nashville has become a product, not a place. By the late 1970s (Figure 7.30), no American with access to radio was denied exposure to electric guitar and melancholy lyric. Fad motion pictures are simultaneously released throughout the country; the same children's toys and

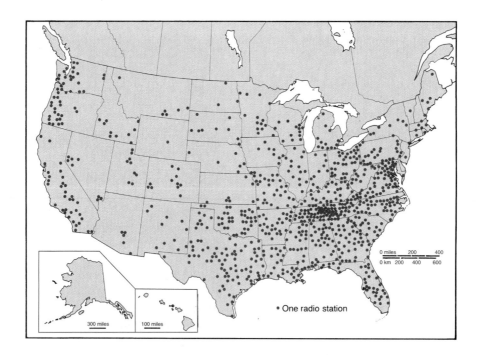

Figure 7.30 **Country music radio stations.** Although still most heavily concentrated in the Upland South, radio stations playing only country music had become a national commonplace by the late 1970s.

Source: Redrawn by permission from George O. Carney, "From Down Home to Uptown," Journal of Geography, 76 (Indiana, PA.: National Council for Geographic Education, 1977), p. 107.

adults' games are everywhere instantly available to satisfy the generated wants.

Wilbur Zelinsky reported on the speed of diffusion of a manufactured desire:

In August, 1958, I drove from Santa Monica, California, to Detroit at an average rate of about 400 miles (650 km) per day; and display windows in almost every drugstore and variety store along the way were being hastily stocked with hula hoops just off the delivery trucks from Southern California. A national television program the week before had roused instant cravings. It was an eerie sensation, surfing along a pseudo-innovation wave.[2]

Diffusion Tracks

Popular culture is marked by the nearly simultaneous adoption over wide areas of an artifact or a nonmaterial element. Knowledge of an innovation is widely and quickly available; television and the national press inform without distance constraints. Mass manufacturing and imitative production place desired items, as Zelinsky discovered, in every store in a remarkably short time. For fads particularly, the tracing of diffusion tracks is difficult and probably meaningless. Recognizable culture hearths and migration paths are not clearly defined by the myriad introductions into the ever-changing popular culture pool. It is not particularly revealing to know that the origin of the Frisbee is apocryphally traced to the pie tins of the old Frisbie Baking Company of Bridgeport, Connecticut, manufactured in plastic and carried as a game toy by college students throughout the United States, or that the Rubik's Cube puzzle was an invention of a Hungarian architect. The one has nothing to do with a New England culture hearth nor the other with East European ethnic influences.

Some more lasting popular changes have been recorded and do provide useful insight into the nature of diffusion and the sequence of acceptance and adoption of new cultural elements. The New Orleans origin of jazz, its upriver movement, and the gradual acceptance by white sophisticates of a new black musical form trace for us the origin, the diffusion path, and the adoption sequence of a major introduction. Cricket as a popular sport followed the spread of empire as British influences were implanted across the world. The names settlers and town founders gave to their communities provide another expression of popular culture and its diffusion (see Figure 5.17). Professor Zelinsky has investigated the origin and spread of classical town names in the United States, documenting on the map America's 19th-century attraction to the Greco-Roman world. Neoclassical (resembling Greek and Roman temples) public architecture, Latin state mottoes, Latin and Greek personal names, and classical town names such as Rome, Ithaca and Syracuse were all part of that Classical Revival. Figure 7.31 summarizes the patterns of innovation and dispersal that he discovered.

---

[2]*The Cultural Geography of the United States* Rev. ed. (Englewood Cliffs, NJ: Prentice Hall, 1992), p. 80, fn 18.

Regional Emphasis

The uneven distribution of classical place names suggests that not all expressions of popular culture are spatially uniform. Areal variations do exist in the extent to which particular elements in the general cultural pool are adopted. These variations impart an aspect of regional differentiation of interest to geographers.

Spatial patterns in sports, for example, reveal that the games played, the migration paths of their fans and players, and the landscape evidence of organized sports constitute regional variables, part of the areal diversity of North American—and world—life. Figure 7.32a, for example, shows that television interest in professional baseball is not universal despite the sport's reputation as "the national pastime." Studies and maps of many encountered regional differences in food and drink preferences, leisure activities, and personal and political tastes—a sampling is presented in Figure 7.32—are suggestive of the growing interest in how people behave and respond, not as echoes of the distant past but as participants in a vibrant and changing contemporary world that still retains evidence of regional contrast along with the commonalities of popular culture.

Vernacular Regions

Ordinary people have a clear view of space. They are aware of variations from place to place in the mix of phenomena, both physical and cultural. They use and recognize as meaningful such common regional names as Corn Belt, Sunbelt, and "the Coast." More important, people individually and collectively agree on where they live. They occupy regions that have reality in their minds and that are reflected in regional journals, in regional museums, and in regionally based names employed in businesses, by sports teams, or in advertising slogans.

These are **vernacular,** or **popular, regions**; they have reality as part of folk or popular culture rather than as political impositions or scholarly constructs. Geographers are increasingly recognizing that vernacular regions are significant concepts affecting the way people view space, assign their loyalties, and interpret their world. One geographer has drawn the boundaries of the large popular regions of North America on the basis of place names and locational identities found in the white pages of central-city telephone directories (Figure 7.33). The 14 large but subnational vernacular regions recognized accord reasonably well with cultural regions defined by more rigorous methods. However, particularly in the West, that accordance is not clearly demonstrated by comparison of the vernacular regions with Figure 6.30. Another, more subjective cultural regionalization of the United States is offered in Figure 7.34. The generalized "consensus" or vernacular regions suggested are based on an understood "sense of place" derived from current population and landscape characteristics as well as on historical differences that impart distinctive regional behaviors and attitudes.

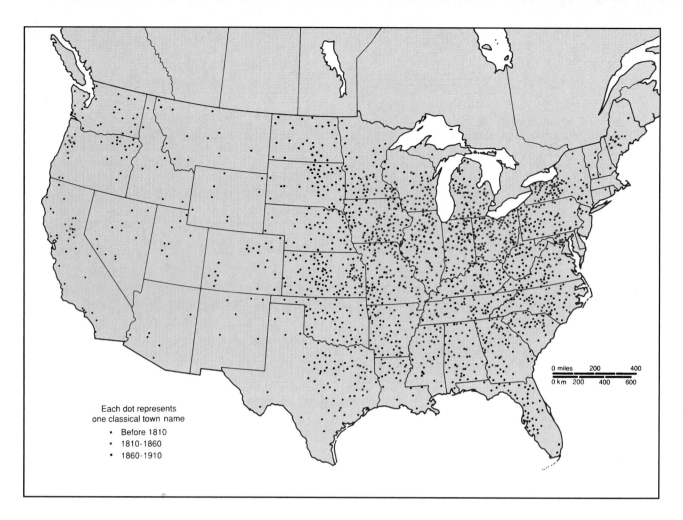

Each dot represents
one classical town name
• Before 1810
• 1810-1860
• 1860-1910

Figure 7.31 **Classical town names.** A permanent reminder of popular culture: the diffusion of classical town names in the United States to 1910.

Source: Redrawn by permission from Wilbur Zelinsky, "Classical Town Names in the United States," Geographical Review *57 (New York: American Geographical Society, 1967).*

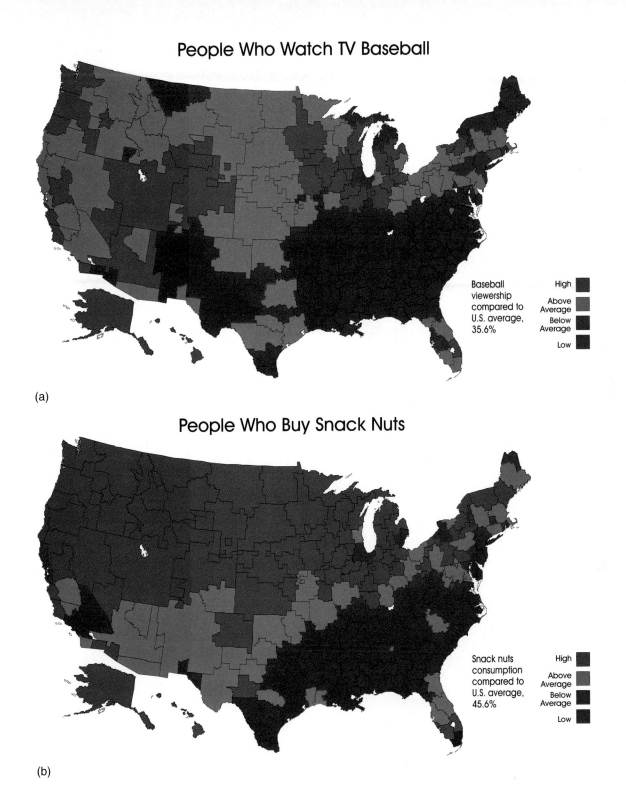

People Who Watch TV Baseball

Baseball viewership compared to U.S. average, 35.6%

High
Above Average
Below Average
Low

(a)

People Who Buy Snack Nuts

Snack nuts consumption compared to U.S. average, 45.6%

High
Above Average
Below Average
Low

(b)

Figure 7.32 **Regional variations in expressions of popular culture.** (*a*) Part of the regional variation in television viewing of baseball reflects the game's lack of appeal in the African American community and, therefore, its low viewership in the Southeast and in metropolitan centers where, additionally, attendance at games is an alternative to watching TV. (*b*) The sharp regional contrasts in snack nut consumption have been attributed to the presumed greater incidence of cocktail parties in the North and West and the higher incidence of religious conservatives (not usual cocktail party hosts or guests) in the Southeast. A map of Americans who habitually listen to religious radio broadcasts is essentially the reverse of this nut map.

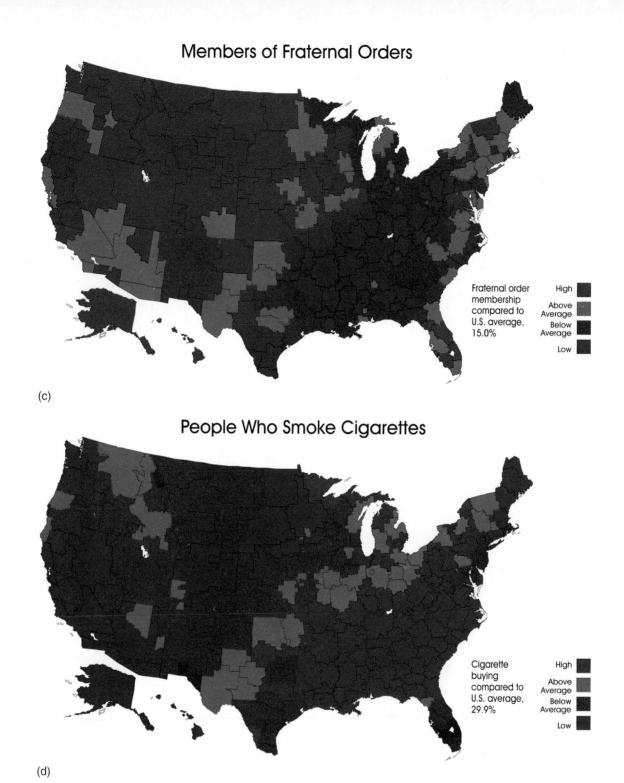

Members of Fraternal Orders

(c)

Fraternal order membership compared to U.S. average, 15.0%

High
Above Average
Below Average
Low

People Who Smoke Cigarettes

(d)

Cigarette buying compared to U.S. average, 29.9%

High
Above Average
Below Average
Low

Figure 7.32 (*continued*) (*c*) Members of fraternal orders tend to concentrate in the rural west, where the clubs provide a social opportunity for rural or small town residents, and in retirement communities, for the younger generation seems less inclined to "join the lodge." (*d*) Even bad habits regionalize. The country's cigarette belt includes notably many of the rural areas where tobacco is grown.

From Michael J. Weiss, Latitudes and Attitudes: An Atlas of American Tastes, Trends, Politics, and Passions, *Boston: Little, Brown and Company, 1994. Used by permission of the author.*

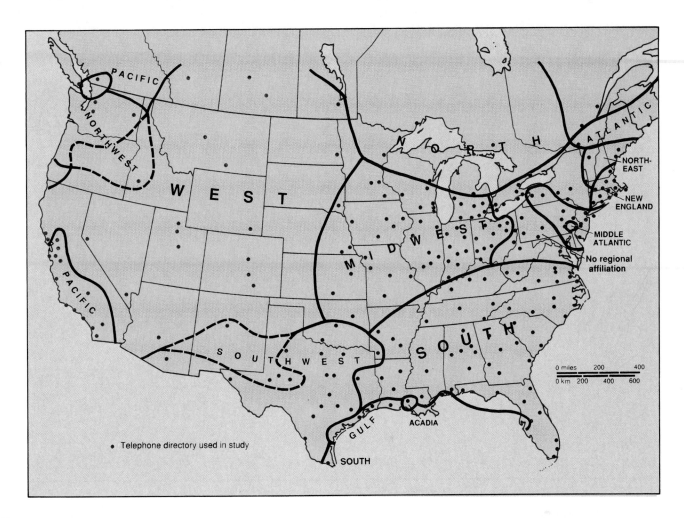

Figure 7.33 **Vernacular regions of North America** as determined by names of enterprises listed in central-city telephone directories. Regions are those in which a given term or a cluster of closely related terms (e.g., Southern, Southland, Dixie) outnumber all other regional or locational references.

Source: Redrawn by permission from Annals of the Association of American Geographers, *W. Zelinsky, Vol. 70, p. 14, Fig. 9, Association of American Geographers, 1980.*

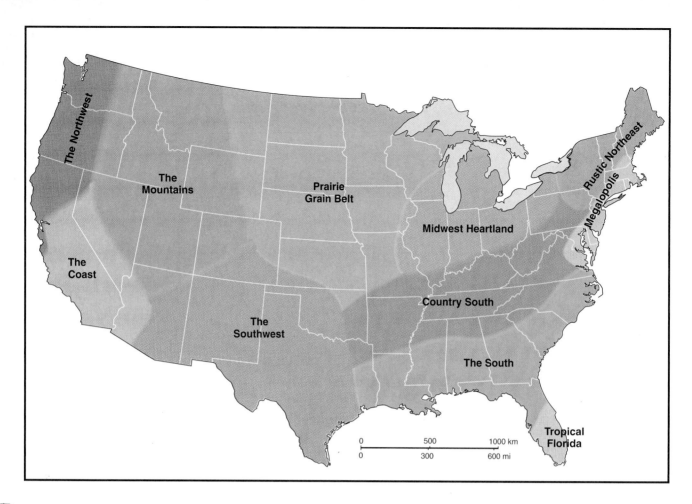

 *Summary*

In the population mix that is the Anglo American, particularly U.S. society, we may recognize two culture-based sources of separation and one of unification. Ethnic culture and folk culture tend to create distinctions between peoples and to impart a special character to the areas in which their influences are dominant. Popular culture implies behavioral unification and the reduction of territorial distinctiveness.

Early arriving ethnic groups were soon Americanized, and their imported cultures were converted from the distinctly ethnic traits of foreigners to the folk cultures of the New World. The foothold settlements of first colonists became separate culture hearths in which imported architectural styles, food preferences, music, and other elements of material and nonmaterial culture were mixed, modified, abandoned, or disseminated along clearly traceable diffusion paths into the continental interior. Ethnic culture was transmuted to folk culture when nurtured in isolated areas and made part of traditional America by long retention and by modification to accommodate local circumstances. Distinctive and bearing the stamp of restricted sections of the nation, those folk cultures contributed both to eastern regional diversity and to the diffusion streams affecting midwestern and western cultural amalgams.

The territorial and social diversities implied by the concepts of *ethnic* and *folk* are modified by the general unifying forces of *popular* culture. Fads, foods, music, dress, toys, games, and other introduced tastes tend to be adopted within a larger society, irrespective of the ethnic or folk distinctions of its parts. Sport interests and rivalries, as we saw at the opening of this chapter, become regional and national—and, increasingly, international—surmounting older limitations of distance. The more modernized and urbanized a country, the more it is uniformly subjected to the mass media and to common entertainment sources, and the more will popular culture subdue the remnant social distinctions still imparted by folk and ethnic customs. At the same time, the greater will be the choices made available to individuals newly freed from the limiting constraints of folk and ethnic traditions.

The threads of diversity traced in Chapter 6 and in this chapter are those of the traits and characteristics of small groups as revealed by behavior and artifact. The reference has been, for the most part, to a single cultural realm and particularly to a single country, the United States, within that realm. The aim has been to impart a clearer understanding of the role of cultural differentiation and unification in the human geography of a region best known to most readers. With different emphases of topic, we could, of course, present similar discussions of cultural mosaics for all other culture realms and sizable countries of the world.

If we were to do so, one recurring condition of material and nonmaterial cultural differentiation would become evident. Much of human activity, including tools made and used, structures built, migrations undertaken, and social controls adopted, is related to economic life. Culture is conditioned by the necessities of production of food and material objects and, increasingly, of their exchange over space and between groups. Many of the threads in the fabric of cultural diversity—and, as well, many of its patterns of uniformity—are those woven by the traditions and technologies of livelihood. We must therefore next turn our attention to a consideration of the varied world of work and to the regional patterns of production and exchange revealed by that encompassing branch of human geography called *economic geography*.

 Key Words

built environment 231
custom 231
folk culture 230
folklore 248

folkways 249
material culture 231
nonmaterial culture 231
popular culture 251

popular region 257
vernacular house 236
vernacular region 257

 For Review

1. What contrasts can you draw between *folk culture* and *popular culture?* What different sorts of material and nonmaterial elements identify them?

2. How many of the early settlement cultural hearths of North America can you name? Did early immigrants create uniform *built environments* within them? If not, why not?

3. When and under what circumstances did popular culture begin to erode the folk and ethnic cultural differences between Americans? Thinking only of your own life and habits, what traces of folk culture do you carry? To what degree does popular culture affect your decisions on dress? On reading material? On recreation?

4. How are we able to recognize hearths and trace diffusions of folk cultural elements? Do items of popular culture have hearths and diffusion paths that are equally traceable? Why or why not?

5. What kinds of connections can you discern between the nature of the physical environment and the characteristics of different *vernacular house* styles in North America? In other parts of the world?

6. If, as some have observed, there is a close relationship between the natural environment and the artifacts of folk culture, is there likely to be a similar causal connection between the environment and expressions of popular culture? Why or why not?

 Focus Follow-up

1. **What is folk culture, and what folk culture hearths and building traditions are found in Anglo America?** pp. 230–242.

Folk culture, often based on ethnic backgrounds, tends to be localized by population groups and areas. It acts to distinguish groups within mixed-culture societies. In Anglo America, diversified immigrant groups settling different, particularly eastern, regions brought their own building traditions to "hearth" regions: those of the *North* in the northeastern United States and southeastern Canada, the *Middle Atlantic,* and the *South.*

2. **What elements and patterns of nonmaterial folk culture can we observe in Anglo America?** pp. 242–249.

The universal elements of nonmaterial folk culture include food and drink preferences and ingredients, music, herbal and customary medicines and cures, recreations, and folkloric oral traditions. In Anglo America, many of the nonmaterial elements of folk culture are associated with the recognized "hearths" of house-building traditions.

3. **What folk culture regions are recognized for the eastern United States?** pp. 249–251.

   Together, elements of material and nonmaterial folk and ethnic cultural distinctions gave rise to a small set of folk culture regions. Dispersions from the *North, Mid-Atlantic,* and *Lowland South* into the *Midwest* helped form a now-disappearing U.S. cultural regionalism.

4. **What is popular culture, and what are its universal and Anglo American evidences?** pp. 251–257.

   Popular culture implies the tastes and habits of the general mass of a society rather than of its small group components. Popular culture is based on changing fads and features of clothing, foods, services, sports, entertainments, and the like and embodies the perhaps temporary dominating "way of life" of a society. Shopping malls, standardized national chains of restaurants, motels, and retail stores, many oriented toward automobile mobility, characterize the contemporary Anglo American landscape and lifestyle.

5. **Are there regional differences and emphases in our mass popular culture?** pp. 257–262.

   Despite national—and increasingly international—uniformities in popular culture, regional differences can be recognized. In Anglo America, these include spatial patterns of sports interests and emphases, food and drink preferences, leisure activities, religious and political affiliations and loyalties, and the like. Those differences give rise to commonly recognized and accepted *vernacular,* or *popular,* regions that overcome even the strong tendencies of Anglo American cultural homogenization.

 Selected References

Bale, John. *Sports Geography.* 2d. ed. New York: Routledge, 2003.

Barer-Stein, Thelma. *You Eat What You Are: A Study of Ethnic Food Traditions.* Toronto: McClelland and Stewart, 1979.

Carney, George O., ed. *The Sounds of People and Places: Readings in the Geography of American Folk and Popular Music.* 3d ed. Lanham, Md.: Rowman & Littlefield, 1994.

Carney, George O., ed. *Fast Food, Stock Cars, and Rock-N-Roll: Space and Place in American Pop Culture.* Lanham, Md.: Rowman & Littlefield, 1995.

Carney, George O., ed. *Baseball, Barns, and Bluegrass: A Geography of American Folklife.* Lanham, Md.: Rowman & Littlefield, 1998.

Carney, George O., ed. *The Sounds of People and Places: A Geography of American Music from Country to Classical and Blues to Bop.* 4th ed. Lanham, Md.: Rowman and Littlefield, 2002.

Coe, Sophie D. *America's First Cuisines.* Austin: University of Texas Press, 1994.

Conzen, Michael P., ed. *The Making of the American Landscape.* Boston: Unwin Hyman, 1990.

Dorson, Richard M., ed. *Folklore and Folklife: An Introduction.* Chicago: University of Chicago Press, 1972.

Ensminger, Robert F. *The Pennsylvania Barn: Its Origin, Evolution, and Distribution in North America.* 2d ed. Baltimore, Md.: Johns Hopkins University Press, 2003.

Fowke, Edith. *Canadian Folklore.* Toronto: Oxford University Press, 1988.

Francaviglia, Richard V. *The Mormon Landscape.* New York: AMS Press, 1978.

Glassie, Henry. *Pattern in the Material Folk Culture of the Eastern United States.* Philadelphia: University of Pennsylvania Press, 1968.

Gordon, Jean, and Jan McArthur. "Popular Culture, Magazines and American Domestic Interiors, 1898–1940." *Journal of Popular Culture* 22, no. 4 (1989): 35–60.

Hart, John Fraser. *The Rural Landscape.* Baltimore, Md.: Johns Hopkins University Press, 1998.

Hart, John Fraser, and Eugene Cotton Mather. "The American Fence." *Landscape* 5, no. 3 (1957): 4–9.

Jackson, John B. *Discovering the Vernacular Landscape.* New Haven, Conn.: Yale University Press, 1984.

Jakle, John A., Robert W. Bastian, and Douglas K. Meyer. *Common Houses in America's Small Towns: The Atlantic Seaboard to the Mississippi Valley.* Athens: University of Georgia Press, 1989.

Johnson, Paul. *The Birth of the Modern: World Society 1815–1830.* New York: HarperCollins, 1991.

Jordan, Terry G., and Matti Kaups. "Folk Architecture in Cultural and Ecological Context." *Geographical Review* 77 (1987): 52–75.

Jordan, Terry G., and Matti Kaups. *The American Backwoods Frontier.* Baltimore and London: Johns Hopkins University Press, 1989.

Jordan, Terry G., Jon T. Kilpinen, and Charles F. Gritzner. *The Mountain West: Interpreting the Folk Landscape.* Baltimore, Md.: The Johns Hopkins University Press, 1996.

Kimber, Clarissa T. "Plants in the Folk Medicine of the Texas-Mexico Borderlands." *Proceedings of the Association of American Geographers* 5 (1973): 130–133.

Kniffen, Fred B. "Folk Housing: Key to Diffusion." *Annals of the Association of American Geographers* 55 (1965): 549–577.

Kniffen, Fred B. "American Cultural Geography and Folklife." In *American*

Folklife, edited by Don Yoder, pp. 51–70. Austin: University of Texas Press, 1976.

Kniffen, Fred B., and Henry Glassie. "Building in Wood in the Eastern United States: A Time-Place Perspective." *Geographical Review* 56 (1966): 40–66.

Lewis, Peirce F. "Common Houses, Cultural Spoor." *Landscape* 19, no. 2 (January 1975): 1–22.

Mather, Eugene Cotton, and John Fraser Hart. "Fences and Farms." *Geographical Review* 44 (1954): 201–223.

Miller, E. Joan Wilson. "The Ozark Culture Region as Revealed by Traditional Materials." *Annals of the Association of American Geographers* 58 (1968): 51–77.

Mitchell, Robert D., ed. *Appalachian Frontiers: Settlement, Society, and Development in the Preindustrial Era.* Lexington: University Press of Kentucky, 1991.

Noble, Allen G. *Wood, Brick, and Stone: The North American Settlement Landscape.* Vol. 1: *Houses*; Vol. 2: *Barns and Farm Structures.* Amherst: University of Massachusetts Press, 1984.

Noble, Allen G., and Richard K. Cleek. *The Old Barn Book.* New Brunswick, N.J.: Rutgers University Press, 1995.

Oliver, Paul. *Dwellings: The House across the World.* Austin: University of Texas Press, 1987.

Paredes, Américo. *Folklore and Culture on the Texas-Mexican Border.* Austin: University of Texas Press, 1995.

Peterson, Fred W. *Homes in the Heartland: Balloon Frame Farmhouses of the Upper Midwest, 1850–1920.* Lawrence: University Press of Kansas, 1992.

Pillsbury, Richard. *No Foreign Food: The American Diet in Time and Place.* Boulder, Colo.: Westview Press, 1998.

Price, Edward T. "Root Digging in the Appalachians: The Geography of Botanical Drugs." *Geographical Review* 50 (1960): 1–20.

Purvis, Thomas L. "The Pennsylvania Dutch and the German-American Diaspora in 1790." *Journal of Cultural Geography* 6, no. 2 (1986): 81–99.

Rapaport, Amos. *House Form and Culture.* Englewood Cliffs, N.J.: Prentice Hall, 1969.

Rooney, John F., Jr., and Richard Pillsbury. *Atlas of American Sport.* New York: Macmillan, 1992.

Rooney, John F., Jr., and Richard Pillsbury. "Sports Regions." *American Demographics* 14 (November 1993): 30–39.

Santelli, Robert, ed., et al. *American Roots Music.* New York: Abrams, 2001.

Shortridge, Barbara G., and James R. Shortridge. *The Taste of American Place: A Reader on Regional and Ethnic Foods.* Lanham, Md.: Rowman & Littlefield, 1998.

Tomlinson, John. *Globalization and Culture.* Chicago: University of Chicago Press, 1999.

Ulrich, Laurel Thatcher. *The Age of Homespun: Objects and Stories in the Creation of an American Myth.* New York: Alfred Knopf, 2001.

Upton, Dell. *America's Architectural Roots: Ethnic Groups That Built America.* National Trust for Historic Preservation, Building Watchers Series. Washington, D.C.: Preservation Press, 1986.

Upton, Dell, and John Michael Vlach, eds. *Common Places: Readings in American Vernacular Architecture.* Athens: University of Georgia Press, 1986.

"Urban Consumption." Review issue of *Urban Studies,* vol. 35, no. 5/6 (May, 1998).

Vermeer, Donald E., and Dennis A. Frate, "Geophagy in a Mississippi County."

Annals of the Association of American Geographers 65 (1975): 414–424.

Vlach, John Michael. "The Shotgun House: An African Architectural Legacy." Parts 1, 2. *Pioneer America* 8 (January 1976): 47–56 (July 1976): 57–70.

Voeks, Robert. "African Medicine and Magic in the Americas." *Geographical Review* 83, no. 1 (1993): 66–78.

Vogel, Virgil J. *American Indian Medicine.* Norman: University of Oklahoma Press, 1970.

Ward, Geoffrey C., et al., *Jazz: A History of America's Music.* New York: Knopf, 2000.

Waterman, Thomas T. *The Dwellings of Colonial America.* Chapel Hill: University of North Carolina Press, 1950.

Weatherford, Jack. *Indian Givers: How the Indians of the Americas Transformed the World.* New York: Ballantine Books, 1990.

Welsch, Roger L. "Sod Construction on the Plains." *Pioneer America* 1, no. 2 (July 1969): 13–17.

Zelinsky, Wilbur. "North America's Vernacular Regions." *Annals of the Association of American Geographers* 70 (1980): 1–16.

Zelinsky, Wilbur. *The Cultural Geography of the United States.* Rev. ed. Englewood Cliffs, N.J.: Prentice Hall, 1992.

Websites: The World Wide Web has a tremendous number and variety of sites pertaining to geography. Websites relevant to the subject matter of this chapter appear in the "Web Links" section of the On-line Learning Center associated with this book. Access it at **www.mhhe.com/fellmann8e**

PART *Three*

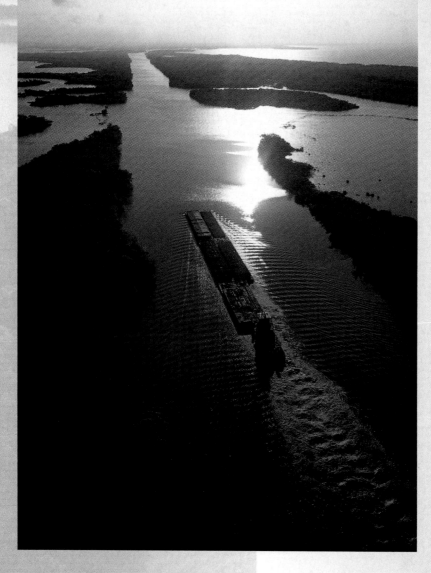

A barge tow on the Intercoastal Waterway in Louisiana.

266

Dynamic Patterns
of the Space Economy

he preceding chapters of Part II focused on prominent elements of unity and diversity of distinctive societies and culture realms of the world. Their theme stressed tradition and conservatism, for the patterns of language, religion, ethnicity, and folk beliefs and customs that were recognized represent long-established and faithfully transmitted expressions of cultural identity. They endure even in the face of profound changes affecting group life and activity. In the following chapters of Part III of our study of human geography, the theme switches from conservative cultural retentions to dynamic economic innovations. Change, as we have seen, is the recurring theme of human societies, and in few other regards is change so pervasive and eagerly sought as it is in the economic orientations and activities of countries and societies throughout the modern world.

Since many—though by no means all—of the material and nonmaterial expressions of cultural differentiation are rooted in the necessities of subsistence and livelihood, basic changes in spatial economic patterns also affect culture groups' established noneconomic sociofacts and mentifacts. Traditional preindustrial societies developed tools and other artifacts, evolved sequences of behavior, and created institutions and practices that were related to their continuing need for finding or producing food and essential materials. That need remains, but the culturally distinctive solutions of traditional societies are being steadily replaced by the shared technologies of modern economies.

Food and raw material production still dominate some economies and areas, but increasingly the world's people are engaging in activities—labeled secondary and tertiary—that involve the processing and exchange of produced primary materials and the provision of personal, business, and professional services to an increasingly interdependent world economy. Simultaneously, economically advanced societies are becoming "postindustrial" while those less advanced seek the prosperity and benefits that are promised by "development." Cultural convergence through shared technologies and intertwined economies has served to reduce the distinguishing contrasts that formerly were preserved in economic isolation. Increasingly, economic development of peoples in all parts of the world has imposed new pressures on the environment to accommodate new patterns of production and exploitation.

Our concern in chapters 8, 9, and 10 will be with the dynamic changes in economic patterns and orientations of regions and societies that have emerged with the development and integration of the world's space economy. These patterns of livelihood, production, and exchange augment and complement the understandings of regional cultural differentiation we have already explored. They add an awareness of the economic unity, diversity, and change so prominent in modern society and so essential to our fuller appreciation of the spatial mosaic of culture.

Eight

Livelihood and Economy:
Primary Activities

Focus Preview

Opposite: Harvesting grapes near Nuits St. George in central Burgundy, France.

The crop bloomed luxuriantly that summer of 1846. The disaster of the preceding year seemed over, and the potato, the sole sustenance of some 8 million Irish peasants, would again yield in the bounty needed. Yet within a week, wrote Father Mathew, "I beheld one wide waste of putrefying vegetation. The wretched people were seated on the fences of their decaying gardens . . . bewailing bitterly the destruction that had left them foodless." Colonel Gore found that "every field was black," and Father O'Sullivan noted that "the fields . . . appeared blasted, withered, blackened, and . . . sprinkled with vitriol. . . ." The potato was irretrievably gone for a second year; famine and pestilence were inevitable.

Within 5 years, the settlement geography of the most densely populated country in Europe was forever altered. The United States received a million immigrants, who provided the cheap labor needed for the canals, railroads, and mines that it was creating in its rush to economic development. New patterns of commodity flows were initiated as American maize for the first time found an Anglo-Irish market—as part of Poor Relief—and then entered a wider European market that had also suffered general crop failure in that bitter year. Within days, a microscopic organism, the cause of the potato blight, had altered the economic and human geography of two continents.

Although the Irish famine of the 1840s was a spatially localized tragedy, it dramatically demonstrated how widespread and intricate are the interrelations between widely separated peoples and areas of the earth. It made vividly clear how fundamental to all human activity patterns are those rooted in economy and subsistence. These are the patterns that, within the broader context of human geography, economic geography isolates for special study.

Simply stated, **economic geography** is the study of how people earn their living, how livelihood systems vary by area, and how economic activities are spatially interrelated and linked. It applies geography's general concern with spatial variation to the special circumstances of the production, exchange, and consumption of goods and services. In reality, of course, we cannot really comprehend the totality of the economic pursuits of more than 6 billion human beings. We cannot examine the infinite variety of productive and service activities found everywhere on the earth's surface, nor can we trace all their innumerable interrelationships, linkages, and flows. Even if that level of understanding were possible, it would be valid for only a fleeting instant of time, for economic activities are constantly undergoing change.

Economic geographers seek consistencies. They attempt to develop generalizations that will aid in the comprehension of the maze of economic variations characterizing human existence. From their studies emerges a deeper awareness of the dynamic, interlocking diversity of human enterprise and of the impact of economic activity on all other facets of human life and culture. From them, too, comes appreciation of the increasing interdependence of differing national and regional economic systems. The potato blight, although it struck only one small island, ultimately affected the economies of continents. In like fashion, the depletion of America's natural resources and the "deindustrialization" of its economy and conversion to postindustrial service and knowledge activities are altering the relative wealth of countries, flows of international trade, domestic employment and income patterns, and more (Figure 8.1).

The Classification of Economic Activity and Economies

The search for understanding of livelihood patterns is made more difficult by the complex environmental and cultural realities controlling the economic activities of humans. Many production patterns are rooted in the spatially variable circumstances of the *physical environment.* The staple crops of the humid tropics, for example, are not part of the agricultural systems of the midlatitudes; livestock types that thrive in American feedlots or on western ranges are not adapted to the Arctic tundra or to the margins of the Saharan desert. The unequal distribution of useful mineral deposits gives some regions and countries economic prospects and employment opportunities that are denied to others. Forestry and fishing depend on still other natural resources unequal in occurrence, type, and value.

Within the bounds of the environmentally possible, *cultural considerations* may condition economic or production decisions. For example, culturally based food preferences rather than environmental limitations may dictate the choice of crops or livestock. Maize is a preferred grain in Africa and the Americas; wheat in North America, Australia, Argentina, southern Europe and Ukraine; and rice in much of Asia. Pigs are not produced in Muslim areas, where religious belief prohibits pork consumption.

Level of *technological development* of a culture will affect its recognition of resources or its ability to exploit them. **Technology** refers to the totality of tools and methods available to and used by a culture group in producing items essential to its subsistence and comfort. Preindustrial societies have no knowledge of or need for the iron ore or coking coal underlying their hunting, gathering, or gardening grounds. *Political decisions* may encourage or discourage—through subsidies, protective tariffs, or production restrictions—patterns of economic activity. And, ultimately, production is controlled by *economic factors* of demand, whether that demand is expressed through a free market mechanism,

Figure 8.1 These Japanese cars unloading at Seattle were forerunners of a continuing flow of imported goods capturing an important share of the domestic market traditionally held by American manufacturers. Established patterns of production and exchange are constantly subject to change in a world of increasing economic and cultural interdependence and of changing relative competitive strengths.

government instruction, or the consumption requirements of a single family producing for its own needs.

Categories of Activity

Regionally varying environmental, cultural, technological, political, and market conditions add spatial details to more generalized ways of classifying the world's productive work. One approach to that categorization is to view economic activity as ranged along a continuum of both increasing complexity of product or service and increasing distance from the natural environment. Seen from that perspective, a small number of distinctive stages of production and service activities may be distinguished (Figure 8.2).

Primary activities are those that harvest or extract something from the earth. They are at the beginning of the production cycle, where humans are in closest contact with the resources and potentialities of the environment. Such activities involve basic foodstuff and raw material production. Hunting and gathering, grazing, agriculture, fishing, forestry, and mining and quarrying are examples. **Secondary activities** are those that add value to materials by changing their form or combining them into more useful—therefore more valuable—commodities. That provision of *form utility* may range from simple handicraft production of pottery or woodenware to the delicate assembly of electronic

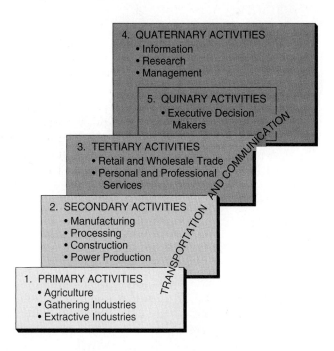

Figure 8.2 **The categories of economic activity.** The five main sectors of the economy do not stand alone. They are connected and integrated by transportation and communication services and facilities not assigned to any single sector but common to all.

goods or space vehicles (Figure 8.3). Copper smelting, steel making, metalworking, automobile production, textile and chemical industries—indeed, the full array of *manufacturing and processing industries*—are included in this phase of the production process. Also included are the production of *energy* (the "power company") and the *construction* industry.

Tertiary activities consist of those business and labor specializations that provide *services* to the primary and secondary sectors and goods and services to the general community and to the individual. These include financial, business, professional, clerical, and personal services. They constitute the vital link between producer and consumer, for tertiary occupations importantly include the wholesale and retail *trade* activities—including "dot-com" Internet sales—necessary in highly interdependent societies. Tertiary activities also provide essential information to manufacturers: the knowledge of market demand without which economically justifiable production decisions are impossible.

In economically advanced societies a growing number of individuals and entire organizations are engaged in the processing and dissemination of information and in the administration and control of their own or other enterprises. The term **quaternary** is applied to this fourth class of economic activities, which is composed entirely of services rendered by "white collar" professionals working in education, government, management, information processing, and research. Sometimes, a subdivision of these management functions—**quinary activities**—is distinguished to recognize high-level decision-making roles in all types of large organizations, public or private. (The distinctions between tertiary, quaternary, and quinary activities are more fully developed in Chapter 9.) As Figure 8.2 suggests, transportation and communication services cut across the general categories of economic activity, unite them, and make possible the spatial interactions that all human enterprise requires (discussed in Chapter 3).

The term *industry*—in addition to its common meaning as a branch of manufacturing activity—is frequently employed as a substitute identical in meaning to *activity* as a designation of these categories of economic enterprise. That is, we can speak of the steel, or automobile, or textile "industry" with all the impressions of factories, mills, raw materials, and products each type of enterprise implies. But with equal logic we can refer in a more generalized way to the "entertainment" or the "travel" industries or, in the present context, to "primary," "secondary," and "tertiary" industries.

Figure 8.3 These logs entering a lumber mill are products of *primary production*. Processing them into boards, plywood, or prefabricated houses is a *secondary activity* that increases their value by altering their form. The products of many secondary industries—sheet steel from steel mills, for example—constitute "raw materials" for other manufacturers.

These categories of production and service activities or industries help us to see an underlying structure to the nearly infinite variety of things people do to earn a living and to sustain themselves. But by themselves they tell us little about the organization of the larger economy of which the individual worker or establishment is a part. For that broader organizational understanding we look to *systems* rather than *components* of economies.

Types of Economic Systems

Broadly viewed, national economies in the early 21st century fall into one of three major types of system: *subsistence, commercial,* or *planned.* None of these economic systems is "pure." That is, none exists in isolation in an increasingly interdependent world. Each, however, displays certain underlying characteristics based on its distinctive forms of resource management and economic control.

In a **subsistence economy,** goods and services are created for the use of the producers and their kinship groups. Therefore, there is little exchange of goods and only limited need for markets. In the **commercial economies** that have become dominant in nearly all parts of the world, producers or their agents, in theory, freely market their goods and services, the laws of supply and demand determine price and quantity, and market competition is the primary force shaping production decisions and distributions. In the extreme form of **planned economies** associated with the communist-controlled societies that have now collapsed in nearly every country where they were formerly created or imposed, producers or their agents disposed of goods and services through government agencies that controlled both supply and price. The quantities produced and the locational patterns of production were strictly programmed by central planning departments.

Rigidly planned economies no longer exist in their classical form; they have been modified or dismantled now in favor of free market structures or only partially retained in a lesser degree of economic control associated with governmental supervision or ownership of selected sectors of increasingly market-oriented economies. Nevertheless, their landscape evidence lives on. The physical structures, patterns of production, and imposed regional interdependencies they created remain to influence the economic decisions of successor societies.

In actuality, few people are members of only one of these systems, although one may be dominant. A farmer in India may produce rice and vegetables privately for the family's consumption but also save some of the produce to sell. In addition, members of the family may market cloth or other handicrafts they make. With the money derived from those sales, the Indian peasant is able to buy, among other things, clothes for the family, tools, and fuel. Thus, that Indian farmer is a member of at least two systems: subsistence and commercial.

In the United States, government controls on the production of various types of goods and services (such as growing wheat or tobacco, producing alcohol, constructing and operating nuclear power plants, and engaging in licensed personal and professional services) mean that the country does not have a purely commercial economy. To a limited extent, its citizens participate in a controlled and planned as well as in a free market environment. Many African, Asian, and Latin American market economies have been decisively shaped by government policies that encourage or demand production of export commodities rather than domestic foodstuffs, or promote through import restrictions the development of domestic industries not readily supported by the national market alone. Example after example would show that there are very few people in the world who are members of only one type of economic system.

Inevitably, spatial patterns, including those of economic activities and systems, are subject to change. For example, the commercial economies of Western European countries, some with sizeable infusions of planned economy controls, are being restructured by both increased free market competition and supranational regulation under the World Trade Organization and the European Union (see pp. 467–468). Many of the countries of Latin America, Africa, Asia, and the Middle East that traditionally were dominated by subsistence economies are now benefiting from technology transfer from advanced economies and integration into expanding global production and exchange patterns.

No matter what economic system may locally prevail, in all systems transportation is a key variable. No advanced economy can flourish without a well-connected transport network. All subsistence societies—or subsistence areas of developing countries—are characterized by their isolation from regional and world routeways (Figure 8.4). That isolation restricts their progression to more advanced forms of economic structure.

Former sharp contrasts in economic organization are becoming blurred and national economic orientations are changing as globalization reduces structural contrasts in national economies. Still, both approaches to economic classification—by types of activities and by organization of economies—help us to visualize and understand changing world economic geographic patterns. In the remainder of this chapter, we will center our attention on the primary industries. In Chapter 9, we will consider secondary through quinary activity patterns.

Primary Activities: Agriculture

Before there was farming, *hunting* and *gathering* were the universal forms of primary production. These preagricultural pursuits are now practiced by at most a few thousands of persons worldwide, primarily in isolated and remote pockets within the low latitudes and among the sparse populations of very high latitudes. The interior of New Guinea, rugged areas of interior Southeast Asia, diminishing segments of the Amazon rain forest, a few districts of tropical Africa and northern Australia, and parts of the Arctic regions still contain such preagricultural people. Their numbers are few and declining, and wherever they are brought into contact with more advanced cultures, their way of life is eroded or lost.

Agriculture, defined as the growing of crops and the tending of livestock whether for the subsistence of the producers or for sale or exchange, has replaced hunting and gathering as economically the most significant of the primary activities. It is spatially the most

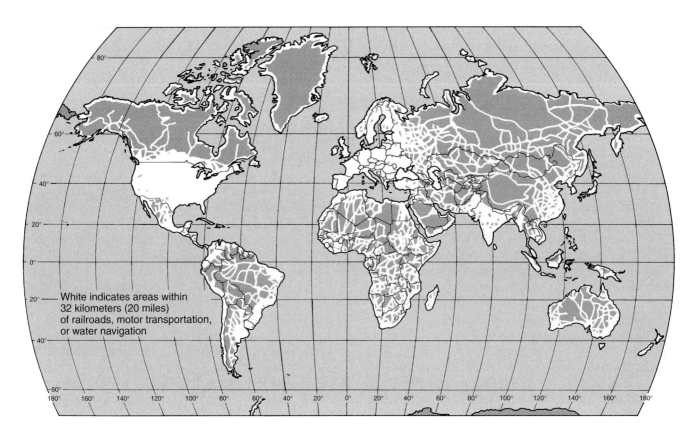

Figure 8.4 **Patterns of access and isolation.** Accessibility is a key measure of economic development and of the degree to which a world region can participate in interconnected market activities. Isolated areas of countries with advanced economies suffer a price disadvantage because of high transportation costs. Lack of accessibility in subsistence economic areas slows their modernization and hinders their participation in the world market.

Sources: Hammond Comparative World Atlas, *New Revised and Expanded Edition, Hammond Inc., Maplewood, N.J.;* Goode's World Atlas, *19th edition, Rand McNally & Company, Chicago, Ill., 1995.*

widespread, found in all world regions where environmental circumstances permit (Figure 8.5). More than one-third of the world's land area (excluding Greenland and Antarctica) is in some form of agricultural use, including permanent pastureland, and crop farming alone covers about 11%. In many developing economies, at least two-thirds of the labor force is directly involved in farming and herding. In some, such as Bhutan in Asia or Burkina Faso and Burundi in Africa, the figure is more than 90%. Overall, however, employment in agriculture is steadily declining in developing economies (Figure 8.6).

Comparable or greater relative reductions in the agricultural labor force have occurred in highly developed commercial economies where farm work involves only a small fraction of the labor force: 8% in most of Western Europe, below 5% in Canada, and less than 3% in the United States. (For the world pattern of the agricultural labor force early in this century, see Figure 10.11.) Indeed, a declining number or proportion of farm workers, along with farm consolidation and increasing output, are typical in all present-day highly developed commercial agricultural systems. On the other hand, agriculture remains a major component in the economies of many of the world's developing countries, producing for domestic markets and providing a major source of national income through exports (Figure 8.7).

It has been customary to classify agricultural societies on the twin bases of the importance of off-farm sales and the level of mechanization and technological advancement. *Subsistence, traditional* (or *intermediate*), and *advanced* (or *modern*) are usual terms employed to recognize both aspects. These are not mutually exclusive but rather are recognized stages along a continuum of farm economy variants. At one end lies production solely for family sustenance, using rudimentory tools and native plants. At the other is the specialized, highly capitalized, near-industrialized agriculture for off-farm delivery that marks advanced economies. Between these extremes is the middle ground of traditional agriculture, where farm production is in part destined for home consumption and in part oriented toward off-farm sale either locally or in national and international markets. We can most clearly see the variety of agricultural activities and the diversity of controls on their spatial patterns by examining the "subsistence" and "advanced" ends of the agricultural continuum.

Subsistence Agriculture

By definition, a *subsistence* economic system involves nearly total self-sufficiency on the part of its members. Production for exchange is minimal, and each family or close-knit social group

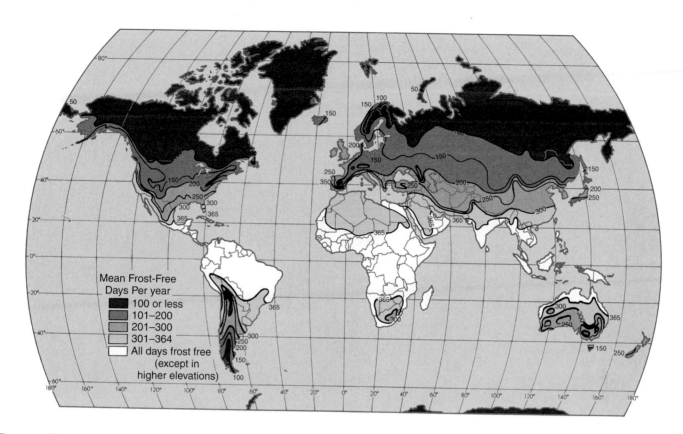

Figure 8.5 **Average length of growing season.** The number of frost-free days is an important environmental control on agriculture, as is the availability of precipitation sufficient in amount and reliability for crop production. Since agriculture is not usually practicable with less than a 90-day growing season, large parts of Russia and Canada have only limited cropping potential. Except where irrigation water is available, arid regions are similarly outside of the margins of regular crop production.

Courtesy Wayne M. Wendland.

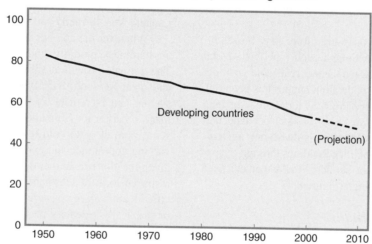

Figure 8.6 In the developing economies worldwide, the percentage of the labor force in agriculture has been steadily declining—and is projected to decrease to even lower levels.

Sources: FAO and World Bank.

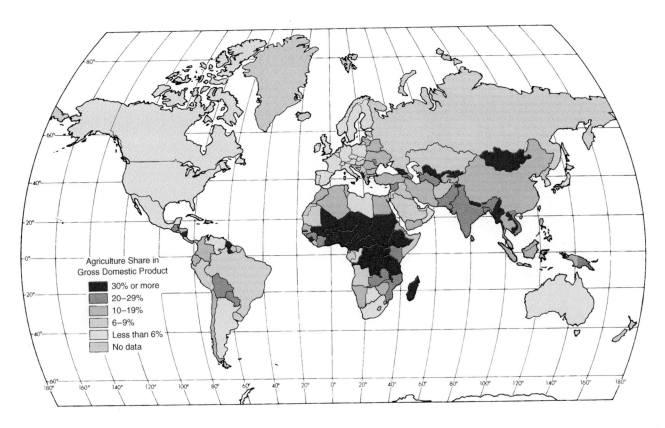

Figure 8.7 **Share of agriculture in gross domestic product.** Agriculture contributed 30% or more of gross domestic product (the total monetary output of goods and services of an economy) of over 30 countries at the end of the 20th century. Most were small, developing economies with less than US $500 in annual per capita income. Together, they held 11% of world population, far less than the 30% in 40 countries with similar agricultural importance in the early 1990s.

Source: The World Bank.

relies on itself for its food and other most essential requirements. Farming for the immediate needs of the family is, even today, the predominant occupation of humankind. In most of Africa, south and east Asia, and much of Latin America, a large percentage of people are primarily concerned with feeding themselves from their own land and livestock.

Two chief types of subsistence agriculture may be recognized: *extensive* and *intensive*. Although each type has several variants, the essential contrast between them is realizable yield per unit of area used and, therefore, population-supporting potential. **Extensive subsistence agriculture** involves large areas of land and minimal labor input per hectare. Both product per land unit and population densities are low. **Intensive subsistence agriculture** involves the cultivation of small landholdings through the expenditure of great amounts of labor per acre. Yields per unit area and population densities are both high (Figure 8.8).

Extensive Subsistence Agriculture

Of the several types of *extensive subsistence* agriculture—varying one from another in their intensities of land use—two are of particular interest.

Nomadic herding, the wandering but controlled movement of livestock solely dependent on natural forage, is the most extensive type of land use system (Figure 8.8). That is, it requires the greatest amount of land area per person sustained. Over large portions of

the Asian semidesert and desert areas, in certain highland zones, and on the fringes of and within the Sahara, a relatively small number of people graze animals for consumption by the herder group, not for market sale. Sheep, goats, and camels are most common, while cattle, horses, and yaks are locally important. The reindeer of Lapland were formerly part of the same system.

Whatever the animals involved, their common characteristics are hardiness, mobility, and an ability to subsist on sparse forage. The animals provide a variety of products: milk, cheese, blood, and meat for food; hair, wool, and skins for clothing; skins for shelter; and excrement for fuel. For the herder, they represent primary subsistence. Nomadic movement is tied to sparse and seasonal rainfall or to cold temperature regimes and to the areally varying appearance and exhaustion of forage. Extended stays in a given location are neither desirable nor possible. *Transhumance* is a special form of seasonal movement of livestock to exploit specific locally varying pasture conditions. Employed by permanently or seasonally sedentary pastoralists amd pastoral farmers, transhumance may involve the regular vertical alteration from mountain to valley pastures between summer and winter months or horizontal movement between established lowland grazing areas to reach pastures temporarily lush from monsoonal (seasonal) rains.

As a type of economic system, nomadic herding is declining. Many economic, social, and cultural changes are causing nomadic groups to alter their way of life or to disappear entirely. On

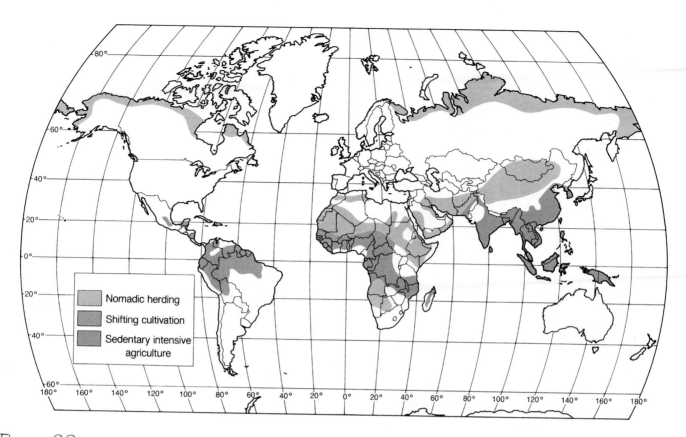

Figure 8.8 **Subsistence agricultural areas of the world.** Nomadic herding, supporting relatively few people, was the age-old way of life in large parts of the dry and cold world. Shifting or swidden agriculture maintains soil fertility by tested traditional practices in tropical wet and wet-and-dry climates. Large parts of Asia support millions of people engaged in sedentary intensive cultivation, with rice and wheat the chief crops.

the Arctic fringe of Russia, herders under communism were made members of state or collective herding enterprises; in post-Soviet years, extensive oil and natural gas exploration and extraction are damaging or destroying large portions of tundra habitat. In northern Scandinavia, Lapps (Saami) are engaged in commercial more than in subsistence livestock farming. In the Sahel region of Africa on the margins of the Sahara, oases formerly controlled by herders have been taken over by farmers, and the great droughts of recent decades have forever altered the formerly nomadic way of life of thousands.

A much differently based and distributed form of extensive subsistence agriculture is found in all of the warm, moist, low-latitude areas of the world. There, many people engage in a kind of nomadic farming. Through clearing and use, the soils of those areas lose many of their nutrients (as soil chemicals are dissolved and removed by surface and groundwater—leaching—or nutrients are removed from the land in the vegetables picked and eaten), and farmers cultivating them need to move on after harvesting several crops. In a sense, they rotate fields rather than crops to maintain productivity. This type of **shifting cultivation** has a number of names, the most common of which are *swidden* (an English localism for "burned clearing") and *slash-and-burn.* Each region of its practice has its own name—for example, *milpa* in Middle and South America, *chitemene* in Africa, *ladang* in Southeast Asia.

Characteristically, the farmers hack down the natural vegetation, burn the cuttings, and then plant such crops as maize (corn), millet (a cereal grain), rice, manioc or cassava, yams, and sugar-cane (Figure 8.9). Increasingly included in many of the crop combinations are such high-value, labor-intensive commercial crops as coffee, which provide the cash income that is evidence of the growing integration of all peoples into exchange economies. Initial yields—the first and second crops—may be very high, but they quickly become lower with each successive planting on the same plot. As that occurs, cropping ceases, native vegetation is allowed to reclaim the clearing, and gardening shifts to another newly prepared site. The first clearing will ideally not be used again for crops until, after many years, natural fallowing replenishes its fertility (see "Swidden Agriculture").

Less than 3% of the world's people are still predominantly engaged in tropical shifting cultivation on some one-sixth of the world's land area (see Figure 8.8). Since the essential characteristic of the system is the intermittent cultivation of the land, each family requires a total occupance area equivalent to the garden plot in current use plus all land left fallow for regeneration. Population densities are traditionally low, for much land is needed to support few people. Here as elsewhere, however, population density must be considered a relative term. In actuality, although crude (arithmetic) density is low, people per unit area of *cultivated* land may be high.

Figure 8.9 An African swidden plot being fired. Stumps and trees left in the clearing will remain after the burn.

Shifting cultivation is one of the oldest and most widely spread agricultural systems of the world. It is found on the islands of Borneo, New Guinea, and Sumatra but is now retained only in small parts of the uplands of Southeast Asia in Vietnam, Thailand, Myanmar, and the Philippines. Nearly the whole of Central and West Africa away from the coasts, Brazil's Amazon basin, and large portions of Central America were formerly all known for this type of extensive subsistence agriculture.

It may be argued that shifting cultivation is a highly efficient cultural adaptation where land is abundant in relation to population and levels of technology and capital availability are low. As those conditions change, the system becomes less viable. The basic change, as noted in Chapter 4, is that land is no longer abundant in relation to population in many of the less developed wet, tropical countries. Their growing populations have cleared and settled the forestlands formerly only intermittently used in swidden cultivation. The **Boserup thesis,** proposed by the economist Ester Boserup, is based on the observation that population increases necessitate increased inputs of labor and technology to compensate for reductions in the natural yields of swidden farming. It holds that population growth independently forces an increased use of technology in farming and—in a reversal of the Malthusian idea (p. 133) that the supply of essential foodstuffs is basically fixed or only slowly expandable—requires a conversion from extensive to intensive subsistence agriculture.

Intensive Subsistence Systems

Fewer than one-half of the people of the world are engaged in intensive subsistence agriculture, which predominates in areas shown in Figure 8.8. As a descriptive term, *intensive subsistence* is no longer fully applicable to a changing way of life and economy in which the distinction between subsistence and commercial is decreasingly valid. While families may still be fed primarily with the produce of their individual plots, the exchange of farm commodities within the system is considerable. Production of foodstuffs for sale in rapidly growing urban markets is increasingly vital for the rural economies of "subsistence farming" areas and for the sustenance of the growing proportion of national and regional populations no longer themselves engaged in farming. Nevertheless, hundreds of millions of Indians, Chinese, Pakistanis, Bangladeshis, and Indonesians plus further millions in other Asian, African, and Latin American countries remain small-plot, mainly subsistence producers of rice, wheat, maize, millet, or pulses (peas, beans, and other legumes). Most live in monsoon Asia, and we will devote our attention to that area.

Intensive subsistence farmers are concentrated in such major river valleys and deltas as the Ganges and the Chang Jiang (Yangtze) and in smaller valleys close to coasts—level areas with fertile alluvial soils. These warm, moist districts are well suited to the production of rice, a crop that under ideal conditions can provide large amounts of food per unit of land. Rice also requires a great deal of time and attention, for planting rice shoots by hand in standing fresh water is a tedious art (Figure 8.10). In the cooler and drier portions of Asia, wheat is grown intensively, along with millet and, less commonly, upland rice.

Rice is known to have been cultivated in parts of China and India for more than 7000 years. Today, wet, or lowland, rice is the mainstay of subsistence agriculture and diets of populations from Sri Lanka and India to Taiwan, Japan, and Korea. It is grown on more than 80% of the planted area in Bangladesh, Thailand, and Malaysia and on more than 50% in six other Asian countries. Almost exclusively used as a human food, rice provides 25% to 80% of the calories in the daily diet of over 2.8 billion Asians, or half the world's population. Its successful cultivation depends on the controlled management of water, relatively easy in humid tropical river valleys with heavy, impermeable, water-retaining soils though more difficult in upland and seasonally dry districts. Throughout Asia, the necessary water management systems have left their distinctive marks on the landscape. Permanently diked fields to contain and control water, levees against unwanted water, and reservoirs, canals, and drainage

Swidden Agriculture

The following account describes shifting cultivation among the Hanunóo people of the Philippines. Nearly identical procedures are followed in all swidden farming regions.

When a garden site of about one-half hectare (a little over one acre) has been selected, the swidden farmer begins to remove unwanted vegetation. The first phase of this process consists of slashing and cutting the undergrowth and smaller trees with bush knives. The principal aim is to cover the entire site with highly inflammable dead vegetation so that the later stage of burning will be most effective. Because of the threat of soil erosion the ground must not be exposed directly to the elements at any time during the cutting stage. During the first months of the agricultural year, activities connected with cutting take priority over all others. It is estimated that the time required ranges from 25 to 100 hours for the average-sized swidden plot.

Once most of the undergrowth has been slashed, chopped to hasten drying, and spread to protect the soil and assure an even burn, the larger trees must be felled or killed by girdling (cutting a complete ring of bark) so that unwanted shade will be removed. The successful felling of a real forest giant is a dangerous activity and requires great skill. Felling in second growth is usually less dangerous and less arduous. Some trees are merely trimmed but not killed or cut, both to reduce the amount of labor and to leave trees to reseed the swidden during the subsequent fallow period.

The crucial and most important single event in the agricultural cycle is swidden burning. The main firing of a swidden is the culmination of many weeks of preparation in spreading and leveling chopped vegetation, preparing firebreaks to prevent flames escaping into the jungle, and allowing time for the drying process. An ideal burn rapidly consumes every bit of litter; in no more than an hour or an hour and a half, only smoldering remains are left.

The Hanunóo, swidden farmers of the Philippines, note the following as the benefits of a good burn: 1) removal of unwanted vegetation, resulting in a cleared swidden; 2) extermination of many animal and some weed pests; 3) preparation of the soil for dibble (any small hand tool or stick to make a hole) planting by making it softer and more friable; 4) provision of an evenly distributed cover of wood ashes, good for young crop plants and protective of newly-planted grain seed. Within the first year of the swidden cycle, an average of between 40 and 50 different types of crop plants have been planted and harvested.

The most critical feature of swidden agriculture is the maintenance of soil fertility and structure. The solution is to pursue a system of rotation of 1 to 3 years in crop and 10 to 20 in woody or bush fallow regeneration. When population pressures mandate a reduction in the length of fallow period, productivity of the region tends to drop as soil fertility is lowered, marginal land is utilized, and environmental degradation occurs. The balance is delicate.

Source: Based on Harold C. Conklin, *Hanunóo Agriculture*, FAO Forestry Development Paper No. 12.

---

channels to control its availability and flow are common sights. Terraces to extend level land to valley slopes are occasionally encountered as well (see Figure 4.24).

Intensive subsistence farming is characterized by large inputs of labor per unit of land, by small plots, by the intensive use of fertilizers, mostly animal manure, and by the promise of high yields in good years (see "The Economy of a Chinese Village"). For food security and dietary custom, some other products are also grown. Vegetables and some livestock are part of the agricultural system, and fish may be reared in rice paddies and ponds. Cattle are a source of labor and of food. Food animals include swine, ducks, and chickens, but since Muslims eat no pork, hogs are absent in their areas of settlement. Hindus generally eat little meat, mainly goat and lamb but not pork or beef. The large number of cattle in India are vital for labor, as a source of milk and cheese, and as producers of fertilizer and fuel.

Urban Subsistence Farming

Not all of the world's subsistence farming is based in rural areas. Urban agriculture is a rapidly growing activity, with some 800 million city farmers worldwide providing, according to United Nations figures, one-seventh of the world's total food production. Occurring in all regions of the world, developed and underdeveloped, but most prevalent in Asia, urban agriculture activities range from small garden plots, to backyard livestock breeding, to raising fish in ponds and streams. Using the garbage dumps of Jakarta, the rooftops of Mexico City, and meager dirt strips along roadways in Kolkata (Calcutta) or Kinshasa, millions of people are feeding their own families and supplying local markets with vegetables, fruit, fish, and even meat—all produced within the cities themselves and all without the expense and spoilage of storage or long-distance transportation.

In Africa where, for example, 2 of 3 Kenyan and Tanzanian urban families engage in farming, a reported 20% of urban nutritional requirement is produced in the towns and cities; in Accra, Ghana's capital, urban farming provides the city with 90% of its fresh vegetables. At the end of the 20th century, city farming in Cuba produced 65% of the country's rice, 43% of its fruits and vegetables, and 12% of roots and fibers; altogether, some 165,000 urban Cubans produced 800,000 tons of fresh produce in 1999.

Urban agriculture occupies city land as well as city residents: in Bangkok, Thailand for example, some 60% of the metropolitan area is cultivated. A similar inclusion of adjacent rural land within urban boundaries is characteristic of China. There, based on an earlier mandate that socialist cities be self-sufficient, municipal boundaries were set to include large areas of rural land

Figure 8.10 Transplanting rice seedlings requires arduous hand labor by all members of the family. The newly flooded diked fields, previously plowed and fertilized, will have their water level maintained until the grain is ripe. This photograph was taken in Indonesia. The scene is repeated wherever subsistence wet-rice agriculture is practiced.

now worked intensively to supply the fruits, vegetables, fish, and the like consumed within the city proper. Chinese urban agriculture—by UN estimates providing 90% of the vegetable supply of cities—is, in reality, periurban (suburban) farming within city administrative control. Little or no backyard (or rooftop) land is available for food production within the densely developed Chinese city proper. In whatever form urban farming efforts are expressed, not all its area or yield is solely for local subsistence. An estimated 200 million global urban dwellers also produce food for sale to others.

In all parts of the developing world, urban-origin foodstuffs have reduced the incidence of adult and child malnutrition in cities rapidly expanding by their own birth rates and by the growing influx of displaced rural folk. City farming is, as well, a significant outlet for underemployed residents. In some cities, as many as one-fifth to two-thirds of all families are engaged in agriculture, a United Nations Development Programme study reports, with as many as one-third of them having no other source of income.

There are both positive and negative environmental consequences of urban agricultural activities. On the plus side, urban farming helps convert waste from a problem to a resource by reducing runoff and erosion from open dumps and by avoiding costs of wastewater treatment and solid waste disposal. In Khartoum, Sudan, for example, about 25% of the city's garbage is consumed by farm animals; in Kolkata (Calcutta), India, city sewage is used to feed some 3000 hectares (7400 acres) of lagoons which, in turn, produce some 6000 tons of fish annually. Additionally, some 20,000 Kolkata residents diligently farm on the city's garbage dumps, converting waste area and rotting refuse to nutrition. Nearly everywhere, human and animal wastes, vegetable debris, and table scraps are composted or applied to garden areas, and nearly everywhere, vegetable gardens and interspersed fruit trees, ornamental plants, and flowers enhance the often drab urban scene.

Negative consequences also attend urban agriculture and frequently evoke restrictive governmental regulations and prohibitions. The widespread use of untreated human waste as fertilizers exposes both producers and consumers to infectious diseases such as cholera and hepatitis. When pesticides and chemical fertilizers are available and indiscriminately used by untrained gardeners, local water supplies may become contaminated. In some instances, limited supplies of drinking water may be severely depleted through diversion for illegal irrigation or watering of subsistence gardens and livestock. In response to its estimate of 35% of fresh drinking water lost through leakage and illegal tapping by urban farmers, Tanzania's National Urban Water Agency imposed severe penalty fees and prohibitions on urban agricultural use. Other municipal water agencies elsewhere have reacted similarly.

The village of Nanching is in subtropical southern China on the Zhu River delta near Guangzhou (Canton). Its pre-communist subsistence agricultural system was described by a field investigator, whose account is here condensed. The system is found in its essentials in other rice-oriented societies.

In this double-crop region, rice was planted in March and August and harvested in late June or July and again in November. March to November was the major farming season. Early in March the earth was turned with an iron-tipped wooden plow pulled by a water buffalo. The very poor who could not afford a buffalo used a large iron-tipped wooden hoe for the same purpose.

The plowed soil was raked smooth, fertilizer was applied, and water was let into the field, which was then ready for the transplanting of rice seedlings. Seedlings were raised in a seedbed, a tiny patch fenced off on the side or corner of the field. Beginning from the middle of March, the transplanting of seedlings took place. The whole family was on the scene. Each took the seedlings by the bunch, ten to fifteen plants, and pushed them into the soft inundated soil. For the first thirty or forty days the emerald green crop demanded little attention except keeping the water at a proper level. But

after this period came the first weeding; the second weeding followed a month later. This was done by hand, and everyone old enough for such work participated. With the second weeding went the job of adding fertilizer. The grain was now allowed to stand to "draw starch" to fill the hull of the kernels. When the kernels had "drawn enough starch," water was let out of the field, and both the soil and the stalks were allowed to dry under the hot sun.

Then came the harvest, when all the rice plants were cut off a few inches above the ground with a sickle. Threshing was done on a threshing board. Then the grain and the stalks and leaves were taken home with a carrying pole on the peasant's shoulder. The plant was used as fuel at home.

As soon as the exhausting harvest work was done, no time could be lost before starting the chores of plowing, fertilizing, pumping water into the fields, and transplanting seedlings for the second crop. The slack season of the rice crop was taken up by chores required for the vegetables which demanded continuous attention, since every peasant family devoted a part of the farm to vegetable gardening. In the hot and damp period of late spring and summer, eggplant and

several varieties of squash and beans were grown. The green-leafed vegetables thrived in the cooler and drier period of fall, winter, and early spring. Leeks grew the year round.

When one crop of vegetables was harvested, the soil was turned and the clods broken up by a digging hoe and leveled with an iron rake. Fertilizer was applied, and seeds or seedlings of a new crop were planted. Hand weeding was a constant job; watering with the long-handled wooden dipper had to be done an average of three times a day, and in the very hot season when evaporation was rapid, as frequently as six times a day. The soil had to be cultivated with the hoe frequently as the heavy tropical rains packed the earth continuously. Instead of the two applications of fertilizer common with the rice crop, fertilizing was much more frequent for vegetables. Besides the heavy fertilizing of the soil at the beginning of a crop, usually with city garbage, additional fertilizer, usually diluted urine or a mixture of diluted urine and excreta, was given every ten days or so to most vegetables.

Source: Adapted from C. K. Yang, *A Chinese Village in Early Communist Transition* (Cambridge, Mass.: Massachusetts Institute of Technology, 1959).

Costs of Territorial Extension

Improved health care in the 20th century lowered infant and crude death rates and accelerated population growth rates in countries of intensive subsistence agriculture. The rising population, of course, puts increasing pressure on the land and, following the Boserup thesis, the response has been to increase further the intensity of agricultural production. Lands formerly considered unsuitable for farming by reason of low fertility, inadequate moisture, difficulty of clearing and preparation, isolation from settlement areas, and other factors have been brought into cultivation.

To till those additional lands, a price must be paid. Any economic activity incurs an additional (called *marginal*) cost in labor, capital, or other unit of expenditure to bring into existence an added unit of production. When the value of the added (marginal) production at least equals the added cost, the effort may be

undertaken. In past periods of lower population pressure, there was no incentive to extend cultivation to less productive or more expensive unneeded lands. Now circumstances are different. In many intensive subsistence agricultural economies, however, possibilities for land conversion to agriculture are limited. More than 60% of the population of the developing world lives in countries in which some three-quarters of possible arable land is already under cultivation and where undeveloped cultivable land has low potential for settlement and use.

When population pressures dictate land conversion, serious environmental deterioration may result. Clearing of wet tropical forests in the Philippines, the Amazon Basin, and Indonesia has converted dense woodland to barren desolation within a very few years as soil erosion and nutrient loss have followed forest destruction. In Southeast Asia, some 10 million hectares (25 million acres) of former forestland are now wasteland, covered by useless saw grasses that supply neither forage, food, nor fuel.

Intensification and the Green Revolution

Increased productivity of existing cropland rather than expansion of cultivated area has accounted for most of the growth of agricultural production over the past few decades. Two interrelated approaches to those yield increases mark recent farming practices.

First, throughout much of the developing world production inputs such as water, fertilizer, pesticides, and labor have been increased to expand yields on a relatively constant supply of cultivable land. Irrigated area, for example, nearly doubled between 1960 and 2000 to comprise by the latter year some 18% of the world's cropland. Global consumption of fertilizers has dramatically increased since the 1950s, and inputs of pesticides and herbicides have similarly grown. Traditional practices of leaving land fallow (uncultivated) to renew its fertility have been largely abandoned, and double and triple cropping of land where climate permits has increased in Asia and even in Africa, where marginal land is put to near-continuous use to meet growing food demands.

Many of these intensification practices are part of the second approach, linked to the **Green Revolution**—the shorthand reference to a complex of seed and management improvements adapted to the needs of intensive agriculture and designed to bring larger harvests from a given area of farmland. Between 1972 and 2002, world total grain production rose 60%. Despite nearly 2.5 billion more people in the world, averaged grain production per capita for the period 1998–2002 was 3% above the 1972 level. More than three-quarters of that production growth was due to increases in yields rather than expansions in cropland. For Asia as a whole, cereal yields grew by more than 40% between 1980 and the end of the century, accounted for largely by increases in China and India; they increased by over 35% in

South America. These Green Revolution yield increases and the improved food supplies they represent have been particularly important in densely populated, subsistence farming areas heavily dependent on rice and wheat cultivation. Chinese rice harvests grew by two-thirds and India's wheat yields doubled between 1970 and the end of the 1990s.

Genetic improvements in rice and wheat have formed the basis of the Green Revolution. Dwarfed varieties have been developed that respond dramatically to heavy applications of fertilizer, that resist plant diseases, and that can tolerate much shorter growing seasons than traditional native varieties. Adopting the new varieties and applying the irrigation, mechanization, fertilization, and pesticide practices they require have created a new "high-input, high-yield" agriculture. Most poor farmers on marginal and rain-fed (nonirrigated) lands, however, have not benefited from the new plant varieties requiring irrigation and high chemical inputs.

Expanded food production made possible through the Green Revolution has helped alleviate some of the shortages and famines predicted for subsistence agricultural regions since the early 1960s (Figure 8.11), saving an estimated 1 billion people from starvation. Despite rapidly growing population numbers since 1961, agricultural production in the developing world has increased by 52% per person and daily food intake grew from 1932 calories, barely enough to sustain life, to over 2700 calories in 2000; it is expected to continue to rise. In the same period, real food prices have declined by more than 50%. According to World Bank calculations, more than 80% of people in developing countries now have adequate diets, versus 55% in 1950. Although the *number* of undernourished people remains near the

Index of Total and Per Capita Food Production, 1961–2001

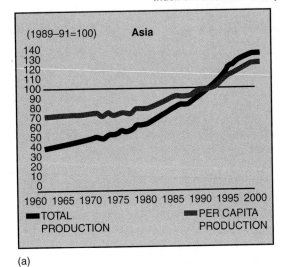

(a)

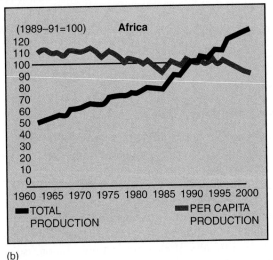

(b)

Figure 8.11 **Trends in food production, 1961–2001.** Globally, production of food crops increased over the 40 year span shown, but the average annual increase dropped from 3% during the 1960s to 2.4% in the 1970s, 2.2% in the 1980s, and to 1% or less in the 1990s. Although *total* food production expanded in nearly all world regions (the area of the former Soviet Union was a notable exception), that expansion has not in all cases been reflected in improved *per capita* availability. (a) Intensification and expansion of farming in Asia resulted in both greatly increased food production and, despite continuing population growth, expanding per capita availability. (b) Population growth presented a different picture in Africa, where total production of food steadily grew over the graphed period, but per capita food supplies persistently declined.

Source: Data from Food and Agriculture Organization, World Resources Institute, and U.S. Department of Agriculture.

900 million mark because of population growth, total world food supply has increased even faster than population and will continue to do so, the UN predicts, through at least 2030. However, a series of years beginning in 2000 in which production of all major grains fell substantially below consumption levels, materially reducing carry-over stocks and, therefore, endangering world food security, called the UN's prediction into question.

But a price has been paid for Green Revolution successes. Irrigation, responsible for an important part of increased crop yields, has destroyed large tracts of land; excessive salinity of soils resulting from poor irrigation practices is estimated to have a serious effect on the productivity of 20 million to 30 million hectares (80,000–120,000 sq mi) of land around the world, out of a world total of some 220 million hectares of irrigated land. And the huge amount of water required for Green Revolution irrigation has led to serious groundwater depletion, conflict between agricultural and growing urban and industrial water needs in developing countries—many of which are in subhumid climates—and to worries about scarcity and future wars over water.

And very serious genetic consequences are feared from the loss of traditional and subsistence agriculture. With it is lost the

food security that distinctive locally adapted native crop varieties (*land races*) provided and the nutritional diversity and balance that multiple-crop intensive gardening assured. Subsistence farming, wherever practiced, was oriented toward risk minimization. Many differentially hardy varieties of a single crop guaranteed some yield whatever adverse weather, disease, or pest problems might occur.

Commercial agriculture, however, aims at profit maximization, not minimal food security. Poor farmers unable to afford the capital investment the Green Revolution demands have been displaced by a commercial monoculture, one often oriented toward specialty and industrial crops designed for export rather than to food production for a domestic market. Traditional rural society has been disrupted, and landless peasants have been added to the urbanizing populations of affected countries. To the extent that land races are lost to monoculture, varietal distinction in food crops is reduced. "Seed banks" rather than native cultivation are increasingly needed to preserve genetic diversity for future plant breeding and as insurance against catastrophic pest or disease susceptibility of inbred varieties (Figure 8.12).

The presumed benefits of the Green Revolution are not available to all subsistence agricultural areas or advantageous to

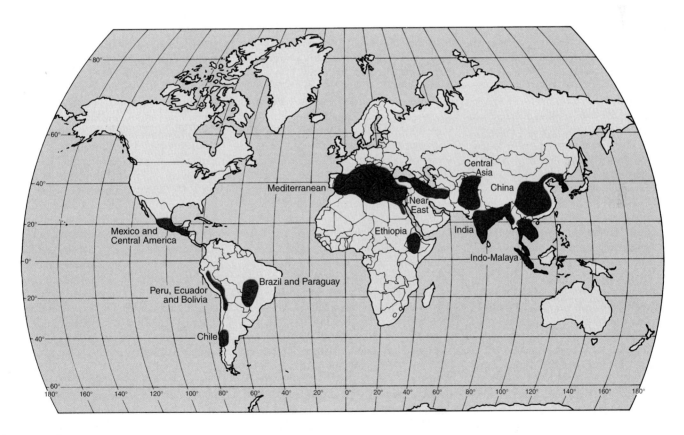

Figure 8.12 **Areas with high current genetic diversity of crop varieties.** Loss of crop varieties characterizes the commercial agriculture of much of the developed world. In place of the many thousands of species and subspecies (varieties) of food plants grown since the development of agriculture 12,000 or more years ago, fewer than 100 species now provide most of the world's food supply. Most of the diversity loss has occurred in the last 100 years. In the United States, for example, 96% of commercial vegetable varieties listed by the Department of Agriculture in 1903 are now extinct. Crop breeders, however, require genetic diversity to develop new varieties that are resistant to evolving plant pest and disease perils. That need necessitates the protection of plant stocks and environments in those temperate and subtropical zones where food plants were first domesticated and are home to the wild relatives of our current food crops.

Sources: J. G. Hawkes, The Diversity of Crop Plants *(Cambridge, Mass.: Harvard University Press, 1983); and Walter V. Reid and Kenton R. Miller,* Keeping Options Alive: The Scientific Basis for Conserving Biodiversity. *(Washington, D.C.: World Resources Institute, 1989), fig. 5, p. 24.*

everyone engaged in farming (see "Women and the Green Revolution"). Africa is a case in point (see Figure 8.11). Green Revolution crop improvements have concentrated on wheat, rice, and maize. Of these, only maize is important in Africa, where principal food crops include millet, sorghum, cassava, manioc, yams, cowpeas, and peanuts. Although new varieties of maize resistant to the drought and acidic soils common in Africa were announced in the middle 1990s, belated research efforts directed to other African crops, the continent's great range of growing conditions, and its abundance of yield-destroying pests and viruses have denied it the dramatic regionwide increases in food production experienced elsewhere in the developing world.

Cassava is a hopeful early exception. The most widely cultivated tuber and second most important food staple in sub-Saharan Africa, cassava has been transformed from a low-yielding subsistence hedge against famine to a high-yielding cash crop. Between 1980 and 2001, total output nearly doubled to 94 million tons thanks to the introduction of new varieties developed by the International Institute of Tropical Agriculture. More recent experimental successes with a variety of genetically modified (GM) crops promise other important yield improvements. Virus-resistant varieties of sweet potatoes and both white and yellow maize and faster-growing bananas are already available though not yet widespread, and other food and fiber crops are receiving attention from African biotechnology scientists in Kenya, South Africa, and Egypt, with contributions from American and other Western investigators.

Women and the Green Revolution

Women farmers grow at least half of the world's food and up to 80% in some African countries. They are responsible for an even larger share of food consumed by their own families: 80% in sub-Saharan Africa, 65% in Asia, and 45% in Latin America and the Caribbean. Further, women comprise between one-third and one-half of all agricultural laborers in developing countries. For example, African women perform about 90% of the work of processing food crops and 80% of the work of harvesting and marketing.

Women's agricultural dominance in developing states is increasing, in fact, as male family members continue to leave for cities in search of paid urban work. In Mozambique, for example, for every 100 men working in agriculture, there are now 153 women. In nearly all other sub-Saharan countries the female component runs between 120 and 150 per 100 men. The departure of men for near or distant cities means, in addition, that women must assume effective management of their families' total farm operations.

Despite their fundamental role, however, women do not share equally with men in the rewards from agriculture, nor are they always beneficiaries of presumed improvements in agricultural technologies and practices. Often, they cannot own or inherit the land on which they work, and they frequently have difficulty in obtaining improved seeds or fertilizers available to male farmers.

As a rule, women farmers work longer hours and have lower incomes than do male farmers. This is not because they are less educated or competent. Rather, it is due to restricting cultural and economic factors. First, most women farmers are involved in subsistence farming and food production for the local market that yields little cash return. Second, they have far less access than men to credit at bank or government-subsidized rates that would make it possible for them to acquire the Green Revolution technology, such as hybrid seeds and fertilizers. Third, in some cultures women cannot own land and so are excluded from agricultural improvement programs and projects aimed at landowners. For example, many African agricultural development programs are based on the conversion of communal land, to which women have access, to private holdings, from which they are excluded. In Asia, inheritance laws favor male over female heirs, and female-inherited land is managed by husbands; in Latin America, discrimination results from the more limited status held by women under the law.

At the same time, the Green Revolution and its greater commercialization of crops has generally required an increase in labor per hectare, particularly in tasks typically reserved for women, such as weeding, harvesting, and postharvest work. If women are provided no relief from their other daily tasks, the Green Revolution for them may be more burden than blessing. But when mechanization is added to the new farming system, women tend to be losers. Frequently, such predominantly female tasks as harvesting or dehusking and polishing of grain—all traditionally done by hand—are given over to machinery, displacing rather than employing women. Even the application of chemical fertilizers (a man's task) instead of cow dung (women's work) has reduced the female role in agricultural development programs. The loss of those traditional female wage jobs means that already poor rural women and their families have insufficient income to improve their diets even in the light of substantial increases in food availability through Green Revolution improvements.

If women are to benefit from the Green Revolution, new cultural norms—or culturally acceptable accommodations within traditional household, gender, and customary legal relations—will be required. These must permit or recognize women's land-owning and other legal rights not now clearly theirs, access to credit at favorable rates, and admission on equal footing with males to government assistance programs. Recognition of those realities fostered the Food and Agriculture Organization of the United Nations' "FAO Plan of Action for Women in Development (1996–2001)" and its "Gender and Development Plan (2002–2007)." Both are aimed at stimulating and facilitating efforts to enhance the role of women as contributors and beneficiaries of economic, social, and political development. Objectives of the plan include promoting gender-based equity in access to, and control of, productive resources; enhancing women's participation in decision- and policy-making processes at all levels, local and national; and encouraging actions to reduce rural women's workload while enhancing their opportunities for paid employment and income.

In many areas showing the greatest past successes, Green Revolution gains are falling off. Recent cereal yields in Asia, for example, are growing at only two-thirds of their 1970s rate; the UN's Food and Agriculture Organization now considers Green Revolution technologies "almost exhausted" of any further productivity gains in Asian rice cultivation. Little prime land and even less water remain to expand farming in many developing countries, and the adverse ecological and social consequences of industrial farming techniques arouse growing resistance. Nor does biotechnology—which many have hailed as a promising new Green Revolution approach—seem likely to fill the gap. Consumer resistance to the genetic modification (GM) of crops, fear of the ecological consequences of such modification, and the high cost and restrictions on the new biotechnologies imposed by their corporate developers all conspire to inhibit the widespread adoption of the new technologies.

Even in those world regions favorable for Green Revolution introductions, its advent has not always improved diets or reduced dependency on imported basic foodstuffs. Often, the displacement of native agriculture involves a net loss of domestic food availability. In many instances, through governmental directive, foreign ownership or management, or domestic market realities, the new commercial agriculture is oriented toward food and industrial crops for the export market or toward specialty crop and livestock production for the expanding urban market rather than food production for the rural population.

Commercial Agriculture

Few people or areas still retain the isolation and self-containment characteristic of pure subsistence economies. Nearly all have been touched by a modern world of trade and exchange and have adjusted their traditional economies in response. Modifications of subsistence agricultural systems have inevitably made them more complex by imparting to them at least some of the diversity and linkages of activity that mark the advanced economic systems of the more developed world. Farmers in those systems produce not for their own subsistence but primarily for a market off the farm itself. They are part of integrated exchange economies in which agriculture is but one element in a complex structure that includes mining, manufacturing, processing, and the service activities of the tertiary, quaternary and quinary sectors. In those economies, farming activities presumably mark production responses to market demand expressed through price and are related to the consumption requirements of the larger society rather than to the immediate needs of farmers themselves.

Production Controls

Agriculture within modern, developed economies is characterized by *specialization*—by enterprise (farm), by area, and even by country; by *off-farm sale* rather than subsistence production; and by *interdependence* of producers and buyers linked through markets. Farmers in a free market economy supposedly produce those crops that their estimates of market price and production cost indicate will yield the greatest return. Theoretically, farm products in short supply will command an increased market price.

That, in turn, should induce increased production to meet the demand with a consequent reduction of market price to a level of equilibrium with production costs. In some developing countries, that equation between production costs and market price is broken and the farm economy distorted when government policy requires uneconomically low food prices for urban workers. It may also suffer material distortion under governmental programs protecting local producers by inhibiting farm product imports or subsidizing production by guaranteeing prices for selected commodities.

Where free market conditions prevail, however, the crop or the mix of crops and livestock that individual commercial farmers produce is a result of an appraisal of profit possibilities. Farmers must assess and predict prices, evaluate the physical nature of farmland, and factor in the possible weather conditions. The costs of production (fuel, fertilizer, capital equipment, labor) must be reckoned. A number of unpredictable conditions may thwart farmers' aspirations for profit. Among them are the uncertainties of growing season conditions that follow the original planting decision, the total volume of output that will be achieved (and therefore the unit cost of production), and the supply and price situation that will exist months or years in the future, when crops are ready for market.

Beginning in the 1950s in the United States, specialist farmers and corporate purchasers developed strategies for minimizing those uncertainties. Processors sought uniformity of product quality and timing of delivery. Vegetable canners—of tomatoes, sweet corn, and the like—required volume delivery of raw products of uniform size, color, and ingredient content on dates that accorded with cannery and labor schedules. And farmers wanted the support of a guaranteed market at an assured price to minimize the uncertainties of their specialization and stabilize the return on their investment.

The solution was contractual arrangements or vertical integrations uniting contracted farmer with purchaser-processor. Broiler chickens of specified age and weight, cattle fed to an exact weight and finish, wheat with a minimum protein content, popping corn with prescribed characteristics, potatoes of the kind and quality demanded by particular fast-food chains, and similar product specification became part of production contracts between farmer and buyer-processor. In the United States, the percentage of total farm output produced under contractual arrangements or by vertical integration (where production, processing, and sales are all coordinated within one firm) rose from 19% in 1960 to well over one-third during the 1990s. The term *agribusiness* is applied to the growing merging of the older, farm-centered crop economy and newer patterns of more integrated production and marketing systems.

Contract farming is spreading as well to developing countries, though it is often criticized as another adverse expression of globalization subjecting small-size farmers to exploitation by powerful Western agribusiness. The UN's FAO, however, argues that well-managed contract arrangements are effective in linking the small farmers of emerging economies with both foreign and local sources of advanced extension advice, seeds, fertilizers, machinery, and profitable markets at stable prices. The agency cites successful examples of contract farming in northern India,

Sri Lanka, Nepal, Indonesia, Thailand, and the Philippines and sees in the arrangements a most promising approach to market-oriented production in areas still dominated by subsistence agriculture.

Even for family farmers not bound by contractual arrangements to suppliers and purchasers, the older assumption that supply, demand, and the market price mechanism are the effective controls on agricultural production is not wholly valid. In reality, those theoretical controls are joined by a number of nonmarket governmental influences that may be as decisive as market forces in shaping farmers' options and spatial production patterns. If there is a glut of wheat on the market, for example, the price per ton will come down and the area sown to it should diminish. It will also diminish regardless of supply if governments, responding to economic or political considerations, impose acreage controls.

Distortions of market control may also be introduced to favor certain crops or commodities through subsidies, price supports, market protections, and the like. The political power of farmers in the European Union (EU), for example, secured for them generous product subsidies and for the EU immense unsold stores of butter, wine, and grains until 1992, when reforms began to reduce the unsold stockpiles even while increasing total farm spending. In Japan, the home market for rice is largely protected and reserved for Japanese rice farmers even though their production ef-

ficiencies are low and their selling price is high by world market standards. In the United States, programs of farm price supports, acreage controls, financial assistance, and other governmental involvements in agriculture have been of recurring and equally distorting effect (Figure 8.13).

A Model of Agricultural Location

Early in the 19th century, before such governmental influences were the norm, Johann Heinrich von Thünen (1783–1850) observed that lands of apparently identical physical properties were utilized for different agricultural purposes. Around each major urban market center, he noted, there developed a set of concentric land use rings of different farm products (Figure 8.14). The ring closest to the market specialized in perishable commodities that were both expensive to ship and in high demand. The high prices they could command in the urban market made their production an appropriate use of high-valued land near the city. Surrounding rings of farmlands farther away from the city were used for less perishable commodities with lower transport costs, reduced demand, and lower market prices. General farming and grain farming replaced the market gardening of the inner ring. At the outer margins of profitable agriculture, farthest from the single central market, livestock grazing and similar extensive land uses were found.

Figure 8.13 Open storage of 1 million bushels of Iowa corn. In the world of commercial agriculture, supply and demand are not always in balance. Both the bounty of nature in favorable crop years and the intervention of governmental programs that distort production decisions can create surpluses for which no market is readily available.

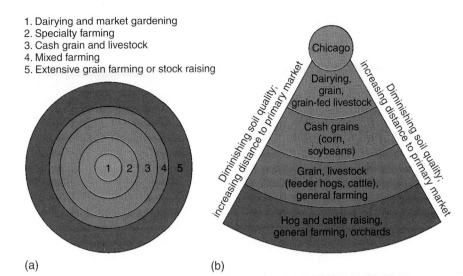

1. Dairying and market gardening
2. Specialty farming
3. Cash grain and livestock
4. Mixed farming
5. Extensive grain farming or stock raising

Chicago

Dairying, grain, grain-fed livestock

Cash grains (corn, soybeans)

Grain, livestock (feeder hogs, cattle), general farming

Hog and cattle raising, general farming, orchards

Diminishing soil quality; increasing distance to primary market

Diminishing soil quality; increasing distance to primary market

(a)   (b)

Figure 8.14 (a) **von Thünen's model.** Recognizing that as distance from the market increases, the value of land decreases, von Thünen developed a descriptive model of intensity of land use that holds up reasonably well in practice. The most intensively produced crops are found on land close to the market; the less intensively produced commodities are located at more distant points. The numbered zones of the diagram represent modern equivalents of the theoretical land use sequence von Thünen suggested over 150 years ago. As the metropolitan area at the center increases in size, the agricultural specialty areas are displaced outward, but the relative position of each is retained. Compare this diagram with Figure 8.18. (b) **A schematic view of the von Thünen zones** in the sector south of Chicago. There, farmland quality decreases southward as the boundary of recent glaciation is passed and hill lands are encountered in southern Illinois. On the margins of the city near the market, dairying competes for space with livestock feeding and suburbanization. Southward into flat, fertile central Illinois, cash grains dominate. In southern Illinois, livestock rearing and fattening, general farming, and some orchard crops are the rule.

Source: (b) Modified with permission from Bernd Andreae, Farming Development and Space: A World Agricultural Geography, *translated by Howard F. Gregor (Berlin; Hawthorne, N.Y.: Walter de Gruyter and Co., 1981).*

To explain why this should be so, von Thünen constructed a formal spatial model—the **von Thünen model**—perhaps the first developed to analyze human activity patterns. He concluded that the uses to which parcels were put was a function of the differing "rent" values placed on seemingly identical lands. Those differences, he claimed, reflected the cost of overcoming the distance separating a given farm from a central market town ("A portion of each crop is eaten by the wheels," he observed). The greater the distance, the higher was the operating cost to the farmer, since transport charges had to be added to other expenses. When a commodity's production costs plus its transport costs just equaled its value at the market, a farmer was at the economic margin of its cultivation. A simple exchange relationship ensued: the greater the transportation costs, the lower the rent that could be paid for land if the crop produced was to remain competitive in the market.

Since in the simplest form of the model, transport costs are the only variable, the relationship between land rent and distance from market can be easily calculated by reference to each competing crop's *transport gradient*. Perishable commodities such as fruits and vegetables would encounter high transport rates per unit of distance; other items such as grain would have lower rates. Land rent for any farm commodity decreases with increasing distance from the central market, and the rate of decline is determined by the transport gradient for that commodity. Crops that have both the highest market price and the highest transport costs will be grown nearest to the market. Less perishable crops with lower production and transport costs will be grown at greater dis-

tances away (Figure 8.15). Since in this model transport costs are uniform in all directions away from the center, a concentric zonal pattern of land use called the *von Thünen rings* results.

The von Thünen model may be modified by introducing ideas of differential transport costs (Figure 8.16), variations in topography or soil fertility, or changes in commodity demand and market price. With or without such modifications, von Thünen's analysis helps explain the changing crop patterns and farm sizes evident on the landscape at increasing distance from major cities, particularly in regions dominantly agricultural in economy. Farmland close to markets takes on high value, is used *intensively* for high-value crops, and is subdivided into relatively small units. Land far from markets is used *extensively* and in larger units.

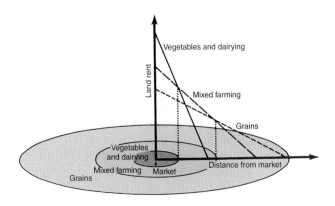

Figure 8.15 **Transport gradients and agricultural zones.**

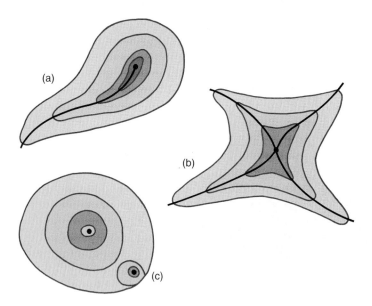

Figure 8.16 **Ring modifications.** Modifications of controlling conditions will alter the details but not change the underlying pattern of the *von Thünen rings.* For example, a change in demand and therefore market price of a commodity would merely expand its ring of production. An increase in transport costs would contract the production area, while reductions in freight rates would extend it. (*a*) If transport costs are reduced in one direction, the circularity—but not the sequence—of the rings will be affected. (*b*) If several roads are constructed or improved, land use sequences assume a star-shaped or digitate outline. (*c*) The addition of a smaller outlying market results in the emergence of a set of von Thünen rings subordinate to it.

In dominantly industrial and postindustrial economies, it has been suggested, the basic forces determining agricultural land use near cities are those associated with urban expansion itself, and von Thünen regularities are less predictable. Rather, irregularities and uncertainties of peripheral city growth, the encroachment on agricultural land by expansion from two or more cities, and the withholding of land from farming in anticipation of subdivision may locally reverse or invert the von Thünen intensity rings. Where those urbanizing forces dominate, the agricultural pattern often may be one of increasing—rather than decreasing—intensity with distance from the city.

Intensive Commercial Agriculture

Following World War II, agriculture in the developed world's market economies turned increasingly to concentrated methods of production. Machinery, chemicals, irrigation, and dependence on a restricted range of carefully selected and bred plant varieties and animal breeds all were employed in a concerted effort to wring more production from each unit of farmland. In that sense, all modern commercial agriculture is "intensive." There are, however, significant differences among the several types of farm specializations and practices in the relative ratios of capital inputs per hectare of farmed land. Those differences underlie generalized distinctions made between intensive and extensive commercial agriculture.

Farmers who apply large amounts of capital (for machinery and fertilizers, for example) and/or labor per unit of land engage in **intensive commercial agriculture.** The crops that justify such costly inputs are characterized by high yields and high market value per unit of land. They include fruits, vegetables, and dairy products, all of which are highly perishable. Near most medium-sized and large cities, dairy farms and **truck farms** (horticultural or "market garden" farms) produce a wide range of vegetables and fruits. Since the produce is perishable, transport costs increase because of the required special handling, such as use of refrigerated trucks and custom packaging. This is another reason for locations close to market. Note the distribution of truck and fruit farming in Figure 8.17.

Livestock-grain farming involves the growing of grain to be fed on the producing farm to livestock, which constitute the farm's cash product. In Western Europe, three-fourths of cropland is devoted to production for animal consumption; in Denmark, 90% of all grains are fed to livestock for conversion not only into meat but also into butter, cheese, and milk. Although livestock-grain farmers work their land intensively, the value of their product per unit of land is usually less than that of the truck farm. Consequently, in North America at least, livestock-grain farms are farther from the main markets than are horticultural and dairy farms.

Normally the profits for marketing livestock (chiefly hogs and beef cattle in the United States) are greater per pound than those for selling corn or other feed, such as alfalfa and clover. As a result, farmers convert their corn into meat on the farm by feeding it to the livestock, efficiently avoiding the cost of buying grain. They may also convert farm grain at local feed mills to the more balanced feed modern livestock rearing requires. Where land is too expensive to be used to grow feed, especially near cities, feed must be shipped to the farm. The grain-livestock belts of the world are close to the great coastal and industrial zone markets. The Corn Belt of the United States and the livestock region of Western Europe are two examples.

Extensive Commercial Agriculture

Farther from the market, on less expensive land, there is less need to use the land intensively. Cheaper land gives rise to larger farm units. **Extensive commercial agriculture** is typified by large wheat farms and livestock ranching.

There are, of course, limits to the land use explanations attributable to von Thünen's model. While it is evident from Figure 8.18 that farmland values decline westward with increasing distance from the northeastern market of the United States, they show no corresponding increase with increasing proximity to the massive West Coast market region until the specialty agricultural areas of the coastal states themselves are reached. The western states are characterized by extensive agriculture, but as a consequence of environmental, not distance, considerations. Climatic conditions obviously affect the productivity and the potential agricultural use of an area, as do associated soils regions and topography. In Anglo America, of course, increasing distance westward from eastern markets is by chance associated with increasing aridity and the beginning of mountainous terrain. In

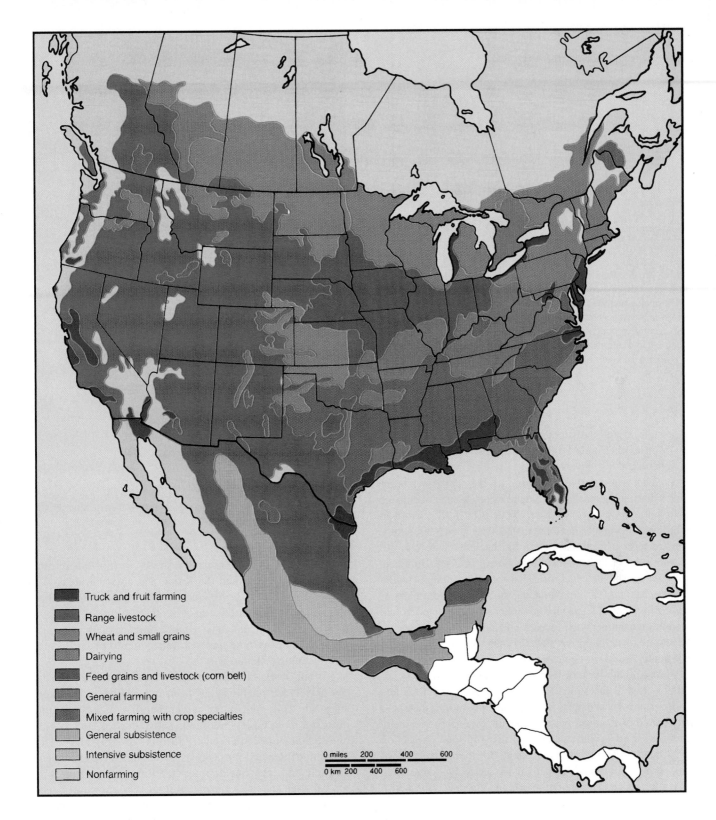

Legend

- Truck and fruit farming
- Range livestock
- Wheat and small grains
- Dairying
- Feed grains and livestock (corn belt)
- General farming
- Mixed farming with crop specialties
- General subsistence
- Intensive subsistence
- Nonfarming

0 miles 200 400 600

0 km 200 400 600

Figure 8.17 **Generalized agricultural regions of North America.**

Sources: U.S. Bureau of Agricultural Economics; Agriculture Canada; and Secretaría de Agricultura y Recursos Hidráulicos, Mexico.

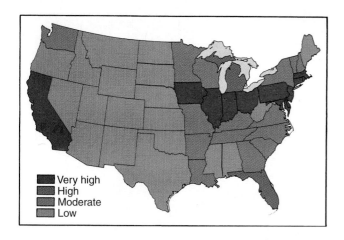

Figure 8.18 **Relative value per acre of farmland and buildings.** In a generalized way, per acre valuations support von Thünen's model. The major metropolitan markets of the Northeast, the Midwest, and California are in part reflected by high rural property valuations, and fruit and vegetable production along the Gulf Coast increases land values there. National and international markets for agricultural goods, soil productivity, climate, and terrain characteristics are also reflected in the map patterns.

Source: Statistical Abstract of the United States.

general, rough terrain and subhumid climates rather than simple distance from market underlie the widespread occurrence of extensive agriculture.

Large-scale wheat farming requires sizable capital inputs for planting and harvesting machinery, but the inputs per unit of land are low; wheat farms are very large. Nearly half the farms in Saskatchewan, for example, are more than 400 hectares (1000 acres). The average farm in Kansas is over 400 hectares, and in North Dakota, more than 525 hectares (1300 acres). In North America, the spring wheat (planted in spring, harvested in autumn) region includes the Dakotas, eastern Montana, and the southern parts of the Prairie Provinces of Canada. The winter wheat (planted in fall, harvested in midsummer) belt focuses on Kansas and includes adjacent sections of neighboring states (Figure 8.19). Argentina is the only South American country to have comparable large-scale wheat farming. In the Eastern Hemisphere, the system is fully developed only east of the Volga River in northern Kazakhstan and the southern part of Western Siberia, and in southeastern and western Australia. Because wheat is an important crop in many agricultural systems—today, wheat ranks first in total production among all the world's grains and accounts for more than 20% of the total calories consumed by humans collectively—large-scale wheat farms face competition from commercial and subsistence producers throughout the world (Figure 8.20).

Livestock ranching differs significantly from livestock-grain farming and, by its commercial orientation and distribution, from the nomadism it superficially resembles. A product of the 19th-century growth of urban markets for beef and wool in Western Europe and the northeastern United States, ranching has been primarily confined to areas of European settlement. It is found in the western United States and adjacent sections of Mexico and Canada (see Figure 8.17); the grasslands of Argentina, Brazil, Uruguay,

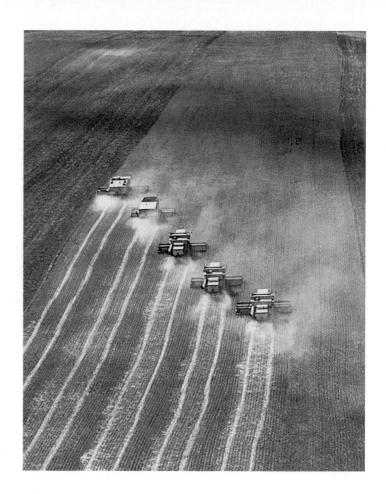

Figure 8.19 Contract harvesters follow the ripening wheat northward through the plains of the United States and Canada.

and Venezuela; the interior of Australia; the uplands of South Island, New Zealand; and the Karoo and adjacent areas of South Africa (Figure 8.21). All except New Zealand and the humid pampas of South America have semiarid climates. All, even the most remote from markets, were a product of improvements in transportation by land and sea, refrigeration of carriers, and of meat-canning technology.

In all of the ranching regions, livestock range (and the area exclusively in ranching) has been reduced as crop farming has encroached on their more humid margins, as pasture improvement has replaced less nutritious native grasses, and as grain fattening has supplemented traditional grazing. Recently, the midlatitude demand for beef has been blamed for expanded cattle ranching and extensive destruction of tropical rain forests in Central America and the Amazon basin.

In areas of livestock ranching, young cattle or sheep are allowed to graze over thousands of acres. In the United States, when the cattle have gained enough weight so that weight loss in shipping will not be a problem, they are sent to livestock-grain farms or to feedlots near slaughterhouses for accelerated fattening. Since ranching can be an economic activity only where alternative land uses are nonexistent and land quality is low, ranching regions of the world characteristically have low population densities, low capitalizations per land unit, and relatively low labor requirements.

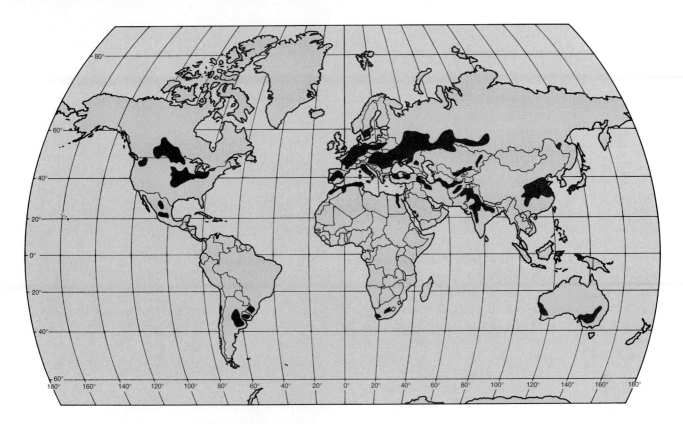

Figure 8.20 **Principal wheat-growing areas.** Only part of the world's wheat production comes from large-scale farming enterprises. In western and southern Europe, eastern and southern Asia, and North Africa, wheat growing is part of general or intensive subsistence farming. Recently, developing country successes with the Green Revolution and subsidized surpluses of the grain in Europe have altered traditional patterns of production and world trade in wheat.

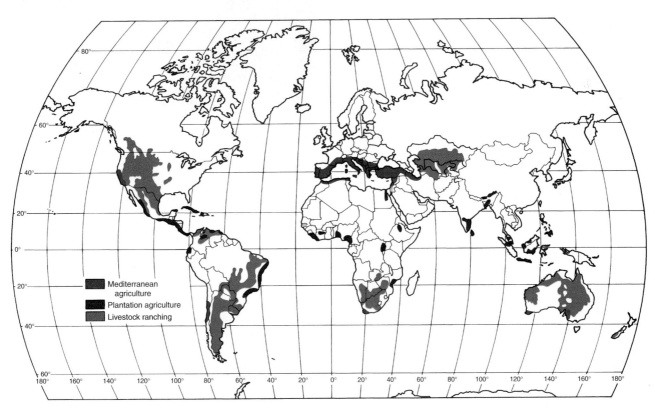

Mediterranean
agriculture
Plantation agriculture
Livestock ranching

Figure 8.21 **Livestock ranching and special crop agriculture.** Livestock ranching is primarily a midlatitude enterprise catering to the urban markets of industrialized countries. Mediterranean and plantation agriculture are similarly oriented to the markets provided by advanced economies of western Europe and North America. Areas of Mediterranean agriculture—all of roughly comparable climatic conditions—specialize in similar commodities, such as grapes, oranges, olives, peaches, and vegetables. The specialized crops of plantation agriculture are influenced by both physical geographic conditions and present or, particularly, former colonial control of the area.

Special Crops

Proximity to the market does not guarantee the intensive production of high-value crops should terrain or climatic circumstances hinder it. Nor does great distance from the market inevitably determine that extensive farming on low-priced land will be the sole agricultural option. Special circumstances, most often climatic, make some places far from markets intensively developed agricultural areas. Two special cases are agriculture in Mediterranean climates and in plantation areas (Figure 8.21).

Most of the arable land in the Mediterranean basin itself is planted to grains, and much of the agricultural area is used for grazing. *Mediterranean agriculture* as a specialized farming economy, however, is known for grapes, olives, oranges, figs, vegetables, and similar commodities. These crops need warm temperatures all year round and a great deal of sunshine in the summer. The Mediterranean agricultural lands indicated in Figure 8.21 are among the most productive in the world. Farmers can regulate their output in sunny areas such as these because storms and other inclement weather problems are infrequent. Also, the precipitation regime of Mediterranean climate areas—winter rain and summer drought—lends itself to the controlled use of water. Of course, much capital must be spent for the irrigation systems. This is another reason for the intensive use of the land for high-value crops that are, for the most part, destined for export to industrialized countries or areas outside the Mediterranean climatic zone and even, in the case of Southern Hemisphere locations, to markets north of the equator.

Climate is also considered the vital element in the production of what are commonly, but imprecisely, known as *plantation crops*. The implication of **plantation** is the introduction of a foreign element—investment, management, and marketing—into an indigenous culture and economy, often employing an introduced alien labor force. The plantation itself is an estate whose resident workers produce one or two specialized crops. Those crops, although native to the tropics, were frequently foreign to the areas of plantation establishment: African coffee and Asian sugar in the Western Hemisphere and American cacao, tobacco, and rubber in Southeast Asia and Africa are examples (Figure 8.22). Entrepreneurs in Western countries such as England, France, the Netherlands, and the United States became interested in the tropics partly because they afforded them the opportunity to satisfy a demand in temperate lands for agricultural commodities not producible in the market areas. Custom and convenience usually retain the term "plantation" even where native producers of local crops dominate, as they do in cola nut production in Guinea, spice growing in India or Sri Lanka, or sisal production in the Yucatán.

The major plantation crops and the areas where they are produced include tea (India and Sri Lanka), jute (India and Bangladesh), rubber (Malaysia and Indonesia), cacao (Ghana and Nigeria), cane sugar (Cuba and the Caribbean area, Brazil, Mexico, India, and the Philippines), coffee (Brazil and Colombia), and bananas (Central America). As Figure 8.21 suggests, for ease of access to shipping, most plantation crops are cultivated along or near coasts since production for export rather than for local consumption is the rule.

Figure 8.22 An Indonesian rubber plantation worker collects latex in a small cup attached to the tree and cuts a new tap just above the previous one. The scene typifies classical plantation agriculture in general. The plantation was established by foreign capital (Dutch) to produce a nonnative (American) commercial crop for a distant, midlatitude market using nonnative (Chinese) labor supervised by foreign (Dutch) managers. Present-day ownership, management, and labor may have changed, but the nature and market orientation of the enterprise remain.

Agriculture in Planned Economies

As their name implies, planned economies have a degree of centrally directed control of resources and of key sectors of the economy that permits the pursuit of governmentally determined objectives. When that control is extended to the agricultural sector—as it was during particularly the latter part of the 20th century in communist-controlled Soviet Union, Eastern Europe, mainland China, and elsewhere—state and collective farms and agricultural communes replace private farms or subsistence gardens, crop production is divorced from market control or family need, and prices are established by plan rather than by demand or production cost.

Such extremes of rural control have in recent years been relaxed or abandoned in the formerly strictly planned economies. Wherever past centralized control of agriculture was imposed and long endured, however, traditional rural landscapes were altered and the organization of rural society was disrupted. The programs set in motion by Stalin and his successors in the former Soviet Union, for example, fundamentally restructured the geography of agriculture of that country (Figure 8.23), transforming the Soviet countryside from millions of small farm holdings to a consolidated pattern of fewer than 50,000 centrally controlled operating units. Reestablishment of private agriculture was undertaken quickly in Russia following the USSR's collapse in late 1991, but not until 2002 was a legal framework established for the sale and purchase of farmland. However, absence of adequate farm registry and boundary descriptions and lack of clear ownership rights hindered outright sale of land, and even in that year the remnants of the old Soviet collective farm system still operated more than 90% of the country's farmlands.

A different progression from private and peasant agriculture, through collectivization, and back to what is virtually a private farming system took place in the planned economy of the People's Republic of China. After its assumption of power in 1949, the communist regime redistributed all farmlands to some 350 million peasants in inefficiently small (0.2 hectare, or 0.5 acre) subsistence holdings that were totally inadequate for the growing food needs of the country. By the end of 1957, 90% of peasant households were collectivized into about 700,000 communes, a number further reduced in the 1970s to 50,000 communes averaging some 13,000 members.

After the death of Chairman Mao Zedong in 1976, what became effectively a private farming system was reintroduced when 180 million new farms were allocated for unrestricted use to peasant families under rent-free leases. Most staple crops are still sold under enforced contracts at fixed prices to government purchasers, but increasingly vegetables and meat are sold on the free market (Figure 8.24), and per capita food production and availability have increased dramatically. Disturbingly, China with 20% of the world's population and only 7% of the world's arable land is losing farmland rapidly to industry, urban development, and environmental deterioration. Any extensive, permanent loss of farmland—economic growth is now swallowing more than 400,000 hectares (some 1 million acres) per year for new factories

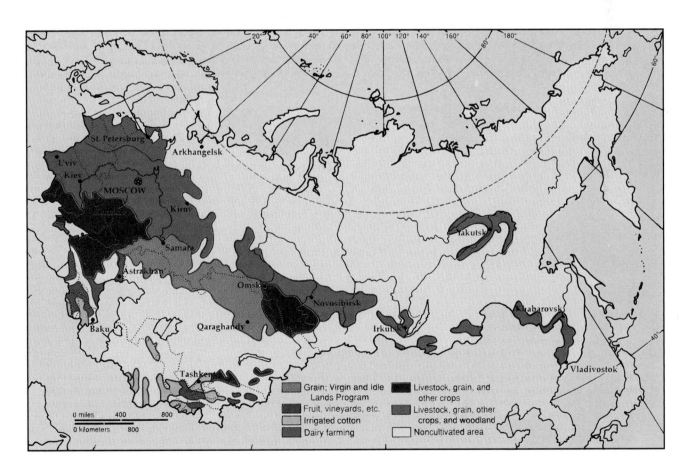

Figure 8.23 **Stalin's Virgin and Idle Lands** program extended grain production, primarily spring wheat, eastward from its traditional European Russian and Ukrainian focus onto marginal and arid land. Wheat constituted nearly 90% of total Soviet food-grain production and 50% of all grains grown at the time of the state's collapse. Sown land totaled some 10% of the USSR, most of it in a "fertile triangle" wedged between the frigid northern and dry southern limits of farming. The eastward expansion into dryer grassland areas released older western grain districts for vegetables, dairy products, and livestock production. But an ecological price has been paid. Soil erosion has forced Kazakhstan to abandon half its cropland since 1980.

Figure 8.24 Independent street merchants, shop owners, and peddlers in modern China are members of both a planned and market system. Free markets and private vendors multiplied after government price controls on most food items were removed in May, 1985. Increasingly, nonfood trade and manufacturing, too, are being freed of central government control and thriving in the private sector. As state-run companies shrank and laid off workers, privately owned businesses in 2002 accounted for over half of China's gross domestic product and employed 130 million workers—only one fifth of the total work force but a large majority of the industrial workers. The photo shows a row of outdoor poultry merchants in Wanxian, Sichuan Province. *Photo by Jon C. Malinowski.*

and real estate developments—that is not compensated by yield increases would again raise the prospect of shortages of domestically produced food.

Primary Activities: Resource Exploitation

In addition to agriculture, primary economic activities include fishing, forestry, and the mining and quarrying of minerals. These industries involve the direct exploitation of natural resources that are unequally available in the environment and differentially evaluated by different societies. Their development, therefore, depends on the occurrence of perceived resources, the technology to exploit their natural availability, and the cultural awareness of their value.

Fishing, forestry, and fur trapping are **gathering industries** based on harvesting the natural bounty of renewable resources that are in serious danger of depletion through overexploitation. Livelihoods based on these resources are areally widespread and involve both subsistence and market-oriented components. Mining and quarrying are **extractive industries,** removing nonrenewable metallic and nonmetallic minerals, including the mineral fuels, from the earth's crust. They are the initial raw material phase of modern industrial economies.

Resource Terminology

Resources or **natural resources** are the naturally occurring materials that a human population, at any given state of economic development and technological awareness, perceives to be necessary and useful to its economic and material well-being. Their occurrence and distribution in the environment are the result of physical processes over which people have little or no direct control. The fact that things exist, however, does not mean that they are resources. To be considered such, a given substance must be *understood* to be a resource—and this is a cultural, not purely a physical, circumstance. Native Americans may have viewed the resource base of Pennsylvania, West Virginia, or Kentucky as composed of forests for shelter and fuel and as the habitat of the game animals (another resource) on which they depended for food. European settlers viewed the forests as the unwanted covering of the resource that *they* perceived to be of value: soil for agriculture. Still later, industrialists appraised the underlying coal deposits, ignored or unrecognized as a resource by earlier occupants, as the item of value for exploitation (Figure 8.25).

Resources may be classified as *renewable* or *nonrenewable.* **Renewable resources** are materials that can be consumed and then replenished relatively quickly by natural or by human-assisted processes. Food crops are renewable resources, for example, as are forests, grasslands, animals and fish, and other living things. Even renewable resources can be exhausted if exploited to extinction or destruction. Soils can be totally eroded,

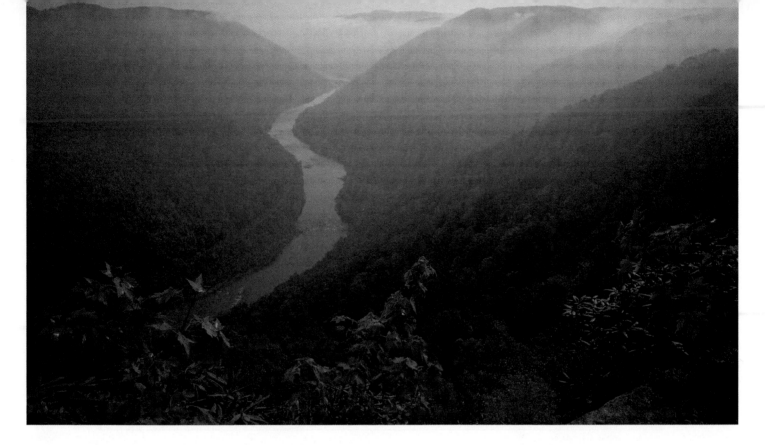

Figure 8.25 The original hardwood forest covering these West Virginia hills was removed by settlers who saw greater resource value in the underlying soils. The soils, in their turn, were stripped away for access to the still more valuable coal deposits below. Resources are as a culture perceives them, though their exploitation may consume them and destroy the potential of an area for alternate uses.

an animal species may be completely eliminated. That is, some resources are renewable only if carefully managed. The **maximum sustainable yield** of a resource is the maximum volume or rate of use that will not impair its ability to be renewed or to maintain the same future productivity. For fishing and forestry, for example, that level is marked by a catch or harvest equal to the net growth of the replacement stock. If that maximum exploitation level is exceeded, the renewable resource becomes a nonrenewable one—an outcome increasingly likely in the case of Atlantic cod and some other food fish species. **Nonrenewable resources** exist in finite amounts and either are not replaced by natural processes—at least not within any time frame of interest to the exploiting society—or are replaced at a rate slower than the rate of use.

Both types of resource are exploited by the nonagricultural primary industries. Fish as a food resource and forests as a source for building materials, cellulose, and fuel are heavily exploited renewable resources. Mining and quarrying extract from nature the nonrenewable minerals essential to industrialized economies.

Fishing

Although fish and shellfish account for less than 20% of all human protein consumption, an estimated 1 billion people depend on fish as their primary source of protein. Reliance on fish is greatest in developing countries of eastern and southeastern Asia, Africa, and parts of Latin America. Fish are also very important in the diets of some advanced states with well-developed fishing industries—Russia, Norway, Iceland, and Japan, for example. While about 75% of the world annual fish harvest is consumed by humans, up to 25% is processed into fish meal to be fed to livestock or used as fertilizer. Those two quite different markets have increased both the demand for and annual harvest of fish.

Three sources are tapped for the annual fish supply:

1. the *marine catch,* comprised of all fish harvested in coastal waters or on the high seas;

2. the *inland catch,* from ponds, lakes, and rivers;

3. *fish farm* production, from inland or coastal controlled containments.

At the start of the 21st century, about two-thirds of the annual fish supply came from *marine fisheries.* And most of that is from coastal wetlands, estuaries, and the relatively shallow coastal waters above the *continental shelf*—the gently sloping extension of submerged land bordering most coastlines and reaching seaward for varying distances up to 150 kilometers (about 100 miles) or more (Figure 8.26). Near shore, shallow embayments and marshes provide spawning grounds, and river waters supply nutrients to an environment highly productive of fish. Offshore, ocean currents and upwelling water move great amounts of nutritive materials from the ocean floor through the sunlit surface waters, nourishing *plankton*—minute plant and animal life forming the base of the marine food chain. Increasingly these are, as well,

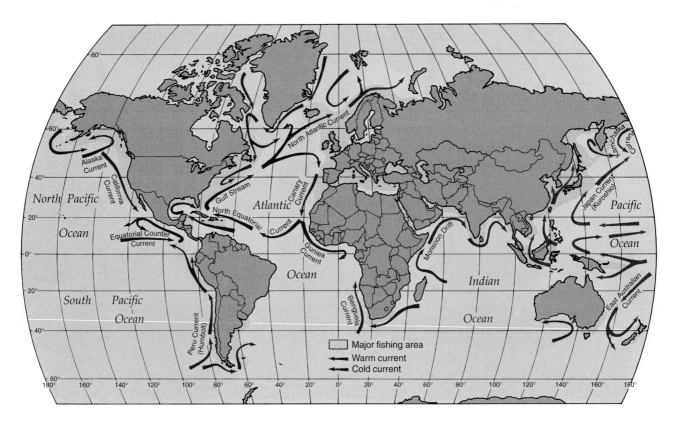

Figure 8.26 **The major commercial marine fisheries of the world.** The waters within 325 kilometers (200 miles) of the United States coastline account for almost one-fifth of the world's annual fish and shellfish harvests. Overfishing, urban development, and the contamination of bays, estuaries, and wetlands have contributed to the depletion of the fish stocks in those coastal waters.

areas seriously affected by pollution from runoff and ocean dumping, an environmental assault so devastating in some areas that fish and shellfish stocks have been destroyed with little hope of revival.

Commercial marine fishing is largely concentrated in northern waters, where warm and cold currents join and mix and where such familiar food species as herring, cod, mackerel, haddock, and flounder congregate or "school" on the broad continental shelves and *banks*—extensive elevated portions of the shelf where environmental conditions are most favorable for fish production. Two of the most heavily fished regions are the Northeast Pacific and Northwest Atlantic, which together yield 40% of the marine catch total.

Tropical fish species tend not to school and, because of their high oil content and unfamiliarity, are less acceptable in the commercial market. They are, however, of great importance for local consumption. Traditional or "artisan" fishermen, nearly all working in inshore waters within sight of land, are estimated to number between 8 and 10 million worldwide, harvesting some 24 million metric tons of fish and shellfish a year—a catch usually not included in world fishery totals. Since each coastal fisherman provides employment for two or three onshore workers, more than 25 million persons are involved in small-scale fisheries. That number is declining year by year as commercial trawlers have materially depleted the fish stocks on which artisanal fishermen depend.

Only a very small percentage of total marine catch comes from the open seas that make up more than 90% of the world's

oceans. Fishing in these comparatively barren waters is an expensive form of maritime hunting and gathering. An accepted equation in distant water fishing is that about 1 ton of diesel oil is burned for every ton of fish caught. In fact, all commercial marine fishing is costly. The Food and Agriculture Organization (FAO) reports that governments subsidize fishers with over $50 billion annually of low-cost loans and direct grants, in effect encouraging uneconomic overexploitation of a decreasing resource.

As Figure 8.27 indicates, world fish catch doubled between 1970 and 2001 as modern technology was increasingly applied to harvesting food fish. That technology included use of sonar, radar, helicopters, and satellite communications to locate schools of fish; more efficient nets and tackle; and factory trawlers to follow fishing fleets to prepare and freeze the catch (Figure 8.28). The rapid rate of increase led to inflated projections of continuing or growing fisheries' productivity and optimism that the resources of the oceans were inexhaustible. Quite the opposite has proved to be true.

In recent years, the productivity of marine fisheries has declined as *overfishing* (catches above reproduction rates) and pollution of coastal waters seriously endangered the supplies of traditional and desirable food species. Adjusted world total catch figures indicate that rather than the steady increase in capture rate shown on Figure 8.27, there has really been a decline of over 660,000 tons per year since the late 1980s. Coupled with growing world population, there has been a serious decline in the average per capita marine catch. The UN reports that all 17 of the world's

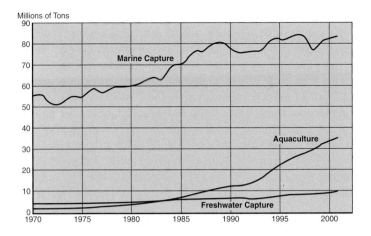

Millions of Tons

Figure 8.27 **Officially recorded annual fish harvests, 1970–2001** rose irregularly from 64 million tons in 1970 to 131 million tons in 2001. On the basis of individual country reports, the FAO recorded fluctuating but slowly growing harvest totals up to 2001; the 1993 and 1998 dips are associated with El Niño-produced ocean temperature changes. Chinese admission of regular over-reporting of, particularly, their marine capture suggests that world marine catch and composite harvest totals actually registered an irregular downward trend each year since 1988. A compensating adjustment to this graph would reduce 2001 total harvest and marine catch figures by some 9 million tons. The FAO estimates that 20 to 40 million tons per year of unintended marine capture of juvenile or undersized fish and nontarget species are discarded each year.

Source: Food and Agriculture Organization (FAO).

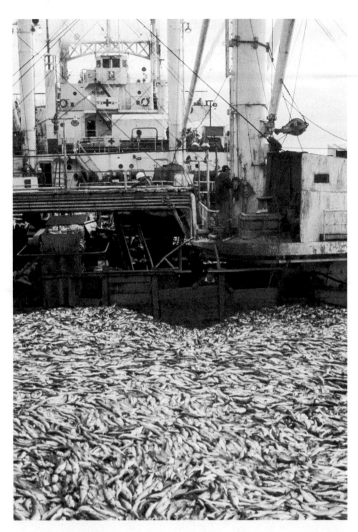

Figure 8.28 A Japanese factory ship in the Bering Sea. Factory ships prepare and freeze the catch of smaller vessels. This ship can process 25 to 30 tons of fish per hour. The increased efficiency of commercial fishing has led to serious depletion of a food source once thought to be inexhaustible.

major oceanic fishing areas are being fished at or beyond capacity; 13 are in decline. The plundering of Anglo American coastal waters has imperiled a number of the most desirable fish species, including haddock, flounder, and cod in New England and eastern Canadian waters; Spanish mackerel, grouper, and red snapper off the Gulf of Mexico; halibut and striped bass off California; and salmon and steelhead in the Pacific Northwest and western Canada. In 1993, Canada shut down its cod industry to allow stocks to recover, and U.S. authorities report that 67 species are overfished and 61 harvested to capacity.

Two fishing control measures hold hope that the problem of oceanic overfishing is being addressed. First, since 1976 coastal states have been claiming a 200 nautical mile (370-km) exclusive economic zone (EEZ) within which they can regulate or prohibit foreign fishing fleets. Since most commercially attractive fish live in coastal waters, these claims, part of the United Nations Convention on the Law of the Sea treaty reviewed in Chapter 12 (p. 465), brought many fisheries under control of the nearest country. Unfortunately, many governments have failed to act on scientific management recommendations and have permitted domestic fleets to expand to replace banned or restricted foreign fishing in territorial waters. Second, in 1995 more than 100 countries adopted a treaty—to become legally binding when ratified by 30 nations—to regulate fishing on the open oceans outside territorial waters. Applying to such species as cod, pollock, and tuna—that is, to migratory and high-seas species—the treaty requires fishermen to report the size of their catches to regional or-

ganizations that would set quotas and subject vessels to boarding to check for violations. These and other fishing control measures could provide the framework of future sustainability of important food fish stocks. Recent research reveals, however, that prolonged overfishing not only depletes stocks but may irrevocably alter the whole ecosystem in which desirable species formerly predominated. Cod catches that suddenly collapsed in the early 1990s off the east coast of Canada, for example, have not been revived by a fishing moratorium; adult cod, once the top predator in the region, now seem to have no place in the restructured ocean fish community. A similar cod depletion in Europe's North Sea also threatens or assures their reduction below the number of mature fish necessary to repopulate those overfished waters.

Overfishing is partly the result of the accepted view that the world's oceans are common property, a resource open to anyone's use with no one responsible for its maintenance, protection, or improvement. The result of this "open seas" principle is but one expression of the so-called **tragedy of the commons**[1]—the economic reality that when a resource is available to all, each user, in the absence of collective controls, thinks he or she is best served by exploiting the resource to the maximum even though this means its eventual depletion.

Usual estimates are that about 10% of the annual fish harvest now comes from the *inland catch* from lakes, rivers, and farm ponds. Although it is apparent that inland fisheries landings have been increasing since the early 1980s, data on the freshwater harvest are poor, especially in the developing countries where the activity is relatively most important in domestic food supply; the FAO reckons that the inland fisheries harvest is greatly underreported—by a factor of two or three. Most of the production increase has occurred in Asia (which produces almost two-thirds of the world's inland fish catch), Africa, and Latin America. In Anglo America, Europe, and the former Soviet Union, landings have declined. Production increases have been supported in many regions by stocking and by introducing nonnative species, but the estimated 8 or 9 million recorded tons of fish caught worldwide from lakes, rivers, and wetlands are also believed to be at or above maximum sustainable yield for these systems.

One approach to increasing the fish supply is through *fish farming* or **aquaculture,** the breeding of fish in freshwater ponds, lakes, and canals or in fenced-off coastal bays and estuaries or enclosures (Figure 8.29). Aquaculture production totaled 36 mil-

lion tons in 2000—27% of the total fish harvest in that year and an increase of nearly 50% over the preceding 5 years. Fish farming's contribution to human food supply is even greater than raw production figures suggest; whereas one-quarter of the conventional fish catch is used to make fish meal and fish oil, virtually all farmed fish are used as human food. Its rapid and continuing production increase makes fish farming the fastest growing sector of the world food economy with promise of overtaking cattle ranching as a human food source by 2010.

Fish farming has long been practiced in Asia, which produces four-fifths of the world's farmed fish, shrimp, and shellfish; China is the leading source, alone contributing over 50% of the world total. Although Asia dominates, aquaculture is practiced today on every continent. Fish farms in the United States, for example, produce significant amounts of catfish, crawfish, trout, mussels, and oysters; Japan specializes in high-value scallops, oysters, and yellowtail; and Norway concentrates on salmon. Aquaculture products are of two distinct types: high-valued species such as shrimp and salmon grown for market and frequently for export, and lower-valued species such as carp and tilapia primarily for local consumption. Small-scale fish aquaculture—until recently, the standard pattern—offers farmers a ready source of both subsistence food and a saleable crop, benefits that promote its expansion. Carp, tilapia, and many other subsistence species are herbivores and, properly managed, impose few adverse environmental impacts, although in China concern over loss of arable land has led to restrictions on any further conversions of farmland to fish farming ponds.

---

[1]The *commons* refers to undivided land available for the use of everyone; usually, it meant the open land of a village that all used as pasture. The *Boston Common* originally had this meaning.

Figure 8.29 Harvesting fish at an aquaculture farm in Thailand.

The development of large-scale intensive aquaculture has contributed significantly to the recent dramatic increase in farmed fish production, particularly of high-value commercial varieties. That growth raised concerns over the resource use diversions of the newer commercial production systems. For example, nearly half the land now used for shrimp farms in Thailand was formerly used for rice paddies; that is, a local subsistence grain crop has been replaced with a cash crop destined solely for export. More serious still is the apparent adverse environmental impact of the intensive, large-scale production operations devoted to raising high-value shrimp, salmon, and other premium commercial species. Shrimp farming, for example, has taken a heavy toll on coastal habitats, with mangrove swamps in Africa and Southeast Asia being extensively and rapidly cleared to make room for shrimp ponds. Mangrove destruction, in turn, has exposed coastal areas to erosion and flooding, altered natural drainage patterns, increased salt intrusions, and destroyed critical habitat for many aquatic species. Whatever the species or producing area, intensive aquaculture operations also lead to water pollution through the release of heavy concentrations of fish feces and organic debris flushed into nearby coastal or river waters. And aquaculture often puts increased pressure on ocean fish stocks rather than reducing it. Shrimp and salmon, among other high-value commercially farmed fish, are carnivorous and depend on high-protein feed formulated from fish meal from low-value marine species. Since some 10% to 15% of all fish meal goes to aquaculture feed and about 2 kilograms of meal are needed to produce a kilogram of farmed fish or shrimp, commercial aquaculture implies a net loss of fish protein.

Forestry

After the retreat of continental glaciers some 12,000 years ago and before the rise of agriculture, the world's forests and woodlands probably covered some 45% of the earth's land area exclusive of Antarctica. They were a sheltered and productive environment for earlier societies that subsisted on gathered fruits, nuts, berries, leaves, roots, and fibers collected from trees and woody plants. Few such cultures remain, though the gathering of forest products is still an important supplemental activity, particularly among subsistence agricultural societies.

Even after millennia of land clearance for agriculture and, more recently, commercial lumbering, cattle ranching, and fuelwood gathering, forests still cover roughly 30% of the world's land area. As an industrial raw material source, however, forests are more restricted in area. Although forests of some type reach discontinuously from the equator northward to beyond the Arctic Circle and southward to the tips of the southern continents, *commercial forests* are restricted to two very large global belts. One, nearly continuous, is found in upper-middle latitudes of the Northern Hemisphere; the second is located in the equatorial zones of South and Central America, Central Africa, and Southeast Asia (Figure 8.30). These forest belts differ in the types of trees they contain and in the type of market or use they serve.

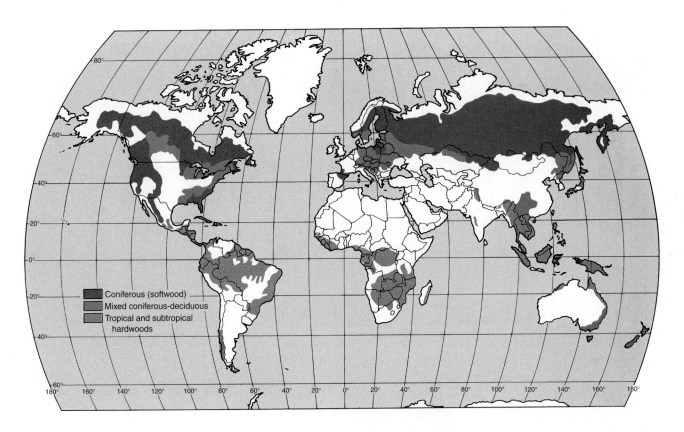

Figure 8.30 **Major commercial forest regions.** Much of the original forest, particularly in midlatitude regions, has been cut over. Many treed landscapes that remain do not contain commercial stands. Significant portions of the northern forest are not readily accessible and at current prices cannot be considered commercial. Deforestation of tropical hardwood stands involves more clearing for agriculture and firewood than for roundwood production.

The northern coniferous, or softwood, forest is the largest and most continuous stand, extending around the globe from Scandinavia across Siberia to North America, then eastward to the Atlantic and southward along the Pacific Coast. The pine, spruce, fir, and other conifers are used for construction lumber and to produce pulp for paper, rayon, and other cellulose products. On the south side of the northern midlatitude forest region are the deciduous hardwoods: oak, hickory, maple, birch, and the like. These and the trees of the mixed forest lying between the hardwood and softwood belts have been greatly reduced in areal extent by centuries of agricultural and urban settlement and development. In both Europe and North America, however, although they—like northern softwoods—have lately been seriously threatened by acid rain and atmospheric pollution, their area has been held constant through conservation, protection, and reforestation. They still are commercially important for hardwood applications: furniture, veneers, railroad ties, and the like.

The tropical lowland hardwood forests are exploited primarily for fuelwood and charcoal, although an increasing quantity of special quality woods are cut for export as lumber. In fact, developing—particularly tropical—countries account for 90% of the world's hardwood log exports (Figure 8.31); some two-thirds of these in the 1990s came from Malaysia alone, with the Malaysian state of Sarawak (on the island of Borneo and about the size of Mississippi) the source then of one-half of the world's hardwood logs.

These contrasting uses document *roundwood* (log) production as a primary economic activity. About 47% of the world's annual logging harvest is for industrial consumption, some 75% of it the output of industrialized countries from the temporal and boreal forest belt. Half of all production of industrial wood is from the United States, Canada, and Russia. Chiefly because of their distance from major industrial wood markets, the developing countries as a group accounted for less than one-quarter of industrial wood production in 2000. The logic of von Thünen's analysis of transportation costs and market accessibility helps explain the pattern.

The other half (53%) of roundwood production is for fuelwood and charcoal; 90% of world fuelwood production comes from the forests of Africa, Asia, Oceania, and Latin America, and demand for fuelwood grows by more than 1.2% per year. Since the populations of developing countries are heavily dependent on fuelwood and charcoal (see "The Energy Crisis in LDCs," p. 367), their growing numbers have resulted in serious depletion of tropical forest stands. Indeed, about 60% (some 1.5 billion people) of those who depend upon fuelwood as their principal energy source are cutting wood at a rate well above the maximum

Figure 8.31 Teak logs for export stacked near Mandalay, Myanmar.

sustainable yield. In tropical areas as a whole, deforestation rates exceed reforestation by 10 to 15 times. During the 1990s, tropical forest and woodlands were converted to agricultural lands at a rate of 10 to 12 million hectares (25 to 30 million acres) annually. Additional millions of hectares, particularly in South and Central America, have been cleared for pasture for beef cattle destined primarily for the North American market.

These uses and conversions have serious implications not only ecologically but also economically. Forest removal without replenishment for whatever reason converts the renewable resource of a gathering industry into a destructively exploited nonrenewable one. Regional economies, patterns of international trade, and prospects of industrial development are all adversely affected. Some world and regional ecological consequences of deforestation are discussed in Chapter 13.

Fur Trapping and Trade

The fur trade is ancient, with references and records dating to early Mediterranean and Far Eastern civilizations. In more recent centuries, demand for fine skins (including sable) lay behind Russia's 17th- and 18th-century expansion eastward to Siberia, Alaska, and down to northern California. Elsewhere in Anglo America, early history was closely tied to a commercial need to satisfy European demand, initially for beaver. Worldwide, the fur trade was and is one of the few sectors of the European and Western economy in which aboriginal people could and can participate and still retain their traditional lifestyles, values, and skills. Particularly for First People of Canada and Alaska, however, those lifestyles and economies have been devastated by the ardent anti-fur campaigns that began in the late 1960s and led to European Union bans on imported seal pelts and fur imports from countries permitting leg-hold traps. To replace their lost income, Canadian trapping communities have increasingly permitted environment-threatening logging, mining, oil drilling, pipeline construction, and forest-flooding hydroelectric projects they formerly rejected. Elsewhere, even today, fur trapping and hunting are still important or essential in the livelihoods of many people in Siberia and South America and support thousands of others worldwide from Mongolia to Australia.

Trapping and hunting are currently only minor parts of the fur trade scene, and a traditional gathering economy is now largely lost, replaced by fur farming as the primary supplier of market demand. Farmed furs—often produced as an income supplement to other farming activities—now account for some 85% of the industry's income generation. Most fur farming takes place in northern Europe (64%) and Anglo America (11%), with the remainder found widely spread, from Argentina to Ukraine, European Russia, and Siberia.

Mining and Quarrying

Societies at all stages of economic development can and do engage in agriculture, fishing, forestry, and trapping. The extractive industries—mining and drilling for nonrenewable mineral wealth—emerged only when cultural advancement and economic necessity made possible a broader understanding of the earth's

resources. Now those industries provide the raw material and energy base for the way of life experienced by people in the advanced economies and are the basis for a major part of the international trade connecting the developed and developing countries of the world.

The extractive industries depend on the exploitation of minerals unevenly distributed in amounts and concentrations determined by past geologic events, not by contemporary market demand. In physically workable and economically usable deposits, minerals constitute only a tiny fraction of the earth's crust—far less than 1%. That industrialization has proceeded so rapidly and so cheaply is the direct result of an earlier ready availability of rich and accessible deposits of the requisite materials. Economies grew fat by skimming the cream. It has been suggested that should some catastrophe occur to return human cultural levels to a preagricultural state, it would be extremely unlikely that humankind ever again could move along the road of industrialization with the resources available for its use.

Our successes in exploiting mineral resources have been achieved, that is, at the expense of depleting the most easily extractable world reserves and with the penalty of increasing monetary costs as the highest-grade deposits are removed (Figure 8.32). Costs increase as more advanced energy-consuming technologies must be applied to extract the desired materials from ever greater depths in the earth's crust or from new deposits of smaller mineral content. That observation states a physical and

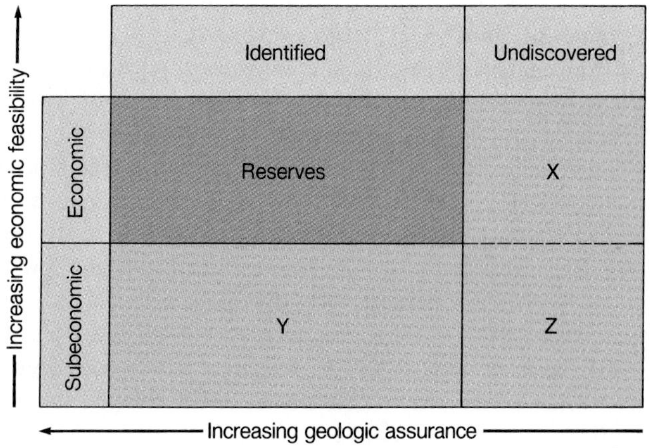

Figure 8.22 **The variable definition of reserves.** Assume the large rectangle includes the total world stock of a particular resource. Some deposits of that resource have been discovered and are shown in the left column as "identified." Deposits not yet known are "undiscovered reserves." Deposits that are economically recoverable with current technology are at the top of the diagram. Those below, labeled "subeconomic" reserves, are not attractive for any of several reasons of mineral content, accessibility, cost of extraction, and so on. Only the pink area can be properly referred to as **usable reserves.** These are deposits that have been identified and can be recovered at current prices and with current technology. X denotes reserves that would be attractive economically but are not yet discovered. Identified but not economically attractive reserves are labeled Y. Z represents undiscovered deposits that would not now be attractive even if they were known.

Source: U.S. Geological Survey.

economic reality relevant particularly to the exploitation of both the *metallic minerals* and the *mineral fuels.* It is less applicable to the third main category of extractive industry, the *nonmetallic minerals.* In few cases, however, does the observation imply that natural scarcity is a limit on resource availability. In fact, as a consequence of modern exploration technologies and extraction efficiencies, known reserves of all fossil fuels and of most commercially important metals are now larger than they were in the middle of the 20th century. That increasing abundance of at least nonfuel resources is reflected in the steady decrease in raw material prices since the 1950s that has so adversely affected some export-oriented developing world economies.

Metallic Minerals

Because usable mineral deposits are the result of geologic accident, it follows that the larger the country, the more probable it is that such accidents will have occurred within the national territory. And in fact, Russia, Canada, China, the United States, Brazil, and Australia possess abundant and diverse mineral resources. It is also true, however, that many smaller developing countries are major sources of one or more critical raw materials and become, therefore, important participants in the growing international trade in minerals.

The production of most metallic minerals, such as copper, lead, and iron ore, is affected by a balance of three forces: the quantity available, the richness of the ore, and the distance to markets. A fourth factor, land acquisition and royalty costs, may equal or exceed other considerations in mine development decisions (see "Public Land, Private Profit"). Even if these conditions are favorable, mines may not be developed or even remain operating if supplies from competing sources are more cheaply available in the market. In the 1980s, more than 25 million tons of iron ore-producing capacity was permanently shut down in the United States and Canada. Similar declines occurred in North American copper, nickel, zinc, lead, and molybdenum mining as market prices fell below domestic production costs. Beginning in the early 1990s, as a result of both resource depletion and low cost imports, the United States became a net importer of nonfuel minerals for the first time. Of course, increases in mineral prices may be reflected in opening or reopening mines that, at lower returns, were deemed unprofitable. However, the developed industrial countries of market economies, whatever their former or even present mineral endowment, find themselves at a competitive disadvantage against developing country producers with lower-cost labor and state-owned mines with abundant, rich reserves.

When the ore is rich in metallic content (in the case of iron and aluminum ores), it is profitable to ship it directly to the market for refining. But, of course, the highest-grade ores tend to be mined first. Consequently, the demand for low-grade ores has been increasing in recent years as richer deposits have been depleted (Figure 8.33). Low-grade ores are often upgraded by various types of separation treatments at the mine site to avoid the cost of transporting waste materials not wanted at the market. Concentration of copper is nearly always mine oriented (Figure 8.34); refining takes place near areas of consumption. The large amount of waste in copper (98% to 99% or more of the ore) and in most other industrially significant ores should not be considered the mark of

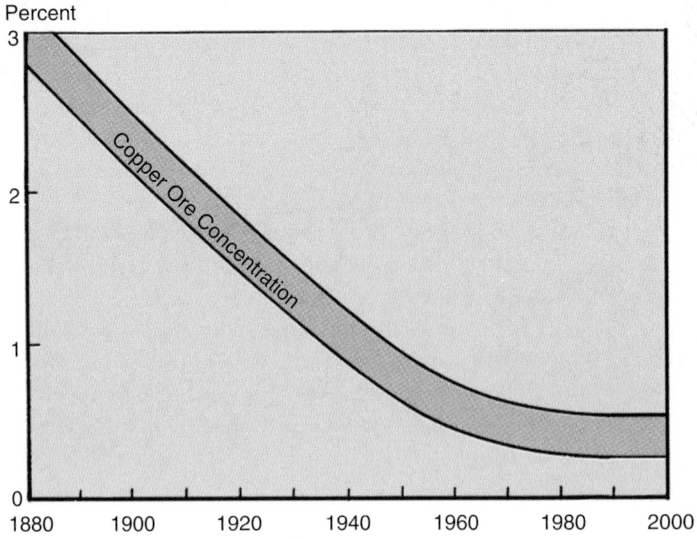

Figure 8.33 **Needed metal content of copper ore for profitable mining.** In 1830, 3% copper ore rock was needed to justify its mining; today, rock with 0.5% ore content is mined. As the supply of a metal decreases and its price increases, the concentration needed for economic recovery goes down. It also goes down as improved and more cost-effective technologies of rock mining and ore extraction come into play.

Source: Data from the U.S. Bureau of Mines.

an unattractive deposit. Indeed, the opposite may be true. Because of the cost of extraction or the smallness of the reserves, many higher-content ores are left unexploited in favor of the utilization of large deposits of even very low-grade ore. The attraction of the latter is a size of reserve sufficient to justify the long-term commitment of development capital and, simultaneously, to assure a long-term source of supply.

At one time, high-grade magnetite iron ore was mined and shipped from the Mesabi area of Minnesota. Those deposits are now exhausted. Yet immense amounts of capital have been invested in the mining and processing into high-grade iron ore pellets of the virtually unlimited supplies of low-grade iron-bearing rock (taconite) still remaining. Such investments do not assure the profitable exploitation of the resource. The metals market is highly volatile. Rapidly and widely fluctuating prices can quickly change profitable mining and refining ventures to losing undertakings. Marginal gold and silver deposits are opened or closed in reaction to trends in precious metals prices. Taconite *beneficiation* (waste material removal) in the Lake Superior region has virtually ceased in response to the decline of the U.S. steel industry. In market economies, cost and market controls dominate economic decisions. In planned economies, cost may be a less important consideration than other concerns such as goals of national development and resources independence.

Nonmetallic Minerals

From the standpoint of volume and weight of material removed, the extraction of nonmetallic earth materials is the most important branch of the extractive industries. The minerals mined are

Geography and Public Policy

Public Land, Private Profit

When U.S. President Ulysses S. Grant signed the Mining Act of 1872, the presidential and congressional goal was to encourage Western settlement and development by allowing any "hard-rock" miners (including prospectors for silver, gold, copper, and other metals) to mine federally owned land without royalty payment. It further permitted mining companies to gain clear title to publicly owned land and all subsurface minerals for no more than $12 a hectare ($5 an acre). Under those liberal provisions, mining firms have bought 1.3 million hectares (3.2 million acres) of federal land since 1872 and each year remove some $1.2 billion worth of minerals from government property. In contrast to the royalty-free extraction privileges granted to metal miners, oil, gas, and coal companies pay royalties of as much as 12.5% of their gross revenues for exploiting federal lands.

Whatever the merits of the 1872 law in encouraging economic development of lands otherwise unattractive to homesteaders, modern-day mining companies throughout the Western states have secured enormous actual and potential profits from the law's generous provisions. In Montana, a company claim to 810 hectares (2000 acres) of land would cost it less than $10,000 for an estimated $4 billion worth of platinum and palladium; in California, a gold mining company in 1994 sought title to 93 hectares (230 acres) of federal land containing a potential of $320 million of gold for less than $1200. Foreign as well as domestic firms may be beneficiaries of the 1872 law. In 1994, a South African firm arranged to buy 411 hectares (1016 acres) of Nevada land with a prospective $1.1 billion in gold from the government for $5100. A Canadian firm in 1994 received title to 800 hectares (nearly 2000 acres) near Elko, Nevada, that cover a likely $10 billion worth of gold—a transfer that Interior Secretary Bruce Babbitt dubbed "the biggest gold heist since the days of Butch Cassidy." And in 1995, Mr. Babbitt conveyed about $1 billion worth of travertine (a mineral used in whitening paper) under 45 hectares (110 acres) of Idaho to a Danish-owned company for $275.

The "gold heist" characterization summarized a growing administration and congressional feeling that what was good in 1872 and today for metal mining companies was not necessarily beneficial to the American public that owns the land. In part, that feeling results from the fact that mining companies commit environmental sins that require public funding to repair or public tolerance to accept. The mining firms may destroy whole mountains to gain access to low-grade ores and leave toxic mine tailings, surface water contamination, and open-pit scarring of the landscape as they move on or disappear. Since under the law mining companies are not required to clean up after themselves, projected public cleanup costs of more than 500,000 abandoned mine sites, 10,000 miles of damaged or dead streams, and 50 billion tons of contaminated waste are estimated at a minimum of $35 billion.

A congressional proposal introduced in 1993 would require mining companies to pay royalties of 8% on gross revenues for all hard-rock ores extracted and prohibit them from outright purchase of federal land. The royalty provision alone would have yielded nearly $100 million annually at 1994 levels of company income. Mining firms claim that imposition of royalties might well destroy America's mining industry. They stress both the high levels of investment they must make to extract and process frequently low-grade ores and the large number of high-wage jobs they provide as their sufficient contribution to the nation. The Canadian company involved in the Elko site, for example, reports that since it acquired the claims in 1987 from their previous owner, it has expended over $1 billion, and in addition has made donations for town sewer lines and schools and created 1700 jobs. The American Mining Congress estimates the proposed 8% royalty charge would cost 47,000 jobs out of 140,000, and even the U.S. Bureau of Mines assumes a loss of 1100 jobs.

Public resistance to Western mining activities is taking its toll. State and federal regulatory procedures, many dragging on for a decade or more, have discouraged opening new mines; newly enacted environmental regulations restricting current mining operations (for example, banning the use of cyanide in gold and silver refining) reduce their economic viability. In consequence, both investment and employment in U.S. mining is in steady decline, eroding the economic base of many Western communities.

Questions to Consider

1. Do you believe the 1872 Mining Law should be repealed or amended? If not, what are your reasons for arguing for retention? If so, would you advocate the imposition of royalties on mining company revenues? At what levels, if any, should royalties be assessed? Should hard-rock and energy companies be treated equally for access to public land resources? Why or why not?

2. Would you propose to prohibit outright land sales to mining companies? If not, should sales prices be determined by surface value of the land or by the estimated (but unrealized) value of mineral deposits it contains?

3. Do you think that cleanup and other charges now borne by the public are acceptable in view of the capital investments and job creation of hard-rock companies? Do you accept the industry's claim that imposition of royalties would destroy American metal mining? Why or why not?

4. Do you favor continued state and federal restrictions on mining operations, even at the cost of jobs and community economies? Why or why not?

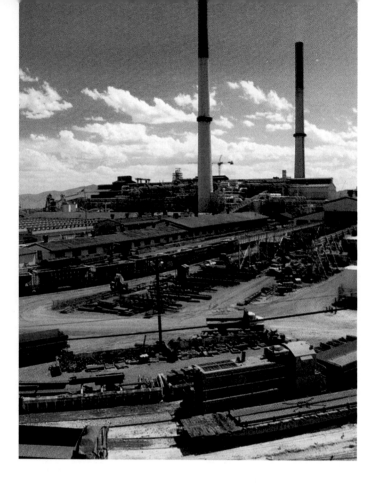

Figure 8.24 Copper ore concentrating and smelting facilities at the Phelps-Dodge mine in Morenci, Arizona. Concentrating mills crush the ore, separating copper-bearing material from the rocky mass containing it. The great volume of waste material removed assures that most concentrating operations are found near the ore bodies. Smelters separate concentrated copper from other, unwanted, minerals such as oxygen and sulfur. Because smelting is also a "weight-reducing" (and, therefore, transportation-cost reducing) activity, it is frequently—though not invariably—located close to the mine as well.

usually classified by their end use. Of widest distribution, greatest use, and least long-distance movement are those used for *construction:* sand and gravel, building stone, and the gypsum and limestone that are the ingredients of cement. Transportation costs play a great role in determining where low-value minerals will be mined. Minerals such as gravel, limestone for cement, and aggregate are found in such abundance that they have value only when they are near the site where they are to be used. For example, gravel for road building has value if it is at or near the road-building project, not otherwise. Transporting gravel hundreds of miles is an unprofitable activity (Figure 8.35).

The mined *fertilizer* minerals include potash and phosphate, which do move in international trade because of their unequal distribution and market value. *Precious* and *semiprecious* stones are also important in the trade of some countries, including South Africa and Sri Lanka.

Mineral Fuels

The advanced economies have reached that status through their control and use of energy. By the application of energy, the conversion of materials into commodities and the performance of services far beyond the capabilities of any single individual are made possible. Energy consumption goes hand in hand with industrial production and with increases in personal wealth. In general, the greater the level of energy consumption, the higher the gross national income per capita. Further, the application of energy can overcome deficiencies in the material world that humans exploit. High-quality iron ore may be depleted, but by massive applications of energy, the iron contained in rocks of very low iron content, such as taconite, can be extracted and concentrated for industrial uses.

Because of the association of energy and economic development, a basic disparity between societies is made clear. Countries that can afford high levels of energy consumption through production or purchase continue to expand their economies and to increase their levels of living. Those without access to energy

Figure 8.35 The Vancouver, British Columbia, municipal gravel quarry and storage yard. Proximity to market gives utility to low-value minerals unable to bear high transportation charges.

or those unable to afford it see the gap between their economic prospects and those of the developed states growing ever greater.

Except for the brief and localized importance of waterpower at the outset of the Industrial Revolution, modern economic advancement has been heavily dependent on the *mineral fuels:* coal, petroleum, and natural gas. Also known as *fossil fuels,* these nonrenewable energy sources represent the capture of the sun's energy by plants and animals in earlier geologic time and its storage in the form of hydrocarbon compounds in sedimentary rocks within the earth's crust.

Coal was the earliest in importance and is still the most plentiful of the mineral fuels. As the first of the major industrial energy sources, coal deposits—as we shall see in Chapter 9—were formerly very important in attracting manufacturing and urbanization in industrializing countries. Although coal is a nonrenewable resource, world supplies are so great—on the order of 10,000 billion (10^{13}) tons—that its resource life expectancy is measured in centuries, not in the much shorter spans usually cited for oil and natural gas. Worldwide, the most extensive deposits are concentrated in the industrialized middle latitudes of the Northern Hemisphere (Table 8.1). Two countries, the United States and China, accounted in roughly equal shares for half of total world coal output in 2002; Russia and Germany, both with large domestic reserves, together produced less than 8%.

Coal is not a resource of constant quality, varying in *rank* (a measure—from lignite to anthracite—of increasing carbon content and fuel quality) and *grade* (a measure of its waste material content, particularly ash and sulfur). The value of a coal deposit depends on these measures and on its accessibility, which is a function of the thickness, depth, and continuity of the coal seam. Much coal can be mined relatively cheaply by open-pit (surface) techniques, in which huge shovels strip off surface material and remove the exposed coal (see Figure 13.21). Much coal, however, is available only by expensive and more dangerous shaft mining, as in Appalachia and most of Europe. In spite of their generally lower heating value, western U.S. coals are attractive because of their low sulfur content. They do, however, require expensive transportation to market or high-cost transmission lines if they are used to generate electricity for distant consumers (Figure 8.36)

Petroleum, first extracted commercially in the 1860s in both the United States and Azerbaijan, became a major power source and a primary component of the extractive industries only early in the 20th century. The rapidity of its adoption as both a favored energy resource and a raw material important in a number of industries from plastics to fertilizers, along with the limited size and the speed of depletion of known and probable reserves, suggest that petroleum cannot continually retain its present position of importance in the energy budget of countries. No one has more than a vague notion of how much oil (or natural gas) remains in the world or how long it will last. Cautious year 2002 estimates were that slightly more than 1000 billion barrels could be classified as proved reserves; another 900 billion may exist in undiscovered

Table 8.1

Proved Petroleum, Natural Gas, and Coal Reserves, January 1, 2003

	Share of Total Petroleum (%)	Share of Total Natural Gas (%)	Share of Total Coal (%)
North America[a]	4.8	4.6	26.2
Europe	1.8	3.6	12.2
Former Soviet Union	7.5	35.6	23.9
Of which Russian Fed.	5.7	30.5	15.9
Others	1.8	5.1	8.0
Central and South America	9.4	4.5	2.2
Africa	7.4	7.6	5.6
Middle East[b]	65.4	36.0	0.2
Australia/New Zealand	0.3	1.6	8.4
Japan	–.–	–.–	0.1
China	1.7	1.0	11.6
Other Asia Pacific	1.7	5.5	9.6
Total World	100.0	100.0	100.0
Of which OPEC[c]	78.2	46.4	NA

[a]Includes Canada, Mexico, U.S.A.
[b]Middle East includes Arabian Peninsula, Iran, Iraq, Israel, Jordan, Lebanon, Syria.
[c]OPEC: Organization of Petroleum Exporting Countries. Member nations are, by world region:
  South America: Venezuela
  Middle East: Iran, Iraq, Kuwait, Qatar, Saudi Arabia, United Arab Emirates (Abu Dhabi, Dubai, Ras-al-Khaimah, and Sharjah)
  North Africa: Algeria, Libya
  West Africa: Nigeria
  Asia Pacific: Indonesia

Source: The BP Amoco, *BP Amoco Statistical Review of World Energy, 2002.* Reprinted by permission.

Figure 8.36 Long-distance transportation to eastern markets adds significantly to the cost of the low-sulfur western coal useful in meeting federal environmental protection standards. To minimize these costs, unit trains carrying only coal engage in a continuous shuttle movement between western strip mines and eastern utility companies.

reservoirs. Assuming total extraction from known reserves and a constant end-of-century rate of extraction, proved reserves at these estimates would last only about 40 years.

More optimistic assessments assure us that petroleum reserves that could be extracted at acceptably competitive prices would last for about 150 years at present consumption rates. Respected experts, however, are convinced that price volatility, carbon dioxide emissions concerns, and the steady drop in price of solar and other alternative energy sources, will reduce oil demand long before supply becomes an issue. On a world basis, petroleum accounted for 47% of commercial energy consumption in 1973, but had dropped to 37% by 2002 as a reflection of its increasing cost and of conservation measures to offset those increases.

Petroleum is among the most unevenly distributed of the major resources. Seventy-five percent of proved reserves are concentrated in just 7 countries, and 82% in only 10. Iran and the Arab states of the Middle East alone control nearly two-thirds of the world total (Table 8.1). The distribution of petroleum supplies differs markedly from that of the coal deposits on which the urban-industrial markets developed, but the substitution of petroleum for coal did little to alter earlier patterns of manufacturing

and population concentration. Because oil is easier and cheaper to transport than coal, it was moved in enormous volumes to the existing centers of consumption via intricate and extensive national and international systems of transportation, a textbook example of spatial interaction, complementarity, and transferability (see Chapter 3 and Figure 3.2).

The uninterrupted international flow of oil is vital to the economic health of the United States and such other advanced industrial economies as those of Europe and Japan (see "A Costly Habit"). That dependence on imported oil gives the oil-exporting states tremendous power, as reflected in the periodic oil "shocks" that reflect the supply and selling-price control exerted by the Organization of Petroleum Exporting Countries (OPEC). Their expressions can be worldwide recessions and large net trade deficits for some importers and a reorientation of international monetary wealth.

Natural gas has been called the nearly perfect energy resource. It is a highly efficient, versatile fuel that requires little processing and is environmentally benign. Geologists estimate that world recoverable gas reserves are sufficient to last to near the last third of the century at 2000 levels of consumption. *Ultimately recoverable reserves,* those that may be found and recovered at very much higher prices, might last another 200 years.

The United States is a crude oil junkie, dependent on daily fixes of petroleum. On average, Americans consumed some 19 million barrels per day at the start of the 21st century. That was the equivalent of 2.9 gallons per person per day, or over 1000 gallons per person per year. What are some of the implications of this dependence? Consider the data in the table.

Notice the imbalance between production and consumption for the United States. In contrast to its hemispheric neighbors, Canada and Mexico, the United States consumes far more oil than it produces. At current rates of consumption, assuming no imports, and barring new discoveries or increased recovery rates from existing fields, the proved reserves would meet domestic demand for only 5 years. Americans continue to drive their cars, and their manufacturing plants continue to turn out a wide range of petroleum-based products, only because the country imports between 10.5 and 11 million barrels a day. That is, the United States relies on foreign sources to meet nearly 60% of its crude-oil needs.

American dependence on imports should ease, and U.S. proved reserves increase significantly as deepwater fields in the Gulf of Mexico are tapped. They are estimated to hold about 15 billion barrels of oil, considerably more oil than the giant Prudhoe Bay fields in Alaska, which are currently among the largest sources of domestic oil. The Gulf fields lie beyond the continental shelf and until recently were considered too deep—deeper than 456 meters (1500 feet)—to reach economically. New deep and lateral drilling techniques coupled with vastly enhanced computerized seismic mapping to reveal geologic structures at great depth and underneath thick salt deposits have improved discovery and recovery prospects for Gulf and other oil fields.

Several petroleum companies have built oil production platforms anchored one-half mile (.8 km) or more below the surface of the water. One company, Shell Oil, has produced from two of these platforms in its Auger and Mars fields since the early and mid-1990s. Two additional fields—Ram-Powell and Ursa—began producing in the late 1990s. Unlike oil rigs in shallower water, which are rigid towers built on fixed platforms attached to the sea bottom, the new platforms float on the surface, tethered by steel tendons to enormous anchors. Each platform cost between $1 billion and $2 billion, reflecting the high price of dependence on oil.

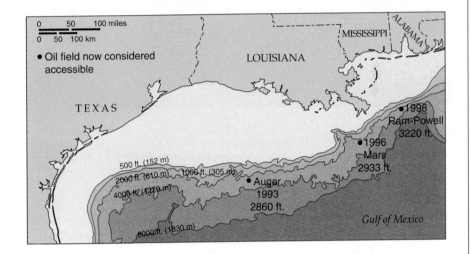

Country	2000 Proved Reserves (billion barrels)	2000 Production (billion bbls)	2000 Consumption (billion bbls)
United States	28.6	2.6	6.6
Canada	6.8	.9	.6
Mexico	28.4	1.2	.6

As we saw for coal and petroleum, reserves of natural gas are very unevenly distributed (Table 8.1). In the case of gas, however, inequalities of supply are not so readily accommodated by massive international movements. Like oil, natural gas flows easily and cheaply by pipeline, but unlike petroleum it does not move freely in international trade by sea. Transoceanic shipment involves costly equipment for liquefaction and for special vessels to contain the liquid under appropriate temperature conditions. Where the fuel can be moved, even internationally, by pipeline, its consumption has increased dramatically. For the world as a whole, gas consumption rose more than 60% between 1974 and 2002, to almost 25% of global energy consumption.

Trade in Primary Products

International trade expanded by some 6% a year between 1965 and the end of the century; it grew by nearly 13% in 2000 alone to account for about 17% of all economic activity. Primary commodities—agricultural goods, minerals, and fuels—made up one-quarter of the total dollar value of that international trade. In the past, the world distribution of supply and demand for those items in general resulted in an understandable pattern of commodity flow: from raw material producers located within less developed countries to processors, manufacturers, and consumers of the more developed ones (Figure 8.37). The reverse flow carried

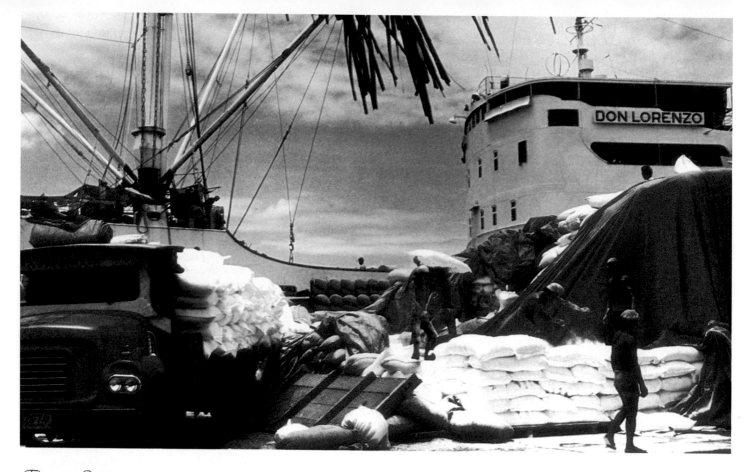

Figure 8.37 Sugar being loaded for export at the port of Cebu in the Philippines. Much of the developing world depends on exports of mineral and agricultural products to the developed economies for the major portion of its income. Fluctuations in market demand and price of some of those commodities can have serious and unexpected consequences.

manufactured goods processed in the industrialized states back to the developing countries. That two-way trade presumably benefited the developed states by providing access to a continuing supply of industrial raw materials and foodstuffs not available domestically and gave less developed countries needed capital to invest in their own development or to expend on the importation of manufactured goods, food supplies, or commodities—such as petroleum—they did not themselves produce.

By the end of the century, however, world trade flows and export patterns of the emerging economies were radically changing. Raw materials greatly decreased and manufactured goods correspondingly increased in the export flows from developing states as a group. In 1990, nonmanufactured (unprocessed) goods accounted for 60% of their exports; by 2000, that share had been cut in half and, in a reversal, manufactured goods made up 60% of the export flows from the developing to the industrialized world. Even with that overall decline in raw material exports, however, trade in unprocessed goods remains dominant in the economic well-being of many of the world's poorer economies. Increasingly, the terms of the traditional trade flows they depend on have been criticized as unequal and damaging to commodity-exporting countries.

Commodity prices are volatile; they may rise sharply in periods of product shortage or international economic growth. Recent commodity price movements, however, have been downward, to the great detriment of material-exporting economies. Prices for agricultural raw materials, for example, dropped by 30% between 1975 and 2000, and those for metals and minerals decreased by almost 40%. Such price declines cut deeply into the export earnings of many emerging economies. As a group, raw-material-exporting states express resentment at what they perceive as commodity price manipulation by rich countries and corporations to ensure low-cost supplies.

Although collusive price-fixing has not been demonstrated, other price-depressing agencies are evident. Technology, for example, has provided industries in advanced countries with a vast array of materials that now can and do substitute for the ores and metals produced by developing states. Glass fibers replace copper wire in telecommunication applications; synthetic rubber replaces natural rubber; glass and carbon fibers provide the raw material for rods, tubes, sheet panels, and other products superior in performance and strength to the metals they replace; and a vast and enlarging array of plastics become the accepted raw materials for commodities and uses for which natural rivals are not even considered. That is, even as the world industrial economy expands, demands and prices for traditional raw materials remain depressed.

While prices paid for developing country commodities tend to be low, prices charged for the manufactured goods offered in exchange by the developed countries tend to be high. To capture processing and manufacturing profits for themselves, some developing states have placed restrictions on the export of unprocessed commodities. Malaysia, the Philippines, and Cameroon, for example, have limited the export of logs in favor of increased domestic processing of sawlogs and exports of lumber. Some developing countries have also encouraged domestic manufacturing to reduce imports and to diversify their exports. Frequently, however, such exports meet with tariffs and quotas protecting the home markets of the industrialized states.

Many developing regions heavily dependent on commodity sales saw their share of global trade fall materially between 1970 and the early 21st century: sub-Saharan Africa from 3.8% to 1%, Latin America from 5.6% to 3.3%, and the least developed states as a group from 0.8% to 0.4%. Those relative declines are understandable in the light of greatly expanding international trade in manufactured goods from China, Korea, Mexico, and other rapidly industrializing states and from the expansion of trade in both manufactured goods and primary products between the industrialized countries themselves within newly established regional free-trade zones. For example, the developed countries acquire some three-quarters by value of their agricultural imports and 70% of their industrial raw materials from one another, diminishing the prospects for developing country exports.

In reaction to the whole range of perceived trade inequities, developing states promoted, in 1964, the establishment of the United Nations Conference on Trade and Development (UNCTAD). Its central constituency—the "Group of 77," expanded by 2003 to 133 developing states—continues to press for a new world economic order based in part on an increase in the prices and values of exports from developing countries, a system of import preferences for their manufactured goods, and a restructuring of international cooperation to stress trade promotion and recognition of the special needs of poor countries. The Word Trade Organization, established in 1995 (and discussed in detail in Chapter 9) was designed in part to reduce trade barriers and inequities. It has, however, been judged by its detractors as ineffective on issues of importance to developing countries. Chief among the complaints is the continuing failure of the industrial countries significantly (or at all) to reduce protections for their own agricultural and mineral industries. The World Bank has calculated, for example, that agricultural trade barriers and subsidies in rich countries reduce incomes in developing countries by at least $20 billion a year. Solutions are yet to be achieved to debated questions of the fair treatment of international trade in primary products.

Summary

How people earn their living and how the diversified resources of the earth are employed by different peoples and cultures are of fundamental concern in human geography. The economic activities that support us and our society are constant preoccupations that color our perception of the world and its opportunities. At the same time, the totality of our culture—technology, religion, customary behavior—and the circumstances of our natural environment influence the economic choices we discern and the livelihood decisions we make.

In seeking spatial and activity regularities in the nearly infinite diversity of human economic activity, it is useful to generalize about systems of economic organization and control and about classes of productive effort and labor specialization. We can observe, for example, that, broadly speaking, there are three types of economic systems: subsistence, commercial, and planned. The first is concerned with production for the immediate consumption of individual producers and family members. In the second, economic decisions ideally respond to impersonal market forces and reasoned assessments of monetary gain. In the third, at least some nonmonetary social rather than personal goals influence production decisions. The three system forms are not mutually exclusive; all societies contain some intermixture of features of at least two of the three pure types, and some economies have elements of all three. Recognition of each type's respective features and controls, however, helps us to understand the forces shaping economic decisions and patterns in different cultural and regional settings.

Our search for regularities is furthered by a classification of economic activities according to the stages of production and the degree of specialization they represent. We can, for example, decide all productive activity is arranged along a continuum of increasing technology, labor specialization, value of product, or sophistication of service. With that assumption, we can divide our continuum into primary activities (food and raw material production), secondary production (processing and manufacturing), tertiary activities (distribution and general professional and personal service), and the quaternary and quinary activities (administrative, informational, and technical specializations) that mark highly advanced societies of either planned or commercial systems.

Agriculture, the most extensively practiced of the primary industries, is part of the spatial economy of both subsistence and advanced societies. In the first instance—whether it takes the form of extensive or intensive, shifting or sedentary production— it is responsive to the immediate consumption needs of the producer group and reflective of the environmental conditions under which it is practiced. Agriculture in advanced economies involves the application of capital and technology to the productive enterprises; as one sector of an integrated economy, it is responsive to consumption requirements expressed through free or controlled markets. Its spatial expression reflects assessments of profitability and the dictates of social and economic planning.

Agriculture, fishing, forestry, trapping, and the extractive (mining) industries are closely tied to the uneven distribution of earth resources. Their spatial patterns reflect those resource potentials, but they are influenced as well by the integration of all societies and economies through the medium of international trade and mutual dependence. The flows of primary products and of manufactured goods suggest the hierarchy of production, marketing, and service activities, which will be the subject of Chapter 9.

Key Words

agriculture 273
aquaculture 298
Boserup thesis 278
commercial economy 273
economic geography 270
extensive agriculture 276, 288
extractive industry 294
gathering industry 294
Green Revolution 282
intensive agriculture 276, 288

maximum sustainable yield 295
natural resource 294
nomadic herding 276
nonrenewable resource 295
planned economy 273
plantation 292
primary activity 271
quaternary activity 272
quinary activity 272
renewable resource 294

resource 294
secondary activity 271
shifting cultivation 277
subsistence economy 273
technology 270
tertiary activity 272
tragedy of the commons 298
truck farm 288
usable reserves 301
von Thünen model 287

For Review

1. What are the distinguishing characteristics of the economic systems labeled *subsistence, commercial,* and *planned*? Are they mutually exclusive, or can they coexist within a single political unit?

2. What are the ecological consequences of the different forms of *extensive subsistence* land use? In what world regions are such systems found? What, in your opinion, are the prospects for these land uses and for the way of life they embody?

3. How is *intensive subsistence* agriculture distinguished from *extensive subsistence* cropping? Why, in your opinion, have such different

land use forms developed in separate areas of the warm, moist tropics?

4. Briefly summarize the assumptions and dictates of von Thünen's agricultural model. How might the land use patterns predicted by the model be altered by an increase in the market price of a single crop? A decrease in the transportation costs of one crop but not of all crops?

5. What is the basic distinction between a *renewable* and a *nonrenewable* resource? Under what circumstances might the distinction between the two be blurred or obliterated?

6. What economic and ecological problems can you cite that do or might affect the viability and productivity of the *gathering industries* of forestry and fishing? What is meant by the *tragedy of the commons?* How is that concept related to the problems you discerned?

7. Why have the mineral fuels been so important in economic development? What are the mineral fuels, and what are the prospects for their continued availability? What economic and social consequences might you anticipate if the price of mineral fuels should double? If it should be cut in half?

Focus Follow-up

1. **How are economic activities and national economies classified?** pp. 270–273.

The innumerable economically productive activities of humans are influenced by regionally varying

environmental, cultural, technological, political, and market conditions. Understanding the world's work is simplified by thinking of economic activity as arranged along a continuum of increasing complexity of product or

service and increasing distance from nature. Primary industries (activities) harvest or extract something from the earth. Secondary industries change the form of those harvested items. Tertiary activities render services, and quaternary efforts reflect professional or managerial talents. Those activity stages are carried out within national economies grouped as subsistence, commercial, or planned.

2. **What are the types and prospects of subsistence agriculture?** pp. 273–285.

Subsistence farming—food production primarily or exclusively for the producers' family needs—still remains the predominant occupation of humans on a worldwide basis. Nomadic herding and shifting ("swidden") cultivation are extensive subsistence systems. Intensive subsistence farming involves large inputs of labor and fertilizer on small plots of land. Both rural and urban subsistence efforts are increasingly marked by some production for market; they have also benefited from Green Revolution crop improvements.

3. **What characterizes commercial agriculture, and what are its controls and special forms?** pp. 285–294.

The modern integrated world of exchange and trade increasingly implies farming efforts that reflect broader market requirements, not purely local or family needs. Commercial agriculture is characterized by specialization, off-farm sale, and interdependence of farmers and buyers linked through complex markets. The von Thünen model of agricultural location suggests that intensive forms of commercial farming—fruits, vegetables, dairy products, and livestock-grain production—should be located close to markets. More extensive commercial agriculture, including large-scale wheat farms and livestock ranches, are by model and reality at more distant locations. Special crops may by value or uniqueness defy these spatial determinants; Mediterranean and plantation agriculture are examples.

4. **What are the special characteristics and problems of nonagricultural primary industries?** pp. 294–307.

The "gathering" industries of fishing, forestry, and trapping and the "extractive" industries of mining and quarrying involve the direct exploitation of areally variable natural resources. Resources are natural materials that humans perceive as necessary and useful. They may be renewable—replenished—by natural processes or nonrenewable once extracted and used. Overexploitation can exceed the maximum sustainable yield of fisheries and forests and eventually destroy the resource. Such destruction is assured in the case of nonrenewable minerals and fuels when their total or economically feasible supply is exhausted.

5. **What is the status and nature of world trade in primary products?** pp. 307–309.

The primary commodities of agricultural goods, fish, forest products, furs, and minerals and fuels account for nearly one-third of the dollar value of international trade. Traditional exchange flows of raw materials outward from developing states that then imported manufactured goods from advanced economies have changed in recent years. Increasingly, the share of manufactured goods in developing world exports is growing, and dependence on income from raw material sales is dropping. However, material-exporting states argue that current international trade agreements are unfavorable to exporters of agricultural products and ores and minerals.

 Selected References

Boserup, Ester. *The Conditions of Agricultural Change: The Economics of Agrarian Change under Population Pressure.* London, England: Allyn & Unwin, 1965.

Bowler, Ian R. *The Geography of Agriculture in Developed Market Economies.* New York: John Wiley & Sons, 1993.

Chang, Claudia, and Harold A. Kostner. *Pastoralists at the Periphery: Herders in a Capitalist World.* Tucson: University of Arizona Press, 1994.

Clark, Gordon L., Maryann P. Feldman, and Meric S. Gertler, eds. *The Oxford Handbook of Economic Geography.* New York: Oxford University Press, 2001.

Courtenay, Percy P. *Plantation Agriculture.* 2d ed. London: Bell and Hyman, 1980.

French, Hilary F. *Costly Tradeoffs: Reconciling Trade and the Environment.* Worldwatch Paper 113. Washington, D.C.: Worldwatch Institute, 1993.

Galaty, John G., and Douglas L. Johnson, eds. *The World of Pastoralism: Herding Systems in Comparative Perspective.* New York: Guilford Press, 1990.

Graham, Edgar, and Ingrid Floering, eds. *The Modern Plantations in the Third World.* New York: St. Martin's Press, 1984.

Grigg, David. *An Introduction to Agricultural Geography.* 2d ed. New York: Routledge, 1995.

Grigg, David. "The Starchy Staples in World Food Consumption." *Annals of the Association of American Geographers* 86, no. 3 (1996): 412–431.

Hamilton, Ian. *Resources and Industry.* The Illustrated Encyclopedia of World Geography. New York: Oxford University Press, 1992.

Hardin, Garrett. "The Tragedy of the Commons." *Science* 162 (1968): 1243–1248.

Harris, David R. "The Ecology of Swidden Agriculture in the Upper Orinoco Rain Forest, Venezuela."

Geographical Review 61 (1971): 475–495.

Kane, Hal. "Growing Fish in Fields." *WorldWatch* 6, no. 5 (September–October, 1993): 20–27.

Lomborg, Bjorn. *The Skeptical Environmentalist: Measuring the Real State of the World.* Cambridge, England: Cambridge University Press, 2001.

McGinn, Anne Platt. *Rocking the Boat: Conserving Fisheries and Protecting Jobs.* Worldwatch Paper 142. Washington, D.C.: Worldwatch Institute, 1998.

Peters, William J., and Leon F. Neuenschwander. *Slash and Burn: Farming in the Third World Forest.* Moscow: University of Idaho Press, 1988.

Sinclair, Robert. "Von Thünen and Urban Sprawl." *Annals of the Association of American Geographers* 57, no. 1 (1967): 72–87.

Smil, Vaclav. *Feeding the World: A Challenge for the Twenty-First Century.* Cambridge, Mass.: MIT Press, 2000.

de Souza, Anthony R., and Stutz, Frederick P. *The World Economy: Resources, Location, Trade and Development.*
3d ed. Upper Saddle River, N.J.: Prentice-Hall, 1998.

Tarrant, John, ed. *Farming and Food.* The Illustrated Encyclopedia of World Geography. New York: Oxford University Press, 1991.

Tuxill, John. *Nature's Cornucopia: Our Stake in Plant Diversity.* Worldwatch Paper 148. Washington, D.C.: Worldwatch Institute, 1999.

Vital Signs: The Trends that Are Shaping Our Future. Washington, D.C.: The Worldwatch Institute, annual.

Weber, Peter. *Net Loss: Fish, Jobs, and the Marine Environment.* Worldwatch Paper 120. Washington, D.C.: Worldwatch Institute, 1994.

Wheeler, James O., and Peter Muller. *Economic Geography.* 3d ed. New York: John Wiley & Sons, 1998.

World Resources. . . . A Report by the World Resources Institute in collaboration with the United Nations Environment Programme and the United Nations Development Program. Washington, D.C.: World Resources Institute, biennial.

Young, John E. *Mining the Earth.* Worldwatch Paper 109. Washington, D.C.: Worldwatch Institute, 1992.

Websites: The World Wide Web has a tremendous number and variety of sites pertaining to geography. Websites relevant to the subject matter of this chapter appear in the "Web Links" section of the On-line Learning Center associated with this book. Access it at **www.mhhe.com/fellmann8e**

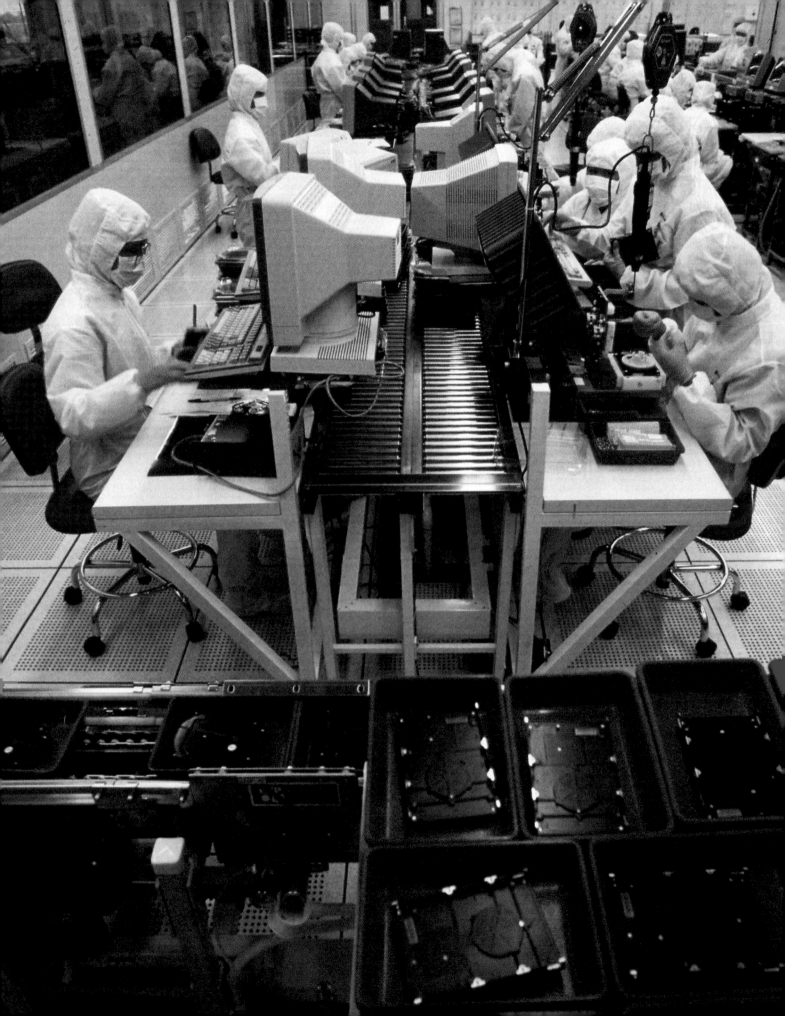

CHAPTER
Nine

Livelihood and Economy:
From Blue Collar to Gold Collar

Focus Preview

1. What principles or considerations guide manufacturing locational decisions, pp. 316–324, and how those considerations have been selectively incorporated in different industrial location theories, pp. 324–327.

2. How other nontheoretical considerations including transnational ownership affect, distort, or reinforce classical locational controls, pp. 327–333.

3. The older world patterns of manufacturing regions, pp. 333–339 and how they have been affected by the special locational characteristics of high-tech industries, pp. 340–343.

4. What the identifying characteristics of tertiary, quaternary, and quinary service activities are, pp. 343–346, and how their recent development impacted world economic patterns and international trade, pp. 346–348.

Opposite: Assembling computer hard drives in Singapore for an American firm. Foreign direct investment, international outsourcing of production, and advanced technology transfer are all features of the globalized eonomy.

*R*oute 837 connects the four U.S. Steel plants stretched out along the Monongahela River south of Pittsburgh. Once, in the late 1960s, 50,000 workers labored in those mills, and Route 837 was choked with the traffic of their cars and of steel haulers' trucks. By 1979, fires were going out in the furnaces of the aging mills as steel imports from Asia and Europe flowed unchecked into domestic markets long controlled by American producers. By the mid-1980s, with employment in the steel plants of the "Mon" Valley well below 5000, the highway was only lightly traveled and only occasionally did anyone turn at the traffic lights into the closed and deserted mills.

At the same time, traffic was building along many highways in the northeastern part of the country. Four-lane Route 1 was clogged with traffic along the 42 kilometers (26 miles) of the "Princeton Corridor" in central New Jersey as that stretch of road in the 1980s had more office space, research laboratories, hotels, conference centers, and residential subdivisions planned and under construction than anywhere else between Washington, D.C. and Boston. Farther to the south, around Washington itself, traffic grew heavy along the Capital Beltway in Virginia, where vast office building complexes, defense-related industries, and commercial centers were converting rural land to urban uses. And east of New York City, traffic jams were monumental around Stamford, Connecticut, in Fairfield County, as it became a leading corporate headquarters town with 150,000 daily in-commuters.

By the early 1990s, traffic in Fairfield County had thinned as corporate takeovers, leveraged buyouts, and "downsizing" reorganizations reduced the number and size of companies and their need for both employees and office space. Vacancies exceeded 25% among the office buildings and research parks so enthusiastically built during the 1970s and 1980s, and vacant "corporate campuses" lined stretches of formerly clogged highways. But soon traffic was building elsewhere in the country as millions of Americans during the 1990s gained technology-related jobs in California's "Silicon Valley," and a whole series of other widely spaced emerging "high-tech" hot spots clustered around such industries as computers, lasers, software, medical devices, and biotechnology. And by the late 1990s, all sections of the United States again were experiencing the traffic volumes that economic prosperity induces, only once more to endure job losses, office vacancies, economic reversals, and altered traffic flows following the "dot-com" bubble collapse of the early 21st century.

These contrasting and fluctuating patterns of traffic flow symbolize the ever-changing nature and structure of the Anglo American space economy. The smokestack industries of the 19th and early 20th centuries have declined, replaced by research park industries, shopping centers, and office building complexes that in their turn experience variable prosperity and adversity. The continent's economic landscape and employment structure are inconstant at best (Figure 9.1). And North America is not alone. Change is the ever-present condition of contemporary economies, whether of the already industrialized, advanced countries or of those newly developing in an integrated world marketplace. Resources are exploited and exhausted; markets grow and decline; patterns of economic advantage, of labor skills, of industrial investment and productive capacity undergo alteration as countries and regions differentially develop, prosper, or experience reversals and decline. Such changes have profound impact on the spatial structure and processes of economic activity.

Components of the Space Economy

All human activity has spatial expression. In the economic sphere we recognize regions of industrial concentration, areas of employment and functional specialization, and specific factory sites and store locations. As geographers, we assume an underlying logic to those spatial economic patterns and seek, through observation and theory, an understanding and explanation of them. In a very preliminary fashion, that understanding has begun through classification of economic activity into *primary, secondary, tertiary, quaternary,* and *quinary* industries. (Remember, the term *industry* may be used in the narrow sense of type of manufacturing activity or enterprise as well as in the broader meaning of category of economic orientation.)

Primary industries, you will recall from Chapter 8, are tied to the natural resources they gather or exploit. Location is therefore predetermined by the distribution of minerals, fuels, forests, fisheries, or natural conditions affecting agriculture and herding. The later (beyond primary) stages of economic activity, however, are increasingly divorced from the conditions of the physical environment. In them, processing, distribution, communication, and management permit enterprise location in response to cultural and economic rather than physical influences. They are movable, rather than spatially tied activities. The locational decisions made and the economic patternings that result differ with the type or level of economic activity in question. Secondary industries involved in material processing and goods production have different spatial constraints than do the retailing activities of tertiary industry or the research parks or office complexes of quaternary and quinary activities. At every industrial or activity level, however, it is assumed that a recurring set of economic controls may be identified.

$\mathscr{F}igure\ 9.1$ This idled Pennsylvania steel mill typifies the structural changes occurring in "post-industrial" America—and in other advanced economies where comparable dislocations and changes are taking place. For heavy industrial jobs lost, replacement employment must increasingly be found in the service industries of the tertiary and quaternary sectors. In adaptive advanced economies, that restructuring is normal. The site of the U.S. Steel Homestead works along the Monongahela River near Pittsburgh, closed in 1986, has been rebuilt, for example, as the mixed-use Waterfront development with more than 50 retail shops, a 22-screen movie theater, and 500 apartments.

Concepts and Controls

The controls that are assumed to exist are rooted in observations about human spatial behavior in general and economic behavior in particular. We have already explored some of those assumptions in earlier discussions. We noted, for example, that the intensity of spatial interaction decreases with increasing separation of places—distance decay, we called it. We observed the importance of complementarity and transferability in the assessment of resource value and trade potential. Von Thünen's model of agricultural land use, you will recall, was rooted in conjectures about transportation cost and land value relationships.

Such simplifying assumptions help us to understand a presumed common set of controls and motivations guiding human economic behavior. We assume, for example, that people are *economically rational;* that is, given the information at their disposal, they make locational, production, or purchasing decisions in light of a perception of what is most cost-effective and advantageous. Behavioral research concludes that while people are not truly rational in the theoretical economic sense, neither are they insane or incompetent. The acceptance of rationality, they conclude, is proper if one also accepts the reality that individuals respond to behavioral traits—envy, rivalry, impulsiveness, forgetfulness of past mistakes, positive wishful thinking, and the like—at odds with purely rational actions or decisions. With those appreciations of behavioral human nature, economic rationality is still the accepted theoretical starting point.

From the standpoint of producers or sellers of goods or services, it is assumed each is intent on *maximizing profit.* To reach that objective, each may consider a host of production and marketing costs and political, competitive, and other limiting factors—and, perhaps, respond to individual behavioral quirks—but the ultimate goal of profit-seeking remains clear. Finally, we assume that in commercial economies the best measure of the correctness of economic decisions is afforded by the *market mechanism.*

At root, that market control mechanism is measured by *price*—the price of land (rent), of labor (wages), of a college course (tuition), or of goods at the store. In turn, price is seen as a function of *supply* and *demand*. In large, complex economies where there are many producers, sellers, and buyers, and many alternative products competing in the marketplace, price is the neutral measure of comparative value and profitability. The theoretical relationship between supply, demand, and price is simple. If demand for a good or service exceeds its available supply, scarcity will drive up the price it can command in the marketplace. That increased price will enhance the profitability of the sale, which will encourage existing producers to increase output or induce new producers or sellers to enter the market (Figure 9.2a). That is, *the higher the price of a good, the more of it will be offered in the market.*

When the price is very high, however, relatively few people are inclined to buy. To dispose of their increased output, old and new producers of the commodity are forced to reduce prices to enlarge the market by making the good affordable to a larger number of potential customers. That is, *at lower prices, more of a good will be purchased* (Figure 9.2b). If the price falls too low, production or sale becomes unprofitable and inefficient suppliers are forced out of business, reducing supply. **Market equilibrium** is marked by the price at which supply equals demand, satisfying the needs of consumers and the profit motivation of suppliers (Figure 9.2c).

These and other modifying concepts and controls of the economist treat supply, demand, and price as if all production, buying, and selling occurred at a single point. But as geographers, we know that human activities have specific locational settings and that neither people, nor resources, nor opportunities are uniformly distributed over the earth. We appreciate that the place or places of production may differ from the locations of demand. We understand that there are spatial relations and interactions based on supply, demand, and equilibrium price. We realize there is a *geography* of supply, a *geography* of demand, and a *geography* of cost.

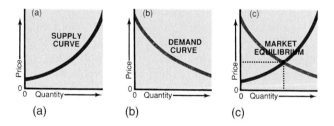

Figure 9.2 **Supply, demand, and market equilibrium.** The regulating mechanism of the market may be visualized graphically. (*a*) The *supply curve* tells us that as the price of a good increases, more of that good will be made available for sale. Countering any tendency for prices to rise to infinity is the market reality that the higher the price, the smaller the demand as potential customers find other purchases or products more cost-effective. (*b*) The *demand curve* shows how the market will expand as prices are lowered and goods are made more affordable and attractive to more customers. (*c*) *Market equilibrium* is marked by the point of intersection of the supply and demand curves and determines the price of goods, the total demand, and the quantity bought and sold.

Secondary Activities: Manufacturing

If we assume free markets, rational producers, and informed consumers, then locational production and marketing decisions should be based on careful consideration of spatially differing costs and opportunities. In the case of primary industries—those tied to the environment—points or areas of possible production are naturally fixed. The only decision is whether or not to exploit known resources. In the instance of secondary and higher levels of economic activity, however, the locational decision is more complex. It involves the weighing of the locational "pulls" of a number of cost considerations and profit prospects.

On the *demand* side, the distribution of populations and of purchasing power defines general areas of marketing opportunities. The regional location of tertiary—sales and service—activities may be nearly as fixed as are primary industries, though specific site decisions are more complex. On the *supply* side, decision making for manufacturers involves a more intricate set of equations. Manufacturers must consider costs of raw materials, distance from them and from markets, wages of labor, outlays for fuel, capital availability and rates, and a host of other inputs to the production and distribution process. It is assumed that the nature and the spatial variability of those myriad costs are known and that rational location decisions leading to profit maximization are based on that knowledge. For market economies, both observation and theory tend to support that assumption.

Locational Decisions in Manufacturing

Secondary activities involve transforming raw materials into usable products, giving them *form utility*. Dominant among them is manufacturing in all of its aspects, from pouring iron and steel to stamping out plastic toys, assembling computer components, or sewing dresses. In every case, the common characteristics are the application of power and specialized labor to the production of standardized commodities in factory settings: in short, the characteristics of industrialization.

Manufacturing poses a different locational problem than does the gathering of primary commodities. It involves the assembly and the processing of inputs and the distribution of the output to other points and therefore presents the question of where the processing should take place. The answer may require multiple spatial levels of consideration. The first is regional and addresses the comparative attractions for different types of industry of different sections of the country or even of different countries at the international scale. Later decision stages become more focused, localized, and specific to an individual enterprise. They involve assessment of the special production and marketing requirements of particular industrialists and of the degree to which those requirements can or will be met at different subregional scales—at the state (in the U.S.), community, and individual site levels. That is, we can ask at one level why northeastern United States-southeastern Canada exerted an earlier pull on industry in general and, at other decision stages, why specific sites along the

Monongahela Valley to the south of Pittsburgh in Pennsylvania were chosen by U.S. Steel Corporation for its mills.

In framing responses, one needs to consider a wide range of industrial pulls and attractions and the modifying influence of a number of physical, political, economic, and cultural constraints. For a great many searches, two or several alternate locations would be equally satisfactory. In very practical financial terms, locational decisions at the state, community, and site levels may ultimately be based on the value of inducements that are offered by rival areas and agencies competing for the new or relocated manufacturing plant (See "Contests and Bribery," p. 332). In both practice and theory, locational factors recognized and analyzed are complexly interrelated, change over time in their relative significance, and differ between industries and regions. But all of them are tied to *principles of location* that are assumed to operate under all economic systems, though to be determinant, perhaps, only in free market, or commercial, economies.

Principles of Location

The principles, or "ground rules," of location are simply stated.

1. Certain input costs of manufacturing are **spatially fixed costs,** that is, are relatively unaffected in their amount or relative importance no matter where the industry is located within a generalized regional or national setting. Wage rates set by national or areawide labor contracts are an example. Fixed costs have no implication for comparative locational advantage.

2. Other input costs of manufacturing are **spatially variable costs;** that is, they show significant differences from place to place in both their amount and their relative contribution to the total cost of manufacturing (Figure 9.3). These will influence locational choices.

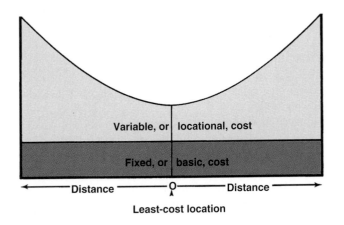

Figure 9.3 **The spatial implications of fixed and variable costs.** *Spatially fixed* (or *basic*) costs represent the minimum price that must be paid at any location for the necessary inputs of production of a given item. Here, for simplicity, a single raw material is assumed and priced at its cheapest source. *Spatially variable* (*locational*) costs are the additional costs incurred at alternate locations in overcoming distance, attracting labor, purchasing the plant site, and so forth. In the example, only the transportation cost of the single material away from its cheapest (source) location is diagrammed to determine *O*, the optimal or least-cost location.

3. The ultimate aim of the economic activity is *profit maximization.* In an economic environment of full and perfect competition, the profit objective is most likely to be achieved if the manufacturing enterprise is situated at the *least total cost* location. Under conditions of imperfect competition, considerations of sales and market may be more important than production costs in fixing "best" locations.

4. Since among the totality of production costs some inputs are approximately the same irrespective of location, fixed costs are not of major importance in determining optimum, or least-cost, locations. Rather, the industrialist bases the locational search on the minimization of variable costs. The locational determinant is apt to be the cost that is both an important component of total costs and shows the greatest spatial variation.

5. Transportation charges—the costs of accumulating inputs and of distributing products—are highly variable costs. As such, they (rather than the commodity transported) may become the locational determinant, imparting an unmistakable *orientation*—a term describing locational tendencies—to the plant siting decision.

6. Individual establishments rarely stand alone; they are part of integrated manufacturing sequences and environments in which *interdependence* increases as the complexity of industrial processes increases. The economies of structural and spatial interdependence may be decisive locational determinants for some industries. *Linkages* between firms may localize manufacturing in areas of industrial agglomeration where common resources—such as skilled labor—or multiple suppliers of product inputs—such as automobile component manufacturers—are found.

These principles are generalized statements about locational tendencies of industries. Their relative weight, of course, varies among industries and firms. Their significance also varies depending on the extent to which purely economic considerations—as opposed, say, to political or environmental constraints—dictate locational decisions.

Raw Materials

All manufactured goods have their origins in the processing of raw materials, but only a few industries at the early stages of the production cycle use raw materials directly from farms or mines. Most manufacturing is based on the further processing and shaping of materials already treated in some fashion by an earlier stage of manufacturing located elsewhere. In general, the more advanced the industrial economy of a nation, the smaller is the role played by truly *raw* materials in its economic structure.

For those industries in which unprocessed commodities are a primary input, however, the source and characteristics of the raw materials upon which they are based are important indeed. The quality, amount, or ease of mining or gathering of a resource may be a locational determinant if cost of raw material is the major variable and multiple sources of the primary material are available. Raw materials may attract the industries that process them when they are bulky, undergo great weight loss in the processing, or are highly perishable. Copper smelting and iron ore beneficiation are examples of weight- (impurity-) reducing industries localized by their ore supplies (see pp. 302 to 304). Pulp, paper, and sawmills are, logically, found in areas within or accessible to timber. Fruit and vegetable canning in California, midwestern meat packing, and Florida orange juice concentration and freezing are different but comparable examples of raw **material orientation.** The reason is simple; it is cheaper and easier to transport to market a refined or stabilized product than one filled with waste material or subject to spoilage and loss.

Multiple raw materials might dictate an intermediate plant location. Least cost may be determined not by a single raw material input but by the spatially differing costs of accumulating several inputs. Steel mills at Gary, Indiana, or Cleveland, Ohio, for example, were not based on local raw material sources but on the minimization of the total cost of collecting at a point the necessary ore, coking coal, and fluxing material inputs for the production process (Figure 9.4). Steel mills along the U.S. East Coast—at Sparrows Point, Maryland, or the Fairless Works near Philadelphia—were localized where imported ores were unloaded from ocean carriers, avoiding expensive transshipment costs. In this latter avoidance, both the Great Lakes and the coastal locations are similar.

Power Supply

For some industries, power supplies that are immobile or of low transferability may serve to attract the activities dependent upon them. Such was the case early in the Industrial Revolution when water power sites localized textile mills and fuel (initially charcoal, later coking coal) drew the iron and steel industry. Metallurgical industries became concentrated in such coal-rich regions as the Midlands of England, the Ruhr district of Germany, and the Donets Basin of Ukraine.

Massive charges of electricity are required to extract aluminum from its processed raw material, *alumina* (aluminum oxide). Electrical power accounts for between 30% and 40% of the cost of producing the aluminum and is the major variable cost influencing plant location in the industry. The Kitimat plant on the west coast of Canada or the Bratsk plant near Lake Baikal in eastern Siberia are examples of industry placed far from raw material sources or market but close to vast supplies of cheap power—in these instances, hydroelectricity.

Labor

Labor also is a spatial variable affecting location decisions and industrial development. Traditionally, three different considerations—price, skill, and amount—of labor were considered to be determinant singly or in combination. For many manufacturers today, an increasingly important consideration is *labor flexibility,* implying more highly educated workers able to apply themselves to a wide variety of tasks and functions. For some activities, a cheap labor supply is a necessity (see "Wage Rates and the Cloth Trades"). For others, labor skills may constitute the locational attraction and regional advantage. Machine tools in Sweden, precision instruments in Switzerland, optical and electronic goods in Japan are examples of industries that have created and depend on localized labor skills. In an increasingly high-tech world of automation, electronics, and industrial robots, labor skills—even at high unit costs—are often more in demand than an unskilled, uneducated work force.

In some world areas, of course, labor of any skill level may be poorly distributed to satisfy the developmental objectives of government planners or private entrepreneurs. In the former Soviet Union, for example, long-standing economic plans called for the fuller exploitation of the vast resources of sparsely populated Siberia, an area generally unattractive to a labor force more attuned to the milder climates and greater amenities of the settled European portion of the country. At the same time, labor surpluses were growing in Soviet Central Asia, where resources were few and rates of natural population increase were high, but whose Muslim populations resisted resettlement outside of their homeland areas.

Market

Goods are produced to supply a market demand. Therefore, the size, nature, and distribution of markets may be as important in industrial location decisions as are raw material, energy, labor, or

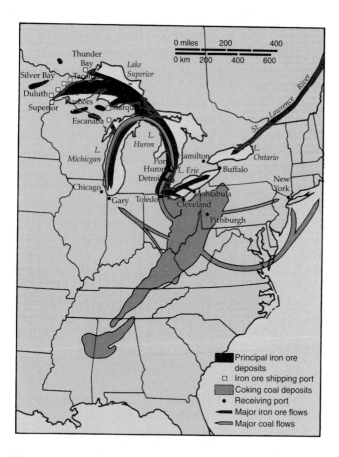

Figure 9.4 **Material flows in the steel industry.** When an industrial process requires the combination of several heavy or bulky ingredients, an intermediate point of assembly of materials is often a least-cost location. In the earlier 20th century, the iron and steel industry of the eastern United States showed this kind of localization—not at the source of any single input but where coking coal, iron ore, and limestone could be brought together at the lowest price.

In the free market world, few industries have been as spatially responsive to labor costs as have textiles and garment making. While a great deal of labor is necessary to turn natural or artificial fibers into a finished product, it need not be skilled labor. In the United States, for example, the cotton and woolen cloth industry established in New England during the 1820s and 1830s drew extensively on a labor pool made up largely of underemployed daughters of the many families working small farms of the area. Later, immigrants provided the surplus unskilled labor supply in the cities and towns of the area. With time and further industrialization, wage rates began to increase in New England to levels unacceptable to the highly competitive textile industry. When (among other reasons) it became clear that a surplus of cheap labor was available in the Piedmont district of the Southeast, the industry quickly moved in that direction. Today, the U.S. textile industry that still remains in the face of foreign competition is concentrated in that region.

Much of the recent competition comes from *newly industrializing countries* (NICs) of Asia where the textile industry is seen as a near-ideal employer of an abundant labor force. Among them earlier were the major suppliers of the American market during the 1970s and 1980s: Taiwan, South Korea, and Hong Kong. As industrialization grew and wage rates increased in these producing areas, however, their relative labor advantage was eroded. The textile industry began to shift to still lower-cost producers, including China, Bangladesh, Indonesia, Mexico, and Thailand. By the early 21st century, labor cost advantages were shifting to still other Asian, African, Central American, and Caribbean countries. The trends of the textile industry confirm the observation that in commercial economies "best location" is only a transient advantage of specific sites, regions, or countries.

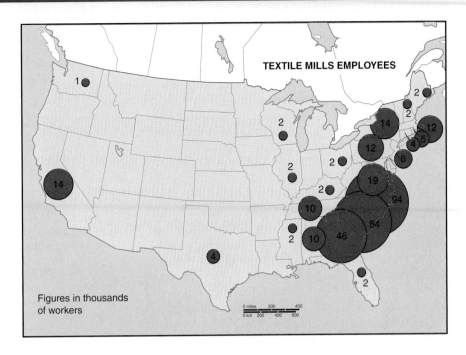

TEXTILE MILLS EMPLOYEES

Figures in thousands of workers

0 miles 200 400
0 km 200 400 600

The point is emphasized in the experience of the apparel industry, which converts finished textiles into clothing. Like textile manufacturing, the apparel industry finds its most profitable production locations in areas of cheap labor. Repetitious, limited-skill, assembly-line operations are necessary for volume production of clothing for a mass market that is highly price competitive; labor costs are the chief locational determinant. Even the lowest-wage areas of the United States have difficulty competing in the manufacture of standardized women's and children's clothing. Wage rates for apparel production workers in the 20 major exporters of garments to the United States range from a low of 2% (Bangladesh) to a high of 25% (Mexico) of the level of their American counterparts. Those 20 low-wage countries accounted for nearly 90% of imported apparel, which, in turn, captured 60% of the U.S. domestic apparel market in the early 1990s. That percentage began to rise in the later 1990s as requirements of the World Trade Organization reduced American import restrictions and quotas and accelerated the apparel industry job loss that already totaled nearly 360,000 between 1972 and 1992 and shed another 475,000 jobs by 2002.

Source: Map data from *Annual Survey of Manufacturers*, 2002.

other inputs. Market pull, like raw material attraction, is at root an expression of the cost of commodity movement. When the transportation charges for sending finished goods to market are a relatively high proportion of the total value of the good (or can be significantly reduced by proximity to market), then the attraction of location near to the consumer is obvious and **market orientation** results.

The consumer may be either another firm or the general public. When a factory is but one stage in a larger manufacturing process—firms making wheels, tires, windshields, bumpers, and the like in the assembly of automobiles, for example—location near the next stage of production is an obvious advantage. The advantage is increased if that final stage of production is also near the ultimate consumer market. To continue our example, automobile assembly plants have been scattered throughout the North American realm in response to the existence of large regional markets and the cost of distribution of the finished automobile. This market orientation is further reflected by the location in North America of auto manufacturing or assembly plants of Asian and European motor vehicle companies, although both foreign and domestic firms again appear to be reconcentrating the industry in the east central part of the United States.

People themselves, of course, are the ultimate consumers. Large urban concentrations represent markets, and major cities have always attracted producers of goods consumed by city dwellers. Admittedly, it is impossible to distinguish clearly between urbanites as market and urbanites as labor force. In either case, many manufacturing activities are drawn to major metropolitan centers. Certain producers are, in fact, inseparable from the immediate markets they serve and are so widely distributed that they are known as **ubiquitous industries.** Newspaper publishing, bakeries, and dairies, all of which produce a highly perishable commodity designed for immediate consumption, are examples.

Transportation

Transportation has been so much the unifying thread of all of these references to "factors" of industrial location that it is difficult to isolate its separate role. In fact, some of the earlier observations about manufacturing plant orientations can be restated in purely transportation cost terms. For example, copper smelting or iron ore beneficiation—described earlier as examples of raw material orientation—may also be seen as industries engaged in *weight reduction* designed to minimize transportation costs by removal of waste material prior to shipment. Some market orientation is of the opposite nature, reflecting *weight-gaining* production. Soft drink bottlers, for example, add large amounts of water to small amounts of concentrated syrup to produce a bulky product of relatively low value. All transport costs are reduced if only the concentrate is shipped to local bottlers, who add the water that is available everywhere and distribute only to local dealers. The frequency of this practice suggests the inclusion of soft drink bottlers among the ubiquitous industries.

No matter the specific characterization of attraction, modern industry is intimately and inseparably tied to transportation systems. The Industrial Revolution is usefully seen as initially and simultaneously a transportation revolution as successive improvements in the technology of movement of peoples and commodities enlarged the effective areas of spatial interaction and made integrated economic development and areal specialization possible. All advanced economies are well served by a diversity of transport media (see Figure 8.4); without them, all that is possible is local subsistence activity. All major industrial agglomerations are simultaneously important nodes of different transportation media, each with its own characteristic advantages and limitations.

Water transportation is the cheapest means of long-distance freight movement (Figure 9.5). Little motive power is required, right-of-way costs are low or absent, and operating costs per unit of freight are low when high-capacity vessels are used. Inland waterway improvement and canal construction marked the first phase of the Industrial Revolution in Europe and was the first stage of modern transport development in the United States. Because the ton-mile costs of water movement remain so relatively low, river ports and seaports have locational attractiveness for industry unmatched by alternative centers not served by water carriers. Although the disadvantages of water carriage of freight are

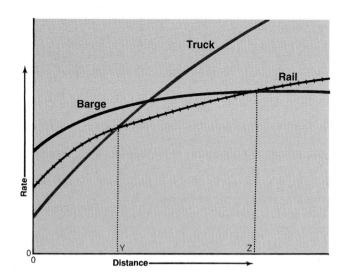

Figure 9.5 **The pattern of carrier efficiency.** Different transport media have cost advantages over differing distances. The usual generalization is that when all three media are available for a given shipment, trucks are most efficient and economical over short hauls of up to about 500 kilometers (about 300 miles), railroads have the cost advantage over intermediate hauls of 500 to 3200 kilometers (about 300 to 2000 miles), and water (ship or barge) movement over longer distances (and, often, over shorter distances where speed of delivery of nonperishable commodities is not a consideration). The differing cost curves represent the differing amounts of fixed or variable costs incurred by each transport medium, as further illustrated in Figure 9.8.

serious, where water routes are in place, as in northwestern Europe or the Great Lakes-Mississippi systems of the United States, they are vital elements in regional industrial economies.

Railroads efficiently move large volumes of freight over long distances at low fuel and labor costs (see Figure 8.36). They are, however, inflexible in route, slow to respond to changing industrial locational patterns, and expensive to construct and maintain. They require high volumes of traffic to be cost-effective. When for any reason traffic declines below minimum revenue levels, rail service may be uneconomic and the lines abandoned—a response of American railroads that abandoned over 125,000 miles of line between 1930 and 2000.

High-volume, high-speed *motor trucks* operating on modern roadway and expressway systems have altered the competitive picture to favor highways over railways in many intercity movements in modern economies. Road systems provide great flexibility of service and are more quickly responsive than railroads to new traffic demands and changing origin and destination points. Intervening opportunities are more easily created and regional integration more cheaply achieved by highway than by railroad (or waterway systems). Disadvantages of highway transport include high maintenance costs of vehicles (and roads) and low efficiency in the long-distance, high-volume movement of bulky commodities.

Increasingly in the United States and elsewhere, greater transport cost efficiencies are achieved by combining short-haul motor carriage with longer-haul rail or ship movement of the same freight containers. Hauling a truck trailer on a railroad flat-car ("piggybacking") or on ship deck serves to minimize total freight rates and transport times. Such *multimodal* freight movements seek the advantages of the most efficient carrier for each stage of the journey from cargo origin point to final destination through the use of prepacked internationally standardized shipping containers (usually 8 × 8 × 20, 30, or 40 feet). The containers with undisturbed content may be transferred to ships for international ocean carriage, to railroads for long-haul land movement, and to truck trailers for shorter-haul distances and pickup and delivery. Their use is increasingly common: on long "trailer-on-flat-car" trains and in the growing volume of international ocean trade (Figue 9.6).

Pipelines provide efficient, speedy, and dependable transportation specifically suited to the movement of a variety of liquids and gases. They serve to localize along their routes the industries—particularly fertilizer and petrochemical plants—that use the transported commodity as raw material. In contrast, *air transport* has little locational significance for most industries despite its growing importance in long-distance passenger and high-value package freight movement. It contributes, of course, to the range of transport alternatives available to large population centers in industrially advanced nations and may increase the attractiveness of airport sites for high-tech and other industries shipping or receiving high-value, low-bulk commodities. Further, air transport may serve as the only effective connection with a larger national economy in the development of outposts of mining or manufacturing—as, for example, in Arctic regions or in interior Siberia. It is not, however, an effective competitor in the usual patterns of freight flow (see "A Comparison of Transport Media").

Transportation and Location

Figure 9.7 indicates the general pattern of industrial orientation related to variable transportation costs. In their turn, those costs are more than a simple function of the distance that goods are carried. Rather, they represent the application of differing **freight rates,** charges made for loading, transporting, and unloading of goods. Freight rates are said to *discriminate* between commodities on the basis of their assumed ability to bear transport costs in relation to their value. In general, manufactured goods have higher value, greater fragility, require more special handling, and can bear higher freight charges than can unprocessed bulk commodities. The higher transport costs for finished goods are therefore seen as a major reason for the increasing market orientation of industry in advanced economies with high-value manufacturing.

In addition to these forms of rate discrimination, each shipment of whatever nature must bear a share of the **fixed costs** of the company's investment in land, plant, and equipment and the assigned *terminal* and *line-haul* costs of the shipment. **Terminal costs** are charges associated with loading, packing, and unloading of a shipment and of the paperwork and shipping documents it entails. **Line-haul** or *over-the-road* **costs** vary with the individual shipments and are the expenses involved in the actual movement of commodities once they have been loaded. They are allocated to each shipment according to equipment used and distance traveled. Total transport costs represent a combination of all pertinent charges and are curvilinear rather than linear functions of distance. That is, carrier costs have a tendency to decline as the length of haul increases because scale economies in long-haul movement permit the averaging of total costs over a greater number of miles. The result is the *tapering principle* diagrammed in Figure 9.8.

Figure 9.6 The Kwai Chung container port, Hong Kong, China. In 2002, containerization accounted for over 70% of all international shipments and newer, larger, and more efficient ships able to carry over 7000 containers were entering service. At the start of the century, Hong Kong was the world's leader in container traffic, handling over 18 million containers per year.

Mode	Uses	Advantages	Disadvantages
Railroad	Intercity medium- to long-haul bulk and general cargo transport.	Fast, reliable service on separate rights-of-way; essentially nonpolluting; energy efficient; adapted to steady flow of single commodities between two points; routes and nodes provide intervening development opportunities.	High construction and operating costs; inflexibility of routes; underutilized lines cause economic drain.
Highway carrier	Local and intercity movement of general cargo and merchandise; pickup and delivery services; feeder to other carriers.	Highly flexible in routes, origins, and destinations; individualized service; maximum accessibility; unlimited intervening opportunity; high speed and low terminal costs.	Low energy efficiency; contributes to air pollution; adds congestion to public roads; high maintenance costs; inefficient for large-volume freight.
Inland waterway	Low-speed haulage of bulk, nonperishable commodities.	High energy efficiency; low per mile costs; large cargo capacity.	High terminal costs; low route flexibility; not suited for short haul; possible delays from ice or low water levels.
Pipelines	Continuous flows of liquids, gases, or suspended solids where volumes are high and continuity is required.	Fast, efficient, dependable; low per mile costs over long distances; maximum safety.	Highly inflexible in route and cargo type; high development cost.
Airways	Medium- and long-haul of high-value, low-bulk cargo where delivery speed is important.	High speed and efficiency; adapted to goods that are perishable, packaged, of a size and quantity unsuited to other modes; high route flexibility; access to areas otherwise inaccessible.	Very expensive; high mileage costs; some weather-related unreliability; inconvenient terminal locations; no intervening opportunities between airports.
Intermodal containerization	Employs standardized closed containers to move a shipment by any combination of water, rail, and truck without unpacking between origin and final destination.	Speed and efficiency of transit and lower shipping costs when multiple carriers are needed; reduced labor charges and pilferage losses.	Requires special terminals and handling machinery to load, off-load, and transfer containers.

One consequence of the necessary assignment of fixed and terminal costs to *every* shipment regardless of distance moved is that factory locations intermediate between sources of materials and final markets are less attractive than location at either end of a single long haul. That is, two short hauls cost more than a single continuous haul over the same distance (Figure 9.9).

Two exceptions to this locational generalization are of practical interest. **Break-of-bulk points** are sites where goods have to be transferred or transshipped from one carrier to another—at ports, for example, where barge or ocean vessel must be unloaded and cargo reloaded to railcar or truck, or between railroad and truck line. When such transfer occurs, an additional fixed or terminal cost is levied against the shipment, perhaps significantly increasing its total transport costs (piggyback transfers reduce, but do not eliminate, those handling charges). There is a tendency for manufacturing to concentrate at such points to avoid the additional charges. As a traffic-generating inducement, **in-transit privilege** may be granted to a manufacturer by a transportation agency through the quotation of a special single rate from material source to market for a movement that may be interrupted for processing or manufacturing *en route*. Such a special rate obviously removes the cost disadvantage of two short hauls and, by equalizing shipping costs between locations, tends to reduce the otherwise dominant attractions of either material or market locations.

Industrial Location Theories

In practice, enterprise locational decisions are based not on the impact of a single selected industrial factor but on the interplay and balance of a number of considerations. Implicit in our review has been the understanding that each type or branch of industry has its own specific set of significant plant siting conditions. For secondary activities as a whole, therefore, a truly bewildering complex of locational determinants exists. Theorists beginning in the first third of the 20th century set themselves the task of sorting through that complex in the attempt to define its underlying structure. The economic world they surveyed at that time, dominated by railroads, based on heavy industry and ideas of national industrial self-sufficiency, no longer is fully consistent with a globalized economy reflecting political, competitive, and social decisions of, for example, the World Trade Organization, transnational corporations, environmental protection agencies,

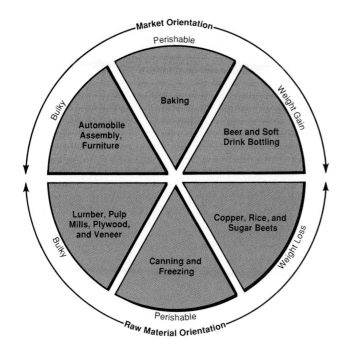

Figure 9.7 **Spatial orientation tendencies.** *Raw material orientation* is presumed to exist when there are limited alternative material sources, when the material is perishable, or when—in its natural state—it contains a large proportion of impurities or nonmarketable components. *Market orientation* represents the least-cost solution when manufacturing uses commonly available materials that add weight to the finished product, when the manufacturing process produces a commodity much bulkier or more expensive to ship than its separate components, or when the perishable nature of the product demands processing at individual market points.

Adapted with permission from Truman A. Hartshorn, Interpreting the City. *Copyright © 1980 John Wiley & Sons, Inc.*

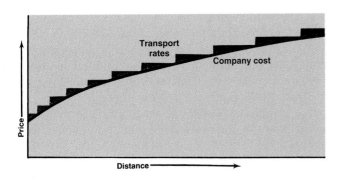

Figure 9.8 **The tapering principle.** The actual costs of transport, including terminal charges and line costs, increase at a decreasing rate as fixed costs are spread over longer hauls. The "tapering" of company cost is differently expressed among media because their mixes of fixed and variable costs are different, as Figure 9.5 diagrams. Note that actual rates charged move in stepwise increments to match the general pattern and level of company costs.

and the like. Nevertheless, the logical systems and concepts they developed and the spatial conclusions they reached still are relevant in understanding present-day industrial locational decisions. Although a full review of all of their contributions is beyond our

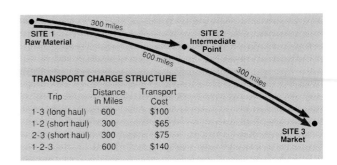

Figure 9.9 **The short-haul penalty.** Plant locations intermediate between material and market are generally avoided because of the realities of transportation pricing that are shown here. Two short hauls simply cost more than a single long haul because two sets of fixed costs must be assigned to the interrupted movement.

scope and interest, it is useful to survey briefly the three fundamental approaches to the problem of plant location those theorists proposed—*least-cost theory, locational interdependence theory,* and *profit-maximization approaches*—and the different conclusions they reach.

Least-Cost Theory

The classical model of industrial location theory, the **least-cost theory,** is based on the work of Alfred Weber (1868–1958) and sometimes called **Weberian analysis.** It explains the optimum location of a manufacturing establishment in terms of minimization of three basic expenses: relative transport costs, labor costs, and agglomeration costs. **Agglomeration** refers to the clustering of productive activities and people for mutual advantage. Such clustering can produce "agglomeration economies" through shared facilities and services. Diseconomies such as higher rents or wage levels resulting from competition for these resources may also occur.

Weber concluded that transport costs are the major consideration determining location. That is, the optimum location will be found where the costs of transporting raw materials to the factory and finished goods to the market are at their lowest. He noted, however, if variations in labor or agglomeration costs are sufficiently great, a location determined solely on the basis of transportation costs may not in fact be the optimum one.

Weber made five controlling assumptions: (1) An area is completely uniform physically, politically, culturally, and technologically. This is known as the **uniform,** or **isotropic, plain** assumption. (2) Manufacturing involves a single product to be shipped to a single market whose location is known. (3) Inputs involve raw materials from more than one known source location. (4) Labor is infinitely available but immobile in location. (5) Transportation routes are not fixed but connect origin and destination by the shortest path; and transport costs directly reflect the weight of items shipped and the distance they are moved.

Given these assumptions, Weber derived the least transport cost location by means of the *locational triangle* (Figure 9.10). It diagrams the cost consequences of fixed locations of materials and market and of movement in any direction of a given weight

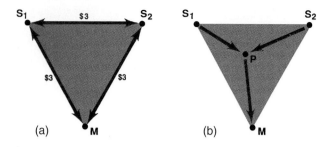

(a)   (b)

Figure 9.10 **Weber's locational triangle** with differing assumptions. (*a*) With one market, two raw material sources, and a finished product reflecting a 50% material weight loss, production could appropriately be located at S_1, S_2, or M since each length of haul is the same. In (*b*) the optimum production point, P, is seen to lie within the triangle, where total transport costs would be less than at corner locations. The exact location of P would depend on the weight-loss characteristics of the two material inputs if only transport charges were involved. P would, of course, be pulled toward the material whose weight is most reduced.

of commodity at a uniform cost per unit of distance. In Figure 9.10a, S_1 and S_2 are the two material sources for a product consumed at M. The problem is to locate the *optimum point of production* where the total ton-distance involved in assembling materials and distributing the product is at a minimum. Each corner of the triangle exerts its pull; each has a defined cost of production should it be chosen as the plant site. If we assume that the material weights are cut in half during manufacturing (so that the finished product weighs the same as each of the original raw materials), then location at either S_1 or S_2 on the diagram would involve a $3 shipping charge from the other raw material source plus $3 to move the product, for a total delivered cost at market of $6. If the market were selected as the plant site, two raw material shipments—again totaling $6—would be involved.

Weberian analysis, however, aims at the least transport cost location, which most likely will be an intermediate point somewhere within the locational triangle. Its exact position will depend on distances, the respective weights of the raw material inputs, and the final weight of the finished product, and may be either material or market oriented (Figure 9.10b). Material orientation reflects a sizable weight loss during the production process; market orientation indicates a weight gain. The optimum placement of P can be found by different analytical means, but the easiest to visualize is by way of a mechanical model of weights and strings (Figure 9.11).

Locational Interdependence Theory

When the locational decision of one firm is influenced by locations chosen by its competitors, a condition of **locational interdependence** exists. It influences the manner in which competitive firms with identical cost structures arrange themselves in space to assure themselves a measure of *spatial monopoly* in their combined market. In locational interdependence theory, the concern is with *variable revenue analysis* rather than, as in the Weber model, with variable costs.

The simplest case concerns the locational decisions of two firms in competition with each other to supply identical goods to

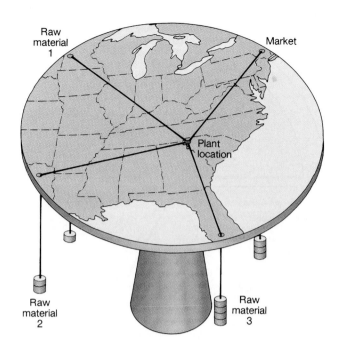

Figure 9.11 **Plane table solution to a plant location problem.** This mechanical model, suggested by Alfred Weber, uses weights to demonstrate the least transport cost point where there are several sources of raw materials. When a weight is allowed to represent the "pull" of raw material and market locations, an equilibrium point is found on the plane table. That point is the location at which all forces balance each other and represents the least-cost plant location.

customers evenly spaced along a linear market. The usual example cited is of two ice cream vendors, each selling the same brand at the same price along a stretch of beach having a uniform distribution of people. All will purchase the same amount of ice cream (that is, demand is *inelastic*—is not sensitive to a change in the price) and will patronize the seller nearer to them. Figure 9.12 suggests that the two sellers would eventually cluster at the midpoint of the linear market (the beach) so that each vendor could supply customers at the extremities of the market without yielding locational advantage to the single competitor.

This is a spatial solution that maximizes return but does not minimize costs. The lowest total cost location for each of the two vendors would be at the midpoint of his or her half of the beach, as shown at the top of Figure 9.12, where the total effort expended by customers walking to the ice cream stands (or cost by sellers delivering the product) is least. To maximize market share, however, one seller might decide to relocate immediately next to the competitor (Figure 9.12b), dominating now three-fourths of the entire beach market. The logical retaliation would be for the second vendor to jump back over the first to recapture market share. Ultimately, side-by-side location at the center line of the beach is inevitable and a stable placement is achieved since neither seller can gain any further advantage from moving. But now the customers collectively have to walk farther to satisfy their ice cream hunger than they did initially; that is, total acquisition cost or delivered price (ice cream purchase plus effort expended) has increased.

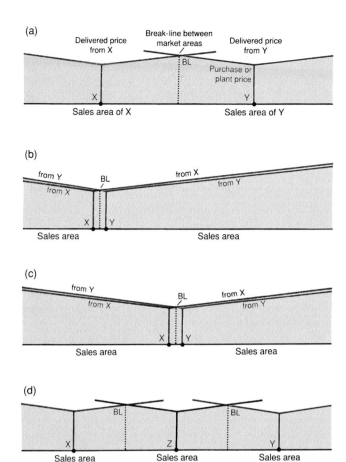

<figure>*Figure 9.12* **Competitive locations in a linear market** (Hotelling model). The initial *socially optimal* locations (*a*) that minimize total distribution costs will be vacated in the search for market advantage (*b*), eventually resulting in *competitive equilibrium* at the center of the market (*c*). Spatial dispersion will again occur if two or more competitors either encounter elasticity of demand or subdivide the market by agreement (*d*).</figure>

The economist Harold Hotelling (1895–1973), who is usually associated with the locational interdependence approach, expanded the conclusion about clustered ice cream sellers to a more generalized statement explaining industrial concentration by multiple producers under conditions of identical production costs and inelastic market demand. However, if the market becomes sensitive to price, sales to more distant customers will be discouraged and producers seeking to maximize sales will again separate rather than aggregate. The conclusion then is that price sensitivity (elasticity of demand) will encourage industrial dispersion.

Profit-Maximization Approaches

For many theorists, the simplicities and rigidities of the least-cost and the locational interdependence explanations are unrealistically restrictive. Ultimately, they maintain, the correct location of a production facility is where the net profit is greatest. They propose employing a **substitution principle** that recognizes that in many industrial processes it is possible to replace a declining amount of one input (e.g., labor) with an increase in another (e.g., capital for automated equipment) or to increase transportation

costs while simultaneously reducing land rent. With substitution, a number of different points may be appropriate manufacturing locations. Further, they suggest, a whole series of points may exist where total revenue of an enterprise just equals its total cost of producing a given output. These points, connected, mark the **spatial margin of profitability** and define the larger area within which profitable operation is possible (Figure 9.13). Location anywhere within the margin assures some profit and tolerates both imperfect knowledge and personal (rather than economic) considerations. Such less-than-optimal, but still acceptable, sites are considered **satisficing locations.**

For some firms, spatial margins may be very broad because transport costs are a negligible factor in production and marketing. Such firms are said to be **footloose**—that is, neither resource nor market oriented. For example, both the raw materials and the finished product in the manufacture of computers are so valuable, light, and compact that transportation costs have little bearing on where production takes place.

Other Locational Considerations and Controls

The behavior of individual firms seeking specific production sites under competitive commercial conditions forms the basis of most classical industrial location theory. But such theory no longer fully explains world or regional patterns of industrial localization or specialization. Moreover, it does not account for locational behavior that is uncontrolled by objective "factors," influenced by new production technologies and corporate structures, or directed by noncapitalistic planning goals.

Traditional theories (including many variants not reviewed here) sought to explain location decisions for plants engaged in

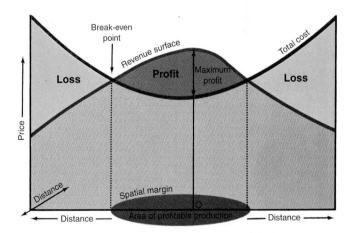

<figure>*Figure 9.13* **The spatial margin of profitability.** In the diagram, *O* is the single optimal profit-maximizing location, but location anywhere within the area defined by the intersects of the total cost and total revenue surfaces will permit profitable operation. Some industries will have wide margins; others will be more spatially constricted. Skilled entrepreneurs may be able to expand the margins farther than less able industrialists. Importantly, a *satisficing* location may be selected by reasonable estimate even in the absence of the totality of information required for an *optimal* decision.</figure>

mass production for mass markets where transportation lines were fixed and transport costs relatively high. Both conditions began to change significantly during the last years of the 20th century. Assembly line production of identical commodities by a rigidly controlled and specialized labor force for generalized mass markets—known as "**Fordism**" to recognize Henry Ford's pioneering development of the system—became less realistic in both market and technology terms. In its place, post-Fordist *flexible manufacturing* processes based on smaller production runs of a greater variety of goods aimed at smaller, niche markets than were catered to by traditional manufacturing have become common. At the same time, information technology applied to machines and operations, increasing flexibility of labor, and declining costs for transportation services that were increasingly viewed from a cost-time rather than a cost-distance standpoint have materially altered underlying assumptions of the classical theories.

Agglomeration Economies

Geographical concentration of economic, including industrial, activities is the norm at the local or regional scale. The cumulative and reinforcing attractions of industrial concentration and urban growth are recognized locational factors, but ones not easily quantified. Both cost-minimizing and profit-maximizing theories, as we have seen, make provision for *agglomeration,* the spatial concentration of people and activities for mutual benefit. That is, both recognize that areal grouping of industrial activities may produce benefits for individual firms that they could not experience in isolation. Those benefits—**agglomeration economies,** or **external economies**—accrue in the form of savings from shared transport facilities, social services, public utilities, communication facilities, and the like. Collectively, these and other installations and services needed to facilitate industrial and other forms of economic development are called **infrastructure.**

Areal concentration may also create pools of skilled and ordinary labor, of capital, ancillary business services, and, of course, a market built of other industries and urban populations. New firms, particularly, may find significant advantages in locating near other firms engaged in the same activity, for labor specializations and support services specific to that activity are already in place. Some may find profit in being near other firms with which they are linked either as customers or suppliers.

A concentration of capital, labor, management skills, customer base, and all that is implied by the term *infrastructure* will tend to attract still more industries from other locations to the agglomeration. In Weber's terms, that is, economies of association distort or alter locational decisions that otherwise would be based solely on transportation and labor costs, and once in existence, agglomerations will tend to grow (Figure 9.14). Through a **multiplier effect,** each new firm added to the agglomeration will lead to the further development of infrastructure and linkages. As we shall see in Chapter 11, the "multiplier effect" also implies total (urban) population growth and thus the expansion of the labor pool and the localized market that are part of agglomeration economies.

Agglomeration—concentration—of like industries in small areas dates from the early industrial age and continues with many of the newest industries. Familiar examples include the town of Dalton, Georgia, in or near which are found all but one of the top 20 United States carpet makers, and Akron, Ohio, which, before 1930, held almost the entire 100 or so tire manufacturers of the country. Silicon Valley dating from the 1960s and other more recent high-tech specialized concentrations simply continue the tradition.

Admittedly, agglomeration can yield disadvantages as well as benefits. Overconcentration can result in diseconomies of congestion, high land values, pollution, increased governmental regulation, and the like. When the costs of aggregation exceed the benefits, a firm will actually profit by relocating to a more isolated position, a process called **deglomeration.** It is a process expressed in the suburbanization of industry within metropolitan areas or the relocation of firms to nonmetropolitan locations.

Just-in-Time and Flexible Production

Agglomeration economies and tendencies are also encouraged by newer manufacturing policies practiced by both older, established industries and by newer post-Fordist plants.

Traditional Fordist industries required the on-site storage of large lots of materials and supplies ordered and delivered well in advance of their actual need in production. That practice permitted cost savings through infrequent ordering and reduced transportation charges and made allowances for delayed deliveries and for inspection of received goods and components. The assurance of supplies on hand for long production runs of standardized outputs was achieved at high inventory and storage costs.

Just-in-time (JIT) manufacturing, in contrast, seeks to reduce inventories for the production process by purchasing inputs for arrival just in time to use and producing output just in time to sell. Rather than costly accumulation and storage of supplies, JIT requires frequent ordering of small lots of goods for precisely timed arrival and immediate deployment to the factory floor. Such "lean manufacturing" based on frequent purchasing of immediately needed goods demands rapid delivery by suppliers and encourages them to locate near the buyer. Recent manufacturing innovations thus reinforce and augment the spatial agglomeration tendencies evident in the older industrial landscape and deemphasize the applicability of older single-plant location theories.

JIT is one expression of a transition from mass-production Fordism to more *flexible production systems*. That flexibility is designed to allow producers to shift quickly and easily between different levels of output and, importantly, to move from one factory process or product to another as market demand dictates. Flexibility of that type is made possible by new technologies of easily reprogrammed computerized machine tools and by computer-aided design and computer-aided manufacturing systems. These technologies permit small-batch, just-in-time production and distribution responsive to current market demand as monitored by computer-based information systems.

Flexible production to a large extent requires significant acquisition of components and services from outside suppliers rather than from in-house production. For example, modular assembly, where many subsystems of a complex final product enter the plant already assembled, reduces factory space and worker requirements. The premium that flexibility places on proximity to

Figure 9.14 On a small scale, the planned industrial park furnishes its tenants external agglomeration economies similar to those offered by large urban concentrations to industry in general. An industrial park provides a subdivided tract of land developed according to a comprehensive plan for the use of (frequently) otherwise unconnected firms. Since the park developers, whether private companies or public agencies, supply the basic infrastructure of streets, water, sewage, power, transport facilities, and perhaps private police and fire protection, park tenants are spared the additional cost of providing these services themselves. In some instances, factory buildings are available for rent, still further reducing firm capital outlays. Counterparts of industrial parks for manufacturers are the office parks, research parks, science parks, and the like for "high-tech" firms and for enterprises in tertiary and quaternary services.

component suppliers adds still another dimension to industrial agglomeration tendencies. "Flexible production regions" have, according to some observers, emerged in response to the new flexible production strategies and interfirm dependencies. Those regions, it is claimed, are usually some distance—spatially or socially—from established concentrations of Fordist industrialization.

Comparative Advantage

The principle of **comparative advantage** is of growing international importance in industrial location and specialization decisions. It tells us that areas tend to specialize in the production of those items for which they have the greatest relative advantage over other areas or for which they have the least relative disadvantage, as long as free trade exists. This principle, basic to the understanding of regional specializations, applies as long as areas have different relative advantages for two or more goods.

Assume that two countries both have a need for and are domestically able to produce two commodities. Further assume that there is no transport cost consideration. No matter what its cost of production of either commodity, Country A will choose to specialize in only one of them if by that specialization and through exchange with Country B for the other, Country A stands to gain more than it loses. The key to comparative advantage is the utilization of resources in such a fashion as to gain, by specialization, a volume of production and a selling price that permit exchange for a needed commodity at a cost level that is below that of the domestic production of both.

At first glance, the concept of comparative advantage may at times seem to defy logic. For example, Japan may be able to produce airplanes and home appliances more cheaply than the United States, thereby giving it an apparent advantage in both goods. But it benefits both countries if they specialize in the good in which they have a comparative advantage. In this instance, Japan's manufacturing cost structure makes it more profitable for Japan to specialize in the volume production of appliances and to buy airplanes from the United States, where large civilian and military markets encourage aircraft manufacturing specialization and efficiency.

When other countries' comparative advantages reflect lower labor, land, raw material, and capital costs, manufacturing activities may voluntarily relocate from higher-cost market locations to lower-cost foreign production sites. Such voluntary **outsourcing**—producing parts or products abroad for domestic sale—by American manufacturers has employment and areal economic consequences no different from those resulting from successful competition by foreign companies or from industrial locational decisions favoring one section of the country over others.[1] When comparative advantage is exploited by individual corporations, one expression of flexible production systems is evident in the erosion of the rigid spatial concentration of production assumed by classical location theory.

Outsourcing has also assumed the meaning of subcontracting production work to outside, often nonunion, domestic companies. It therefore becomes an important element in just-in-time acquisition of preassembled components for snap-together fabrication of finished products, often built only to orders actually received from customers. Reducing parts inventories and introducing build-to-order production demands a high level of flexible freight movement increasingly supplied by "logistics" firms that themselves may become involved in packaging, labeling, and even manufacturing products for client companies. When comparative advantage and outsourcing are exploited by individual corporations, one expression of flexible production systems is evident in the erosion of the rigid spatial concentration of manufacturing assumed by classical location theory.

A clear example of that impact is seen in the changing nature of automobile manufacturing. Formerly, motor vehicle companies were largely self-contained production entities controlling raw material inputs through their own steel and glass plants and producing themselves all parts and components required in the assembly of their products. Since late 1992, that self-containment has been abandoned as car companies have divested themselves of raw material production facilities and have in large part sold off their in-house parts production. Increasingly, they are purchasing parts and subassemblies from independent, often distant, suppliers. In fact, some observers of the changing vehicle production scene predict that established automobile companies will eventually convert themselves into "vehicle brand owners," retaining for themselves only such essential tasks as vehicle design, engineering, and marketing. All else, including final product assembly, is projected to be done through outsourcing to parts suppliers. Similar trends are already evident in consumer electronics where a third of manufacturing is (2004) estimated to be outsourced.

A distinctive regional illustration of more diversified industrial deconcentration is found along the northern border of Mexico. In the 1960s, Mexico enacted legislation permitting foreign (specifically, American) companies to establish "sister" plants, called *maquiladoras,* within 20 kilometers (12 miles) of the U.S. border for the duty-free assembly of products destined for reexport. By 2000, more than 3000 such assembly and manufacturing plants had been established to produce a diversity of goods including electronic products, textiles, furniture, leather goods,

toys, and automotive parts. By 2000, the plants generated direct and indirect employment for more than 1 million Mexican workers (Figure 9.15) and for large numbers of U.S. citizens, employees of growing numbers of American-side *maquila* suppliers and of diverse service-oriented businesses spawned by the "multiplier effect." The North American Free Trade Agreement (NAFTA) creating a single Canadian–United States–Mexican production and marketing community turns "outsourcing" in the North American context from a search abroad for low-cost production sites to a review of best locations within a broadened unified economic environment.

The United States also benefits from outsourcing by other countries. Japanese and European companies have established

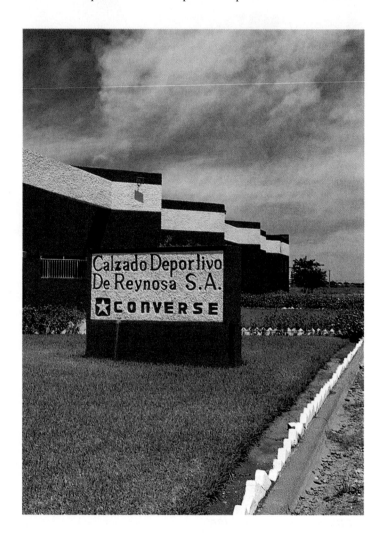

Figure 9.15 American manufacturers, seeking lower labor costs, began in the 1960s to establish light manufacturing, component production, and assembly operations along the international border in Mexico. *Outsourcing* to such plants as this Converse Sport Shoe factory at Reynosa has moved a large proportion of American electronics, small appliance, toy, and garment industries to off-shore subsidiaries or contractors in Asia and Latin America. More than a quarter-million Mexican *maquilador* jobs and 350 plants were lost between 2000 and late 2002, however, largely to competition from lower cost, more efficient Chinese and other Asian producers. Comparative advantage is not a permanent condition.

---

[1]*Outsourcing* has also assumed the meaning of subcontracting production work to outside, particularly nonunion, domestic companies.

automobile and other manufacturing plants in part to take advantage of lower American production and labor costs and in part in response to American protectionist policies and automobile import quotas. A strong yen in relation to the value of the dollar during the 1980s, by raising the relative cost of imported cars, also encouraged Japanese auto plant location in the United States followed by Japanese auto parts manufacturers. At least a portion of the products of both automobile assemblers and parts suppliers were available for reexport to other national markets.

Comparative advantage is not a fixed, unchanging relationship. As we shall see in Chapter 10, technology transfer from economically advanced to underdeveloped economies is transforming the world economy by introducing a new *international division of labor*. In the 19th and first half of the 20th centuries, that division invariably involved exports of manufactured goods from the "industrial" countries and of raw materials from the "colonial" or "undeveloped" economies. Roles have now altered. Manufacturing no longer is the mainstay of the economy and employment structure of Europe or Anglo America and, as the NAFTA example showed, the world pattern of industrial production is shifting to reflect the growing dominance of countries formerly regarded as subsistence peasant societies but now emerging as the source areas for manufactured goods of all types.

Imposed Considerations

Locational theories dictate that in a pure, competitive economy, the costs of material, transportation, labor, and plant should be controlling in locational decisions. Obviously, neither in the United States nor in any other market economy do the idealized conditions exist. Other constraints—some representing cost considerations, others political or social impositions—also affect, perhaps decisively, the locational decision process. Land use and zoning controls, environmental quality standards and regulations, governmental area-development inducements, local tax abatement provisions or developmental bond authorizations, noneconomic pressures on quasi-governmental corporations, and other considerations constitute attractions or repulsions for industry outside of the context and consideration of pure theory (see "Contests and Bribery"). If these noneconomic forces become compelling, the assumptions of the commercial economy classification no longer apply, and locational controls reminiscent of those enforced by centrally planned economies become determining.

No other imposed considerations were as pervasive as those governing industrial location in planned economies. The theoretical controls on plant location decisions that apply in commercial economies were not, by definition, determinant in the centrally planned Marxist economies of Eastern Europe and the former Soviet Union. In those economies, plant locational decisions were made by government agencies rather than by individual firms.

Bureaucratic rather than company decision making did not mean that location assessments based on factor cost were ignored; it meant that central planners were more concerned with other than purely economic considerations in the creation of new industrial plants and concentrations. Important in the former Soviet Union, for example, was a controlling policy of the *rationalization* of industry through full development of the resources of the country wherever they were found and without regard to the cost or competitiveness of such development. Inevitably, although the factors of industrial production are identical in capitalist and noncapitalist economies, the philosophies and patterns of industrial location and areal development will differ between them. Since major capital investments are relatively permanent additions to the landscape, the results of their often noneconomic political or philosophical decisions are fixed and will long remain to influence industrial regionalism and competitive efficiencies into the post-communist present and future. Those same decisions and rigidities continue to inhibit the transition by the formerly fully planned economies to modern capitalist industrial techniques and flexibilities.

Transnational Corporations

Outsourcing is but one small expression of the growing international structure of modern manufacturing and service enterprises. Business and industry are increasingly stateless and economies borderless as giant **transnational corporations (TNCs)**—private firms that have established branch operations in nations foreign to their headquarters' country—become ever-more important in the globalizing world space economy. Early in the 21st century, there were some 65,000 transnational (or *multinational*) companies controlling at least 850,000 foreign affiliates employing about 54 million workers. Excluding the parent companies themselves, the TNC affiliates accounted for almost $19 trillion in sales, one-tenth of world GDP and one-third of world exports.

Measured by value added (not total sales), 29 of the world's top 100 economic entities in 2002 were corporations, not countries. The great majority of them, all TNCs, are engaged in secondary industries. That is, except for a few resource-based firms, they are principally involved in producing and selling manufactured goods. Although as we shall see, tertiary and quaternary activities have also become international in scope and transnational in corporate structure, the locational and operational advantages of multicountry operations were first discerned and exploited by manufacturers.

TNCs are increasingly international in origin and administrative home, based in a growing number of economically advanced countries. At the start of the 21st century, about 90% of the world's 100 largest TNCs had home offices in the European Union, United States, and Japan; only two developing countries' firms were on the list. Because of their outsourced purchases of raw materials, parts and components, and services, the total number of worldwide jobs associated with TNCs in 2001 reached 150 million or more.

Their foreign impact, however, is limited to relatively few countries and regions. Less than 20% (2000) of inflows of foreign direct investment (FDI) went to developing countries and the majority of that was concentrated in 10 to 15 states, mainly in South, Southeast, and East Asia (China is the largest developing country recipient) and in Latin America and the Caribbean. The 49 least developed countries as a group—including nearly all African states—received only 0.3% of world foreign direct investment in 2000. Despite poor countries' hopes for foreign investment to spur their economic growth, the vast majority of FDI flows not to the poor or developing worlds but to the rich.

Geography and Public Policy

Contests and Bribery

In 1985, it cost Kentucky over $140 million in incentives—some $47,000 a job—to induce Toyota to locate an automobile assembly plant in Georgetown, Kentucky. That was cheap. By 1993, Alabama spent $169,000 per job to lure Mercedes-Benz to that state, Kentucky bid $350,000 per job in tax credits to bring a Canadian steel mill there, Mississippi agreed to $400 million in spending and tax rebates to Nissan in 2001, and in 2002, Georgia gave DaimlerChrysler $320 million in incentives in successful competition with South Carolina to secure the company's proposed new factory.

The spirited auction for jobs is not confined to manufacturing. A University of Minnesota economist calculates that his state will have spent $500,000 for each of the 1500 or more permanent jobs created by Northwest Airlines at two new maintenance facilities. Illinois gave $240 million in incentives ($44,000 per job) to keep 5400 Sears, Roebuck employees within the state, and New York City awarded $184 million to the New York Mercantile Exchange and more than $30 million each to financial firms Morgan Stanley and Kidder, Peabody to induce them to stay in the city. For some, the bidding between states and locales to attract new employers and employment gets too fierce. Kentucky withdrew from competition for a United Airlines maintenance facility, letting Indianapolis have it when Indiana's offered package exceeded $450 million.

Inducements to lure companies are not just in cash and loans—though both figure in some offers. For manufacturers, incentives may include workforce training, property tax abatement, subsidized costs of land and building or their outright gifts, below-market financing of bonds, and the like. Similar offers are regularly made by states, counties, and cities to wholesalers, retailers, major office worker and other service activity employers. The objective, of course, is not just to secure the new jobs represented by the attracted firm but to benefit from the general economic stimulus and employment growth that those jobs—and their companies—generate. Auto parts manufacturers are presumably attracted to new assembly plant locations; cities grow and service industries of all kinds—doctors, department stores, restaurants, food stores—prosper from the investments made to attract new basic employment.

Not everyone is convinced that those investments are wise, however. A poll of Minnesotans showed a majority opposed the generous offer made by the state to Northwest Airlines. In the late 1980s, the governor of Indiana, a candidate for Kentucky's governorship, and the mayor of Flat Rock, Michigan, were all defeated by challengers who charged that too much had been spent in luring the Suburu-Isuzu, Toyota, and Mazda plants, respectively. Established businesses resent what often seems neglect of their interests in favor of spending their tax money on favors to newcomers. The Council for Urban Economic Development, surveying the escalating bidding wars, has actively lobbied against incentives, and many academic observers note that industrial attraction amounts to a zero-sum game: unless the attracted newcomer is a foreign firm, whatever one state achieves in attracting an expanding U.S. company comes at the expense of another state.

Some doubt that inducements matter much, anyway. Although, sensibly, companies seeking new locations will shop around and solicit the lowest-cost, best deal possible, their site choices are apt to be determined by more realistic business considerations: access to labor, suppliers, and markets; transportation and utility costs; weather; the nature of the work force; and overall costs of living. Only when two or more similarly attractive locations have essentially equal cost structures might such special inducements as tax reductions or abatements be determinants in a locational decision.

Questions to Consider

1. As citizen and taxpayer, do you think it is appropriate to spend public money to attract new employment to your state or community?
2. If not, why not? If yes, what kinds of inducements and what total amount offered per job seem appropriate to you? What reasons support your opinion?
3. If you believe that "best locations" for the economy as a whole are those determined by pure location theory, what arguments would you propose to discourage locales and states from making financial offers designed to circumvent decisions clearly justified on abstract theoretical grounds?

The advanced-country destination of those capital flows is understandable: TNCs are actively engaged in merging with or purchasing competitive established firms in already-developed foreign market areas, and cross-border mergers and acquisitions have been the main stimuli behind FDI. Between 1980 and 2000, some 225,000 mergers or purchases were announced worldwide; merger activity continued but at a markedly slower pace during the worldwide economic slowdown beginning in 2001. Because most transnational corporations operate in only a few industries—computers, electronics, petroleum and mining, motor vehicles, chemicals, and pharmaceuticals—the worldwide impact of their consolidations is significant. Some dominate the marketing and distribution of basic and specialized commodities. In raw materials, a few TNCs account for 85% or more of world trade in wheat, maize, coffee, cotton, iron ore, and timber, for example. In manufactures, the highly concentrated world pharmaceutical industry is dominated by just six firms, and the world's 15 major automobile producers at the start of the century, it has been predicted, will fall to 5 or 10 by 2010.

Because they are international in operation with multiple markets, plants, and raw material sources, TNCs actively exploit the principle of comparative advantage. In manufacturing they have internationalized the plant-siting decision process and multiplied the number of locationally separated operations that must be assessed. TNCs produce in that country or region where costs of materials, labor, or other production inputs are minimized, while maintaining operational control and declaring taxes in localities where the economic climate is most favorable. Research and development, accounting, and other corporate activities are placed wherever economical and convenient.

TNCs have become global entities because global communications make it possible (Figure 9.16). Often, they have lost their original national identities and are no longer closely associated with or controlled by the cultures, societies, and legal systems of a nominal home country. At the same time, their multiplication of economic activities has reduced any earlier identifications with single products or processes and given rise to "transnational integral conglomerates" that span a large spectrum of both service and industrial sectors.

World Manufacturing Patterns and Trends

Whether locational decisions are made by private entrepreneurs or central planners—and on whatever considerations those decisions are based—the results over many years have produced a distinctive world pattern of manufacturing. Figure 9.17 suggests the striking prominence of a relatively small number of major industrial concentrations localized within relatively few countries primarily but not exclusively parts of the "industrialized" or "developed" world. These may be roughly grouped into four commonly recognized major manufacturing regions: *Eastern Anglo America, Western and Central Europe, Eastern Europe,* and *Eastern Asia.* Together, the industrial plants within these established regional clusters account for an estimated three-fifths of the world's manufacturing output by volume and value.

Their continuing dominance is by no means assured. The first three—those of Anglo America and Europe—were the beneficiaries of an earlier phase in the development and spread of manufacturing following the Industrial Revolution of the 18th century and lasting until after World War II. The countries within them now are increasingly "postindustrial" and traditional manufacturing and processing are of declining relative importance.

The fourth—the East Asian district—is part of the wider, newer pattern of world industrialization that has emerged in recent years, the result of massive international *cultural convergence* (p. 51) and technology transfers in the latter half of the 20th century. The older rigid economic split between the developed and developing worlds has rapidly weakened as the full range of industrial activities from primary metal processing (e.g., the iron and steel industries) through advanced electronic assembly has been dispersed from, or separately established within, an ever-expanding list of countries.

Such states as Mexico, Brazil, China, and others of the developing world have created industrial regions of international significance, and the contribution to world manufacturing activity of the smaller newly industrializing countries (NICs) has been growing significantly. The spreading use of efficient and secure containerized shipment of high-value goods to Anglo American and European markets has been a major contributor to their competitive success. Even economies that until recently were overwhelmingly subsistence or dominated by agricultural or mineral exports have become important players in the changing world manufacturing scene. Foreign branch plant investment in low-wage Asian, African, and Latin American states has not only created there an industrial infrastructure but has as well increased their gross national incomes and per capita incomes sufficiently to permit expanded production for growing domestic—not just export—markets.

Nevertheless, those countries separately and collectively figure less prominently in world manufacturing volumes and values than do the Anglo American and European industrial regions that still remain major components on the world economic landscape and are now matched by the Eastern Asian region of more recent origin. Because of either their traditional or newly emerging world significance, each of those Western and Eastern Hemisphere industrial regions warrants a closer look.

Anglo America

The importance of manufacturing in Anglo America has been steadily declining. In 1960, the 28% of the labor force engaged in manufacturing generated nearly one-third of the region's wealth. In the early 21st century, manufacturing employment had dropped to about 16% of a much larger labor force, and manufacturing contributed less than a fifth of the gross domestic product of the Anglo American realm.

Manufacturing is found particularly in the urbanized sections of North America, but is not uniformly distributed. Its primary concentration is in the northeastern part of the United States and adjacent sections of southeastern Canada, the *Anglo American Manufacturing Belt* (Figure 9.18). That district contains the majority of the urban population of the two countries, their densest and best-developed transportation network, the largest number of their manufacturing establishments, and the preponderance of heavy industry.

Anglo American manufacturing began early in the 19th century in southern New England, where waterpowered textile mills, iron plants, and other small-scale industries began to free Canada and the new United States from total dependence on European—particularly English—sources. The eastern portion of the manufacturing belt contained early population centers, a growing canal and railroad network, a steady influx of immigrant skilled and unskilled labor, and concentrations of investment capital to invest in new manufacturing enterprises. The U.S. eastern seaboard remains an important producer of consumer goods, light industrial, and high-technology products on the basis of its market and developed labor skills. Its core is *Megalopolis,*[2] a 1000-kilometer- (600-mile-) long city system stretching from southern Maine to Norfolk, Virginia, with a great array of market-oriented industries and thousands of individual industrial plants.

The heart of the Anglo American manufacturing belt developed across the Appalachians in the interior of the continent. The Ohio River system and the Great Lakes provided the early—and

---

[2]*Megalopolis* or *conurbation* is an extended urbanized area formed by the gradual merger of several individual cities.

Nokia—Finland
(Pakistan)

Nestlé—Switzerland (Egypt)

IBM—United States (Vietnam)

Sony—Japan (Vietnam)

British Petroleum—U.K. (China)

Ford—United States (China)

Figure 9.16 The world's transnational corporations increased in number from some 7,000 in 1970 to nearly 65,000 in 2003. Ninety of the top 100 TNCs are headquartered in the *Triad*—the European Union, the United States, and Japan. Their recognition and impact, however, are global, as suggested by these billboards advertising just a sample of leading TNCs in distant settings. Corporate names and headquarters countries are followed by billboard locations in parentheses.

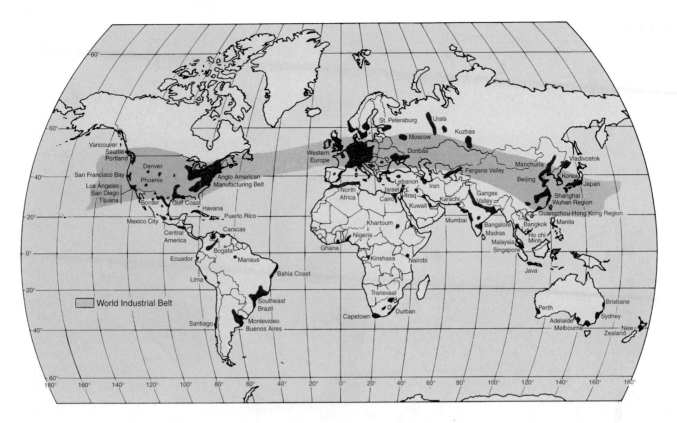

Figure 9.17 **World industrial regions.** Industrial districts are not as continuous or "solid" as the map suggests. Manufacturing is a relatively minor user of land even in areas of greatest concentration. There is a loose spatial association of major industrial districts in an "industrial belt" extending from Western Europe eastward to the Ural Mountains and, through outliers in Siberia, to the Far East. The belt picks up again on the west coast of North America, though its major Anglo American concentration lies east of the Mississippi River. The former overwhelming production dominance of that belt is being steadily and increasingly eroded by the expanding industrialization of countries throughout the developing world.

still important—"highways" of the interior (Figure 9.19), supplemented later by canals and, after the 1850s, by the railroads that tied together the agricultural and industrial raw materials, the growing cities, and multiplying manufacturing plants of the interior with markets and materials throughout the country. The early heavy metallurgical emphasis—the U.S. Steel plants of the Monongahela Valley are an example—has declined and been succeeded early in the 21st century by advanced material processing and fabrication plus high-tech manufacturing, creating a renewed and modernized diversified industrial base.

The Canadian portion of the Anglo American manufacturing belt lies close to neighboring U.S. industrial districts. About one-half of Canada's manufacturing labor force is localized in southern Ontario. With Toronto as the hub, the industrial belt extends westward to Windsor, across from Detroit. Another third of Canadian manufacturing employment is found in Quebec, with Montreal as the obvious core but with energy-intensive industries—particularly aluminum plants and paper mills—along the St. Lawrence River.

By the 1990s, manufacturing employment and volume was declining everywhere in the Anglo American economy. What remained showed a pattern of relocation to Western and Southern zones reflecting national population shifts and changing material and product orientations.

In the *Southeast,* textiles, tobacco, food processing, wood products, furniture, and a Birmingham-based iron and steel industry became important users of local resources. In the *Gulf Coast–Texas district,* petroleum and natural gas provide wealth, energy, and raw materials for a vast petrochemical industry; sulfur and salt support other branches of chemical production. Farther west, *Denver* and *Salt Lake City* have become major, though isolated, industrial centers with important "high-tech" orientations. On the West Coast, three distinctive industrial subregions have emerged. In the *Northwest,* from Vancouver to Portland, orientation to both a regional and a broader Asian-Pacific market is of greater significance than are the primary domestic markets of Canada and the United States. Seattle's aircraft production and the software industry of the Northwest are, by their high-value products, largely unaffected by transport costs to world markets. The *San Francisco Bay district* is home to Silicon Valley and the electronics/computer/high-tech manufacturing that name implies. Food specializations there (wine, for example) for a national market have their counterpart farther south in the *Los Angeles–San Diego corridor,* where fruits and vegetables are grown and packed. More important, however, is diversified, particularly consumer-goods, production for the rapidly growing California and western market.

Livelihood and Economy: From Blue Collar to Gold Collar   **335**

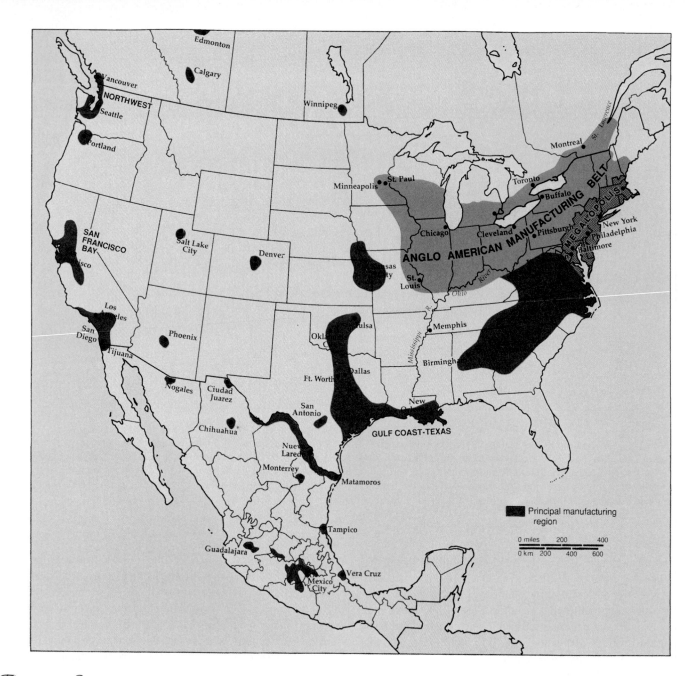

Figure 9.18 **North American manufacturing districts.** Although the preponderance of North American industry is still concentrated in Anglo America, Mexican manufacturing activity is rapidly growing and diversifying—for both expanding domestic and export markets. While Mexico City alone yields nearly half of the country's manufacturing output volume, industrial plants are also localized in the Central Plateau area and along the northern border with the United States, where most *maquiladoras* have been established.

North America's fastest growing industrial region lies along the U.S.-Mexican border. Called *la frontera* by its Mexican workers and extending 2100 miles from the Pacific Ocean to the Gulf of Mexico, this subregion served us earlier as an example of "outsourcing" and comparative advantage (p. 330).

Western and Central Europe

The Industrial Revolution that began in England in the late 1700s and spread to the continent during the 19th century established Western and Central Europe as the world's premier manufacturing region and the source area for the diffusion of industrialization across the globe. By 1900, Europe accounted for 80% of the world's industrial output though, of course, its relative position has since eroded, particularly after World War II. Although industry is part of the economic structure of every section and every metropolitan complex of Europe, the majority of manufacturing output is concentrated in a set of distinctive districts stretching from the Midlands of England in the west to the Ural Mountains in the east (Figure 9.20).

Figure 9.19 **A barge "tow" passing St. Louis on the Mississippi River.** About 15% of the total ton-miles of freight movement in the United States is by inland water carriers. Crude and refined petroleum accounts for three-fifths of the tonnage. Farm products, chemicals (including fertilizers), and nonmetallic minerals (sand, rock, and gravel) make up much of the rest.

Waterpowered mechanical spinning and weaving in the textile industry of England began the Industrial Revolution, but it was steam power, not waterpower, that provided the impetus for the full industrialization of that country and of Europe. Consequently, coal fields, not rivers, were the sites of the new manufacturing districts in England. London, although remote from coal deposits, became the largest single manufacturing center of the United Kingdom, its consumers and labor force potent magnets for new industry.

Technologies developed in Britain spread to the continent. The coal fields distributed in a band across northern France, Belgium, central Germany, the northern Czech Republic, southern Poland, and eastward to southern Ukraine, as well as iron ore deposits, localize the metallurgical industries to the present day. Other pronounced industrial concentrations focus on the major metropolitan districts and capital cities of the countries of Europe.

The largest and most important single industrial area of Europe extends from the French–Belgian border to western Germany. Its core is Germany's Ruhr, a compact, highly urbanized industrial concentration of more than 50 major cities housing iron and steel, textiles, automobiles, chemicals, and all the metal-forming and metal-using industries of modern economies. In France, heavy industry located near the iron ore of Nancy and the coal of Lille also specialized in textile production. Like London, Paris lacks raw materials, but with easy access to the sea and to the domestic market, it became the major manufacturing center of France. Farther east, the Saxony district began to industrialize as early as the 1600s, in part benefiting from labor skills brought by

immigrant artisans from France and Holland. Those skills have been preserved in a district noted for the quality of its manufactured goods.

Western Europe is experiencing a deindustrialization accompanied by massive layoffs of workers in coal mining because of declining demand and in iron mining because of ore depletion. Iron and steel, textiles, and shipbuilding—the core industries of the Industrial Revolution—have been particularly hard hit. As in the Anglo American Manufacturing Belt, a restructuring of the Western European economy is introducing new industrial and service orientations and employment patterns.

Eastern Europe

Between the end of World War II and 1990, Eastern European industrial concentrations, such as that of Silesia in Poland and the Czech portion of the Bohemian Basin (Figure 9.20), were largely cut off from their earlier connections with the larger European market and economy. Instead, they were controlled by centralized industrial planning and tied to the regional economic plans imposed by the Soviet Union. Since its fall, Eastern European states have struggled with a generally poorly conceived, technologically antiquated, uneconomic industrial structure that, in its creation and operation, was unresponsive to market realities.

Farther east, in Russia and Ukraine, two distinctly different industrial orientations predominate, both dating from Czarist times and strengthened under Soviet-era planning. One emphasis is on light industrial, market-oriented production primarily

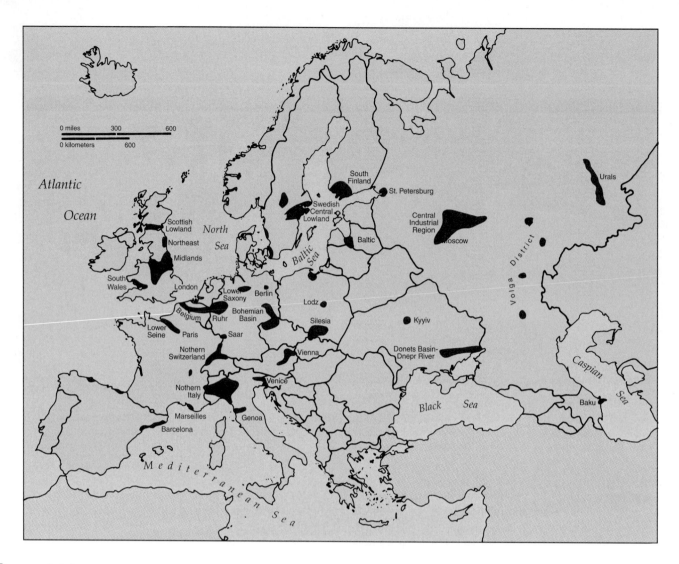

Figure 9.20 **The industrial regions of Europe.**

focused on Russia's Central Industrial Region of Greater Moscow and surrounding areas (Figure 9.21). The other orientation is heavy industrial. Its Czarist beginnings were localized in the southern Ukrainian Donets Basin-Dnepr River district where coking coal, iron ore, fluxing materials, and iron alloys are found near at hand. Under the Stalinist Five-year plans, with their emphasis on creation of multiple sources of supply of essential industrial goods, heavy industry was also developed elsewhere in the Soviet Union. The industrial districts of Russia's Volga, Urals, Kuznetsk Basin, Baikal, and Far East regions, and the industrial complexes of the Caucasus, Kazakhstan, and Central Asia resulted from those Soviet programs first launched in 1928.

Eastern Asia

The Eastern Asian sphere is rapidly becoming the most productive of the world's industrial regions (Figure 9.22). Japan has emerged as the overall second-ranked manufacturing nation. China—building on a rich resource base, massive labor force, and nearly insatiable market demand—is industrializing rapidly

and ranks among the top 10 producers of a number of major industrial commodities. South Korea, Taiwan, Singapore, and (before its inclusion in China's mainland economy in 1997) Hong Kong, were recognized as "the four tigers," swiftly industrializing Asian economies that have become major presences in markets around the world.

Japanese industry was rebuilt from near total destruction during World War II to its present leading position in some areas of electronics and other high-tech production. That recovery was accomplished largely without a domestic raw material base and primarily with the export market in mind. Dependence on imports of materials and exports of product has encouraged a coastal location for most factories. The industrial core of modern Japan is the heavily urbanized belt from Tokyo to northern Kyushu (Figure 9.22).

When the communists assumed control of China's still war-damaged economy in 1949, that country was essentially unindustrialized. Most manufacturing was small-scale production geared to local subsistence needs. A massive industrialization program initiated by the new regime greatly increased the volume, diversity, and dispersion of manufacturing in China. Until 1976,

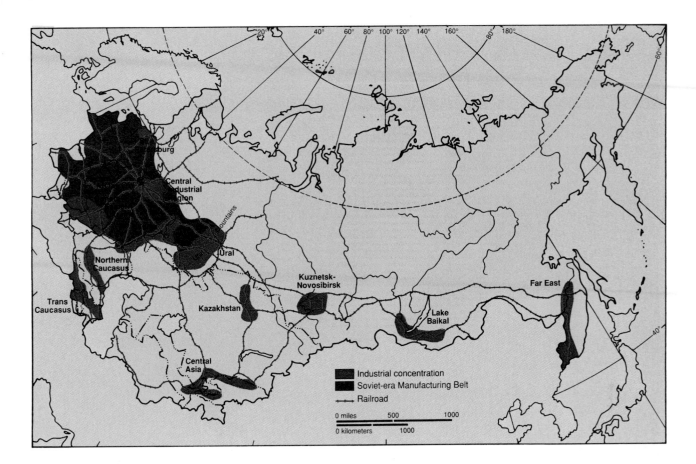

Figure 9.21 **Industrial regions under central planning in the former Soviet Union.** The Volga, the Central Industrial, and the St. Petersburg (Leningrad) concentrations within the former Soviet manufacturing belt were dependent on transportation, labor, and market pulls. All the other planned industrial regions had a strong orientation to materials and were developed despite their distance from the population centers and markets of the west.

domestic needs rather than foreign markets were the principal concern of an industrial development totally controlled by the state and the communist party. From the late 1970s, manufacturing activities were freed from absolute state control and industrial output grew rapidly with most dramatic gains coming not from state enterprises but from quickly multiplying rural collectives.

By the start of the 21st century, however, it was foreign direct investment, foreign firm outsourcing of production to low-cost Chinese suppliers, and relocation to China of manufacturing and assemblage of a host of consumer electronics, clothing, toys, and industrial equipment from other Asian countries that propelled China to the forefront among East Asian and developing countries worldwide as an industrial powerhouse. Its admission in 2001 to the World Trade Organization (see Chapter 12) further enhanced the competitive position of an economy already accounting for 8% of total world exports of manufactured goods in 2000, a share predicted to grow to 14% by 2007.

Unlike Japan, China possesses a relatively rich and diversified domestic raw material base of ores and fuels. The pattern of resource distribution in part accounts for the spatial pattern of industry, though coastal locations, urban agglomerations, and market orientations are equally important (Figure 9.22).

Three smaller East Asian economies—Taiwan, South Korea, and Singapore—have outgrown their former "developing country" status to become advanced industrialized states. Their rise to prominence has been rapid, and their share of market in those branches of industry in which they have chosen to specialize has increased dramatically (Figure 9.23). Although the specifics of their industrial successes have differed, in each case an educated, trainable labor force; economic and social systems encouraging industrial enterprise; and national programs directed at capital accumulation, industrial development, and export orientation fueled the programs.

Their ranks have recently been joined by an expanded list of other industrial "tigers"—nations demonstrating the capacity for rapid, sustained economic growth. At the least, the new Asian tiger group includes Malaysia and Thailand and may soon be joined by the Philippines, Indonesia, and Vietnam. Other Asian manufacturing concentrations are also emerging as important participants in the world's industrial economy. India, for example, benefits from expanding industrial bases centered in metropolitan Bangalore, Mumbai (Bombay), Delhi, Kolkata (Calcutta), and elsewhere, each with its own developing specializations.

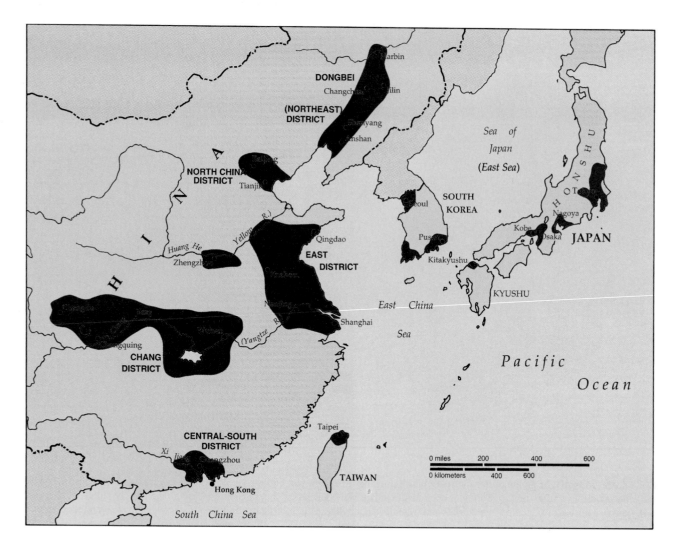

$\mathcal{F}igure\ 9.22$ **The industrial regions of Eastern Asia.**

High-Tech Patterns

Major industrial districts of the world developed over time as entrepreneurs and planners established traditional secondary industries according to the pulls and orientations predicted by classical location theories. Those theories are less applicable in explaining the location of the latest generation of manufacturing activities: the high-technology—or *high-tech*—processing and production that is increasingly part of the advanced economies. For these firms, new and different patterns of locational orientation and advantage have emerged based on other than the traditional regional and site attractions.

High technology is more a concept than a precise definition. It probably is best understood as the application of intensive research and development efforts to the creation and manufacture of new products of an advanced scientific and engineering character. Professional—"white collar"—workers make up a large share of the total workforce. They include research scientists, engineers, and skilled technicians. When these high-skill specialists are added to administrative, supervisory, marketing, and other

professional staffs, they may greatly outnumber actual production workers in a firm's employment structure. In the world of high-tech, that is, the distinction between secondary (manufacturing) and quaternary (knowledge) activities and workers is increasingly blurred.

Although only a few types of industrial activity are generally reckoned as exclusively high-tech—electronics, communication, computers, software, pharmaceuticals and biotechnology, aerospace, and the like—advanced technology is increasingly a part of the structure and processes of all forms of industry. Robotics on the assembly line, computer-aided design and manufacturing, electronic controls of smelting and refining processes, and the constant development of new products of the chemical industries are cases in point.

The impact of high-tech industries on patterns of economic geography is expressed in at least three different ways. First, high-tech activities are becoming major factors in employment growth and manufacturing output in the advanced and newly industrializing economies. In the United States, for example, between 1986 and 2000, the largest five high-tech industry groups ployment shifts, while many of the newly industrializing economies of East and Southeast Asia have registered high-tech employment growth of similar or greater proportions.

Many of the confident generalizations concerning, particularly, the computer and software segments of high-technology industry and employment and the prospects for their continuing expansion were called into question by the abrupt dot-com bubble collapse of the first years of the century. Tens of thousands of software jobs lost, scores of innovative, visionary, or simply hopeful software and hardware ventures bankrupt, billions of dollars of venture capital and stock valuations erased, and lavish offices and plants left vacant marked the end of a period of exuberant growth and expectations that shaped high-tech locational patterns and bolstered national economies throughout the world. Although details of the software and computer hardware distributions, specializations, and employment structures of the end of the 20th century were severely disturbed or destroyed, the new century saw both a continuation of high-technology applications to established general manufacturing activities and a gradual revival of investment and employment in software development. Locational concentrations and specializations may change but high-tech's fundamental alteration of older economic structures and patterns endures.

The products of high-technology activity represent an increasing share of total industrial output of individual countries and of the trade between them. As early as 1995, high-tech manufactures represented 15% of manufacturing output in both the United States and Japan, 14% in the United Kingdom, some 10% in Germany and France; among the emerging Asian countries, 12.5% of China's and 15% of South Korea's total output came from high-tech manufacturing. In all these and other countries, high-tech production increased in volume and percentage by the start of the century. The 2000 U.S. high-tech share increased to near 20%. Global data are incomplete, but Figure 9.24 suggests the great disparity between countries in the importance of high-tech products in their exports of manufactured goods.

A second impact is more clearly spatial. High-tech industries have tended to become regionally concentrated in their countries of development, and within those regions, they frequently form self-sustaining, highly specialized agglomerations. California, for example, has a share of U.S. high-tech employment far in excess of its share of American population. Along with California, the Pacific Northwest (including British Columbia), New England, New Jersey, Texas, and Colorado all have proportions of their workers in high-tech industries above the national average. And within these and other states or regions of high-tech concentration, specific locales have achieved prominence: "Silicon Valley" of Santa Clara County near San Francisco; Irvine and Orange County south of Los Angeles; the "Silicon Forest" near Seattle; North Carolina's Research Triangle; Utah's "Software Valley"; Routes 128 and 495 around Boston; "Silicon Swamp" of the Washington, D.C. area; Ottawa, Canada's "Silicon Valley North"; or the Canadian Technology Triangle west of Toronto are familiar Anglo American examples.

Within such concentrations, specialization is often the rule: medical technologies in Minneapolis and Philadelphia, biotechnology around San Antonio, computers and semiconductors in

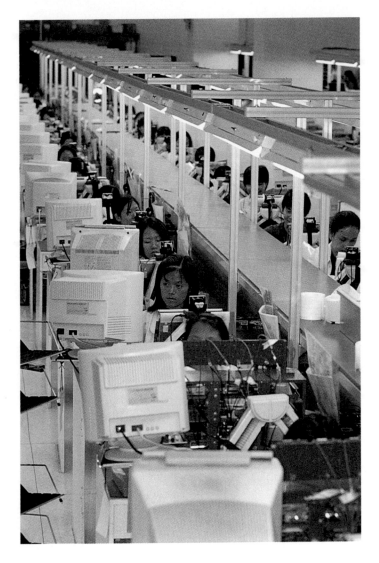

Figure 9.23 Industrialization in Taiwan began in the 1950s with low-skill textiles, plastic toys, simple appliances, and housewares production for export. In the 1960s, televisions, bicycles, refrigerators, radios, electrical goods, and the like represented more advanced manufacturing processes and products. In the 1970s, under governmental planning, factories for intermediate goods and producers goods—petrochemicals, machine tools, heavy machinery, and the like—were developed to support an expanding industrial base. The 1980s and 1990s saw the rise of science-based high-technology industries fully competitive in the world market. By 2000, Taiwan was becoming a contract manufacturer to the world, designing and assembling a growing array of, particularly, consumer electronic, equipment, though an increasing volume of its production was "outsourced," particularly to mainland China.

alone added more than 8 million new jobs to the secondary sector of the economy, helping to replace many thousands of other workers who lost jobs to outsourcing, foreign competition, changing markets, and deindustrialization. In 2000, total U.S. high-technology employment equaled more than 16% of all non-farm workers, while even workers in more strictly defined "high-technology industries" totaled 7% of all employment. The United Kingdom, Germany, Japan, and other advanced countries—though not yet those of Eastern Europe—have had similar em-

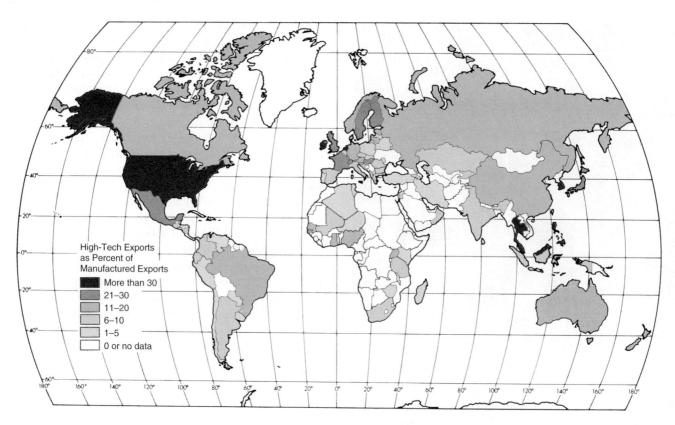

Figure 9.24 This map of **high-tech exports** clearly suggests the importance of the industrialized countries—particularly the United States and Western Europe—in high-tech manufacturing and exports. Less evident is the relative role of high-tech in the manufactured exports of a few smaller, developing states: 60% for Singapore; 54% for Malaysia; and 71% for the Philippines in the late 1990s. By 1999, in fact, high-tech goods made up a larger proportion of the exports of developing countries than of advanced industrial ones, and the disparity continues to grow. The map, of course, does not report a country's ranking in volumes or values of high-tech manufactured goods exports.

Source: The World Bank.

eastern Virginia and at Austin, Texas, biotechnology and telecommunications in New Jersey's Princeton Corridor, telecommunications and Internet industries near Washington, D.C. Elsewhere, Scotland's Silicon Glen, England's Sunrise Strip and Silicon Fen, Wireless Valley in Stockholm, Zhong Guancum in suburban Beijing, and Bangalore, India, are other examples of industrial landscapes characterized by low, modern, dispersed office-plant-laboratory buildings rather than by massive factories, mills, or assembly structures, freight facilities, and storage areas. Planned business parks catering to the needs of smaller companies are increasingly a part of regional and local economic planning. Irvine, California's Spectrum, for example, housed 44,000 employees and 2200 companies, most of them high-tech start-ups.

The older distributional patterns of high-tech industries suggest they respond to different localizing forces than those controlling traditional manufacturing industries. At least five locational tendencies have been recognized: (1) Proximity to major universities or research facilities and to a large pool of scientific and technical labor skills; (2) avoidance of areas with strong labor unionization where contract rigidities might slow process innovation and work force flexibility; (3) locally available venture capital and entrepreneurial daring; (4) location in regions and major metropolitan areas with favorable "quality of life" reputations—

climate, scenery, recreation, good universities, and an employment base sufficiently large to supply needed workers and provide job opportunities for professionally trained spouses; (5) availability of first-quality communication and transportation facilities to unite separated stages of research, development, and manufacturing and to connect the firm with suppliers, markets, finances, and the government agencies so important in supporting research. Essentially all of the major high-tech agglomerations have developed on the semirural peripheries of metropolitan areas but far from inner-city problems and disadvantages. Many have emerged as self-sufficient areas of subdivisions, shopping centers, schools, and parks in close proximity to company locations and business parks that form their core. While the New York metropolitan area is a major high-tech concentration, most of the technology jobs are suburban, not in Manhattan; the periphery's share of computer-related employment in the region amounted to 80% at the end of the 20th century.

Agglomerating forces are also important in this new industrial locational model. The formation of new firms is frequent and rapid in industries where discoveries are constant and innovation is continuous. Since many are "spin-off" firms founded by employees leaving established local companies, areas of existing high-tech concentration tend to spawn new entrants and to provide necessary labor skills. Agglomeration, therefore, is both a product and a cause of spatial associations.

Not all phases of high-tech production must be concentrated, however. The spatial attractions affecting the professional, scientific, and knowledge-intensive aspects of high tech have little meaning for many of the component manufacturing and assembly operations, which may be highly automated or require little in the way of labor skills. These tasks, in our earlier locational terminology, are "footloose"; they require highly mobile capital and technology investments but may be advantageously performed by young women in low-wage areas at home or—more likely—in countries such as Taiwan, Singapore, Malaysia, or Mexico. Contract manufacturers totally divorced spatially and managerially from the companies whose products they produce accounted for an estimated 15% to 20% of the output of electronics hardware. Most often the same factory produces similar or identical products under a number of different brand names. Through such manufacturing transfers of technology and outsourcing, therefore, high-tech activities are spread to newly industrializing countries—from the center to the periphery, in the developmental terms we will explore in Chapter 10. This globalization through areal transfer and dispersion represents a third impact of high-tech activities on world economic geographic patterns already undergoing significant but variable change in response to the new technologies.

Tertiary and Beyond

Primary activities, you will recall, gather, extract, or grow things. *Secondary* industries, we have seen in this chapter, give form utility to the products of primary industry through manufacturing and processing efforts. A major and growing segment of both domestic and international economic activity, however, involves *services* rather than the production of commodities. These **tertiary activities** consist of business and labor specializations that provide services to the primary and secondary sectors, to the general community, and to the individual. They imply pursuits other than the actual production of tangible commodities.

As we have seen in Chapters 8 and 9, regional and national economies undergo fundamental changes in emphasis in the course of their development. Subsistence societies exclusively dependent on primary industries may progress to secondary stage processing and manufacturing activities. In that progression, the importance of agriculture, for example, as an employer of labor or contributor to national income declines as that of manufacturing expands. Many parts of the formerly underdeveloped world have made or are making that developmental transition, as we shall review in Chapter 10.

The advanced countries that originally dominated the world manufacturing scene, in contrast, saw their former industrial primacy reduced or lost during the last third of the 20th century. Rising energy and labor costs, the growth of transnational corporations, transfer of technology to developing countries, and outsourcing of processing or assembly have all changed the structure and pattern of the world economy. The earlier competitive manufacturing advantages of the developed countries could no longer

be maintained and new economic orientations emphasizing service and information activities became the replacement. Advanced economies that have most completely made that transition are often referred to as "postindustrial."

Perhaps more than any other economy, the United States has reached postindustrial status. Its primary sector component fell from 66% of the labor force in 1850 to 2% in 2000, and the service sector rose from 18% to 80% (Figure 9.25). Of the 22 million new jobs created in the United States between 1990 and 2000, more than half occurred in services. Comparable changes are found in other countries. At the start of the 21st century, between 65% and 80% of jobs in such economies as Japan, Australia, Canada, Israel, and all major Western European countries were also in the services sector; Russia and Eastern Europe averaged rather less.

The significance of tertiary activities to national economies and the contrast between more developed and less developed states are made clear not just by employment but also by the differential contribution of services to the gross domestic products of states. The relative importance of services displayed in Figure 9.26 shows a marked contrast between advanced and subsistence societies. The greater the service share of an economy, the greater is the integration and interdependence of that society. That share has grown over time among most regions, and all national income categories as all economies have shared to some degree in world developmental growth (Table 9.1). Indeed, the expansion of the tertiary sector in modernizing East Asia, South Asia, and the Pacific was three times the world average in the 1990s. In Latin America and the Caribbean, services accounted for more than 60% of total output in 2000.

"Tertiary" and "service," however, are broad and imprecise terms that cover a range of activities from neighborhood barber to World Bank president. The designations are equally applicable both to traditional low-order personal and retail activities and, importantly, to higher-order knowledge-based professional services performed primarily for other businesses, not for individual consumption.

Logically, the composite tertiary category should be subdivided to distinguish between those activities answering to the daily living and support needs of individuals and local communities and

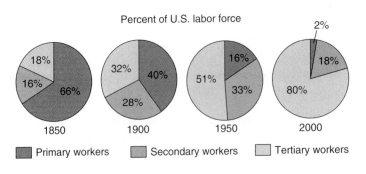

Percent of U.S. labor force

Figure 9.25 The changing sectoral allocation of the U.S. labor force is a measure of the economic development of the country. Its progression from a largely agricultural to postindustrial status is clearly evident.

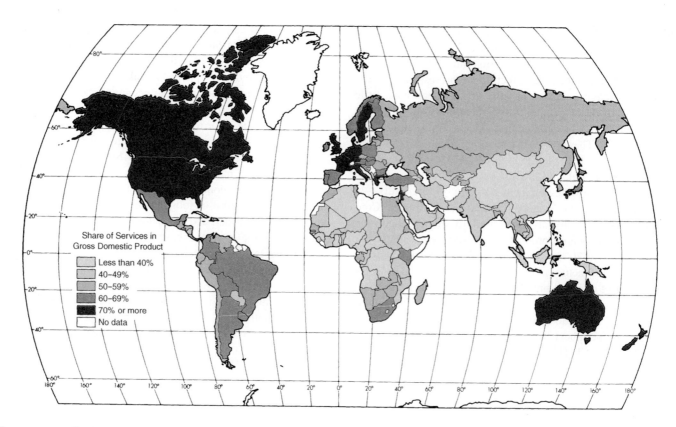

Figure 9.26 Services accounted for nearly two-thirds of global GDP in 2000, up sharply from about 50% 20 years earlier. As the map documents, the contribution of services to individual national economies varied greatly, while Table 9.1 indicates that all national income categories shared to some degree in the expansion of service activities.

Table 9.1

Contribution of the Service Sector to Gross Domestic Product

Country Group	Percentage of GDP		
	1960	1980	2000
Low income	32	30	44
Middle income	47	46	55
High income	54	59	64
United States	58	63	73
World		55	63

Source: World Bank.

those involving professional, administrative, or financial management tasks at regional, national, and international scales. Those differing levels and scope of activity represent different locational principles and quite different roles in their contribution to domestic and world economies.

To recognize such fundamental contrasts, we may usefully restrict the term "tertiary" specifically to those lower-level services largely related to day-to-day needs of people and to the usual range of functions found in smaller towns and cities worldwide. We can then assign higher-level, more specialized information research, and management activities to distinctive "quaternary" and "quinary" categories (see Figure 8.2) with quite different and distinctive characteristics and significance.

Tertiary Services

Some services are concerned with the wholesaling or retailing of goods, providing what economists call *place utility* to items produced elsewhere. They fulfill the exchange function of advanced economies and provide the market transactions necessary in highly interdependent societies. In commercial economies, tertiary activities also provide vitally needed information about market demand, without which economically justifiable production decisions are impossible.

The locational controls for tertiary enterprises are rather simpler than those for the manufacturing sector. Service activities are by definition market oriented. Those dealing with transportation and communication are concerned with the placements of people and commodities to be connected or moved; their locational determinants are therefore the patterns of population distribution and the spatial structure of production and consumption.

Most tertiary activities, however, are concerned with personal and business services performed in shops, restaurants, and company and governmental offices that cluster in cities large and small. The supply of those kinds of low-level services of necessity must be identical to the spatial distribution of *effective demand*—that is, wants made meaningful through purchasing power. Retail and personal services are localized by their markets, because the production of the service and its consumption are simultaneous occurrences. Retailers and personal service providers tend to locate, therefore, where market density is greatest and multiple service demands are concentrated (Figure 9.27). Their locational patterns and the employment support they imply are important aspects of urban economic structure and are dealt with in Chapter 11.

In all of the world's increasingly interdependent postindustrial societies, the growth of the service component reflects not only the development of ever more complex social, economic, and administrative structures. It also indicates changes made possible by growing personal incomes and alterations in family structure and individual lifestyle. For example, in subsistence economies families produce, prepare, and consume food within the household. Urbanizing industrial societies have increasing dependence on specialized farmers growing food and wholesalers and retailers selling food to households that largely prepare and consume it at home. Postindustrial America increasingly opts to purchase prepared foods in restaurants, fast-food, or carry-out establishments with accelerating growth of the tertiary food service workers that change demands. People are still fed, but the employment structure has altered.

Part of the growth in the tertiary component is statistical, rather than functional. We saw in our discussion of modern in-

Figure 9.27 Low-level services are most efficiently and effectively performed where demand and purchasing power are concentrated, as this garment repairman in an Ecuador city marketplace demonstrates. Such "informal sector" employment, further illustrated in Figure 10.6, usually escapes governmental registration and is not included in official service employment totals. One informed source estimates that 60% of Latin America's retailing is conducted in the unofficial informal sector.

dustry that "outsourcing" was increasingly employed as a device to reduce costs and enhance manufacturing and assembly efficiencies. In the same way, outsourcing of services formerly provided in-house is also characteristic of current business practice. Cleaning and maintenance of factories, shops, and offices—formerly done by the company itself as part of internal operations—now are subcontracted to specialized service providers. The jobs are still done, perhaps even by the same personnel, but worker status has changed from "secondary" (as employees of a manufacturing plant, for example) to "tertiary" (as employees of a service company).

Special note should be made of *tourism*—travel undertaken for purposes of recreation rather than business. It has become not only the most important single tertiary sector activity but is, as well, the world's largest industry in jobs and total value generated. On a worldwide basis, tourism accounts for some 250 million recorded jobs and untold additional numbers in the informal economy. Altogether, 15% or more of the world's work force is engaged in providing services to recreational travelers, and the total economic value of tourism goods and services at the end of the 20th century reached about $4 trillion, or some 14% of the world's gross domestic product. In middle- and high-income countries, tourism supports a diversified share of domestic expenditures through transportation-related costs, roadside services, entertainment, national park visits, and the like. International tourism, on the other hand, generates new income and jobs of growing importance in developing states since one-fifth of all international tourists now travel from an industrial country to a developing one. Altogether, worldwide international tourist visits numbered about 700 million in 2000, more than a quarter of them destined for the less developed low- and middle-income countries. That inbound flow produced over 8% of all foreign earnings of developing states in 2000 and—in the form of goods and services such as meals, lodging, and transport consumed by foreign travelers—comprised 44% of their total "service exports." For half of the world's 49 poorest countries, tourism became by 2000 the leading service export sector.

Whatever the origins of tertiary employment growth, the social and structural consequences are comparable. The process of development leads to increasing labor specialization and economic interdependence within a country. That was true during the latter 20th century for all economies, as Table 9.1 attests. Carried to the postindustrial stage of advanced technology-based economies and high per capita income, the service component of both the gross domestic product (see Figure 9.26) and the employed labor force rises to dominance.

Beyond Tertiary

Available statistics unfortunately do not always permit a clear distinction between *tertiary* service employment that is a reflection of daily lifestyle or corporate structural changes and the more specialized, higher-level *quaternary* and *quinary* activities.

The **quaternary** sector may be seen realistically as an advanced form of services involving specialized knowledge, technical skills, communication ability, or administrative competence. These are the tasks carried on in office buildings, elementary and university classrooms, hospitals and doctors' offices, theaters, accounting and brokerage firms, and the like. With the explosive growth in demand for and consumption of information-based services—mutual fund managers, tax consultants, software developers, statisticians, and more at near-infinite length—the quaternary sector in the most highly developed economies has replaced all primary and secondary employment as the basis for economic growth. In fact, over half of all workers in rich economies are in the "knowledge sector" alone—in the production, storage, retrieval, or distribution of information.

Quaternary activities performed for other business organizations often embody "externalization" of specialized services similar to the outsourcing of low-level tertiary functions. The distinction between them lies in the fact that knowledge and skill-based free-standing quaternary service establishments can be spatially divorced from their clients; they are not tied to resources, affected by the environment, or necessarily localized by market. They can realize cost reductions through serving multiple clients in highly technical areas and permit client firms to utilize specialized skills and efficiencies to achieve competitive advantage without the expense of adding to their own labor force.

Often, of course, when high-level personal contacts are required, the close functional association of client and service firms within a country encourages quaternary locations and employment patterns similar to those of the headquarters distribution of the primary and secondary industries served. But the transportability of quaternary services also means that many of them can be spatially isolated from their client base. In the United States, at least, these combined trends have resulted both in the concentration of certain specialized services—merchant banking or bond underwriting, for example—in major metropolitan areas and, as well, in a regional diffusion of the quaternary sector to accompany a growing regional deconcentration of the client firm base. Similar locational tendencies have been noted even for the spatially more restricted advanced economies of, for example, England and France.

Information, administration, and the "knowledge" activities in their broadest sense are dependent on communication. Their spatial dispersion has been facilitated by the underlying technological base of most quaternary activities: electronic digital processing and telecommunication transfer of data. That technology permits many "back-office" tasks to be spatially far distant from the home office locations of either the service or client firms. Insurance claims, credit card charges, mutual fund and stock market transactions, and the like, are more efficiently and economically recorded or processed in low-rent, low-labor cost locations—often in suburbs or small towns and in rural states—than in the financial districts of major cities. Production and consumption of such services can be spatially separated in a way not feasible for tertiary, face-to-face activities.

And finally there are the **quinary activities,** the "gold collar" professions of the chapter title, another separately recognized subdivision of the tertiary sector representing the special and highly paid skills of top business executives, government officials, research scientists, financial and legal consultants, and the like. These people find their place of business in major metropolitan centers, in and near major universities and research parks, at first-rank medical centers, and in cabinet and department-level

offices of political capitals. Within their cities of concentration they may be highly localized by prestigious street addresses (Park Avenue, Wall Street) or post offices (Princeton, New Jersey), or by notable "signature" office buildings (Transamerica Building, Seagram Building). Their importance in the structure of advanced economies far outweighs their numbers.

The list of tertiary, quaternary, and quinary employment is long. Its diversity and familiarity remind us of the complexity of modern life and of how far removed we are from the subsistence economies. As societies advance economically, the share of employment and national income generated by the primary, secondary, and composite tertiary sectors continually changes; the spatial patterns of human activity reflect those changes. The shift is steadily away from production and processing, and toward the trade, personal, and professional services of the tertiary sector and the information and control activities of the quaternary and quinary. That transition is the essence of the now-familiar term *postindustrial.*

Services in World Trade

Just as service activities have been major engines of national economic growth, so too have they become an increasing factor in international trade flows and economic interdependence. Between 1980 and 2000, services increased from 15% of total world trade to nearly 25%. The fastest growing segment of that increase was in such private services as financial, brokerage, and leasing activities, which grew to 50% of all commercial services trade by the end of the 20th century. As in the domestic arena, rapid advances and reduced costs in information technology and electronic data transmission have been central elements in the internationalization of services (Figure 9.28). Many services considered nontradable even during the 1980s are now actively exchanged at long distance.

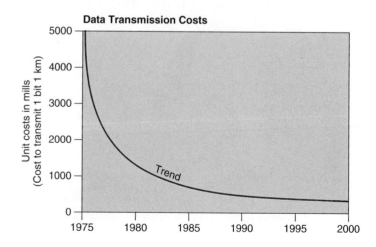

Figure 9.28 The costs of transmitting data electronically have plummeted since the mid-1970s and promise to be lower still in coming years as fiber optics and wireless transmission become widely available. Technologies developed by the early 21st century have the capability to transmit the equivalent of 90,000 volumes of an encyclopedia per second.

Sources: Probe Research, Inc.; Telcordia (Bellcore); Progressive Policy Institute.

Developing countries have been particular beneficiaries of the new technologies. Their exports of services—valued at $260 billion in 2000—grew at an annual 15% rate in the 1990s, twice as fast as service exports from industrial regions. The increasing tradability of services has expanded the international comparative advantage of developing states in relatively labor-intensive long-distance service activities such as mass data processing, computer software development, and the like. At the same time, they have benefited from increased access to efficient, state-of-the-art equipment and techniques transferred from advanced economies. The concentration of computer software development around Bangalore and Hyderabad has made India a major world player in software innovation, for example, while there and elsewhere in that country, increasing volumes of back-office work for Western insurance, financial, and accounting companies and airlines are being performed. Customer interaction services ("call centers") formerly based in the United States are now increasingly relocated to India, employing workers trained to speak to callers in perfect American English. Claims processing for life and health insurance firms have become concentrated in English-speaking Caribbean states to take advantage of the lower wages and availability of a large pool of educated workers there. In all such cases, the result is an acceleration in the transfer rate of technology in such expanding areas as information and telecommunications services and an increase in the rate of developing-country integration in the world economy.

Most of the current developing-country gains in international quaternary services are the result of increased foreign direct investment (FDI) in the services sector. Those flows accounted for three-fifths of all FDI at the start of 2000. The majority of such investment, however, is transferred among the advanced countries themselves rather than between industrial and developing states. In either case, as transnational corporations employ mainframe computers around the clock for data processing, they can exploit or eliminate time zone differences between home office countries and host countries of their affiliates. Such cross-border intrafirm service transactions are not usually recorded in balance of payment or trade statistics, but materially increase the volume of international services flows.

Despite the increasing share of global services trade held by developing countries, world trade—imports plus exports—in services is still overwhelmingly dominated by a very few of the most advanced states (Table 9.2). The country and category contrasts are great, as a comparison of the "high-income" and "low-income" groups documents. At a different level, the single small island state of Singapore had more than twice the 2000 share (1.7%) of world services trade as all of sub-Saharan Africa (0.7%).

The same cost and skill advantages that enhance the growth and service range of quaternary firms and quinary activities on the domestic scene also operate internationally. Principal banks of all advanced countries have established foreign branches, and the world's leading banks have become major presences in the primary financial capitals. In turn, a relatively few world cities have emerged as international business and financial centers whose operations and influence are continuous and borderless, while a host of offshore banking havens have emerged to exploit gaps in regulatory controls and tax laws (Figure 9.29). Accounting firms, advertising agencies, management consulting companies, and similar establishments of primarily North American or European origin have increasingly established their international presence, with main branches located in principal business centers worldwide. Those advanced and specialized service components help swell the dominating role of the United States and European Union in the structure of world trade in services.

Table 9.2

Shares of World Trade in Services (exports plus imports, 2000)

Country or Category	% of World
United States	16.7
Germany	7.5
United Kingdom	7.3
Japan	6.5
France	5.0
China	4.7
Italy	3.9
Netherlands	3.6
Spain	3.0
Belgium	2.9
Canada	2.7
Austria	2.1
Total	**65.9**
High-Income States	80.3
Low-Income States	3.1
Sub-Saharan Africa	0.7

Source: Data from World Bank.

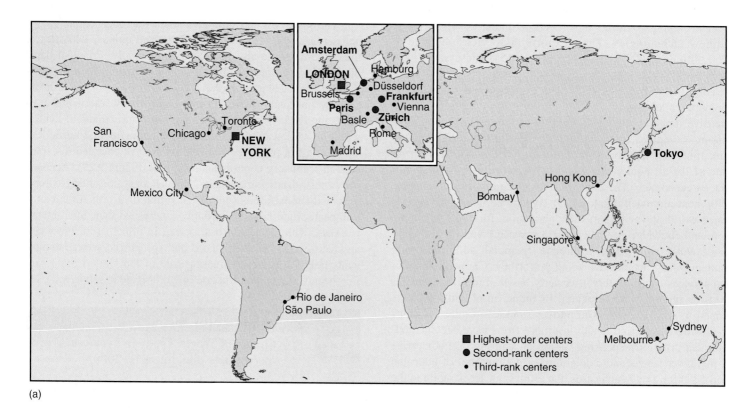

(a)

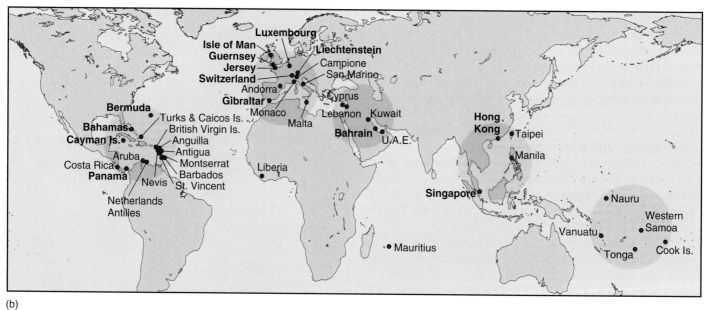

(b)

Figure 9.29 (*a*) **The hierarchy of international financial centers,** topped by New York and London, indicates the tendency of highest-order quaternary activities to concentrate in a few world and national centers. (*b*) At the same time, the multiplication of offshore locations—estimated at 35 or more in 2000—where "furtive money" avoiding regulatory control and national taxes finds refuge, suggests that dispersed convenience sites also serve the international financial community. In 2002, under international pressure, all but seven of the tax havens agreed to greater openness and less protective secrecy.

Source: Peter Dicken. Global Shift, *3d. ed. New York: Guilford Press, 1998, Figures 12.9 and 12.10.*

Summary

The spatial patterns of the world's manufacturing regions represent the landscape evidence of industrial location theories. Those theories are based on assumed regularities of human economic behavior that is responsive to profit and price motivations and on simplifying assumptions about fixed and variable costs of manufacturing and distribution. In commercial economies, market mechanisms and market prices guide investment and production decisions.

Industrial cost components considered theoretically important are raw materials, power, labor, market accessibility, and transportation. Weberian analysis argues that least-cost locations are optimal and are strongly or exclusively influenced by transportation charges. Locational interdependence theory suggests that firms situate themselves to assure a degree of market monopoly in response to the location of competitors. Profit maximization concepts accept the possibility of multiple satisficing locations within a spatial margin of profitability. Agglomeration economies and the multiplier effect may make attractive locations not otherwise predicted for individual firms, while comparative advantage may dictate production, if not location, decisions of entrepreneurs. Location concepts developed to explain industrial distributions under Fordist production constraints have been challenged as new just-in-time and flexible production systems introduce different locational considerations.

Major industrial districts of Eastern Anglo America, Western and Central Europe, Eastern Europe, and Eastern Asia are part of a world-girdling "industrial belt" in which the vast majority of global secondary industrial activity occurs. The most advanced countries within that belt, however, are undergoing deindustrialization as newly industrializing countries with more favorable cost structures compete for markets. In the advanced economies, tertiary, quaternary, and quinary activities become more important as secondary-sector employment and share of gross national income decline. The new high-tech and postindustrial spatial patterns are not necessarily identical to those developed in response to theoretical and practical determinants of manufacturing success.

The nearly empty highway of the Monongahela Valley and the crowded expressways of high-tech and office park corridors are symbols of those changes in North America. As economic activity becomes less concerned with raw materials and freight rates, it becomes freer of the locational constraints of an older industrial society. Increasingly, skills, knowledge, communication, and population concentrations are what attract and hold the newer economic sectors in the most advanced economies. At the same time, much of the less developed world is striving for the transfer of manufacturing technology from developed economies and for the industrial growth seen as the path to their future prosperity. Those aspirations for economic development and the contrasts they imply in the technological subsystems of the countries of the world are topics of concern in Chapter 10.

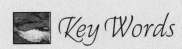

Key Words

agglomeration 325
agglomeration (external) economies 328
break-of-bulk point 324
comparative advantage 329
deglomeration 328
fixed cost 323
footloose firm 327
Fordism 328
freight rates 323
infrastructure 328
in-transit privilege 324
least-cost theory 325

line-haul (over-the-road) costs 323
locational interdependence 326
market equilibrium 318
market orientation 321
material orientation 320
multiplier effect 328
outsourcing 330
quaternary activities 345
quinary activities 346
satisficing location 327
secondary activities 318

spatially fixed costs 319
spatially variable costs 319
spatial margin of profitability 327
substitution principle 327
terminal costs 323
tertiary activities 343
transnational (multinational) corporation (TNC) 331
ubiquitous industry 322
uniform (isotropic) plain 325
Weberian analysis 325

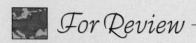
1. What are the six *principles of location* outlined in this chapter? Briefly explain each and note its contribution to an entrepreneur's spatial search.

2. What is the difference between *fixed* and *variable* costs? Which of the two is of interest in the plant locational decision? What kinds of variable costs are generally reckoned as most important in locational theory?

3. What role do prices play in the allocation of resources in commercial economies? Are prices a factor in resource allocation in planned economies? What differences in locational patterns of industry are implicit in the different treatment of costs in the two economies?

4. *Raw materials, power, labor, market,* and *transportation* are "factors of location" usually considered important in industrial placement decisions. Summarize the role of each, and cite examples of where each could be decisive in a firm's location.

5. What were Weber's controlling assumptions in his theory of plant location? What "distortions" did he recognize that might alter the locational decision?

6. With respect to plant siting, in what ways do the concepts and conclusions of *locational interdependence theory* differ from those of *least-cost theory*?

7. What is the *spatial margin of profitability?* What is its significance in plant location practice?

8. How have the concepts or practices of *comparative advantage* and *outsourcing* affected the industrial structure of advanced and developing countries?

9. In what ways are the locational constraints for *high-tech* industries significantly different from those of more basic secondary activities?

10. As high-tech industries and *quaternary* and *quinary* employment become more important in the economic structure of advanced nations, what consequences for economic geographic patterns do you anticipate? Explain.

 Focus Follow-up

1. **What are the principal elements of locational theory, and how do different classical theories employ them?** pp. 316–327.

   Costs of raw materials, power, labor, market access, and transportation are the assumed controls governing industrial location decisions. They receive different emphases and imply different conclusions in the theories considered here. *Least-cost* (Weber) analysis concludes transport costs are the fundamental consideration; *locational interdependence* (Hotelling) considers that location of competitors determines a firm's siting decision; *profit maximization* maintains a firm should locate where profit is maximized by utilizing the substitution principle.

2. **How do agglomeration, just-in-time, comparative advantage, and TNC control affect traditional location theory outcomes?** pp. 327–333.

   By sharing infrastructure, agglomerating companies may reduce their individual total costs, while JIT supply flows reduce their inventory

capital and storage charges. Comparative advantage recognizes that different regions or nations have different industrial cost structures. Companies utilize *outsourcing* of part of their production to exploit those differences. Transnational corporations distribute their operations based on comparative advantage: manufacturing in countries where production costs are lowest; performing research, accounting, and other service components where economical or convenient; and maintaining headquarters in locations that minimize taxes. Outsourcing and TNC practices evade the single location implications of classical location theories.

3. **What influences high-tech activity location, and what is the impact of high-tech growth on established world manufacturing regions?** pp. 333–343.

   Long-established industrial regions of Eastern Anglo America and of Western, Central, and Eastern Europe developed over time in response to

predications of classical location analysis. Eastern Asia, the most recently developed major industrial region, has been influenced by both classical locational pulls and outsourcing and high-tech locational needs. High-tech industries tend to create regionally specialized agglomerations reflecting proximity to scientific research centers, technically skilled labor pools, venture capital availability, quality of life environments, and superior transport and communication facilities. Their emergence has altered traditional industrial emphases and distributional patterns.

4. **What are the functional and locational characteristics of tertiary, quaternary, and quinary service activities** (pp. 343–346), **and how are they reflected in world trade patterns?** (pp. 346–348).

   Tertiary industries include all nongoods production activities and provide services to goods producers, the general community, and individuals. Subdivided for easier recognition, the general tertiary category contains: (*a*) low-level personal and professional services, retailing of goods, and the like involved in daily life and market-oriented functions. This subcategory is also called *tertiary*. (*b*) The *quaternary* sector comprises advanced forms of services suggested by the term "knowledge industries" that are performed in classrooms, hospitals, accounting and brokerage firms, corporate office buildings, etc. (*c*) *Quinary* activities include highly specialized and advanced services of research scientists, highest-level corporate executives and governmental officials, and the like. The growing world trade in services, made possible by plummeting costs of information transmission, has altered international economic relations and encouraged cultural and functional integration.

 Selected References

Corbridge, Stuart. *World Economy.* The Illustrated Encyclopedia of World Geography. New York: Oxford University Press, 1993.

Daniels, Peter, John Bryson, and Barney Warf. *Service Industries in the New Economy.* New York: Routledge, 2003.

Daniels, Peter, and W. F. Lever, eds. *The Global Economy in Transition.* White Plains, New York: Longman, 1996.

Dicken, Peter. *Global Shift: Reshaping the Global Economic Map in the 21st Century.* 4th ed. New York: Guilford Press, 2003.

Ettlinger, Nancy. "The Roots of Competitive Advantage in California and Japan." *Annals of the Association of American Geographers* 81, no. 3 (1991): 391–407.

Greenhut, Melvin L. *Plant Location in Theory and in Practice.* Chapel Hill: University of North Carolina Press, 1956. Reprint. Westport, Conn.: Greenwood Press, 1982.

Hall, Colin M., and Stephen Page. *Geography of Tourism and Recreation.* 2d ed. New York: Routledge, 2002.

Hamilton, Ian, ed. *Resources and Industry.* The Illustrated Encyclopedia of World Geography. New York: Oxford University Press, 1992.

Harrington, James W., and Barney Warf. *Industrial Location: Principles, Practice, and Policy.* London and New York: Routledge, 1995.

Hayter, Roger. *The Dynamics of Industrial Location.* New York: John Wiley & Sons, 1997.

Hoover, Edgar M. *The Location of Economic Activity.* New York: McGraw-Hill, 1948.

Hudman, Lloyd, and Richard Jackson. *Geography of Travel and Tourism.* 3d ed. Albany, N.Y.: Delmar Publishers, 1999.

International Bank for Reconstruction and Development/The World Bank. *World Development Indicators.* Washington, D.C.: The World Bank, annual.

International Bank for Reconstruction and Development/The World Bank. *World Development Report.* Published annually for the World Bank by Oxford University Press, New York.

Knox, Paul L., and John A. Agnew. *The Geography of the World Economy.* 3d ed. New York: John Wiley & Sons, 1998.

Kotkin, Joel. *The New Geography: How the Digital Revolution Is Reshaping the American Landscape.* New York: Random House, 2000.

Mastny, Lisa. *Traveling Light: New Paths for International Tourism.* Worldwatch Paper 159. Washington, D.C.: Worldwatch Institute, 2001.

South, Robert B. "Transnational 'Maquiladora' Location." *Annals of the Association of American Geographers* 80, no. 4 (1990): 549–570.

de Souza, Anthony R., and Frederick P. Stutz. *The World Economy: Resources, Location, Trade, and Development.* 3d ed. Upper Saddle River, N.J.: Prentice-Hall, 1998.

United Nations Conference on Trade and Development (UNCTAD). *World Investment Report.* New York and Geneva: United Nations, annual.

Weber, Alfred. *Theory of the Location of Industries.* Translated by Carl J. Friedrich. Chicago: University of Chicago Press, 1929. Reissue. New York: Russell & Russell, 1971.

Wheeler, James O., Peter Muller, Grant Thrall, and Timothy Fik. *Economic Geography.* 3d ed. New York: John Wiley & Sons, 1998.

Williams, Stephen. *Tourism Geography.* New York: Routledge, 1998.

Websites: The World Wide Web has a tremendous number and variety of sites pertaining to geography. Websites relevant to the subject matter of this chapter appear in the "Web Links" section of the On-line Learning Center associated with this book. Access it at **www.mhhe.com/fellmann8e**

CHAPTER
Ten

Patterns of Development
and Change

Focus Preview

Opposite: A study in contrasts: low-tech transportation beneath a billboard advertising China's leading Web portal.

The Hindu funeral pyres burned day and night; Muslims were buried five and more together in common graves. Countless dead cattle, buffalo, and dogs were hastily gathered and dumped in pits. In a sense, on that unseasonably cold December night in central India, all had died for economic development (Figure 10.1). Some 40% of the Indian population exists in poverty. Eager to attract modern industry to its less developed states, to create additional industrial and urban employment, and to produce domestically the chemicals essential to its drive for agricultural self-sufficiency, the Indian government in 1969 granted Union Carbide Corporation a license to manufacture pesticides at a new plant built on vacant land on the outskirts of Bhopal. A principal ingredient was deadly methyl isocyanate gas, the silent killer that escaped from its storage tank that winter night of 1984 after a sudden and unexplained buildup of its temperature and pressure.

To assure the plant's success, Union Carbide had been exempted from many local taxes, and land, water, and power costs were heavily subsidized. To yield maximum benefit to the local economy and maximum transfer of technology and skills, 50% ownership in the enterprise was retained for Indian investors along with total local control of construction and operation of the plant. The 1000 jobs were considered so important by the state and local governments that despite six accidents and one death in the years before the night of disaster, reports critical of plant safety and operation were shelved and ignored. A local official who had called for the removal of the factory to a more isolated area was himself removed from office.

By the time of the fatal accident, Bhopal had grown from 300,000 to over 900,000 people. More than 130,000 residents lived in the slums and shantytowns they built for themselves just across the street from a factory they thought produced "plant medicine" to keep crops healthy; they were the principal victims. Before the week was over, almost 3000 people had died. Another 300,000 had been affected by exposure to the deadly poison, and perhaps 150,000 of those suffer long-term permanent disabilities—blindness, sterility, kidney and liver infections, and brain damage.

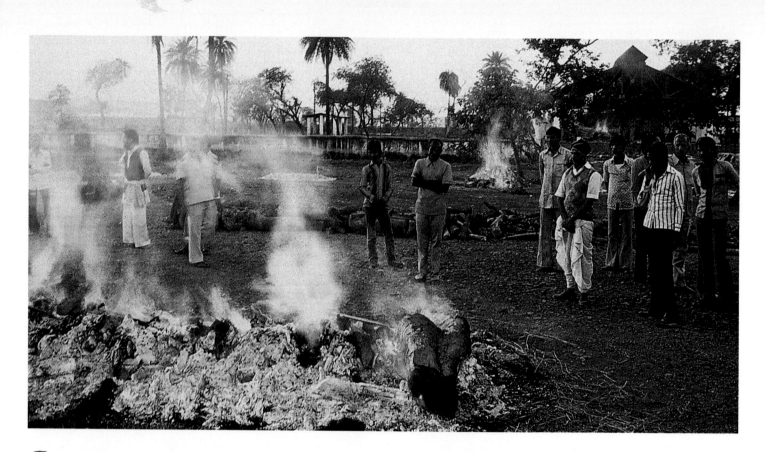

Figure 10.1 Burning the dead at Bhopal. At the time of the tragedy, India was more prepared than many developing countries to accept the transfer of advanced technology. In 1984, it ranked among the top 15 countries in manufacturing output and supplied most basic domestic needs from its own industry. India still sought modern plants and processes and, particularly, industry supporting agricultural improvement and expansion.

Development as a Cultural Variable

Whatever its immediate cause of equipment failure or operator error, the tragedy of Bhopal—seen by opponents of globalization as an emblem of the evils of multinationalism—is witness to the lure of economic development so eagerly sought that safety and caution are sacrificed to achieve it. That lure is nearly irresistible for those countries and regions that look to industrialization and urban employment as their deliverance from traditional economies no longer able to support their growing populations or to satisfy their hopes for an improved quality of life.

Any view of the contemporary world quickly shows great—almost unbelievable—contrasts from place to place in levels of economic development and people's material well-being. Variations in these are indicative of the tools, energy sources, and other *artifacts* (p. 51) differing societies employ in production and the kinds of economic activities in which they engage, and underlie the social organizations and behavior patterns they have developed. A look around tells us that these interrelated economic and social structures are not shared by all societies; they vary between cultures and countries. The ready distinction that we make between the "Gold Coast" and the "slum" indicates that different groups have differential access to the wealth, tools, and resources of the global and national societies of which they are a part.

At an international scale, we distinguish between "advanced" or "rich" nations, such as Canada or Switzerland, and "less developed" or "poor" countries, like Bangladesh or Burkina Faso, though neither class of states may wish those adjectives applied to its circumstances. Hunter-gatherers of southwest Africa or Papua New Guinea, shifting gardeners of Amazonia, or subsistence farmers of southeastern Asia may be largely untouched by the modernization, industrialization, or urbanization of society commonplace elsewhere. Development differentials exist within countries, too. The poverty of drought- and hunger-plagued northeastern Brazil stands in sharp contrast to the prosperous, industrialized modernity of São Paulo state or city (Figure 10.2), while in the United States, farmers of the hillsides of Appalachia live in a different economic and cultural reality than do midwestern cash grain farmers.

Dividing the Continuum: Definitions of Development

Countries display different levels of development. **Development** in that comparative sense means simply the extent to which the resources of an area or country have been brought into full productive use. It may also carry in common usage the implications of economic growth, modernization, and improvement in levels of material production and consumption. For some, it also suggests changes in traditional social, cultural, and political structures to resemble more nearly those displayed in countries and economies deemed "advanced." Many of the attributes of development under an economic definition can be quantified by reference to statistics of national production, per capita income, energy consumption, nutritional levels, labor force characteristics, and the like. Taken together, such variables might calibrate a scale of achievement against which the level of development of a single country may be compared.

Such a scale would reveal that countries lie along a continuum from the least advanced in technology or industrialization to the most developed in those and similar characteristics. Geographers (and others) attempt to classify and group countries along the continuum in ways that are conceptually revealing and spatially informative. The extremes are easy; the middle ground is less clear-cut, and the terminology referring to it is mixed.

In the broadest view, "developed" countries stand in easy contrast to the "underdeveloped," "less developed," or "developing" world. (*Developing* was the term introduced by President Harry S. Truman in 1949 as a replacement for *backward,* the unsatisfactory and unflattering reference then in use.) **Underdevelopment** from a

(a)

(b)

Figure 10.2 The modern high-rise office and apartment buildings of prosperous São Paulo (*a*), a city that generates over one-third of Brazil's national income, stand a world apart from the poverty and peasant housing of northeastern Brazil (*b*). The evidences and benefits of "development" are not equally shared by all segments of any country or society.

strictly economic point of view suggests the possibility or desirability of applying additional capital, labor, or technology to the resource base of an area to permit the present population to improve its material well-being or to allow populations to increase without a deterioration in their quality of life.

The catch-all category of *underdeveloped,* however, does not tell us in which countries such efforts at improvement have occurred or been effective. With time, therefore, more refined subdivisions of development have been introduced, including such indistinctly relative terms as *moderately, less,* or *least developed* countries.[1] Since development is commonly understood to imply industrialization and to be reflected in improvements in national and personal income, the additional terms *newly industrializing countries* (NICs) (which we encountered in Chapter 9) and *middle income countries* have been employed. More recently, *emerging economy* has become a common designation, providing a more positive image than "underdeveloped." In a corruption of its

original meaning, the term *Third World* is often applied to the developing countries as a group, though when first used that designation was a purely political reference to nations not formally aligned with a "First World" of industrialized free market (capitalist) nations or a "Second World" of centrally controlled (communist bloc) economies. And increasingly, the name *Fourth World* has been attached to the UN-recognized group of least developed states.

In 1980 the contrasting terms *North* and *South* were introduced (by the Independent Commission on International Development Issues, commonly called the Brandt Report[2]) as a broad and not wholly accurate generalization to emphasize the distinctions between the rich, advanced, developed countries of the Northern Hemisphere (to which Southern Hemisphere Australia and New Zealand are added)—the *North*—and, roughly, all the rest of the world—the *South* (Figure 10.3). This split agreed with the United Nations classification that placed all of Europe and

---

[1]In 1971, the General Assembly of the UN listed 24 "least developed" countries identified by per capita gross domestic product, share of manufacturing in GDP, and adult literacy. In later years the criteria were expanded to include a quality of life index, economic diversification index, and population size; at the same time the number of countries included in the least developed group rose steadily to 49 in 2003. Over the years, only one country, Botswana, has ever "graduated" from the list. See Figure 10.3.

[2]*North-South: A Programme for Survival.* The Commission was established in 1977 at the suggestion of the chairman of the World Bank. Under its charge, "global issues arising from economic and social disparities of the world community" were to be studied and "ways of promoting adequate solutions to the problems involved in development" were to be proposed. The former Soviet Union was at that time included within the North, and its successor states retain that association.

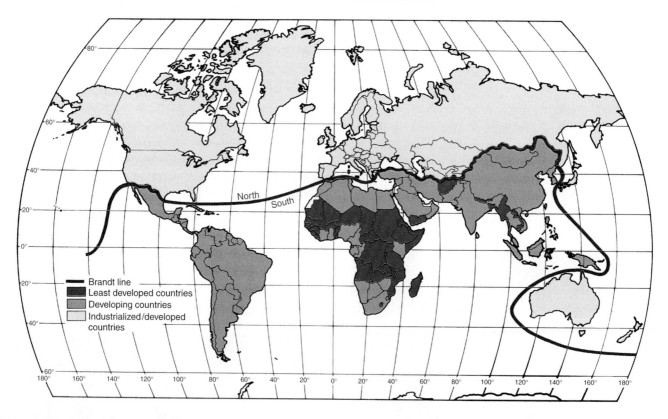

Figure 10.3 **Comparative development levels.** The "North–South" line of the 1980 *Brandt Report* suggested a simplified world contrast of development and underdevelopment based largely on degree of industrialization and per capita wealth recorded then. In 2003, the United Nations Economic and Social Council and the UN Conference and Trade and Development (UNCTAD) recognized 49 "least developed countries." That recognition reflects low ratings in three criteria: gross domestic product per capita; human resources as measured by a series of "quality of life" indicators; and level of economic diversification. The inclusive category of "developing countries" ignores recent significant economic and social gains in several Asian and Latin American states, raising them now to "industrialized/developed" status. Some "least developed" states are small island countries not shown at this map scale.

Sources: UNCTAD and United Nations Development Programme.

North America, plus Australia, Japan, New Zealand, and the former USSR in a *more developed country* category, with all other states classed as *less developed countries* (LDCs).

The variety of terms devised—not all of them accurately descriptive or acceptable to those countries designated—represent honest efforts to categorize countries whose developmental circumstances are defined by a variety of economic and social measures along a continuum of specific or composite characteristics. In the remainder of this chapter, broad developmental contrasts between countries or regions will conform to the "North-South" and the UN "more developed-less developed" categorizations. Our primary attention in maps and text, however, will be given to the developing countries of the "South."

The terminology of development is usually applied to country units, but it is equally meaningful at the regional and local levels within them, for few countries are uniformly highly developed or totally undeveloped. Many emerging economies contain pockets—frequently the major urban centers—of productivity, wealth, and modernity not shared by the rest of the state. For example, Mexico is a leading NIC, but more than 50% of its industrial workers and over 60% of value of manufacturing are located in metropolitan Mexico City. Many other parts of the country and, particularly, its Indian population, remain untouched by the development concentrated in the capital city. Even within the most advanced societies, some areas and populations remain outside the mainstream of progress and prosperity enjoyed by the majority. Fourth World deprivation is not just a whole country concept.

Explanations of Underdevelopment

It is one thing to devise categories of relative development and to assign countries to them; it is quite another to see in those categories an explanation of their spatial pattern. Why are different countries arranged as they are along the continuum of advancement? What conditions underlie their relative degrees of development? Are those conditions common to all countries at the same level of technology? And do those conditions have spatial expression and spatial explanation?

The Brandt Report hints at one frequent but simplistic spatial explanation: Development is a characteristic of the rich "North"—the midlatitudes, more precisely; poverty and underdevelopment are tropical conditions. Proponents of the latitudinal explanation support their conviction not only by reference to such topical maps as Figures 10.3, 10.8, or 10.11, but by noting that rich countries—some 30 in number—have 93% of their population resident in temperate or "snow belt" zones; 42 of the world's poorest states have 56% of their people in tropical latitudes and 18% in arid zones. They also note that observable differences in development and wealth exist within individual countries. Brazilians of the southeastern temperate highlands, for example, have average incomes several times higher than their compatriots of tropical Amazonia. Annual average incomes of Mexicans of the temperate north far exceed those of southern Yucatán. Australians of the tropical north are poorer than Australians of the temperate south. Unfortunately for the search for easy explanation, many of the poorer nations of the "South" lie partially or wholly within the midlatitudes or at temperate elevations—

Afghanistan, North Korea, and Mongolia are examples—while equatorial Singapore and Malaysia prosper. Geography, many argue forcefully, is not destiny, although tropical regions admittedly face the major ecological handicaps of low agricultural productivity and high incidence of plant, animal, and human disease.

Other generalizations seem similarly inconclusive: (1) Resource poverty is cited as a limit to developmental possibilities. Although some developing countries are deficient in raw materials, others are major world suppliers of both industrial minerals and agricultural goods—bauxite, cacao, and coffee, for example. Admittedly, a Third World complaint is that their materials are underpriced in the developed world markets to which they flow or are restricted in that flow by tariffs and subsidized destination country competitors. Those, however, are matters of marketing, politics, and economics, not of resources. Further, economists have long held that reliance on natural resource wealth and exports by less developed countries undermines their prospects for growth by interfering with their development of industry and export-oriented manufacturing. (2) Overpopulation and overcrowding are frequently noted as common denominators of national underdevelopment, but Singapore prospers with 6800 per square kilometer (17,000 per sq mi) while impoverished Mali is empty with 9 per square kilometer (24 per sq mi) (Figure 10.4). (3) Former colonial status is often blamed for present underdevelopment. The accusation is arguably valid for countries where—as in

Figure 10.4 Landlocked and subject to severe droughts, Mali is one of the poorest of the "least developed" countries. Low densities of population are not necessarily related to prosperity, or high densities to poverty. Mali has only 9 people per square kilometer (24 per sq mi); Japan has 337 per square kilometer (873 per sq mi). These Dogon women crossing a parched millet field near Sanga are on their way to get water—a time- and energy-consuming daily task for many least-developed-country women. Even in more humid South Africa, rural women on average spend 3 hours and 10 minutes each day fetching water, according to a government survey.

sub-Saharan Africa and southern Asia—colonizers temporarily ruled but left largely intact the indigenous population. It may, however, be a surprise to the ex-colonies of Australia, New Zealand, Canada, or the United States where colonists replaced the aboriginal occupants—and to Ethiopia, which never was colonized but is a case study in underdevelopment.

Although there appears to be no single, simple explanation of Third World status, just as there is no single measure of underdevelopment that accounts for every Third World case, the Harvard Institute for International Development did attempt to quantify differences in national economic development. It argued that "physical geography" is one of four factors influencing global patterns of growth; the least developed countries are almost without exception located in ecological zones that pose serious health conditions—including much shorter life spans—not found in the midlatitudes and have agricultural limitations that are very different from those of wealthy states. The other three factors are initial economic level, government policy, and demographic change. The Institute's conclusions were that landlocked countries grew more slowly than coastal economies, that—because of poor health and unproductive farming—tropical states were slower to develop than temperate zone ones, and that sparse natural resources and transport isolation inhibited growth possibilities and rates.

These physical differences and environmental limitations, the Institute found, were far less explanatory of national growth rates than were market economies, prudent fiscal policies, and the rule of laws prohibiting corruption, breach of contract, expropriation of property, and the like. These are circumstances and controls independent of locational or resource differentials. That conclusion is buttressed by a United Nations report concluding that "good government," including protection of property rights under a stable political and legal system, is the top priority in poverty-fighting and the key to sustainable development. These are, however, conditions reflective of western market economy standards that are not necessarily acceptable to all cultures.

The Core-Periphery Argument

Core-periphery models are based on the observation that within many spatial systems sharp territorial contrasts exist in wealth, economic advancement, and growth—in "development"—between economic heartlands and outlying subordinate zones. Wealthy urban cores and depressed rural peripheries, or prospering "high-tech" concentrations and declining "rust belts," are contrasts found in many developed countries. On the international scene, core-periphery contrasts are discerned between, particularly, Western Europe, Japan, and the United States as prosperous cores and the Fourth World as underdeveloped periphery. At all spatial scales, the models assume that at least partially and temporarily the growth and prosperity of core regions is at the expense of exploited peripheral zones.

That conclusion is drawn from the observation that linkages and interactions exist between the contrasting parts of the system. As one variant of the model suggests, if for any reason (perhaps a new industrial process or product) one section of a country experiences accelerated economic development, that section by its expanding prosperity becomes increasingly attractive for investors and other entrepreneurs. Assuming national investment capital is limited, growth in the developing core must come at the expense of the peripheries of the country.

A process of **circular and cumulative causation** thus set in motion continues to polarize development and, according to economist Gunnar Myrdal, leads to a permanent division between prosperous (and dominating) cores and depressed (and exploited) peripheral districts that are milked of surplus labor, raw materials, and profits. In its *dependency theory* form (p. 371), this version of the core-periphery argument sees the developing world as effectively held captive by the leading industrial nations. It is drained of wealth and deprived of growth by remaining largely a food and raw material exporter and an importer of manufactured commodities—and frequently suffering price discrimination in both their sales and purchases. A condition of *neocolonialism* is said to exist in which economic and even political control is exercised by developed states over the economies and societies of legally independent countries of the underdeveloped world.

A more hopeful variant of the model observes that regional income inequalities exist within all countries but that they tend to be greater in less developed countries than in the developed ones. That is, within market economies, income disparities tend to be reduced as developmental levels increase. Eventually, it is argued, income convergence will occur as **trickle-down effects,** or **spread effects,** work to diffuse benefits outward from the center in the form of higher prices paid for needed materials or through the dispersion of technology to branch plants or contract suppliers in lower-cost regions of production. On the international scale, such spread effects should work to reduce the dominance of formerly exploitative cores and equalize incomes between world regions. The increasing wealth of the newly industrializing economies and the penetration of European and American markets by, for example, Asian-produced goods ranging from cheap textiles to expensive automobiles and high-technology electronics are cited in support of this model variant.

Core-periphery models stress economic relationships and spatial patterns of control over production and trade. Indeed, the usual measures and comparisons of development are stated in economic terms (Figure 10.5). As we shall see later in this chapter, noneconomic measures may also be employed, though usually not without reference to their relationships to national or per capita income or to other measures of wealth and productivity. We shall also see that composite measures of developmental level are perhaps more useful and meaningful than those restricted to single factors or solely to matters of either economy or social welfare.

And finally, we should remember that "development" is a culturally relative term. It is usually interpreted in western, democratic, market economy terms that presumably can be generalized to apply to all societies. Others insist that it must be seen against the background of diverse social, material, and environmental conditions that differently shape cultural and economic aspirations of different peoples, many of whom specifically reject those western cultural and economic standards.

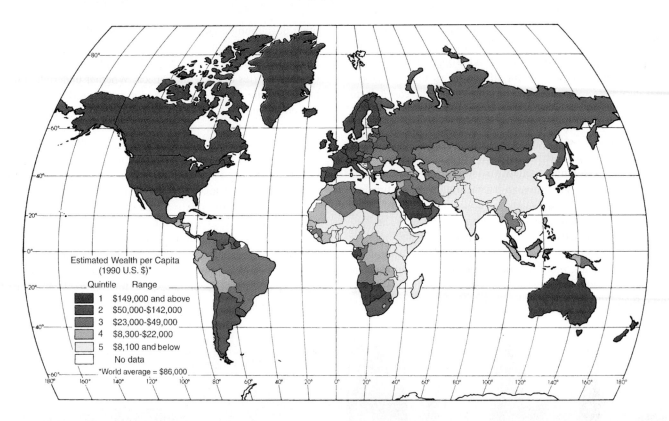

Estimated Wealth per Capita
(1990 U.S. $)*

Quintile	Range
1	$149,000 and above
2	$50,000–$142,000
3	$23,000–$49,000
4	$8,300–$22,000
5	$8,100 and below
	No data

*World average = $86,000

Figure 10.5 **The wealth of nations.** Sustainable development, the World Bank observes, implies providing future generations with as much or more capital *per capita* as we ourselves possess. Usual measures of national wealth emphasize brick, mortar, machinery, and infrastructural physical capital and try to recognize less tangible national assets of education, health, and social organizations. But, the Bank suggests, a national wealth account must also be credited with the value of possessed natural and mineral resources and adjusted downward by explicit recognition of the costs of environmental degradation and mineral and agricultural resource depletion.

By that accounting system, a preliminary 1996 Bank ranking of per capita wealth of countries put Australia at $835,000 per person and Canada ($704,000) on top because their great natural wealth (capital) is owned by relatively small populations. "Developed countries" of the North rate highly on the combined basis of their human, social, and produced capital, even though their natural capital may be modest. At the bottom (Ethiopia, with a per capita national wealth of $1400, was lowest) are states with poor natural endowments and low levels of human, social, and produced capital development and those where large and growing populations further subdivide a limited composite capital account. The Bank's "environmental accounting" method also recognized dozens of states such as Kenya, Libya, Nigeria, Venezuela, and others where the accumulation of human and produced capital has been offset by the depletion of raw materials and fertile farm land. Although the dollar values are now dated, other sources indicate the relative rankings are still valid.

Source: Data from The World Bank.

Economic Measures of Development

The developing countries as a group have made significant progress along the continuum of economic development. Between 1990 and 2000, their economies collectively grew at an average annual rate of over 3% compared to less than 2% per year for the industrial states. As a result of those growth differentials and of overall changes in the composition of their gross domestic products, the less developed states were in a decisively different relative position at the end of the 20th century than they were at its start. In 1913, on the eve of World War I, the 20 or so countries now known as the rich industrial economies produced almost 80% of world manufacturing. In 1950, the United States alone accounted for around half of world output, about the share produced by all the developed economies together in the late 1990s.

Particularly after midcentury, the spread effects of technology transfer, industrialization, and expanding world trade substantially reduced—though have not yet eliminated—the core-periphery contrasts in productivity and structure of gross domestic product that formerly seemed insurmountably great. Significantly, manufactured goods at the end of the century accounted for more than 80% of all merchandise exports of developing countries as a group—including the rapidly industrializing and exporting economies of the East Asian "tigers" (fast-growing economies, such as Singapore, South Korea, and Taiwan)—up from only 5% in 1955. Their growing importance in manufacturing exports clearly marks the restructuring of the economies of many developing states away from subsistence agriculture and primary commodity production. Not all developing regions have experienced that restructuring, of course; the share of manufacturing in exports for sub-Saharan Africa was still less than 30% in 2000.

In 1965, 30% of the developing countries' combined income came from agriculture and another 30% from industry. By the end of the century, agriculture's share had dropped to about 12% while industry grew to account for some 37% of their collective national earnings; over the same span, services increased from less than 40% to 53% of earnings. Not all of the employment shifts and structural changes within developing countries, however, are accounted for by official statistics. In all countries, at least a portion of goods and services are produced and workers supported by the **informal** (shadow, underground) **economy.** For emerging economies as a group, an international study estimates, underground activity is equivalent to around one-third of gross domestic product (and to about 15% in rich countries). The proportion of economic activity escaping official notice and thus not part of published gross national income (GNI) totals varies greatly between states, the 1999 study reports. Nigeria and Thailand have the world's largest shadow economies, both accounting for more than 70% of official gross domestic product (GDP); for

other developing countries, the proportion, though lower, is in all cases still significant. The World Bank estimates that the informal sector accounts for between 25% and 40% of GDP in developing countries of Asia and Africa. Whether undertaken to avoid taxes, to hide illegal enterprises, or simply reflecting the efforts and employment of those unable to find jobs with registered businesses, informal economic activity obviously distorts government statistics on total employment and real GDP and their agricultural, industrial, and service components (Figure 10.6).

The world and regional impacts of continuing shifts in global economic balance are obscured by traditional market exchange rate estimates of the relative size of national economies. Those estimates measure gross national income or per capita share of gross domestic product in U.S. dollar equivalents, making no assessment of relative price levels in different countries or of the value of nontraded goods and services. Seeking a more realistic measure of national economies, the International Monetary Fund and the World Bank now use **purchasing power parity (PPP),**

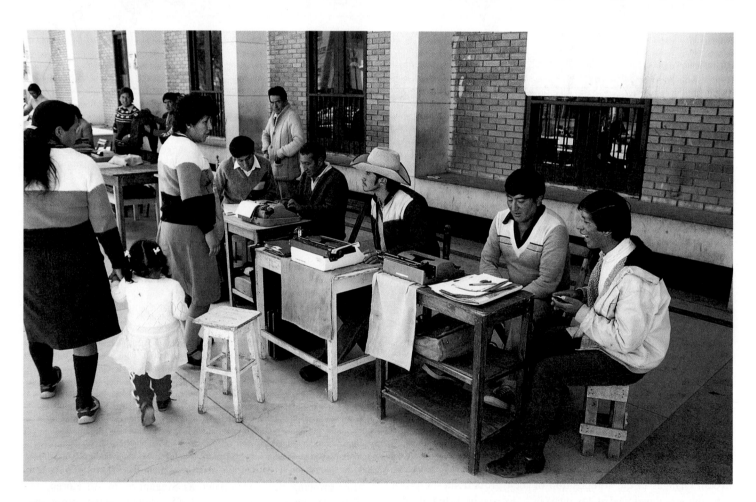

Figure 10.6 "Informal sector" initiative by street typists for hire in Huancayo, Peru. Between 1965 and 2000, the percentage of the labor force in agriculture dropped precipitously in all countries of Latin America. That decline was not matched by a proportional increase in jobs in manufacturing and other industries. In Peru, employment in agriculture fell from 50% to less than 8% of the total, but the share of workers in industry remained constant at about 20%. Many of the former rural workers found urban work in the informal or shadow economy sector, the generator of 60% of all new Latin American jobs in the 1990s. They became errand runners, street vendors, odd-job handymen, open-air dispensers of such personal services as barbering, shoe shining, clothes mending, letter-writing, and the like, as well as unregistered workers in small-scale construction and repair shops. For Latin America as a whole, the UN reports, 74% of the female and 55% of the male labor force worked in the informal economy at the start of the century. For developing countries as a group, informal sector employment makes up 37% of total jobs and reaches 45% in Africa.

which takes account of what money actually buys in each country. (PPP is based on the idea that an identical basket of traded goods should cost the same in all countries. If, for example, a loaf of bread costs $1 in the United States but only 50 cents in Thailand, the Thai citizen's assigned per capita dollar income should be adjusted upward to reflect what can actually be bought with a dollar's equivalent in Thai currency.) When the new PPP measure was introduced in 1993, it gave a clearer picture of the world economy and radically changed traditional assessments of it.

Immediately, the relative economic importance of the Third World doubled. Using purchasing power parity, East Asia's share of world 2000 output jumped from 6% to 17%, and the revised weight for all developing states vaulted from 20% to 45%. That is, the rich industrial economies in 2000 accounted for just 55% of total global output using PPPs, a marked come-down from their assumed 80% contribution under the old exchange rate system. If present trends hold, the World Bank forecasts, by 2020 the current "rich world's" share of global output could shrink to less than two-fifths.

Impressive as the shifts in global economic balance between the industrial and developing countries may be, they do little to reduce the contrasts that remain between the richest of the North and the poorest of the South. While as a group the developing countries—127 of them as usually defined—are catching up to the developed world in total productivity, a good part of the aggregate gain is lost in per capita terms (see "Poverty and Development"). With the South's first-of-century population increasing

Poverty and Development

According to the World Bank and the United Nations Development Programme (UNDP), of the world's start of-century 6 billion people, 2.8 billion—almost half—lived on less than $2 per day. Over 1.2 billion— about a fifth of the world total— experience "absolute poverty." These folk exist on an income of less than $1 per day ($440 per year in purchasing power parity), a figure below which, it is usually reckoned, people are unable to buy adequate food, shelter, and other necessities. This is the same number, though a slightly smaller percentage of the world's population, as were poor in 1990. About 44% of the very poor live in South Asia and another 25% in sub-Saharan Africa. East Asia and the Pacific account for 23% of those in absolute poverty and Latin America and the Caribbean add another 6.5%.

Although the dollar definition of poverty is applied as if it were a worldwide constant, in reality poverty is comprised of two separate elements that are regional variables. One of these is the reasonably objective observation that you are impoverished if you can't afford a minimum standard of nutrition. The other element is more subjective and equates poverty with inability to buy basic goods that other citizens of your country regard as necessities. The UNDP has attempted to combine these separate elements in devising a human poverty index (HPI) that, first, identifies poverty populations by the simple income test and then concentrates on measuring deprivation in three essential dimensions of human life: longevity (percentage of people not expected to live to age 40); knowledge (percentage of adults who are il-

literate); and deprivation in living standards (measured by access to safe water, access to health services, and percentage of malnourished children under 5).

By these standards, the world made significant progress in reducing human poverty in the 1990s. In developing countries, the percentage of newly born people not expected to survive to age 40 declined from 20% to 14% during the decade. Adult illiteracy dropped from 35% to 28% and access to safe water increased from 68% to 72%. The purchasing power income poverty rate, even at the $1-a-day standard, dropped from 17% of world population in 1970 to about 7% in 2000. Income poverty declined during the 1990s in every developing region, though the decline was not uniform and ranged from 11 percentage points in East Asia to only 0.3 percentage points in sub-Saharan Africa.

The UNDP reports that some countries have made notable progress. Malaysia reduced income poverty from 60% in 1960 to 14% in 1993, China from 33% in 1978 to 18% in 1998, and India from 54% in 1974 to 44% in 1997. In contrast, in Niger, the African country lowest on the human poverty index scale at the start of the century, some two-thirds of residents live in conditions of poverty, the highest figure for any of the world's developing countries. In the Western Hemisphere, Haiti has the highest HPI, and Nicaragua is the Latin American poverty leader. In Asia, Nepal is worst off among South Asian states and Yemen leads the Arab world.

Poverty is an areal variable within countries. In Burkina Faso and Zambia, rural poverty fell and urban poverty rose during the 1990s. In Mexico between 1989 and 1994, overall poverty declined modestly, but there were large variations across regions within the country. In Thailand, the incidence of poverty in the rural northeast was almost double the national average in 1992. The World Bank notes that, in general, poverty tends to be associated with distance from cities and the seacoast.

Africa is a continuing problem region. Because of its rapid population growth, stagnation or decline in per capita food production, weakness in infrastructure and facilities systems, periodic drought, and devastating civil wars, in sub-Saharan Africa the number of poor people increased from 217 million in 1987 to more than 291 million in 1999. The World Bank projects that the sub-Saharan region will have 360 million people or 40% of its population in poverty in 2010.

One key to improving both the economic and social lot of the "poorest of the poor," the World Bank and United Nations argue, is to target public spending on their special needs of education and health care and to pursue patterns of investment and economic growth that can productively employ that underutilized and growing labor force so abundant in the least developed countries. Another needed solution, many argue, is for rich creditor countries and international lending agencies to provide relief from the massive burden of interest and principal repayments owed by the group of 40 or more highly indebted poor countries (HIPCs).[a]

---

[a]See "Does Foreign Aid Help?", p. 372.

at about a 2% per annum rate, growing national prosperity has to be—at least statistically—divided among an increasing number of claimants. In the early 1990s, the South had to apportion its recalculated one-third share of *gross global product* (the total value of goods and services produced by the world economy) among more than three-fourths of world population. By 2000, the South's population had increased by more than 600 million to over 80% of the world total. Because of that growth, the number of persons worldwide living in poverty hardly changed during the 1990s. And the very rich, in contrast, continue to get richer. Average income in the richest 20 countries was at the end of the 1990s 37 times the average in the poorest 20—a gap that had doubled over the preceding 40 years.

As core-periphery theorists suggest, the industrial economies still account for three-quarters of world exports and dominate the international financial markets. Further, the South's economies are far more (though decreasingly) oriented toward raw material production than are those of the developed world. For example, in 1965, more than 80% of the exports from developing countries—but less than one-third of those from industrial market countries—were minerals and agricultural products. By 2000, the disparity had materially lessened; minerals and agricultural goods represented only 23% by value of the developing countries'[3] exports and 14% of those from industrial market economies.

The Diffusion of Technology

Composite figures mask the disparities that exist within the ranks of developing countries. The world's 49 "least developed" states in 2001 produced less than 1% of global wealth, measured at market exchange rates. In contrast, the small, industrialized "four Tigers"—Hong Kong, Singapore, South Korea, and Taiwan—alone contributed 3.5% to gross world product. Obviously, there are differences within the developing world in the successful application of technology to the creation of wealth.

Technology refers to the totality of tools and methods available to and used by a culture group in producing items essential to its subsistence and comfort. We saw in Chapter 2 how in antiquity there emerged *culture hearths*—centers of technological innovation, of new ideas and techniques that diffused or were carried out from the core region. Innovation is rarely a single event; as cultures advance, needs multiply and different solutions develop to meet expanding requirements. The ancient hearths (see Figure 2.15) were locales of such multiple invention and innovation. Their modern counterparts are the highly urbanized, industrialized advanced nations whose creativity is recorded by patent registrations and product and process introductions. The changing rate of innovation over time is suggested by Figure 2.20.

In all periods there has existed between hearths and outlying regions a **technology gap,** a contrast in the range and productivity of artifacts introduced at the core and those known or employed at the periphery. That gap widened at an accelerating rate as technology moved farther away from the shared knowledge of earlier periods. During the Industrial Revolution, the technological distance between cottage hand looms of 18th-century English villagers and the power looms of their neighboring factories was

one of only moderate degree (Figure 10.7). Far greater is the gap between the range of traditional crafts known throughout the world and the modern technologies of the most advanced societies. It is much more difficult now for a less developed society to advance to the "state of the art" by its own efforts than it was for British colonies or the rest of Europe to re-create the textile or iron-making industries first developed in England.

The persistence and expansion of the technology gap suggest that the idea of **cultural convergence**—the increasing similarity in technologies and ways of life among societies at the same levels of development—does not as well unite the most and the least advanced economies. In the modern world, as we saw in Chapter 2 (p. 51), there is a widespread sharing of technologies, organizational forms, and developed cultural traits. But not all countries are at the same developmental state. Not all are equally able to draw on advanced technology to create the same products with identical efficiency and quality, although there is increasing awareness of the existence of those products and the benefits of their use.

The technology gap matters. At any given level of technology, the resources of an area will have a limited population supporting capacity. As population growth approaches or exceeds those limits, as it has in many less developed areas of the globe, poverty, famine, and political and social upheaval can result. Understandably, all countries aspire to expand their resource base, increase its support levels through application of improved technologies, or enter more fully into an income-producing exchange relationship with other world regions through economic development. Their objective is a **technology transfer,** placing in their own territory and under their own control the productive plants and processes marking the more advanced countries. The chemical plant of Bhopal was one item in a technology transfer sought

Figure 10.7 Early in the Industrial Revolution, new techniques that diffused most readily from the English hearth were those close to handicraft production processes. In some industries, the important innovation was the adoption of power and volume production, not radically new machines or products. For textiles and similar light industries, capital requirements were low and workers required little training in new skills. The picture shows carding, drawing, and spinning machinery built by memory in the United States in 1790 by Samuel Slater, an Englishman who introduced the new technology despite British prohibitions on exports of drawings or models of it.

---

by the state of Madhya Pradesh and the government of India, one step in the process of moving the region and country further along the continuum from less developed to more developed status.

Not all technology, of course, is equally transferable. Computers, information management techniques, cell phones, and the like easily make the move between advanced and emerging economies. Other technologies, particularly in the life sciences, materials innovation, and energy, are more specific to the markets, monetary resources, and needs of the rich countries and not adapted to those of the less developed states. Even where transfer is feasible, imported innovations may require domestic markets sufficient to justify their costs, markets that poor countries will not possess at their current national income levels. And the purchase of technology presumes recipient country export earnings sufficient to pay for it, again a condition not met by the poorest states.

Developing countries have, in a form of reverse flow, contributed to scientific and engineering innovation. Advanced technologies and scientific breakthroughs depend on public and private research institutions and corporate research and development departments common in the rich states. Many of the advances produced by those agencies have been made by poor-country scientists working in the rich-country laboratories. Indian and Chinese technologists and engineers, for example, are major components in the work force of all the high-tech concentrations discussed in Chapter 9 (p. 340).

The Complex of Development

Technology transfer is only one aspect of economic development. The process as a whole intricately affects all facets of social and economic life. The terms *level of living* and *standard of living* bring to mind some of the ways in which economic advancement implies both technological and societal change, including amount of personal income, levels of education, food consumption, life expectancy, and the availability of health care. The complexity of the occupational structure, the degree of specialization in jobs, the ways in which natural resources are used, and the level of industrialization are also measures of development and of the innovation or adoption of technology within a society.

The generalized summary table "Relative Characteristics of Development" outlines some of the many implications and attributes of development. It also makes clear why no single measure

Relative Characteristics of Development

Less Developed

1. Per capita incomes are low, and capital is scarce.
2. Wealth is unevenly distributed within individual countries (e.g., in Colombia 2.6% of population owns 40% of the national wealth, and in Gabon 1% owns 56% of total wealth).
3. Primary industries (farming, forestry, quarrying, mining, fishing) dominate national economies.
4. High proportion (over 50%) of population is engaged in agriculture.
5. Farming is mostly at the subsistence level and is characterized by hand labor methods and underemployment. Farm holdings are small, mechanization is limited, and crop yields are low.
6. Populations are dominantly rural, though impoverished urban numbers are growing.
7. Birth and death rates are high, and life expectancy is low. There tends to be a high proportion of children. Rates of natural increase are high.
8. Inadequate or unbalanced diets resulting from a relatively low consumption of protein; hunger and malnutrition are common.
9. Infectious, respiratory, and parasitic diseases are common; medical services are poor.
10. Overcrowding, poor housing, few public services, and bad sanitation yield poor social conditions.
11. Poor educational facilities and high levels of illiteracy hinder scientific and technological advancement. (In sub-Saharan Africa, over 40% of population is illiterate.)
12. Women may be held in inferior position in society.

Developed

1. Per capita incomes are high, and capital is readily available.
2. Wealth within individual countries is comparatively evenly distributed (e.g., in Canada, 10% of the population owns 24% of national wealth).
3. Manufacturing and service industries dominate national economies.
4. Very small proportion (under 10%) of population is engaged in agriculture.
5. Farming is mostly commercial, efficient, and highly mechanized. Farm holdings are generally large, and crop yields are high.
6. Populations are predominantly urban, with near 80% living in towns and cities.
7. Birth and death rates are low, and life expectancy is high. There is often a high proportion of people over 65 years of age. Rates of natural increase are low.
8. Generally adequate supplies of food and balanced diets; overeating is sometimes a problem.
9. Primary diseases are related to age and lifestyle; good medical services are available.
10. Social conditions are generally good, with adequate housing space and a high level of public health facilities and sanitation.
11. Highly developed educational facilities and low levels of illiteracy are the norm. Technical proficiency is advanced.
12. Women are increasingly treated on equal terms with men.

Adapted with permission from Charles Whynne-Hammond, *Elements of Human Geography*. 2d ed., p. 171 (London: Unwin Hyman Limited, 1985).

is sufficient to assess the comparative stage of economic development or level of living of a society. We might, for example, simplistically assume that national contrasts in average per capita income would serve to measure the level of living in all of its aspects; but personal income figures are particularly hard to compare across national borders. An income of (U.S.) $50,000 in Sweden is taxed at a much higher rate than a similar income in the United States. But social welfare programs, higher education, and medicine receive greater central governmental funding in Sweden; the American family must set aside a larger portion of its income for such services.

Further, identical incomes will be spent on different amounts and types of goods and services in different countries. Americans, because of lower prices, spend a smaller proportion of their personal income on food than do the residents of most European states or Japan; those living in upper latitudes must buy fuel and heavy clothing not necessary in tropical household budgets; and price levels may vary widely in different economies for similar essential goods. Of course, national average personal income figures do not indicate how earnings are distributed among the citizenry. In some countries, the wealthiest 5% of the population control over 50% of the income, whereas in others, revenues are more uniformly distributed.

To broaden the limited view afforded by per capita income figures, therefore, a variety of more specific and descriptive mea-

sures has been employed to suggest national levels of development. Each such measure can present only part of a total picture of developmental status. Taken together, however, the comparative criteria tend to show a high, but not perfect, correlation that collectively supports the accepted North-South global split. The primarily economic measures discussed here are among those commonly accepted as most revealing of the relative progress of countries of the "South" along the scale of development. They are (1) gross national income and purchasing power parity (PPP) per capita, (2) per capita energy consumption, (3) percent of the workforce engaged in agriculture, and (4) calorie intake per capita.

Gross National Income and PPP per Capita

Gross national income (GNI) is a commonly available statistic (formerly known as gross national product) that reports the total market value of goods and services[4] produced within an economy within a given time period, usually a year. Expressed in per capita terms (see Appendix B), GNI is the most frequently used indicator of a country's economic performance (Figure 10.8).

[4]Adjusted by deducting income earned by foreign interests and adding income accruing to residents from foreign investments or remittances. Without these adjustments, the statistic is called **gross domestic product,** itself subject to different adjustments.

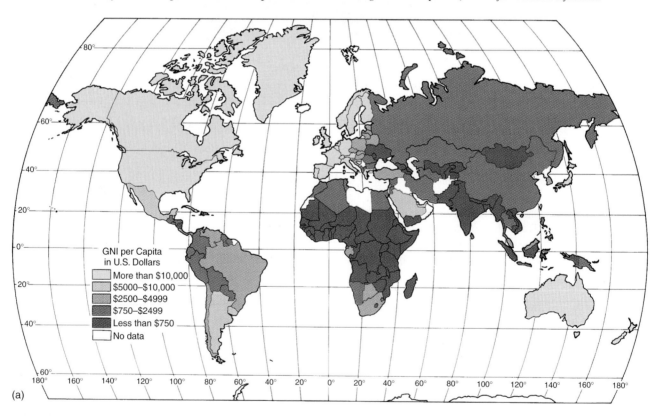

(a)

Figure 10.8 **Gross national income per capita.** GNI per capita is a frequently employed summary of degree of economic advancement, though high incomes in sparsely populated, oil-rich countries may not have the same meaning in developmental terms as do comparable per capita values in industrially advanced states. The map implies an unrealistic precision. For many states, when uncertain GNI is divided by unreliable population totals, the resulting GNI per capita is at best a rough approximation that varies between reporting agencies. A comparison of this map and Figure 10.11 presents an interesting study in regional contrasts.

Source: Data from World Bank and Population Reference Bureau.

Like any other single index of development, gross national income tells only part of a complex story. Indeed, its concept, and that of the related gross domestic product, is under increasing attack for its assumed distortions of reality. One group, including environmentalists, argues that the GNI overstates the wealth of a society by ignoring the cost of ecological damage and the drain modern economies place on natural resources (see Figure 10.5). An opposing group holds that GNI understates the strength of economic growth by overlooking much of the quality and productivity improvements brought by technology (safer automobiles, faster, more powerful computers, etc.).

Of course, gross national income per capita is not a personal income figure, but simply a calculated assignment of each individual's share of a national total. Change in total population or in total national income will alter the average per capita figure but need have no impact on the personal finances of any individual citizen. Nor is per capita GNI a totally realistic summary of developmental status. It tends to distort a more inclusive picture of underdevelopment by overemphasizing the purely monetary circumstances of countries and not accurately representing the economic circumstances of countries with dominantly subsistence economies, for example, many of the nations of Asia and Africa with low income figures.

As expected, the countries with the highest GNI per capita are those in northwestern Europe, where the Industrial Revolution began, and in the midlatitude colonial areas—North America, Australia, and New Zealand—to which the new technologies were first transplanted. In the middle position are found many of the countries of Latin America and of southern and eastern Europe. Large sections of Africa and Asia, in contrast, are at the low end of average income figures, since the money value of the nontraded goods and services that subsistence farmers provide for themselves and their communities goes unrecorded in the GNI. That problem is partly resolved, you will recall, by calculating what are sometimes called "real per capita gross domestic products," but more usually summarized as *purchasing power parities* (see Figure 10.9 and p. 360).

Energy Consumption per Capita

Per capita energy consumption is a common measure of technological advancement of nations because it loosely correlates with per capita income, degree of industrialization, and use of advanced technology. In fact, the industrialized countries use about 10 times more energy on a per capita basis than developing economies do. The consumption rather than the production of

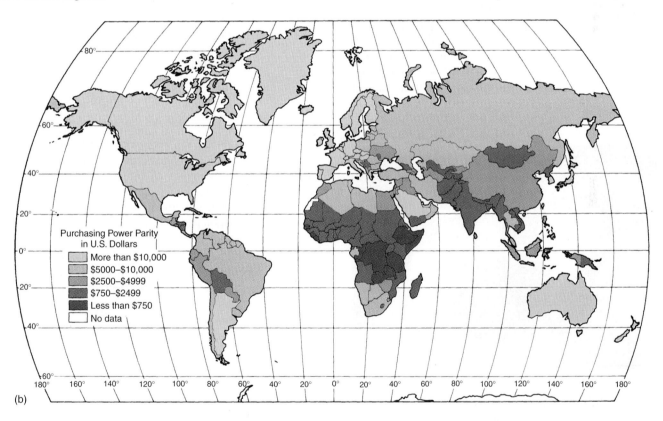

Figure 10.9 **Purchasing power parity (PPP).** When local currency measures of gross national income or gross domestic product are converted into purchasing power parities, there is a twofold revision of the usual view of world economic status. The first result is a sharp increase in developing countries' share of total world output. By the PPP calculation, China has the world's second largest economy, and India, Mexico, and the Russian Federation all emerge as bigger than Canada. Second, the abject poverty suggested by per capita gross national income or gross domestic product is seen to be much reduced in many developing countries. India, for example, showed a 2001 gross national income per capita at market exchange rates of $460; in purchasing power parity, the figure rose to $2820, and the Central African Republic's people jumped from $260 to $1300. Compare this map with Figure 10.8 to see how PPP changes our impressions of some countries' economic status.

Sources: Data from World Bank and United Nations.

energy is the concern. Many of the highly developed countries consume large amounts of energy but produce relatively little of it. Japan, for example, must import from abroad the energy supplies its domestic resource base lacks. In contrast, many less developed countries have very high per capita or total energy production figures but primarily export the resource (petroleum). Libya, Nigeria, and Brunei are cases in point. Most of the less and least developed countries depend less on commercial forms of inanimate energy (petroleum, coal, lignite, natural gas, hydropower, etc.) than they do on animate energy (human and animal labor) and the firewood, crop residues, dung, peat, and other domestic fuels on which subsistence populations must depend. Both rudimentary and some advanced technologies are locally and gradually improving that picture as solar stoves, waste matter converters (Figure 10.10), solar photovoltaic panels, and the like come into use.

The advanced countries developed their economic strength through the use of cheap energy and its application to industrial processes. But energy is cheap only if immense capital investment is made to produce it at a low cost per unit. The less advanced nations, unable to make those necessary investments or lacking domestic energy resources, use expensive animate energy or such decreasingly available fuels as firewood (see "The Energy Crisis in LDCs"), and they must forgo energy-intensive industrial development. Anything that increases the cost of energy further removes it from easy acquisition by less developed countries. Periodic surges in petroleum prices beginning in the 1970s and the consequent increase in the price of all purchased energy supplies served to widen further the gulf between the technological subsystems of the rich and the poor countries of the world.

𝒻𝒾𝑔𝓊𝓇𝑒 10.10 An anaerobic digester in Pakistan. Human, animal, and vegetable wastes are significant energy sources in developing economies such as Pakistan, India, Thailand, and China. There, such wastes are fermented to produce methane gas (*biogas*) as a fuel for cooking, lighting, and heating. The simple technology involves only a stone fermentation tank (foreground, above) fed with wastes—straw and other crop residues, manure, human waste, kitchen scraps, and the like. These are left to decompose and ferment; the emitted methane gas passes into a large collection chamber (background tank) and later is drawn through a hose into the farm kitchen. After the gas is spent, the remaining waste is pumped out and used for fertilizer in the fields.

Percentage of the Workforce Engaged in Agriculture

A high percentage of employment in agriculture (Figure 10.11) is almost invariably associated with low per capita gross national income and low energy consumption, that is, with underdevelopment. Economic development always means a range of occupational choices far greater than those available in a subsistence agricultural society. Mechanization of agriculture increases the productivity of a decreasing farm labor force; surplus rural workers are made available for urban industrial and service employment, and if jobs are found, national and personal prosperity increases. When a labor force is primarily engaged in agriculture, on the other hand, subsistence farming, low capital accumulation, and limited national economic development are usually indicated.

Landlessness

Developing region economies devoid of adequate urban industrial or service employment opportunities can no longer accommodate population growth by bringing new agricultural land into cultivation. In the most densely settled portions of the developing world, rural population expansion increasingly means that new entrants to the labor force are denied access to land either through ownership or tenancy. The problem is most acute in southern Asia, particularly on the Indian subcontinent, where the landless rural population is estimated to number some 290 million—as large as the total population of the United States. Additional millions have access to parcels too small to adequately feed the average household. A landless agricultural labor force is also of increasing concern in Africa and Latin America (Figure 10.12).

Landlessness is in part a function of an imbalance between the size of the agricultural labor force and the arable land resource. It is also frequently a reflection of concentration of ownership by a few and consequent landlessness for many. Restricted ownership of large tracts of rural land appears to affect not just the economic fortunes of the agricultural labor force itself but also to depress national economic growth through inefficient utilization of a valuable but limited resource. Large estates are often farmed carelessly, are devoted to production of crops for export with little benefit for low-paid farm workers, or even left idle. In some societies, governments concerned about undue concentration of ownership have imposed restrictions on total farm size—though not always effectively.

In Latin America, where farms are often huge and most peasants landless, land reform—that is, redistribution of arable land to farm workers—has had limited effect. The Mexican revolution early in the 20th century resulted in the redistribution of nearly half the country's agricultural land over the succeeding 60 years, but the rural discord in Chiapas beginning in the 1990s reflects the persistence there of underutilized large estates and peasant landlessness. The Bolivian revolution of 1952 was followed by a redivision of 83% of the land. Some 40% of Peru's farming area was redistributed by the government during the 1970s. In other Latin American countries, however, land reform movements have been less successful. In Guatemala, for example, 85% of rural households are landless or nearly so, and the top 1% of landowners control 34% of arable land.

The Energy Crisis in LDCs

"The poor man's energy crisis" is a phrase increasingly applicable to the rising demand for and the decreasing supply of traditional fuels—wood, crop residue, dung, and the like—in the developing countries. It is a different kind of crisis from that faced by industrialized countries encountering rising prices and diminishing supplies of petroleum and natural gas. The crisis of the less developed societies involves cooking food and keeping warm, not running machines, cooling theaters, or burning lights.

More than 2.5 billion people in developing countries depend on the traditional fuels, primarily wood. The UN Food and Agriculture Organization, indeed, estimates that wood accounts for at least 60% of the fuel used in the developing countries and exceeds 90% in the poorest countries such as Ethiopia and Nepal. The agency reports that wood accounts for nearly two-thirds of all energy consumed in Africa (excluding Egypt and South Africa), more than 40% in the Far East (excluding China), 20% in Latin America, and 14% in the Near East. Demand for fuelwood, the main or sole source of domestic energy for two-fifths of the world's population, continues to grow by well above 1% per year, and declining supplies are having serious human and natural consequences. More than 100 million people consume amounts of energy—mainly fuelwood— "below minimum requirements" for cooking, heating, and other domestic purposes. Another 1.3 billion people meet their needs only by serious depletion of the wood reserves upon which they totally depend. Some two-thirds of those people live in Asia. The most serious shortages and depletions are in the drier areas of Africa (more than 50 million Africans face acute fuelwood shortages), in the mountainous districts of Asia—the Himalayas are particularly affected—and in the Andean uplands of Latin America.

As a result of shortages and deforestations in such widely scattered areas as Nepal and Haiti, families have been forced to change their diets to primary dependence on less nutritious foods that need no cooking. Reports of whole villages reduced to only one cooked meal a day are common. With the average villager requiring a ton of wood per year, an increasing proportion of labor must be expended to secure even minimal supplies of fuel, to the detriment of food- or income-producing activities. In parts of Tanzania in East Africa, because of time involved in traveling to and from forestlands and gathering the wood itself, between 250 and 300 workdays are needed to fill the yearly firewood needs of a single household. The figure is 230 person-days in the highlands of Nepal. Growing populations assure that the problem of fuel shortages will continue to plague developing countries even though recently introduced improved stoves, solar reflector ovens, and backyard fermentation tanks to convert human and animal excreta and organic wastes into methane gas (biodigesters) for cooking, lighting, and heating fuel have begun to lower per capita fuelwood use in many regions.

In India, where two-thirds of rural families either have no land at all or own less than 2 hectares (5 acres), a government regulation limits ownership of "good" land to 7 hectares (18 acres). That limitation has been effectively circumvented by owners distributing title to the excess land to their relatives. Population growth has reduced the amount of land available to the average farmer on the Indonesian island of Java to only 0.3 hectares (three-quarters of an acre), and the central government reports that over half of Java's farmers now work plots too small to support them.

The rural landless are the most disadvantaged segment of the poorest countries of the least developed regions of the world. They have far higher levels of malnutrition and incidence of disease and lower life expectancies than other segments of their societies. In Bangladesh, for example, the rural landless consume only some 80% of the daily caloric intake of their landholding neighbors. To survive, many there and in other countries where landlessness is a growing rural problem leave the agricultural labor force and migrate to urban areas, swelling the number of shantytown residents but not necessarily improving their fortunes.

Poverty, Calories, and Nutrition

Poverty is the most apparent common characteristic of countries, regions, communities, or households afflicted by malnourishment. Availability of urban employment or rural access to arable land is far more important in determining national levels of undernourishment than is a country's aggregate per capita food production. During the Bangladesh famine of 1974, for example, total food availability per capita was at a long-term peak; starvation, according to World Bank reports, was the result of declines in real wages and employment in the rural sector and short-term speculative increases in the price of rice. In India in 2002, huge stockpiles of government-owned wheat purchased at high subsidy costs rotted in storage while held for sale at prices beyond the reach of malnourished or starving but impoverished citizens.

Nourishment levels, therefore, are as truly an indicator of economic development of a country as are any of the dollar-indexed measures of production and income or summary statements about the structure of national employment. Indeed, no other economic measure of national prosperity or development level can be as meaningful as the availability of food supplies sufficient in caloric content to meet individual daily energy requirements and so balanced as to satisfy normal nutritional needs. Food, as the essential universal consumption necessity and the objective of the majority of human productive activity, is the ultimate indicator of economic well-being.

Calorie requirements to maintain moderate activity vary according to a person's type of occupation, age, sex, and size, and to climate conditions. The Food and Agriculture Organization (FAO) of the UN specifies 2360 calories as the minimum necessary daily consumption level, but that figure has doubtful universal applicability. By way of a benchmark, per capita daily calorie availability in the United States is nearly 3700. Despite the

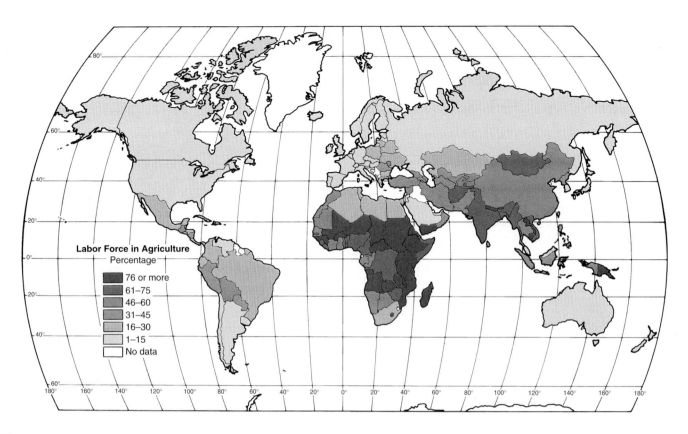

Figure 10.11 **Percentage of labor force engaged in agriculture.** For the world as a whole, agricultural workers make up slightly less than half of the total labor force. Highly developed economies usually have relatively low proportions of their labor forces in the agricultural sector, but the contrast between advanced and underdeveloped countries in the agricultural labor force measure is diminishing. Rapid Third World population growth has resulted in increased rural landlessness and poverty from which escape is sought by migration to cities. The resulting reduction in the agricultural labor force percentage is an expression of relocation of poverty and unemployment, not of economic advancement.

Sources: Data from C.I.A., The World Factbook 2002 *and United Nations Development Programme.*

Figure 10.12 Throughout much of the developing world, growing numbers and proportions of rural populations are either landless agricultural laborers with, at best, tiny garden plots to provide basic food needs or independent holders of parcels inadequate in size or quality to provide food security for the family. In either case, size of land holding and poverty of farmer restrict the operator to rudimentary agricultural implements and practices. In this photo, a Nuer woman of Sudan cultivates corn with a simple hand tool.

limitations of the FAO standards, Figure 10.13 uses them to report calorie intake as a percentage of the minimum daily requirements.

Like other national indicators, caloric intake figures must be viewed with suspicion; the dietary levels reported by some states may more reflect self-serving estimates or fervent hopes than actual food availability. Even if accurate, of course, they report national averages, which may seriously obscure the food deprivation of large segments of a population. But even the data reported on Figure 10.13 show nearly one-seventh of the world's population, almost all in the less developed countries, to be inadequately supplied with food energy (Figure 10.14). Areal or household incidences of hunger or malnutrition that exist in the more developed and affluent economies are masked within high national averages.

Low caloric intake is usually coupled with lack of dietary balance, reflecting an inadequate supply of the range and amounts of carbohydrates, proteins, fats, vitamins, and minerals needed for optimum physical and mental development and maintenance of health. The World Health Organization estimates that more than 2 billion people worldwide suffer from some form of micronutrient malnutrition that leads to high infant and child mortality, impaired physical and mental development, and weakened immune responses. As Figure 10.13 indicates, dietary insufficiencies—with inevitable adverse consequences for life expectancy, physical vigor, and intellectual acuity—are most likely to be encountered in those developing countries that have large proportions of their populations in the young age groups (see Figure 4.11). Indeed, undernourishment is damaging and widespread throughout the developing world where, collectively, 30% of children under 5 years are moderately to severely underweight and one-third are stunted. South Asia shows the highest incidence of childhood nutritional problems measured by standardized weight-for-age and weight-for-height measures. There, of the under-5 age group 49% are moderately or severely underweight, 48% show stunting, and half the world's undernourished children are found. Malnutrition among young Indians, for example, is proportionately nearly twice as high as in sub-Saharan Africa.

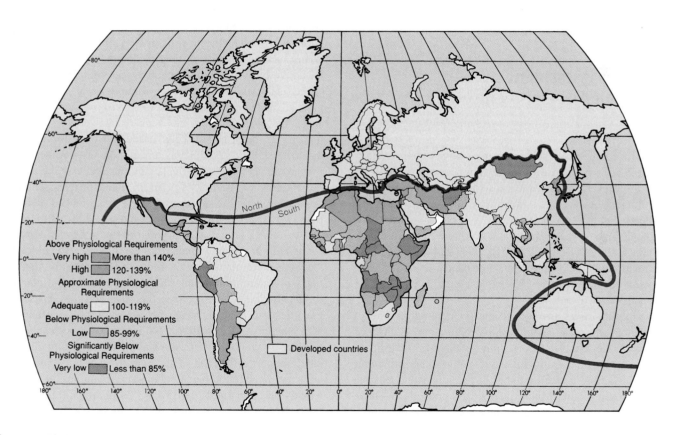

Figure 10.13 **The South: Percentage of required dietary energy supply received daily.** If the world's food supply were evenly divided, all would have an adequate diet; each person's share would be between 2600 and 2700 calories. Even developing world populations *as a group* would be adequately fed with between 2400 and 2500 calories each. In reality, in 2002, there were around 840 million undernourished people worldwide facing chronic hunger or starvation, undernutrition, and deficiencies of essential iron, iodine, Vitamin A, and other micronutrients. Further, for many, sickness and parasites take the nutritive value from what little food is eaten.

Yet, progress in feeding the world's people has been made. In 1970, more than one-third of people living in poor countries were undernourished; by 2000 the figure had been lowered to 16%. Only in Africa has the hunger problem remained largely the same, a product of the continent's continuing poverty and progressive drop in per capita food production since the 1960s (see figure 8.11). In contrast to the regional variation shown on the map of the South, all countries of the North have average daily per capita caloric intake above 110% of physiological requirements.

Sources: Data from Bread for the World Institute and United Nations Development Programme.

Figure 10.14 Malnourished Sudanese children at an aid center. The FAO estimates that early in the 21st century, some 200 million children under 10 years of age were among the over 800 million people chronically undernourished in the developing world alone. As a result of hunger, 6 million children under the age of 5 die each year. The occasional and uncertain supplies of food dispensed by foreign aid programs and private charities are not sufficient to assure them of life, health, vigor, or normal development.

Composite Assessment of Economic Development

Although single-factor evaluations of technological development tend to identify the same set of countries as "less developed," the correspondence is not exact. For each measure selected for comparison, each country finds itself in the company of a slightly different set of peers. As a consequence, no one individual measure of technological development, wealth, or economic well-being fully reflects the diversity of characteristics of individual countries, though revealing summary indexes have been prepared. Using data from the 1950s, Brian Berry (1934–) compressed through factor analysis, 43 different, dominantly economic measurements of development for 95 separate countries into a single index of technological status.

More recent similar attempts at country ranking on an economic development scale show the earlier situation has changed very little with the passage of time. Some shifts have occurred, of course. Thanks to oil wealth, some petroleum exporting countries, like Libya, and a small number of recently industrializing countries—South Korea, Malaysia, Taiwan, and others—have moved out of the ranks of the "least developed." India placed relatively well in the 1950 rankings, but 50 years later its low GNI per capita put it among the poorer countries of the world. Such relatively modest changes emphasize rather than contradict the conclusion that improvements in a country's relative technological position and national wealth are difficult to achieve.

A Model for Economic Development

The realization that economic growth is not automatic and inevitable has been a discouraging reversal of an earlier commonly held and optimistic belief: that there was an inevitable process of development that all countries could reasonably expect to experience and that progress toward development would be marked by recognizable stages of achievement.

A widely cited model for economic advancement was proposed in 1960 by W. W. Rostow (1916–2003). Generalizing on the "sweep of modern history," Rostow theorized that all developing economies may pass through five successive stages of growth and advancement. *Traditional societies* of subsistence agriculture, low technology levels, and poorly developed commercial economies can have only low productivity per capita. The *preconditions for takeoff* are established when those societies, led by an enterprising elite, begin to organize as political rather than kinship units and to invest in transportation systems and other productive and supportive infrastructure.

The *takeoff* to sustained growth is the critical developmental stage, lasting perhaps 20 to 30 years, during which rates of investment increase, new industries are established, resources are exploited, and growth becomes the expected norm. The *drive to maturity* sees the application of modern technology to all phases of economic activity; diversification carries the economy beyond the industrial emphases first triggering growth, and the economy becomes increasingly self-sufficient. Finally, when consumer goods and services begin to rival heavy industry as leading economic sectors and most of the population has consumption levels far above basic needs, the economy has completed its transition to the *age of mass consumption.* More recently—and referring to most advanced economies (and discussed in Chapter 9)—a sixth stage, the *postindustrial,* has been recognized. Services replace industry as the principal sector of the economy, professional and technical skills assume preeminence in the labor force, and information replaces energy as the key productive resource.

Rostow's expectations of an inevitable progression of development proved illusory. Many LDCs remain locked in one of the first two stages of his model, unable to achieve the takeoff to self-sustained growth despite importation of technology and of foreign aid investment funds from the more developed world (see "Does Foreign Aid Help?"). Indeed, it has become apparent to many observers that despite the efforts of the world community, the development gap between the most and the least advanced countries widens rather than narrows over time. A case in point is sub-Saharan Africa; between 1975 and 2000, per capita income declined by almost 1% a year, leaving all but a tiny elite significantly poorer at the end of the period. Over the same years, income per head in the industrial market economies grew at a 1.8% annual rate. The 1960s, 1970s, and 1980s were all proclaimed by United Nations resolutions as "Development Decades." They proved instead to be decades of dashed hopes—at least by economic measures—for many of the world's least developed states and for those who believed in definable and achievable "stages of development."

Other development theories and models try to address these realities. The concept of the "Big Push" concludes that underdeveloped economies can break out of their poverty trap by coordinated investment in both basic—but high wage—industries and infrastructure, creating simultaneously an expanding consumer base and steadily falling costs and rising volumes of production. These, in turn, encourage creation of backward- and forward-linked industries, further cost reductions, faster growth, and perhaps the industrial specializations that foster agglomeration economies and trade expansion.

Another viewpoint holds that national growth rate differentials are rooted in differing investments in "human capital," an ill-defined composite of skills, habits, schooling, and knowledge that—more importantly than labor force numbers or capital availability—contributes to successful economic development and sustained growth. Technological progress in recent decades, it is pointed out, has been notably dependent on more educated work forces equipped with high levels of capital investment. The current deep global imbalance in literate and technically trained people has been called the most potent force of divergence in well-being between the rich world and the poor.

A corollary concept concludes that for least developed or newly industrializing countries, incentives encouraging foreign direct investment and technology transfer are the most important policy. When imported ideas and technology that help create "human capital" labor and intellectual skills are combined with domestic industrial control, encouragement of education, and local research and development, there will certainly follow industrial specializations, massive exports, and rising levels of living—as presumably they did for Taiwan, Singapore, and the other Asian "tigers."

Obviously, not all less developed countries have been able to follow that same path to success. Indeed, those countries where the poorest 20% of the world's people live were, in the late 1990s, 60 times worse off than those where the richest fifth live, and the gap between the two groups had doubled since the early 1960s. *Dependency theory* holds that these differentials are not accidental but the logical result of the ability and necessity of developed countries and power elites to exploit and subjugate other populations and regions to secure for themselves a continual source of new capital. Transnational corporations, the theory contends, tend to dominate through their investments key areas of developing state economies. They introduce technologies and production facilities to further their own corporate goals, not to further the balanced development of the recipient economies. Development aid where proffered, dependency theory holds, involves a forced economic reliance on donor countries and economies that continues an imposed cycle in which, in a sense, selective industrialization leads not to independent growth but to further dependent underdevelopment—a negative consequence of circular and cumulative causation (see p. 358).

Balancing these polar extremes, we have seen, is the emerging world reality of accelerating economic growth for the developing countries as a group. Their present status as collective generators of nearly one-half of world output and 25% of world trade clearly suggests that, development theories aside, the best stimulus for economic development has been the widespread relaxation of restrictive economic and political controls on all economies and on international trade flows in the past generation. Transnational corporations, technology transfer, pro-development national policies, trade restriction relaxations, and selective foreign aid and lending have all played a part. But the major impetus to the transition to Rostow's *takeoff* to sustained growth appears to be near worldwide conversion from controlled to free market economies, a conversion to economic and cultural globalization resented, rejected, and resisted by many individuals and groups.

Noneconomic Measures of Development

Development is measured by more than economic standards, though income and national wealth strongly affect the degree to which societies can invest in education, sanitation, health services, and other components of individual and group well-being. Indeed, the relationship between economic and social measures of development is direct and proportional. The higher the per capita gross national income is, for example, the higher the national ranking tends to be in such matters as access to safe drinking water, prevalence of sanitary waste treatment, availability of physicians and hospital beds, and educational and literacy levels.

In contrast, the relationship between social-economic and demographic variables is usually inverse. Higher educational or income levels, that is, are usually associated with lower infant mortality rates, birth and death rates, rates of natural increase, and the like. However it is measured, the gap between the most and least developed countries in noneconomic characteristics is at least as great as it is in their economic-technological circumstances. Table 10.1 suggests that the South as a whole has made progress in reducing its disadvantages in some human well-being measures. In others, however, the gap between rich and poor remains or is increasing, and disparities still persist after the three UN "development decades."

Education

A literate, educated labor force is essential for the effective transfer of advanced technology from the developed to developing countries. Yet in the poorest societies two-thirds or more of adults are illiterate; for the richest, the figure is 1% or less (Figure 10.15). The problem in part stems from a national poverty that denies funds sufficient for teachers, school buildings, books, and other necessities of the educational program. In part it reflects the lack of a trained pool of teachers and the inability to expand their number rapidly enough to keep up with the ever-increasing size of school-age populations. In African countries worst hit by the AIDS epidemic, deaths among established teachers exceeded the supply of new teachers entering the profession beginning in the late 1990s. For the same number of potential pupils, the richest countries may have 20 to 25 times as many teachers as do the poorest countries. In Denmark in the late 1990s, there was 1 teacher for every 12 children of school age; in Burkina Faso the ratio was 1 to more than 270. Both wealth and commitment appear important in the student-teacher ratios. Oil-rich Qatar had 11 students per teacher; in similarly rich Saudi Arabia, the figure was nearly 30. Israel had more teachers per 1000 students than did wealthier Switzerland or the United States.

Lack of facilities and teachers, family poverty that makes tuition fees prohibitive and keeps millions of school-age children in full-time work, and national poverty that underfunds all levels of

Geography and Public Policy

Does Foreign Aid Help?

A 1998 World Bank report on "Assessing Aid" concluded that the raw correlation between rich-country aid and developing country growth is near zero. Simply put, more aid does not mean more growth, certainly not for countries with "bad" economic policies (high inflation, large budget deficits, corrupt bureaucracy); for them, the report claims, aid actually retards growth and does nothing to reduce poverty. Other studies similarly have found no clear link between aid and faster economic development. The $1 trillion rich countries and international agencies gave and loaned to poor ones between 1950 and 2000 did not have the hoped-for result of eliminating poverty and reducing economic and social disparities between the rich and poor countries of the world.

In part, that was because economic growth was not necessarily a donor country's first priority. During the Cold War, billions flowed from both the Soviet Union and the United States to prop up countries whose leaders favored the donor state agendas. Even today, strategic considerations may outweigh charitable or developmental aims. Israel gets a major share of American aid for historical reasons; Egypt, Pakistan, and Colombia get sizeable portions for political and strategic reasons; and Russia and Ukraine receive billions for not selling nuclear warheads. America, in fact, spends only 40% of its modest foreign aid budget on assistance to poorer states; the rest goes to middle-income countries.

About one-quarter of all aid from whatever source has been tied to purchases that must be made in the donor country, and additional large shares flow, regardless of need or merit, to former colonies of donor countries. In part, a World Bank report admits, aid failures reflect the fact that the Bank and its sister agencies have wasted billions on ill-conceived projects.

More optimistic conclusions are drawn by other observers who note that (*a*) foreign aid tends to reduce poverty in countries with market-based economic policies but is ineffective where those policies do not exist; (*b*) aid is most effective in lowering poverty if it is given to poor, rather than less poor, countries; and (*c*) aid targeted to specific objectives—eradication of disease or Green Revolution crop improvement, for example—are often remarkably successful, though spending on food aid or on aid tied to purchases from donor countries are of little use.

Although some countries—Botswana, the Republic of Korea, China, different Southeast Asian states—made great progress thanks to development assistance, a large number of others have seen their prospects worsen and their economies decline. Slow growth and rising populations have lowered per capita incomes, poor use of aid and loans has failed to improve their infrastructures and social service levels. Most critically for the economic and social development prospects of those countries is that the financing offered to them over the years in the hopes of stimulating new growth has become a burden of unmanageable debt.

So great and intractable has their debt problem become that the international community has now recognized a whole class of countries distinguished by their high-debt condition: Heavily Indebted Poor Countries (HIPCs) that are so far in debt that many of them are paying more in interest and loan payments to industrialized countries and international agencies than they are receiving in exports to or aid from those sources. Gradually, the rich world has accepted that debt relief, not lectures on capitalism, is the correct approach to helping the world's poor countries and people. In 1996 the World Bank, International Monetary Fund, and other agencies launched the first HIPC initiative, identifying 41 very poor countries and acknowledging that their total debt burden (including the share owed to international institutions) must be reduced to sustainable levels. In the years since, differing definitions of

Table 10.1

The Narrowing North–South Disparity in Human Development, 1960–2000

	North		South		Absolute Disparity	
	1960	2000	1960	2000	1960	2000
Life expectancy (years)	69	75	46	63	23	12
Adult literacy (%)	95	98	46	72	49	26
Nutrition (daily calorie supply as % of requirement)	124	141	90	113	34	26
Infant mortality (per 1000 live births)	37	8	149	63	112	55
Child mortality (under age 5)	46	8	216	93	170	85
Access to safe water (% of population)	100	100	40	72	60	26

Sources: United Nations Development Programme, World Bank, and UNICEF.

"sustainable" and criteria for debt relief have been adopted but remain rooted in the requirement that benefiting countries must face an unsustainable debt burden, maintain good economic policies, and prepare a "Poverty Reduction Strategy Paper," which is a blueprint laying out how a country will fight poverty and promote health and educational programs and how savings from debt relief will help.

International meetings in Cologne, Germany, in 1999 and Prague, Czech Republic, in 2000 formalized the details of the assistance proposals. Approval of the applications and Strategy Papers of the first 20 countries, it was announced, resulted in the commitment of $30 billion worth of debt relief. In addition, during 1999 and 2000, the richest countries pledged 100% forgiveness of bilateral debts owed them by HIPCs. To fulfill part of its agreements, the United States in November, 2000, provided $435 million in HIPC debt relief and gave approval for International Monetary Fund plans to use the proceeds from some limited sales of gold reserves for further debt relief. In March of 2002, President Bush pledged to increase America's foreign aid budget—part destined for HIPC programs—by 50% to $15 billion by 2006.

By late 2002, 26 countries had been accepted into the HIPC debt relief program, and 6 more applications were pending. The debt-reduction packages approved for the 26 states—22 of them in Africa—will remove over $40 billion in debt, about half of what the countries owe. Funds freed from debt service are now being spent on social services. Overall, additional expenditures for health and education are projected to absorb about two-thirds of total debt relief, with increased spending on HIV/AIDS programs, rural development and water supply, and road construction figuring prominently in most of the approved "Strategy Papers."

The expressed hope of the international community after the Prague meeting was that the answer to the question "Does foreign aid help?" would finally be "Yes." In a reconsideration of its former pessimism, the World Bank in 2002 concluded that, indeed, the answer was affirmative. It claimed that foreign aid in all forms had been instrumental in increasing life expectancy at birth in developing countries by 20 years since 1960, cutting adult illiteracy in half since 1970, reducing the number of people in abject poverty by 200 million since 1980 even as world population increased by 1.6 billion, and more than doubling the per capita income in developing countries since 1965.

Questions to Consider

1. In light of World Bank and other studies concluding aid does not correlate with development or poverty reduction in recipient countries, do you think the rich world and international agencies should halt all further monetary assistance to developing states? Why or why not?

2. Do you think donor countries such as the United States should completely ignore all self-interest including, for example, extra generosity toward friendly or politically compatible states, in making aid decisions? Why or why not?

3. Do you think international programs of forgiveness of debts contracted by sovereign states is appropriate or fair to lending countries and their citizens? Why or why not?

4. One widely held opinion is that money now spent on direct and indirect foreign aid more properly should be spent on domestic programs dealing with poverty, unemployment, homelessness, inner-city decay, inequality, and the like. An equally strongly held contrary view is that foreign aid should take priority, for it is needed to address world and regional problems of overpopulation, hunger, disease, destruction of the environment, and civil and ethnic strife those conditions foster. Assuming you had to choose one of the two polar positions, which view would you support, and why?

education together combine to restrict school enrollment in poor countries to a fraction of normal rich country expectations. In the least developed states in the late 1990s, only 60% of primary-age children and 20% of secondary school-age students were actually in school. Whatever the enrollment percentages were in individual countries, girls were less apt to be in school than were boys. Again for least developed countries, female primary school enrollment was only at 85% of the rate for males; at the secondary level, only 14% of girls were enrolled, and that rate was only 60% of the rate for boys.

Public Services

Development implies more than industrial expansion or agricultural improvement. The quality of public services and the creation of facilities to assure the health of the labor force are equally significant evidences of national advancement. Safe drinking water and the sanitary disposal of human waste are particularly important in maintaining human health (see Figure 4.19). As Table 10.1 notes, disparities in access to safe water are being steadily reduced between developed and developing countries, but sanitation statistics are getting worse worldwide. In 2000, according to the UN, 2.6 billion of the world's poor people—some 44% of total world population—lacked access to basic hygienic sanitation.

The accepted presence of pure water and sanitary toilets in the North and their general absence in, particularly, rural areas and urban slums in the less developed world present a profound contrast between the two realms. Less than half of the rural populations of the predominantly rural least developed states have access to water safe to drink. Within the expanding cities of the developing countries, nearly a quarter-billion people live in shantytowns and slums devoid of adequate water supply or sanitary disposal facilities (Figure 10.16). Worldwide, more than 1 billion people in the developing countries lack a dependable sanitary supply of water (Figure 10.17) and water-related diseases kill

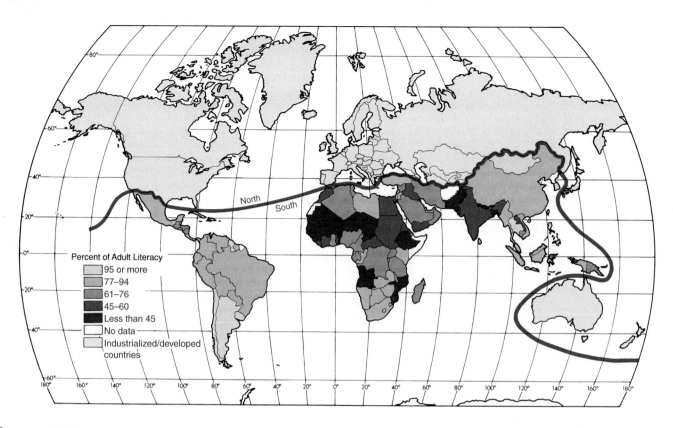

Figure 10.15 **The South: Adult literacy rate,** as a percentage of the adult population (over 15 years of age) able to read and write short, simple statements relating to their everyday life. With almost no exceptions, adult literacy was 95% or more in countries of the North at the start of the 21st century. With only a few exceptions, literacy rates in all countries of the South improved dramatically during the 1990s. For developing countries as a group, 74% of adults were literate in 2000 compared to 64% in 1990; for least developed countries, the improvement was from 45% literate at the start of the 1990s to 53% literate in 2000.

Source: Data from UNESCO.

approximately 10 million people every year. Yet, significant progress has been made; during the 1980s (the UN designated it the International Drinking Water Supply and Sanitation Decade), 1.2 billion people worldwide were added to the ranks of those with access to potable water; nearly another 1 billion were supplied during the 1990s.

Health

Access to medical facilities and personnel is another spatial variable with profound implications for the health and well-being of populations. Within the less developed world, vast numbers of people are effectively denied the services of physicians. While in industrial countries, on average, one physician serves 350 people, the figure for developing countries is over 5800. For sub-Saharan Africa as a whole, the ratio is about 18,500 to 1. In the developing world, there are simply too few trained health professionals to serve the needs of expanding populations. Those few who are in practice tend to congregate in urban areas, particularly in the capital cities. Rural clinics are few in number and the distance to them so great that many rural populations are effectively denied medical treatment of even the most rudimentary nature.

Increasingly, those sorts of health-related contrasts between advanced and developing countries have become matters of international concern and attention (see "Poverty and Development" p. 361). We saw in Chapter 4 how important for developing states population growth is the transfer of advanced technologies of medicine and public health: insecticides, antibiotics, and immunization, for example. Most recently, childhood diseases and deaths in developing countries have come under coordinated attack by the World Health Organization under the Task Force for Child Survival program (Figure 10.18). Gains have been impressive. If the 1960 worldwide infant mortality rate had remained in 2000, 15 million more children would have died than in fact did. Yet stark contrasts between most developed and least developed societies remain. Based on the mortality levels for children under 5 in industrialized countries in 2000 (7 per thousand), the United Nations estimated that more than 90% of the approximately 15 million infant and child deaths in developing countries (115 per thousand) in that year were preventable.

Taken at their extremes, advanced and developing countries occupy two distinct worlds of disease and health. One is affluent; its death rates are low, and the chief killers of its mature populations are cancers, heart attacks, and strokes. The other world is

Figure 10.16 Because they have no access to safe drinking water or sanitary waste disposal, impoverished populations of a developing country's unserved rural districts and urban slums—like this one in Capetown, South Africa—are subject to water-borne and sanitation-related diseases: 900 million annual cases of diarrhea including 2 million childhood deaths, 900 million cases of roundworm, 200 million of schistosomiasis, and additional millions of other similarly related infections and deaths.

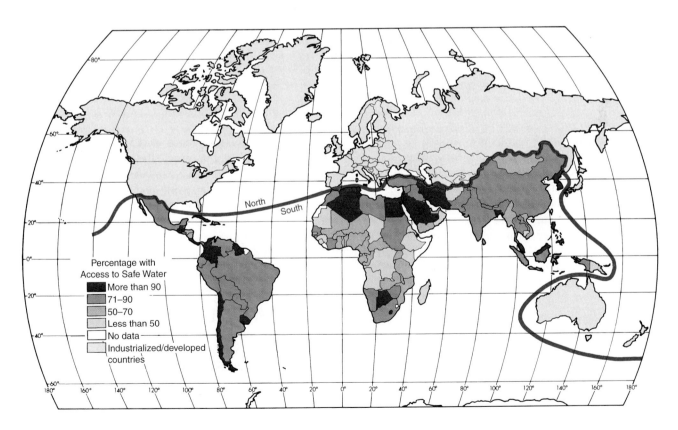

Percentage with Access to Safe Water
- More than 90
- 71–90
- 50–70
- Less than 50
- No data
- Industrialized/developed countries

Figure 10.17 **Percentage of population with access to safe drinking water.** Between 1975 and 2000, access to safe water increased by more than two-thirds to make potable water available to some 90% of urban residents in developing countries and 71% of rural folk (though only to 63% of total populations in the least developed states). By the start of the 21st century, thanks to WHO, United Nations, and World Bank programs and to targeted foreign aid, much of developing world was approaching the levels of safe water availability formerly found only in industrialized states of the North.

Sources: Data from United Nations Development Programme and World Health Organization.

Figure 10.18   The World Health Organization (WHO) is the agency of the United Nations that helps bring modern preventive health care, safe water, and sanitation to the less developed world. WHO workers help to fight certain diseases, advise on nutrition and living conditions, and aid developing countries in strengthening their health services. When the organization launched its Expanded Programme on Immunization in 1974, only 5% of the world's children were immunized against measles, diphtheria, polio, tetanus, whooping cough, and tuberculosis—diseases claiming 7 million young lives annually. In the early 1990s, more than 70% of children in developing countries were vaccinated against basic childhood diseases following an accelerated campaign by Unicef and WHO. But by 2002, poor countries reported that only 56% of their children received such immunization because of a falloff in financial support from wealthy nations. In response to the slowdown, in 1999, a Global Alliance for Vaccines and Immunization was formed by national and international agencies, philanthropies, and pharmaceutical companies to revive the earlier efforts. The UN Children's Fund reported at the end of the century that more than 90% of the children of the developing world lived in countries again making significant progress toward reducing malnutrition and preventing diseases. Pictured is preventive health care in a Micronesian clinic.

impoverished, often crowded, and prone to disease. The deadly dangers of its youthful populations are infectious, respiratory, and parasitic diseases made more serious by malnutrition.

In 1978, the World Health Organization endorsed preventive health care as an attainable goal and adopted "health for all by the year 2000" as its official target. It was to be reached through primary health care: low technologies aimed at disease prevention in poorer nations. Although substantial improvements in global health were made by the target year and disparities between the developed and developing worlds had been reduced, gaps had actually widened between the developing world as a whole and its "least developed" components, and health gains have actually been reversed in some states. The World Health Assembly of 1998, recognizing the continuing challenges, renewed the global commitment to "health for all" and established new targets for the early 21st century.

The general determinants of health are well known: enough purchasing power to secure the food, housing, and medical care essential to it; a healthful physical environment that is both sanitary and free from infectious disease; and a particularly female educational level sufficient to comprehend the essentials of nutrition and hygiene. Family planning, health, and infrastructure and economic developmental programs have begun to increase the numbers in the developing world that now have access to at least rudimentary health services.

Unfortunately, resurgence of old diseases and emergence of new ones may disrupt or reverse the hoped-for transition to better health in many world areas (See "Our Delicate State of Health," (p. 121). Almost 10% of world population now suffer from one or more tropical diseases, many of which—malaria, affecting 200 to 300 million people with up to 3 million deaths annually, is an example—were formerly thought to be eradicable but now are spreading in drug-resistant form. One such scourge, tuberculosis, is appearing as a major concern among particularly poorer populations outside tropical regions. Low income countries are also hard hit by the spread of AIDS (acquired immune deficiency syndrome). In 2002, the UN Global Program on AIDS reports, over 90% of a worldwide estimated 35 million adult and 10 million child cases of HIV infection were found in the developing world.

The high and rising costs of modern medications place unbearable burdens on strained budgets of developing states. Those costs increasingly must include health care for the rapidly growing number of their elderly citizens and for those exposed to the health risks that come with economic development and industrialization: higher consumption of alcohol, tobacco, and fatty foods, pollution, motor vehicle accidents, and the like. The World

Health Organization is concerned that health services in poor developing countries may be overwhelmed by the twin burdens of poverty-related illness and health problems of industrialization and urbanization; heart disease and cancer now claim as many developing world as industrial world lives.

Aggregate Measures of Development and Well-Being

As we have seen, no single measure adequately summarizes the different facets of national development or gives a definitive comparison of countries on the continuum of development. Composite measures to achieve that summary aim can, of course, be devised from the growing body of comparative statistics regularly published by United Nations agencies, the World Bank, and other sources. Many of those—Figure 10.3 is an example—have been criticized for being based too strongly on economic and infrastructural indicators: gross national income, per capita income, sectoral structure of national economies, import and export data, miles of railroad or paved highways, and the like.

Development, it is maintained, is more than the purely economic and physical, and personal development may have little or nothing to do with objective statistical measures. The achievement of development must also be seen in terms of individual and collective well-being: a safe environment, freedom from want, opportunity for personal growth and enrichment, and access to goods and services beyond the absolute minimum to sustain life (see "Measuring Happiness"). Health, safety, educational and

Measuring Happiness

In an article in *The Times* (London) of 26 May 1975, "Introducing the Hedonometer, a New Way of Assessing National Performance *or* Why We Should Measure Happiness Instead of Income," Geraldine Norman compares England and Botswana under six headings.

Three years ago I spent my honeymoon in the eastern highlands of Rhodesia [now Zimbabwe] trying to construct a hedonometer, a means of measuring happiness per head of the population. I did not have a thermometer type of thing in mind: it was to be a statistical structure, on the lines of Keynesian national accounting, that would end up by measuring gross national happiness instead of gross national product. The unit of measurement would be psychological satisfaction rather than money. I envisaged my hedonometer as a tool for political policy making of such power that boring chat about economic growth would be ousted, forgotten and interred.

The motivation for this eccentric undertaking was a growing conviction that the accretion of wealth and/or the expansion of income, whether at national or individual level, was not necessarily a recipe for happiness. Indeed, I suspected that in some cases greater wealth reduced the likelihood of happiness—that in certain circumstances there could be a negative correlation between happiness and money. . . .

From my psychology book . . . I learnt that the two primary needs of a human being are security and achievement, achievement being the positive satisfaction of individual and species drives. Conning [the] book with attention—and using a bit of imagination—I arrived at the six principal factors which contribute to a happy life, the basis of my hedonometer. The list is as follows (the first two factors provide security and the next four require satisfaction for an adequate level of achievement):

1. Understanding of your environment and how to control it.

2. Social support from family and friends.

3. Species drive satisfaction (sex and parental drives).

4. Satisfaction of drives contributing to physical well-being (hunger, sleep, etc.).

5. Satisfaction of aesthetic and sensory drives.

6. Satisfaction of the exploratory drive (creativity, discovery, etc.).

You might like to modify the table by changing the relative importance of different factors, the scores, the countries, even the factors themselves. On the purely subjective judgment of Geraldine Norman, Botswana is a "happier" place than England. The measurement of happiness is indeed difficult and very subjective. The exercise is worthwhile, however, because it shows how the conventional measures of development such as energy consumption and GNI per inhabitant are really very subjective too.

©Times Newspapers Limited, 1975.

| | Importance | England | | Botswana | |
		Satisfaction (%)	Product (score)	Satisfaction (%)	Product (score)
Understanding	15	50	750	70	1050
Social support	20	40	800	80	1600
Species satisfaction	10	70	700	70	700
Physical well-being	35	92	3220	72	2520
Aesthetic	5	40	200	60	300
Exploratory	15	30	450	60	900
Total	**100**		**6120**		**7070**

cultural development, security in old age, political freedom, and similar noneconomic criteria are among the evidences of comparative developmental level that are sought in composite statistics. Also sought is a summary statistic of development that is value free; that is, the input data should not measure development by expenditure patterns or performance standards that are ethnocentric or colored by political agendas. The values of one culture—for example, in housing space per person, in educational levels achieved, or in distribution of national income—are not necessarily universally applicable or acceptable, and a true comparative statistic should not imply that they are.

Seeking a value-free measure of the extent to which minimum human needs are being satisfied among the world's countries, the Overseas Development Council devised a Physical Quality of Life Index (PQLI). Three indicators—infant mortality, life expectancy, and literacy—are each scored 0–100, with 0 an explicitly "worst" performance. A national achievement level is calculated by averaging the three indicators. The PQLI is but one of many attempts to recognize that national development and human welfare are complex achievements not measurable by a single indicator. Each approach has attempted to integrate into a composite index a larger or smaller number of national variables detailing physical, economic, political, and social conditions specific to country units. On the basis of the national rankings they derived, each has explicitly or implicitly ranked the countries of the world on a continuum from least to most developed.

One such ranking gaining increasing recognition is employed by the United Nations Development Programme. Its "human development index" (HDI) combines purchasing power (not just dollar amount of per capita GNI), life expectancy, and literacy (Figure 10.19). The HDI reflects the Programme's conviction that the important human aspirations are leading a long and healthy life, receiving adequate education, and having access to assets and income sufficient for a decent quality of life. The arbitrary weighting of the three input variables—longevity (measured by life expectancy at birth), knowledge (indicated by weighted measures of adult literacy and mean years of schooling), and income (based on a poverty-adjusted statistic of gross domestic product per capita)—makes the derived national rankings subjective rather than fully objective. The HDI, like all attempts at measuring developmental levels of countries and categorizing their variations in qualities of life and human welfare, is a recognition both of the complexity of the economic and social structures involved and of the need to focus developmental efforts.

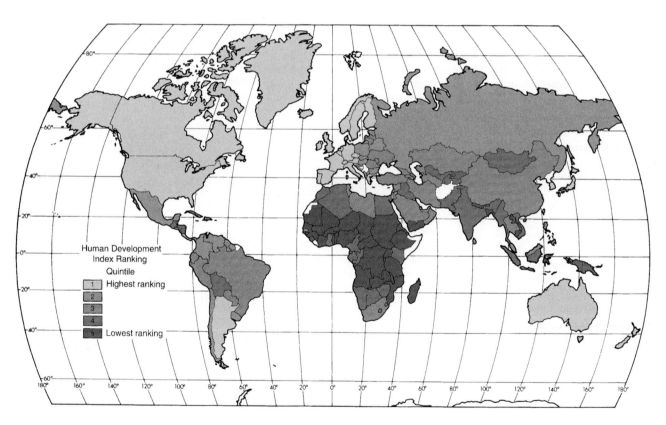

Figure 10.19 **Country rankings according to the Human Development Index** of the United Nations Development Programme. Since the index is intended to measure the absence of deprivation, it discounts incomes higher than needed to achieve an acceptable level of living and therefore is uninformative in comparing the levels of development of the richest countries. The four measures that are used by the UNDP—life expectancy, adult literacy, combined school enrollment ratios, and real (PPP) income—are highly correlated with one another. For that reason, it has been noted, the rankings derived by the HDI differ only slightly from income rankings adjusted for purchasing power parity; the Indian minister for human resources in 2002 objected that HDI ignored "spiritual happiness" and "intellectual advances." Fifth quintile countries, at the bottom of the Human Development Index, closely match the "least developed" countries recognized by the UN and shown on Figure 10.3.

Source: "Human Development Index," country rankings are made and reported by United Nations Development Programme in its annual Human Development Report.

The UN Development Programme has also developed a reverse image of poverty in its Human Poverty Index (HPI). While the HDI measures average *achievement,* the HPI measures *deprivation* in the same three measures of development underlying the HDI. For the poverty index, the benchmarks of concern are probability of not surviving to age 40; exclusion from full social intercourse because of illiteracy; and deprivation of a decent level of living as measured by lack of safe water access and percentage of underweight small children. The Human Poverty Index is discussed in more detail in "Poverty and Development" (p. 361).

The Role of Women

Many of the common measures of development and change within and between countries take no account of the sex and age structures of the societies examined. Gross national income per capita, literacy rates, percentage of labor force in agriculture, and the like are statistics that treat all members (or all adult members) of the society uniformly. Yet among the most prominent strands in the fabric of culture are the social structures (*sociofacts*) and relationships that establish distinctions between males and females in the duties assigned and the rewards afforded to each.

Because gender relationships and role assignments vary among societies, the status of women is a cultural spatial variable. Because so much of that variation is related to the way economic roles and production and reward assignments are allocated by sex, we might well assume a close tie between the status of women in different societies and their level and type of economic development. Further, it would be logical to believe that advancement in the technological sense would be reflected in an enhancement of the status and rewards of both men and women in developing countries. Should that prove true, it would logically follow that contrasts between the developed and developing world in gender relationships and role assignments would steadily diminish.

The pattern that we actually observe is not quite that simple or straightforward, for gender relationships and role assignments are only partially under the control of the technological subsystem. **Gender** in the cultural sense refers to socially created—not biologically based—distinctions between femininity and masculinity. Therefore, religion and custom play their own important roles. Further, it appears that at least in the earlier phases of technological change and development, women generally lose rather than gain in status and rewards. Only recently and only in the most developed countries have gender-related contrasts been reduced within and between societies.

Hunting and gathering cultures observed a general egalitarianism; each sex had a respected, productive role in the kinship group (see Figure 2.11). Gender is more involved and changeable in agricultural societies (see "Women and the Green Revolution," p. 284). The Agricultural Revolution—a major change in the technological subsystem—altered the earlier structure of gender-related responsibilities. In the hoe agriculture found in much of sub-Saharan Africa and in South and Southeast Asia, women became responsible for most of the actual field work, while still retaining their traditional duties in child rearing, food preparation, and the like.

Plow agriculture, on the other hand, tended to subordinate the role of women and diminish their level of equality. Women may have hoed, but men plowed, and female participation in farm work was drastically reduced. This is the case today in Latin America and, increasingly, in sub-Saharan Africa where women are often more visibly productive in the market than in the field (Figure 10.20). As women's agricultural productive role declined, they were afforded less domestic authority, less control over their own lives, and few if any property rights independent of male family members.

Western industrial—"developed"—society emerged directly from the agricultural tradition of the subordinate female who was not considered an important element in the economically active population, no matter how arduous or essential the domestic tasks assigned, and who was not afforded full access to education or similar amenities of an advancing society. European colonial powers introduced that attitude along with economic development into Third World cultures. Only within the later 20th century, and then largely in the more developed countries, has that subordinate role pattern changed.

The rate and extent of women's participation in the labor force has expanded everywhere in recent years. Between 1970 and 1997, both the percentage of the total labor force who are women and the percentage of women who are economically active[5] increased in nearly every world region—developed and developing (Figure 10.21). Women's increased participation in the work force reflects several changing conditions. Women have gained greater control over their fertility, thus increasing their opportunities for education and employment. Further, attitudes toward employed women have changed and public policies on, for example, child care, maternity benefits, and the like, are more favorable. Economic growth, including the expansion of service sector jobs open to women, was also important in many regions. Permissive attitudes and policies with regard to micro and small enterprises, including financing and credit programs, have in some areas played a major role in encouraging women entrepreneurs (see "Empowering Women Financially").

Considering all work—paid and unpaid economic activity and unpaid housework—women spend more hours per day working than do men in all developing and developed regions except Anglo America and Australia. In developing countries, the UN estimates, when unpaid agricultural work and housework are considered along with wage labor, women's work hours exceed men's by 30% and may involve at least as arduous—or heavier—physical labor. The FAO reports "rural women in the developing world carry 80 tons or more of fuel, water and farm produce for a distance of 1 km during the course of a year. Men carry much less. . ." Everywhere women are paid less than men for comparable employment, but in most world regions the percentage of

[5]The International Labour Office defines "economically active" work as that "producing significant amounts of 'economic' (that is, marketable) goods, or of visible income." Included in the "economically active population" are all employed and unemployed persons seeking employment and all wage earners, unpaid family workers, and members of producers' cooperatives.

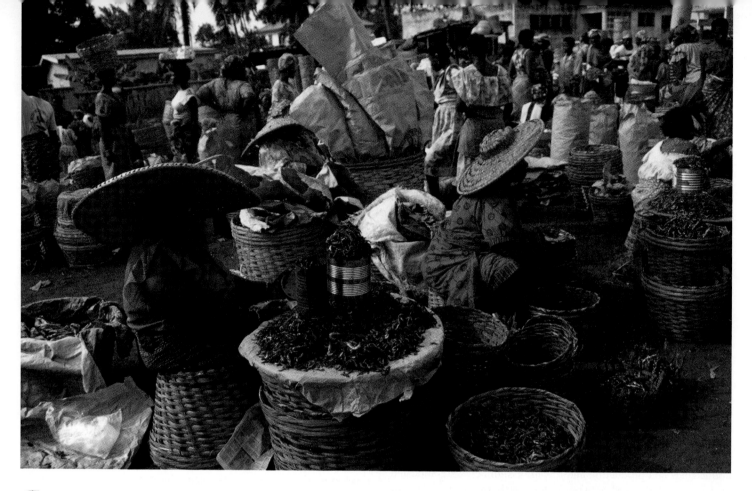

Figure 10.20 Women dominate the once-a-week *periodic* markets in nearly all developing countries. Here they sell produce from their gardens or the family farm and often offer processed goods for sale (to which their labor has added value)—oil pressed from seeds or, in Niger, for example, from peanuts grown on their own fields; cooked, dried, or preserved foods; simple pottery and baskets, or decorated gourds. In West Africa, the Caribbean, and Asia, between 70% and 90% of all farm and marine produce is traded by women. The market shown here is in the West African country of Ghana. More than half of the economically active women in sub-Saharan Africa and southern Asia and about one-third in northern Africa and the rest of Asia are self-employed, working primarily in the informal sector. In the developed world, only about 14% of active women are self-employed.

economically active women holding wage or salaried positions is about equal to the rate for men. Exceptions are Latin America, where a higher proportion of active women than men are wage earners, and Africa, where wage-earning opportunities for women are few; in several African states, less than 10% of economically active women are wage earners.

Despite these and similar widely applicable generalizations, the present world pattern of gender-related institutional and economic role assignments is varied. It is influenced by a country's level of economic development, by the persistence of the religious and customary restrictions its culture imposes on women, and by the specific nature of its economic—particularly agricultural—base. The first control is reflected in contrasts between the developed and developing world; the second and third are evidenced in variations within the developing world itself.

The differential impact of these and other conditions is evident in Figure 10.22. The pattern shows a distinct gender-specific regionalization among the countries of the developing world. Among the Arab or Arab-influenced Muslim areas of western Asia and North Africa, the recorded proportion of the female population that is economically active is low. Religious tradition restricts women's acceptance in economic activities outside of the home, a tradition that results in probable under-reporting of female employment by the countries involved. The same cultural limitations do not apply under the different rural economic conditions of Muslims in southern and southeastern

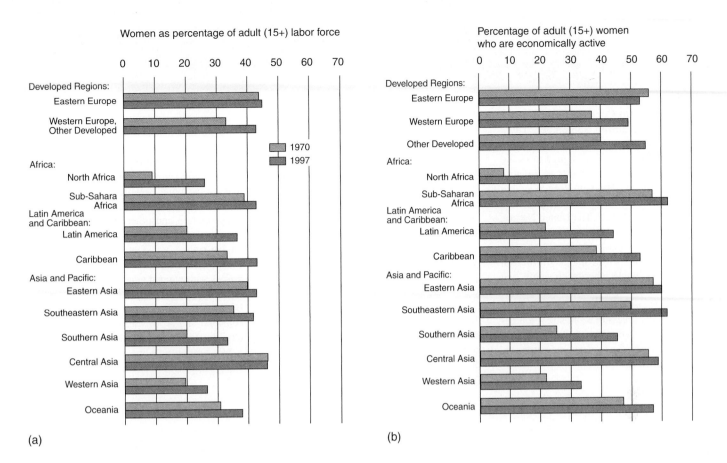

Women as percentage of adult (15+) labor force

Percentage of adult (15+) women who are economically active

(a)

(b)

Figure 10.21 (*a*) **Women's share of the labor force** increased in almost all world regions between 1970 and 1997. Worldwide, women were recorded by the World Bank at 41% of the total labor force in 2000 and comprised at least one-third of the workers in all areas except North Africa and Western Asia according to 2000 World Bank figures. (*b*) **Women's economic activity rates** showed a mixed pattern of change between and within many world regions. More than half of the world's female labor force lived in Asia and the Pacific area in 1997, a proportion also registered in 2000. Although the regional share of economically active women varies widely, the UN estimates that women made up half the labor force in most countries and regions shortly after 2000.

Sources: Based on charts 5.1 and 5.2 of United Nations, The World's Women 2000: Trends and Statistics. *Social Statistics and Indicators, Series K., No. 16 (New York: United Nations, 2000) and on International Labour Office surveys.*

Asia, where labor force participation by women in Indonesia and Bangladesh, for example, is much higher than it is among the western Muslims.

In Latin America, women have been overcoming cultural restrictions on their employment outside the home and their active economic participation has been increasing. That participation is occurring almost entirely outside of the agricultural realm, where the high degree of farm labor tenancy as well as custom limits the role of females. Sub-Saharan Africa, highly diverse culturally and economically, in general is very dependent on female farm labor and market income. The historical role of strongly independent,

property-owning females formerly encountered under traditional agricultural and village systems, however, has increasingly been replaced by subordination of women with the modernization of agricultural techniques and the introduction of formal, male-dominated financial and administrative farm-sector institutions.

A "gender empowerment measure" devised by the United Nations Development Programme emphasizes female participation in national economic, political, and professional affairs and clearly displays areal differentials in the position of women in different cultures and world regions (Figure 10.23).

Empowering Women Financially

The Fourth World Conference on Women held in Beijing during September, 1995, called on all governments to formulate strategies, programs, and laws designed to assure women their full human rights to equality and development. The Conference's final declaration, reinforced at the "Beijing Plus Five" Conference held at The United Nations in June, 2000, detailed recommended policies in the areas of sexuality and child-bearing, violence against women, discrimination against girls, female inheritance rights, and family protection. Its particular emphasis, however, focused on efforts to "ensure women's equal access to economic resources including land, credit, . . . and markets as a means to further advancement and empowerment of women and girls."

Two-thirds of the total amount of work women do is unpaid, but that unpaid work amounts to an $11 trillion addition to the total world economy. The Beijing Conference declaration was a recognition that women's economic contribution would be even greater—and of more social and personal benefit—were governments to grant them equal opportunity through financial support to engage as owners in small-scale manufacturing, trade, or service enterprises. In fact, both the model and proof of success in granting women access to credit were already in place.

In 1976, a Bangladeshi economist, Muhammad Yunus, wandered into a poor village and got an idea that has captured international interest and changed accepted beliefs and practices of banking in developing

countries. The concept behind the Grameen Bank he established is simple: if individual borrowers are given access to credit, they will be able to identify and engage in viable income-producing activities such as pottery making, weaving, sewing, buying and marketing simple consumer goods, or providing transportation and other basic services. Declaring that "Access to credit should be a human right," Mr. Yunus was a pioneer in extending "microcredit" for "microenterprises" with women emerging as the primary borrowers and beneficiaries of Grameen Bank's practice of lending money without collateral and at low rates of interest. To be eligible for the average loan of about U.S. $160, women without assets must join or form a "cell" of five unrelated women, of

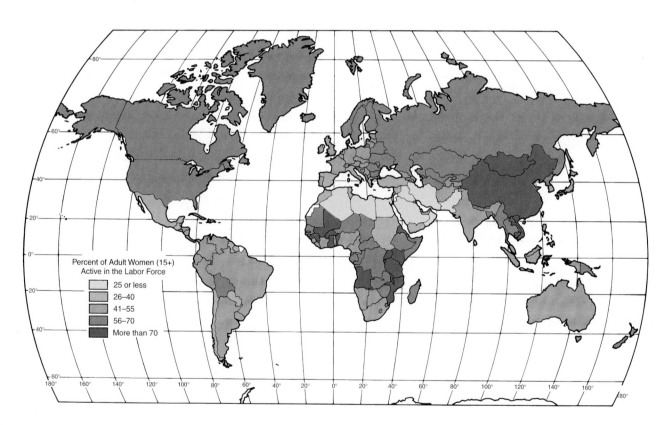

Figure 10.22 **Economically active women.** Since female participation in the labor force is reported by individual countries with differing definitions of "economically active," international comparisons may be misleading. The International Labor Office definition is given as a footnote on page 379. Since a higher proportion of the female than the male labor force is engaged in the "informal" sector, their recorded presence in the workplace is officially understated. Worldwide, 61% of all women over 15 years of age were in the labor force in 2000, and women comprise more than 40% of the labor force in developing countries. The ILO maintains that "in many developing areas . . . the number of women in the labor force . . . is much larger than that given in official statistics."

Sources: United Nations and International Labour Office.

whom only two can borrow at first though all five are responsible for repayment. When the first two begin to repay, two more can borrow, and so on. As a condition of the loan, clients must also agree to increase their savings, observe sound nutritional practices, and educate their children.

By 2000, the bank had made over 2 million loans in 40,000 villages in Bangladesh. More than 94% of the borrowers are women, and repayment rates reach above 95%. The average household income of Grameen Bank members has risen to about 59% higher than that of nonmembers in the same villages, with the landless benefiting most and marginal landowner families following closely. Because of enterprise incomes resulting from the lending program, there has been a sharp reduction in the number of Grameen Bank members living below the poverty line—to 20% compared to nearly 60% for nonmembers. There has also been a marked shift from low-status agricultural labor to self-employment in simple manufacturing and trading.

The Grameen concept has spread from its Bangladesh origins to elsewhere in Asia and to Latin America and Africa. By 2002, some 10,000 microfinance institutions worldwide had reached nearly 20 million clients, among the world's poorest people. With loans of as little as $40, increasing by some 37% a year, the Grameen concept—according to Mr. Yunus—was, in 2002, on course to offer microcredit to 100 million families by 2005. But the women of those families still would represent only a faction of the estimated (2000) 500 to 600 million women worldwide who have virtually no access to credit—or to the economic, social, educational, and nutritional benefits that come from its availability. It is that globally enormous number of women now effectively denied credit equality that the resolutions of the Fourth World Conference on Women sought to benefit.

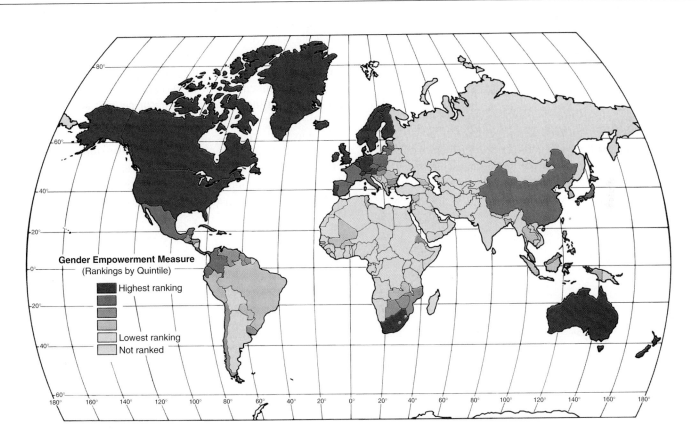

Figure 10.23 **The gender empowerment measure (GEM)** summarizes women's access to political and economic power based on three variables: female share of parliamentary seats; share of professional and technical jobs; and share of administrative and managerial positions. The GEM rankings show that gender equality in political, economic, and professional activities is not necessarily related to level of national wealth or development. According to this measure, some developing countries—China, for example, where women are afforded a large share of legislative seats and political administrative positions—outperform industrialized South Korea. Only 102 countries are ranked; in this report from the end of the 20th century; in most, women are in a distinct minority in the exercise of economic power and decision making.

Source: Rankings from United Nations Development Programme.

Summary

Development as a concept and process implies change and improvement. It suggests the fuller and more productive use of the resources of an area through the application of advanced levels of technology. The result is presumed to be improved conditions of life and well-being for constant or growing populations and, for the society undergoing development, a fuller integration into—and more equal share of—the world space economy.

Development in that light can be seen as a cultural variable with a distinctive spatial patterning. No two countries have exactly the same position on the continuum of development in all of its many different possible economic and noneconomic measures. For this reason, precise classification of countries by developmental level is impossible, and a variety of general descriptive terms has been introduced, including the following: developed, developing, underdeveloped, least (or less) developed, Third or Fourth World, and the like. Whatever the terms, the overall world pattern of development is clear: The advanced and relatively wealthy countries of the economic core are those of Europe, North America, Japan, Australia, and New Zealand and a small but growing number of newly industrialized countries with high incomes and quality of life—Taiwan, South Korea, Singapore, and the like. The rest of the world is considered to be "developing" on the economic periphery, where individual countries are progressing at different rates and with different degrees of success.

A variety of comparative economic and noneconomic data are available to help identify the relative position of individual countries. *Gross national income* and *purchasing power parity per capita* document the basic core-periphery pattern while making clear the diversity among the developing countries in the monetary success of their economies. *Per capita consumption of commercial energy* reveals the immense size of the technology gap between most and least developed states, for energy use may be loosely equated with modern industrial plant and transportation facilities. A high percentage of a country's *workforce in agriculture* is associated with less developed subsistence economies with low labor productivity and low levels of national wealth. The price of underdevelopment—and of the relative poverty it implies—is malnutrition. Although the correlation is not exact, countries registering *average caloric intake* below daily requirements are also countries registering poorly on all purely economic measures of development.

Earlier hopes that underdevelopment was simply the common starting point in a series of expected and inevitable stages of advancement have been dashed. Many countries appear unable to accumulate the capital, develop the skills, or achieve the technology transfer necessary to carry them along the path to fuller economic development and prosperity. Without that development, countries score poorly on noneconomic measures such as literacy, safe water, and conditions of health. With it, they can—as the experience of newly industrializing countries demonstrates—experience growing cultural and technological convergence with the most advanced states. That convergence, in fact, is increasing, and the share of the *gross world product* attributable to what is still called the "developing" world continues to grow and amounted to over 45% at the start of the 21st century.

Development implies pervasive changes in the organizational and institutional structuring of peoples and space. Urbanization of populations and employment has invariably accompanied economic development, as has a more complete and rigorous political organization of space. We turn our attention in Chapters 11 and 12 to these two important expressions of human geographic variation, beginning first with an examination of city systems and of the spatial variations observable in the structure of urban units.

Key Words

circular and cumulative causation 358
core-periphery model 358
cultural convergence 362
development 355
gender 379

gross domestic product 364
gross national income (GNI) 364
informal economy 360
purchasing power parity (PPP) 360
spread effect 358

technology 362
technology gap 362
technology transfer 362
trickle-down effect 358
underdevelopment 355

384

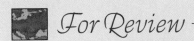

For Review

1. How does the *core-periphery* model help us understand observed contrasts between developed and developing countries? In what way is *circular and cumulative causation* linked either to the perpetuation or the reduction of those contrasts? How does the concept of *trickle-down effects,* or *spread effects,* explain the equalization of development and incomes on a regional or international scale?

2. What are some of the reasons that have been given to explain why some countries are *developed* and others are *underdeveloped?*

3. What different ways and measures do we have to indicate degrees of development of particular countries or regions? Do you think these measures can be used to place countries or regions into uniform *stages of development?*

4. Why should any country or society concern itself with *technology transfer* or with the *technology gap?* What do these concepts have to do with either development or societal well-being?

5. What kinds of material and nonmaterial economic and noneconomic contrasts can you cite that differentiate more developed from less developed societies?

6. Assume you are requested to devise a composite index of national development and well-being. What *kinds* of characteristics would you like to include in your composite? Why?

What specific *measures* of those characteristics would you like to cite?

7. Why is energy *consumption* per capita considered a reliable measure of level of national economic development? If a country has a large per capita *production* of energy, can we assume that it also has a high level of development? Why or why not?

8. Have both males and females shared equally in the benefits of economic development in its early stages? What are the principal contrasts in the status of women between the developed and developing worlds? What regional contrasts within the developing world are evident in the economic roles assigned to women?

Focus Follow-up

1. **How do we define development and explain the occurrence or persistence of underdevelopment?** pp. 355–358.

   Development implies improvement in economic and quality-of-life aspects of a society. It presumably results from technology transfer from advanced to developing states and, through consequent cultural convergence, promises the full integration of the developing society into the larger modern world order. When that stage of advancement is reached, transition from the world economic and social "periphery" to its "core" has been achieved. Persistence of underdevelopment is usually attributed to failure of a culture or region to accumulate capital, develop skills, or achieve technology transfers to improve its prosperity or quality of life.

2. **What economic measures mark a country's stage of development or its progress from underdevelopment?** pp. 359–371.

   Gross national income and purchasing power parity per capita, per capita commercial energy consumption,

percentage of labor force in agriculture, and average daily caloric intake are common, accepted measures of development. Attempts to model the process of development have led to inconclusive and contrasting theories of inevitable "stages of growth," optimistic "Big Push" ideas of coordinated investment, and pessimistic "dependency theory" concepts of perpetual exploitation of underdeveloped regions.

3. **What are noneconomic aspects of development, and how are they related to measures of economic growth?** pp. 371–379.

   Education, sanitation, and health services are among many noneconomic indices of development that are strongly related to income and national wealth. The higher a country's ranking on purely economic measures, the more it can and does spend on improvement of quality-of-life conditions for its citizens. Similarly, the higher those expenditures are, the lower on average are national rates of infant mortality,

births and deaths, rates of natural increase, and the like. "Happiness" or satisfaction of such cultural wants as social support, aesthetic and sensory needs, creativity outlets, etc., also figure as importantly into well-being assessments as do gross domestic product or energy consumption.

4. **What conditions underlie the varying world pattern of women's roles, status, and rewards?** pp. 379–383.

   The status of women is a cultural spatial variable reflecting gender relationships characteristic of different societies. The world pattern of gender-related institutional and economic role assignments and rewards appears strongly influenced by national levels of economic development and by the persistence of customary and religious restrictions on women. With few exceptions, women worldwide spend more hours per day working than do men; everywhere they are paid less for comparable work. A general world trend is toward greater equality for women in political and economic opportunities and status.

Selected References

Armstrong, R. Warwick, and Jerome D. Fellmann. "Health: One World or Two." In *The New Third World,* 2d ed., edited by Alonso Gonzalez and Jim Norwine, pp. 75–92. Boulder, Colo.: Westview Press, 1998.

Berry, Brian J. L. "An Inductive Approach to the Regionalization of Economic Development." In *Essays on Geography and Economic Development,* edited by Norton Ginsburg, pp. 78–107. Chicago: University of Chicago, Department of Geography, Research Paper no. 62, 1960.

Boserup, Ester. "Development Theory: An Analytical Framework and Selected Applications." *Population and Development Review* 22, no. 3 (1996): 505–515.

Boserup, Ester. *Woman's Role in Economic Development.* London: Allen and Unwin, 1970.

Butz, David, Steven Lonergan, and Barry Smit. "Why International Development Neglects Indigenous Social Reality." *Canadian Journal of Development Studies* 12, no. 1 (1991): 143–157.

Dickenson, John, ed. *Geography of the Third World.* 2d ed. New York: Routledge, 1996.

Gardner, Gary, and Brian Halweil. *Underfed and Overfed: The Global Epidemic of Malnutrition.* Worldwatch Paper 150. Washington, D.C.: Worldwatch Institute, 2000.

Hodder, Rupert. *Development Geography* New York: Routledge, 2000.

Independent Commission on International Development Issues. *North-South: A Programme for Survival.* Cambridge, Mass.: MIT Press, 1980.

International Bank for Reconstruction and Development/The World Bank. *World Development Indictors.* Washington, D.C.: The World Bank, annual.

International Bank for Reconstruction and Development/The World Bank. *World Development Report.* Published annually for the World Bank by Oxford University Press, New York.

Jacobson, Jodi L. *Gender Bias: Roadblock to Sustainable Development.* Worldwatch Paper 110. Washington, D.C.: Worldwatch Institute, 1992.

Knox, Paul L., John Agnew, and Linda McCarthy, *The Geography of the World Economy.* 4th ed. New York: Oxford University Press, 2004.

Landes, David S. *The Wealth and Poverty of Nations: Why Some Are So Rich and Some So Poor.* New York: Norton, 1998.

Lenssen, Nicholas. *Empowering Development: The New Energy Equation.* Worldwatch Paper 111. Washington, D.C.: Worldwatch Institute, 1992.

Momsen, Janet H. *Women and Development in the Third World.* New York: Routledge, 1991.

Momsen, Janet H. "Gender Bias in Development." In *The New Third World,* 2d ed., edited by Alonso Gonzalez and Jim Norwine, pp. 93–111. Bounder, Colo.: Westview Press, 1998.

Momsen, Janet H., and Vivian Kinnaird. *Different Places, Different Voices: Gender and Development in Africa, Asia and Latin America.* New York: Routledge, 1993.

Porter, Philip W., and Eric S. Sheppard. *A World of Difference: Society, Nature, Development.* New York: Guilford, 1998.

Riley, Nancy E. "Gender, Power, and Population Change." *Population Bulletin* 52, no. 1. Washington, D.C.: Population Reference Bureau, 1997.

Roodman, David Malin. *Still Waiting for the Jubilee; Pragmatic Solutions for the Third World Debt Crisis.* Worldwatch Paper 155. Washington, D.C.: Worldwatch Institute, 2001.

Rostow, Walter W. *The Stages of Economic Growth.* London: Cambridge University Press, 1960, 1971.

Sachs, J. D., et al. "The Geography of Poverty and Wealth." *Scientific American* vol. 284 (March 2001): 70–75.

Seager, Joni. *The Penguin Atlas of Women in the World.* 3d ed. New York: Penguin USA, 2003.

Simpson, Edward S. *The Developing World: An Introduction.* 2d ed. Essex, England: Addison-Wesley Longman, 1994.

Tata, Robert J., and Ronald R. Schultz. "World Variation in Human Welfare: A New Index of Development Status." *Annals of the Association of American Geographers* 78, no. 4 (1988): 580–593.

UNICEF. *The State of the World's Children.* New York: United Nations, annual.

UNIFEM/United Nations Development Fund for Women. *Progress of the World's Women.* United Nations, biennial.

United Nations. *The World's Women 2000: Trends and Statistics.* Social Statistics and Indicators, Series K, no. 16. New York: United Nations, 2000.

United Nations Development Programme. *Human Development Report.* New York: Oxford University Press, annual.

Websites: The World Wide Web has a tremendous number and variety of sites pertaining to geography. Websites relevant to the subject matter of this chapter appear in the "Web Links" section of the On-line Learning Center associated with this book. Access it at **www.mhhe.com/fellmann8e**

PART *Four*

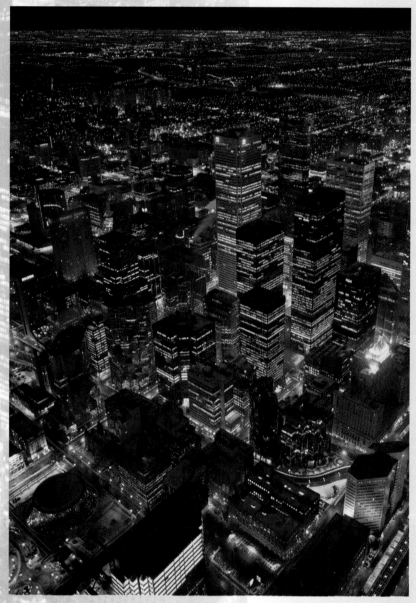

Nighttime Toronto, Ontario, Canada.

Landscapes of
Functional Organization

The more advanced a region's economy, the more highly developed and integrated are its organizational controls. Subsistence economies of hunting, gathering, and shifting agriculture were locally based, self-sufficient, small-group ways of life. As the preceding chapters make clear, they have nearly everywhere been replaced by more advanced and spatially integrated economies based on secondary, tertiary, and "postindustrial" activities. These newer forms of economic orientation require more concentrated and structured systems of production and a more formal organization of society and territory than any previously experienced. They have inevitably given rise to distinctive urban and political landscapes.

Urbanization has always accompanied economic advancement. Manufacturing and trade imply concentrations of workers, managers, merchants, and supporting populations and institutions. They require a functional organization and control of space far more pervasive and extensive than any guiding the subsistence economies that went before. Chapter 11, the first of the two chapters of this section of our study of human geography, looks at the systems of cities that contain an increasing proportion of the population of all culture realms. In them, too, are concentrated a growing share of the world's productive activities.

Cities, we shall see, are functional entities producing and exchanging goods, performing services, and administering the space economy for the larger society of which they are the operational focus. The functions they house and the kind of economy that has shaped them help determine the size and spatial patterns of the urban systems that have developed in different parts of the world. Within those systems, each city is a separate landscape entity, distinct from surrounding nonurban areas. Internally, each displays a complex of land use arrangements and social geographies, in part unique to it but in part, as well, influenced by the cultural, economic, and ideological setting that it reflects. Both city systems and internal city structures are fundamental features of spatial organization and of cultural differentiation.

Advancement in cultural and economic development beyond the local and the self-sufficient also implies hierarchies of formal territorial control and administration. Structured political landscapes have been created as the accompaniment of the economic development and urbanization of societies and regions. In Chapter 12, the second of the chapters of this section, we examine the variety of forms and levels—from local to international—of the political control of space as another key element in the contemporary human geographic mosaic.

CHAPTER
Eleven

Urban Systems and Urban Structures

Focus Preview

Opposite: A residential suburb of Meknes, Morocco.

In the 1930s, Mexico City was described as perhaps the handsomest city in North America and the most exotic capital city of the hemisphere, essentially unchanged over the years and timeless in its atmosphere. It was praised as beautifully laid out, with wide streets and avenues, still the "city of palaces" that Baron von Humboldt called it in the 19th century. The 70-meter- (200-ft-) wide Paseo de la Reforma, often noted as "one of the most beautiful avenues in the world," was shaded by a double row of trees and lined with luxurious residences.

By the 1950s, with a population of over 2 million and an area of 52 square kilometers (20 sq mi), Mexico City was no longer unchanged. The old, rich families who formerly resided along the Paseo de la Reforma had fled from the noise and crowding. Their "palaces" were being replaced by tall blocks of apartments and hotels. Industry was expanding and multiplying,

tens of thousands of rural folk were flocking in from the countryside every year. By 2000, with its population estimated at more than 18 million and its area at over 3000 square kilometers (1160 sq mi), the Mexico City metropolitan area was among the world's largest urban complexes.

The toll exacted by its growth has been heavy. Each year the city pours more than 5 million tons of pollutants into the air. Some 80% comes from unburned gas leaked from residents' stoves and heaters and from the exhausts of their estimated 4 million motor vehicles; the rest is produced by nearly 35,000 industrial plants. More than 5 million people citywide have no access to tap water; in many squatter neighborhoods less than half do. Some 4 million residents have no access to the sewage system. About one-third of all families—and they average five people—live in but a single room, and that room generally is in a hovel in one of the largest slums in the world.

The changes in Mexico City since the 1930s have been profound (Figure 11.1). Already one of the world's most populous centers, Mexico City is a worst-case scenario of an urban explosion that sees an increasing proportion of the world's population housed within a growing number of immense cities. While growth rates have declined or even reversed in recent years among some of the world's largest cities, urban population overall is growing more rapidly than the population as a whole and, by most estimates, by larger annual increments than ever before.

The Urbanizing Century

Figure 11.2 gives evidence that the growth of major metropolitan areas was astounding in the 20th century. Some 411 metropolitan areas each had in excess of 1 million people in 2000; in 1900, there were only 12. Expectations are for 564 "million cities" in 2015. Nineteen metropolises had populations of more than 10 million people in 2000, earning them the title of *megacities* (Figure 11.3). In 1900, none was of that size. It follows, of course, that since the world's total population greatly increased (Chapter 4), so too would its urban component. Importantly, the urban share of the total has everywhere increased greatly as urbanization has spread to all parts of the globe.

The amount of urban growth differs from continent to continent and from region to region (Figure 11.4), but nearly all countries have two things in common: the proportion of their people living in cities is rising, and the cities themselves are large and growing. In consequence, most of the world's people will soon be city dwellers. The UN projection is for world urban populations to become the majority by 2005, and soon after that to become

dominant in essentially all regions of the world (Figure 11.5). Urban population is increasing much more rapidly in developing countries than in the more developed economies. In 1970, urban dwellers were about evenly divided between the less- and the more-developed portions of the world. At the start of the 21st century, that ratio was slightly more than 2 to 1; it is projected to pass 3 to 1 by 2015 and approach 4 to 1 by 2025. Even though large parts of the developing world (Southeast and South Asia, for example) still have relatively low proportions of people in cities, their absolute number of people in urban areas is extraordinarily high. Given the huge populations in Asia and the relatively heavy emphasis on agriculture (excluding Japan and Korea), it sometimes escapes us that many very large cities exist throughout parts of the world where most people are still subsistence farmers.

Megacities and Merging Metropolises

The emergence of megacities—a term originally coined by the UN in the 1970s—aroused dire predictions by the early 1980s that supercities were destined to dominate the world urban structure and distort the economies and city hierarchies of countries everywhere. Predictions based on observed or projected growth rates envisioned there would soon be cities of totally unmanageable size—20 million inhabitants or more. Fears as well were voiced of catastrophic human poverty and unbearable environmental deterioration thought certain to accompany such megacity growth. Various estimates claimed that ten or more metropolises—including Mumbai (Bombay) and Delhi, India; Tokyo, Japan; Lagos, Nigeria; Dhaka, Bangladesh; São Paulo, Brazil; Karachi, Pakistan; Mexico City; and Jakarta, Indonesia—but none in Anglo America or Europe—would exceed 20 million population by early in the 21st century.

Figure 11.1 Sprawling Mexico City, with some 18 million residents, is one of the world's largest urban concentrations. Ringed by mountains, the metropolis endures frequent temperature inversions during the winter that result in the health-endangering smog visible in the background of this photograph.

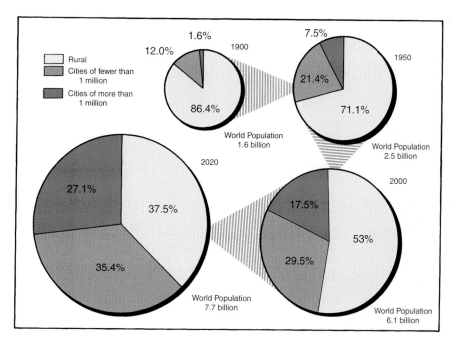

Figure 11.2 **Trends of world urbanization.** Reflecting the steady decline in rural population proportions throughout the 20th century, the United Nations estimates that virtually all the population growth expected during 2000–2020 will be concentrated in the urban areas of the world.

Estimates and projections from Population Reference Bureau, United Nations, and other sources.

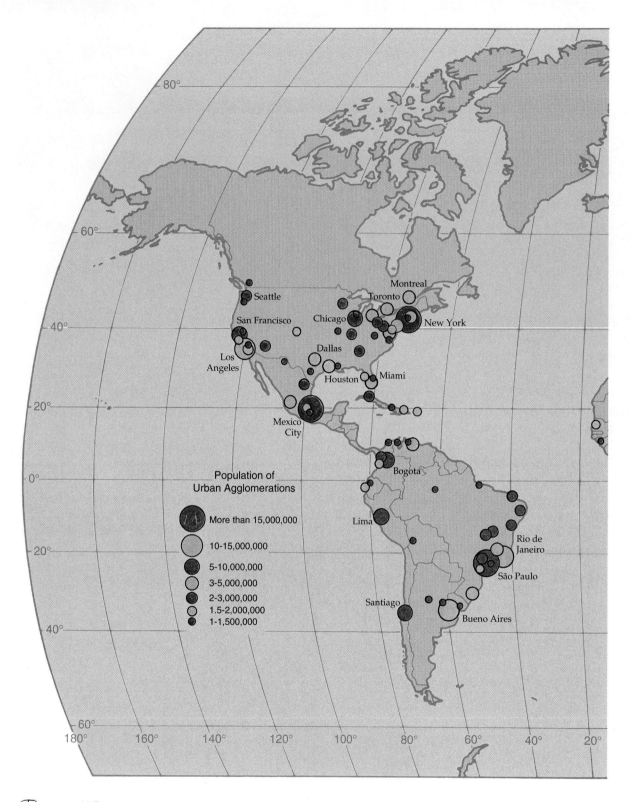

Figure 11.3 **Metropolitan areas of 1 million or more.** Massive urbanized districts are no longer characteristic only of the industrialized, developed nations. They are now found on every continent, in all latitudes, as part of most economies and societies. Not all cities in congested areas are shown.

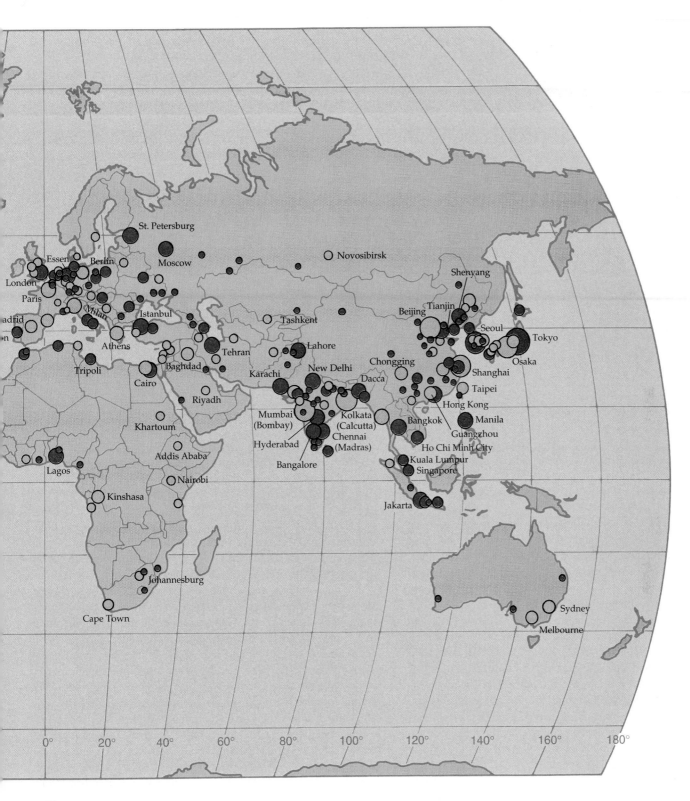

St. Petersburg
Essen · Berlin · Moscow · Novosibirsk
London · Shenyang
Paris · Milan · Istanbul · Tashkent · Beijing · Tianjin · Seoul
Madrid · Athens · Tehran · Lahore · Chongqing · Shanghai · Tokyo
adn · Baghdad · Karachi · New Delhi · Dacca · Taipei · Osaka
Tripoli · Cairo · Riyadh · Mumbai (Bombay) · Kolkata (Calcutta) · Bangkok · Hong Kong · Manila
Khartoum · Hyderabad · Chennai (Madras) · Ho Chi Minh City · Guangzhou
Lagos · Addis Ababa · Bangalore · Kuala Lumpur · Singapore
Kinshasa · Nairobi · Jakarta
Johannesburg
Cape Town · Sydney · Melbourne

0° 20° 40° 60° 80° 100° 120° 140° 160° 180°

Figure 11.3 (continued)

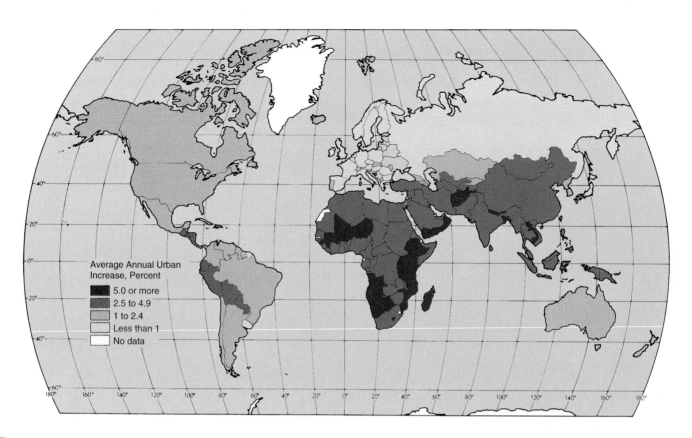

Figure 11.4 **Average annual urban population growth rates, 1995–2000.** In general, developing countries show the highest percentage increases in their urban populations, and the already highly urbanized and industrialized countries have the lowest—less than 1% per year in most of Europe.

Source: United Nations Population Fund.

Those size predictions now appear to have been overblown in many cases, reflecting simple projections of estimated past percentage and numerical increases. In reality, both personal migration choices and corporate and government investment decisions have resulted in growth rates and city sizes below those earlier anticipated. Although data are not totally conclusive, the rapid expansion of many megacities seems to be slowing, and some of the largest may now, in fact, be stabilized or even losing population. In 1984, Mexico City was said to already hold 17 million people and a decade later the U.S. Census Bureau was guessing it housed 24 million. More realistic UN figures, however, put Mexico City's 2000 population at about 18 million. Kolkata (Calcutta) was projected to have 15 million inhabitants by the end of the 1990s; it appears to have had less than 13 million for its entire metropolitan area in 2000. While São Paulo and Mexico City may, indeed, have stabilized, mid-size cities such as Curitiba (Brazil) and Monterrey (Mexico) within the same countries are growing, with at least part of their growth representing outmigrants from the megacities and government programs encouraging investment and population retention within smaller towns and mid-sized cities.

While it is certain that growth rates have slowed for most developing country megacities in recent years, even those lower rates represent more new residents each year than during the middle of the 20th century because the rates are applied to an expanding population base. In consequence, the world's population is increasingly housed in megacities. In 1975, less than 2% of global population lived in cities of 10 million or more. In 2000, the proportion exceeded 4% and is projected to top 5% by 2015.

When separate major metropolitan complexes of whatever size expand along the superior transportation facilities connecting them, they may eventually meet, bind together at their outer margins, and create the extensive metropolitan regions or **conurbations** suggested on Figure 11.3. Where this increasingly common pattern has emerged, the urban landscape can no longer be described in simple terms. No longer is there a single city with a single downtown area set off by open countryside from any other urban unit in its vicinity. Rather, we must now recognize extensive regions of continuous urbanization made up of multiple centers that have come together at their edges.

Megalopolis, already encountered in Chapter 9, is the major conurbation of North America, a nearly continuous urban string that stretches from north of Boston (southern Maine) to south of Washington, D.C. (southern Virginia). Other North American present or emerging conurbations include

- the southern Great Lakes region stretching from north of Milwaukee through Chicago and eastward to Detroit, Cleveland, and Pittsburgh;
- the Coastal California zone of San Francisco—Los Angeles—San Diego—Tijuana, Mexico;

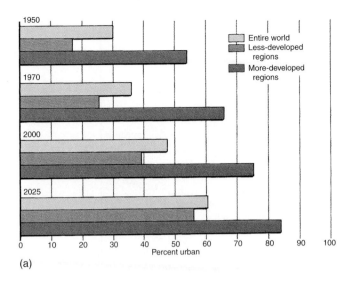

(a)

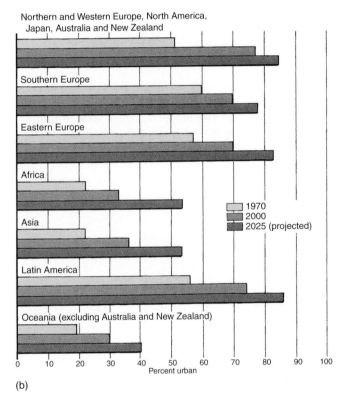

(b)

Figure 11.5 (a) **Percentage of population that is urban.** In general, developing countries show the highest percentage increases in their urban populations, and the already highly urbanized and industrialized countries have the lowest—less than 1% per annum in most of Europe. The UN projects that the less developed regions will have accounted for 93% of global urban population increase that occurs between 1970 and 2020. (b) **World regional urbanization levels.** Within the larger continental summaries shown, regional differences in urbanization levels may be pronounced. Within Asia, for example, national levels range (2000) from about 11% in Nepal to over 90% in Israel and Qatar and 100% for Kuwait and Singapore. See also Figure 4.27.

Source: Data from Population Division, United Nations.

- the Canadian "core region" conurbation from Montreal to Windsor, opposite Detroit, Michigan, where it connects with the southern Great Lakes region;
- the Vancouver—Willamette strip ("Cascadia") in the West, and the Gulf Coast and the Coastal Florida zones in the Southeast (Figure 11.6).

Outside North America, examples of conurbations are numerous and growing, still primarily in the most industrialized European and East Asian (Japanese) districts, but forming as well in other world regions where urban clusters and megacities emerged in developing countries still primarily rural in residential pattern (see Figure 11.3).

Settlement Roots

The major cities of today had humbler origins, their roots lying in the clustered dwellings which everywhere have been the rule of human settlement. People are gregarious and cooperative. Even Stone Age hunters and gatherers lived and worked in groups, not as single individuals or isolated families. Primitive cultures are communal for protection, cooperative effort, sharing of tasks by age and sex, and for more subtle psychological and social reasons. Communal dwelling became the near-universal rule with the advent of sedentary agriculture wherever it developed, and the village became the norm of human society.

In most of the world still, most rural people live in nucleated settlements, that is, in villages or hamlets, rather than in dispersed dwellings or isolated farmsteads. Only in Anglo America, parts of northern and western Europe, and in Australia and New Zealand do rural folk tend to live apart, with houses and farm buildings located on land that is individually worked. In those regions, farmsteads tend to be spatially separate one from another, and the farm village is a much less common settlement form. Communal settlements were not, of course, unknown in Anglo America. Mormon Utah, Mennonite Manitoba, and other districts of cluster migration (see p. 202) were frequently village-centered, as were such cooperative and utopian communities as Oneida, New York; Amana, Iowa; New Harmony, Indiana; the various Shaker settlements; and others of the 19th century. Elsewhere in the world, villages and hamlets were the settlement norm, though with size and form that varied by region and culture. Intensity of agricultural land use, density of population, complexity and specialization of life and livelihood, and addition of functions other than the purely residential affected the size, distribution, external form, and internal structure of settlements (Figures 11.7 and 11.8).

Rural settlements in developing countries are often considered as expressions of subsistence economic systems in which farming and fishing cultures produce no more than their individual families can consume. That clearly is not always the case. Even in the poorest farm settlements of India or Bangladesh, for example, there is a good deal of trading, buying, and selling of farm goods and family crafts for other needed commodities, and at least some village land is used for other than residential purposes (Figure 11.9). The farming or fishing settlement itself, however, may

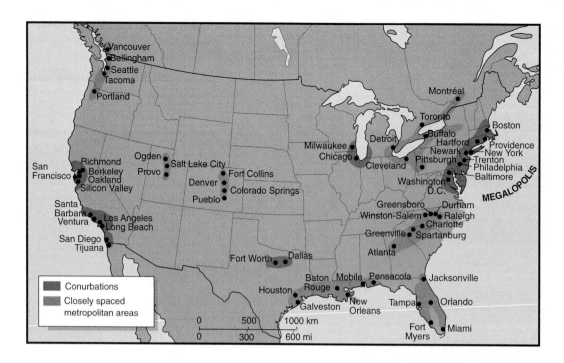

Figure 11.6 **Megalopolis and other Anglo American conurbations.** The northeast U.S. Boston-to-Norfolk urban corridor comprises the original and largest *Megalopolis* and contains the economic, political, and administrative core of the United States. A Canadian counterpart core region anchored by Montreal and Toronto connects with U.S. conurbations through Buffalo, New York, and Detroit, Michigan. For some of their extent, conurbations fulfill their classic definition of continuous built-up urban areas. In other portions, they are more statistical than landscape entities, marked by counties that qualify as "urban" or "metropolitan" even though land uses may appear dominantly rural.

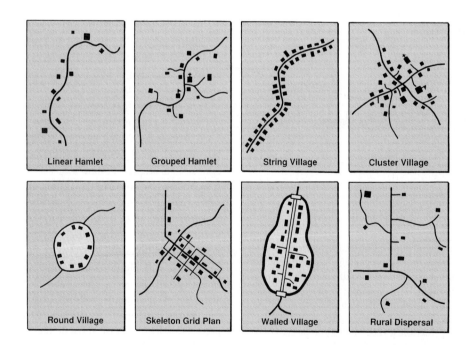

Figure 11.7 **Basic settlement forms.** The smallest organized rural clusters of houses and nonresidential structures are commonly called *hamlets*, and may contain only 10–15 buildings. *Villages* are larger agglomerations, although not as sizable or functionally complex as urban *towns*. The distinction between village and town is usually a statistical definition that varies by country.

From Introducing Cultural Geography, *2d ed., by Joseph E. Spencer and William L. Thomas. Copyright © 1978 John Wiley & Sons, Inc. Reprinted by permission of John Wiley & Sons, Inc.*

(a)

(b)

(c)

Figure 11.8 Rural settlements in largely subsistence economies vary from the rather small populations characteristic of compact African villages, such as the Zulu village, or kraal, in South Africa shown in (*a*), to more dispersed and populous settlements such as the Nepalese high pasture summer village of Konar seen in (*b*), to the very large, densely populated Indian rural communities like that seen in (*c*).

Source: (b) Nepalese village of Konar courtesy of Colin Thorn.

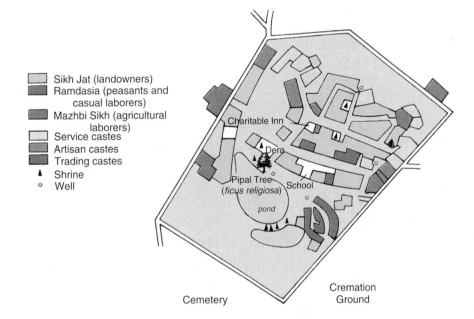

Sikh Jat (landowners)
Ramdasia (peasants and casual laborers)
Mazhbi Sikh (agricultural laborers)
Service castes
Artisan castes
Trading castes
▲ Shrine
○ Well

Charitable Inn
Dera
Pipal Tree
(*ficus religiosa*)
School
pond

Cemetery

Cremation Ground

Figure 11.9 **A village in the Punjab region, India.** In the 1960s, Kunran village had some 1000 inhabitants of several different occupational castes. Most numerous were the Sikh Jat (landowners: 76 households), Ramdasia (peasants and casual laborers: 27 households), and Mazhbi Sikh (agricultural laborers: 12 households). Other castes (and occupations) included Tarkhan (carpenter), Bazigar (acrobat), Jhiwar (water carrier), Sunar (goldsmith), Nai (barber), and Bania (shopkeeper). The trades, crafts, and services they (and others) pursued created a more complex land use pattern than is implied by the generalized village forms depicted on Figure 11.7.

Redrawn with permission from Jan O. M. Broek and John M. Webb, A Geography of Mankind, *copyright © 1968 McGraw-Hill, Inc.*

be nearly self-contained, with little commercial exchange with neighboring villages or between villages and distant cities.

When trade does develop between two or more rural settlements, they begin to take on new physical characteristics as their inhabitants engage in additional types of occupations. The villages lose the purely social and residential character of subsistence agricultural settlements and assume urban features. There is a tendency for the houses to cluster along the main road or roads, creating a linear, cross, or starlike pattern. No longer are the settlements nearly completely self-contained; they become part of a system of communities. The beginnings of urbanization are seen in the types of buildings that are erected and in the heightened importance of the main streets and of the roads leading to other settlements. The location of villages relative to one another becomes significant as the once self-sufficient rural settlements become towns and cities engaged in urban activities and interchange.

The Nature of Cities

Cities are among the oldest marks of civilization. Dating from at least 6000 years or more ago, they originated in—or diffused from—the culture hearths that first developed sedentary agriculture. They are as well among the newest experiences of a growing share of the world's population. Whether ancient or modern, all cities show recurring themes and regularities appropriate to their time and place of existence.

First, all of them perform functions—have an economic base—generating the income necessary to support themselves and their contained population. Second, none exists in a vacuum; each is part of a larger society and economy with which it has essential reciprocal connections. That is, each is a unit in a system of cities and a focus for a surrounding nonurban area. Third, each urban unit has a more or less orderly internal arrangement of land uses, social groups, and economic functions. These arrangements may be partially planned and controlled and partially determined by individual decisions and market forces. Finally, all cities, large or small, ancient or modern, have experienced problems of land use, social conflict, and environmental concern. Yet cities, though flawed, remain the capstone of our cultures, the organizing focuses of modern societies and economies, the magnet of people everywhere.

Whatever their size, age, or location, urban settlements exist for the efficient performance of functions required by the society that creates them. They reflect the saving of time, energy, and money that the agglomeration of people and activities implies. The more accessible the producer to the consumer, the worker to the workplace, the citizen to the town hall, the worshiper to the church, or the lawyer or doctor to the client, the more efficient is the performance of their separate activities, and the more effective is the integration of urban functions.

Urban areas may provide all or some of the following types of functions: retailing, wholesaling, manufacturing, professional and personal services, entertainment, business and political administration, military defense, educational and religious functions, and transportation and communication services. Because all urban functions and people cannot be located at a single point, cities themselves must take up space, and land uses and populations must have room within them. Because interconnection is essential, the nature of the transportation system will have an enormous bearing on the total number of services that can be performed and the efficiency with which they can be carried out. The totality of people and functions of a city constitutes a distinctive cultural landscape whose similarities and differences from place to place are the subjects for urban geographic analysis.

Some Definitions

Urban units are not of a single type, structure, or size. Their common characteristic is that they are nucleated, nonagricultural settlements. At one end of the size scale, urban areas are hamlets or small towns with at most a single short main street of shops; at the opposite end, they are complex multifunctional metropolitan areas or supercities (Figure 11.10). The word *urban* is often used to describe such places as a town, city, suburb, and metropolitan area, but it is a general term, not used to specify a particular type or size of settlement. Although the terms designating the different types of urban settlement, like *city,* are employed in common speech, not everyone uses them in the same way. What is recognized as a city by a resident of rural Vermont or West Virginia might not at all be afforded that name and status by an inhabitant of California or New Jersey. It is necessary in this chapter to agree on the meanings of terms commonly employed but varyingly interpreted.

The words **city** and **town** denote nucleated settlements, multifunctional in character, including an established central business district and both residential and nonresidential land uses. Towns are smaller in size and have less functional complexity than cities, but they still have a nuclear business concentration. **Suburb** implies a subsidiary area, a functionally specialized segment of a larger urban complex. It may be dominantly or exclusively residential, industrial, or commercial, but by the specialization of its land uses and functions, a suburb is not self-sufficient. It depends on and is integrated with urban areas outside of its boundaries. Suburbs can, however, be independent political entities. For large cities having many suburbs, it is common to call that part of the urban area contained within the official boundaries of the main city around which the suburbs have been built the **central city.**

Some or all of these urban types may be associated into larger landscape units. The **urbanized area** refers to a continuously built-up landscape defined by building and population densities with no reference to political boundaries. It may be viewed as the *physical city* and may contain a central city and many contiguous cities, towns, suburbs, and other urban tracts. A **metropolitan area,** on the other hand, refers to a large-scale *functional* entity, perhaps containing several urbanized areas, discontinuously built-up but nonetheless operating as an integrated economic whole. Figure 11.11 shows these areas in a hypothetical American county.

The Bureau of the Census has redefined the concept of "metropolitan" from time to time to summarize the realities of the changing population, physical size, and functions of urban regions. The current *metropolitan statistical areas* are comprised of

(a)

(b)

Figure 11.10 The differences in size, density, and land use complexity are immediately apparent between (a) New York City and (b) a small Ohio town. Clearly, one is a city, one is a town, but both are *urban*.

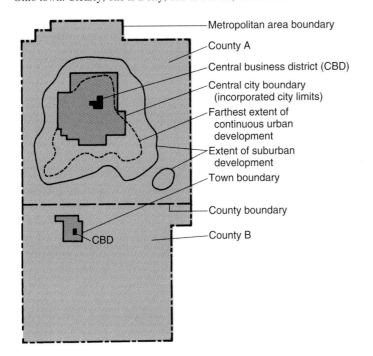

— Metropolitan area boundary
— County A
— Central business district (CBD)
— Central city boundary (incorporated city limits)
— Farthest extent of continuous urban development
— Extent of suburban development
— Town boundary
— County boundary
— County B
CBD

Figure 11.11 **A hypothetical spatial arrangement of urban units within a metropolitan area.** Sometimes official limits of the central city are very extensive and contain areas commonly considered suburban or even rural. On the other hand, older eastern U.S. cities (and some such as San Francisco in the West) more often have restricted limits and contain only part of the high-density land uses and populations of their metropolitan or urbanized areas.

a central county or counties with at least one urbanized area of at least 50,000 population, plus adjacent outlying counties with a high degree of social and economic integration with the central county as measured by commuting volumes. A *micropolitan statistical* area is a similar but smaller version of the metropolitan concept. It is based on a central city county with at least one urban cluster of between 10,000 and 50,000 population plus outlying counties with considerable social and economic integration with it.

The Location of Urban Settlements

Urban centers are functionally connected to other cities and to rural areas. In fact, the reason for the existence of an urban unit is not only to provide services for itself, but for others outside of it. The urban center is a consumer of food, a processor of materials, and an accumulator and dispenser of goods and services. But it must depend on outside areas for its essential supplies and as a market for its products and activities.

In order to adequately perform the tasks that support it and to add new functions as demanded by the larger economy, the city must be efficiently located. That efficiency may be marked by centrality to the area served. It may derive from the physical characteristics of its site. Or placement may be related to the resources, productive regions, and transportation network of the country so that the effective performance of a wide array of activities is possible.

In discussing urban settlement location, geographers usually mention the significance of site and situation, concepts already introduced in Chapter 1 (see p. 10 and Figures 1.6 and 1.7). You will recall that *site* refers to the exact terrain features associated with the city, as well as—less usefully—to its absolute (globe grid) location. Classifications of cities according to site characteristics have been proposed, recognizing special placement circumstances. These include *break-of-bulk* locations such as river crossing points where cargoes and people must interrupt a journey; *head-of-navigation* or *bay head* locations where the limits of water transportation are reached; and *railhead* locations where the railroad ended. In Europe, security and defense—island locations or elevated sites—were considerations in earlier settlement locations. Waterpower sites of earlier stages and coalfield sites of later phases of the Industrial Revolution were noted in Chapters 8 and 9 and represent a union of environmental and cultural-economic considerations.

If site suggests absolute location, *situation* indicates relative location that places a settlement in relation to the physical and cultural characteristics of surrounding areas. Very often it is important to know what kinds of possibilities and activities exist in the area near a settlement, such as the distribution of raw materials, market areas, agricultural regions, mountains, and oceans.

Although in many ways more important than site in understanding the functions and growth potentials of cities, situation is more nearly unique to each settlement and does not lend itself to easy generalization.

The site or situation that originally gave rise to an urban unit may not remain the essential ingredient for its growth and development for very long. Agglomerations originally successful for whatever reason may by their success attract people and activities totally unrelated to the initial localizing forces. By what has been called a process of "circular and cumulative causation" (see p. 358), a successful urban unit may acquire new populations and functions attracted by the already existing markets, labor force, and urban facilities.

The Functions of Cities

The key concept is *function*—what cities actually do within the larger society and economy that established them. No city stands alone. Each is linked to other towns and cities in an interconnected city system; each provides services and products for its surrounding tributary region—its *hinterland* or trade area. Those linkages reflect complementarity and the processes of spatial interaction that we explored in Chapter 3; they are rooted in the different functions performed by different units within the urban system. However, not all of the activities carried on within a city are intended to connect that city with the outside world. Some are necessary simply to support the city itself. Together, these two levels of activity make up the **economic base** of an urban settlement.

The Economic Base

Part of the employed population of an urban unit is engaged either in the production of goods or the performance of services for areas and people outside the city itself. They are workers engaged in "export" activities, whose efforts result in money flowing into the community. Collectively, they constitute the **basic sector** of the city's total economic structure. Other workers support themselves by producing goods or services for residents of the urban unit itself. Their efforts, necessary to the well-being and the successful operation of the settlement, do not generate new money for it but comprise a **service** or **nonbasic sector** of its economy. These people are responsible for the internal functioning of the urban unit. They are crucial to the continued operation of its stores, professional offices, city government, local transit, and school systems.

The total economic structure of an urban area equals the sum of its basic and nonbasic activities. In actuality, it is the rare urbanite who can be classified as belonging entirely to one sector or another. Some part of the work of most people involves financial interaction with residents of other areas. Doctors, for example, may have mainly local patients and thus are members of the nonbasic sector, but the moment they provide a service to someone from outside the community, they bring new money into the city and become part of the basic sector.

Variations in basic employment structure among urban units characterize the specific functional role played by individual cities. Most centers perform many export functions, and the larger the urban unit, the more multifunctional it becomes. Nonetheless, even in cities with a diversified economic base, one or a very small number of export activities tends to dominate the structure of the community and to identify its operational purpose within a system of cities. Figure 11.12 indicates the functional specializations of some large U.S. metropolitan areas.

Such functional specialization permits the classification of cities into categories: manufacturing, retailing, wholesaling, transportation, government, and so on. Such specialization may also evoke images when the city is named: Detroit, Michigan, or Tokyo, Japan, as manufacturing centers; Tulsa, Oklahoma, for oil production; Nice, France, as a resort; Ottawa, Canada, in government; and so on. Certain large regional, national, or world capitals—as befits major multifunctional concentrations—call up a whole series of mental associations, such as New York with banking, the stock exchange, entertainment, the fashion industry, port activities, and others.

Base Ratios

Assuming it were possible to divide with complete accuracy the employed population of a city into totally separate basic and service (nonbasic) components, a ratio between the two employment groups could be established. With exception for some high-income communities, this *basic/nonbasic ratio* is roughly similar for urban units of similar size irrespective of their functional specializations. Further, as a settlement increases in size, the number of nonbasic personnel grows faster than the number of new basic workers. Thus, in cities with a population of 1 million, the ratio is about 2 nonbasic workers for every basic worker. The addition of ten new basic employees implies the expansion of the labor force by 30 (10 basic, 20 nonbasic) and an increase in total population equal to the added workers plus their dependents. A **multiplier effect** thus exists, associated with economic growth. The term implies the addition of nonbasic workers and dependents to a city's total employment and population as a supplement to new basic employment. The size of the multiplier effect is determined by the community's basic/nonbasic ratio (Figure 11.13).

The changing numerical relationships shown in Figure 11.13 are understandable when we consider how settlements add functions and grow in population. A new industry selling services to other communities requires new workers who thus increase the basic workforce. These new employees, in turn, demand certain goods and services, such as clothing, food, and medical assistance, which are provided locally. Those who perform such services must themselves have services available to them. For example, a grocery clerk must also buy groceries. The more nonbasic workers a city has, the more nonbasic workers are needed to support them, and the application of the multiplier effect becomes obvious.

The growth of cities may be self-generating—"circular and cumulative"—in a way related not to the development of

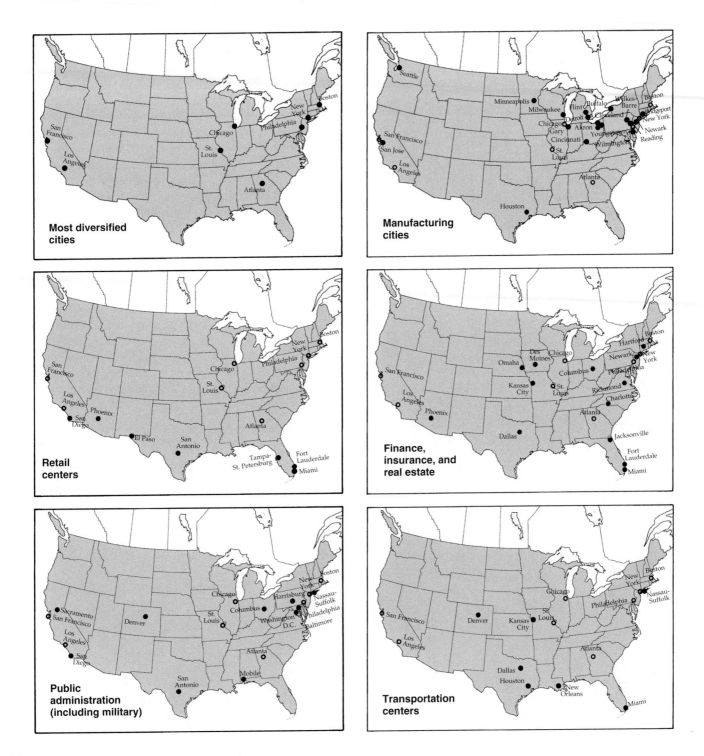

Figure 11.12 **Functional specialization of selected U.S. metropolitan areas.** Five categories of employment were chosen to show patterns of specialization for some U.S. metropolitan areas. In addition, the category "Most Diversified" includes representative examples of cities with a generally balanced employment distribution. Since their "balance" implies performance of a variety of functions, the diversified cities are included as open circles on the other specialization maps. Note that the most diversified urban areas tend to be the largest.

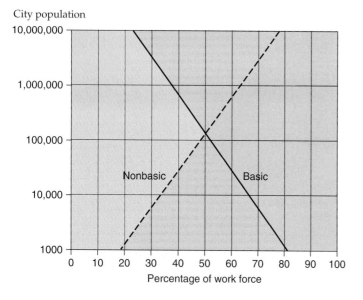

Figure 11.13 **A generalized representation of the proportion of the work force engaged in basic and nonbasic activities.** As settlements become larger, a greater proportion of the work force is employed inn nonbasic activities. Larger centers are therefore more self-contained.

industries that specialize in the production of material objects for export, like automobiles and paper products, but to the attraction of what would be classified as *service* activity. Banking and legal services, a sizable market, a diversified labor force, extensive public services, and the like may generate additions to the labor force not basic by definition, but nonbasic. In recent years, service industries have developed to the point where new service activities serve older ones. For example, computer systems firms aid banks in developing more efficient computer-driven banking systems.

In much the same way as settlements grow in size and complexity, so do they decline. When the demand for the goods and services of an urban unit falls, obviously fewer workers are needed, and both the basic and the services components of a settlement system are affected. There is, however, a resistance to decline that impedes the process and delays its impact. That is, settlements can grow rapidly as migrants respond quickly to the need for more workers, but under conditions of decline, those that have developed roots in the community are hesitant to leave or may be financially unable to move to another locale. Figure 11.14 and Table 11.1 show that in recent years urban areas in the South and West of the United States have been growing, while some decline is evident in the Northeast and the Midwest regions.

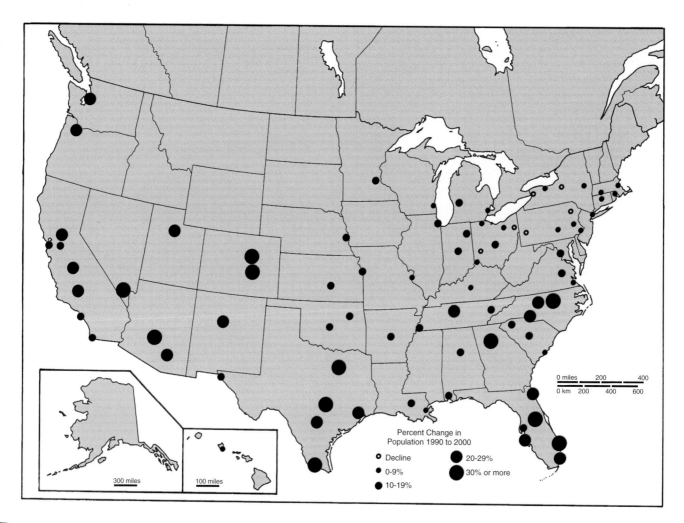

Figure 11.14 **The pattern of metropolitan growth and decline in the United States, 1990–2000.** Shown are metropolitan areas with 600,000 or more population in 2000. The cities of the Southern and Southwestern Sun Belt showed the greatest relative growth; only modest growth, stability, or decline generally marked the Northeast and Midwest.

Data from U.S. Bureau of the Census.

Table 11.1

Fasting Growing U.S. Metropolitan Statistical Areas, 1990–2000, with More than 1 Million Population

Rank	MSA	% Growth
1	Las Vegas, NV–AZ	83.3
2	Austin–San Marcos, TX	48.7
3	Phoenix–Mesa, AZ	45.3
4	Atlanta, GA	38.9
5	Raleigh–Durham-Chapel Hill, NC	38.9
6	Orlando, FL	34.3
7	West Palm Beach–Boca Raton, FL	31.0
8	Denver–Boulder–Greeley, CO	30.4
9	Dallas–Fort Worth, TX	29.3
10	Charlotte, NC	29.0

Slowest Growing or Declining U.S. Metropolitan Statistical Areas, 1990–2000, with More than 1 Million Population

Rank	MSA	% Growth
1	Buffalo–Niagara Falls, NY	–1.6
2	Pittsburgh, PA	–1.5
3	Hartford, CT	2.2
4	Cleveland, OH	3.0
5	Rochester, NY	3.4
6	New Orleans, LA	4.1
7	St. Louis, MO	4.5
8	Providence–Fall River, RI–MA	4.8
9	Philadelphia–Wilmington, PA–DE–NJ	5.0
10	Milwaukee, WI	5.1

Biggest Population Gains, 1990–2000, U.S. MSAs and Consolidated Metropolitan Statistical Areas (CMSA)

Rank	MSA/CMSA	Gain (in thousands)
1	Los Angeles–Riverside–Orange County	1842
2	New York–Northern New Jersey–Long Island	1650
3	Dallas–Fort Worth	1185
4	Atlanta	1152
5	Phoenix–Mesa	1013
6	Houston–Galveston–Brazoria	938
7	Chicago–Gary–Kenosha	918
8	Washington–Baltimore	881
9	San Francisco–Oakland–San Jose	786
10	Las Vegas	711

Source: U.S. Bureau of the Census.

Systems of Urban Settlements

The functional structure of a settlement affects its current size and growth prospects. At the same time, its functions are a reflection of that community's location and its relationships with other urban units in the larger city system of which all are a part. A simple but revealing threefold functional classification of urban settlements recognizes them as either transportation centers, special-function cities, or central places. Each class has its own characteristic spatial arrangement; together, the three classes help explain the distributional pattern and the size and functional hierarchies of the entire city system.

The spatial pattern of *transportation centers* is that of alignment— along seacoasts, major and minor rivers, canals, or railways. Routes of communication form the orienting axes along which cities developed and on which at least their initial functional success depended (Figure 11.15). *Special-function cities* are those engaged in mining, manufacturing, or other activities the localization of which is related to raw material occurrence, agglomeration economies, or the circular and cumulative attractions of constantly growing market and labor concentrations. Special-function cities show a pattern of urban clustering—as the mining and manufacturing cities of the Ruhr district of Germany, the Midlands of England, or the Donets Basin in Ukraine, for example. More familiarly, they appear in the form of the multifunctional metropolitan concentrations recognized by the Metropolitan Statistical Areas of the United States, the Census Metropolitan Areas of Canada, or in such massive urbanized complexes as metropolitan Tokyo, Moscow, Paris, London, Buenos Aires, and others worldwide.

A common property of all settlements is centrality, no matter what their recognized functional specializations. Every urban unit provides goods and services for a surrounding area tributary to it. For many, including mining or major manufacturing centers, service to tributary areas is only a very minor part of their economic base. Some settlements, however, have that rural service and trade function as their dominant role, and these make up the third simplified category of cities: *central places.*

The Urban Hierarchy

Perhaps the most effective way to recognize how systems of cities are organized is to consider the **urban hierarchy,** a ranking of cities based on their size and functional complexity. One can measure the numbers and kinds of functions each city or metropolitan area provides. The hierarchy is then like a pyramid; a few large and complex cities are at the top and many smaller, simpler ones are at the bottom. There are always more smaller cities than larger ones.

When a spatial dimension is added to the hierarchy as in Figure 11.16, it becomes clear that an areal system of metropolitan centers, large cities, small cities, and towns exists. Goods, services, communications, and people flow up and down the hierarchy. The few high-level metropolitan areas provide specialized

Together, all centers at all levels in the hierarchy constitute an urban system.

Rank-Size and Primacy

The observation that there are many more small than large cities within an urban system ("the larger the fewer") is a statement about hierarchy. For many large countries of great regional diversity and, usually but not always, advanced economy, the city size hierarchy is summarized by the **rank-size rule.** It tells us that the nth largest city of a national system of cities will be $\frac{1}{n}$ the size of the largest city. That is, the second largest settlement will be $\frac{1}{2}$ the size of the largest, the 10th biggest will be $\frac{1}{10}$ the size of the first-ranked city, and so on.

The rank-size ordering may describe the urban-size patterning in complex economies where urban history is long and urbanizing forces are many and widely distributed. Although no national city system exactly meets the requirements of the rank-size rule, that of the United States closely approximates it. It is less applicable to countries with developing economies or where the urban size hierarchy has been distorted through concentration of functions in a single, paramount center.

In some countries the urban system is dominated by a **primate city,** one that is far more than twice the size of the second-ranked city. In fact, there may be no obvious "second city" at all, for a characteristic of a primate city hierarchy is one very large city, few or no intermediate-sized cities, and many subordinate smaller settlements. For example, metropolitan Seoul contains over 40% of the total population and one-half of the urban population of South Korea, and Luanda has almost two-thirds of Angola's urban folk. The capital cities of many developing countries display that kind of overwhelming primacy. In part, their primate city pattern is a heritage of their colonial past, when economic development, colonial administration, and transportation and trade activities were concentrated at a single point (Figure 11.17); Kenya (Nairobi is the primate city) and many other African countries are examples.

In other instances—Egypt (Cairo) or Mexico (Mexico City), for example—development and population growth have tended to concentrate disproportionately in a capital city whose very size attracts further development and growth. Many European countries—Austria, the United Kingdom, and France are familiar examples—also show a primate structure, often ascribed to the former concentration of economic and political power around the royal court in a capital city that was, perhaps, also the administrative and trade center of a larger colonial empire.

World Cities

Standing at the top of national systems of cities are a relatively few cities that may be called **world cities.** These are urban centers that are control points for international production and marketing and for international finance. Three world cities—New York, London, and Tokyo—dominate commerce in their respective parts of the world. Each has a number of other secondary-level world cities (Osaka, Rhine-Ruhr, Chicago, Paris, Frankfurt, Sydney, and Zurich are often cited) directly linked to it. All are

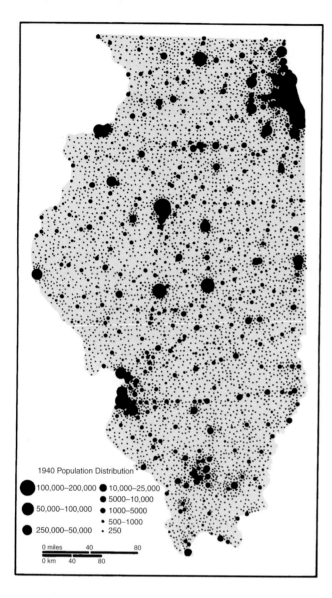

Figure 11.15 **Urban alignments in Illinois.** Railroads preceded settlement in much of the Anglo American continental interior, and urban centers were developed—frequently by the railroad companies themselves—as collecting and distributing points expected to grow as the farm populations increased. Located at constant 8- to 10-kilometer (5- to 6-mile) intervals in Illinois, the rail towns were the focal points of an expanding commercial agriculture. The linearity of the town pattern in 1940, at the peak of railroad influence, unmistakably marks the rail routes. Also evident are such special-function clusterings as the Chicago and St. Louis metropolitan districts and the mining towns of Southern Illinois.

1940 Population Distribution

- ⬤ 100,000–200,000
- ⬤ 50,000–100,000
- ⬤ 250,000–50,000
- ● 10,000–25,000
- ● 5000–10,000
- • 1000–5000
- · 500–1000
- · 250

0 miles 40 80
0 km 40 80

functions for large regions, while the smaller cities serve smaller districts. The separate centers interact with the areas around them, but since cities of the same level provide roughly the same services, those of the same size tend not to serve each other unless they provide some very specialized activity, such as housing a political capital of a region or a major university. Thus, the settlements of a given level in the hierarchy are not independent but interrelated with communities of other levels in that hierarchy.

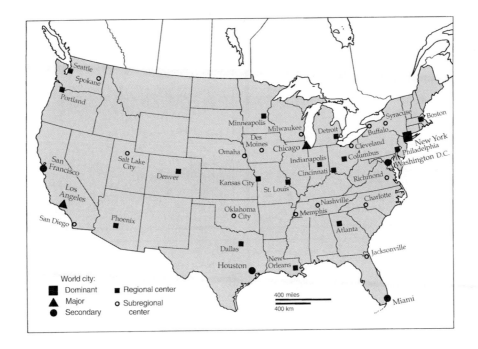

Figure 11.16 **A functional hierarchy of U.S. metropolitan areas.** Only the major metropolitan areas are shown. The hierarchy includes smaller urban districts (not shown) that depend on or serve the larger centers.

Redrawn from P. L. Knox, ed., The United States: A Contemporary Human Geography. *Harlow, England. Longman, 1988, Fig. 5.5, p. 144.*

bound together in complex networks that control the organization and management of the global system of finance, manufacturing, and trade. Figure 11.18 shows the links between the dominant centers and the suggested major and secondary world cities. These are all interconnected mainly by advanced communication systems between governments, major corporations, stock and futures exchanges, securities and commodity markets, major banks, and international organizations.

Major international corporations themselves spur world city development and dominance. The growing size and complexity of transnational corporations dictate their need to outsource central managerial functions to specialized service firms to minimize the complexity of control over dispersed operations. Those specialized service agencies—legal, accounting, financial, and so on—in their turn need to draw on the very large pools of expertise, information, and talent available only in very large world-class cities.

Urban Influence Zones

Whatever its position in its particular urban hierarchy, every urban settlement exerts an influence upon its immediately surrounding area. A small city may influence a local region of some 1000+ square kilometers (400 sq mi) if, for example, its newspaper is delivered to that district. Beyond that area, another city may be the dominant influence. **Urban influence zones** are the areas outside of a city that are still affected by it. As the distance away from a community increases, its influence on the surrounding countryside decreases (recall the idea of distance decay discussed in Chapter 3, p. 69). The sphere of influence of an urban unit is usually proportional to its size.

A large city located, for example, 100 kilometers (62 miles) away from a small city may influence that and other small cities through its banking services, its TV stations, and its large shopping malls. There is an overlapping hierarchical arrangement, and the influence of the largest cities is felt over the widest areas, a "market area" dominance basic to central place theory, discussed in the following text.

Intricate relationships and hierarchies are common. Consider Grand Forks, North Dakota, which for local market purposes dominates the rural area immediately surrounding it. However, Grand Forks is influenced by political decisions made in the state capital, Bismarck. For a variety of cultural, commercial, and banking activities, Grand Forks is influenced by Minneapolis. As a center of wheat production, Grand Forks and Minneapolis are subordinate to the grain market in Chicago. Of course, the pervasive agricultural and other political controls exerted from Washington, D.C., on Grand Forks, Minneapolis, and Chicago indicate how large and complex are urban zones of influence.

Central Places

An effective way to realize the meaning of influence zones and to grasp how cities and towns are interrelated is to consider urban settlements as **central places,** that is, as centers for the distribution of economic goods and services to surrounding nonurban populations. They are at the same time essential links in a system of interdependent urban settlements. Central places show size and spacing regularities not seen where special function or transportation cities predominate. That is, instead of showing patterns of clustering or alignment, central places display a regularity of distribution, with towns of about the same size and performing

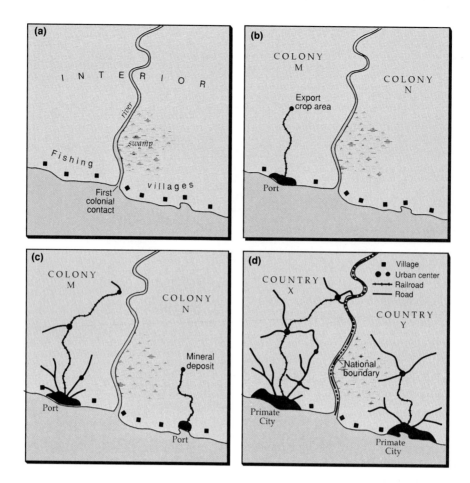

Figure 11.17 **Primate city evolution.** At first colonial contact (*a*), settlements are coastal and unconnected with each other. Joining a newly productive hinterland by European-built railroad to a new colonial port (*b*) begins to create a pattern of core-periphery relations and to focus European administration, trade, and settlement at the port. Mineral discoveries and another rail line in a neighboring colony across the river (*c*) mark the beginnings of a new set of core-periphery relationships and of a new multifunctional colonial capital nearby but unconnected by land with its neighbor. With the passage of time and further transport and economic development, two newly independent nations (*d*) display *primate city* structures in which further economic and population growth flows to the single dominating centers of countries lacking balanced regional transport networks, resource development, and urban structures. Both populations and new functions continue to seek locations in the primate city where their prospects for success are greatest.

Adapted from E. S. Simpson, The Developing World: An Introduction. *(Harlow, Essex, England: Longman Group UK Limited, 1987).*

about the same number and kind of functions located about the same distance from each other.

In 1933, the German geographer **Walter Christaller** (1893–1969) attempted to explain those observed regularities of size, location, and interdependence of settlements. He recognized that his **central place theory** could best be visualized in rather idealized, simplified circumstances. Christaller assumed that the following propositions were true:

1. Towns that provide the surrounding countryside with such fundamental goods as groceries and clothing would develop on a uniform plain with no topographic barriers, channelization of traffic, or variations in farm productivity.

2. The farm population would be dispersed in an even pattern across that plain.

3. The characteristics of the people would be uniform; that is, they would possess similar tastes, demands, and incomes.

4. Each kind of product or service available to the dispersed population would have its own *threshold*, or minimum number of consumers needed to support its supply. Because such goods as sports cars or fur coats are either expensive or not in great demand, they would have a high threshold, while a fewer number of customers within smaller tributary areas would be needed to support a small grocery store.

5. Consumers would purchase goods and services from the nearest opportunity (store or supplier).

When all of Christaller's assumptions are considered simultaneously, they yield the following results:

1. Since each customer patronizes the nearest center offering the needed goods, the agricultural plain is automatically divided into noncompeting market areas—*complementary regions*—where each individual town (and its merchants) has a sales monopoly.

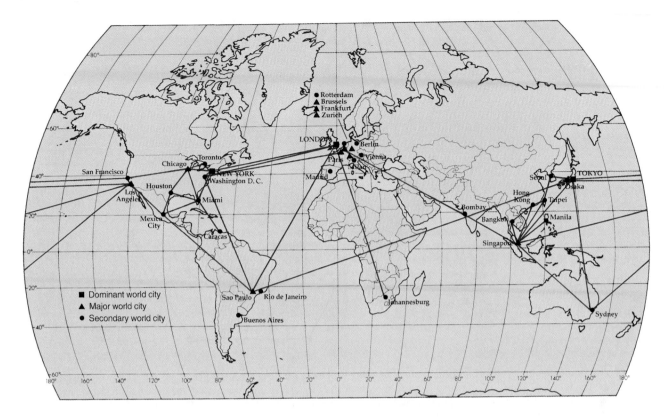

Figure 11.18 **A classification of world cities.** The ties between the cities represent the flow of financial and economic information.

2. Those market areas will take the form of a series of hexagons that cover the entire plain, as shown in Figure 11.19. Since the hypothetical plain must be completely subdivided, no area can be unserved and none can have equal service from two competing centers (Figure 11.20).

3. There will be a central place at the center of each of the hexagonal market areas.

4. The largest central places (with the largest market areas) will supply all the goods and services the consumers in that area demand and can afford.

5. The size of the market area of a central place will be proportional to the number of goods and services offered from that place.

6. Contained within or at the edge of the largest market areas are central places serving a smaller population and offering fewer goods and services. As Figure 11.19 indicates, the central place pattern shows a "nesting" of complementary regions in which part or all of multiple lower-order service areas are contained within the market area of a higher-order center.

In addition, Christaller reached two important conclusions. First, towns at the same size (functional level) in the central place system will be evenly spaced, and larger towns (higher-order places) will be farther apart than smaller ones. This means that many more small than large towns will exist. In Figure 11.19, the ratio of the number of small towns to towns of the next larger size is 3 to 1. This distinct, steplike series of towns in size classes differentiated by both size and function is called a *hierarchy of central places.*

Second, the system of towns is interdependent. If one central place were eliminated, the entire system would have to readjust in its spatial pattern, its offered goods, or both. Consumers need a variety of products, each of which has a different minimum number of customers required to support it. The towns containing many goods and services become regional retailing centers, while the smaller central places serve just the people immediately in their vicinity. Customers are willing to travel great distances to buy expensive luxury items, but not basic foodstuffs like bread or milk. The higher the threshold of a desired product, the farther, on average, the consumer must travel to purchase it.

These conclusions have been shown to be generally valid in widely differing areas of the world. When varying incomes, cultures, landscapes, and transportation systems are taken into account, the results, though altered to some extent, hold up rather well. They are particularly applicable to agricultural areas. One has to stretch things a bit to see the model operating in highly industrialized areas where cities are more than just retailing centers. Therefore, in an increasingly industrializing and modernizing world, pure central place theory has decreasing applicability as the sole explanation of observed urban spatial patterns. However, if we combine a Christaller-type approach with the ideas that help us understand the cluster patterns of special-function cities and the alignments of transportation-based cities, we have a fairly good understanding of the distribution of the majority of cities and towns.

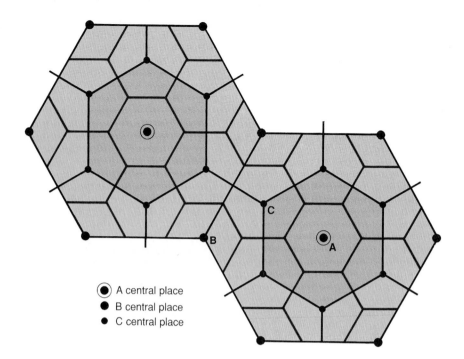

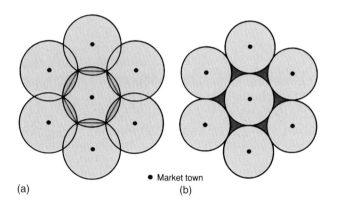

⊙ A central place
● B central place
· C central place

Figure 11.19 **Complementary regions and the pattern of central places.** The two A central places are the largest on this diagram of one of Christaller's models. The B central places offer fewer goods and services for sale and serve only the areas of the intermediate-sized hexagons. The many C central places, which are considerably smaller and more closely spaced, serve still smaller market areas. The goods offered in the C places are also offered in the B and A places, but the latter offer considerably more and more specialized goods. Notice that places of the same size are equally spaced.

Arthur Getis and Judith Getis, "Christaller's Central Place Theory." Journal of Geography, 1966. Used with permission of the National Council for Geographic Education, Indiana, PA.

(a)   (b)
● Market town

Figure 11.20 **The derivation of complementary regions.** (*a*) If the hypothetical region were totally covered by circular complementary regions, areas of overlap would occur. Since Christaller's assumption was that people will only shop at the nearest center, areas of overlap must be divided so that those on each side of the boundary are directed to their nearest service point. (*b*) Circular areas too small to cover the region completely result in impermissible unserved populations.

Network Cities

In recent years, a new kind of urban spatial pattern has begun to appear. A **network city** evolves when two or more previously independent nearby cities, potentially complementary in functions, strive to cooperate by developing between them high-speed transportation corridors and communications infrastructure. For example, with the reunion of Hong Kong and China proper, an infrastructure of highway and rail lines and of communications improvements has been developed to help integrate Hong Kong with Guangzhou, the huge, rapidly growing industrial and economic hub on the mainland. In Japan, three distinctive, nearby cities—Kyoto, Osaka, and Kobe—are joining together to compete with the Tokyo region as a major center of commerce. Kyoto, with its temples and artistic treasures, is the cultural capital of Japan; Osaka is a primary commercial and industrial center; and Kobe is a leading port. Their complementary functional strengths are reinforced by high-speed rail transport connecting the cities and by an airport (Kansai) designed to serve the entire region.

In Europe, the major cities of Amsterdam, Rotterdam, and The Hague, together with intermediate cities such as Delft, Utrecht, and Zaanstad, are connected by high-speed rail lines and contain a major airport that serves them all collectively. Each of these cities has special functions not duplicated in the others, and there is no intention of developing competition between them. In a sense, this region—called the Randstad—is in a strong position to rival London for dominant world city status.

No similar network city has yet developed in Anglo America. The New York-Philadelphia, the Chicago-Milwaukee, the San Francisco-Oakland, or the Los Angeles-San Diego city pairings do not yet qualify for network city status since there has been no concerted effort to bring their competing interests together into a single structure of complementary activities.

Inside the City

The structure, patterns, and spatial interactions of systems of cities make up only half of the story of urban settlements. The other half involves the distinctive cultural landscapes that are the cities themselves. An understanding of the nature of cities is incomplete without a knowledge of their internal characteristics. So far, we have explored the location, the size, and the growth and decline tendencies of cities within hierarchical urban systems. Now we look into the city itself in order to better understand how its land uses are distributed, how social areas are formed, and how institutional controls such as zoning regulations affect its structure. We will begin on familiar ground and focus our discussion primarily on U.S. cities. Later in this chapter we will review urban land use patterns and social geographies in different world settings.

It is a common observation that recurring patterns of land use arrangements and population densities exist within urban areas. There is a certain sameness to the ways cities are internally organized, especially within one particular culture sphere like Anglo America or Western Europe. The major variables shaping those Anglo American regularities were accessibility, a competitive market in land, the transportation technologies available during the periods of urban growth, and the collective consequences of individual residential, commercial, and industrial locational decisions.

The Competitive Bidding for Land

For its effective operation, the city requires close spatial association of its functions and people. As long as those functions were few and the population small, pedestrian movement and pack-animal haulage were sufficient for the effective integration of the urban community. With the advent of large-scale manufacturing and the accelerated urbanization of the economy during the 19th century, however, functions and populations—and therefore city areas—grew beyond the interaction capabilities of pedestrian movement alone. Increasingly efficient and costly mass transit systems were installed. Even with their introduction, however, only land within walking distance of the mass transit routes or terminals could successfully be incorporated into the expanding urban structure.

Usable—because accessible—land, therefore, was a scarce commodity, and by its scarcity it assumed high market value and demanded intensive, high-density utilization. Because of its limited supply of usable land, the industrial city of the mass transit era (the late 19th and early 20th centuries) was compact, was characterized by high residential and structural densities (Figure 11.21), and showed a sharp break on its margins between urban and nonurban uses. The older central cities of, particularly, the northeastern United States and southeastern Canada were of that vintage and pattern.

Within the mass transit city, parcels of land were allocated among alternate potential users on the basis of the relative ability of those users to outbid their competitors for a chosen site. There was, in gross generalization, a continuous open auction in land in which users would locate, relocate, or be displaced in accordance with "rent-paying ability." The attractiveness of a parcel, and therefore the price that it could command, was a function of its accessibility. Ideally, the most desirable and efficient location for all the functions and the people of a city would be at the single point at which the maximum possible interchange could be achieved. Such total coalescence of activity is obviously impossible.

Because uses must therefore arrange themselves spatially, the attractiveness of a parcel is rated by its relative accessibility to all other land uses of the city. Store owners wish to locate where they can easily be reached by potential customers; factories need a convenient assembling of their workers and materials; residents desire easy connection with jobs, stores, and schools. Within the older central city, the radiating mass transit lines established the elements of the urban land use structure by freezing in the landscape a clear-cut pattern of differential accessibility. The convergence of that system on the city core gave that location the highest accessibility, the highest desirability, and hence, the highest land values of the entire built-up area. Similarly, transit junction points were accessible to larger segments of the city than locations along single traffic routes; the latter were more desirable than parcels lying between the radiating lines (Figure 11.22).

Society deems certain functions desirable without regard to their economic competitiveness. Schools, parks, and public buildings are assigned space without being participants in the auction for land. Other uses, through the process of that auction, are granted spaces by market forces. The merchants with the widest variety and highest order of goods and the largest threshold requirements bid most for and occupy parcels within the **central business district (CBD),** which became localized at the convergence of mass transit lines. The successful bidders for slightly less accessible CBD parcels were the developers of tall office buildings of major cities, the principal hotels, and similar land uses that help produce the distinctive *skylines* of high-order commercial centers.

Comparable, but lower-order, commercial aggregations developed at the outlying intersections—transfer points—of the mass transit system. With time, a distinctive retailing hierarchy emerged within the urban settlement, an intracity central place pattern based on the purchasing-power thresholds and complementary regions of city populations themselves. Industry took control of parcels adjacent to essential cargo routes: rail lines, waterfronts, rivers, or canals. Strings of stores, light industries, and high-density apartment structures could afford and benefit from location along high-volume transit routes. The least accessible locations within the city were left for the least competitive uses: low-density residences. A diagrammatic summary of this repetitive allocation of space among competitors for urban sites in American mass transit cities is shown in Figure 11.23. Compare it to the generalized land use map of Calgary, Alberta, in Figure 11.27.

The land use regularities of the older, eastern mass transit central cities were not fully replicated in the 20th-century urban centers of western United States. The density and land use structures of those newer cities have been influenced more by the automobile than by mass transit systems. They spread more readily, evolved at lower densities, and therefore display less tightly

Figure 11.21 Duplexes, apartment buildings, and row houses like these in the Crown Heights district of Brooklyn were characteristic 19th-century residential responses to the price and scarcity of developable urban land. Where detached single-family dwellings were built, they were usually placed on far smaller lots than became the rule during the middle 20th century.

structured and standardized land use patterns than their eastern predecessors.

Land Values and Population Density

Theoretically, the open land auction should yield two separate although interconnected distance decay patterns, one related to land values and the other to population density (as distance increases away from the CBD, population density decreases). If one thinks of the land value surface of the older central city as a topographic map with hills representing high valuations and depressions showing low prices, a series of peaks, ridges, and valleys would reflect the differentials in accessibility marked by the pattern of mass transit lines, their intersections, and the unserved interstitial areas.

Dominating these local variations, however, is an overall decline of valuations with increasing distance away from the *peak land value intersection,* the most accessible (by mass transit) and costly location of the central business district. As would be expected in a distance-decay pattern, the drop in valuation is precipitous within a short linear distance from that point, and

then the valuation declines at a lesser rate to the margins of the built-up area.

With one important variation, the population density pattern of the central city shows a comparable distance decay arrangement, as suggested by Figure 11.24. The exception is the tendency to form a hollow *at the center,* the CBD, which represents the inability of all but the most costly apartment houses to compete for space against alternative users desiring these supremely accessible parcels. Yet accessibility is attractive to a number of residential users and brings its penalty in high land prices. The result is the high-density residential occupancy of parcels *near the center* of the city—by those who are too poor to afford a long-distance journey to work; are consigned by their poverty to overcrowding in obsolescent slum tenements near the heart of the inner city; or are self-selected occupants of luxury apartments whose high rents are made necessary by the price of land. Other urbanites, if financially able, may opt to trade off higher commuting costs for lower-priced land and may reside on larger parcels away from high-accessibility, high-congestion locations. Residential density declines with increasing distance from the city center as this option is exercised.

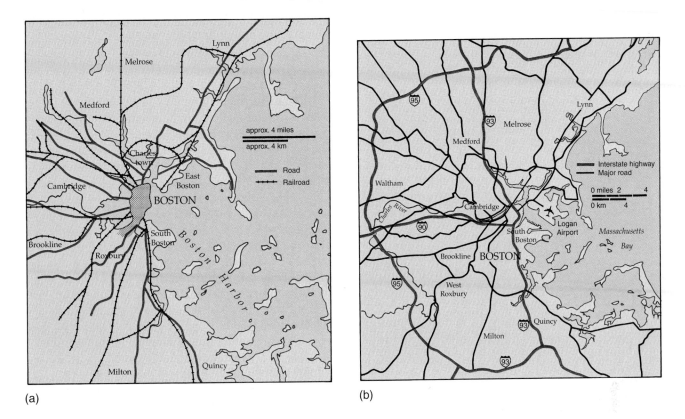

(a)   (b)

Figure 11.22 **Major access lines in Boston in 1872 and 1994.** (*a*) The convergence of mass transit lines in the 19th century gave to the central city and its downtown core a centrality reduced or lost with (*b*) the freeway pattern and motor vehicle dominance in Boston of the 1990s. See also Figure 11.36.

As a city grows in population, the peak densities no longer increase, and the pattern of population distribution becomes more uniform. Secondary centers begin to compete with the CBD for customers and industry, and the residential areas become less associated with the city center and more dependent on high-speed transportation arteries. Peak densities in the inner city decline, and peripheral areas increase in population concentration. The validity of these generalizations may be seen on Figure 11.25, a time series graph of population density patterns for Cleveland, Ohio, over a 50-year period. The peak density was 2.8 miles from the CBD in 1940, but by 1990, it was at 5.8 miles. As the city expanded, density close to the center decreased.

Models of Urban Land Use Structure

Generalized models of urban growth and land use patterns were proposed during the 1920s and 1930s describing the results of these controls on the observed structure of the central city. The models were simplified graphic summaries of U.S. mass transit city growth processes as interpreted by different observers. Although the culture, society, economy, and technology they summarized have now been superseded, the physical patterns they explained or summarized still remain as vestiges and controls on the current landscape. A review of their propositions and conclusions still helps our understanding of the modern U.S. urban complex.

The common starting point of the classical models is the distinctive central business district found in every older central city. The core of this area displays intensive land development: tall buildings, many stores and offices, and crowded streets. Framing the core is a fringe area of wholesaling activities, transportation terminals, warehouses, new car dealers, furniture stores, and even light industries. Just beyond the central business district frame is the beginning of residential land uses.

The land use models shown in Figure 11.26 differ in their explanation of patterns outside the CBD. The **concentric zone (or zonal) model** (Figure 11.26a), developed to explain the sociological patterning of American cities in the 1920s, sees the urban community as a set of nested rings. It recognizes four concentric circles of mostly residential diversity at increasing distance in all directions from the wholesaling, warehousing, and light industry border of the high-density CBD core:

* A zone in transition marked by the deterioration of old residential structures abandoned, as the city expanded, by the former wealthier occupants and now containing high-density, low-income slums, rooming houses, and perhaps ethnic ghettos.

* A zone of "independent working people's homes" occupied by industrial workers, perhaps second-generation Americans able to afford modest but older homes on small lots.

* A zone of better residences, single-family homes, or high-rent apartments occupied by those wealthy enough to exercise

choice in housing location and to afford the longer, more costly journey to CBD employment.

• A commuters' zone of low-density, isolated residential suburbs, just beginning to emerge when this model was proposed.

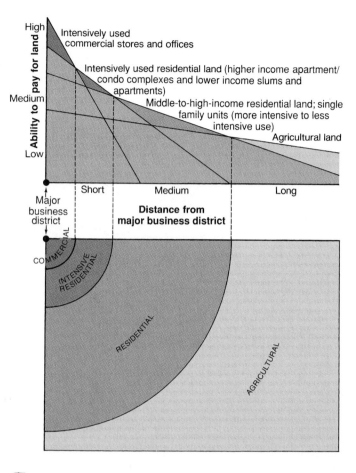

Figure 11.23 **Generalized urban land use pattern.** The model depicts the location of various land uses in an idealized city where the highest bidder gets the most accessible land.

The model is dynamic; it imagines the continuous expansion of inner zones at the expense of the next outer developed circles and suggests a ceaseless process of *invasion* and *succession* that yields a restructured land use pattern and population segregation by income level.

The **sector model** (Figure 11.26b) also concerns itself with patterns of housing and wealth, but it arrives at the conclusion that high-rent residential areas are dominant in city expansion and grow outward from the center of the city along major arterials. New housing for the wealthy, the model concludes, is added in an outward extension of existing high-rent axes as the city grows. Middle-income housing sectors lie adjacent to the high-rent areas, and low-income residents occupy the remaining sectors of growth. There tends to be a *filtering down* process as older areas are abandoned by the outward movement of their original inhabitants, with the lowest-income populations (closest to the center of the city and farthest from the current location of the wealthy) becoming the dubious beneficiaries of the least desirable vacated areas. The accordance of the sector model with the actual pattern that developed in Calgary, Canada, is suggested in Figure 11.27 and for Chicago in Figure 11.29.

The concentric circle and sector models assume urban growth and development outward from a single central core, the site of original urban settlement that later developed into the central business district. These "single-node" models are countered by a **multiple-nuclei model** (see Figure 11.26c), which maintains that large cities develop by peripheral spread from several nodes of growth, not just one. Individual nodes of special function—commercial, industrial, port, residential—are originally developed in response to the benefits accruing from the spatial association of like activities. Peripheral expansion of the separate nuclei eventually leads to coalescence and the meeting of incompatible land uses along the lines of juncture. The urban land use pattern, therefore, is not regularly structured from a single center in a sequence of circles or a series of sectors but based on separately expanding clusters of contrasting activities. The metropolitan consequences of that pattern may be glimpsed in Figures 11.31 and 11.32.

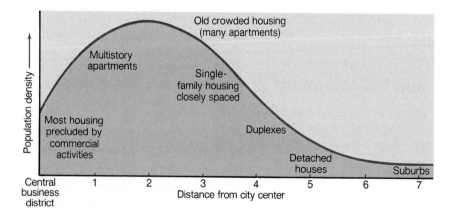

Figure 11.24 **A summary population density curve.** As distance from the area of multistory apartment buildings increases, the population density declines.

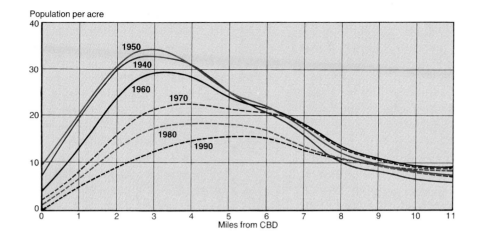

Figure 11.25 **Population density gradients for Cleveland, Ohio, 1940–1990.** The progressive depopulation of the central core and flattening of the density gradient over time to the city margin is clearly seen as Cleveland passed from mass transit to automobile domination. The Cleveland pattern is consistent with conclusions drawn from other urban density studies: density gradients tend to flatten over time, and the larger the city, the flatter the gradient.

Anupa Mukhopadhyay and Ashok K. Dutt, "Population Density Gradient Changes of a Postindustrial City—Cleveland, Ohio 1940–1990," GeoJournal 34:517, no. 4, 1994. Redrawn by permission of Kluwer Academic Publishers and Ashok K. Dutt.

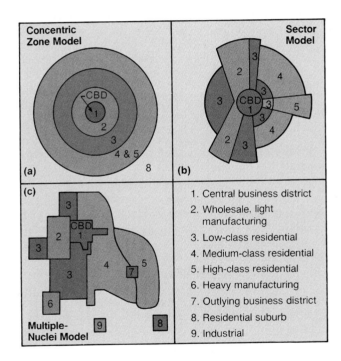

Figure 11.26 **Three classic models of the internal structure of the cities.**

Redrawn from "The Nature of Cities" by C. D. Harris and E. L. Ullman in Volume No. 242 of The Annals of the American Academy of Political and Social Science. Copyright © 1945 The American Academy of Political and Social Science, Philadelphia, PA. Used by permission of the publisher and authors.

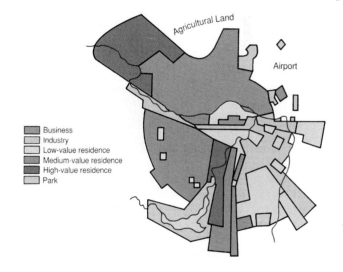

Figure 11.27 **The land use pattern in and around Calgary, Alberta, in 1981.** The circular arrangement of uses suggested by the concentric zone theory (Figure 11.26a) might result if a city developed on a flat surface. In reality, hills, rivers, railroads, and highways affect land uses in uneven ways. Physical and cultural barriers and the evolution of cities over time tend to result in a sectoral pattern of similar land uses. Calgary's central business district was the focus for many of the sectors.

Revised and redrawn with permission from P. J. Smith, "Calgary: A Study in Urban Patterns" in Economic Geography Vol. 38, p. 328. Copyright © 1962 Clark University, Worcester, MA.

Social Areas of Cities

Vestiges of these classical models of American city layout can be seen in modern interpretations of urban structure based on observed social segregation within urban areas. The larger and more economically and socially complex cities are, the stronger is the tendency for their residents to segregate themselves into groups based on *social status, family status,* and *ethnicity.* In a large metropolitan region with a diversified population, this territorial behavior may be a defense against the unknown or the unwanted, a desire to be among similar kinds of people, a response to income constraints, or a result of social and institutional barriers. Most people feel more secure when they are near those with whom they can easily identify. In traditional societies, these groups are the families and tribes. In modern society, people group according to income or occupation (social status), stages in the life cycle (family status), and language or race (ethnic characteristics). Awareness of such social and economic clustering is important in the commercial world (see "Birds of a Feather").

Many of these social area groupings are fostered by the size and the value of available housing. Land developers, especially in cities, produce homes of similar quality in specific areas. The social sorting process, then, takes place in relation to existing land uses, themselves the product of older generations of urban growth. Of course, as time elapses, there is a change in the condition and quality of that housing, and new groups may replace previous tenants. In any case, neighborhoods of similar social characteristics evolve.

Social Status

The social status of an individual or a family is determined by income, education, occupation, and home value, though it may be measured differently in different cultures. In the United States, high income, a college education, a professional or managerial position, and high home value constitute high status. High home value can mean an expensive rental apartment as well as a large house with extensive grounds.

A good housing indicator of social status is persons per room. A low number of persons per room tends to indicate high status. Low status characterizes people with low-income jobs living in low-value housing. There are many levels of status, and people tend to filter out into neighborhoods where most of the heads of households and household incomes are of similar rank.

Social status patterning agrees with the sector model, and in most cities, people of similar status are grouped in sectors which fan out from the innermost urban residential areas (Figure 11.28). The pattern in Chicago is illustrated in Figure 11.29. If the number of people within a given social group increases, they tend to move away from the central city along an arterial connecting them with the old neighborhood. Major transport routes leading to the city center are the usual migration routes out from the center.

Family Status

As the distance from the city center increases, the average age of the adult residents declines, or the size of their family increases, or both. Within a particular sector—say, that of high status—older people whose children do not live with them or young professionals, unmarried or without families, tend to live close to the city center. Between these are the older families who lived at the

Birds of a Feather . . . or, Who Are the People in Your Neighborhood?

How does a McDonald's or a Burger King decide which 99-cent menu items to promote at a certain site, or whether it can profitably offer salads at that franchise? Are there enough families with children to justify building a play area? On what basis does a Starbucks or an Easy Lube determine in what neighborhood to seek a new store location?

Many businesses, large and small, base their sales and locational decisions on a marketing analysis system developed by Claritas, Inc., that uses ZIP codes and census data to categorize Americans by the social and economic characteristics they share with their neighborhoods. People tend to cluster together in roughly homogeneous areas based on social status, family status, ethnicity, and other cultural markers. People in any one cluster tend to have or adopt similar lifestyles. As Claritas puts it, "You are where you live." Residents of a cluster tend to read the same kinds of books, subscribe to the same magazines and newspapers, watch the same movies and television shows. They exhibit similar preferences in food and drink,

clothes, furniture, cars, and all the other goods a consumer society offers.

Claritas uses a number of variables to classify areas of the country: household density per square mile; area type (city, suburb, town, farm); degree of ethnic diversity; family type (married with children, single, and so on); predominant age group; extent of education; type of employment; housing type; and neighborhood quality. After analyzing the data, the firm characterizes each ZIP code as belonging to from one to five of 62 possible neighborhood lifestyle categories. Catchy names have been assigned to these clusters, ranging from "Blueblood Estates" (elite, super-rich families) to "Hard Scrabble" (older families in poor, isolated areas). Some of the others: "Winner's Circle" (executive suburban families); "Pools and Patios" (established empty nesters); "Upward Bound" (young, upscale white collar families); and "Big City Blend" (middle income immigrant families).

The company realizes that the designations don't define the tastes and habits of every single person in a community. But despite critical commentary from some social science researchers, it maintains that the identified clusters summarize the behavior that most people within them are apt to follow. In the "Towns and Gowns" (college-town singles aged 18–34) ZIP code area, for example, residents are likely to be college basketball fans, own a computer, have a school loan, watch television shows advertisers gear to their age and income group, and read magazines similarly focused. But in the "Money and Brains" (older, sophisticated townhouse couples predominantly employed in white-collar jobs) ZIP code, residents are most likely to have a passport, own bonds, shop at Nordstroms, own or lease a European luxury car, watch Public Television shows, and to have purchased, in the past year, a business suit priced at more than $250.

If you would like to know how Claritas marketing analysts have categorized your neighborhood and to judge how closely their summary agrees with your own observations, go to **www.dellvader.claritas.com/YAWYL/ ziplookup.wjsp** and enter your ZIP code.

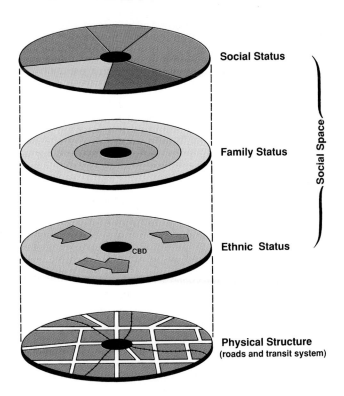

Figure 11.28 **The social geography of American and Canadian cities.**

Redrawn with permission from Robert A. Murdie, Factorial Ecology of Metropolitan Toronto. Research Paper 116. *Department of Geography Research Series, University of Chicago, 1969.*

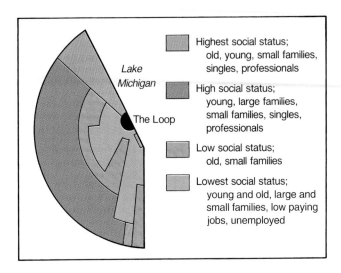

Figure 11.29 **A diagrammatic representation of the major social areas of the Chicago region.** The central business district of Chicago is known as the "Loop."

Redrawn with permission from Phillip Rees, "The Factorial Ecology of Metropolitan Chicago" M.A. Thesis, University of Chicago, 1968.

outskirts of the city in an earlier period before expansion moved beyond them. The young families seek space for child rearing, and older people may covet more the accessibility of the cultural and business life of the city. However, where inner-city life is unpleasant, there is a tendency for older people to migrate to the suburbs or to retirement communities.

Within lower-status sectors, the same pattern tends to emerge. Transients and single people are housed in the inner city, and families, if they find it possible or desirable, live farther from the center. The arrangement that emerges is a concentric circle patterning according to family status, as Figure 11.28 suggests. In general, inner-city areas house older people and outer-city areas house younger populations.

Ethnicity

For some groups, ethnicity is a more important residential location determinant than is social or family status. Areas of homogeneous ethnic identification appear in the social geography of cities as separate clusters or nuclei, reminiscent of the multiple-nuclei concept of urban structure. For some ethnic groups, cultural segregation is both sought and vigorously defended, even in the face of pressures for neighborhood change exerted by potential competitors for housing space, as we saw in Chapter 6. The durability of "Little Italys" and "Chinatowns" and of Polish, Greek, Armenian, and other ethnic neighborhoods in many American cities is evidence of the persistence of self-maintained segregation.

Certain ethnic or racial groups, especially African Americans, have had segregation in nuclear communities forced on them. Every city in the United States has one or more black areas that in many respects may be considered cities within a city, with their own self-contained social geographies of social status, income, and housing quality. Social and economic barriers to movement outside the area have always been high, as they also have been for Hispanics and other non-English-speaking minorities.

As whites and Asians increase their household incomes, they tend to move to neighborhoods that match their economic standing. Due to persistent residential segregation, census data disclose, blacks with similar income growth are less able to move to less segregated neighborhood settings. Although segregation in 2000 was slightly lower nationally than in 1990, at the start of the 21st century the average city black lived in a census tract that was more than 75% minority and three-fifths black in racial composition. Figure 6.16 (p. 210) illustrates the concentration of blacks, Hispanics, and other ethnic groups in Los Angeles. Elsewhere, black segregation varies by region. Black-white separation is highest in metropolitan areas in the Northeast and Midwest; greatest integration is found in the metropolitan South and West and, notably, in military towns such as Norfolk, Virginia, and San Diego, California.

Of the three social geographic patterns depicted on Figure 11.28, family status has undergone the most widespread change in recent years. Today, the suburbs house large numbers of singles and childless couples, as well as two-parent families. Areas near the central business district have become popular for young professionals. Much of this is a result of changes in family structure and the advent of large numbers of new jobs for professionals in the suburbs and central business districts, but not in between. With more women in the workforce than ever before, and as a result of multiple-earner families, residential site selection has become a more complex undertaking.

Institutional Controls

Over the past century, and particularly since World War II, institutional controls have strongly influenced the land use arrangements and growth patterns of most U.S. cities. Indeed, the governments—local and national—of most Western urbanized societies have instituted myriad laws to control all aspects of urban life with particular emphasis on the ways in which individual property and city areas can be developed and used. In the United States, emphasis has been on land use planning, subdivision control and zoning ordinances, and building, health, and safety codes. All have been designed to assure a legally acceptable manner and pattern of urban development, and all are based on broad applications of the police powers of municipalities to assure public health, safety, and well-being even when private property rights are infringed.

These nonmarket controls on land use are designed to minimize incompatibilities (residences adjacent to heavy industry, for example), provide for the creation in appropriate locations of public uses (the transportation system, waste disposal facilities, government buildings, parks), and private uses (colleges, shopping centers, housing) needed for and conducive to a balanced, orderly community. In theory, such careful planning should prevent the emergence of slums, so often the result of undesirable adjacent uses, and should stabilize neighborhoods by reducing market-induced pressures for land use change.

Zoning ordinances and land use planning have sometimes been criticized as being unduly restrictive and unresponsive to changing land use needs and patterns of economic development. Zoning and subdivision control regulations that specify large lot sizes for residential buildings and large house-floor areas have been particularly criticized as devices to exclude from upper-income areas lower-income populations or those who would choose to build or occupy other forms of residences: apartments, special housing for the aged, and so forth. Bitter court battles have been waged, with mixed results, over "exclusionary" zoning practices that in the view of some serve to separate rather than to unify the total urban structure and to maintain or increase diseconomies of land use development. All institutional controls, of course, interfere with the market allocation of urban land, as do the actions of real estate agents who "steer" people of certain racial and ethnic groups into neighborhoods that the agent thinks are appropriate.

In most of Asia there is no zoning, and it is quite common to have small-scale industrial activities operating in residential areas. Even in Japan, a house may contain living space and several people doing piecework for a local industry. In both Europe and Japan, neighborhoods have been built and rebuilt gradually over time to contain a wide variety of building types from several eras intermixed on the same street. In Anglo America, such mixing is much rarer and is often viewed as a temporary condition in a process of transition to total redevelopment.

Suburbanization in the United States

The 20 years before World War II (1939–1945) saw the creation of a technological, physical, and institutional structure that resulted after that war in a sudden and massive alteration of past urban forms. The improvement of the automobile increased its reliability and range, freeing its owner from dependence on fixed-route public transit for travel to home, work, or shopping. The new transport flexibility opened up vast new acreages of rural land for urban development. The acceptance of a maximum 40-hour work week guaranteed millions of Americans the time for a commuting journey not possible when workdays of 10 or more hours were common.

Finally, to stimulate the economy by the ripple effect associated with expanding home construction, the Federal Housing Administration was established as part of the New Deal programs under President Franklin D. Roosevelt. It guaranteed creditors the security of their mortgage loans, thus reducing down-payment requirements and lengthening mortgage repayment periods, and made owned rather than rented housing possible for working-class Americans. In addition, veterans of World War II were granted generous terms on new housing.

Demand for housing, pent up by years of economic depression and wartime restrictions, was loosed in a flood after 1945, and a massive suburbanization of people and functions altered the existing pattern of urban America. Between 1950 and 1970, the two most prominent patterns of population growth were the *metropolitanization* of people and, within metropolitan areas, their *suburbanization*. During the 1970s, the interstate highway system was substantially completed and major metropolitan expressways put in place, allowing sites 30 to 45 or more kilometers (20 to 30 or more miles) from workplaces to be within acceptable commuting distance from home. The major metropolitan areas rapidly expanded in area and population. Growth patterns for the Chicago area, reflecting those developments, are shown in Figure 11.30.

Suburban expansion reached its maximum pace during the decade of the 1970s when developers were converting open land to urban uses at the rate of 80 hectares (200 acres) an hour. High energy prices of the 1970s slowed the rush to the suburbs, but during the 1980s, suburbanization again proceeded apace. In much of the recent outward flow, the tendency has been as much for "filling in" as for continued sprawl.

Residential land uses led the initial rush to the suburbs. Typically uniform, spatially discontinuous housing developments were built beyond the boundaries of most older central cities. The new design was an unfocused sprawl because it was not tied to mass transit lines and the channelized pattern of nodes and links they imposed. Further, it represented a massive relocation of purchasing power to which retail merchants were quick to respond. The planned major regional shopping center became the suburban

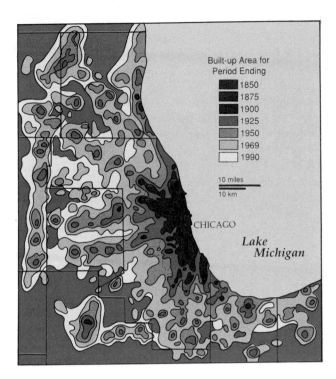

Figure 11.30 **A history of urban sprawl.** In Chicago, as in most larger and older U.S. cities, the slow peripheral expansion recorded during the late 19th and early 20th centuries suddenly accelerated as the automobile suburbs began developing after 1945.

Revised with permission from B. J. L. Berry, Chicago: Transformation of an Urban System, *1976, with additions from other sources.*

counterpart to higher-order central places and the outlying commercial districts of the central city. Smaller shopping malls and strip shopping centers gradually completed the retailing hierarchy.

Faced with a newly suburbanized labor force, industry followed the outward move, drawn as well by the economies derived from modern single-story plants with plenty of parking space for employees and the new freedom from railroad access made possible by the motor truck and the expanding interstate highway system. Service industries were also attracted by suburbanizing purchasing power and labor force, and along with the new shopping malls, office building complexes began to be developed. Like the malls, these localized at freeway intersections and along freeway frontage roads and major connecting highways.

In time, in the United States, new metropolitan land use and functional patterns emerged that could no longer be satisfactorily explained by the classic ring, sector, or multiple-nuclei models. Yet traces of the older generation concepts seemingly remained applicable. Multiple nuclei of specialized land uses appeared, expanded, and coalesced. Sectors of high-income residential use continued their outward extension beyond the central city limits, usurping the most scenic and most desirable suburban areas and

segregating them by price and zoning restrictions. As shown in Figure 11.31, middle-, lower-middle-, and lower-income groups found their own income-segregated portions of the fringe. Ethnic minorities were frequently relegated to the inner city and to some older industrial suburbs, although increasingly immigrants are locating in suburbs as their first step on arrival in the country, and a growing share of native-born minorities are also suburbanizing. In the country's largest metropolitan areas, minorities accounted for 27% of suburbanites in 2000, up from 19% in 1990. In all metropolitan fringe areas blacks and other minorities, irrespective of income levels, are apt to live in suburban racial and ethnic enclaves. Unlike the depressed segregated neighborhoods of central cities, however, suburban segregation in *ethnoburbs* (see p. 214) appears to be a residential choice of more affluent minorities.

By the 1990s, a new urban America had emerged on the perimeters of the major metropolitan areas. With increasing sprawl and the rising costs implicit in the ever-greater spatial separation of the functional segments of the fringe, peripheral

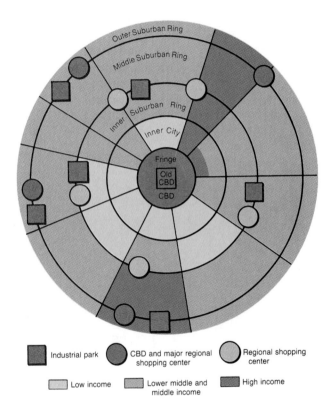

Figure 11.31 **A diagram of the present-day United States metropolitan area.** Note that aspects of the concentric zone, sector, and multiple-nuclei patterns are evident and carried out into the suburban fringe. The "major regional shopping centers" of this earlier, mid-1970s model are increasingly the cores of newly developing "outer cities."

Figure 4.10 (redrawn) from The North American City, *4th ed. by Maurice Yeates. Copyright © 1997. Reprinted by permission of Pearson Education, Inc., Upper Saddle River, NJ 07458.*

expansion slowed, the supply of developable land was reduced (with corresponding increases in its price), and the intensity of land development grew. No longer dependent on the central city, the suburbs were reborn as vast, collectively self-sufficient outer cities, marked by landscapes of skyscraper office parks, massive retailing complexes, established industrial parks, and a proliferation of apartment and condominium districts and gated communities (see "The Gated Community").

The new suburbia began to rival older central business districts in size and complexity. Collectively, the new centers surpassed the central cities as generators of employment and income. Individually, each of the major suburban complexes established its own particular role in the metropolitan economy. Together with the older CBDs, the suburbs perform the many tertiary and quaternary services that mark the postindustrial metropolis. During the 1980s, more office space was created in the suburbs than in the central cities of America. Tysons Corner, Virginia, for example, became the ninth largest central business district in the United States. Regional and national headquarters of leading corporations, banking, professional services of all kinds, major hotel complexes and recreational centers—all formerly considered immovable keystones of central business districts— became parts of the new outer cities. And these outer cities themselves filled in and made more continuous the urban landscape of all North American conurbations. Recognized as **edge cities,** they are defined by their large nodes of concentrated office and commercial structures and characterized by having more jobs than residents within their boundaries.

Edge cities now exist in all regions of urbanized Anglo America. The South Coast Metro Center in Orange County, California; the City Post Oak-Galleria center on Houston's west side; King of Prussia and the Route 202 corridor northwest of Philadelphia; the Meadowlands, New Jersey, west of New York City; and Schaumburg, Illinois, in the western Chicago suburbs are but a very few examples of the new urban forms. The metropolis has become polynucleated, and urban regions are increasingly "galactic"—that is, galaxies of economic activity nodes organized primarily around the freeway systems (Figure 11.32). Commuting across the galaxy is far more common than journeys-to-work between suburbs and central cities. In recent years, suburban outliers and edge cities are coalescing, creating continuous metropolitan belts on the pattern shown in Figure 11.6.

Central City Change

Continuing urbanization and metropolitanization of America at the end of the 20th century led to two contrasting sets of central city patterns and problems. One is characteristic of older, eastern cities and of their older generation of suburbs, unable to expand and absorb the new growth areas on their margins and to maintain the balanced, profitable economic and social base originally theirs. The other set is more characteristic of western U.S. cities that developed in the automobile, rather than mass transit era. Although they are able to incorporate within their political boundaries new growth areas on their margins, they are faced with

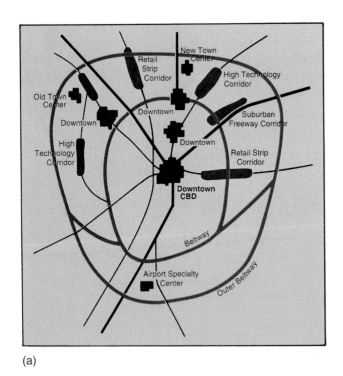

(a)

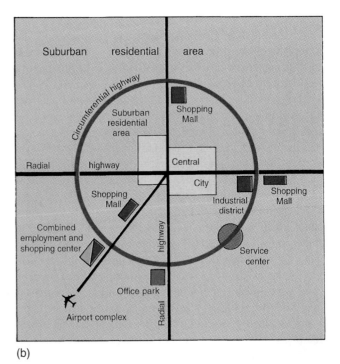

(b)

Figure 11.32 **The galactic city's** multiple downtowns and special function nodes and corridors are linked by the metropolitan expressway systems in these conceptualizations proposed by (*a*) Truman Hartshorn and Peter Muller and (*b*) Chauncy Harris.

(a) *Redrawn with permission from V. H. Winston and Son, Inc., "Suburban Downtowns and the Transformation of Atlanta's Business Landscape," Urban Geography, 10:382 (Silver Springs, Md.: V. H. Winston and Son, Inc., 1989); (b) Redrawn with permission from Winston and Son, Inc., "'The Nature of Cities' and Urban Geography in the Last Half Century." Urban Geography, 18:17 (Silver Springs, Md.: Winston and Son, Inc., 1997).*

The Gated Community

Approximately one in six Americans—some 48 million people—lives in a master-planned community. Particularly characteristic of the fastest growing parts of the country, most of these communities are in the South and West, but they are increasingly common everywhere. In many regions, more than half of all new houses are being built in private developments. Master-planned communities in the United States trace their modern start back to the 1960s, when Irvine, California, and Sun City, Arizona, were built, but their roots can be found much earlier. Tuxedo Park, New York, for example, was planned and built in 1886 as a fully protected, socially exclusive community, and in the 1920s Kansas City's Country Club District was established as a restricted residential development with land use controlled by planning and deed restrictions and a self-governing homeowners association providing a variety of governmental, cultural, and recreational services.

A subset of the master-planned community is the **gated community,** a fenced or walled residential area with checkpoints staffed by guards and access limited to designated individuals and identified guests. By 2002, 9 million Americans were living in these middle- and high-income gated communities within communities. With private security forces, surveillance systems monitoring common recreational areas such as community swimming pools, tennis courts, and health clubs and—often—with individual home security systems, the walled enclaves provide a sense of refuge from high crime rates, drug abuse, and other social problems of urban America.

Gated and sheltered communities are not just an American phenomenon but are increasingly found in all parts of the world. More and more guarded residential enclaves have been sited in such stable Western European states as Spain, Portugal, and France.

Elsewhere, as in Argentina or Venezuela in South America or Lebanon in the Near East, with little urban planning, unstable city administration, and inadequate police protection, not only rich but also middle-class citizens are opting for protected residential districts. In China and Russia, the sudden boom in private and guarded settlements reflects in part a new form of post-communist social class distinction, while in South Africa gated communities serve as effective racial barriers.

Within the United States, the typical developer-created gated community is conceived as a unit and built following a master plan. To preserve the upscale nature of the development and protect land values, community associations enact and enforce a range of conditions and restrictions on private property use. Pervasive and detailed, they may specify such things as the size, construction materials, and design of houses, the color of walls and fences, the size and permitted uses of rear and side yards, even the design of exterior lights and mailboxes and the display of flags and outdoor decorations. Some go so far as to tell residents what trees they can plant and what pets they may have. The comfort of protected living may carry a high price in loss of individuality.

problems of providing infrastructure, services, and environmental protection to an ever-more sprawled residential and functional base.

Constricted Central Cities

The economic base and the financial stability of those central cities unable to expand and absorb new growth areas have been grievously damaged by the process of suburbanization. In earlier periods of growth, as new settlement areas developed beyond the political margins of the city, annexation absorbed new growth within the corporate boundaries of the expanding older city. The additional tax base and employment centers became part of the municipal whole. But in states that recognized the right of separate incorporation for the new growth areas—particularly in the eastern part of the United States—the ability of the city to continue to expand was restricted. Where possible, suburbanites opted for a separation from the central city and for aloofness from the costs, the deterioration, and the adversities associated with it. Their homes, jobs, shopping, schools, and recreation all existed outside the confines of the city from which they had divorced themselves.

The redistribution of population caused by suburbanization resulted not only in the spatial but also in the political segregation of social groups of the metropolitan area. The upwardly mobile resident of the city—younger, wealthier, and better educated—took advantage of the automobile and the freeway to leave the central city. The poorer, older, least-advantaged urbanites were left behind. The central cities and the suburbs became increasingly differentiated. Large areas within those cities now contain only the poor and minority groups, including women (see "Women in the City"), a population little able to pay the rising costs of the social services that their numbers, neighborhoods, and condition require.

The services needed to support the poor include welfare payments, social workers, extra police and fire protection, health delivery systems, and subsidized housing. Central cities, by themselves, are unable to support such an array and intensity of social services since they have lost the tax bases represented by suburbanized commerce, industry, and upper-income residential uses. Lost, too, are the job opportunities that were formerly a part of the central city structure. Increasingly, the poor and minorities are trapped in a central city without the possibility of nearby employment and are isolated by distance, immobility,

Women in the City

Maurice Yeates has noted that women have quite different needs, problems, and patterns than men with respect to urban social space.

In the first place, women are more numerous in large central cities than are men. Washington, D.C. probably is the most female-dominant (numerically) of any city in North America, with a "sex ratio" of eighty-seven (or 115 females for every 100 males). In Minneapolis it is eighty-four. The preponderance of women in central cities is related to an above-average number of household units headed by women, and to the larger numbers of women among the elderly.

A second characteristic is that women, along with their children, constitute the bulk of the poor. This feminization of poverty among all races is a consequence of the low wage rates, part-time work, and lack of security of employment in many "women's jobs." Central cities, with their low-cost but often low-quality rental housing units, house the vast majority of poor women.

A third spatial characteristic of women in urban areas is that they have shorter

journeys to work and rely more heavily upon public transportation than do men, a reflection of the lower incomes received by women, the differences in location of "female jobs," and the concentration of women in the central cities. Women on the whole simply cannot afford to spend as much on travel costs as men and make greater use of public transportation, which in the United States is usually inferior and often dangerous. The concentration of employment of women in clerical, sales, service jobs, and nursing also influences travel distances because these "women's jobs" are spread around the metropolitan area more than "men's jobs," which tend to be concentrated. It might well be argued that the more widespread location of "women's jobs" helps maintain the relative inaccessibility of many higher-paid "men's jobs" to a large number of women.

Given the allocation of roles, the resulting inequities, and the persistence of these inequities, there are spatial issues that impinge directly upon women. One is that many women find that their spatial range of employment opportunities is limited as

a result of the inadequate availability of child-care facilities within urban areas. A second spatial issue relates to the structure of North American metropolitan areas and to the design of housing in general. North American cities are the outcome of male-dominant traits. Suburbs, in particular, reflect a male-paid work and female-home/children ethos. The suburban structure mitigates against women by confining them to a place and role in which there are very few meaningful choices. It has been argued that suburban women really desire a greater level of accessibility to a variety of conveniences and services, more efficient housing units, and a range of public and private transportation that will assure higher levels of mobility. These requirements imply higher-density urban areas.

Text excerpt from *The North American City*, 4th ed., by Maurice Yeates. Copyright © 1997. Reprinted by permission of Pearson Education, Inc., Upper Saddle River, NJ, 07458.

and unawareness—by *spatial mismatch*—from the few remaining low-skill jobs, which are now largely in the suburbs.

Abandonment of the central city by people and functions has nearly destroyed the traditional active, open auction of urban land, which led to the replacement of obsolescent uses and inefficient structures in a continuing process of urban modernization. In the vacuum left by the departure of private investors, the federal government, particularly after the landmark Housing Act of 1949, initiated urban renewal programs with or without provisions for a partnership with private housing and redevelopment investment. Under a wide array of programs, slum areas were cleared; public housing was built (Figure 11.33); cultural complexes and industrial parks were created; and city centers have been reconstructed.

With the continuing erosion of the urban economic base and the disadvantageous restructuring of the central city population mix, the hard-fought governmental battle to maintain or revive the central city is frequently judged to be a losing one. Public assistance programs have not reduced the central city burden of thousands of homeless people (see "The Homeless"), and central city economies, with their high land and housing values, limited

unskilled job opportunities, and inadequate resources for social services, appeared to many observers to offer few or no prospects for change.

That pessimistic outlook, however, began to change dramatically in the 1990s. Although their death was widely reported, central cities by 2000 were showing many positive signs of revival and renewed centrality both in the expanded metropolitan districts they anchored and in the larger national economy. That revival reflects at least two rediscovered attractions of core cities: as centers of economic opportunity and employment and as competitive and attractive residential locations for educated and affluent home-seekers.

Central cities were often dismissed in the 1980s as anachronisms in the coming age of fax machines, the Internet, mobile phones, and the like that would eliminate the need for the face-to-face interaction intrinsic to cities. Instead, communications have become centralizing concerns of knowledge-based industries and activities such as finance, entertainment, health care, and corporate management that depend on dense, capital-intensive information technologies concentrated in geographically centralized markets. Cities—particularly large metropolitan cores—provide

Figure. 11.33 **Faulty towers.** Many elaborate—and massive—public housing projects have been failures. Chicago's Robert Taylor Homes, shown here, consisted of 28 identical 16-story buildings, the largest public housing unit in the world and the biggest concentration of poverty in America. Many of the 4400 apartments were abandoned—victims of soaring vandalism and crime rates—before the first of several of the project's buildings to be razed was demolished in May, 1997; 16 are slated for demolition by 2006, part of the city's intent to flatten all of its 58 "family high-rises." The growing awareness that public high-rise buildings intended to revive the central city do not meet the housing and social needs of their inhabitants led to razing nearly 100,000 of the more than 1.3 million public housing units in cities around the country during the 1990s, many replaced by low-rise apartments or mixed-use developments.

Geography and Public Policy

The Homeless

In recent decades, the number of homeless people in the United States rose dramatically: to anywhere between 600,000 and 3 million in the early 21st century, according to various "official" counts. Their existence and persistence raise a multitude of questions—with the answers yet to be agreed upon by public officials and private Americans. Who are the homeless, and why do their numbers increase? Who should be responsible for coping with the problems they present? Are there ways to eliminate homelessness?

Some people believe the homeless are primarily the impoverished victims of a rich and uncaring society. They view them as ordinary people, but ones who have had a bad break and been forced from their homes by job loss, divorce, domestic violence, or incapacitating illness. They point to the increasing numbers of families, women, and children among the homeless, less visible than the "loners" (primarily men) because they tend to live in cars, emergency shelters, or doubled-up in substandard buildings. Advocates of the homeless argue that government policies of the 1980s and 1990s that led to a dire shortage of affordable housing are partly to blame. Federal outlays for building low-income and subsidized housing were more than $30 billion in 1980 but dropped to $7.5 billion a decade later. Simultaneously, local governments pursued policies of destruction of low-income housing, especially single-room-occupancy hotels, and encouraged gentrification. In addition, federal regulations and reduced state funding for mental hospitals cast institutionalized patients onto the streets to join people displaced by gentrification, job loss, or rising rents.

A contrary view is presented by those who see the homeless chiefly as people responsible for their own plight, not unlike the skid row denizens of former years. In the words of one commentator, the homeless are "deranged, pathological predators who spoil neighborhoods, terrorize passersby, and threaten the commonweal." They point to studies showing that nationally between 66% and 85% of all homeless suffer from alcoholism, drug abuse, or mental illness and argue that people are responsible for the alcohol and drugs they ingest; they are not helpless victims of disease.

Communities have tried a number of strategies to cope with their homeless populations. Some set up temporary shelters, especially in cold weather; some subsidize permanent housing and/or group homes. They encourage private, nonprofit groups to establish soup kitchens and food banks. Others attempt to drive the homeless out of town or at least to parts of town where they will be less visible. They forbid loitering in city parks or on beaches after midnight, install sleep-proof seats on park benches and bus stations, and outlaw aggressive panhandling.

Neither point of view appeals to those who believe that homelessness is more than simply a lack of shelter, that it is a matter of a mostly disturbed or distressed population with severe physical, emotional, or financial problems. What the homeless need, they say, is a "continuum of care"—an entire range of services that includes education; treatment for drug and alcohol abuse and mental illness; and job training.

A homeless man finds shelter on a bench near the White House in Washington, D.C.

Questions to Consider

1. What is the nature of the homeless problem in the community where you live or with which you are most familiar?

2. Where should responsibility for the homeless lie: at the federal, state, or local governmental level? Is it best left to private groups such as churches and charities? Or is it ultimately best recognized as a personal matter to be handled by homeless individuals themselves? What reasons form or support your response?

3. Some people argue that giving money, food, or housing but no therapy to street people makes one an "enabler" or accomplice of addicts. Do you agree? Why or why not?

4. One columnist has proposed quarantining male street people on military bases and compelling them to accept medical treatment. Those who resist would be charged with crimes of violence and turned over to the criminal justice system. Do you believe the homeless should be forced into treatment programs or institutionalized against their will? If so, under what conditions?

the first-rate telecommunications and fiber optics infrastructures and the access to skilled workers, customers, investors, research, educational, and cultural institutions needed by the modern, postindustrial economy. As a reflection of their renewed attractions, employment and gross domestic product in the country's 50 largest urban areas grew significantly in the 1990s, reversing stagnation and decline in the preceding decade. Demand for downtown office space was met by extensive new construction and urban renewal, and even manufacturing revived in the form of small and midsize companies providing high-tech equipment and processes. These, in turn, support a growing network of suppliers and specialized services, with "circular and cumulative" growth the result.

Part of the new vigor of central cities comes from its new residents. Between 1980 and 2000, some 15 million immigrants arrived in the United States, most concentrating in "gateway" cities where they have become deeply rooted in their new communities by buying and renovating homes in inner-city areas, spending money in neighborhood stores, and most importantly establishing their own businesses. They also are important additions to the general urban labor force, providing the skilled and unskilled workers needed in expanding office-work, service, and manufacturing sectors.

Another part of central city residential revival is found in **gentrification,** the rehabilitation of housing in the oldest and now deteriorated inner-city areas by middle- and high-income groups (Figure 11.34). Welcomed by many as a positive, privately financed force in the renewal of depressed urban neighborhoods, gentrification also has presumed serious negative social and housing impacts on the low-income, frequently minority families

displaced. Gentrification, it is argued, is simply another expression of the continuous remaking of urban land use and social patterns in accordance with the rent-paying abilities of alternate potential occupants. Yet the rehabilitation and replacement of housing stock that it implies yield inflated rents and prices that may push out established residents, disrupt the social networks they have created, and totally alter the characteristics and services of their home areas.

The extent of resident displacement, of course, depends on the city and community involved and the economic climate encouraging gentrification. Government and some academic studies, however, suggest that perhaps some 500,000 households (approximately 2 million people) are annually displaced, perhaps forced to move to worse, more expensive, or overcrowded accommodations. Although some research suggests that current residents may tend to remain in gentrifying neighborhoods, induced to stay by the improvements they recognize, gentrification inevitably results in significant change in the established population mix and in the physical and economic character of the gentrified community.

The city districts usually targeted for gentrification are those close to downtown jobs, with easy access to transit, and often of interesting older architecture. The replacement population of younger, wealthier professionals has helped revitalize and repopulate inner city zones already the destination of growing numbers of urbanites. A study of 26 cities nationwide found each expecting its downtown population to grow by 2010, some by double-digit percentages.

The reason for that expected growth lies in demographics. Young professionals are marrying and having children later or,

Figure 11.34 Gentrified housing in the Georgetown section of Washington, D.C. Gentrification is especially active in the major urban centers of the eastern United States, from Boston south along the Atlantic Coast to Charleston, South Carolina, and Savannah, Georgia; it is also increasingly a part of the regeneration of older, deteriorated, first-generation residential districts in major central cities across the country.

often, are divorced or never-married. For them—a growing proportion of Americans—suburban life and shopping malls hold few attractions, while central city residence offers high-tech and executive jobs within walking or biking distance and cultural, entertainment, and boutique shopping opportunities close at hand. The younger group has been joined by "empty-nesters," couples who no longer have children living at home and who find big houses on suburban lots no longer desirable. By their interests and efforts, these two groups have largely or completely remade and upgraded such old city neighborhoods as the Mill District of Minneapolis; the Armory District of Providence, Rhode Island; the Denny Regrade and Belltown of Seattle; Main St./Market Square district of Houston; and many others throughout the country.

Individual home buyers and rehabbers opened the way; commercial developers followed, greatly increasing the stock of quality housing in downtown areas—but often only after local, state, or federal government made the first investments in slum clearance, park development, cultural center construction, and the like. Milwaukee built a riverside walk and attracted $50 million in private investment, for example. Indianapolis city officials have been emptying housing projects in the Chatham Arch neighborhood and selling them to developers for conversion into apartments and condominiums. Renovation of an old cotton mill in the Cabbagetown district of Atlanta produced 500 new apartments in a building recycling project common to many older cities. And as whole areas are gentrified or redeveloped residentially, other investment flows into nearby commercial activities. For example, Denver's LoDo district, once a skid row, has been wholly transformed into a thriving area of shops, restaurants, and sports bars along with residential lofts.

Expanding Central Cities

During the latter part of the 20th century, the most dynamic U.S. urban growth areas were in the 13 states of the Mountain and Pacific West. In 1940, little more than half of all Westerners lived in cities; by 2000, nearly 90% were urbanites. Arizona, California, Nevada, and Utah all have a higher percentage of city dwellers than New York, and 6 of the 10 U.S. metropolitan areas with the biggest population gains from 1990 to 2000 were in the West (see Table 11.1). For the most part, these newer "automobile" metropolises were able to expand physically to keep within the central city boundaries the new growth areas on their peripheries. Nearly without exception they placed few restrictions on physical expansion. That unrestricted growth has often resulted in the coalescence of separate cities into ever-larger metropolitan complexes.

The speed and volume of growth has spawned a complex of concerns, some reminiscent of older eastern cities and others specific to areas of rapid urban expansion as in the West. As in the East, the oldest parts of western central cities tend to be pockets of poverty, racial conflict, and abandonment. In addition, western central city governments face all the economic, social, and environmental consequences of unrestricted marginal expansion. Scottsdale, Arizona, for example, covered a single square mile in 1950; by 2000, it grew to nearly 200 square miles, four times the physical size of San Francisco. Phoenix, with which Scottsdale has now coalesced, surpasses in sprawl Los Angeles, which has three times as many people. The phenomenal growth of Las Vegas, Nevada, has similarly converted vast areas of desert landscape to low-density urban use (Figure 11.35).

Figure 11.35 Urban sprawl in the Las Vegas, Nevada, metropolitan area. Like many western U.S. cities, Las Vegas has spread out over great expanses of desert in order to keep pace with its rapidly growing population. The fastest growing metropolitan area of the United States in the 1990s, Las Vegas, increased from a little more than 850,000 people in 1990 to more than 1.5 million by 2000, an increase of over 83%.

Such unrestricted central city expansion has introduced its own fiscal crises. In many instances limited by state law or constitution from raising taxes, central cities have been unable to provide the infrastructural improvements and social services their far-flung new populations require. Schools remain unbuilt and underfunded, water supplies are increasingly difficult and expensive to obtain, open space requirements are ignored, street and highway improvements and repairs are inadequate even as demand for them increases. In short, each additional unit of unrestricted growth costs the municipality more than the additional development generates in tax revenue.

Increasingly, central cities and metropolitan areas of both East and West are seeking to restrain rather than encourage physical growth. Portland, Oregon, drew a "do not pass" line around itself in the late 1970s, prohibiting urban conversion of surrounding forests, farmlands, and open space. Even with its vigorously enforced "Smart growth" policies, however, Portland has been unable to stop some suburbanization of rural land in the face of continuing population growth, although at least part of that growth was accommodated by higher densities per square mile rather than by low-density sprawl.

Faced with the realization that the United States loses two acres of mostly prime farmland every minute to development, other cities, metropolitan areas, and states are also beginning to resist and restrict urban expansion. "Smart growth" programs have been adopted by such states as Colorado, Delaware, Minnesota, and Washington, spurred by 1991 federal legislation (the Intermodal Surface Transportation Efficiency Act) that gave local planners increased say in the expenditure of the highway trust fund monies. Diverting a portion of them from highways to mass transit support, and resisting plans for expansion or extension of freeways and other roadways in order to prevent further traffic generation and urban construction, cities of both the West and East are beginning to tighten controls on unrestricted and uneconomic expansion.

World Urban Diversity

The city, Figure 11.3 reminds us, is a global phenomenon. It is also a regional and cultural variable. The descriptions and models that we have used to study the functions, land use arrangements, suburbanization trends, and other aspects of the U.S. city would not in all—or even many—instances help us understand the structures and patterns of cities in other parts of the world. Those cities have been created under different historical, cultural, and technological circumstances. They have developed different functional and structural patterns, some so radically different from our U.S. model that we would find them unfamiliar and uncharted landscapes indeed. The city is universal; its characteristics are cultural and regional.

The Anglo American City

Even within the seemingly homogeneous culture realm of the United States and Canada, the city shows subtle but significant differences—not only between older eastern and newer western U.S. cities, but between cities of Canada and those of the United States. Although the urban expression is similar in the two countries, it is not identical. The Canadian city, for example, is more compact than its U.S. counterpart of equal population size, with a higher density of buildings and people and a lesser degree of suburbanization of populations and functions (Figure 11.36).

Space-saving multiple-family housing units are more the rule in Canada, so a similar population size is housed on a smaller land area with much higher densities, on average, within the central area of cities. The Canadian city is better served by and more dependent on mass transportation than is the U.S. city. Since Canadian metropolitan areas have only one-quarter the number of miles of expressway lanes per capita as U.S. metropolises—and at least as much resistance to constructing more—suburbanization of peoples and functions is less extensive north of the border than south.

In social as well as physical structure, Canadian–United States contrasts are apparent. While cities in both countries are ethnically diverse—Canadian communities, in fact, have the higher proportion of foreign born—U.S. central cities exhibit far greater internal distinctions in race, income, and social status and more pronounced contrasts between central city and suburban residents. That is, there has been much less "flight to the suburbs" by middle-income Canadians. As a result, the Canadian city shows greater social stability, higher per capita average income, more retention of shopping facilities, and more employment opportunities and urban amenities than its U.S. central city counterpart. In particular, it does not have the rivalry from well-defined competitive "edge cities" of suburbia that so spread and fragment U.S. metropolitan complexes.

The West European City

If such significant urban differences are found even within the tightly knit Anglo American region, we can only expect still greater divergences from the U.S. model at greater linear and cultural distance and in countries with long urban traditions and mature cities of their own. The political history of France, for example, has given to Paris an overwhelmingly primate position in its system of cities. Political, economic, and colonial history has done the same for London in the United Kingdom. On the other hand, Germany and Italy came late to nationhood, and no overwhelmingly dominant cities developed in their systems.

Nonetheless, a generally common heritage of medieval origins, Renaissance restructurings, and industrial period extensions has given to the cities of Western Europe features distinctly different from those of cities in other regions founded and settled by European immigrants. Despite wartime destructions and postwar redevelopments, many still bear the impress of past occupants and technologies, even back to Roman times in some cases. An irregular system of narrow streets may be retained from the random street pattern developed in medieval times of pedestrian and pack-animal movement. Main streets radiating from the city center and cut by circumferential "ring roads" tell us the location of primary roads leading into town through the gates in city walls now gone and replaced by circular

Figure 11.36 The central and outlying business districts of Toronto, easily visible in this photo, are still rooted firmly by mass transit convergence and mass transit usage. On a per capita basis, Canadian urbanites are two and a half times more dependent on public transportation than are American city dwellers. That reliance gives form, structure, and coherence to the Canadian central city, qualities now irretrievably lost in the sprawled and fragmented U.S. metropolis.

boulevards. Broad thoroughfares, public parks, and plazas mark Renaissance ideals of city beautification and the esthetic need felt for processional avenues and promenades.

Although each is unique historically and culturally, West European cities as a group share certain common features that set them off from the U.S. model, though they are less removed from the Canadian norm. Cities of Western Europe have, for example, a much more compact form and occupy less total area than American cities of comparable population; most of their residents are apartment dwellers. Residential streets of the older sections tend to be narrow, and front, side, or rear yards or gardens are rare.

European cities were developed for pedestrians and still retain the compactness appropriate to walking distances. The sprawl of American peripheral or suburban zones is generally absent. At the same time, compactness and high density do not mean skyscraper skylines. Much of urban Europe predates the steel frame building and the elevator. City skylines tend to be low, three to five stories in height, sometimes (as in central Paris) held down by building ordinance (Figure 11.37), or by prohibitions on private structures exceeding the height of a major public building, often the central cathedral. Those older restrictions are increasingly relaxed as taller office buildings and blocks are developed in London and other commercial centers.

Compactness, high densities, and apartment dwelling encouraged the development and continued importance of public transportation, including well-developed subway systems. The private automobile has become much more common of late, though most central city areas have not yet been significantly restructured with wider streets and parking facilities to accommo-

date it. The automobile is not the universal need in Europe that it has become in American cities. Home and work are generally more closely spaced in Europe—often within walking or bicycling distance—while most sections of towns have first-floor retail and business establishments (below upper-story apartments), bringing both places of employment and retail shops within convenient distance of residences.

A very generalized model of the social geography of the West European city has been proposed (Figure 11.38). Its exact counterpart can be found nowhere, but many of its general features are part of the spatial social structure of most major European cities. In the historic core, now increasingly gentrified, residential units for the middle class, the self-employed, and the older generation of skilled artisans share limited space with preserved historic buildings, monuments, and tourist attractions.

The old city fortifications may mark the boundary between the core and the surrounding transitional zone of substandard housing, 19th-century industry, and recent immigrants. The waterfront has similar older industry; newer plants are found on the periphery. Public housing and some immigrant concentrations may be near that newer industry, while other urban socioeconomic groups aggregate themselves in distinctive social areas within the body of the city.

The West European city is not characterized by inner-city deterioration and out-migration. Its core areas tend to be stable in population and attract, rather than repel, the successful middle class and upward mobile. Nor does it always feature the ethnic neighborhoods of U.S. cities although some, like London, do (see "The Caribbean Map in London," page 210), particularly for immigrants of non-European origin. Similar segregation, though

Figure 11.37 Even in their central areas, many European cities show a low profile, like that of Paris seen here from the Eiffel Tower. Although taller buildings—20, 30, even 50 or more stories in height—have become more common in major cities since World War II, they are not the universal mark of central business districts that they have become in the United States, nor the generally welcomed symbols of city progress and pride.

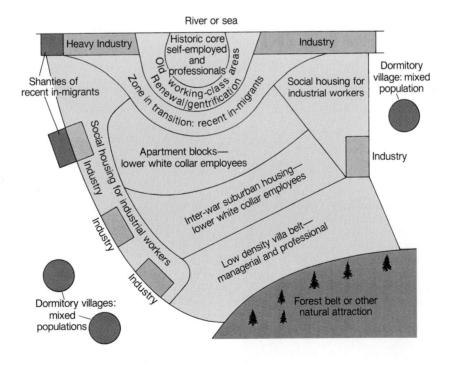

Figure 11.38 **A diagrammatic representation of the West European city.**

Redrawn from Paul White, The West European City: A Social Geography. *Copyright © Longman Group Limited 1984, reprinted by permission of Pearson Education Limited.*

there in suburban apartment clusters, is found on the margins of French cities, particularly for North African ethnics.

The East European City

Cities of Eastern Europe, including Russia and the former European republics of the Soviet Union, make up a separate urban class—the East European city. It is an urban form that shares many of the traditions and practices of West European cities, but it differs from them in the centrally administered planning principles that were in the communist period (1945–1990) designed to shape and control both new and older settlements. For reasons both ideological and practical, the particular concerns were, first, limitation on size of cities to avoid supercity growth and metropolitan sprawl; second, assurance of an internal structure of neighborhood equality and self-sufficiency; and third, strict land use segregation. The planned East European city fully achieved none of these objectives, but by attempting them it emerged as a distinctive urban form.

In general structural terms, the city is compact, with relatively high building and population densities reflecting the nearly universal apartment dwelling, and with a sharp break between urban and rural land uses on its margins. Like the older generation West European city, the East European city depended nearly exclusively on public transportation.

During the communist period, it differed from its Western counterpart in its purely governmental rather than market control of land use and functional patterns. That control dictated that the central area of cities (the Central Cultural District or CCD) should be reserved for public use, not occupied by retail establishments or office buildings on the Western, capitalist model. A large central square ringed by administrative and cultural buildings was the preferred pattern. Nearby, space was provided for a large recreational and commemorative park. In the Russian prototype, neither a central business district nor major outlying business districts were required or provided. Residential areas were expected to be largely self-contained in the provision of at least low-order goods and services, minimizing the need for a journey to centralized shopping locations.

Residential areas are made up of *microdistricts,* assemblages of uniform apartment blocks housing perhaps 10,000 to 15,000 persons, surrounded by broad boulevards, and containing centrally sited nursery and grade schools, grocery and department stores, theaters, clinics, and similar neighborhood necessities and amenities (Figure 11.39a). Plans called for effective separation of residential quarters from industrial districts by landscaped buffer zones, but in practice many microdistricts were built by factories for their own workers and were located immediately adjacent to the workplace. Since microdistricts were most easily and rapidly constructed on open land at the margins of expanding cities, high residential densities have been carried to the outskirts of town (Figure 11.39b).

These characteristic patterns will change in the decades to come as market principles of land allocation are adopted. Now

(a)

(b)

Figure 11.39 (a) This scene from Bucharest, Romania, clearly shows important recurring characteristics of the East European socialist-era city design: mass transit service to boulevard-bordered "superblocks" of self-contained apartment-house microdistricts with their own shopping, schools, and other facilities. (b) High-density apartment houses bordered by wheat fields mark the urban margin of Poprad, Slovakia; the Tatra Mountains are in the background.

that private interests can own land and buildings, the urban areas may take on forms more similar to those of the West European city. In Moscow, the prototypical communist-era East European city, a recent spate of building is rapidly remaking a landscape and skyline dominated by Soviet-style drab gray concrete monoliths. Glass and metal apartment buildings, modernistic Western-style shopping malls, gated communities of luxury apartments and individual houses, and commercial and residential redesign and redevelopment within existing older structures are the 21st-century trend experienced as well in other major Russian and East European cities. Currently, a prominent trend in all principal East European cities is to construct more spacious privately owned apartments and single-family houses for the newly rich.

Cities in the Developing World

Still farther removed from the U.S. urban model are the cities of Africa, Asia, and Latin America. Industrialization has come to them only recently, modern technologies in transportation and public facilities are sparsely available, and the structures of cities and the cultures of their inhabitants are far different from the urban world familiar to North Americans. The developing world is vast in extent and diverse in physical and social content; generalizations about it or its urban landscapes lack certainty and universality. Islamic cities of North Africa, for example, are entities sharply distinct from the sub-Saharan African, the Southeast Asian, or the Latin American city.

Yet, by observation and consensus, some common features of developing-world cities are recognizable. All, for example, have endured massive in-migrations from rural areas, and most have had even faster rates of natural increase than of immigration. As a result, most are ringed by vast squatter settlements high in density and low in public facilities and services (see "The Informal Housing Problem"). All, apparently, have populations greater than their formal functions and employment bases can support. In all, large numbers support themselves in the "informal" sector—as snack-food vendors, peddlers of cigarettes or trinkets, streetside barbers or tailors, errand-runners or package carriers, and the like outside the usual forms of wage labor (see Figure 10.6). All of the large cities have modern centers of commerce, not unlike their Western counterparts, and many are undergoing significant change in their physical and employment structure in response to an influx of foreign investment.

But the extent of acceptable generalization is limited, for the backgrounds, developmental histories, and current economies and administrations of developing-world cities vary so greatly. Some are still preindustrial, with only a modest central commercial core or central bazaar; they lack industrial districts, public transportation, or any meaningful degree of land use separation. Some are the product of Western colonialism, established as ports or outposts of administration and exploitation, built by Europeans on the Western model, though increasingly engulfed by later, indigenous urban forms. In some, Western-style skyscraper central areas and commercial cores have been newly constructed; in others, commerce is conducted in different forums and formats (Figure 11.40). Urban structure is a function of the role the city plays in its own cultural milieu. Some may be religious centers,

as Mecca in Saudi Arabia and Varanasi in India; others may be traditional market centers for a wide area, as are Timbuktu in Mali and Lahore in Pakistan, or serve as cultural capitals such as Addis Ababa in Ethiopia and Cuzco in Peru.

Wherever the automobile or modern transport systems are an integral part of the growth of developing-world cities, the metropolis begins to take on Western characteristics. But in places like Mumbai (India), Lagos (Nigeria), Jakarta (Indonesia), Kinshasa (Congo), and Bangkok (Thailand), where the public transport system is limited, the result has been overcrowded cities centered on a single major business district in the old tradition.

The developing countries, emerging from formerly dominant subsistence economies, have experienced disproportionate population concentrations, particularly in their national and regional capitals. Lacking or relatively undeveloped is the substructure of maturing, functionally complex smaller and medium-sized centers characteristic of more advanced and diversified economies. The primate city dominates their urban systems (Figure 11.17). Nearly a quarter of all Nicaraguans live in metropolitan Managua, and Libreville contains 40% of the populace of Gabon. Vast numbers of surplus, low-income rural populations have been attracted to these developed seats of wealth and political centrality in the hope of finding a job.

Whatever their relative or absolute size within their respective states, large cities of the developing world typically produce a significant share of the gross domestic income (GDI) of their countries. Within Latin America, for example, Lima contributes 44% of Peru's GDI and São Paulo yields 37% of Brazil's. Major cities of Asia show comparable relative economic importance: Bangkok is credited with 38% of Thailand's gross domestic income, and Manila, Philippines; Karachi, Pakistan; and Shanghai, China, contributed 25%, 18%, and 12%, respectively, to the GDI of their countries, the UN reported in 2001.

Although attention may be lavished on creating urban cores on the skyscraper model of Western cities (Figure 11.41), most of the new urban multitudes have little choice but to pack themselves into squatter shanty communities on the fringes of the city, isolated from the sanitary facilities, the public utilities, and the job opportunities that are found only at the center. As many as half of Nairobi, Kenya's, 3 million residents live in slums, most without electricity, running water, or sewers; in that city's sprawling slum district, called Mathare Valley, some 250,000 people are squeezed into 15 square kilometers (6 sq mi) and are increasing by 10,000 inhabitants per year. Such impoverished squatter districts exist around most major cities in Africa, Asia, and Latin America (Figure 11.42), creating an *inverse concentric zone* pattern where the elite and upper class reside in central areas and social status declines with increasing distance from the center. Proposed models of urban structures in the developing world help define some of the regional and cultural contrasts that distinguish those cities (Figure 11.43).

The Asian City and African City

Many large cities of Asia and Africa were founded and developed by European colonialists. For example, the British built Kolkata (Calcutta) and Mumbai (Bombay) in India, and Nairobi and

Between one-third and two-thirds of the population of most developing world cities is crowded into shantytowns and squatter settlements built by the inhabitants, often in defiance of officialdom. These unofficial communities usually have little or no access to publicly provided services such as water supply, sewerage and drainage, paved roads, and garbage removal. In such megacities as Rio de Janeiro, São Paulo, Mexico City, Bangkok, Chennai (Madras), Cairo, or Lagos, millions find refuge in the shacks and slums of the "informal housing sector." Crumbling tenements house additional tens of thousands, many of whom are eventually forced into shantytowns by the conversion of tenements into commercial property or high-income apartments.

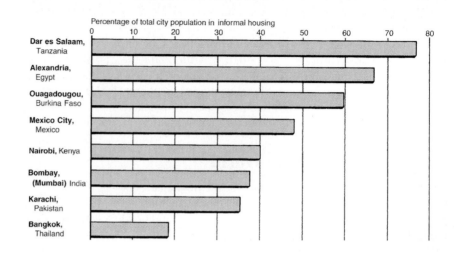

No more than 20% of the new housing in Third World cities is produced by the formal housing sector; the rest develops informally, ignoring building codes, zoning restrictions, property rights, and infrastructure standards. As the graph indicates, informal settlements house varying percentages of these populations, but for low-income developing countries as a whole during the 1980s and 1990s, only one formal housing unit was added for every nine new households, and between 70% and 90% of all new households found shelter in shanties or slums. Peripheral squatter settlements, though densely built, may provide adequate household space and even water, sewers, and defined traffic lanes through the efforts of the residents. More usually, however, overcrowding transforms these settlements into vast zones of disease and squalor subject to constant danger from landslides, fire, and flooding. The informality and often illegality of the squatter housing solution means that those who improvise and build their own shelters lack registration and recognized ownership of their domiciles or the land on which they stand. Without such legal documentation, no capital accumulation based on housing assets is possible, and no collateral for home improvement loans or other purposes is created.

Ssource: Graph data from United Nations Development Programme and other sources.

Harare in Africa; the French developed Ho Chi Minh City (Saigon) in Vietnam, Dakar in Senegal, and Bangui in the Central African Republic. The Dutch had as their main outpost Jarkarta in Indonesia, and many colonial countries established Shanghai. These and many other cities in Asia and Africa have certain important similarities derived in part from their colonial heritage and the imprint of alien cultures which they still bear.

The large *Southeast Asian city* is shown in Figure 11.43a as a composite. The port and its associated areas were colonial creations, retained and strengthened in independence. Around them are found a Western-style central business district with European shops, hotels, and restaurants; one or more "alien commercial zones" where merchants of the Chinese and, perhaps, Indian communities have established themselves; and the more widespread zone of mixed residential, light industrial, and indigenous commercial uses. Central slums and peripheral squatter settlements house up to two-thirds of the total city population. Market gardening and recent industrial development mark the outer metropolitan limits.

The *South Asian city* appears in two forms. Figure 11.43b summarizes the internal structure of the colonial-based city, making clear the spatial separation of local and European residential areas, the mixed-race enclave between them, and the 20th-century new growth areas housing the wealthier local elites. Figure 11.43c depicts the traditional bazaar city, its city center focused on a crossroads around which are found the houses of the wealthier residents. Merchants live above or behind their shops, and the entire city center is characterized by mixed residential, commercial, manufacturing land uses. Beyond the inner core is, first, an upper-income residential area shared (but not in the same structures) with poorer servants. Still farther out are the slums and squatter communities, generally sharply segregated according to ethnic, religious, caste, or native village of their inhabitants.

Asia's past and projected urban growth is explosive. From 1960 to 1990, some 45% of the continent's total population growth came within its urban areas. The pace of urbanization increased during the 1990s, and the United Nations estimates that essentially all of Asia's net population increase between 2000 and 2020 will be in cities, raising Asia's urban population from 1.2 billion to 2.3 billion. That annual average growth of 55 million new city dwellers will exaggerate the already considerable problems of environmental degradation posed by urban growth

Figure 11.40 The Grand Bazaar of Istanbul, Turkey, with its miles of crowded streets, was built in the mid-17th century and now houses more than 3000 shops. It is a vibrant reminder that a successful, thriving commercial economy need not be housed in Western-style business districts.

unsupported by adequate infrastructure development in water, sewer, and other facilities.

Most Asian governments, recognizing the problems of substandard housing, inadequate public services, and environmental deterioration their dominantly primate city population concentrations create, have adopted policies encouraging the establishment of intermediate-sized cities to disperse urbanization and its developmental benefits more widely across their territories. China has achieved more success in this regard than have, for example, India or Pakistan.

With nearly 490 million officially recorded urban residents in 2002, plus unnumbered millions of unregistered migrants from rural areas, China had nearly tripled its urban population since the late 1970s. The number of cities also tripled, and existing centers received massive investment in physical redevelopment and infrastructure improvement. In rapidly expanding older cities, vast areas of inner-city overcrowded tenement mazes have been cleared and replaced with modern high-rise office, luxury residential, and commercial buildings. The former residents, relocated to outer zones and satellite suburbs, have been rehoused in concrete high-rise apartments. Older inner-city factories similarly have been moved to new industrial zones on the outskirts, and new urban transit systems are being installed. By the early 21st century, light rail surface and subway systems were built or under construction in 20 of China's 34 cities of more than 1 million in population. Despite massive sums spent on infrastructural improvement and modernization, China's cities face continuing problems associated with their burgeoning numbers and size. Some 110 large cities, for example, suffer from severe water shortages, and inadequacies of sewage systems and dangerously high levels of air pollution are among widespread Chinese urban problems.

The *African city* is less easily generalized. Sub-Saharan Africa, with little more than one-quarter of its people living in cities, is the least urbanized segment of the developing world. It has, however, the fastest urban growth rates. No more than half of their growth reflects the natural increase of populations already in the cities, and future African urban expansion will largely come from rural to urban migration and the incorporation of villages into spreading metropolitan complexes.

As they did in Asia, European colonialists created new centers of administration and exploitation. Many were designed with spread-out, tree-lined European residential districts separated by open land from the barracks built for African laborers. Disregarding local climate, building materials, and wisdom, British colonists imposed English building codes more concerned with snow load than tropical heat. Since independence, these former colonial outposts have grown apace, with the largest cities expanding at rates

Figure 11.41 Downtown Nairobi, Kenya, is a busy, modern urban core, complete with high-rise commercial buildings.

Figure 11.42 Millions of people of the developing world live in shantytown settlements on the fringes of large cities, without benefit of running water, electricity, sewage systems, or other public services. The hillside slum pictured here is one of the many *favelas* that are home for nearly half of Rio de Janeiro's 11 million residents.

(a) SOUTHEAST ASIAN CITY

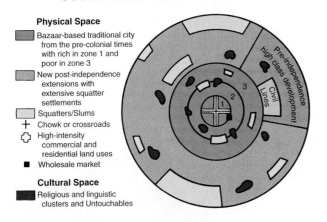

1 Alien commercial zone
2 Alien commercial zone
3 Western commercial zone
a Squatter area
b Suburb

(b) COLONIAL-BASED SOUTH ASIAN CITY

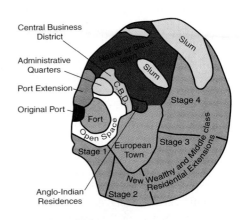

(c) BAZAAR-BASED (Traditional) SOUTH ASIAN CITY

Physical Space

▨ Bazaar-based traditional city from the pre-colonial times with rich in zone 1 and poor in zone 3

▦ New post-independence extensions with extensive squatter settlements

☐ Squatters/Slums

+ Chowk or crossroads

⚑ High-intensity commercial and residential land uses

■ Wholesale market

Cultural Space

▨ Religious and linguistic clusters and Untouchables

(d) LATIN AMERICAN CITY

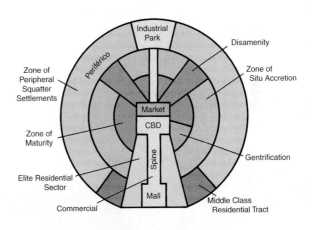

Figure 11.43 **Developing world urban models.**

Sources: (a) Redrawing of top half of figure 25, p. 128 in Terence G. McGee, The Southeast Asian City: A Social Geography of the Primate Cities of Southeast Asia. Copyright © 1967 by Terence G. McGee. Reprinted by permission of the author. (b) Figures 9.7 and 9.9 Ashok K. Dutt, "Cities of South Asia," from Cities of the World: World Regional Development 2d ed. by Stanley D. Brunn and Jack F. Williams. Copyright © 1993 by Harper & Row, Publishers, Inc. Redrawn by permission of the author. (c) Figure 9, page 25 of Global Urganization: Trends, Form and Density Gradients by Ashok K. Dutt. Copyright © Professor R. N. Dubey Foundation. Reprinted by permission of the author. (d) Redrawn by permission from Larry Ford, "A New and Improved Model of Latin American City Structure," in Geographical Review 86 (1996) American Geographical Society.

upwards of 10% per year in some countries. That explosive growth reflects the centralization of government and the concentration of wealth and power in single cities that the small urban elites view as symbols of their countries' economic growth and modernity.

Many, like Lagos, Nigeria, present a confused landscape of teeming, dirt-street shanty developments, unserved by running water or sewerage lines surrounding a modern urban core of high-rise buildings, paved streets and expressways, and modern facilities, and the older, lower building commercial, governmental, and residential district near the harbor. In contrast, others, like Abidjan in the Ivory Coast, are clean, well-designed, and orderly cities nearly Western in appearance.

In all African cities, however, spatial contrasts in social geography are great, and in most sub-Saharan cities socioeconomic divisions are coupled with a partition of squatter slum areas into ethnically based subdivisions. Former greenbelts have been densely filled with cardboard and sheet metal shacks of the poor who are still denied access to the spacious suburbs of the well-to-

do and influential. The richest 10% of Nairobi's population, for example, occupy two-thirds of the city's residential land.

The Latin American City

"City life" is the cultural norm in Latin America. The vast majority of the residents of Mexico, Venezuela, Brazil, Argentina, Chile, and other countries live in cities, and very often in the primate city. The urbanization process is rapidly making Latin American cities among the largest in the world. Analysts predict that by the year 2015, 6 of the largest 28 cities will be in Latin America, and Rio de Janeiro and São Paulo, Brazil, will have merged into a continuous megalopolis 350 miles long.

Latin cities (Figure 11.43d) still retain the focus on their central areas which has been so largely lost in their Anglo American counterparts. The entire transportation system focuses on the downtown, where the vast majority of jobs are found. The city centers are lively and modern with many tall office buildings, clubs, restaurants, and stores of every variety. Condominium apartments house the well-to-do who prefer living in the center because of its convenience to workplaces, theaters, museums, friends, specialty shops, and restaurants (Figure 11.44). Thousands of commuters pour into the urban core each day, some coming from the outer edge of the city (perhaps an hour or two commuting time) where the poorest people live. The mixed usages of the city center are reflected in its increasing segregation into two parts: the modernizing CBD of self-contained newer high-rise office, hotel, and department store buildings and the older traditional "market" segment of small, street-oriented businesses and shops.

Two features of the Latin American city pattern are noteworthy. One is the *spine,* which is a continuation of the features of the city center outward along the main wide boulevard. Here one finds the upper-middle-class housing stock, which is again apartments and town houses. A mall or developing competitive major suburban business node often lies at the end of the elite commercial spine. A ring highway (*periférico*) is becoming common in most large Latin American cities, serving to connect the mall and developing industrial parks and to ensure access for the growing

number of outlying elite residential communities and middle-class housing tracts. It also marks the separation of the better inner-city residential areas and the peripheral squatter settlements and slums.

The second feature is the residential districts arranged in concentric rings around the core and housing ever poorer people as distance increases from the center. This social patterning is just the opposite of many U.S. cities. The slums and squatter settlements (*barrios, favelas*) are on the outskirts of the city. In rapidly growing centers like Mexico City, the barrios are found in the farthest concentric ring, which is several kilometers wide. Many people within these areas find a meager living in the informal sector, selling goods and services to other slum dwellers.

Once Latin residents establish themselves in the city, they tend to remain at their original site, and as income permits they improve their homes. When times are good, there is a great deal of house repair and upgrading activity in this middle zone of *in situ accretion.* Those in the city for the longest time are generally the most prosperous. As a result, the quality of housing continually improves inward toward the city center. The homes in the *zone of maturity* closest to the center are substantial and need lit-

Figure 11.44 Buildings along the Paseo de la Reforma in Mexico City. Part of the central business district, this area contains apartment houses, theaters and nightclubs, and commercial high-rises.

tle upgrading; the zone may also contain a small, elite sector of gentrification and historic preservation.

Each of the idealized land use models in Figure 11.43 presents a variant of the developing world's collective urban dilemma: an urban structure not fully capable of housing the peoples so rapidly thrust upon it. The great increases in city populations exceed urban support capabilities, and unemployment rates are nearly everywhere disastrously high. There is little chance to reduce them as additional millions continue to swell cities already overwhelmed by poverty. The problems, cultures, environments, and economies of developing-world cities are tragically unique to them. The urban models that give us understanding of U.S. cities are of little assistance or guidance in such vastly different culture realms.

Summary

The city is the essential activity focus of every society advanced beyond the subsistence level. Although they are among the oldest marks of civilization, only in the past century have cities become the home of the majority of the people in the industrialized countries and both the commercial crossroads and place of refuge for uncounted millions in the developing world.

All settlements growing beyond their village origins take on functions uniting them to the countryside and to a larger system of settlements. As they grow, they become functionally complex. Their economic base, composed of both *basic* and *service* activities, may become diverse. Basic activities represent the functions performed for the larger economy and urban system, while service (nonbasic) activities satisfy the needs of the urban residents themselves. Functional classifications distinguish the economic roles of urban centers, while simple classification of them as transportation and special-function cities or as central places helps define and explain their functional and size hierarchies and the spatial patterns they display within a system of cities.

As Anglo American urban centers expanded in population size and diversity, they developed structured land use and social patterns based on market allocations of urban space, channelization of traffic, and socioeconomic aggregation. The observed regularity of land use arrangements has been summarized for U.S. cities by the concentric circle, sector, and multiple-nuclei models. Social area counterparts of land use specializations are based on social status, family status, and ethnicity. Since 1945, these older models of land uses and social areas have been modified by the suburbanization of people and functions that has led to the creation of new and complex outer urban areas and "edge" cities and to the deterioration of the older central city itself. Recent economic trends and gentrification have, however, enhanced the employment and residential importance of central city downtown areas.

Urbanization is a global phenomenon, and the Anglo American models of city systems, land use, and social area patterns are not necessarily or usually applicable to other cultural contexts. In Europe, stringent land use regulations have brought about a compact urban form ringed by greenbelts. Although rapidly changing, the East European urban areas still show a pattern of density and land use reflecting recent communist principles of city structure. Models descriptive of developing-world cities do little to convey the fact that those settlements are currently growing faster than it is possible to provide employment, housing, safe water, sanitation, and other minimally essential services and facilities.

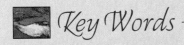

Key Words

basic sector 402
central business district (CBD) 411
central city 400
central place 407
central place theory 408
Christaller, Walter 408
city 400
concentric zone (zonal) model 413
conurbation 396
economic base 402

edge city 420
gated community 421
gentrification 425
metropolitan area 400
multiple-nuclei model 414
multiplier effect 402
network city 410
nonbasic (service) sector 402
primate city 406

rank-size rule 406
sector model 414
service (nonbasic) sector 402
suburb 400
town 400
urban hierarchy 405
urban influence zone 407
urbanized area 400
world city 406

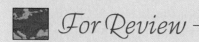 *For Review*

1. Consider the city or town in which you live, attend school, or with which you are most familiar. In a brief paragraph, discuss that community's *site* and *situation*. Point out the connection, if any, between its site and situation and the basic functions that it earlier or now performs.

2. Describe the *multiplier effect* as it relates to the population growth of urban units.

3. What area does a *central place* serve, and what kinds of functions does it perform? If an urban system were composed solely of central places, what summary statements could you make about the spatial distribution and the urban size hierarchy of that system?

4. Is there a hierarchy of retailing activities in the community with which you are most familiar? Of how many and of what kinds of levels is that hierarchy composed? What localizing forces affect the distributional pattern of retailing within that community?

5. Briefly describe the urban land use patterns predicted by the *concentric circle,* the *sector,* and the *multiple-nuclei* models of urban development. Which one, if any, best corresponds to the growth and land use pattern of the community most familiar to you? How well do Figures 11.31 or 11.32 depict the land use patterns in the metropolitan area with which you are most familiar?

6. In what ways do *social status, family status,* and *ethnicity* affect the residential choices of households? What expected distributional patterns of urban social areas are associated with each? Does the social geography of your community conform to the predicted pattern?

7. How has suburbanization damaged the economic base and the financial stability of the U.S. central city?

8. In what ways does the Canadian city differ from the pattern of its U.S. counterpart?

9. What are *primate cities?* Why are primate cities so prevalent in the developing world? How are some governments attempting to reduce their relative importance in their national systems of cities?

10. What are the significant differences in the generalized pattern of land uses of Anglo American, West European, East European, Asian, and African cities?

11. Why are metropolitan areas in developing countries expected to grow larger than Western metropolises by 2020? What do you expect the population density profile for Mexico City to look like in the year 2020?

 Focus Follow-up

1. **What common features define the origin, nature, and locations of cities?** pp. 392–402.

   Cities arose 4000–6000 years ago as distinctive evidence of the growing cultural and economic complexity of early civilizations. Distinct from the farm villages of subsistence economies, true cities provided an increasing range of functions—religious, military, trade, production, etc.—for their developing societies. Their functions and importance were affected by the sites and situations chosen for them. The massive recent increase in number and size of cities worldwide reflects the universality of economic development and total population growth in the latter 20th century.

2. **How are cities structured economically, and how are systems of cities organized?** pp. 402–410.

   The economic base of a city—the functions it performs—is divided between basic and nonbasic (or service) activities. Through a multiplier effect, adding basic workers increases both the number of service workers and the total population of a city. The amount of growth reflects the base ratio characteristic of the city. Cities may be hierarchically ranked by their size and functional complexity. Rank-size, primate, and central place hierarchies are commonly cited but distinctly different.

3. **How are cities structured internally, and how do people distribute themselves within them?** pp. 411–427.

   Cities are themselves distinctive land use and cultural area landscapes. In the United States, older cities show repetitive land use patterns that are largely determined by land value and accessibility considerations. Classical land use models include the concentric circle, sector, and multiple nuclei patterns. Distinct social area arrangements have been equated with those land use models. Newer cities and growing metropolitan areas have created different land use and social area structures with suburbs, edge cities, and galactic metropolises as recognized urban landscape features.

4. **Are there world regional and cultural differences in the land use and population patterns of major cities?** pp. 427–437.

   Cities are regional and cultural variables; their internal land use and social area patterns reflect the differing historical, technological, political, and cultural conditions under which they developed. Although the Anglo American city is the familiar U.S.–Canadian model, we can easily recognize differences among it and West European, East European, Asian, African, and Latin American city types.

Selected References

Batten, David F. "Network Cities: Creative Urban Agglomerations for the 21st Century." *Urban Studies* 32 (1995): 313–327.

Beaverstock, Jonathan V., Richard G. Smith, and Peter J. Taylor. "World-City Network: A New Metageography?" *Annals of the Association of American Geographers* 90, no. 1 (2000): 123–134.

Brockerhoff, Martin P. "An Urbanizing World." *Population Bulletin* 55, no. 3. Washington, D.C.: Population Reference Bureau, 2000.

Brunn, Stanley D., Jack F. Williams, and Donald Ziegler, eds. *Cities of the World: World Regional Urban Development.* 3 rev. ed. Lanham, MD: Rowman and Littlefield, 2003.

Burtenshaw, D., M. Bateman, and G. J. Ashworth. *The European City: A Western Perspective.* New York: John Wiley & Sons, 1991.

Carter, Harold. *The Study of Urban Geography.* 4th ed. London: Arnold, 1995.

Cartier, Carolyn. "Cosmopolitics and the Maritime World City." *Geographical Review* 89, no. 2 (1999): 278–289.

Coffey, William J., and Richard G. Shearmur. "The Growth and Location of High Order Services in the Canadian Urban System, 1971–1991." *Professional Geographer* 49, no. 4 (1997): 404–418.

Drakakis-Smith, David. *Third World Cities.* 2d rev. ed. London and New York: Routledge, 2000.

Dutt, Ashok K. *Global Urbanization: Trends, Form and Density Gradients.* Allahabad, India: Professor R.N. Dubey Foundation, 2001.

Dutt, Ashok K., et al., eds. *The Asian City: Processes of Development, Characteristics and Planning.* Dordrecht/Boston/London: Kluwer Academic Publishers, 1994.

Ewing, Gordon O. "The Bases of Differences between American and Canadian Cities." *The Canadian Geographer/Le Géographie Canadien* 36, no. 3 (1992): 266–279.

Fogelson, Robert M. *Downtown: Its Rise and Fall, 1880–1950.* New Haven, Conn.: Yale University Press, 2001 and 2003.

Ford, Larry R. *America's New Downtowns.* Baltimore: Johns Hopkins University Press, 2003.

Ford, Larry R. *Cities and Buildings: Skyscrapers, Skid Rows, and Suburbs.* Baltimore: Johns Hopkins University Press, 1994.

Gugler, Josef, ed. *The Urban Transformation of the Developing World.* Oxford, England: Oxford University Press, 1996.

Hall, Tim. *Urban Geography.* 2d ed. New York: Routledge, 2001.

Hartshorn, Truman. *Interpreting the City.* 3d ed. New York: John Wiley & Sons, 1998.

Kaplan, David H., and Steven R. Holloway. *Segregation in Cities.* AAG Resource Publication 1998-1. Washington, D.C.: Association of American Geographers, 1998.

Katz, Bruce, and Robert E. Lang, eds. *Defining Urban and Suburban America: Evidence from Census 2000.* Washington, D.C.: Brookings Institution Press, 2003.

Knox, Paul L., and Steven Pinch. *Urban Social Geography: An Introduction.* 4th ed. Upper Saddle River, N.J.: Prentice-Hall, 2000.

Knox, Paul L., and Peter J. Taylor, eds. *World Cities in a World System.* Cambridge, England: Cambridge University Press, 1995.

Lo, Fu-chen, and Yue-Man Yeuny, eds. *Globalization and the World of Large Cities.* New York: United Nations University Press, 1998.

Lowder, Stella. *The Geography of Third World Cities.* Totowa, N.J.: Barnes & Noble, 1986.

Marcuse, Peter, and Ronald van Kampen, eds. *Globalizing Cities: A New Spatial Order?* Oxford, England: Blackwell Publishers, 2000.

Mitchell, John G. "Urban Sprawl." *National Geographic* vol. 200, no. 1 (July, 2001):48–73.

Pacione, Michael. *Urban Geography: A Global Perspective.* New York: Routledge, 2001.

Roseman, Curtis C., Hans Dieter Laux, and Gunther Thieme, eds. *EthniCity: Geographic Perspectives on Ethnic Change in Modern Cities.* Lanham, Md.: Rowman & Littlefield, 1996.

Sassen, Saskia, ed. *Global Networks: Linked Cities.* New York: Routledge, 2002.

Sheehan, Molly O'Meara. *City Limits: Putting the Brakes on Urban Sprawl.* Worldwatch Paper 156. Washington, D.C.: Worldwatch Institute, 2001.

Short, John R. *The Urban Order: An Introduction to Urban Geography.* Cambridge, Mass: Blackwell, 1996.

Short, John R., and Yeong-Hyun Kim. *Globalization and the City.* New York: Addison Wesley Longman, 1999.

Siegel, Fred, and Jan Rosenberg, eds. *Urban Society.* Issued annually. Guilford, Conn.: Dushkin/McGraw-Hill.

Simon, David. *Cities, Capital, and Development: African Cities in the World Economy.* New York: John Wiley & Sons, 1992.

Smith, David W. *Third World Cities.* 2d ed. New York: Routledge, 2000.

Smith, Neil. *The New Urban Frontier: Gentrification and the Revanchist City.* New York: Routledge, 1996.

Teaford, Jon C. *Post-Suburbia: Government and Politics in the Edge Cities.* Baltimore, Md.: Johns Hopkins University Press, 1996.

United Nations Center for Human Settlements (Habitat). *An Urbanizing World: Global Report on Human Settlements 1996.* New York: Oxford University Press, 1996.

United Nations Center for Human Settlements (Habitat). *The State of the World's Cities 2001.* Nairobi, Kenya: United Nations, 2001.

Wilson, David, ed. "Globalization and the Changing U.S. City." A special issue of *Annals of the American Academy of Political and Social Science.* Vol. 551, May, 1997.

Yeates, Maurice. *The North American City.* 5th ed. New York: Longman, 1997.

Websites: The World Wide Web has a tremendous number and variety of sites pertaining to geography. Websites relevant to the subject matter of this chapter appear in the "Web Links" section of the On-line Learning Center associated with this book. Access it at **www.mhhe.com/fellmann8e**

Twelve

The Political
Ordering of Space

Focus Preview

Opposite: The Hungarian National Assembly building on the Danube River in Budapest.

They met together in the cabin of the little ship on the day of the landfall. The journey from England had been long and stormy. Provisions ran out, a man had died, a boy had been born. Although they were grateful to have been delivered to the calm waters off Cape Cod that November day of 1620, their gathering in the cramped cabin was not to offer prayers of thanksgiving but to create a political structure to govern the settlement they were now to establish. The Mayflower Compact was an agreement among themselves to "covenant and combine our selves togeather into a civill Body Politick . . . to enacte, constitute, and frame such just and equall Lawes, Ordinances, Acts, Constitutions, and Offices . . . convenient for ye Generall good of ye Colonie. . . ." They elected one of their company governor, and only after those political acts did they launch a boat and put a party ashore.

The land they sought to colonize had for more than 100 years been claimed by the England they had left. The New World voyage of John Cabot in 1497 had invested their sovereign with title to all of the land of North America and a recognized legal right to govern his subjects dwelling there. That right was delegated by royal patent to colonizers and their sponsors, conferring upon them title to a defined tract and the authority to govern it. Although the Mayflower settlers were originally without a charter or patent, they recognized themselves as part of an established political system. They chose their governor and his executive department annually by vote of the General Court, a legislature composed of all freemen of the settlement.

As the population grew, new towns were established too distant for their voters to attend the General Court. By 1636, the larger towns were sending representatives to cooperate with the executive branch in making laws. Each town became a legal entity, with election of local officials and enactment of local ordinances the prime purpose of the town meetings that are still common in New England today.

The Mayflower Compact, signed by 41 freemen as their initial act in a New World, was the first step in a continuing journey of political development for the settlement and for the larger territory of which it became a part (Figure 12.1). From company patent to crown colony to rebellious commonwealth under the Continental Congress to state in a new country, Massachusetts (and Plimoth Plantation) were part of a continuing process of the political organization of space.

Figure 12.1 Signing the Mayflower Compact, probably the first written plan for self-government in America. Forty-one adult males signed the Compact aboard the *Mayflower* before going ashore.

That process is as old as human history. From clans to kingdoms, human groups have laid claim to territory and have organized themselves and administered their affairs within it. Indeed, the political organizations of society are as fundamental an expression of culture and cultural differences as are forms of economy or religious beliefs. Geographers are interested in that structuring because it is both an expression of the human organization of space and is closely related to other spatial evidences of culture, such as religion, language, and ethnicity.

Political geography is the study of the organization and distribution of political phenomena, including their impact on other spatial components of society and culture. Nationality is a basic element in cultural variation among people, and political geography traditionally has had a primary interest in country units, or *states* (Figure 12.2). Of particular concern have been spatial patterns that reflect the exercise of central governmental control, such as questions of boundary delimitation and effect. Increasingly, however, attention has shifted both upward and downward on the political scale. On the world scene, international alliances, regional compacts, and producer cartels—some requiring the surrender of at least a portion of national sovereignty—have increased in prominence since World War II, representing new forms of spatial interaction. At the local level, voting patterns, constituency boundaries and districting rules, and political

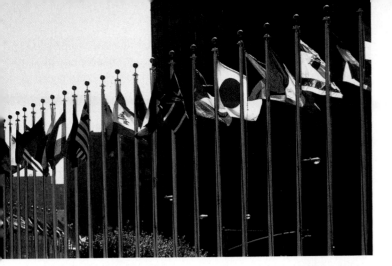

Figure 12.2 These flags, symbols of separate member states, grace the front of the United Nations building in New York City. Although central to political geographic interest, states are only one level of the political organization of space.

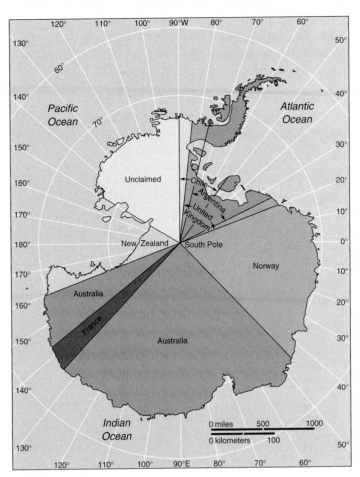

Figure 12.3 **Territorial claims in Antarctica.** Seven countries claim sovereignty over portions of Antarctica, and those of Argentina, Chile, and the United Kingdom overlap. The Antarctic Treaty of 1959 froze those claims for 30 years, banned further land claims, and made scientific research the primary use of the continent. The treaty was extended for 50 years in 1991. Antarctica is neither a sovereign state—it has no permanent inhabitants or local government—nor a part of one.

fragmentation have directed public attention to the significance of area in the domestic political process.

In this chapter, we consider some of the characteristics of political entities, examine the problems involved in defining jurisdictions, seek the elements that lend cohesion to a political entity, explore the implications of partial surrender of sovereignty, and consider the significance of the fragmentation of political power. We begin with states (countries) and end with local political systems.

Emphasis here on political entities should not make us lose sight of the reality that states are rooted in the operations of the economy and society they represent, that social and economic disputes are as significant as border confrontations, and that in some regards transnational corporations and other nongovernmental agencies may exert more influence in international affairs than do the separate states in which they are housed or operate. Some of those expanded political considerations are alluded to in the discussions that follow; others were developed more fully in Chapter 9.

National Political Systems

One of the most significant elements in cultural geography is the nearly complete division of the earth's land surface into separate national units, as shown on the Countries of the World map inside this book's cover. Even Antarctica is subject to the rival territorial claims of seven countries, although these claims have not been pressed because of the Antarctic Treaty of 1959 (Figure 12.3). A second element is that this division into country units is relatively recent. Although countries and empires have existed since the days of early Egypt and Mesopotamia, only in the last century has

the world been almost completely divided into independent governing entities. Now people everywhere accept the idea of the state and its claim to sovereignty within its borders as normal.

States, Nations, and Nation-States

Before we begin our consideration of political systems, we need to clarify some terminology. Geographers use the words *state* and *nation* somewhat differently than the way they are used in everyday speech; sometimes confusion arises because each word has more than one meaning. A state can be defined as either (1) any of the political units forming a federal government (e.g., one of the United States) or as (2) an independent political entity holding sovereignty over a territory (e.g., the United States). In this latter sense, *state* is synonymous with *country* or *nation*. That is, a nation can also be defined as (1) an independent political unit holding sovereignty over a territory (e.g., a member of the United Nations). But it can also be used to describe (2) a community of people with a common culture and territory (e.g., the Kurdish nation). The second definition is *not* synonymous with state or country.

To avoid confusion, we shall define a **state** on the international level as an independent political unit occupying a defined, permanently populated territory and having full sovereign control over its internal and foreign affairs. We will use *country* as a synonym for the territorial and political concept of "state." Not all recognized territorial entities are states. Antarctica, for example, has neither established government nor permanent population, and it is, therefore, not a state. Nor are *colonies* or *protectorates* recognized as states. Although they have defined extent, permanent inhabitants, and some degree of separate governmental structure, they lack full control over all of their internal and external affairs. More importantly, they lack recognition as states by the international community, a decisive consideration in the proper use of the term "state."

We use nation in its second sense, as a reference to people, not to political structure. A **nation** is a group of people with a common culture occupying a particular territory, bound together by a strong sense of unity arising from shared beliefs and customs. Language and religion may be unifying elements, but even more important are an emotional conviction of cultural distinctiveness and a sense of ethnocentrism. The Cree nation exists because of its cultural uniqueness, not by virtue of territorial sovereignty.

The composite term **nation-state** properly refers to a state whose territorial extent coincides with that occupied by a distinct nation or people or, at least, whose population shares a general sense of cohesion and adherence to a set of common values (Figure 12.4a). That is, a nation-state is an entity whose members feel a natural connection with each other by virtue of sharing language, religion, or some other cultural characteristic strong enough both to bind them together and to give them a sense of distinction from all others outside the community. Although all countries strive for consensus values and loyalty to the state, few can claim to be true nation-states since few are or have ever been

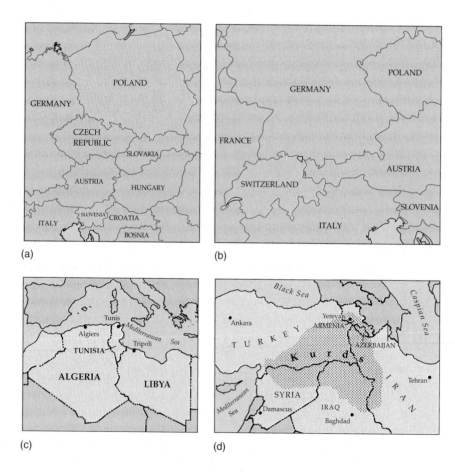

(a)   (b)

(c)   (d)

Figure 12.4 **Types of relationships between "states" and "nations."** (*a*) A **nation-state.** Poland and Slovenia are examples of states occupied by a distinct nation, or people. (*b*) A **multinational state.** Switzerland shows that a common ethnicity, language, or religion is not necessary for a strong sense of nationalism. (*c*) A **part-nation state.** The Arab nation extends across and dominates many states in northern Africa and the Middle East. (*d*) A **stateless nation.** An ancient group with a distinctive language, the Kurds are concentrated in Turkey, Iran, and Iraq. Smaller numbers live in Syria, Armenia, and Azerbaijan.

wholly uniform ethnically. Iceland, Slovenia, Poland, and the two Koreas are often cited as acceptable examples.

A *binational* or *multinational state* is one that contains more than one nation (Figure 12.4b). Often, no single ethnic group dominates the population. In the constitutional structure of the former Soviet Union before 1988, one division of the legislative branch of the government was termed the Soviet of Nationalities. It was composed of representatives from civil divisions of the Soviet Union populated by groups of officially recognized "nations": Ukrainians, Kazakhs, Tatars, Estonians, and others. In this instance, the concept of nationality was territorially less than the extent of the state.

Alternatively, a single nation may be dispersed across and be predominant in two or more states. This is the case with the *part-nation state* (Figure 12.4c). Here, a people's sense of nationality exceeds the areal limits of a single country. An example is the Arab nation, which dominates 17 states.

Finally, there is the special case of the *stateless nation*, a people without a state. The Kurds, for example, are a nation of some 20 million people divided among six states and dominant in none (Figure 12.4d). Kurdish nationalism has survived over the centuries, and many Kurds nurture a vision of an independent Kurdistan. Other stateless nations include Gypsies (Roma), Basques, and Palestinians.

The Evolution of the Modern State

The concept and practice of political organization of space and people arose independently in many parts of the world. Certainly, one of the distinguishing characteristics of very early culture hearths—including those shown on Figure 2.15—was the political organization of their peoples and areas. The larger and more complex the economic structures they developed, the more sophisticated became their mechanisms of political control and territorial administration.

Our Western orientations and biases may incline us to trace ideas of spatial political organization through their Near Eastern, Mediterranean, and Western European expressions. Mesopotamian and classical Greek city states, the Roman Empire, and European colonizing kingdoms and warring principalities were, however, not unique. Southern, southeastern, and eastern Asia had their counterparts, as did sub-Saharan Africa and the Western Hemisphere. Although the Western European models and colonization strongly influenced the forms and structures of modern states in both hemispheres, the cultural roots of statehood run deeper and reach further back in many parts of the world than European example alone suggests.

The now universal idea of the modern state was developed by European political philosophers in the 18th century. Their views advanced the concept that people owe allegiance to a state and the people it represents rather than to its leader, such as a king or feudal lord. The new concept coincided in France with the French Revolution and spread over Western Europe, to England, Spain, and Germany.

Many states are the result of European expansion during the 17th, 18th, and 19th centuries, when much of Africa, Asia, and the Americas was divided into colonies. Usually these colonial claims were given fixed and described boundaries where none had earlier been formally defined. Of course, precolonial native populations had relatively fixed home areas of control within which there was recognized dominance and border defense and from which there were, perhaps, raids of plunder or conquest of neighboring "foreign" districts. Beyond understood tribal territories, great empires arose, again with recognized outer limits of influence or control: Mogul and Chinese; Benin and Zulu; Incan and Aztec. Upon them where they still existed, and upon the less formally organized spatial patterns of effective tribal control, European colonizers imposed their arbitrary new administrative divisions of the land. In fact, groups that had little in common were often joined in the same colony (Figure 12.5). The new divisions, therefore, were not usually based on meaningful cultural or physical lines. Instead, the boundaries simply represented the limits of the colonizing empire's power.

As these former colonies have gained political independence, they have retained the idea of the state. They have generally accepted—in the case of Africa, by a conscious decision to avoid precolonial territorial or ethnic claims that could lead to war—the borders established by their former European rulers (Figure 12.6). The problem that many of the new countries face is "nation building"—developing feelings of loyalty to the state among their arbitrarily associated citizens. For example, the Democratic Republic of the Congo, the former Belgian Congo, contains some 270 frequently antagonistic ethnic groups. Julius Nyerere, president of Tanzania, noted in 1971, "These new countries are artificial units, geographic expressions carved on the map by European imperialists. These are the units we have tried to turn into nations."

The idea of separate statehood grew slowly at first and, more recently, has accelerated rapidly. At the time of the Declaration of Independence of the United States in 1776, there were only some 35 empires, kingdoms, and countries in the entire world. By the beginning of World War II in 1939, their number had only doubled to about 70. Following that war, the end of the colonial era brought a rapid increase in the number of sovereign states. From the former British Empire and Commonwealth, there have come the independent countries of India, Pakistan, Bangladesh, Malaysia, Myanmar (Burma), and Singapore in Asia, and Ghana, Nigeria, Kenya, Uganda, Tanzania, Malawi, Botswana, Zimbabwe, and Zambia in Africa. Even this extensive list is not complete. A similar process has occurred in most of the former overseas possessions of the Netherlands, Spain, Portugal, and France. By 1990, independent states totaled some 180, and their number increased again following—among other political geographic developments—the disintegration during the 1990s of the USSR, Czechoslovakia, and Yugoslavia, which created more than 20 countries where only three had existed before (Figure 12.7).

The proliferation of states means that about half of the world's independent countries had in 2000 smaller populations than the U.S. states of Maryland or Arizona. All told, nearly 90 countries had populations under 5 million, 55 had less than 2.5 million, and 33 had fewer than a half-million population at the start of the century. The great increase in the number of smaller countries is an affirmation of the ideal of nation-state.

Figure 12.5 **The discrepancies between ethnic groups and country boundaries in Africa.** Cultural boundaries were ignored by European colonial powers. The result has been significant ethnic diversity in nearly all African countries.

Redrawn from World Regional Geography: A Question of Place *by Paul Ward English, with James Andrew Miller. Copyright © 1977 Harper & Row. Used by permission of the author.*

Geographic Characteristics of States

Every state has certain geographic characteristics by which it can be described and that set it apart from all other states. A look at the world political map inside the cover of this book confirms that every state is unique. The size, shape, and location of any one state combine to distinguish it from all others. These characteristics are of more than academic interest, because they also affect the power and stability of states.

Size

The area that a state occupies may be large, as is true of China, or small, as is Liechtenstein. The world's largest country, Russia, occupies over 17 million square kilometers (6.5 million sq mi), some 11% of the earth's land surface—nearly as large as the whole continent of South America and more than 1 million times as large as Nauru, one of the *ministates* or *microstates* found in all parts of the world (see "The Ministates").

An easy assumption would be that the larger a state's area, the greater is the chance that it will include the ores, energy sup-

plies, and fertile soils from which it can benefit. In general, that assumption is valid, but much depends on accidents of location. Mineral resources are unevenly distributed, and size alone does not guarantee their presence within a state. Australia, Canada, and Russia, though large in territory, have relatively small areas capable of supporting productive agriculture. Great size, in fact, may be a disadvantage. A very large country may have vast areas that are remote, sparsely populated, and hard to integrate into the mainstream of economy and society. Small states are more apt than large ones to have a culturally homogeneous population. They find it easier to develop transportation and communication systems to link the sections of the country, and, of course, they have shorter boundaries to defend against invasion. Size alone, then, is not critical in determining a country's stability and strength, but it is a contributing factor.

Shape

Like size, a country's shape can affect its well-being as a state by fostering or hindering effective organization. Assuming no major topographical barriers, the most efficient form would be a circle

AFRICA, 2000

TUNISIA 1956
MOROCCO 1956
ALGERIA 1962
LIBYA 1951
ARAB REPUBLIC OF EGYPT 1922
Madeira
Canary Is.
WESTERN SAHARA
MAURITANIA 1960
MALI 1960
NIGER 1960
CHAD 1960
SUDAN 1956
ERITREA 1993
formerly Br. Somaliland
DJIBOUTI 1977
GAMBIA 1965
SENEGAL 1960
GUINEA - BISSAU 1974
GUINEA 1958
BURKINA FASO 1960
IVORY COAST 1960
GHANA 1957
NIGERIA 1960
CAMEROON 1960
CENTRAL AFRICAN REP. 1960
ETHIOPIA 1st Millenium B.C.
SIERRA LEONE 1961
LIBERIA 1847
TOGO 1960
BENIN 1960
EQUAT. GUINEA 1968
SÃO TOMÉ & PRINCIPE 1975
GABON 1960
CONGO 1960
DEM. REP. CONGO 1960
UGANDA 1962
RWANDA 1962
BURUNDI 1962
KENYA 1963
SOMALIA 1960
Equator
CABINDA (Angola)
TANZANIA 1961
Comoro Is.
ANGOLA 1975
ZAMBIA 1964
MALAWI 1964
MOZAMBIQUE 1975
NAMIBIA 1990
BOTSWANA 1966
ZIMBABWE 1965
MADAGASCAR 1960
REPUBLIC OF SOUTH AFRICA 1910; 1961
LESOTHO 1966
SWAZILAND 1968
1000 miles
1000 kilometers

Former Colonies
- British
- French
- Belgian
- Italian
- Spanish
- Portuguese

Dates indicate year of independence

(b)

COLONIAL AFRICA, 1939

TANGIER (International)
SPANISH MOROCCO
MOROCCO
TUNISIA
Madeira
IFNI
Canary Is.
RIO DE ORO
ALGERIA
LIBYA
KINGDOM OF EGYPT
Suez Canal Zone (British occupation)
FRENCH WEST AFRICA
FRENCH EQUATORIAL AFRICA
ANGLO-EGYPTIAN SUDAN
FRENCH SOMALILAND
BRITISH SOMALILAND
GAMBIA
PORT. GUINEA
SIERRA LEONE
LIBERIA
GOLD COAST
TOGOLAND (Br. & Fr. Mandates)
NIGERIA
ITALIAN EAST AFRICA
Equator
CAMEROONS (Br. & Fr. Mandates)
SP. GUINEA
CABINDA
BELGIAN CONGO
UGANDA
KENYA
RUANDA URUNDI (Belgian Mandate)
TANGANYIKA
ANGOLA
NORTHERN RHODESIA
NYASALAND
Comoro Is.
SOUTHERN RHODESIA
MOÇAMBIQUE
MADAGASCAR
SOUTH WEST AFRICA (South African Mandate)
BECHUANALAND
SWAZILAND & BASUTOLAND (British Protectorates)
UNION OF SOUTH AFRICA
1000 miles
1000 kilometers

Alien Rule
- British
- British mandate
- French
- French mandate
- Belgian
- Belgian mandate
- Italian
- Spanish
- Portuguese

(a)

Figure 12.6 **Africa—from colonies to states.** (*a*) Africa in 1939 was a patchwork of foreign claims and alien rule, some dating from the 19th century, others of more recent vintage. For example, Germany lost its claim to South West Africa, Tanganyika, Togoland, and the Cameroons after World War I, and Italy asserted control over Ethiopia during the 1930s. (*b*) **Africa in 2000** was a mosaic of separate states. Their dates of independence are indicated on the map. French West Africa and French Equatorial Africa have been extensively subdivided, and Ethiopia and Somaliland emerged from Italian control. Most of the current countries retain the boundaries of their former colonial existence, though the continent's structure of political influence and regional power changed through civil wars and neighboring state interventions. These marked the decline of earlier African principles of inviolability of borders and noninterference in the internal affairs of other states.

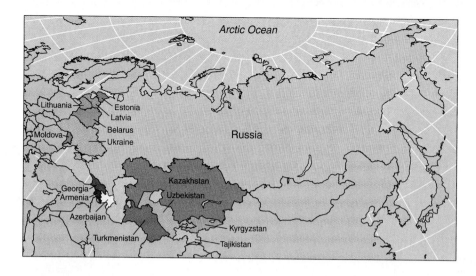

Figure 12.7 By mid-1992, 15 newly independent countries had taken the place of the former USSR.

with the capital located in the center. In such a country, all places could be reached from the center in a minimal amount of time and with the least expenditure for roads, railway lines, and so on. It would also have the shortest possible borders to defend. Uruguay, Zimbabwe, and Poland have roughly circular shapes, forming a **compact state** (Figure 12.8).

Prorupt states are nearly compact but possess one or sometimes two narrow extensions of territory. Proruption may simply reflect peninsular elongations of land area, as in the case of Myanmar and Thailand. In other instances, the extensions have an economic or strategic significance, recording a past history of international negotiation to secure access to resources or water routes or to establish a buffer zone between states that would otherwise adjoin. The proruptions of Afghanistan, Democratic Republic of the Congo, and Namibia fall into this category. The Caprivi Strip of Namibia, for example, which extends eastward from the main part of the country, was designed by the Germans to give what was then their colony of Southwest Africa access to the Zambezi River. Whatever their origin, proruptions tend to isolate a portion of a state.

The least efficient shape administratively is represented by countries like Norway, Vietnam, or Chile, which are long and narrow. In such **elongated states,** the parts of the country far from the capital are likely to be isolated because great expenditures are required to link them to the core. These countries are also likely to encompass more diversity of climate, resources, and peoples than compact states, perhaps to the detriment of national cohesion or, perhaps, to the promotion of economic strength.

A fourth class of **fragmented states** includes countries composed entirely of islands (e.g., the Philippines and Indonesia), countries that are partly on islands and partly on the mainland (Italy and Malaysia), and those that are chiefly on the mainland but whose territory is separated by another state (the United States). Fragmentation makes it harder for the state to impose centralized control over its territory, particularly when the parts of the state are far from one another. This is a problem in the Philippines and Indonesia, the latter made up of over 13,000 is-

lands stretched out along a 5100-kilometer (3200-mi) arc. Fragmentation helped lead to the disintegration of Pakistan. It was created in 1947 as a spatially divided state with East and West Pakistan separated by 1610 kilometers (1000 mi). That distance exacerbated economic and cultural differences between the two, and when the eastern part of the country seceded in 1971 and declared itself the independent state of Bangladesh, West Pakistan was unable to impose its control.

A special case of fragmentation occurs when a territorial outlier of one state, an **exclave,** is located within another country. Before German unification, West Berlin was an outlier of West Germany within the eastern German Democratic Republic. Europe has many such exclaves. Kleinwalsertal, for example, is a patch of Austria accessible only from Germany. Baarle-Hertog is a fragment of Belgium inside Holland; Campione d'Italia is an Italian outlier in Switzerland and Büsingen is a German one; and Llivia is a totally Spanish town just inside France. Exclaves are not limited to Europe, of course. African examples include Cabinda, an exclave of Angola, and Mililla and Ceuta, two Spanish exclaves in Morocco.

The counterpart of an exclave, an **enclave,** helps to define the fifth class of country shapes, the **perforated state.** A perforated state completely surrounds a territory that it does not rule as, for example, the Republic of South Africa surrounds Lesotho. The enclave, the surrounded territory, may be independent or may be part of another country. Two of Europe's smallest independent states, San Marino and Vatican City, are enclaves that perforate Italy. As an *exclave* of former West Germany, West Berlin perforated the national territory of former East Germany and was an *enclave* in it. The stability of the perforated state can be weakened if the enclave is occupied by people whose value systems differ from those of the surrounding country.

Location

The significance of size and shape as factors in national well-being can be modified by a state's location, both absolute and

Totally or partially autonomous political units that are small in area and population pose some intriguing questions. Should size be a criterion for statehood? What is the potential of ministates to cause friction among the major powers? Under what conditions are they entitled to representation in international assemblies like the United Nations?

Of the world's growing number of small countries, more than 40 have under 1 million, the population size adopted by the United Nations as the upper limit defining "small states," though not too small to be members of that organization. Nauru has about 12,000 inhabitants on its 21 square kilometers (8.2 sq mi). Other areally small states like Singapore (580 sq km; 224 sq mi) have populations (4.2 million) well above the UN criterion. Many are island countries located in the Caribbean, the Pacific or Indian Ocean, but Europe (Vatican City and Andorra), Asia (Bahrain and Brunei), Africa (Djibouti and Equatorial Guinea), and Central and Caribbean America (Belize, Dominica, Grenada) have their share.

Many ministates are vestiges of colonial systems that no longer exist. Some of the small countries of West Africa and the Arabian peninsula fall into this category. Others, such as Mauritius, served primarily as refueling stops on transoceanic shipping lanes. However, some occupy strategic locations (such as Bahrain, Malta, and Singapore), and others contain valuable minerals (Kuwait, Nauru, and Trinidad). The possibility of claiming 370-kilometer-wide (200 nautical mile) zones of adjacent seas (see "Specks and Spoils," p. 467) adds to the attraction of yet others.

Their strategic or economic value can expose small islands and territories to unwanted attention from larger neighbors. The 1982 war between Britain and Argentina over the Falkland Islands (claimed as the Islas Malvinas by Argentina) and the Iraqi invasion of Kuwait in 1990 demonstrate the ability of such areas to bring major powers into conflict and to receive world attention that is out of proportion to their size and population.

The proliferation of tiny countries raises the question of their representation and their voting weight in international assemblies. Should there be a minimum size necessary for participation in such bodies? Should countries receive a vote proportional to their population? New members accepted into the United Nations in 1999 and 2000 included four small Pacific island countries, all with populations under 100,000: Nauru, Tonga, Kiribati, and Tuvalu. Within the United Nations, the Alliance of Small Island States (AOSIS) has emerged as a significant power bloc, controlling more than one-fifth of UN General Assembly votes—far more than the combined voting strength of all the countries of South America.

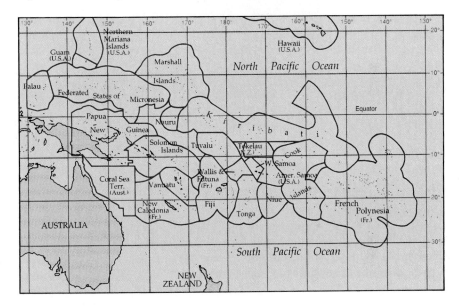

Pacific Ocean Ministates

relative. Although both Canada and Russia are extremely large, their *absolute location* in upper middle latitudes reduces their size advantages when agricultural potential is considered. To take another example, Iceland has a reasonably compact shape, but its location in the North Atlantic Ocean, just south of the Arctic Circle, means that most of the country is barren, with settlement confined to the rim of the island.

A state's *relative location,* its position compared to that of other countries, is as important as its absolute location. *Landlocked* states, those lacking ocean frontage and surrounded by other states, are at a commercial and strategic disadvantage (Figure 12.9). They lack easy access to both maritime (sea-borne) trade and the resources found in coastal waters and submerged lands. Typically, a landlocked country arranges to use facilities at a foreign port plus to have the right to travel to that port. Bolivia, for example, has secured access to the Chilean port of Arica, the Peruvian port of Ilo, and the Argentinian city of Rosario on the Parana River (Figure 12.10). The number of landlocked states—about 40—increased greatly with the dissolution of the Soviet Union and the creation of new, smaller nation-states out of such former multinational countries as Yugoslavia and Czechoslovakia.

In a few instances, a favorable relative location constitutes the primary resource of a state. Singapore, a state of only 580 square kilometers (224 sq mi), is located at a crossroads of world shipping and commerce. Based on its port and commercial activities and buttressed by its more recent industrial development, Singapore has become a notable Southeast Asian economic success. In general, history has shown that countries benefit from

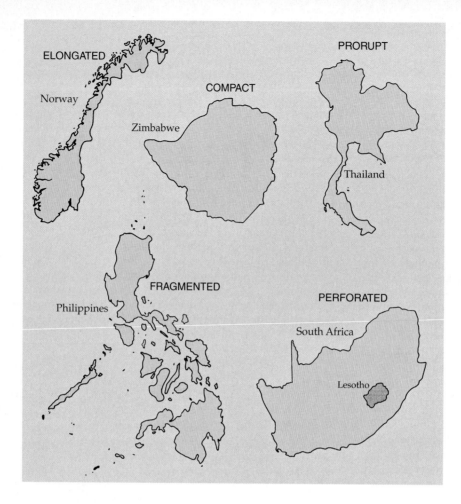

$\mathcal{F}igure\ 12.8$ **Shapes of states.** The sizes of the countries should not be compared. Each is drawn on a different scale.

a location on major trade routes, not only from the economic advantages such a location carries, but also because they are exposed to the diffusion of new ideas and technologies.

Cores and Capitals

Many states have come to assume their present shape, and thus the location they occupy, as a result of growth over centuries. They grew outward from a central region, gradually expanding into surrounding territory. The original nucleus, or **core area,** of a state usually contains its most developed economic base, densest population and largest cities, the most highly developed transportation systems, and—at least formerly if no longer—the resources which sustained it. All of these elements become less intense away from the national core. Transportation networks thin, urbanization ratios and city sizes decline, and economic development is less concentrated on the periphery than in the core. The outlying resource base may be rich but generally is of more recent exploitation with product and benefit tending to flow to the established heartlands. The developed cores of states, then, can be contrasted to their subordinate peripheries just as we saw the *core-periphery* idea applicable in an international developmental context in Chapter 10.

Easily recognized and unmistakably dominant national cores include the Paris Basin of France, London and southeastern England, Moscow and the major cities of European Russia, northeastern United States and southeastern Canada, and the Buenos Aires megalopolis in Argentina. Not all countries have such clearly defined cores—Chad, or Mongolia, or Saudi Arabia, for example—and some may have two or more rival core areas. Ecuador, Nigeria, Democratic Republic of the Congo, and Vietnam are examples of multicore states.

The capital city of a state is usually within its core region and frequently is the very focus of it, dominant not only because it is the seat of central authority but because of the concentration of population and economic functions as well. That is, in many countries the capital city is also the largest or *primate* city, dominating the structure of the entire country. Paris in France, London in the United Kingdom, and Mexico City are examples of that kind of political, cultural, and economic primacy (see p. 406 and Figure 11.17).

This association of capital with core is common in what have been called the *unitary states,* countries with highly centralized governments, relatively few internal cultural contrasts, a strong sense of national identity, and borders that are clearly cultural as well as political boundaries. Most European cores and capitals

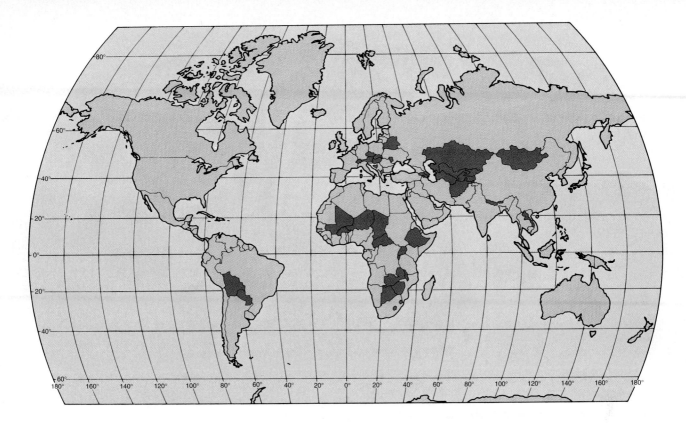

Figure 12.9 **Landlocked states.**

Figure 12.10 Like many other landlocked countries, Bolivia has gained access to the sea through arrangements with neighboring states. Unlike most landlocked countries, however, Bolivia can access ports on two oceans.

are of this type. It is also found in many newly independent countries whose former colonial occupiers established a primary center of exploitation and administration and developed a functioning core in a region that lacked an urban structure or organized government. With independence, the new states retained the established infrastructure, added new functions to the capital, and, through lavish expenditures on governmental, public, and commercial buildings, sought to create prestigious symbols of nationhood.

In *federal states,* associations of more or less equal provinces or states with strong regional governmental responsibilities, the national capital city may have been newly created or selected to serve as the administrative center. Although part of a generalized core region of the country, the designated capital was not its largest city and acquired few of the additional functions to make it so. Ottawa, Canada; Washington, D.C.; and Canberra, Australia, are examples (Figure 12.11).

A new form of state organization, *regional* government or *asymmetric federalism,* is emerging in Europe as formerly strong unitary states acknowledge the autonomy aspirations of their several subdivisions and grant to them varying degrees of local administrative control while retaining in central hands authority over matters of nationwide concern, such as monetary policy, defense, foreign relations, and the like. That new form of federalism involves recognition of regional capitals, legislative assemblies, administrative bureaucracies, and the like. The asymmetric federalism of the United Kingdom, for example, now involves separate status for Scotland, Wales, and Northern Ireland, with their

Figure 12.11 Canberra, the planned capital of Australia, was deliberately sited away from the country's two largest cities, Sydney and Melbourne. Planned capitals are often architectural showcases, providing a focus for national pride.

own capitals at Edinburgh, Cardiff, and Belfast, respectively. That of Spain recognizes Catalonia and the Basque country with capitals in Barcelona and Vitoria, respectively.

All other things being equal, a capital located in the center of the country provides equal access to the government, facilitates communication to and from the political hub, and enables the government to exert its authority easily. Many capital cities, such as Washington, D.C., were centrally located when they were designated as seats of government but lost their centrality as the state expanded.

Some capital cities have been relocated outside of peripheral national core regions, at least in part to achieve the presumed advantages of centrality. Two examples of such relocation are from Karachi inland to Islamabad in Pakistan and from Istanbul to Ankara, in the center of Turkey's territory. A particular type of relocated capital is the *forward-thrust capital* city, one that has been deliberately sited in a state's interior to signal the government's awareness of regions away from an off-center core and its interest in encouraging more uniform development. In the late 1950s, Brazil relocated its capital from Rio de Janeiro to the new city of Brasília to demonstrate its intent to develop the vast interior of the country. The West African country of Nigeria has been building the new capital of Abuja near its geographic center since the late 1970s, with relocation there of government offices and foreign embassies in the early 1990s.

The British colonial government relocated Canada's capital six times between 1841 and 1865, in part seeking centrality to the mid-19th-century population pattern and in part seeking a location that bridged that colony's cultural divide (Figure 12.12). A

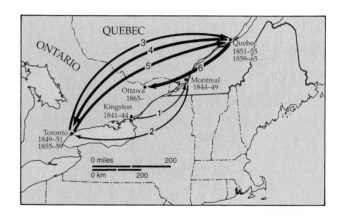

Figure 12.12 **Canada's migratory capital.** Kingston was chosen as the first capital of the united Province of Canada in preference to either Quebec, capital of Lower Canada, or Toronto, that of Upper Canada. In 1844, governmental functions were relocated to Montreal where they remained until 1849, after which they shifted back and forth—as the map indicates—between Toronto and Quebec. An 1865 session of the provincial legislature was held in Ottawa, the city that became the capital of the Confederation of Canada in 1867.

Redrawn with permission from David B. Knight, A Capital for Canada *(Chicago: University of Chicago, Department of Geography, Research Paper No. 182, 1977), Figure 1, p. vii.*

Japanese law of 1997 calling for the relocation out of Tokyo of the parliament building, Supreme Court, and main ministries by 2010 is more related to earthquake fears and a search for seismic safety than to enhanced convenience or governmental efficiency. Putrajaya, the new administrative seat of Malaysia 25 miles south of the present capital, Kuala Lumpur, and Astana, the new national capital of Kazakhstan located on a desolate stretch of Siberian steppe, are other examples of several recent new national capital proposals or creations.

Boundaries: The Limits of the State

We noted earlier that no portion of the earth's land surface is outside the claimed control of a national unit, that even uninhabited Antarctica has had territorial claims imposed upon it (see Figure 12.3). Each of the world's states is separated from its neighbors by *international boundaries,* or lines that establish the limit of each state's jurisdiction and authority. Boundaries indicate where the sovereignty of one state ends and that of another begins.

Within its own bounded territory, a state administers laws, collects taxes, provides for defense, and performs other such governmental functions. Thus, the location of the boundary determines the kind of money people in a given area use, the legal code to which they are subject, the army they may be called upon to join, and the language and perhaps the religion children are taught in school. These examples suggest how boundaries serve as powerful reinforcers of cultural variation over the earth's surface.

Territorial claims of sovereignty, it should be noted, are three-dimensional. International boundaries mark not only the outer limits of a state's claim to land (or water) surface, but are also projected downward to the center of the earth in accordance with international consensus allocating rights to subsurface resources. States also project their sovereignty upward, but with less certainty because of a lack of agreement on the upper limits of territorial airspace. Properly viewed, then, an international boundary is a line without breadth; it is a vertical interface between adjacent state sovereignties.

Before boundaries were delimited, nations or empires were likely to be separated by *frontier zones,* ill-defined and fluctuating areas marking the effective end of a state's authority. Such zones were often uninhabited or only sparsely populated and were liable to change with shifting settlement patterns. Many present-day international boundaries lie in former frontier zones, and in that sense the boundary line has replaced the broader frontier as a marker of a state's authority.

Natural and Geometric Boundaries

Geographers have traditionally distinguished between "natural" and "geometric" boundaries. **Natural** (or **physical**) **boundaries** are those based on recognizable physiographic features, such as mountains, rivers, and lakes. Although they might seem to be attractive as borders because they actually exist in the landscape and are visible dividing elements, many natural boundaries have proved to be unsatisfactory. That is, they do not effectively separate states.

Many international boundaries lie along mountain ranges, for example in the Alps, Himalayas, and Andes, but while some have proved to be stable, others have not. Mountains are rarely total barriers to interaction. Although they do not invite movement, they are crossed by passes, roads, and tunnels. High pastures may be used for seasonal grazing, and the mountain region may be the source of water for irrigation or hydroelectric power. Nor is the definition of a boundary along a mountain range a simple matter. Should it follow the crests of the mountains or the *water divide* (the line dividing two drainage areas)? The two are not always the same. Border disputes between China and India are in part the result of the failure of mountain crests and headwaters of major streams to coincide (Figure 12.13).

Rivers can be even less satisfactory as boundaries. In contrast to mountains, rivers foster interaction. River valleys are likely to be agriculturally or industrially productive and to be densely populated. For example, for hundreds of miles the Rhine River serves as an international boundary in Western Europe. It is also a primary traffic route lined by chemical plants, factories, and power stations and dotted by the castles and cathedrals that make it one of Europe's major tourist attractions. It is more a common intensively used resource than a barrier in the lives of the nations it borders. With any river, it is not clear precisely where the boundary line should lie: along the right or left bank, along the center of the river, or perhaps along the middle of the navigation channel. Even an agreement in accordance with international custom that the boundary be drawn along the main channel may be impermanent if the river changes its course, floods, or dries up.

The alternative to natural boundaries are **geometric** (or **artificial**) **boundaries.** Frequently delimited as segments of parallels of latitude or meridians of longitude, they are found chiefly in Africa, Asia, and the Americas. The western portion of the United States–Canada border, which follows the 49th parallel, is an example of a geometric boundary. Many such were established when the areas in question were colonies, the land was only sparsely settled, and detailed geographic knowledge of the frontier region was lacking.

Figure 12.13 Several international borders run through the jumble of the Himalayas. The mountain boundary between India and China has long been in dispute.

Boundaries Classified by Settlement

Boundaries can also be classified according to whether they were laid out before or after the principal features of the cultural landscape were developed. An **antecedent boundary** is one drawn across an area before it is well populated, that is, before most of the cultural landscape features were put in place. To continue our earlier example, the western portion of the United States–Canada boundary is such an antecedent line, established by a treaty between the United States and Great Britain in 1846.

Boundaries drawn after the development of the cultural landscape are termed **subsequent.** One type of subsequent boundary is **consequent** (also called *ethnographic*), a border drawn to accommodate existing religious, linguistic, ethnic, or economic differences between countries. An example is the boundary drawn between Northern Ireland and Eire (Ireland). Subsequent **superimposed boundaries** may also be forced on existing cultural landscapes, a country, or a people by a conquering or colonizing power that is unconcerned about preexisting cultural patterns. The colonial powers in 19th-century Africa superimposed boundaries upon established African cultures without regard to the tradition, language, religion, or tribal affiliation of those whom they divided (see Figure 12.5).

When Great Britain prepared to leave the Indian subcontinent after World War II, it was decided that two independent states would be established in the region: India and Pakistan. The boundary between the two countries, defined in the partition settlement of 1947, was thus both a *subsequent* and a *superimposed* line. As millions of Hindus migrated from the northwestern portion of the subcontinent to seek homes in India, millions of Muslims left what would become India for Pakistan. In a sense, they were attempting to ensure that the boundary would be *consequent,* that is, that it would coincide with a division based on religion.

If a former boundary line that no longer functions as such is still marked by some landscape features or differences on the two sides, it is termed a **relic boundary** (Figure 12.14). The abandoned castles dotting the former frontier zone between Wales and England are examples of a relic boundary. They are also evidence of the disputes that sometimes attend the process of boundary making.

Boundary Disputes

Boundaries create many possibilities and provocations for conflict. Since World War II, almost half of the world's sovereign states have been involved in border disputes with neighboring countries. Just like householders, states are far more likely to have disputes with their neighbors than with more distant parties. It follows that the more neighbors a state has, the greater the likelihood of conflict. Although the causes of boundary disputes and open conflict are many and varied, they can reasonably be placed into four categories.

1. **Positional disputes** occur when states disagree about the interpretation of documents that define a boundary and/or the way the boundary was delimited. Such disputes typically arise when the boundary is antecedent, preceding effective human settlement in the border region. Once the area becomes populated and gains value, the exact location of the

Figure 12.14 Like Hadrian's Wall in the north of England or the Great Wall of China, the Berlin Wall was a demarcated boundary. Unlike them, it cut across a large city and disrupted established cultural patterns. The Berlin Wall, therefore, was a *subsequent superimposed* boundary. The dismantling of the wall in 1990 marked the reunification of Germany; any of it that remains standing as a historic monument is a *relic* boundary.

boundary becomes important. The boundary between Argentina and Chile, originally defined during Spanish colonial rule and formalized by treaty in 1881, was to follow "the most elevated crests of the Andean Cordillera dividing the waters" between east- and west-flowing rivers. Because the southern Andes had not been adequately explored and mapped, it was not apparent that the crest lines (highest peaks) and the watershed divides do not always coincide. In some places, the water divide is many miles east of the highest peaks, leaving a long, narrow area of some 52,000 square kilometers (20,000 sq mi) in dispute (Figure 12.15). In Latin America as a whole, the 21st century began with at least 10 unresolved border disputes, some dating back to colonial times.

2. **Territorial disputes** over the ownership of a region often, though not always, arise when a boundary that has been superimposed on the landscape divides an ethnically homogeneous population. Each of the two states then has some justification for claiming the territory inhabited by the ethnic group in question. We noted previously that a single nation may be dispersed across several states (see Figure 12.4d). Conflicts can arise if the people of one state want to annex a territory whose population is ethnically related to that of the state but now subject to a foreign government. This type of expansionism is called **irredentism.** In the 1930s, Hitler used the existence of German minorities in Czechoslovakia and Poland to justify German invasion and occupation of

Figure 12.15 **The disputed boundary between Argentina and Chile** in the southern Andes. The treaty establishing the boundary between the two countries preceded adequate exploration and mapping of the area, leaving its precise location in doubt and in contention. After years of friction, the last remaining territorial dispute between Chile and Argentina in the Andes was settled in an accord signed in late 1998.

those countries. More recently, Somalia has had many border clashes with Ethiopia over the rights of Somalis living in that country, and the area of Kashmir has been a cause of dispute and open conflict between India and Pakistan since the creation of the two countries (Figure 12.16).

3. Closely related to territorial conflicts are **resource disputes.** Neighboring states are likely to covet the resources— whether they be valuable mineral deposits, fertile farmland, or rich fishing grounds—lying in border areas and to disagree over their use. In recent years, the United States has been involved in disputes with both its immediate neighbors: with Mexico over the shared resources of the Colorado River and Gulf of Mexico and with Canada over the Georges Bank fishing grounds in the Atlantic Ocean.

One of the causes of the 1990–1991 war in the Persian Gulf was the huge oil reservoir known as the Rumaila field, lying mainly under Iraq with a small extension into Kuwait (Figure 12.17a). Because the two countries were unable to agree on percentages of ownership of the rich reserve, or a formula for sharing production costs and revenues, Kuwait pumped oil from Rumaila without any international agreement. Iraq helped justify its invasion of Kuwait by contending that the latter had been stealing Iraqi oil in what amounted to economic warfare.

4. **Functional disputes** arise when neighboring states disagree over policies to be applied along a boundary. Such policies may concern immigration, the movement of traditionally nomadic groups, customs regulations, or land use. U.S. relations with Mexico, for example, have been affected by the increasing number of illegal aliens and the flow of drugs entering the United States from Mexico (Figure 12.17b).

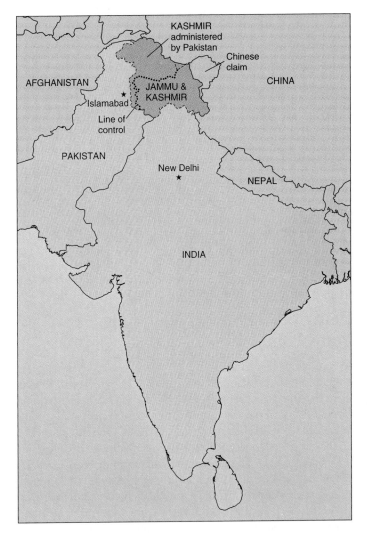

Figure 12.16 **Kashmir,** a disputed area that has been the cause of two wars between India and Pakistan. The resolution of the problem of possession of Kashmir may be a permanent partition along the cease-fire line, though the continuing Chinese claim to a portion of eastern Kashmir clouds the ownership picture.

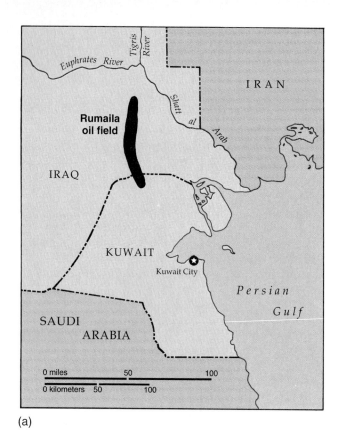

(a)

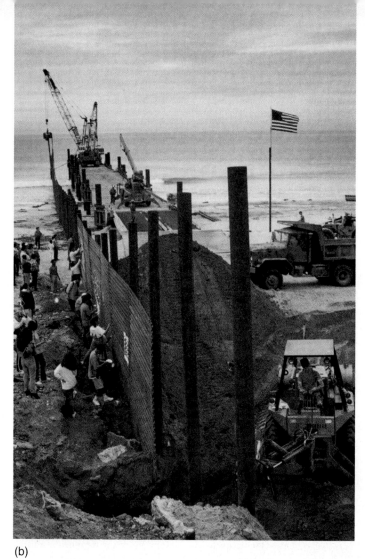

(b)

Figure 12.17 (*a*) **The Rumaila oil field.** One of the world's largest petroleum reservoirs, Rumaila straddles the Iraq-Kuwait border. Iraqi griev-ances over Kuwaiti drilling were partly responsible for Iraq's invasion of Kuwait in 1990. (*b*) To stem the flow of undocumented migrants entering Cali-fornia from Baja California, the United States in 1993 constructed a fence 3 meters (10 feet) high along the border.

Centripetal Forces: Promoting State Cohesion

At any moment in time, a state is characterized by forces that promote unity and national stability and by others that disrupt them. Political geographers refer to the former as **centripetal forces.** These are factors that bind together the people of a state, that enable it to function and give it strength. **Centrifugal forces,** on the other hand, destabilize and weaken a state. If centrifugal forces are stronger than those promoting unity, the very existence of the state will be threatened. In the sections that follow, we ex-amine four centripetal (uniting) forces—nationalism, unifying in-stitutions, effective organization and administration of government, and systems of transportation and communication—to see how they can promote cohesion.

Nationalism

One of the most powerful of the centripetal forces is **nationalism,** an identification with the state and the acceptance of national

goals. Nationalism is based on the concept of allegiance to a sin-gle country and the ideals and the way of life it represents; it is an emotion that provides a sense of identity and loyalty and of col-lective distinction from all other peoples and lands.

States purposely try to instill feelings of allegiance in their citizens, for such feelings give the political system strength. Peo-ple who have such allegiance are likely to accept common rules of action and behavior and to participate in the decision-making process establishing those rules. In light of the divisive forces present in most societies, not everyone, of course, will feel the same degree of commitment or loyalty. The important considera-tion is that the majority of a state's population accepts its ideolo-gies, adheres to its laws, and participates in its effective operation. For many countries, such acceptance and adherence has come only recently and partially; in some, it is frail and endangered.

We noted earlier that true nation-states are rare; in only a few countries do the territory occupied by the people of a particular nation and the territorial limits of the state coincide. Most coun-tries have more than one culture group that considers itself sepa-rate in some important way from other citizens. In a multicultural society, nationalism helps integrate different groups into a unified population. This kind of consensus nationalism has emerged in countries such as the United States and Switzerland, where different culture groups have joined together to create political entities commanding the loyalties of all their citizens.

States promote nationalism in a number of ways. *Iconography* is the study of the symbols that help unite people. National anthems and other patriotic songs; flags, national sports teams, and officially designated or easily identified flowers and animals; and rituals and holidays are all developed by states to promote nationalism and attract allegiance (Figure 12.18). By ensuring that all citizens, no matter how diverse the population may be, will have at least these symbols in common, they impart a sense of belonging to a political entity called, for example, Japan or Canada. In some countries, certain documents, such as the Magna Carta in England or the Declaration of Independence in the United States, serve the same purpose. Royalty may fill the need: in Sweden, Japan, and the United Kingdom, the monarchy functions as the symbolic focus of allegiance. Such symbols are significant, for symbols and beliefs are major components of the ideological subsystem (p. 51) of every culture. When a society is very heterogeneous, composed of people with different customs, religions, and languages, belief in the national unit can help weld them together.

Unifying Institutions

Institutions as well as symbols help to develop the sense of commitment and cohesiveness essential to the state. Schools, particularly elementary schools, are among the most important of these. Children learn the history of their own country and relatively little about other countries. Schools are expected to instill the society's goals, values, and traditions, to teach the common language that conveys them, and to guide youngsters to identify with their country.

Other institutions that advance nationalism are the armed forces and, sometimes, a state church. The armed forces are of necessity taught to identify with the state. They see themselves as protecting the state's welfare from what are perceived to be its enemies. In some countries, the religion of the majority of the people may be designated a state church. In such cases the church sometimes becomes a force for cohesion, helping to unify the population. This is true of Buddhism in Thailand, Hinduism in Nepal, Islam in Pakistan, and Judaism in Israel. In countries like

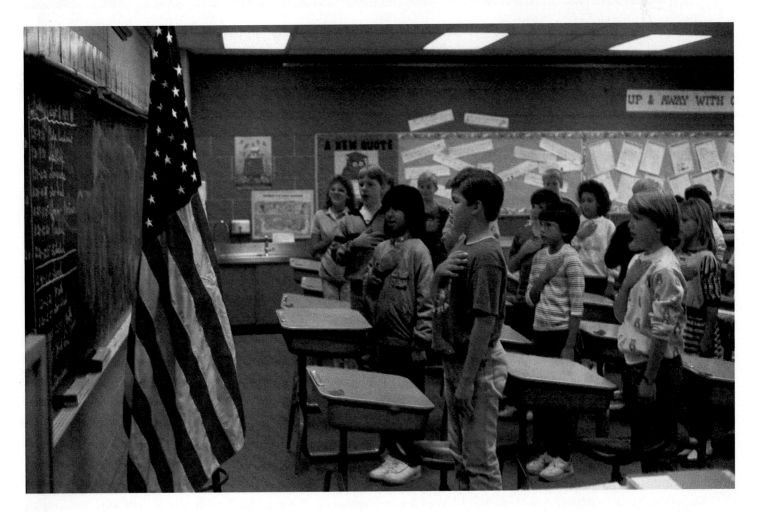

Figure 12.18 The ritual of the pledge of allegiance is just one way in which schools in the United States seek to instill a sense of national identity in students.

these, the religion and the church are so identified with the state that belief in one is transferred to allegiance to the other.

The schools, the armed forces, and the church are just three of the institutions that teach people what it is like to be members of a state. As institutions, they operate primarily on the level of the sociological subsystem of culture, helping to structure the outlooks and behaviors of the society. But by themselves, they are not enough to give cohesion, and thus strength, to a state. Indeed, each of the institutions we have discussed can also be a destabilizing centrifugal force.

Organization and Administration

A further bonding force is public confidence in the effective organization of the state. Can it provide security from external aggression and internal conflict? Are its resources distributed and allocated in such a way as to be perceived to promote the economic welfare of all its citizens? Are all citizens afforded equal opportunity to participate in governmental affairs (see "Legislative Women")? Do institutions that encourage consultation and the peaceful settlement of disputes exist? How firmly established are the rule of law and the power of the courts? Is the system of decision making responsive to the people's needs?

The answers to such questions, and the relative importance of the answers, will vary from country to country, but they and similar ones are implicit in the expectation that the state will, in the words of the Constitution of the United States, "establish justice, insure domestic tranquility, provide for the common defence, [and] promote the general welfare. . . ." If those expectations are not fulfilled, the loyalties promoted by national symbols and unifying institutions may be weakened or lost.

Legislative Women

Women, a majority of the world's population, in general fare poorly in the allocation of such resources as primary and higher education, employment opportunities and income, and health care. That their lot is improving is encouraging. In nearly every developing country, women have been closing the gender gap in literacy, school enrollment, and acceptance in the job market.

But in the political arena—where power ultimately lies—women's share of influence is increasing only slowly and selectively. In 2003, fewer than 15 countries out of a world total of over 200 had women as heads of government: presidents or prime ministers. Nor did they fare much better as members of parliaments. Women in late-2002 held just 15% of all the seats in the world's legislatures.

Only in 23 countries did women in 2002 occupy one-quarter or more of the seats in the lower or single legislative House; in none were women a legislative majority. Nine of them were European—five Nordic plus the Netherlands, Germany, Spain, and Bulgaria. Sweden was the most feminist of any country with 45% female members. The worldwide roster was completed by Mozambique, Namibia, Rwanda, Seychelles, and South Africa in Africa; Argentina, Costa Rica, Cuba, and Grenada in the Americas; Australia and New Zealand in Oceania; and East Timor, Turkmenistan, and Vietnam in Asia.

Europe also counts an additional 14 countries with between 15% and 25% female representatives, though many of the established democracies of Western Europe and virtually all of the countries of Southern and Eastern Europe fell below even that modest 15% share. Although in the Parliament of the European Union women comprised 31% of members in 2002, they held only 9% of the seats in the Greek parliament, 12% of France's National Assembly, and 10% of Italy's Chamber of Deputies. Japan made an even poorer showing with but a 7% female membership. Arab states average only 6%. Nor did the United States show a very significant number of women members. At the start of the 108th Congress (January, 2003), only 14 women served in the Senate and 60 in the House of Representatives, for a 14% share of seats in both Houses. At that time, both numbers were at their highest-ever levels.

In the later 1990s, women's legislative representation began to expand materially in many developed and developing democracies, and their "fair share" of political power began to be formally recognized or enforced. In Western countries, particularly, improvement in female parliamentary participation has become a matter of plan and pride for political parties and, occasionally, for governments themselves.

Political parties from Mexico to China have tried to correct female underrepresentation, usually by setting quotas for women candidates, and a few governments—including Belgium and Italy—have tried to require their political parties to improve their balance. France went further than any other country in acknowledging the right of women to equal access to elective office when in 1999 it passed a constitutional amendment requiring *parité*—parity or equality. A year later, the National Assembly enacted legislation requiring the country's political parties to fill 50% of the candidacies in all elections in the country (municipal, regional, and European Parliament) with women, or lose a corresponding share of their state-provided campaign funding. India similarly proposed to reserve a third of the seats in parliament for women.

Quotas are controversial, however, and often are viewed with disfavor even by avowed feminists, who argue that quotas are demeaning because they imply women cannot match men on merit alone. Others fear that other groups—for example, religious groups or ethnic minorities—would also seek quotas to guarantee their proportionate legislative presence.

A significant presence of women in legislative bodies makes a difference in the kinds of bills that get passed and the kinds of programs that receive governmental emphasis. Regardless of party affiliation, women are more apt than their male counterparts to sponsor bills and vote for measures affecting child care, elderly care, women's health care, medical insurance, and bills affecting women's rights and family law.

Transportation and Communication

A state's transportation network fosters political integration by promoting interaction between areas and by joining them economically and socially. The role of a transportation network in uniting a country has been recognized since ancient times. The saying that all roads lead to Rome had its origin in the impressive system of roads that linked Rome to the rest of its empire. Centuries later, a similar network was built in France, joining Paris to the various departments of the country. Often the capital city is better connected to other cities than the outlying cities are to one another. In France, for example, it can take less time to travel from one city to another by way of Paris than by direct route.

Roads and railroads have played a historically significant role in promoting political integration. In the United States and Canada, they not only opened up new areas for settlement but increased interaction between rural and urban districts. Because transportation systems play a major role in a state's economic development, it follows that the more economically advanced a country is, the more extensive its transport network is likely to be (see Figure 8.4). At the same time, the higher the level of development, the more resources there are to be invested in building transport routes. The two reinforce one another.

Transportation and communication, while encouraged within a state, are frequently curtailed or at least controlled between them as a conscious device for promoting state cohesion through limitation on external spatial interaction (Figure 12.19). The mechanisms of control include restrictions on trade through tariffs or embargoes, legal barriers to immigration and emigration, and limitations on travel through passports and visa requirements.

Centrifugal Forces: Challenges to State Authority

State cohesion is not easily achieved or, once gained, invariably retained. Destabilizing *centrifugal forces* are ever-present, sowing internal discord and challenges to the state's authority. Transportation and communication may be hindered by a country's shape or great size, leaving some parts of the state not well integrated with the rest. A country that is not well organized or administered stands to lose the loyalty of its citizens. Institutions that in some states promote unity can be a divisive force in others.

Organized religion, for example, can be a potent centrifugal force. It may compete with the state for people's allegiance—one reason the former USSR and other communist governments suppressed religion and promoted atheism. Conflict between majority and minority faiths within a country—as between Catholics and Protestants in Northern Ireland or Hindus and Muslims in Kashmir and Gujarat State in India—can destabilize social order. Opposing sectarian views within a single, dominant faith can also promote civil conflict. Recent years have seen particularly Muslim militant groups attempt to overturn official or constitutional policies of secularism or replace a government deemed insufficiently ardent in its imposition of religious laws and regulations. Islamic fundamentalism led to the 1979 overthrow of the Shah of Iran; more recently, Islamic militancy has been a destabilizing force in, among other countries, Afghanistan, Algeria, Tunisia, Egypt, and Saudi Arabia.

Nationalism, in contrast to its role as a powerful centripetal agency, is also a potentially disruptive centrifugal force. The idea

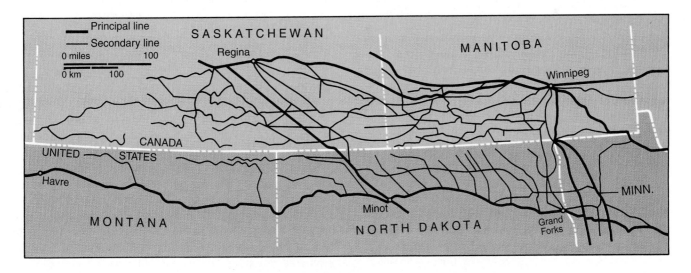

Figure 12.19 **Canadian–U.S. railroad discontinuity.** Canada and the United States developed independent railway systems connecting their respective prairie regions with their separate national cores. Despite extensive rail construction during the 19th and early 20th centuries, the pattern that emerged even before recent track abandonment was one of discontinuity at the border. Note how the political boundary restricted the ease of spatial interaction between adjacent territories. Many branch lines approached the border, but only eight crossed it. In fact, for over 480 kilometers (300 miles), no railway bridged the boundary line. The international border—and the cultural separation it represents—inhibits other expected degrees of interaction. Telephone calls between Canadian and U.S. cities, for example, are far less frequent than would be expected if distance alone were the controlling factor, and research indicates that a Canadian province in the middle 1990s was 12 times more likely to trade merchandise and 40 times more likely to trade services with another Canadian province than with an American state of similar size and distance.

of the nation-state is that states are formed around and coincide with nations. It is a small step from that to the notion that every nation has the right to its own state or territory. Centrifugal forces may be very strong in countries containing multiple nationalities and unassimilated minorities, racial or ethnic conflict, contrasting cultures, and a multiplicity of languages or religions. Such states are susceptible to nationalist challenges from within their borders *if* the minority group has an explicit territorial identification and believes that its right to *self-determination*—the right of a group to govern itself in its own state—has not been satisfied.

A dissident minority that has total or partial secession from the state as its primary goal is said to be guided by **separatism** or **autonomous nationalism.** In recent years, such nationalism has created currents of unrest within many countries, even long-established ones. Canada, for example, houses a powerful secessionist movement in French-speaking Quebec, the country's largest province. In October, 1995, a referendum to secede from Canada and become a sovereign country failed in Quebec by a razor-thin margin. Quebec's nationalism is fueled by strong feelings of collective identity and distinctiveness and by a desire to protect its language and culture, and by the conviction that the province's ample resources and advanced economy would permit it to manage successfully as a separate country.

In Western Europe, five countries (the United Kingdom, France, Belgium, Italy, and Spain) house separatist political movements whose members reject total control by the existing sovereign state and who claim to be the core of a separate national entity (Figure 12.20). Their basic demand is for *regional autonomy,* usually in the form of self-government or "home rule" rather than complete independence. Accommodation of those demands has resulted in some degrees of **devolution**—the transfer of some central powers to regional or local governments—and in the forms of asymmetric federalism discussed earlier (p. 451) with the United Kingdom and Spain as examples.

Separatist movements affect many states outside of Western Europe and indeed are more characteristic of developing countries, especially those formed since the end of World War II and containing disparate groups more motivated by enmity than affinity. The Basques of Spain and the Bretons of France have their counterparts in the Palestinians in Israel, the Sikhs in India, the Moros in the Philippines, the Tamils in Sri Lanka, and many others. Separatist movements are expressions of **regionalism,** minority group self-awareness and identification with a region rather than with the state.

The countries of Eastern Europe and the republics of the former Soviet Union have seen many instances of regionally rooted nationalist feelings. Now that the forces of ethnicity, religion, language, and culture are no longer suppressed by communism, ancient rivalries are more evident than at any time since World War II. The end of the Cold War aroused hopes of decades of peace. Instead, the collapse of communism and the demise of the USSR spawned dozens of smaller wars. Numerous ethnic groups large and small are asserting their identities and what they perceive to be their right to determine their own political status.

The national independence claimed in the early 1990s by the 15 former Soviet constituent republics did not assure the satisfaction of all separatist movements within them. Many of the new

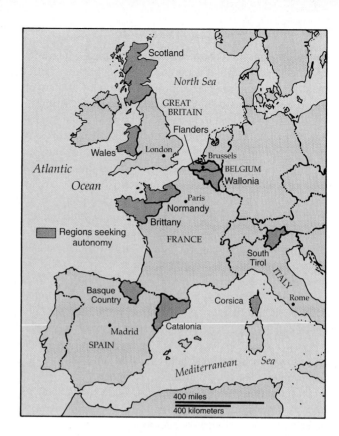

Figure 12.20 **Regions in Western Europe seeking autonomy.** Despite long-standing state attempts to assimilate these historic nations culturally, each contains a political movement that has sought or is seeking a degree of self-rule recognizing its separate identity. Separatists on the island of Corsica, for example, want to secede from France, as do the Basques from Spain. The desires of nationalist parties in both Wales and Scotland were partially accommodated by the creation in 1999 of their own parliaments and a degree of regional autonomy, an outcome labeled "separation but not divorce" from the United Kingdom.

individual countries are themselves subject to strong destabilizing forces that threaten their territorial integrity and survival. The Russian Federation itself, the largest and most powerful remnant of the former USSR, has 89 components, including 21 "ethnic republics" and a number of other nationality regions, many of which are rich in natural resources, have non-Russian majorities, and seek greater autonomy within the federation. Some, indeed, want total independence. One, the predominantly Muslim republic of Chechnya, in 1994 claimed the right of self-determination and attempted to secede from the federation, provoking a bloody civil war that escalated again in 1996 and 1999. Under similar separatist pressures, Yugoslavia shattered into five pieces in 1991–1992; more peacefully, Czechs and Slovaks agreed to split former Czechoslovakia into two ethnically based states in 1993.

Recently, several European governments have moved peacefully in the direction of regional recognition and devolution. In France, 22 regional governments were established in 1986; Spain has a program of devolution for its 17 "autonomous communities," a program that Portugal is beginning to emulate. Italy, Germany, and the Nordic countries have, or are developing, similar recognitions of regional communities with granted powers of local administration and relaxation of central controls.

The two preconditions common to all regional autonomist movements are *territory* and *nationality*. First, the group must be concentrated in a core region that it claims as a national homeland and seek to regain control of land and power that it believes were unjustly taken from it. Second, certain cultural characteristics must provide a basis for the group's perception of separateness, identity, and unity. These might be language, religion, or distinctive group customs and institutions that promote feelings of group identity at the same time that they foster exclusivity. Normally, these cultural differences have persisted over several generations and have survived despite strong pressures toward assimilation.

Other characteristics common to many separatist movements are a *peripheral location* and *social* and *economic inequality*. Troubled regions tend to be peripheral, often isolated in rural pockets, and their location away from the seat of central government engenders feelings of alienation and exclusion. They perhaps sense what has been called the *law of peripheral neglect*, which observes that the concern of the capital for its controlled political space decreases with increasing distance from it. Second, the dominant culture group is often seen as an exploiting class that has suppressed the local language, controlled access to the civil service, and taken more than its share of wealth and power. Poorer regions complain that they have lower incomes and greater unemployment than prevail in the rest of the state and that "outsiders" control key resources and industry. Separatists in relatively rich regions believe that they could exploit their resources for themselves and do better economically without the constraints imposed by the central state.

The Projection of Power

Territorial and political influence or control by a state need not necessarily halt at its recognized land borders. Throughout history, states have projected power beyond their home territories where such power could credibly be applied or asserted. Imperial powers such as Rome, Czarist Russia, and China extended control outward over adjacent peoples and territories through conquest or *suzerainty*—control over vassal states. The former Soviet Union, for example, not only conquered and incorporated such adjacent states as Estonia, Latvia, and Lithuania but also, claiming to be first among equals, asserted the right to intervene militarily to preserve communist regimes wherever they appeared threatened.

Colonial empires such as those of England, France, Spain, and Portugal exerted home state control over noncontiguous territories and frequently retain influence even after their formal colonial dominion has been lost. The Commonwealth (originally, the British Commonwealth of Nations), for example, is a free association of some 50 countries that recognize the British sovereign as head of the Commonwealth and retain use of the English language and legal system. The French Community comprises autonomous states formerly part of the French colonial empire that opted to remain affiliated with the Community, that generally retain the French language and legal system, have various contrac-

tual cooperative arrangements with the former ruling state, and have in the instance of African members occasionally called on France to intervene militarily to protect established regimes.

Geopolitical Assessments

Geopolitics is a branch of political geography that considers the strategic value of land and sea area in the context of national economic and military power and ambitions. In that light, geopolitical concerns and territorial assessments have always influenced the policies of governments. "Manifest Destiny" rationalized the westward territorial spread of the United States; the Monroe Doctrine declared the Western Hemisphere off-limits to further European colonization; creation of a "Greater East Asia Co-Prosperity Sphere" justified Japan's Asian and Pacific aggression before and during World War II.

Modern geopolitics was rooted in the early 20th-century concern of an eminent English geographer, Halford Mackinder (1861–1947), with the world balance of power at a time of British expansion and overseas empire. Believing that the major powers would be those that controlled the land, not the seas, he developed what came to be known as the **heartland theory.** The greatest land power, he argued, would be sited in Eurasia, the "World-Island" containing the world's largest landmass in both area and population. Its interior or heartland, he warned, would provide a base for world conquest, and Eastern Europe was the core of that heartland. Mackinder warned, "Who rules East Europe commands the Heartland, who rules the Heartland commands the World-Island, who rules the World-Island commands the World."[1]

Developed in a century that saw first Germany and then the Soviet Union dominate East Europe and the decline of Britain as a superpower, Mackinder's theory impressed many. Even earlier, Alfred Mahan (1840–1914) recognized the core position of Russia in the Asian landmass and anticipated conflict between Russian land power and British sea power, though Mahan argued that control of the world's sea lanes to protect commerce and isolate an adversary was the key to national strength. Near the end of World War II, Nicholas Spykman (1894–1943) also agreed that Eurasia was the likely base for potential world domination, but argued that the coastal fringes of the landmass, not its heartland, were the key (Figure 12.21). The continental margins, Spykman reasoned, contained dense populations, abundant resources, and had controlling access both to the seas and to the continental interior. His **rimland theory,** published in 1944, stated "Who controls the Rimland rules Eurasia, who rules Eurasia controls the destinies of the world."[2] The rimland has tended throughout history to be politically fragmented, and Spykman concluded that it would be to the advantage of both the United States and the USSR if it remained that way.

By the end of World War II, the Heartland was equated in American eyes with the USSR. To prevent Soviet domination of the World-Island, U.S. foreign policy during the Cold War was

---

[1]Halford J. Mackinder, *Democratic Ideals and Reality* (London: Constable, 1919), p. 150.
[2]Nicholas J. Spykman, *The Geography of the Peace* (New York: Harcourt Brace, 1944), p. 43.

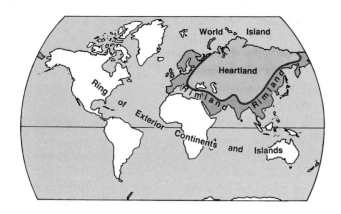

Figure 12.21 **Geopolitical viewpoints.** Both Mackinder and Spykman believed that Eurasia possessed strategic advantages, but they disagreed on whether its heartland or rimland provided the most likely base for world domination. Mahan recognized sea power as the key to national strength, advocating American occupation of the Hawaiian Islands, control of the Caribbean, and construction of an interocean canal through Central America.

based on the notion of **containment,** or confining the USSR within its borders by means of a string of regional alliances in the Rimland: The North Atlantic Treaty Organization (NATO) in Western Europe, the Central Treaty Organization (CENTO) in West Asia, and the Southeast Asia Treaty Organization (SEATO). Military intervention was deemed necessary where communist expansion, whether Soviet or Chinese, was a threat—in Berlin, the Middle East, and Korea, for example.

A simple spatial model, the **domino theory,** was used as an adjunct to the policy of containment. According to this analogy, adjacent countries are lined up like dominoes; if one topples, the rest will fall. In the early 1960s, the domino theory was invoked to explain and justify U.S. intervention in Vietnam, and in the 1980s, the theory was applied to involvement in Central America. The fear that war among the Serbs, Croatians, and Bosnians in Bosnia-Herzegovina would lead to the destruction of that state and spread into other parts of the former Yugoslavia led in 1995 to NATO airstrikes against the Serbs, a peace agreement forged with American help in Dayton, Ohio, and stationing of United Nations peacekeeping forces in Bosnia.

These (and other) models aimed at realistic assessments of national power and foreign policy stand in contrast to "organic state theory" based on the 19th-century idea of German geographer Friedrich Ratzel (1844–1904) that the state acted as if it were an organism conforming to natural laws and forced to grow and expand into new territories (*Lebensraum*) in order to secure the resources needed for survival. Without that growth, the state would wither and die. These ideas, later expanded in the 1920s by the German Karl Haushofer (1869–1946) as *Geopolitik,* were used by the Nazi party as the presumed intellectual basis for wartime Germany's theories of race superiority and need for territorial conquest. Repudiated by events and Germany's defeat, *Geopolitik* for many years gave bad odor to any study of geopoli-

tics, which only recently has again become a serious subfield of political geography.

In a rapidly changing world, many analysts believe the older geopolitical concepts and ideas of geostrategy no longer apply. A number of developments have rendered them obsolete: the dissolution of the USSR and the presumed end of the Cold War; the proliferation of nuclear technology; and the rise of Japan, China, and Western Europe to world power status. Geopolitical reality is now seen less in terms of military advantage and confrontation—the East-West rivalry of the Cold War era—and more as a reflection of other forms of competition.

One of those forms expresses itself in violent assaults carried out by individuals and groups motivated by their total rejection of an established order they despise. Traditional geopolitical theories and assessments have dealt with projections of state power and influence embodied in military and economic strength. As a series of destructive incidents throughout the world before and after the tragic September 11, 2001 assaults on the World Trade Center and the Pentagon—and the aroused response to them from the United States and other governments—made clear, terrorism, including state-sponsored terrorism, must also be recognized as a substantially different form of global geopolitical activity. **Terrorism,** defined as the use or threat of violence to advance a political cause, was usually thought of as dissident individual or small group assault on civilian populations to weaken state authority; force change in governmental policy; or erode the social, cultural, or political organization of a society. The truck bombing of the Oklahoma City Murrah Federal Building in 1995 was of that individual terrorist nature. It caused 168 deaths and led to the conviction and execution of Timothy McVeigh, who attributed his action to rage over lethal federal agency actions against a family at Ruby Ridge, Idaho, in 1992 and a religious group in Waco, Texas, in 1993. But other, broader forms of organized terrorism exist.

On the wider global scene, the State Department in late 2002 listed 28 international terrorist organizations with goals as varied as eliminating Israel; replacing secular Muslim state governments with strict Islamic rule; securing regional separatism or unification in Ireland, India, Spain, Sri Lanka, the Philippines, Indonesia, and elsewhere; and the like. The State Department's annually updated terrorist list cites groups dedicated to changing—by violent action if necessary—current conditions they deem intolerable. Many are subnational ethnic groups (Sri Lanka's Tamil Liberation Tigers, for example) or those with specific ideological objectives, such as the Marxist Shining Path in Peru.

Religious or faith-based groups, dominantly though not exclusively Muslim, are particularly prominent on the late 20th and early 21st centuries' terrorist rosters. In their home countries, the common objective of the Muslim organizations is to replace existing secular or westernized governments with regimes committed to strict enforcement of fundamentalist religious law. On the international scene, however, they have primarily targeted the United States and its citizens. The bomb destruction of Pan Am flight 103 over Scotland in 1988, car bombing of the World Trade Center in New York in 1993, bombing of American embassies in

Kenya and Tanzania in 1998, and the *USS Cole* bombing in the port of Aden, Yemen, in 2000—collectively with the loss of hundreds of lives and many thousands of injuries—were the work of Muslim terrorist individuals and groups. The al-Qaeda network, responsible for the attacks of 9/11, was the creation of Osama bin Laden, who issued a 1998 *fatwa,* or religious judgment, claiming it was the duty of Muslims to conduct holy war against the United States for its support of both the state of Israel and the ruling Saudi Arabian royal family. Governments and citizens of European Union countries have also been subject to Muslim terrorism at home and abroad.

Although individuals and nongovernmental groups have been most active in terrorist attacks directed against Americans and Europeans, state-sponsored and supported terrorism is also evident. The governments of Iraq, Libya, Afghanistan (under Taliban control), Sudan, North Korea, Yemen, and other states have in recent years been judged by the United States to have been secret sponsors, financial supporters, and refuge areas of terrorists or served as their co-conspirators in international hostile acts. The marshalling of United States resistance through the Department of Homeland Security signals an awareness of the serious potential of the terrorist forms of international threat.

Two other types of global competition are also evident on the modern geopolitical scene. One is economic rivalry within the developed world and between economic core countries and emerging peripheral states—the North-South split introduced in Chapter 10 and expressed in the development of international blocs aligned by economic interests. The other is competition rooted in more fundamental and perhaps enduring conflicts between different "civilizations." It has been suggested that the world will increasingly be shaped by the interactions and conflicts among seven or eight major civilizations: Western, Confucian, Japanese, Islamic, Hindu, Slavic, Latin American, and possibly African. The differences among such civilizations, it is thought, are basic and antagonistic, rooted in enduring differences of history, language, culture, tradition, and religion. These differences, the argument runs, are less easily resolved than purely political and economic ones. They underlie such recent clashes as Indian rivalries between Hindus and Muslims, those of Sri Lanka between Hindu Tamils and Buddhist Sinhalese, multicultural conflicts in former Yugoslavia and between Armenians and Azeris in the Caucasus, and among and within other states and areas where "civilizations" come in contact and competition. They also motivate the Muslim faith-based terrorist groups' animosity against Western states and the globalization of their economies, societies, and popular cultures.

International Political Systems

As changing geopolitical theories and outlooks suggest, in many ways individual countries are now weaker than ever before. Many are economically frail, others are politically unstable, and some are both. Strategically, no country is safe from military attack, for technology now enables us to shoot weapons halfway around the world. Some people believe that no national security is possible in the atomic age.

The recognition that a country cannot by itself guarantee either its prosperity or its own security has led to increased cooperation among states. In a sense, these cooperative ventures are replacing the empires of yesterday. They are proliferating quickly, and they involve countries everywhere. They are also adding a new dimension to the concept of "political boundaries" since the associations of states have themselves limits that are marked by borders of a higher spatial order than those between individual states. Such boundaries as the former Iron Curtain, the current division between NATO (North Atlantic Treaty Organization) and non-NATO states, or that between the European Union area and other European countries represent a different scale of the political ordering of space.

Supranationalism

Such associations also represent a new dimension in the ordering of national power and national independence. Recent trends in economic globalization and international cooperation suggest to some that the sovereign state's traditional responsibilities and authorities are being diluted by a combination of forces and partly delegated to higher-order political and economic organizations. Even corporations and nongovernmental economic and communication agencies often operate in controlling ways outside of nation-state jurisdiction.

The rise of transnational corporations dominant in global markets, for example, limits the economic influence of individual countries; cyberspace and the Internet are controlled by no one and are largely immune to the state restrictions on the flow of information exerted by many governments. Those information flows help create and maintain the growing number of international *nongovernmental organizations* (NGOs), estimated at over 20,000 in number and including such well-known groups as Greenpeace, Amnesty International, and Doctors without Borders. NGOs, through petitions, demonstrations, court actions, and educational efforts, have become effective influences on national and international political and economic actions. And increasingly, individual citizens of any country have their lives and actions shaped by decisions not only of local and national authorities, but by those of regional economic associations (the North American Free Trade Agreement, for example), multinational military alliances (NATO), and global political agencies (the United Nations).

The roots of such multistate cooperative systems are ancient—for example, the leagues of city states in the ancient Greek world or the Hanseatic League of free German cities in Europe's medieval period. The creation of new ones has been particularly active since 1945. They represent a world trend toward a **supranationalism** comprised of associations of three or more states created for mutual benefit and to achieve shared objectives. Although many individuals and organizations decry the loss of national independence that supranationalism entails, the many supranational associations in existence early in the 21st century are evidence of their attraction and pervasiveness. Almost all countries, in fact, are members of at least one—and most of many—supranational groupings. All at least are members of the United Nations.

The United Nations and Its Agencies

The United Nations (UN) is the only organization that tries to be universal, and even it is not all-inclusive. With its membership expanded from 51 countries in 1945 to 191 by 2002, the UN is the most ambitious attempt ever undertaken to bring together the world's nations in international assembly and to promote world peace. Stronger and more representative than its predecessor, the League of Nations, it provides a forum where countries may discuss international problems and regional concerns and a mecha-

nism, admittedly weak but still significant, for forestalling disputes or, when necessary, for ending wars (Figure 12.22). The United Nations also sponsors 40 programs and agencies aimed at fostering international cooperation with respect to specific goals. Among these are the World Health Organization (WHO), the Food and Agriculture Organization (FAO), and the United Nations Educational, Scientific, and Cultural Organization (UNESCO). Many other UN agencies and much of the UN budget are committed to assisting member states with matters of economic growth and development.

Member states have not surrendered sovereignty to the UN, and the world body is legally and effectively unable to make or enforce a world law. Nor is there a world police force. Although there is recognized international law adjudicated by the International Court of Justice, rulings by this body are sought only by

Figure 12.22 United Nations peacekeeping forces on duty in East Timor (Timor-Leste). Under the auspices of the UN, soldiers from many different countries staff peacekeeping forces and military observer groups in many world regions in an effort to halt or mitigate conflicts. Demand for peacekeeping and observer operations is indicated by recent deployment of UN forces in Bosnia, Croatia, Cyprus, East Timor, Haiti, Iraq, Israel, Kosovo, Lebanon, Pakistan/India, Sierra Leone, Somalia, and elsewhere.

countries agreeing beforehand to abide by its arbitration. In 2002, a UN treaty was signed (without U.S. participation) creating a permanent International Criminal Court able to investigate and prosecute those individuals accused of crimes against humanity, genocide, and crimes of war. The United Nations has no authority over the military forces of individual countries.

A pronounced change both in the relatively passive role of the United Nations and in traditional ideas of international relations has begun to emerge. Long-established rules of total national sovereignty that allowed governments to act internally as they saw fit, free of outside interference, are fading as the United Nations increasingly applies a concept of "interventionism." The Persian Gulf War of 1991 was UN authorized under the old rules prohibiting one state (Iraq) from violating the sovereignty of another (Kuwait) by attacking it. After the war, the new interventionism sanctioned UN operations within Iraq against the Iraqi government's will to protect Kurds within the country. Later, the UN intervened with troops and relief agencies in Somalia, Bosnia, and elsewhere, invoking an "international jurisdiction over inalienable human rights" that prevails without regard to state frontiers or sovereignty considerations.

Whatever the long-term prospects for interventionism replacing absolute sovereignty, for the short term the United Nations remains the only institution where the vast majority of the world's countries can collectively discuss matters of international political and economic concerns and attempt peacefully to resolve their differences. It has been particularly influential in formulating a law of the sea.

Maritime Boundaries

Boundaries define political jurisdictions and areas of resource control. But claims of national authority are not restricted to land areas alone. Water covers about two-thirds of the earth's surface, and increasingly countries have been projecting their sovereignty seaward to claim adjacent maritime areas and resources. A basic question involves the right of states to control water and the resources that it contains. The inland waters of a country, such as rivers and lakes, have traditionally been considered within the sovereignty of that country. Oceans, however, are not within any country's borders. Are they, then, to be open to all states to use, or may a single country claim sovereignty and limit access and use by other states?

For most of human history, the oceans remained effectively outside individual national control or international jurisdiction. The seas were a common highway for those daring enough to venture on them, an inexhaustible larder for fishermen, and a vast refuse pit for the muck of civilization. By the end of the 19th century, however, most coastal countries claimed sovereignty over a continuous belt 3 or 4 nautical miles wide (a *nautical mile,* or *nm,* equals 1.15 statute miles, or 1.85 km). At the time, the 3-nm limit represented the farthest range of artillery and thus the effective limit of control by the coastal state. Though recognizing the rights of others to innocent passage, such sovereignty permitted the enforcement of quarantine and customs regulations, allowed national protection of coastal fisheries, and made claims of neutrality effective during other people's wars. The primary concern was with security and unrestricted commerce. No separately codified laws of the sea existed, however, and none seemed to be needed until after World War I.

A League of Nations Conference for the Codification of International Law, convened in 1930, inconclusively discussed maritime legal matters and served to identify areas of concern that were to become increasingly pressing after World War II. Important among these was an emerging shift from interest in commerce and national security to a preoccupation with the resources of the seas, an interest fanned by the *Truman Proclamation* of 1945. Motivated by a desire to exploit offshore oil deposits, the federal government under this doctrine laid claim to all resources on the continental shelf contiguous to its coasts. Other states, many claiming even broader areas of control, hurried to annex their own adjacent marine resources. Within a few years, a quarter of the earth's surface was appropriated by individual coastal countries.

Unrestricted extensions of jurisdiction and disputes over conflicting claims to maritime space and resources led to a series of United Nations conferences on the Law of the Sea. Meeting over a period of years, delegates from more than 150 countries attempted to achieve consensus on a treaty that would establish an internationally agreed-upon "convention dealing with all matters relating to the Law of the Sea." The meetings culminated in a draft treaty in 1982, the **United Nations Convention on the Law of the Sea.**

An International Law of the Sea

The convention delimits territorial boundaries and rights by defining four zones of diminishing control (Figure 12.23):

- A *territorial sea* of up to 12 nm (19 km) in breadth over which coastal states have sovereignty, including exclusive fishing rights. Vessels of all types normally have the right of innocent passage through the territorial sea, though under certain circumstances, noncommercial (primarily military and research) vessels can be challenged.

- A *contiguous zone* to 24 nm (38 km). Although a coastal state does not have complete sovereignty in this zone, it can enforce its customs, immigration, and sanitation laws and has the right of hot pursuit out of its territorial waters.

- An **exclusive economic zone (EEZ)** of up to 200 nm (370 km) in which the state has recognized rights to explore, exploit, conserve, and manage the natural resources, both living and nonliving, of the seabed and waters (see Figure 12.24 and "Specks and Spoils"). Countries have exclusive rights to the resources lying within the continental shelf when this extends farther, up to 350 nm (560 km), beyond their coasts. The traditional freedoms of the high seas are to be maintained in this zone.

- The *high seas* beyond the EEZ. Outside any national jurisdiction, they are open to all states, whether coastal or landlocked. Freedom of the high seas includes the right to sail ships, fish, fly over, lay submarine cables and pipelines, and pursue scientific research. Mineral resources in the international deep seabed area beyond national jurisdiction are declared the common heritage of humankind, to be managed for the benefit of all the peoples of the earth.

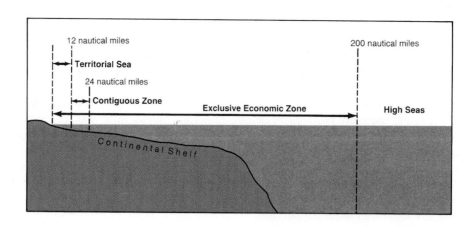

Figure 12.23 **Territorial claims permitted by the 1982 United Nations Convention on the Law of the Sea (UNCLOS).**

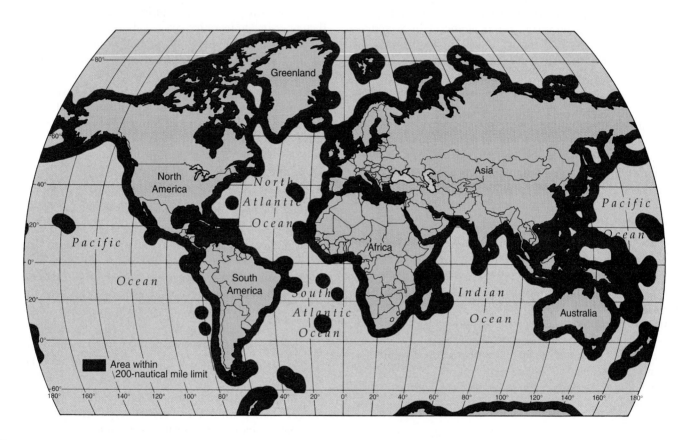

Figure 12.24 **The 200-nautical mile exclusive economic zone (EEZ) claims of coastal states.** The provisions of the Law of the Sea Convention have in effect changed the maritime map of the world. Three important consequences flow from the 200-nm EEZ concept: (1) islands have gained a new significance (see "Specks and Spoils"), (2) countries have a host of new neighbors, and (3) the EEZ lines result in overlapping claims. EEZ lines are drawn around a country's possessions as well as around the country itself. Every island, no matter how small, has its own 200-nm EEZ. This means that while the United States shares continental borders only with Canada and Mexico, it has maritime boundaries with countries in Asia, South America, and Europe. All told, the United States may have to negotiate some 30 maritime boundaries, which is likely to take decades. Other countries, particularly those with many possessions, will have to engage in similar lengthy negotiations.

By the end of the 1980s, most coastal countries, including the United States, had used the UNCLOS provisions to proclaim and reciprocally recognize jurisdiction over 12-nm territorial seas and 200-nm economic zones. Despite reservations held by the United States and a few other industrial countries about the deep seabed mining provisions, the convention received the necessary ratification by 60 states and became international law in 1994.

Other fully or essentially global supranational organizations with influences on the economic, social, and cultural affairs of states and individuals have been created. Most are specialized in-

Specks and Spoils

The Convention on the Law of the Sea gives to owners of islands claims over immense areas of the surrounding sea and, of course, to the fisheries and mineral resources in and under them. Tiny specks of land formerly too insignificant in size or distant in location to arouse the emotions of any nation now are avidly sought and fervently claimed. Remote Rockall, a British islet far west of Scotland, was used by Britain in 1976 to justify extending its fishing rights claim farther into the North Atlantic than otherwise was possible. Argentina nearly went to war with Chile in 1978 over three islands at the tip of South America at the Atlantic end of the Beagle Channel. Chile had lodged its claim of ownership hoping to gain access to known South Atlantic fish resources and hoped-for petroleum deposits. In 1982, Argentina seized the Falkland Islands from Britain, ostensibly to reclaim the Malvinas (their Spanish name) as national territory, but with an underlying economic motive as well. British forces retook the islands and subsequently used sovereignty over them to claim a sea area three times as large as Britain. Japan has encased a disappearing islet in concrete to maintain territorial claims endangered through erosion of the speck of land supporting them.

The Paracel and Spratly Islands, straddling trade routes in the South China Sea, have attracted more attention and claimants than most island groups, thanks to presumed large reserves of oil and gas in their vicinities. The Japanese seized the Paracels from China during World War II and at its end surrendered them to Nationalist Chinese forces that soon retreated to Taiwan. South Vietnam took them over until 1974, when they were forcibly ejected by the mainland Chinese. In 1979, a united Vietnam reasserted its claims, basing them on 17th- and 18th-century maps. China countered with reference to 3rd-century explorations by its geographers and maintained its control.

The location of the Paracels to the north, near China, in the South China Sea places them in a different and less difficult status than that of the Spratlys, whose nearest neighbors are the Philippines and Malaysia.

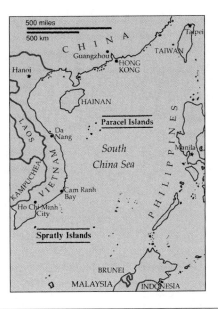

Mere dots in the sea, the largest of the Spratlys is about 100 acres—no more than one-eighth the size of New York's Central Park. But under the Convention on the Law of the Sea, possession of the island group would confer rights to the resources (oil, it is hoped) found beneath about 400,000 square kilometers (150,000 sq mi) of sea. That lure has made rivals of six governments and posed the possibility of conflict. Until early in 1988, Vietnam, the Philippines, Malaysia, Taiwan, and tiny Brunei had all maintained in peaceful coexistence garrisons on separate islets in the Spratly group. Then China landed troops on islands near the Vietnamese holdings, sank two Vietnamese naval ships, and accused Vietnam of seizing "Chinese" territory on the pretext of searching for their missing sailors. Although China agreed in 1992 that ownership disputes in the Spratlys should be resolved without violence, it also, in 1993, passed a law repeating its claims to all the islands and its determination to defend them. In early 1995, China occupied "Mischief Reef," close to—and already claimed by—the Philippines, but in late 2002 agreed with other Southeast Asian countries to avoid future disputes over the islands.

Assertions of past discovery, previous or present occupation, proximity, and simple wishful thinking have all served as the basis for the proliferating claims to seas and seabeds. The world's oceans, once open and freely accessible, are increasingly being closed by the lure of specks of land and the spoils of wealth they command.

---

ternational agencies, autonomous and with their own differing memberships but with affiliated relationships with the United Nations and operating under its auspices. Among them are the Food and Agriculture Organization (FAO), the International Bank for Reconstruction and Development (World Bank), the International Labor Organization (ILO), the United Nations Children's Fund (UNICEF), the World Health Organization (WHO), and—of growing economic importance—the World Trade Organization (WTO).

The WTO, which came into existence at the start of 1995, has become one of the most significant of the global expressions of supranational economic control. It is charged with enforcing the global trade accords that grew out of years of international negotiations under the terms of the General Agreement on Tariffs and Trade (GATT). The basic principle behind the WTO is that the 144-plus (2003) member countries should work to cut tariffs, dismantle nontariff barriers to trade, liberalize trade in services, and treat all other countries uniformly in matters of trade. Any preference granted to one should be available to all. Increasingly, however, regional rather than global trade agreements are being struck, and free trade areas are proliferating. Only a few WTO members are not already part of some other regional trade association. Such areal alliances—some 80 of them by early in the 21st century—it can be argued, make world trade less free by scrapping tariffs on trade among member states but retaining them separately or as a group on exchanges with nonmembers.

Regional Alliances

In addition to their membership in such international agencies, countries have shown themselves willing to relinquish some of their independence to participate in smaller multinational systems. These groupings may be economic, military, or political, and many have been formed since 1945. Cooperation in the economic sphere seems to come more easily to states than does military and political collaboration.

Economic Alliances

Among the oldest, most powerful, and far-reaching of the regional economic alliances are those that have evolved in Europe, particularly the European Union and its several forerunners. Shortly after the end of World War II, the Benelux countries (Belgium, the Netherlands, and Luxembourg) formed an economic union to create a common set of tariffs and to eliminate import licenses and quotas. Formed at about the same time were the Organization for European Cooperation (1948), which coordinated the distribution and use of Marshall Plan funds, and the European Coal and Steel Community (1952), which integrated the development of that industry in the member countries. A few years later, in 1957, the *European Economic Community (EEC),* or *Common Market,* was created, composed at first of only six states: France, Italy, West Germany, and the Benelux countries.

To counteract these Inner Six, as they were called, other countries in 1960 formed the European Free Trade Association (EFTA). Known as the Outer Seven, they were the United Kingdom, Norway, Denmark, Sweden, Switzerland, Austria, and Portugal (Figure 12.25). Between 1973 and 1986, three members (the United Kingdom, Denmark, and Portugal) left EFTA for membership in the Common Market and were replaced by Iceland and Finland. Other Common Market additions were Greece in 1981 and Spain and Portugal in 1986. Austria, Finland, and Sweden became members of the **European Union (EU),** as the organization embracing the Common Market is now called, in 1995. Invitations to preliminary entry negotiations, conditional on continued economic restructuring, were issued to Poland, the Czech Republic, Hungary, Slovenia, and Estonia in mid-1997 (Figure 12.26). At meetings in 2000, the EU pledged they would be ready to receive new members in 2004. That pledge was confirmed in 2002 with the decision to admit—subject to their acceptance by referendum—former Soviet bloc nations from Estonia on the north to Slovenia in the south and adding as well the island states of Malta and Cyprus. These ten additions increase the EU's land mass by 23%, raise its total population to more than 450 million people, and expand its economy to rival and perhaps exceed that of the United States, and make it the world's largest and richest bloc of nation-states.

Over the years, members of the European Union have taken many steps to integrate their economies and coordinate their policies in such areas as transportation, agriculture, and fisheries. A Council of Ministers, a Commission, a European Parliament, and a Court of Justice give the European Union supranational institutions with effective ability to make and enforce laws. By January 1, 1993, the EU had abolished most remnant barriers to free trade and the free movement of capital and people, creating a single European market. In another step toward economic and monetary union, the EU's single currency, the *euro,* replaced separate national currencies in 1999. And all applicant members in 2002 added 80,000 pages of EU law to their own legal systems.

We have traced this European development history, not because the full history of the EU is important to remember, but simply to illustrate the fluid process by which regional alliances are made. Countries come together in an association, some drop out, and others join. New treaties are made, and new coalitions emerge. Indeed, a number of such regional economic and trade associations have been added to the world supranational map. None are as encompassing in power and purpose as the EU, but all represent a cession of national independence to achieve broader regional goals.

NAFTA, the North American Free Trade Agreement launched in 1994 and linking Canada, Mexico, and the United States in an economic community aimed at lowering or removing trade and movement restrictions among the countries, is perhaps the best known to North American students. The Americas as a whole, however, have other similar associations with comparable trade enhancement objectives, though frequently they—in common with other world regional alliances—have social, political, and cultural interests also in mind. CARICOM (Caribbean Community and Common Market), for example, was established in 1974 to further cooperation among its members in economic, health, cultural, and foreign policy arenas. MERCOSUR—the Southern Cone Community Market—which unites Brazil, Argentina, Uruguay, and Paraguay in the proposed creation of a customs union to eliminate levies on goods moving among them, is a South American example.

A similar interest in promoting economic, social, and cultural cooperation and development among its members underpins

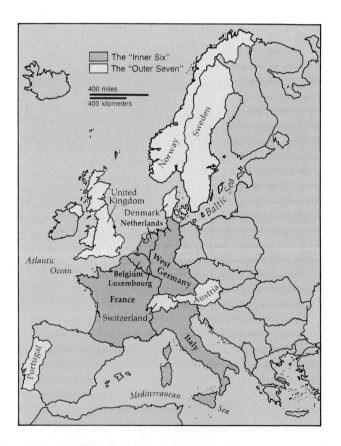

Figure 12.25 **The original Inner Six and Outer Seven of Europe.**

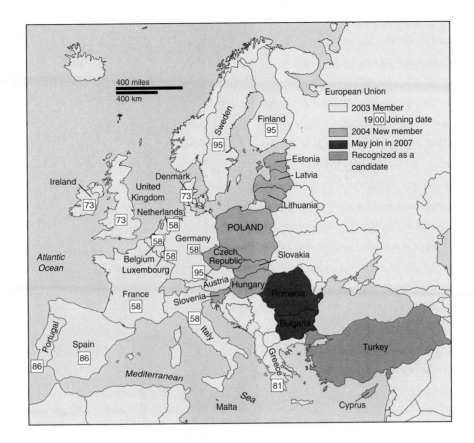

Figure 12.26 **The 15 members of the European Union (EU)** as of January, 2001, when 13 additional states were applicants for membership. The EU has stipulated that in order to join, a country must have stable institutions guaranteeing democracy, the rule of law, human rights and protection of minorities; a functioning market economy; and the ability to accept the obligations of membership, including the aims of political, economic, and monetary union. Ten applicants accepted and met those conditions and, subject to final negotiations, were admitted to the Union by 2004. The EU now spreads from the Mediterranean to the Arctic. In addition, some 70 states in Africa, the Caribbean, and the Pacific have been affiliated with the EU by the Lomé Convention, which provides for developmental aid and favored trade access to EU markets.

the Association of Southeast Asian Nations (ASEAN), formed in 1967. A similar, but much less wealthy African example is ECOWAS, the Economic Community of West African States. The Asia Pacific Economic Cooperation (APEC) forum includes China, Japan, Australia, Canada, and the United States among its 18 members and has a grand plan for "free trade in the Pacific" by 2020. More restricted bilateral and regional preferential trade arrangements have also proliferated, numbering over 400 early in this century and creating a maze of rules, tariffs, and commodity agreements that result in trade restrictions and preferences contrary to the free trade intent of the World Trade Organization.

Some supranational alliances, of course, are more cultural and political in orientation than these cited agencies. The League of Arab States, for example, was established in 1945 primarily to promote social, political, military, and foreign policy cooperation among its 22 members. In the Western Hemisphere, the Organization of American States (OAS) founded in 1948 concerns itself largely with social, cultural, human rights, and security matters affecting the hemisphere. A similar concern with peace and security underlay the Organization of African Unity (OAU) formed in 1963 by 32 African countries and, by 2001, expanded to 53 members and renamed the African Union.

Economic interests, therefore, may motivate the establishment of most international alliances, but political, social, and cul-

tural objectives also figure largely or exclusively in many. Although the alliances themselves may change, the idea of single- and multiple-purpose supranational associations has been permanently added to the national political and global realities of the 21st century. The world map pattern those alliances create must be recognized to understand the current international order.

Three further points about regional international alliances are worth noting. The first is that the formation of a coalition in one area often stimulates the creation of another alliance by countries left out of the first. Thus, the union of the Inner Six gave rise to the treaty among the Outer Seven. Similarly, a counterpart of the Common Market was the Council of Mutual Economic Assistance (CMEA), also known as Comecon, which linked the former communist countries of Eastern Europe and the USSR through trade agreements.

Second, the new supranational unions tend to be composed of contiguous states (Figure 12.27). This was not the case with the recently dissolved empires, which included far-flung territories. Contiguity facilitates the movement of people and goods. Communication and transportation are simpler and more effective among adjoining countries than among those far removed from one another, and common cultural, linguistic, historical, and political traits and interests are more to be expected in spatially proximate countries.

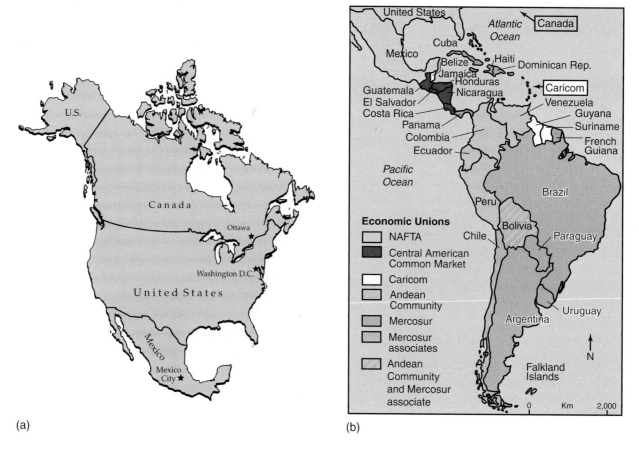

(a)   (b)

Figure 12.27 (*a*) **The North American Free Trade Agreement (NAFTA)** is intended to unite Canada, the United States, and Mexico in a regional free trade zone. Under the terms of the treaty, tariffs on all agricultural products and thousands of other goods were to be eliminated by the end of 1999. In addition, all three countries are to ease restrictions on the movement of business executives and professionals. If fully implemented, the treaty will create one of the world's richest and largest trading blocs. (*b*) **Western Hemisphere economic unions** in 2003. In addition to these subregional alliances, President George H. W. Bush in 1990 proposed a "free trade area of the Americas" to stretch from Alaska to Cape Horn.

Finally, it does not seem to matter whether countries are alike or distinctly different in their economies, as far as joining economic unions is concerned. There are examples of both. If the countries are dissimilar, they may complement each other. This was one basis for the European Common Market. Dairy products and furniture from Denmark are sold in France, freeing that country to specialize in the production of machinery and clothing. On the other hand, countries that produce the same raw materials hope that by joining together in an economic alliance, they might be able to enhance their control of markets and prices for their products. The Organization of Petroleum Exporting Countries (OPEC), mentioned in Chapter 8, is a case in point. Other attempts to form commodity cartels and price agreements between producing and consuming nations include the International Tin Agreement, the International Coffee Agreement, and others.

Military and Political Alliances

Countries form alliances for other than economic reasons. Strategic, political, and cultural considerations may also foster cooperation. *Military alliances* are based on the principle that unity assures strength. Such pacts usually provide for mutual assistance

in the case of aggression. Once again, action breeds reaction when such an association is created. The formation of the North Atlantic Treaty Organization (NATO), a defensive alliance of many European countries and the United States, was countered by the establishment of the Warsaw Treaty Organization, which joined the USSR and its satellite countries of Eastern Europe. Both pacts allowed the member states to base armed forces in one another's territories, a relinquishment of a certain degree of sovereignty uncommon in the past.

Military alliances depend on the perceived common interests and political goodwill of the countries involved. As political realities change, so do the strategic alliances. NATO was created to defend Western Europe and North America against the Soviet military threat. When the dissolution of the USSR and the Warsaw Pact removed that threat, the purpose of the NATO alliance became less clear and, during the 1990s, its members put its relationships with Eastern European states and Russia under review. Most of those countries sought ways to foster cooperation with NATO, and three of them—Poland, Hungary, and the Czech Republic—joined the alliance in 1999.

All international alliances recognize communities of interest. In economic and military associations, common objectives are clearly seen and described, and joint actions are agreed on with respect to the achievement of those objectives. More generalized mutual concerns or appeals to historical interest may be the basis for primarily *political alliances*. Such associations tend to be rather loose, not requiring their members to yield much power to the union. Examples are the Commonwealth of Nations (formerly the British Commonwealth), composed of many former British colonies and dominions, and the Organization of American States, both of which offer economic as well as political benefits.

There are many examples of abortive political unions that have foundered precisely because the individual countries could not agree on questions of policy and were unwilling to subordinate individual interests to make the union succeed. The United Arab Republic, the Central African Federation, the Federation of Malaysia and Singapore, and the Federation of the West Indies fall within this category.

Although many such political associations have failed, observers of the world scene speculate about the possibility that "superstates" will emerge from one or more of the economic or political alliances that now exist. Will a "United States of Europe," for example, under a single government be the logical outcome of the successes of the EU? No one knows, but as long as the individual state is regarded as the highest form of political and social organization (as it is now) and as the body in which sovereignty rests, such total unification does not appear imminent.

Local and Regional Political Organization

The most profound contrasts in cultures tend to occur among, rather than within, states, one reason political geographers traditionally have been primarily interested in country units. The emphasis on the state, however, should not obscure the fact that for most of us it is at that local level that we find our most intimate and immediate contact with government and its influence on the administration of our affairs. In the United States, for example, an individual is subject to the decisions and regulations made by the school board, the municipality, the county, the state, and, perhaps, a host of special-purpose districts—all in addition to the laws and regulations issued by the federal government and its agencies. Among other things, local political entities determine where children go to school, the minimum size lot on which a person can build a house, and where one may legally park a car. Adjacent states of the United States may be characterized by sharply differing personal and business tax rates; differing controls on the sale of firearms, alcohol, and tobacco; variant administrative systems for public services; and different levels of expenditures for them (Figure 12.28).

All of these governmental entities are *spatial systems*. Because they operate within defined geographic areas and because they make behavior-governing decisions, they are topics of interest to political geographers. In the concluding sections of this chapter, we examine two aspects of political organization at the local and regional level. Our emphasis will be on the U.S. and Canadian scene simply because their local political geography is

Figure 12.28 The Four Corners Monument, marking the meeting of Utah, Colorado, Arizona, and New Mexico. Jurisdictional boundaries within countries may be precisely located but are usually not highly visible in the landscape. At the same time, those boundaries may be very significant in citizens' personal affairs and in the conduct of economic activities.

familiar to most of us. We should remember, however, Anglo American structures of municipal governments, minor civil divisions, and special-purpose districts have counterparts in other regions of the world (Figure 12.29).

The Geography of Representation: The Districting Problem

There are more than 85,000 local governmental units in the United States. Slightly more than half of these are municipalities, townships, and counties. The remainder are school districts, water-control districts, airport authorities, sanitary districts, and other special-purpose bodies. Around each of these districts, boundaries have been drawn. Although the number of districts does not change greatly from year to year, many boundary lines are redrawn in any single year. When the size or shape of a district is based on population numbers or distribution, such *redistricting* or *reapportionment* is made necessary by shifts in population, as areas gain or lose people.

For example, every 10 years following the U.S. census, updated figures are used to redistribute the 435 seats in the House of Representatives among the 50 states. Redrawing the Congressional districts to reflect population changes is required by the Constitution, the intention being to make sure that each legislator represents roughly the same number of people. Since 1964, Canadian provinces and territories have entrusted redistricting for federal offices to independent electoral boundaries commissions. Although a few states in the United States also have independent, nonpartisan boards or commissions draw district boundaries, most rely on state legislatures for the task. Across the United States, the decennial census data are also used to redraw the boundaries of legislative districts within each state as well as those for local offices, such as city councils and county boards.

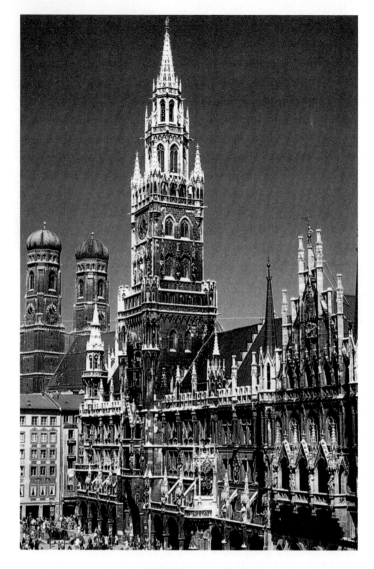

Figure 12.29 The Rathaus—city hall—of Munich, Germany. Before the rise of strong central governments, citizens of wealth and power in medieval and renaissance Europe focused their loyalties on their home cities and created within them municipal buildings and institutions that would reflect their pride and substance. The city hall was frequently the grandest public building of the community, prominently located in the center of town.

The analysis of how boundaries are drawn around voting districts is one aspect of **electoral geography,** which also addresses the spatial patterns yielded by election results and their relationship to the socioeconomic characteristics of voters. In a democracy, it might be assumed that election districts should contain roughly equal numbers of voters, that electoral districts should be reasonably compact, and that the proportion of elected representatives should correspond to the share of votes cast for a given political party. Problems arise because the way in which the electoral boundary lines are drawn can maximize, minimize, or effectively nullify the representational power of a group of people.

Gerrymandering is the practice of drawing the boundaries of voting districts so as to unfairly favor one political party over another, to fragment voting blocs, or to achieve other nondemo-

cratic objectives (Figure 12.30). A number of strategies have been employed over the years for that purpose. *Stacked* gerrymandering involves drawing circuitous boundaries to enclose pockets of strength or weakness of the group in power; it is what we usually think of as "gerrymandering." The *excess vote* technique concentrates the votes of the opposition in a few districts, which they can win easily, but leaves them few potential seats elsewhere. Conversely, the *wasted vote* strategy dilutes the opposition's strength by dividing its votes among a number of districts.

Assume that *X* and *O* represent two groups with an equal number of voters but different policy preferences. Although there are equal numbers of *X*s and *O*s, the way electoral districts are drawn affects voting results. In Figure 12.31a, the *X*s are concentrated in one district and will probably elect only one representative of four. The power of the *X*s is maximized in Figure 12.31b, where they may control three of the four districts. The voters are evenly divided in Figure 12.31c, where the *X*s have the opportunity to elect two of the four representatives. Finally, Figure 12.31d shows how both political parties might agree to delimit the electoral districts to provide "safe seats" for incumbents. Such partitioning offers little chance for change.

Figure 12.31 depicts a hypothetical district, compact in shape with an even population distribution and only two groups competing for representation. In actuality, voting districts are often oddly shaped because of such factors as the city limits, current population distribution, and transportation routes—as well as past gerrymandering. Further, in any large area, many groups vie for power. Each electoral interest group promotes its version of fairness in the way boundaries are delimited. Minority interests, for example, seek representation in proportion to their numbers so that they will be able to elect representatives who are concerned about and responsive to their needs (see "Voting Rights and Race").

In practice, gerrymandering is not always and automatically successful. First, a districting arrangement that appears to be unfair may be appealed to the courts. Further, voters are not unthinking party loyalists; key issues may cut across party lines, scandal may erode, or personal charm increase votes unexpectedly; and the amount of candidate financing or number of campaign workers may determine election outcome if compelling issues are absent.

The Fragmentation of Political Power

Boundary drawing at any electoral level is never easy, particularly when political groups want to maximize their representation and minimize that of opposition groups. Furthermore, the boundaries that we may want for one set of districts may *not* be those that we want for another. For example, sewage districts must take natural drainage features into account, whereas police districts may be based on the distribution of the population or the number of miles of street to be patrolled, and school attendance zones must consider the numbers of school-aged children and the capacities of individual schools.

As these examples suggest, the United States is subdivided into great numbers of political administrative units whose areas of control are spatially limited. The 50 states are partitioned into

THE GERRY-MANDER. (Boston, 1811.)

Figure 12.30 **The original gerrymander.** The term *gerrymander* originated in 1811 from the shape of an electoral district formed in Massachusetts while Elbridge Gerry was governor. When an artist added certain animal features, the district resembled a salamander and quickly came to be called a gerrymander.

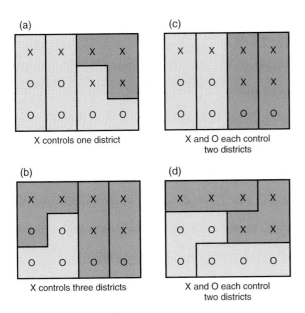

(a)

X controls one district

(b)

X controls three districts

(c)

X and O each control
two districts

(d)

X and O each control
two districts

Figure 12.31 **Alternative districting strategies.** *X*s and *O*s might represent Republicans and Democrats, urban and rural voters, blacks and whites, or any other distinctive groups.

Geography and Public Policy

Voting Rights and Race

Irregularly shaped Congressional voting districts such as those shown here were created by several state legislatures after the 1990 census to make minority representation in Congress more closely resemble minority presence in the state's voting-age population. Most were devised to contain a majority of black voters, but what opponents called racial gerrymandering was in a few cases utilized to accommodate Hispanic majorities.

All represented a deliberate attempt to balance voting rights and race; all were specifically intended to comply with the federal Voting Rights Act of 1965, which provides that members of racial minorities shall not have "less opportunity than other members of the electorate . . . to elect representatives of their choice."

Because at least some of the newly created districts had very contorted boundaries, on appeal by opponents they have at least in part been ruled unconstitutional by the Supreme Court. The state legislatures' attempts at fairness and adherence to Congres-

sional mandate contained in the Voting Rights Act were held not to meet such other standards as rough equality of district population size, reasonably compact shape, and avoidance of disenfranchisement of any class of voters. The conflicts reflected the uncertainty of exactly what were the controlling requirements in voting district creation.

In North Carolina, for example, although 22% of the 1990 population of that state were black, past districting had divided black voters among a number of districts, with the result that blacks had not elected a single Congressional representative in the 20th century. In

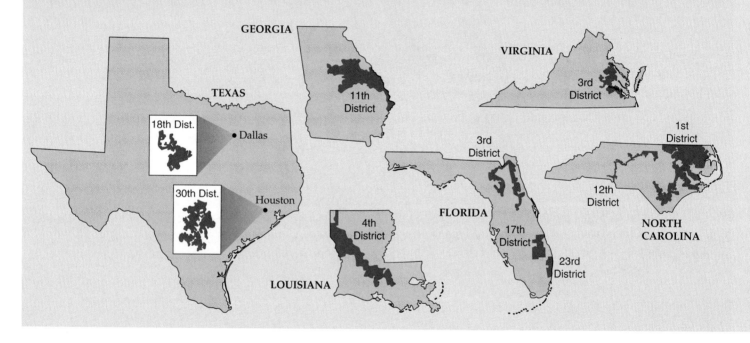

more than 3000 counties ("parishes" in Louisiana), most of which are further subdivided into townships, each with a still lower level of governing power. This political fragmentation is further increased by the existence of nearly innumerable special-purpose districts whose boundaries rarely coincide with the standard major and minor civil divisions of the country or even with each other (Figure 12.32). Each district represents a form of political allocation of territory to achieve a specific aim of local need or legislative intent (see "Too Many Governments").

Canada, a federation of ten provinces and three territories, has a similar pattern of political subdivision. Each of the provinces contains minor civil divisions—municipalities—under provincial control, and all (cities, towns, villages, and rural municipalities) are governed by elected councils. Ontario and Que-

bec also have counties that group smaller municipal units for certain purposes. In general, municipalities are responsible for police and fire protection, local jails, roads and hospitals, water supply and sanitation, and schools, duties that are discharged either by elected agencies or appointed commissions.

Most North Americans live in large and small cities. In the United States these, too, are subdivided, not only into wards or precincts for voting purposes but also into special districts for such functions as fire and police protection, water and electricity supply, education, recreation, and sanitation. These districts almost never coincide with one another, and the larger the urban area, the greater the proliferation of small, special-purpose governing and taxing units. Although no Canadian community has quite the multiplication of governmental entities plaguing many

[Continued]

1991, the Justice Department ordered North Carolina to redistrict so that at least two districts would contain black majorities. Because of the way the black population was distributed, the only way to form black-majority districts was to string together cities, towns, and rural areas in very elongated sinuous belts. The two newly created districts had slim (53%) black majorities.

The redistricting in North Carolina and other states had immediate effects. Black membership in the House of Representatives increased from 26 in 1990 to 39 in 1992; blacks constituted nearly 9% of the House as against 12% of African Americans in the total population. Within a year, those electoral gains were threatened as lawsuits challenging the redistricting were filed in a number of states. The chief contention of the plaintiffs was that the irregular shapes of the districts were a product of racial gerrymandering and amounted to reverse discrimination against whites.

In June, 1993, a sharply divided Supreme Court ruled in *Shaw v. Reno* that North Carolina's 12th Congressional District might violate the constitutional rights of white voters and ordered a district court to review the case. The 5–4 ruling gave evidence that the country had not yet reached agreement on how to comply with the Voting Rights Act. It raised a central question: Should a state maximize the rights of racial minorities or not take racial status into consideration? A divided Court provided answers in 1995, 1996, and 1997 rulings that rejected Congressional

redistricting maps for Georgia, Texas, and North Carolina on the grounds that "race cannot be the predominant factor" in drawing election district boundaries, nor can good-faith efforts to comply with the Voting Rights Act insulate redistricting plans from constitutional attack.

The difficulty in interpreting and complying with the Act is illustrated by the fact that although the Supreme Court in 1996 ruled North Carolina's 12th Congressional District unconstitutional, federal courts in 1998 and 1999 rejected alternate district designs. In *Easley v. Cromartie* (2001) the Court approved both a redrawn 12th District and using race as a redistricting consideration as long as it was not the "dominant and controlling" one. Finally, in *Georgia v. Ashcroft* (2003), the court permitted using minority political influence, not just the number of minority voters, in redrawing districts. Some observers hail these decisions as providing needed guidance in new district creation; others contend the rulings leave unresolved conflicts between the Voting Rights Act and the Court's admonitions against using race as a redistricting determinant.

Questions to Consider

1. Do you believe that race should be a consideration in the electoral process? Why or why not? If so, should voting districts be drawn to increase the likelihood that representatives of racial or ethnic minorities will win elections? If not, how can one be certain that the voting

power of minorities will not be unacceptably diluted?

2. With which of the following arguments in *Shaw v. Reno* do you agree? Why? ". . . Racial gerrymandering, even for remedial purposes, may balkanize us into competing racial factions; it threatens to carry us further from the goal of a political system in which race no longer matters." (Justice Sandra Day O'Connor) ". . . Legislators will have to take race into account in order to avoid dilution of minority voting strength." (Justice David Souter).

3. One of the candidates in North Carolina's 12th Congressional District said, "I love the district because I can drive down I-85 with both car doors open and hit every person in the district." Given a good transportation and communication network, how important is it that voting districts be compact?

4. Blacks face difficult obstacles in being elected in districts that do not have a black majority, as witness their numerical underrepresentation in most legislative bodies. But critics of "racial gerrymandering" contend that blacks have been and can continue to be elected in white-majority districts—as, in 1996, were two black incumbents from abolished majority-black districts in Georgia—and, further, that white politicians can and do adequately represent the needs of all, including black, voters in their districts. Do you agree? Why or why not?

U.S. urban areas, major Canadian cities may find themselves with complex and growing systems of similar nature. Even before its major expansion on January 1, 1998, for example, Metropolitan Toronto had more than 100 identified authorities that could be classified as "local governments."

The existence of such a great number of districts in metropolitan areas may cause inefficiency in public services and hinder the orderly use of space. *Zoning ordinances,* for example, controlling the uses to which land may be put, are determined by each municipality and are a clear example of the effect of political decisions on the division and development of space. Unfortunately, in large urban areas, the efforts of one community may be hindered by the practices of neighboring communities. Thus, land zoned for an industrial park or shopping mall in one city may

abut land zoned for single-family residences in an adjoining municipality. Each community pursues its own interests, which may not coincide with those of its neighbors or the larger region.

Inefficiency and duplication of effort characterize not just zoning but many of the services provided by local governments. The efforts of one community to avert air and water pollution may be, and often are, counteracted by the rules and practices of other towns in the region, although state and national environmental protection standards are now reducing such potential conflicts. Social as well as physical problems spread beyond city boundaries. Thus, nearby suburban communities are affected when a central city lacks the resources to maintain high-quality schools or to attack social ills. The provision of health care facilities, electricity and water, transportation, and recreational space affects the whole

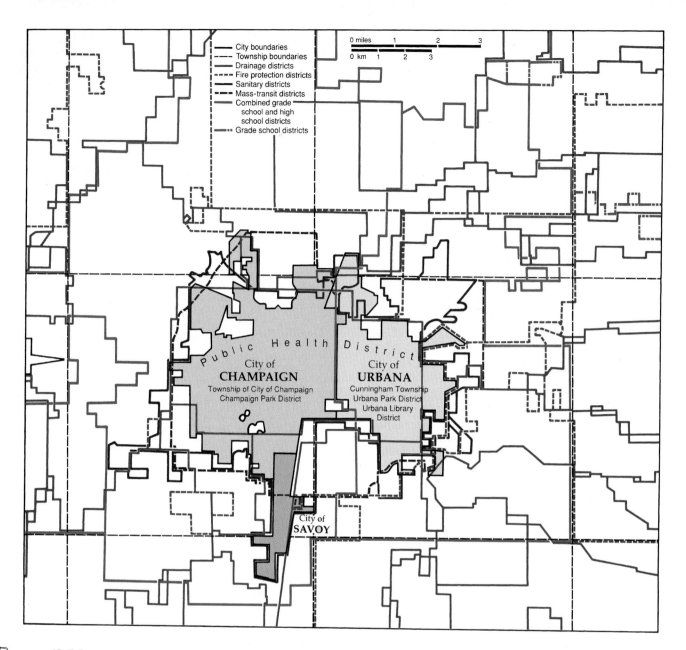

Legend:
— City boundaries
--- Township boundaries
— Drainage districts
-·- Fire protection districts
— Sanitary districts
-·- Mass-transit districts
— Combined grade school and high school districts
-·- Grade school districts

0 miles 1 2 3
0 km 1 2 3

P u b l i c   H e a l t h   D i s t r i c t

City of
CHAMPAIGN
Township of City of Champaign
Champaign Park District

City of
URBANA
Cunningham Township
Urbana Park District
Urbana Library
District

City of
SAVOY

Figure 12.32 **Political fragmentation in Champaign County, Illinois.** The map shows a few of the independent administrative agencies with separate jurisdictions, responsibilities, and taxing powers in a portion of a single Illinois county. Among the other such agencies forming the fragmented political landscape are Champaign County itself, a forest preserve district, a public health district, a mental health district, the county housing authority, and a community college district.

region and, many professionals think, should be under the control of a single consolidated metropolitan government.

To achieve that goal, various forms of **unified government**—often called *metro* or *unigov*—have been introduced in different U.S. and Canadian metropolitan areas. Metropolitan Government of Toronto, Canada, established in 1954, was the prototype followed, with variations in Montreal, Calgary, Vancouver, and elsewhere in Canada. United States examples include (among several others) the Jacksonville, Florida; Portland, Oregon; and Seattle, Washington, urban areas. In a few instances, unigov has involved city-county consolidation: Miami–Dade

County, Florida; Nashville–Davidson County, Tennessee; and Indianapolis–Marion County, Indiana, are familiar examples.

But for many urbanites, secession and not consolidation is the goal. Key Biscayne has seceded from Miami, Florida. Activists of the San Fernando Valley have advocated separation from Los Angeles and put the question on the ballot. Roxbury, Massachusetts, wants to sever its ties to Boston, and West Seattle also seeks independence from its parent city. And secession movements have at least been suggested if not actively pursued in other U.S. metropolitan areas. The issue is usually a desire for "home rule" based on the conviction that the central city has

Too Many Governments

If you are a property owner in the city of Urbana, Cunningham Township, Champaign County, Illinois, here's who takes a bite out of your tax dollar: the city, the township, the county, the public school district, the community college district, the forest preserve district, the park district, the sanitary district, the public health district, the mass transit district, and the library district (see Figure 12.32).

The number of administrative units levying taxes there does not reflect Urbana's population density or the devious tricks of Champaign County politicians. That's just the way things are in Illinois, host to more governmental units than any other state in the country. The Census Bureau reports there are 6835 local administrative entities in Illinois; Pennsylvania is in second place with just 5070 units—1765 fewer than Illinois. The average for all 50 states is only 1750.

In addition to its 102 counties, Illinois houses nearly 1300 municipalities, over 1400 townships, and nearly 1000 school districts. But the biggest share of the government unit total are the single function special districts. These were up from 2600 in 1982 to some 3100 by 2002, with no end to their increase foreseeable. Special districts range from Chicago's massive Metropolitan Sanitary District to the Caseyville Township Street Lighting District. Champaign County alone contains 177 separate governments, including 107 single special function districts. Most of them have property-taxing power; some also impose sales or utility taxes. The Chicago area alone has one unit of government for every 6000 people—five times the ratio in greater Los Angeles and seven times the New York City ratio.

Blame this Illinois proliferation partly on good intentions. Fearful of overtaxation, the framers of the state's 1870 constitution limited the borrowing and taxing power of local governments to 5% of the assessed value of properties within their jurisdictional control. As demands on and functions of local governments increased over the years with the growth in population, officials and voters financially restricted by the Constitution were forced to create new taxing bodies—special function districts to address specific public needs. In addition to circumventing municipal debt limitations, such special districts also could be adapted and shaped to serve users without regard to city, township, or county boundaries.

For some, the multiple special-function districts simply duplicate efforts, assure inefficiencies, and produce both higher costs and higher taxes. For their supporters, however, park, library, public health, sanitary, and other special-purpose units fulfill the ideal of government close to the people and responsive to constituent needs.

---

grown too large to be managed efficiently and, in fact, is managed to the benefit of only the central core, not of the prosperous, populous peripheral districts that are—the argument goes—overtaxed and underserved in police and fire protection, school funding, infrastructure maintenance and improvement, and other municipal services.

The growth in the number and size of metropolitan areas has increased awareness of their administrative and jurisdictional problems. Too much governmental fragmentation and too little local control are both seen as metropolitan problems demanding attention and solution. The one concern is that multiple jurisdictions prevent the pooling of resources to address metropolitan-wide needs. The other is that local community needs and interests are subordinated to social and economic problems of a core city for which outlying communities feel little affinity or concern.

Summary

The sovereign state is the dominant entity in the political subdivision of the world. It constitutes an expression of cultural separation and identity as pervasive as that inherent in language, religion, or ethnicity. A product of 18th-century political philosophy, the idea of the state was diffused globally by colonizing European powers. In most instances, the colonial boundaries they established have been retained as their international boundaries by newly independent countries.

The greatly varying physical characteristics of states contribute to national strength and stability. Size, shape, and relative location influence countries' economies and international roles, while national cores and capitals are the heartlands of states. Boundaries, the legal definition of a state's size and shape, determine the limits of its sovereignty. They may or may not reflect preexisting cultural landscapes and in any given case may or may not prove to be viable. Whatever their nature, boundaries are at the root of many international disputes. Maritime boundary claims, particularly as reflected in the UN Convention on the Law of the Sea, add a new dimension to traditional claims of territorial sovereignty.

State cohesiveness is promoted by a number of centripetal forces. Among these are national symbols, a variety of institutions, and confidence in the aims, organization, and administration of government. Also helping to foster political and economic integration are transportation and communication connections. Destabilizing centrifugal forces, particularly ethnically based separatist movements, threaten the cohesion and stability of many states. Assessments of the possession and projection of national power have always colored international relations. The validity of older geopolitical theories has recently been questioned in light of new developments involving organized terrorism and global competition between conflicting "civilizations."

Although the state remains central to the partitioning of the world, a broadening array of political entities affects people individually and collectively. Recent decades have seen a significant increase in supranationalism, in the form of a number and variety of global and regional alliances to which states have surrendered some sovereign powers. At the other end of the spectrum, expanding Anglo American urban areas and governmental responsibilities raise questions of fairness in districting procedures and of effectiveness when political power is fragmented.

Key Words

antecedent boundary 454
artificial boundary 453
autonomous nationalism 460
centrifugal force 456
centripetal force 456
compact state 448
consequent (ethnographic) boundary 454
containment 462
core area 450
devolution 460
domino theory 462
electoral geography 472
elongated state 448
enclave 448
European Union (EU) 468
exclave 448

exclusive economic zone (EEZ) 465
fragmented state 448
functional dispute 455
geometric boundary 453
geopolitics 461
gerrymandering 472
heartland theory 461
irredentism 454
nation 444
nationalism 456
nation-state 444
natural boundary 453
perforated state 448
physical boundary 453
political geography 442
positional dispute 454

prorupt state 448
regionalism 460
relic boundary 454
resource dispute 455
rimland theory 461
separatism 460
state 444
subsequent boundary 454
superimposed boundary 454
supranationalism 463
territorial dispute 454
terrorism 462
unified government 476
United Nations Convention on the Law of the Sea 465

For Review

1. What are the differences between a *state,* a *nation,* and a *nation-state?* Why is a colony not a state? How can one account for the rapid increase in the number of states since World War II?

2. What attributes differentiate states from one another? How do a country's size and shape affect its power and stability? How can a piece of land be both an *enclave* and an *exclave?*

3. How may boundaries be classified? How do they create opportunities for conflict? Describe and give examples of three types of border disputes.

4. How does the *United Nations Convention on the Law of the Sea* define zones of diminishing national control? What are the consequences of

the concept of the 200-nm *exclusive economic zone?*

5. Distinguish between *centripetal* and *centrifugal* political forces. What are some of the ways national cohesion and identity are achieved?

6. What characteristics are common to all or most regional autonomist movements? Where are some of these movements active? Why do they tend to be on the periphery rather than at the national core?

7. What types of international organizations and alliances can you name? What were the purposes of their establishment? What generalizations can you make regarding economic alliances?

8. How did MacKinder and Spykman differ in their assessments of Eurasia as a likely base for world conquest? What post-1945 developments suggest that there may be no enduring correlation between location and national power?

9. Why does it matter how boundaries are drawn around electoral districts? Theoretically, is it always possible to delimit boundaries "fairly"? Support your answer.

10. What reasons can you suggest for the great political fragmentation of the United States? What problems stem from such fragmentation? Describe two approaches to insuring the more efficient administration of large urban areas.

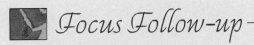 *Focus Follow-up*

1. **What are the types and geographic characteristics of countries and the nature of their boundaries?**
pp. 442–455.

States are internationally recognized independent political entities. When culturally uniform, they may be termed nation-states. Their varying physical characteristics of size, shape, and location have implications for national power and cohesion. Boundaries define the limits of states' authority and underlie many international disputes.

2. **How do states maintain cohesiveness, instill nationalism, and project power internationally?**
pp. 456–463.

Cohesiveness is fostered through unifying institutions, education, and efficient transport and communication systems. It may be eroded by minority group separatist wishes and

tendencies. Older geopolitical theories of state military power projection have been modified by concepts of economic rivalry and conflicting cultural ideals.

3. **Why are international alliances proliferating, and what objectives do they espouse and serve?**
pp. 463–471.

In an economically and technologically changing world, alliances are presumed to increase the security and prosperity of states. The UN claims to represent and promote worldwide cooperation; its Law of the Sea regulates use and claims of the world's oceans. Regional alliances involving some reduction of national independence promote economic, military, or political objectives of groups of states related spatially or ideologically. They are expressions of the growing trend toward *supranationalism* in international affairs.

4. **What problems are evident in defining local political divisions in Anglo America, and what solutions have been proposed or instituted?**
pp. 471–477.

The great political fragmentation within, particularly, the United States reflects the creation of special-purpose units to satisfy a local or administrative need. States, counties, townships, cities, and innumerable special-purpose districts all have defined and often overlapping boundaries and functions. Voting rights, reapportionment, and local political boundary adjustments represent areas of continuing political concern and dispute. In the United States, racial gerrymandering is a current legal issue in voting district definition.

 Selected References

Agnew, John. *Geopolitics: Re-Visioning World Politics.* New York: Routledge, 1998.

Blacksell, Mark. *Political Geography.* New York: Routledge, 2002.

Blake, Gerald H., ed. *Maritime Boundaries.* London and New York: Routledge, 1994.

Blouet, Brian W. *Geopolitics and Globalization in the Twentieth Century.* London: Reaktion Books, 2001.

Boyd, Andrew. *An Atlas of World Affairs.* 10th ed. New York: Routledge, 1998.

Demko, George J., and William B. Wood, eds. *Reordering the World: Geopolitical Perspectives on the Twenty-First Century.* 2d ed. Boulder, Colo.: Westview Press, 1999.

Elbow, Gary S. "Regional Cooperation in the Caribbean: The Association of Caribbean States." *Journal of Geography* 96, no. 1 (1997): 13–22.

Gibb, Richard, and Wieslaw Michalak, eds. *Continental Trading Blocs: The Growth of Regionalism in the World Economy.* New York: John Wiley & Sons, 1994.

Gibb, Richard, and Mark Wise. *The European Union.* London: Edward Arnold, 2000.

Glassner, Martin I. *Political Geography.* 3d ed. New York: John Wiley & Sons, 2003.

Hartshorne, Richard. "The Functional Approach in Political Geography." *Annals of the Association of American Geographers* 40 (1950): 95–130.

Hooson, David, ed. *Geography and National Identity.* Oxford, England: Blackwell Publishers, 1994.

Jones, Martin, Rhys Jones, and Michael Woods. *An Introduction to Political Geography: Space, Place and Politics.* New York: Routledge, 2003.

Michalak, Wieslaw, and Richard Gibb. "Trading Blocs and Multilateralism in the World Economy." *Annals of the Association of American Geographers* 87, no. 2 (1997): 264–279.

Morrill, Richard. "Gerrymandering." *Focus* 41, no. 3 (Fall, 1991): 23–27.

Newhouse, John. "Europe's Rising Regionalism." *Foreign Affairs* 76 (January/February 1997): 67–84.

O'Loughlin, John, ed. *Dictionary of Geopolitics.* Westport, Conn.: Greenwood Publishing Group, 1994.

Pinder, John. *The European Union: A Very Short Introduction.* New York: Oxford University Press, 2001.

"The Rise of Europe's Little Nations." *The Wilson Quarterly* 18, no. 1 (1994):50–81.

Shelley, Fred M., J. Clark Archer, Fiona M. Davidson, and Stanley D. Brunn. *Political Geography of the United States.* New York: Guilford Publications, 1996.

Taylor, Peter J., and Colin Flint. *Political Geography: World-Economy, Nation-State and Locality.* 4th ed. Upper Saddle River, N.J.: Prentice Hall, 2000.

Websites: The World Wide Web has a tremendous number and variety of sites pertaining to geography. Websites relevant to the subject matter of this chapter appear in the "Web Links" section of the On-line Learning Center associated with this book. Access it at **www.mhhe.com/fellmann8e**

PART Five

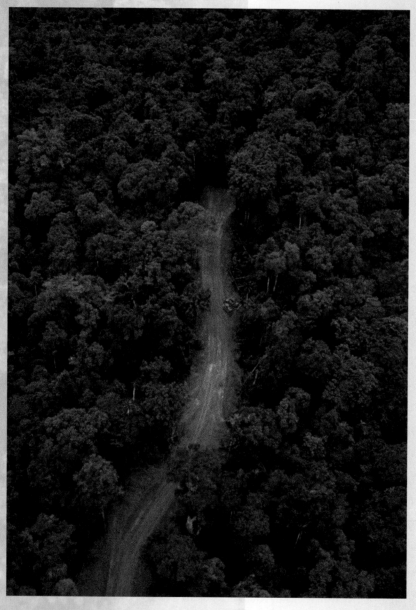

A new logging road in virgin lowland tropical rain forest in Papua New Guinea is the first stage in the exploitation of an untouched natural resource.

Human Actions and Environmental Impacts

*T*he final chapter of our study of human geography brings to the fore the recurring theme of all geographic study and the unifying thread running through each of the preceding chapters: human activities and physical environments in interaction. In those chapters we have come to understand how over the past several thousand years people in their increasing numbers and growing technological skills have placed their mark on the natural landscape, altering it to conform to their needs. In some instances, such as that of modern cities, the human imprint may be so complete that the original landscape of nature has been totally wiped away and replaced by a created cultural environment. People, we now understand, are the dominant agents in the continuing drama of human–environmental interaction. Ecological alteration, damage, or destruction may be the unplanned and unwanted consequences of the power they possess.

We need that understanding, for increasingly evident environmental deterioration has become an ever-present and growing concern of people and governments throughout the world. Ecological damage or change is no longer occasional and localized; it has now become permanent and generalized. Climatic modification, air and water pollution, soil erosion, natural vege-tation destruction, and loss of productive lands to advancing deserts are just part of the testimony of destructive cultural pressures that has aroused widespread public and private discussion. Increasingly, the inseparable interplay of the cultural and physical environments—seen in the imprint of humans on the endowment of nature—is apparent and accepted.

For convenience and focus, geographers may arbitrarily separate the physical and human systems that together comprise the reality of the earth's surface we occupy. We have made such a separation in our study of human geography, for our primary concern has been with the processes and patterns of human spatial organization. In addition to exploring the structure and logic of social spatial systems, that focus has also prepared us to evaluate the relationship of these systems to the physical environment humans occupy and alter. With the insights we have gained, we can now bring a more informed voice to discussion of the interaction of human cause and environmental consequence. Further background for that discussion is offered in our concluding chapter, "Human Impacts on Natural Systems." Its subject matter bridges our temporarily convenient subdivision of the discipline and brings back into focus the inseparable unity of human and physical geography.

Thirteen

Human Impacts
on Natural Systems

Focus Preview

Opposite: Deforested and deeply eroded hills along a heavily silted river in Madagascar mark an irreversible human impact on the natural environment.

hen the daily tides come in, a surge of water high as a person's head moves up the rivers and creeks of the world's largest delta, formed where the Ganges and Brahmaputra rivers meet the Bay of Bengal in the South Asian country of Bangladesh. Within that Wisconsin-sized country that is one-fifth water, millions of people live on thousands of alluvial islands known as "chars." These form from the silt of the rivers and are washed away by their currents and by the force of cyclones that roar upstream from the bay during the annual cyclone period. As the chars are swept away so, too, are thousands and tens of thousands of their land-hungry occupants who fiercely battled each other with knives and clubs to claim and cultivate them.

Late in April of 1991, an atmospheric low-pressure area moved across the Malay Peninsula of Southeast Asia and gained strength in the Bay of Bengal, generating winds of nearly 240 kilometers (150 miles) per hour. As it moved northward, the storm sucked up and drew along with it a wall of water 6 meters (20 feet) high. At 1:00 A.M. on April 30, with a full moon and highest tides,

the cyclone and its battering ram of water slammed across the chars and the deltaic mainland. When it had passed, some of the richest rice fields in Asia were gray with the salt that ruined them, islands totally covered with paddies were left as giant sand dunes, others—densely populated—simply disappeared beneath the swirling waters. An estimated 200,000 lives were lost to the storm and to subsequent starvation, disease, and exposure.

Each year lesser variants of the tragedy are repeated; each year survivors return to rebuild their lives on old land or new still left after the storms or created as the floods ease and some of the annual 2.5 billion tons of river-borne silt is deposited to form new chars. Deforestation in the Himalayan headwaters of the rivers increases erosion there and swells the volume of silt flowing into Bangladesh. Dams on the Ganges River in India alter normal flow patterns, releasing more water during floods and increasing silt deposits during seasonal droughts. And, always, population growth adds to the number of desperate people seeking homes and fields on lands more safely left as the realm of river and sea.

Physical Environments and Cultural Impacts

The people of the chars live with an immediate environmental contact that is not known to most of us in the highly developed, highly urbanized countries of the world. In fact, much of the content of the preceding chapters has detailed ways that humans isolate themselves from the physical environment and how they superimpose cultural landscapes on it to accommodate the growing needs of their growing numbers.

Many cultural landscape changes are minor in themselves. The forest clearing for swidden agriculture or the terracing of hillsides for subsistence farming are modest alterations of nature. Plowing and farming the prairies, harnessing major river systems by dams and reservoirs, building cities and their connecting highways, or opening vast open-pit mines (Figure 13.1) are much more substantial modifications. In some cases the new landscapes are apparently completely divorced from the natural ones which preceded them—as in enclosed, air-conditioned shopping malls and office towers. The original minor modifications have cumulatively become totally new cultural creations.

But suppression of the physical landscape does not mean eradication of human-environmental interactions. They continue, though in altered form, as humans increasingly become the active and dominant agents of environmental change. More often than not, the changes we have set in motion create unplanned cultural landscapes and unwanted environmental conditions. We have altered our climates, polluted our air and water and soil, destroyed natural vegetation and land contours while stripping ores and

fuels from the earth. At the same time, we have found it increasingly difficult and costly to provide with food and resources our growing populations. Such adverse consequences of human impact on the environment are fundamental elements in our human geographic study. They are the unforeseen creations of the landscapes of culture we have been examining and analyzing.

Environment is an overworked word that means the totality of things that in any way affect an organism. Humans exist within a natural environment—the sum of the physical world—that they have modified by their individual and collective actions. Those actions include clearing forests, plowing grasslands, building dams, and constructing cities. On the natural environment, then, we have erected our cultural environment, modifying, altering, or destroying the conditions of nature that existed before human impact was expressed.

Even in the absence of humans, those conditions were marked by constant alteration and adjustment that nonetheless preserved intact the **biosphere** (or **ecosphere**), the thin film of air, water, and earth within which we live. This biosphere is composed of three interrelated parts: (1) the *atmosphere,* a light blanket of air enveloping the earth, with more than half of its mass within 6.5 kilometers (4 miles) of the surface and 98% within 26 km (16 mi); (2) the *hydrosphere,* the surface and subsurface waters in oceans, rivers, lakes, glaciers, and groundwater; and (3) the *lithosphere,* the upper reaches of the earth's crust containing the soils that support plant life, the minerals that plants and animals require for life, and the fossil fuels and ores that humans exploit.

The biosphere is an intricately interlocked system, containing all that is needed for life, all that is available for life to use, and, presumably, all that ever will be available. The ingredients

Figure 13.1 The Bingham Canyon open-pit copper mine in Utah is said to be the largest human-made excavation on earth. Mining operations here have removed approximately 15 billion tons of material, creating a pit more than 800 meters (2600 feet) deep and 4 kilometers (2.5 miles) wide. Giant machinery and intensive application of capital and energy make possible such monumental reshapings of contours and landscapes.

of the thin ecosphere must be and are constantly recycled and renewed in nature: plants purify the air; the air helps to purify the water; plants and animals use the water and the minerals, which are returned to the system for reuse. Anything that upsets the interplay of the ecosphere or diminishes its ability to recycle itself or to sustain life endangers all organisms within it, including humans.

Climates, Biomes, and Change

The structure of the ecosphere is not eternal and unchanging. On the contrary, alteration is the constant rule of the physical environment and would be so even in the absence of humans and their distorting impacts. Climatic change, year-to-year variations in weather patterns, fires, windstorms, floods, diseases, or the unexplained rise and fall of predator and prey populations all call for new environmental configurations and forever prevent the establishment of a single, constant "balance of nature."

Remember that we began to track cultural geographic patterns from the end of the last continental glaciation, some 11,000 years ago. Our starting point, then, was a time of environmental change when humans were too few in number and primitive in technology to have had any impact on the larger structure of the biosphere. Their numbers increased and their technologies became vastly more sophisticated and intrusive with the passage of time, but for nearly all of the period of cultural development to modern times, human impact on the world environment was absorbed and accommodated by it with no more than local distress. The rhythm and the regularity of larger global systems proceeded largely unaffected by people.

Over the millennia since the last glaciation—with a few periods of unusual warming or cooling as the exceptions—a relatively stable pattern of climatic regions emerged, a global system of environmental conditions within which human cultures developed and differentiated. That pattern reflected enduring physical controls and balances: the tilt of the earth's axis; the earth's rotation and its movement about the sun; its receipt of energy from the sun and the seasonal variations in energy effectiveness in the Northern and Southern Hemispheres; the reradiation of some of that received energy back through the atmosphere in the form of heat (Figure 13.2); and, in finer detail, the pattern of land and water distribution and of ocean and atmospheric currents.

In combination these and other controls determine global patterns of temperature and precipitation, the basic variables in world climatic systems. The continual warmth of the tropical (equatorial) regions is replaced by the seasonal temperature variations of the midlatitudes, where land and water contrasts also affect the temperatures recorded even at the same latitude. Summers become cooler and shorter farther towards the poles until, finally, permanent ice cap conditions prevail. Precipitation patterns are more complexly determined than are those of temperature but are important constituents of regional environmental variation.

The pattern of global climates that these physical controls established (Figure 13.3) was, at the same time, a pattern of biomes. **Biomes** are major communities of plants and animals

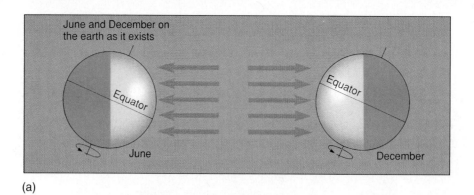

(a)

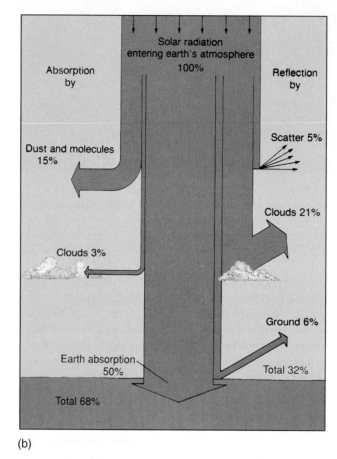

(b)

Figure 13.2 **Incoming solar energy** is indicated by the yellow arrows in (*a*). Because of the tilt of the earth's axis, the most intense of the sun's rays are received north of the equator in June and south of the equator in December. The tilt plus the earth's daily rotation on its axis also means that every point in the Northern (or Southern) Hemisphere summer has more hours of daylight than of darkness each day. The more direct rays received over longer daylight periods assure seasonal differences in hemispheric heating and cooling. (*b*) Consider the incoming solar radiation as 100%. The portion that is absorbed into the earth (50%) is eventually released to the atmosphere and then reradiated into space. Notice that the outgoing radiation is equal to 100%, showing that there is an energy balance on the earth. Percentages shown are estimated averages.

occupying extensive areas of the earth's surface in response to climatic conditions. We know them by such descriptive names as *desert, grassland* or *steppe* or as the *tropical rain forest* and *northern coniferous forest* that we met in Chapter 8. Biomes, in turn, contain smaller, more specialized **ecosystems:** self-contained, self-regulating, and interacting communities adapted to local combinations of climate, topography, soil, and drainage conditions.

Ecosystems were the first to feel the destructive hand of humans and the cultural landscapes they made. We saw in Chapter 2 the results of human abuse of the local environment in the Chaco Canyon and Easter Island deforestations. Forest removal, overgrazing, and ill-considered agriculture turned lush hillsides of the Mediterranean Basin into sterile and impoverished landscapes by the end of the Roman Empire. Other similar local and even regional alterations of natural environmental conditions

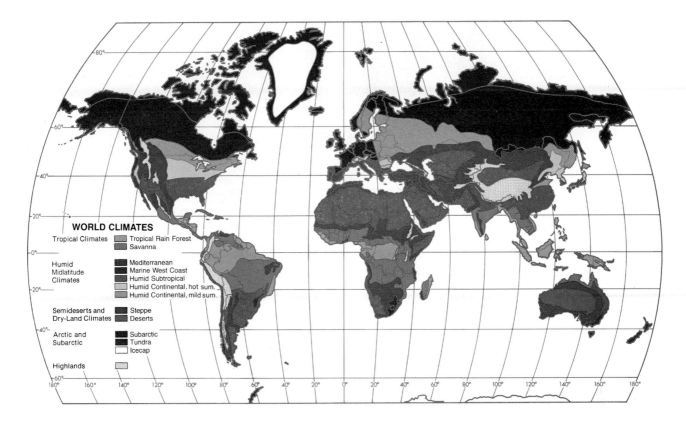

Figure 13.3 **Climates of the world.** Complex interrelationships of latitude, land and water contrasts, ocean currents, topography, and wind circulation make the global pattern of climates more intricate than this generalized map reveals.

occurred and increased as humans multiplied and exerted growing pressure on the resources and food potentials of the areas they occupied.

At a global scale, however, human impact was minimal. Long-term and short-term deviations from average conditions were induced by natural, not cultural, conditions (Figure 13.4; see also "Our Inconstant Climates"). But slowly, unnoticed at first, human activity began to have a global impact, carrying the consequences of cultural abuse of the biosphere far beyond the local scene. The atmosphere, the one part of the biosphere that all the world shares, began to react measurably during the last half of the 20th century to damage that humans had done to it since the beginning of the Industrial Revolution in the 18th century. If those reactions prove permanent and cumulative, then established patterns of climates and the biomes based on them are destined to be altered in fundamental ways.

At first it appeared the danger was from overcooling, an onset of a new glacial stage. The cause is implicit in Figure 13.2b. Part of incoming solar energy is intercepted by clouds and by solid and liquid particles—*aerosols*—and reradiated back to space. An increase in reflectors decreases energy receipts at the earth's surface, and a cooling effect, the **icebox effect,** is inevitable. Aerosols are naturally injected into the atmosphere from such sources as dust storms, forest fires, or volcanos. Indeed, increases in volcanic eruptions are thought by some not only to be triggering events for years-long cooling cycles but possibly even for ice age development itself. The famous "year without a summer,"

1816, when snow fell in June in New England and frost occurred in July, was probably the climatic reflection of the 1815 eruption of the Indonesian volcano, Tambora. That explosion ejected an estimated 200 million tons of gaseous aerosols—water vapor, sulfur dioxide, hydrogen chloride, and others—and upward of 50 cubic kilometers (30 cubic miles) of dust and ash into the atmosphere. The reflective cooling effect lasted for years. (A similar, but less extreme, Northern Hemisphere summer temperature drop in the early 1990s was attributed to the eruption of Mount Pinatubo in the Philippines in 1991.)

Aerosols in solid and gaseous form are products of human activities as well. Ever-increasing amounts of them are ejected from the smokestacks of factories, power plants, and city buildings and from the tailpipes of vehicles and exhaust plumes of jet aircraft. The global cooling that became noticeable by the late 1940s seemed to presage a new ice age, partly the product of natural conditions but hastened and deepened by human pressures upon the atmosphere.

The fears those pressures generated began to be replaced, in the 1980s, by a three-part package of different concerns: (1) a global warming caused by the "greenhouse" effect, (2) acid rain, and (3) ozone depletion. These, too, are presumed threats ascribed to human introduction into the atmosphere of kinds and amounts of materials that natural systems apparently cannot handle or recycle. Some lines of evidence and projections based on them indicate that these changing conditions may well alter the composition of the atmosphere and the pattern of climates established as the

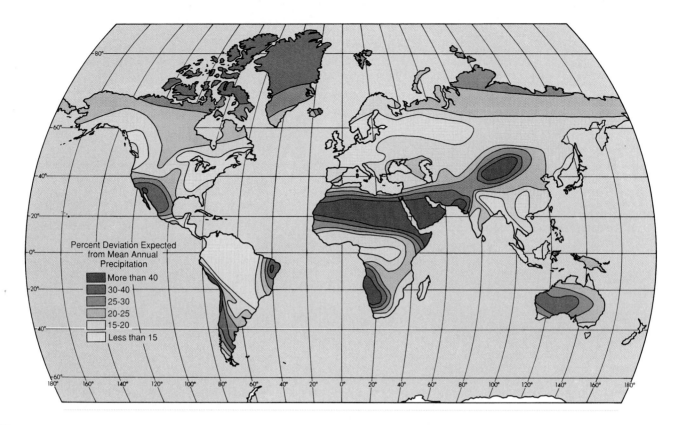

$\mathcal{F}igure\ 13.4$ **The pattern of precipitation variability.** Note that regions of low total precipitation tend to have high variability. In general, the lower the amount of long-term annual precipitation, the lower is the probability that the "average" will be recorded in any single year. Short-run variability and long-term progressive change in climatic conditions are the rule of nature and occur independent of any human influence.

expected norm over the past few decades. Other scientifically plausible evidence, however, suggests that concerns and reactions based on the fear of human-induced global warming, acid precipitation, and ozone destruction are exaggerated and in some measure unwarranted.

Global Warming

To those who fear its reality, the evidence of global atmospheric warming seems compelling and modern civilization's role in its occurrence appears easily traced. Humankind's massive assault on the atmosphere presumably began with the Industrial Revolution. First coal and then increasing amounts of petroleum and natural gas have been burned to power industry, heat and cool cities, and drive vehicles. Their burning has turned fuels into carbon dioxide and water vapor. At the same time, the world's forest lands—most recently its tropical rain forests—have been destroyed wholesale by logging and to clear land for agriculture. With more carbon dioxide in the atmosphere and fewer trees to capture the carbon and produce oxygen, carbon dioxide levels have risen steadily.

The role of trees in managing the carbon cycle is simple: Probably more than half the carbon dioxide put into the atmosphere by burning fossil fuels is absorbed by the earth's oceans, plants, and soil. The rest of the carbon dioxide remains in the atmosphere where it traps earth heat radiation. In theory, atmo-

spheric carbon dioxide could be reduced by expanding plant carbon reservoirs, or "sinks," on land. Under actual circumstances of expanded combustion of fuels and reduction of forest cover, atmospheric carbon dioxide levels now total well over 200% of their amounts at the start of the Industrial Revolution and continue to rise. Yearly carbon emissions that totaled 1.6 billion tons in 1950 reached 6.5 billion tons in 2000, a quadrupling in just 50 years. The International Energy Agency predicts that annual carbon emissions will rise to over 8.8 billion tons in 2010.

That extra carbon dioxide makes the atmosphere just a bit less transparent to the long-wave heat energy radiated back into space from the earth. Along with three other partially man-made gases (methane, nitrous oxides, and chlorofluorocarbons), the carbon dioxide traps the heat before it can escape. That so-called **greenhouse effect** is a natural condition and a necessary element in earth's heat budget. Without the atmospheric heat absorption and retention provided by carbon dioxide and water vapor, energy reradiated by the globe would pass through the atmosphere and be lost in space; earth temperatures would fluctuate widely as they do on airless Mars, and plant and animal life as we know it could not exist. The "greenhouse effect" that is of recent concern is the *increased* absorption of long-wave radiation from the earth's surface induced by the apparent increase in atmospheric carbon dioxide concentrations. *That* greenhouse effect is far less benign and nurturing than the name implies (Figure 13.5). Slowly but inexorably the retained heat raises the average temperature of

Our Inconstant Climates

In 1125, William of Malmesbury favorably compared the number and productivity of the vineyards of England with those of France; England has almost no vineyards today. In the 10th century, the Vikings established successful colonies in Greenland; by 1250, that island was practically cut off by extensive drift ice, and by the early 15th century, its colonies were forgotten and dead. The Medieval Warm Period, which evidence indicates lasted from A.D. 800 to 1200, was marked by the warmest climate that had occurred in the Northern Hemisphere for several thousand years. Glaciers retreated and agricultural settlement spread throughout Europe. Those permissive conditions were not to last, however. Change and fluctuation in environmental conditions are the rules of nature.

In recent years archaeologists and historians have found evidence that ancient seats of power—such as Sumeria in the Middle East, Mycenae in southern Greece, the Maya civilization in Yucatán, the Peruvian Moche culture, and Mali in Africa—may have fallen not to barbarians but to unfavorable alterations in the climates under which they came to power. Climatic change certainly altered the established structure of European society when, between 1550 and 1850, a "little ice age" descended on the Northern Hemisphere. Arctic ice expanded, glaciers advanced, Alpine passes were closed to traffic, and crop failures and starvation were common in much of Europe. Systems of agriculture, patterns of trade, designs of buildings, styles of clothing, and rhythms of life responded to climatic conditions vastly less favorable than those of the preceding centuries.

A new pronounced warming trend began about 1890 and lasted to the early 1940s. During that period, the margin of agriculture was extended northward, the pattern of commercial fishing shifted poleward, and the reliability of crop yields increased. But by the 1950s, natural conditions again seemed poised to change. From the late 1940s to the 1970s, the mean temperature of the globe declined. The growing season in England became 2 weeks shorter; disastrous droughts occurred in Africa and Asia; and unexpected freezes altered crop patterns in Latin America.

This time, humans were aware that changes in the great natural systems were partly traceable to things that they themselves were doing. The problem was, they were doing so many things, with such contrary consequences, that the combined and cumulative impact was unclear. One scenario predicted the planet would slowly cool over the next few centuries and enter a new ice age, helped along that path by the large amount of soot and dust in the atmosphere traceable to human activity that prevents incoming solar energy from reaching and warming the earth. A quite different scenario foretold a planet warming dangerously as different human stresses on the atmosphere overcame a natural cycle of cooling. Instead of nature's refrigerator, a man-made sauna may be the more likely prospect. A third conclusion maintains that fundamental climatic changes result from natural conditions essentially unaffected by human activities.

---

the earth; slowly but unavoidably, if the process continues, new patterns of climates and biomes must result.

During the first century of the Industrial Revolution, from 1780 to 1880, mean global temperature rose 0.3° Celsius (0.5° Fahrenheit). In the next hundred years—even allowing for a slight cooling between 1945 and 1975—average temperatures increased about 0.6° C (a bit over 1° F). They rose another half degree Celsius in the last half of the 1980s alone. Apparently the rate of heating was increasing. The 20th was the warmest century for the past 600 years, and most of its warmest years were concentrated near its end. The pattern of high and increasing average global temperatures has continued into the 21st century.

Because of the time lag in developing the greenhouse effect, temperatures would continue to rise even if carbon dioxide amounts were stabilized at today's levels. If temperatures rise by the 1995 "best estimate" made by the international Intergovernmental Panel on Climate Change of 2° C (3.6° F) over the 21st century, the effects upon world climates could be profound. The panel, a United Nations and World Meteorological Organization group of 2000 scientists from around the world, was established in 1988 to assess the science of climate change, determine the impact of any changes on the environment and society, and formulate strategies to respond. Its "worst-case" scenario concluded that temperatures could rise by 6.3° F by 2100. That same investigative agency in 2001 warned of an even more serious set of possibilities: increases of 2.5° F probable and 10.4° F worst-case over the same time span. But the outcomes under either the earlier or later year's forecast are not clearly foreseeable; climate prediction is not an exact science.

The role of humans in global warming has been disputed. Skeptics note that nearly half the observed atmospheric warming occurred before 1940 even though almost all the increased production of carbon dioxide and other greenhouse gases came after that date. Doubters further note that every millennium since the end of the last Ice Age has had one or two centuries in which temperatures have risen by at least as much as they have in the last century. It is reasonable, they claim, to assume that recent atmospheric temperature increases are part of a natural warming cycle and have nothing to do with carbon dioxide. Further, they point out, increases in temperatures recorded at terrestrial observation stations were inevitable but misleading. Many weather stations originally sited in rural areas have, through urban expansion, now been made part of recognized city "heat islands." The result, the climate-change skeptics maintain, has been a distortion of officially recorded long-term temperature trends and a false suggestion of steadily rising temperatures.

Those arguments were countered by the Intergovernmental Panel's conclusion that the warming of the last century, and especially of the last few years, "is unlikely to be entirely due to natural causes, and . . . a pattern of climatic response to human

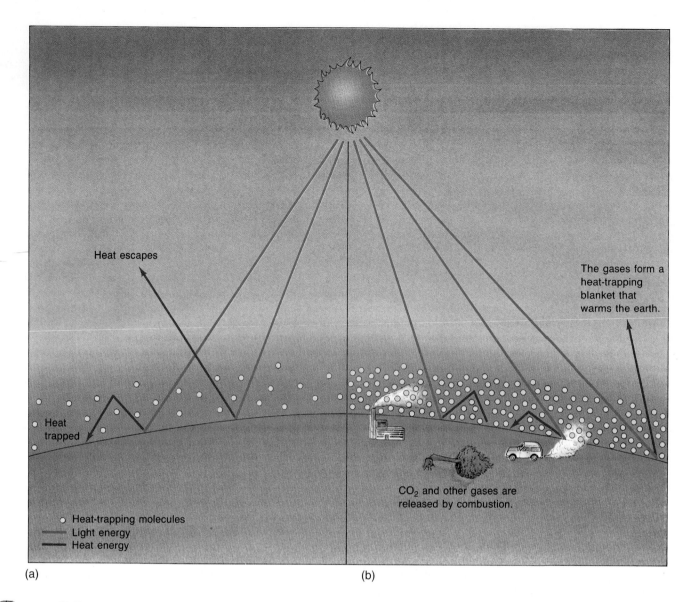

Heat escapes

The gases form a
heat-trapping
blanket that
warms the earth.

Heat
trapped

○ Heat-trapping molecules
— Light energy
— Heat energy

CO_2 and other gases are
released by combustion.

(a)   (b)

Figure 12.5 **Creating the greenhouse effect.** When the level of carbon dioxide (CO_2) in the air is low, as in (*a*), incoming solar radiation strikes the earth's surface, heating it up, and the earth radiates the energy back into space as heat. The greenhouse effect, depicted in (*b*), is the result of the more than 6 billion tons of CO_2 that the burning of fossil fuels adds to the atmosphere each year. The carbon dioxide molecules intercept some of the reradiated energy, deflecting it groundward and preventing it from escaping from the atmosphere.

activities is identifiable in the climatic record." A 2001 U.S. National Academy of Science review of available evidence reached the same conclusion: that "Greenhouse gases are accumulating in Earth's atmosphere as a result of human activities."

Whatever the attributable causes of global heating, climatologists agree on certain of its general consequences should it continue. Increases in sea temperatures would cause ocean waters to expand slightly and the polar ice caps to melt at least a bit. The reality of Arctic ice melt is already clearly evident. Between 1978 and 2000, the coverage of Arctic sea ice in winter decreased by nearly 1.2 million square kilometers (almost a half million square miles)—equivalent to an area twice the size of Texas. Further, the average thickness of Arctic ice over the same period declined by 42%, from 3.1 to 1.8 meters (10.2 to 5.9 ft), and research reports suggest the possibility of its complete summertime disappearance by the middle of the 21st century (open water was observed at the

North Pole in August, 2000). More serious consequences would result from the simultaneously observed melting of the Greenland ice sheet and the rapid retreat or total melting of glaciers throughout the world. Melting sea ice would have no effect on sea levels; water melted from continental sources, however, is added to ocean volumes. Inevitably, sea levels would rise, perhaps 0.5 to 3 meters (1.5 to 10 feet) or more within a hundred years. Even a conservative 1-meter (3-foot) rise would be enough to cover the Maldives and other low-lying island countries. The homes of between 50 and 100 million people would be inundated, a fifth of Egypt's arable land in the Nile Delta would be flooded, and the impact on the people of the Bangladesh chars would be catastrophic.

Other water problems would result from changes in precipitation patterns. Shifts in weather conditions might well increase the aridity of some of the world's already dry areas, such as Africa's Sahel, and increase rainfall in some already wet areas—

although traditionally damp England's extended period of dry weather during the middle and late 1990s has also been attributed to global warming. Almost certainly, much of the continental interiors of middle latitudes would receive less precipitation than they do now and suffer at least periodic drought, if not absolute aridity. Precipitation might decline by as much as 40% in the U.S. corn and wheat belts, drastically reducing agricultural productivity, bringing to near ruin the rural economy, and altering world patterns of food supply and trade (Figure 13.6). That same 40% reduction of rainfall would translate into significantly reduced flows of such western rivers as the Colorado, cutting back the water supply of major southwestern cities and irrigated farming districts.

Climate change may also be expressed as short-term extremes of weather. Temperature and precipitation records from 1980 through the 1990s show the incidence of extreme one-day precipitation, overall precipitation, above-normal temperatures, and drought have risen in many parts of the United States. Those increases in weather extremes—the eastern states' drought of 1999 is a recent example—are directly attributable, scientists conclude, to the increase in greenhouse gases.

On the global and long-term scales, of course, some areas would benefit from general temperature rises. Parts of Russia, Scandinavia, and Canada would get longer growing seasons; indeed, by the end of the 1990s, the growing season north of 45°N already was 12 days longer than it had been earlier in the century, and summer temperatures in Siberia were their warmest in 1000 years. In North America, crop patterns could shift northward, making the northern Great Lakes states and Canada the favored agricultural heartland climatically, though without the soil base supporting the present patterns and volumes of production.

Figure 13.6 Many climatologists noted that the parched cornfields in the U.S. Midwest during the summer of 1988 were a sample of what could be expected on a recurring basis if a worst-case scenario of global warming were to be realized. In actuality, the 1988 drought was a natural climatic fluctuation—much like the abundant rain and floods of 1993 and the varying wet and dry periods during following years—but an event whose probability of recurrence is increased by the long-term accumulation of greenhouse gases.

On average, some climatologists conclude, established middle- and upper-middle-latitude farm districts would be net beneficiaries of global warming through longer growing seasons and faster crop growth resulting from extra atmospheric carbon. Skeptics, however, remind us that a scientific rule of thumb is that a 1 degree Celsius (1.8° F) rise in temperature above the optimum reduces gain yields by 10%. Greater summer warmth, that is, might reduce, not increase, farm productivity.

Certainly, global warming would tend to reduce latitudinal differences in temperature; higher latitudes would become relatively more heated than equatorial regions. Among the consequences of that poleward shift of warmth is an already observed 80 to 100 kilometer (50 to 60 mile) poleward shift of the ranges of many animal and bird species. Similar latitudinal shifts in plant associations can be assumed, though they will be slower to materialize. Shifts in structure and distribution of ecosystems and biomes would be inevitable.

These are generally (but not completely) agreed upon scenarios of change, but local details are highly uncertain. Temperature differences are the engine driving the global circulation of winds and ocean currents and help create conditions inducing or inhibiting winter and summer precipitation and daily weather conditions. Exactly how those vital climatic details would express themselves locally and regionally is uncertain since the best of current climate models are still unable to project those details reliably at those scales. The only realistic certainty is that the patterns of climates and biomes developed since the last glaciation and shown on Figure 13.3 would be drastically altered.

Global warming and climatic change would impact most severely, of course, on developing countries highly dependent on natural, unmanaged environments for their economic support. Agriculture, hunting and gathering, forestry, and coastal fishing have that dependency, but even in those economic sectors the impact of greenhouse warming is not certain. Studies suggest that warming would reduce yields in many crops but also that the associated fertilization effect of higher carbon dioxide content would probably offset the negative impact of warming, at least for the next century. Indeed, the UN's Food and Agriculture Organization observes that global crop productivity could increase by up to 30% if the concentration of carbon dioxide doubles as they foresee over the next 50 years. But certainly, small and poor countries with great dependence on agriculture are potentially most at risk from projected climatic changes. The lower-latitude states would be most vulnerable as increased heat and higher evaporation rates would greatly stress wheat, maize, rice, and soybean crops. Most economic activities in industrialized countries do not have a close dependency on natural ecosystems. The consensus is that the impacts of climate change on diversified developed countries are likely to be small, at least over the next half-century.

Nevertheless, on the world scene, any significant continuing deviation from the present norm would at the very least disrupt existing patterns of economy, productivity, and population-supporting potential. At the worst, severe and pervasive changes could result in a total restructuring of the landscapes of culture and the balances of human–environmental relationships presently established. Nothing, from population distributions to the relative strength of countries, would ever be quite the same again. Such grim predictions were the background for major international conferences and treaty proposals of the 1990s seeking to address and limit the dangers prophesied (see "Climate Change Summits").

Despite the intuitive response that remedial efforts are needed to avert the problems foreseen, economic analyses have suggested that it would be far more expensive to radically cut carbon dioxide emissions than just to pay the costs of adjustment to the predicted temperature increases. And it is not certain that even accepting the very high costs of the Kyoto Protocol would

yield significant results. A model by one of the Climate Change Panel's lead authors, indeed, predicts that the Treaty if fully implemented would only lower the expected temperature increase of 2.1° C in 2100 to an increase of 1.9° C instead—or, that is, to postpone anticipated 2094 temperatures by just 6 years to 2100.

A cautionary note is needed. A scenario of general global warming through a human-induced greenhouse effect may be countered by an equally plausible outline of the ecological and cultural consequences of a replenished Northern Hemisphere gla-

Climate Change Summits

Accumulating evidence of global warming, projections about its long-term effect, and growing public and political determination to address its causes led during the 1990s to two high-level international conferences and treaty proposals.

The first, the "Earth Summit," was held in Rio de Janeiro in June, 1992. The Framework Convention on Climate Change signed by 166 nations called on industrialized countries to try to cap emissions of greenhouse gases at 1990 levels by the year 2000 as a necessary first step to prevent disruption of world agriculture and natural ecosystems. Small island countries, fearing their possible obliteration with rising seas, proposed even more stringent reductions.

The European Union and the United States, agreeing with the overall emissions caps proposed, based their plans for voluntary compliance on hoped-for improved energy efficiencies. Unlike the industrialized countries collectively responsible for most of the present and past production of carbon dioxide, China and other developing economies were not to be bound by any precise targets or timetables; they successfully rejected being subject to treaty provisions that would lower their economic growth prospects by limiting industrialization and the expansion of fossil fuel use such growth implies.

It became apparent in the years after the Rio summit that most advanced economies were not going to meet the voluntary greenhouse gas reductions envisioned there and that the production of such gases by developing states was increasing more rapidly than earlier projected. At subsequent lower-level conferences, notably Berlin in 1995, it was determined that a second Earth Summit was required and that the gas emission targets to be adopted there had to be made mandatory

and binding on all parties concerned. The Kyoto (Japan) Climate Change Summit of 1997 was to be the stage for those binding treaty arrangements.

The world's nations came to Kyoto with different interests and bargaining positions. The European Union, for example, proposed that industrial nations—including its own members—reduce emissions of CO_2 and other heat-trapping gases to 85% of their 1990 levels within 12 years. The United States, in contrast, proposed that emissions be reduced no lower than 1990 levels and not until some time between 2008 and 2012. The developing countries demanded that industrialized countries collectively achieve a 35% emissions reduction by 2020. The Kyoto Protocol, the result of 10 days of intense discussion and bargaining, represents compromises among the various extreme positions originally held and established at least an initial institutional framework and mechanism for addressing the global warming problem in future years.

By the variable goals it set, the adopted climate accord acknowledged the diversity of concerns among and between developed and developing economies. Thirty-eight industrial nations were required collectively to reduce greenhouse emissions by an annual average of 5.2% below 1990 levels from 2008 through 2012—a 30% reduction below what they otherwise would likely be. The actual targets differ among them, however. The European Union's goal was set at an 8% reduction, that of the United States at 7%, and Japan's at 6%. Some industrial states would have smaller reductions and a few would not face any cuts immediately. No specific goals were set on developing countries—though as a group they were asked to set voluntary reduction quotas—and no enforcement provisions for developed country compliance were agreed on.

The basic accord adopted at the Kyoto conference needs ratification by at least 55 countries representing over 55% of 1990 carbon dioxide emissions to take effect. However, the accepted rules of ratification assure that the Protocol cannot be binding if the United States, responsible for nearly 23% of world carbon emissions in the mid-1990s, does not approve it. That approval was effectively denied in March, 2001, when President George W. Bush made clear his view that the Protocol was "fatally flawed," unequal in the obligations it placed on developed and developing states, and potentially unacceptably damaging to the U.S. economy. Other industrialized countries and the world community in general expressed dismay at the American position and in follow-up meetings made plans to proceed with the agreement without U.S. participation if necessary. A November, 2001 conference in Morocco to finalize details of the Protocol, however, accepted compromises and escape clauses that, critics complained, weakened the Kyoto agreement and permitted countries to ratify the treaty without enacting language to make it legally binding.

By late 2002, international scientific and political opinion began to reassess the value of the Kyoto Protocols in light of new studies indicating that even full compliance with them would reduce worst case estimates of global warming by only an insignificant 0.06° Celsius by 2050. In late 2002 at meetings in both Johannesburg, South Africa, and New Delhi, India, summary conference declarations advocated that the global community should realistically focus its attention on adapting to inevitable climate change since natural forces already in motion meant inevitable continued climate warming no matter what now was done to ratify and enforce the Kyoto Protocol.

cial cover. The onset of a new ice age is predicted on solid climatic reasoning that greenhouse-related temperature increases would likely warm upper latitudes sufficiently to permit heavy snowfall to commence in regions—like northern Greenland—where it is now so cold that snow rarely falls. That increase in upper-latitude precipitation—already being recorded in Antarctica—would yield a dramatic influx of fresh water and accumulation of ice subject to melting in polar oceans. That influx would dilute the salt content of ocean currents, making them less dense. Density conferred by salinity affects both current flow and capacity to transport heat between lower latitudes and, particularly, the North Atlantic. When the globe-girdling heat exchange of ocean currents is interrupted by a lowered salt content, polar temperatures can drop abruptly, and significant changes in world climatic patterns can occur. Recent evidence, it is claimed, indicates that polar regions did, in fact, grow slightly warmer before the onset of the last glacial period, just as the greenhouse effect is warming them now. Further, detailed ocean sediment cores provide strong evidence of a recurring 1500-year cycle of alternating cold and warm spells—of global warmings and "little ice ages"—dating back at least 32,000 years and totally divorced from human-caused greenhouse gas accumulations.

One of the greatest concerns about global warming, therefore, is that it would shift Atlantic Ocean currents now warming northern Europe with a possible 11° C (20° F) drop in temperature within a 10-year span. Should that sequence of climatic changes occur rather than the one predicted under the usual global warming hypothesis, an equally profound, but quite different, set of human consequences would ensue—one still triggered by the greenhouse effect, but with a new ice age outcome.

Air Pollution and Acid Rain

Every day thousands of tons of pollutants are discharged into the air by natural events and human actions. Air is polluted when it contains alien substances in sufficient amounts and concentrations to have a harmful effect on living things and human-made objects. Truly clean air has never existed, for atmospheric pollution can and does result in nature from ash from volcanic eruptions, marsh gases, smoke from naturally occurring forest fires, and wind-blow dust. Normally these pollutants are of low volume, are widely dispersed in the atmosphere, and have no significant long-term effect on air quality.

Far more damaging are the substances discharged into the atmosphere by human actions. These pollutants come primarily from burning fossil fuels—coal, oil, and gas—in power plants, factories, furnaces, and vehicles, and from fires deliberately set to clear forests and grasslands for agricultural expansion or swidden garden plot clearing and renewal. Air pollution is a global problem; areas far from the polluting activity may be adversely impacted as atmospheric circulation moves pollutants freely without regard to political or other earth-based boundaries. Where burning for farmland improvement or expansion combines with rapidly expanding urban and industrial development—increasingly common in large areas of the developing world—unbelievable widespread atmospheric contamination results. For example, current full-color satellite cameras regularly reveal a nearly continuous, 2-mile-thick blanket of soot, organic compounds, dust, ash, and other air debris stretching across much of India, Bangladesh, and Southeast Asia, reaching northward to the industrial heart of China. The pollution shroud in and around India, researchers find, reduces sunlight there by 10%, enough to cut rice yields by 3% to 10% across much of the country. The World Health Organization (WHO) reports that of India's 23 cities of more than 1 million people, not one meets the organization's air pollution standards. In southern China and Southeast Asia, as many as 1.4 million people die annually from respiratory ills caused by human-induced pollution that drifts across the Pacific to the Western Hemisphere and beyond.

In addition to the very serious human health consequences of air pollution, the interaction of pollutants with each other or with natural atmospheric constituents such as water vapor may create derivative pollutants highly damaging to vegetation, surface and groundwater, and structures. Among these secondary agents is the acid rain that is the second of the recent trio of environmental concerns.

Unexpectedly, acid precipitation is a condition in part traceable to actions taken in developed countries in past decades to alleviate the smoke and soot that poured into the skies from the chimneys of their power plants, mills, and factories. The urban smoke abatement and clean air programs demanded by environmentalists and the general public reacting to growing health and property damage concerns usually incorporated prohibitions against the discharge of atmospheric pollutants damaging to areas near the discharge point. The response was to raise chimneys to such a height that smoke, soot, and gases were carried far from their origin points by higher elevation winds (Figure 13.7).

But when power plants, smelters, and factories were fitted with tall smokestacks to free local areas from pollution, the sulfur dioxide and nitrogen oxides in the smoke instead of being deposited locally were pumped high into the atmosphere. There they mixed with water and other chemicals and turned into sulfuric and nitric acid that was carried to distant areas. They were joined in their impact by other sources of acid gases. Motor vehicles are particularly prolific producers of nitrogen oxides in their exhausts, and volcanos can add immense amounts of acidic gases, as the Tambora eruption demonstrated.

When acids from all sources are washed out of the air by rain, snow, or fog the result is **acid rain,** though *acid precipitation* is a more precise designation. Acidity levels are described by the *pH factor,* the measure of acidity/alkalinity on a scale of 0 to 14. The average pH of normal rainfall is 5.6, slightly acidic, but acid rainfalls with a pH of 2.4—approximately the acidity of vinegar and lemon juice—have been recorded. Primarily occurring in developed nations, acid rain has become a serious problem in many parts of Europe, North America, and Japan. It expresses itself in several forms, though the most visible are its corrosive effects on marble and limestone sculptures and buildings and on metals such as iron and bronze (Figure 13.8) and in the destruction of forests. Trees at higher elevations are particularly susceptible, with widespread forest loss clearly apparent on the hillsides and mountain tops of New England, Scandinavia, and Germany, where acid rain had apparently degraded much of that country's famous forests by the early 1990s.

Figure 13.7 Before concern with acid rain became widespread, the U.S. Clean Air Act of 1970 set standards for ground-level air quality that could be met most easily by building smokestacks high enough to discharge pollutants into the upper atmosphere. Stacks 300 meters (1000 feet) and more high became a common sight at utility plants and factories, far exceeding the earlier norm of 60–90 meters (200–300 feet). What helped cleanse one area of pollution greatly increased damage elsewhere. The farther and higher the noxious emissions go, the longer they have to combine with other atmospheric components and moisture to form acids. Thus, the taller stacks directly aggravated the acid rain problem. Recognizing this, the Environmental Protection Agency in 1985 issued rules discouraging the use of tall smokestacks to disperse emissions. The Clean Air Act Amendments of 1990 required that sulfur dioxide and nitrogen oxide emissions from smokestacks be cut in half.

Damage to lakes, fish, and soils is less immediately evident, but more widespread and equally serious. Acid rain has been linked to the disappearance of fish in thousands of streams and lakes in New England, Canada, and Scandinavia, and to a decline in fish populations elsewhere. It leaches toxic constituents such as aluminum salts from the soil and kills soil microorganisms that break down organic matter and recycle nutrients through the ecosystem.

Acid deposition can harm and decrease yields of many food crops and increase the content of poisonous heavy metals in drinking water supplies. The culprit acids are borne in the atmosphere and so may wreak their injury far from the power plants or cities (or volcanos) that put them in the air (Figure 13.9). In North America, midwestern coal-burning power stations and industries are blamed for acid rain contamination in New England. They, along with other U.S. industries, power plants, and automobiles, are credited with the widespread acid rain damage in Canada (including some 2000 lakes totally dead to fish life and another 150,000 in danger). Canada itself has major urban and industrial pollution sources contributing to its acid rain incidence.

The Trouble with Ozone

The forest damage usually blamed exclusively on acid rain has, on closer investigation, proved to be at least partially the product of ozone poisoning. **Ozone** is a molecule consisting of three oxygen atoms rather than the two of normal oxygen. Sunlight produces it from standard oxygen, and a continuous but thin layer of ozone accumulates at upper levels in the atmosphere. There it blocks the cancer-causing ultraviolet (UV) light that damages DNA, the molecule of heredity and cell control. That upper atmospheric shield now appears in danger of destruction by chemicals released into the air by humans (see "Depleting the Ozone Layer").

At lower levels, however, the problem is accumulation, not depletion, of ozone. Relatively harmless to humans, ozone is injurious to plants. Exposed to too much of it, their growth may be stunted, their yields reduced (by as much as 30% for wheat), or they may even die. That, apparently, is an important contributor to forest damage and destruction commonly attributed to acid rain. In the lower atmosphere, ozone is produced in *photochemical smogs* by sunlight and pollution, with the main pollutant being motor vehicle exhaust fumes (Figure 13.10). Their nitrogen oxides and hydrocarbons are particularly good at converting oxygen to ozone. The resulting smog, unlike the ozone alone, has serious adverse consequences for human respiratory health. The increasing use of automobiles in Europe, not acid rain, has done the harm to that continent's forests, a fact that explains the rise of forest destruction during the same years that sulfur dioxide emissions from power plants were being significantly reduced.

There is an element of presumed irreversibility in both the greenhouse effect and ozone depletion. Once the processes creating them are launched, they tend to become cumulative and continuous. Even if carbon dioxide levels stayed as they are now, temperatures would continue to climb. Even if all CFCs were immediately banned and no more were released into the atmosphere, it would take more than a century to replenish the ozone already lost. Since population growth, industrial development, and chemical pollution will continue—though perhaps under tighter control—assaults upon the atmosphere will also continue rather than cease. The same disquieting irreversibility seems to characterize three other processes of environmental degradation: *tropical deforestation; desertification* of cropland, grazing areas, and deforested lands; and air, land, and water *pollution*. Each stands alone as an identified problem of global concern, and each is a component part of cumulative human pressures upon the biosphere greater than its recuperative powers can handle.

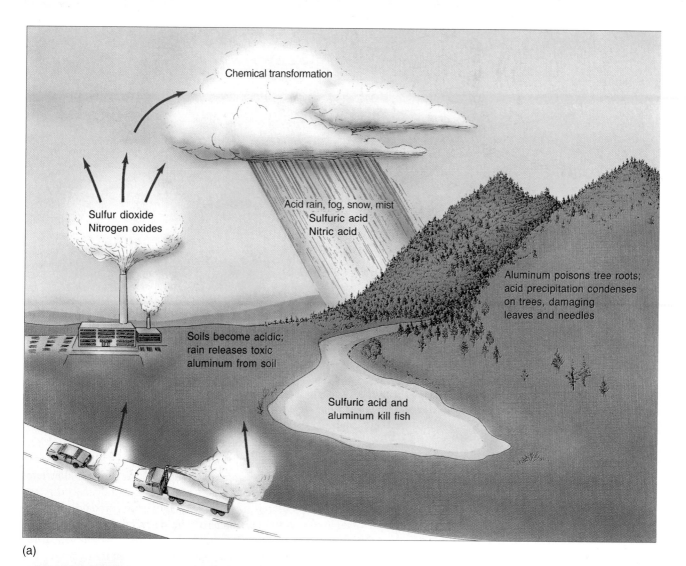

Chemical transformation

Sulfur dioxide
Nitrogen oxides

Acid rain, fog, snow, mist
Sulfuric acid
Nitric acid

Aluminum poisons tree roots;
acid precipitation condenses
on trees, damaging
leaves and needles

Soils become acidic;
rain releases toxic
aluminum from soil

Sulfuric acid and
aluminum kill fish

(a)

(b)

Figure 12.8 **The formation and effects of acid precipitation.** (*a*) Sulfur dioxide and nitrogen oxides produced by the combustion of fossil fuels are transformed into sulfate and nitrate particles; when the particles react with water vapor, they form sulfuric and nitric acids, which then fall to earth. (*b*) The destructive effect of acid rain is evident on this limestone statuary at the cathedral in Reims, France.

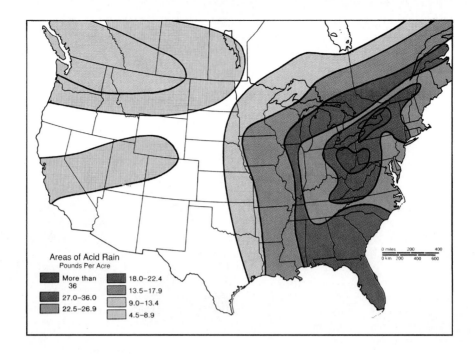

Figure 13.9 **Where acid rain falls.** In general, the areas that receive the most acid rain in the eastern United States and Canada are those that are least able to tolerate it. Their surface waters tend to be acidic rather than alkaline and are unable to neutralize the acids deposited by rain or snow. Ironically, the highly industrialized Ohio Valley and southern Great Lakes districts have high natural resistance to soil and water acidification.

Data from Canadian government.

Land Use and Land Cover

Human-induced alterations in land use and vegetative cover affect the radiation balance of the earth and, therefore, contribute to climatic change. Since the beginning of the 19th century, vast portions of the earth's surface have been modified, whole ecosystems destroyed, and global biomes altered or eliminated. North American and European native forests have largely vanished; the grasslands of interior United States, Canada, and Ukraine have been converted into farmland. Marshes and wetlands have been drained, dams built, and major water impoundments created. Steppe lands have become deserts; deserts have blossomed under irrigation.

At least locally, every such change alters surface reflectivity for solar radiation, land and air temperature conditions, and water balances. In turn, these changes in surface conditions affect the climates of both the local area and of areas downwind. On a global basis their cumulative impact is less clear, but certain generalizations are agreed upon. The generation of methane gas, an important contributor to the greenhouse effect, is almost certainly reduced by drainage of swamps, but it is also greatly increased as a by-product of expanding paddy rice production and of growing herds of cattle on pastureland. Heavy applications of nitrogen fertilizers are thought to be important in increasing the nitrous oxide content of the atmosphere. When they enter streams and lakes through farm runoff, the fertilizers encourage the algal growth that alters water surface reflectivity and evaporation rates.

But unquestionably the most important of the land surface changes has been that of clearing of forests and plowing of grasslands. Both effect drastic environmental changes that alter temperature conditions and water balances and release—through vegetative decomposition—large quantities of carbon dioxide and other gases to the atmosphere. The destruction of both biomes represents, as well, the loss of major "sinks" that extract carbon dioxide from the atmosphere and hold it in plant tissue.

Tropical Deforestation

Forests, we saw in Chapter 8, still cover some 30% of the earth's land surface (see Figure 8.30), though the forest biomes have suffered mightily as human pressures on them have increased. Forest clearing accompanied the development of agriculture and spread of people throughout Europe, Central Asia, the Middle East, and India. European colonization had much the same impact on the temperate forests of eastern North America and Australasia. In most midlatitude developed countries, although original forest cover is largely gone, replanting and reversion of cropland to timber has tended to replenish woodlands at about their rate of cutting.

Now it is the tropical rain forest—also known as the tropical moist forest—biome that is feeling the pressure of growing population numbers, the need for more agricultural land, expanded demand for fuel and commercial wood, and a midlatitude market for beef that can be satisfied profitably by replacing tropical forest with cleared grazing land. These disappearing forests—covering no more than 6% of the planet's land surface—extend across parts of Asia, Africa, and Latin America, and are the world's most diverse and least understood biome. About 45% of their original expanse has already been cleared or degraded. Africa has

In the summer of 1986, scientists for the first time verified that a "hole" had formed in the ozone layer over Antarctica. In fact, the ozone was not entirely absent, but it had been reduced from earlier recorded levels by some 40%. As a result, Antarctic life—particularly the microscopic ocean plants (phytoplankton) at the base of the food chain—that had lived more or less in ultraviolet (UV) darkness was suddenly getting a trillionfold (1 followed by 12 zeros) increase above the natural rate of UV receipt.

The ozone hole typically occurs over Antarctica during late August through early October and breaks up in mid-November. From 1987 when the ozone loss was 60% to 2000 when more than 85% of ozone in the lower stratosphere was destroyed, the hole grew larger, lasted longer, and spread farther outward each year toward South America and Australia. The NASA satellite image shows the pattern of ozone depletion during September, 2000, when the gap at its largest-ever extent measured 44.3 million square kilometers (17.1 million sq mi), an area nearly as large as all of Asia (44.8 million sq km) and an increase of nearly 4 million square kilometers (1.5 million sq mi) over its 1999 extent. The color scale below the image shows the total ozone levels; the dark blues and purples indicate the areas of greatest ozone depletion.

Most observers attribute the ozone decline to pollution from human-made chemicals, particularly *chlorofluorocarbons* (CFCs) used as coolants, cleansing agents, propellants for aerosols, and in insulating foams. In a chain reaction of oxygen destruction, each of the chlorine atoms released can over time destroy upwards of 10,000 ozone molecules. Ozone reduction is an increasing and spreading atmospheric problem. A similar ozone hole about the size of Greenland opens in the Arctic, too, and the ozone shield over the midlatitudes has dropped significantly since 1978.

Why should the hole in the ozone layer have appeared first so prominently over Antarctica? In most parts of the world, horizontal winds tend to keep chemicals in the air well mixed. But circulation patterns are such that the freezing whirlpool of air over the south polar continent in winter is not penetrated by air currents from warmer earth regions. In the absence of sunlight and atmospheric mixing, the CFCs work to destroy the ozone. During the Southern Hemisphere summer, sunlight works to replenish it. A different scenario exists in the Northern Hemisphere: one of accumulating greenhouse gases causing increased upper atmosphere cooling and further ozone loss over the North Pole. In either Hemisphere, ozone depletion has identical adverse effects. Among other things, increased exposure to UV radiation increases the incidence of skin cancer and, by suppressing bodily defense mechanisms, increases risk from a variety of infectious diseases. Many crop plants are sensitive to UV radiation, and the very existence of the microscopic plankton at the base of the marine food chain is threatened by it.

Some scientists dispute both the existence and the cause of ozone layer depletion and claim "there is no observational evidence that man-made chemicals like CFCs are dangerously thinning the ozone layer. . . ." Nevertheless, production and use of CFCs is being phased out under the Montreal Protocol, a 1987 international agreement made effective in 1992, that requires production of CFCs be ended in developed countries after 1995 and to cease in developing states by 2010. Because of those restrictions, ozone depletion is being slowed and even reversed.

In striking contrast to its 2000 maximum, satellite observation of the hole in 2002 showed it at its smallest extent (15.5 million sq km, or 6 million sq mi) since 1988 and split in two. Although federal scientists concluded the reduced size and split of the hole were due to an unusual single year confluence of events, recent research suggests that the peak of Southern Hemisphere ozone loss will likely occur between 2000 and 2008. The layer should then begin to mend and return to normal around 2050 in the Antarctic.

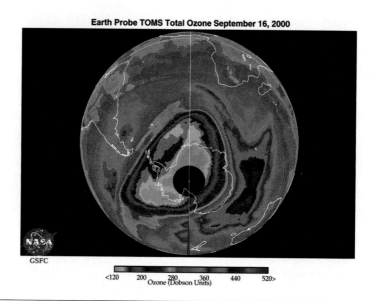

Earth Probe TOMS Total Ozone September 16, 2000

NASA
GSFC

<120   200   280   360   440   520>
Ozone (Dobson Units)

lost more than half of its original rain forest; nearly half of Asia's is gone; 70% of the moist forests of Central America and some 40% of those of South America have disappeared. Every year additional thousands of square kilometers are lost, though recent satellite surveys indicate the overall rate of tropical forest cutting is not as great as estimates of the 1980s and 1990s feared. Even so, FAO data of 2001 suggest tropical deforestation still exceeds 130,000 square kilometers (50,000 sq mi) annually. Tropical forest removal raises three principal global concerns and a host of local ones.

First, on a worldwide basis, all forests play a major role in maintaining the oxygen and carbon balance of the earth. This is particularly true of tropical forests because of their total area and volume. Humans and their industries consume oxygen; vegetation

(a)

(b)

Figure 13.10 (*a*) Photochemical smog in sunny California during the late 1970s. When air remains stagnant over Los Angeles, it can accumulate increasing amounts of automobile and industrial exhausts, reducing afternoon sunlight to a dull haze and sharply lifting ozone levels. Such occurrences are increasingly rare in Los Angeles—where peak levels of ozone have dropped to a quarter of their 1955 levels—and in other major American cities with past serious smog and ozone dangers. Mandates of the Clean Air Act and, particularly, more stringent restrictions on automobile emissions assure continued improvements in metropolitan air quality. Europe is only beginning the same kinds of protection, and summer ozone levels in such cities as Paris today are triple the worst Los Angeles readings. (*b*) The Germans call it *Waldsterben*—forest death—a term now used more widely to summarize the destruction of trees by a combination of ozone; heavy metals; and acidity in clouds, rain, snow, and dust. It first strikes at higher elevations where natural stresses are greatest and acidic clouds most prevalent, but it slowly moves downslope until entire forests are gone. Here at Mount Mitchell in North Carolina, *Waldsterben* is thought to result from pollution traveling eastward from the Ohio and Tennessee valleys. Forests throughout eastern Anglo America from Georgia northward into Canada display evidence of similar pollution-related damage as do forests in the Front Range of Colorado and the San Gabriel Mountains near Los Angeles. Similar impacts are increasingly seen in Europe.

replenishes it through photosynthesis and the release of oxygen back into the atmosphere as a by-product. At the same time, plants extract the carbon from atmospheric carbon dioxide, acting as natural retaining sponges for the gas so important in the greenhouse effect. Each year, each hectare (2.47 acres) of Amazon rain forest absorbs a ton of carbon dioxide. When the tropical rain forest is cleared, not only is its role as a carbon sink lost but the act of destruction itself through decomposition or burning releases as carbon dioxide the vast quantities of carbon the forest had stored.

A second global concern is also climate related. Forest destruction changes surface and air temperatures, moisture content, and reflectivity. Conversion of forest to grassland, for example, increases surface temperature, raises air temperatures above the treeless ground, and therefore increases the water-holding capacity of the warmer air. As winds move the hotter, drier air, it tends to exert a drying effect on adjacent forest and agricultural lands. Trees and crops outside the denuded area experience heat and aridity stresses not normal to their geographical locations. It is

calculated that cutting the forests of South America on a wide scale should raise regional temperatures from 3° C to 5° C (5.5–9° F), which in turn would extend the dry season and greatly disrupt not only regional but global climates.

In some ways, the most serious long-term global consequence of the eradication of tropical rain forests will be the loss of a major part of the biological diversity of the planet. Of the estimated 5–10 million plant and animal species believed to exist on earth, a minimum of 40% to 50%—and possibly 70% or more—are native to the tropical rain forest biome. Many of the plants have become important world staple food crops: rice, millet, cassava, yam, taro, banana, coconut, pineapple, and sugarcane to name but a very few well-known ones. Unknown additional potential food species remain as yet unexploited. Reports from Indonesia suggest that in that country's forests alone, some 4000 plant species have proved useful to native peoples as foodstuffs of one sort or another, though less than one-tenth have come into wide use. The rain forests are, in addition, the world's main storehouse of drug-yielding plants and insects, including thousands with proven or prospective anticancer properties and many widely used as sources of antibiotics, antivirals, analgesics, tranquilizers, diuretics, and laxatives, among a host of other items (see "Tropical Forests and Medical Resources"). The loss of the zoological and botanical storehouse that the rain forests represent would deprive humans of untold potential benefits that might never be realized.

On a more local basis, tropical forests play for their inhabitants and neighbors the same role taken by forests everywhere. They protect watersheds and regulate water flow. After forest cutting, unregulated flow accentuates the problems of high and low water variations, increases the severity of valley flooding, and makes more serious and prolonged the impact of low water flow on irrigation agriculture, navigation, and urban and rural water supply. Accelerated **soil erosion**—the process of removal of soil particles from the ecosystem, usually by wind or running water—quickly removes the always thin, infertile tropical forest soils from deforested areas. Lands cleared for agriculture almost immediately become unsuitable for that use partially because of soil loss (Figure 13.11). The surface material removed is transported and deposited downstream, changing valley contours, extending the area subject to flooding, and filling irrigation and drainage channels. Or it may be deposited in the reservoirs behind the increasing number of major dams on rivers within the tropical rain forests or rising there (see "Dam Trouble in the Tropics").

Desertification

The tropical rain forests can succumb to deliberate massive human assaults and be irretrievably lost. With much less effort, and with no intent to destroy or alter the environment, humans

Figure 13.11 Wholesale destruction of the tropical rain forest guarantees environmental degradation so severe that the forest can never naturally regenerate itself. Exposed soils quickly deteriorate in structure and fertility and are easily eroded, as this growing gully in Amazonia clearly shows.

Tropical Forests and Medical Resources

Tropical forests are biological cornucopias, possessing a stunning array of plant and animal life. Costa Rica, about the size of South Carolina, contains as many bird species as all of North America, more species of insects, and nearly half the number of plant species. One stand of rain forest in Kalimantan (Borneo) contains more than 700 species of tree, as many as exist in North America, and half a square kilometer of Malaysia's forest can feature as many tree and shrub species as the whole of the United States and Canada. Forty-three species of ant inhabit a single tree variety in Peru, dependent on it for food and shelter and providing in return protection from other insects.

The tropical forests yield an abundance of chemical products used to manufacture alkaloids, steroids, anesthetics, and other medicinal agents. Indeed, one quarter to one half of all modern drugs, including strychnine, quinine, curare, and ipecac, come from the tropical forests. A single flower, the Madagascar periwinkle, produces two drugs used to treat leukemia and Hodgkin's disease.

As significant as these and other modern drugs derived from tropical plants are, scientists believe that the medical potential of the tropical forests remains virtually untapped. They fear that deforestation will eradicate medicinal plants and traditional formulas before their uses become known, depriving humans of untold potential benefits that may never be realized. Indigenous peoples make free use of plants of the rain forest for such purposes as treating stings and snakebites, relieving burns and skin fungi, reducing fevers, and curing earaches. Yet botanists have only recently begun to identify tropical plants and study traditional herbal medicines to discover which plants might contain pharmaceutically important compounds.

A second concern is that forest destruction will create shortages of drugs already derived from those plants. Reportedly, as many as 60,000 plants with valuable medical properties are likely to become extinct by 2050. Already endangered is reserpine, an ingredient in certain tranquilizers that is derived from the *Rauwolfia serpentina* plant found in India. Also threatened are cinchona, whose bark produces quinine, and foxglove varieties that are used to make the heart medications digitoxin and acetyldigitoxin.

Dam Trouble in the Tropics

The great tropical river systems have a sizeable percentage of the world's undeveloped power potential. The lure of that power and its promise for economic development and national modernization have proved nearly irresistible. But the tropical rain forests have been a particularly difficult environment for dam builders. The dams (and their reservoirs) often carry a heavy ecological price, and the clearing and development of the areas they are meant to serve often assure the destruction of the dam projects themselves.

The creation of Brokopondo in Suriname in 1964 marked the first large reservoir in a rain forest locale. Without being cleared of their potentially valuable timber, 1480 square kilometers (570 sq mi) of dense tropical forest disappeared underwater. As the trees decomposed, producing hydrogen sulfide, an intolerable stench polluted the atmosphere for scores of miles downwind. For more than

2 years, employees at the dam wore gas masks at work. Decomposition of vegetation produced acids that corroded the dam's cooling system, leading to costly continuing repairs and upkeep.

Water hyacinth spreads rapidly in tropical impoundments, its growth hastened by the rich nutrients released by tree decomposition. Within a year of the reservoir's completion, a 130-square-kilometer (50-sq-mi) blanket of the weed was afloat on Lake Brokopondo, and after another year almost half the reservoir was covered. Another 440 square kilometers (170 sq mi) were claimed by a floating fern, *Ceratopteris*. Identical problems plague most rain forest hydropower projects.

The expense, the disruption of the lives of valley residents whose homes are to be flooded, and the environmental damage of dam projects in the rain forest all may be in vain. Deforestation of river banks and clear-

ing of vegetation for permanent agriculture usually results in accelerated erosion, rapid sedimentation of reservoirs, and drastic reduction of electrical generating capacity. The Ambuklao Reservoir in the Philippines, built with an expected payback period of 60 years, now appears certain to silt up in half that time. The Anchicaya Reservoir in Colombia lost 25% of its storage capacity only 2 years after it was completed and was almost totally filled with silt within 10 years. The Peligre Dam in Haiti was completed in 1956 with a life expectancy of at least 50 years; siltation reduced its usefulness by some 15 years. El Cajon Dam in Honduras, Arenal in Costa Rica, Chixoy in Guatemala and many others—all built to last decades or even centuries—have, because of premature siltation, failed to repay their costs or fulfill their promise. The price of deforestation in wet tropics is high indeed.

are assumed to be similarly affecting the arid and semiarid regions of the world. The process is called **desertification,** the expansion or intensification of areas of degraded or destroyed soil and vegetation cover. While the Earth Summit of 1992 defined desertification broadly as "land degradation in arid, semiarid and dry subhumid areas, resulting from climatic variations and human activities," the process is often charged to increasing human pressures exerted through overgrazing, deforestation for fuel wood, clearing of original vegetation for cultivation, and burning, and implies a continuum of ecological alteration from slight to extreme (Figure 13.12).

Both satellite measurements and paleoclimatological studies, however, are forcing a reassessment of what is really happening to arid and semiarid drylands along the perimeter of the Sahara and other major world deserts. Imagery dating from 1980 indicates that the Sahel drylands region on the southern border of the Sahara, for example, did not move steadily south as usually assumed. Rather, the vegetation line fluctuated back and forth in response to variable rainfall patterns, with year to year shifts ranging between 30 and 150 miles. Since Africa's drylands climate has shifted back and forth between periods of extended drought and higher rainfall for at least 10,000 years, many scientists now believe climate variation keeps the drylands in a continual state of disequilibrium. That, rather than human abuse of the land, is thought to be the major influence on dryland ecology and the shifting margins of deserts.

Certainly much past periodic desertification of vast areas has been induced by nature rather than by humans. Over the past 10,000 years, for example, several prolonged and severe droughts

far more damaging than the "Dust Bowl" period of the 1930s converted vast stretches of the Great Plains from Texas and New Mexico to Nebraska and South Dakota into seas of windblown sand dunes like those of the Sahara. Such conditions were seen most recently in the 18th and 19th centuries, before the region was heavily settled, but after many explorers and travelers noted—as did one in 1796 in present-day Nebraska—"a great desert of drifting sand, without trees, soil, rock, water or animals of any kind." Today, those same areas are covered only thinly by vegetation and could revert to shifting desert—as they almost did in the 1930s—with a prolonged drought of the type that might accompany global warming. The U.S. Bureau of Land Management's warning that almost half of the Great Plains is prone to desertification was underscored by the prolonged drought of the late 1990s and early 21st century that, particularly in the northern plains, was the most severe in more than a century.

Whatever its degree of development, when desertification results from human rather than climatic causes, it begins in the same fashion: the disruption or removal of the native cover of grasses and shrubs through farming or overgrazing (Figure 13.13). If the disruption is severe enough, the original vegetation cannot reestablish itself and the exposed soil is made susceptible to erosion during the brief, heavy rains that dominate precipitation patterns in semiarid regions. Water runs off the land surface instead of seeping in, carrying soil particles with it and leaving behind an *erosion pavement*. When the water is lost through surface flow rather than seepage downward, the water table is lowered. Eventually, even deep-rooted bushes are unable to reach groundwater, and all natural vegetation is lost. The

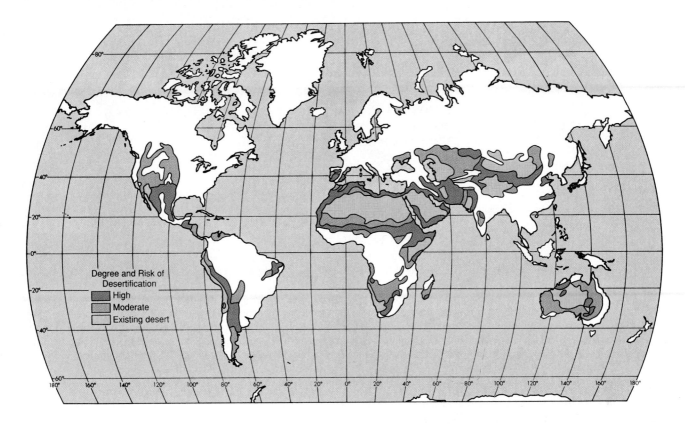

𝓕𝑖𝑔𝑢𝑟𝑒 13.12 **Desertification** is usually understood to imply the steady advance of the margins of the world's deserts into their bordering dry-lands, converting through human mistreatment formerly productive or usable pastures and croplands into barren and sterile landscapes. In reality, the process may result from natural climatic fluctuations as much as from human abuse; local and regional variations in those two causal conditions are reflected in the reversal or stabilization of desertification in some areas. Because of different criteria, areas shown as "desert" here are not identical to desert regions of Figure 13.3. See also Figure 13.15.

Sources: Based on H. E. Dregne, Desertification of Arid Lands, *Figure 1.2, copyright © 1983 Harwood Academic Publishers; and* A World Map of Desertification, *UNESCO/FAO.*

process is accentuated when too many grazing animals pack the earth down with their hooves, blocking the passage of air and water through the soil. When both plant cover and soil moisture are lost, desertification has occurred.

It happens with increasing frequency in many areas of the earth as pressures on the land continue. Worldwide, desertification affects about 1 billion people in 110 countries and impacts about 1.2 billion hectares—about the size of China and India combined. According to the United Nations, between one-quarter and one-third of the planet's land surface now qualifies as degraded semidesert. Africa is most at risk; the United Nations has estimated that 40% of that continent's nondesert land is in danger of human-induced desertification. But nearly a fifth of Latin America's lands and a third of Asia's are similarly endangered. China's Environmental Protection Agency reports, for example, that the Gobi desert of that country grew by 52,000 square kilometers (20,000 sq mi) from 1994 to 1999.

In countries where desertification is particularly extensive and severe (Algeria, Ethiopia, Iraq, Jordan, Lebanon, Mali, and Niger) per capita food production declined by nearly half between 1950 and the late 1990s. The resulting threat of starvation spurs populations of the affected areas to increase their farming and livestock pressures on the denuded land, further contributing to their desertification. It has been suggested that Mali may be the first country in the world rendered uninhabitable by environmental destruction. Many of its more than 11 million inhabitants begin their day by shoveling their doorways clear of the night's accumulation of sand (Figure 13.14).

Soil Erosion

Desertification is but one expression of land deterioration leading to accelerated soil erosion, a worldwide problem of biosphere deterioration. Over much of the earth's surface, the thin layer of topsoil upon which life depends is only a few inches deep, usually less than 30 centimeters (1 ft). Below it, the lithosphere is as lifeless as the surface of the moon. A **soil** is a complex mixture of rock particles, inorganic mineral matter, organic material, living organisms, air, and water. Under natural conditions, soil is constantly being formed by the physical and chemical decomposition of rock material and by the decay of organic matter. It is simultaneously being eroded, for soil erosion is as natural a process as soil formation and occurs even when land is totally covered by forests or grass. Under most natural conditions, however, the rate of soil formation equals or exceeds the rate of soil erosion, so soil depth and fertility tend to increase with time.

When land is cleared and planted to crops or when the vegetative cover is broken by overgrazing, deforestation, or other

Figure 13.13   The margin of the desert. Intensive grazing pressure destroys vegetation, compacts soil, and leads to soil degradation and desertification, as this desert-margin view from Burkina Faso suggests. Elsewhere, the Argentine government reports a vast area in the southern part of that country has become part of an expanding desert mainly due to human activities that have degraded the lands to the point that "their use to man is practically nil" and the damage is "economically irreversible." Similarly, China reports its vast Gobi Desert is encroaching into northern crop and grazing lands at a rate of 2500 square kilometers (950 sq mi) per year as a result of overgrazing and "excessive gathering of firewood."

Figure 13.14   Windblown dust is engulfing the scrub forest in this drought-stricken area of Mali, near Timbuktu. The district is part of the Sahel region of Africa where desertification has been accelerated by both climate and human pressures on the land. From the late 1930s to 2000, some 650,000 square kilometers (250,000 sq mi) were added to the southern Sahara. It has expanded on its northern and eastern margins as well. On an annual basis, marginal fluctuation rather than steady expansion is the rule, and some scientists prefer to speak of an "ebb and flow" of the Sahara margins and of land degradation rather than of permanent conversion to true desert.

disturbances, the process of erosion inevitably accelerates. When its rate exceeds that of soil formation, the life-sustaining veneer of topsoil becomes thinner and eventually disappears, leaving behind only sterile subsoil or barren rock. At that point the renewable soil resource has been converted through human impact into a nonrenewable and dissipated asset. Carried to the extreme of bare rock hillsides or wind-denuded plains, erosion spells the total end of agricultural use of the land. Throughout history, such extreme human-induced destruction has occurred and been observed with dismay.

Any massive destruction of the soil resource could spell the end of the civilization it had supported. For the most part, however, farmers—even those in difficult climatic and topographic circumstances—devised ingenious ways to preserve and even improve the soil resource on which their lives and livelihoods depended. Particularly when farming was carried on outside of fertile, level valley lands, farmers' practices were routinely based on some combination of crop rotation, fallowing, and terracing.

Rotation involves the planting of two or more crops simultaneously or successively on the same area to preserve fertility or to provide a plant cover to protect the soil. **Fallowing** leaves a field idle (uncropped) for 1 or more years to achieve one of two outcomes. In semiarid areas the purpose is to accumulate soil moisture from one year to apply to the next year's crop; in tropical wet regions, as we saw in Chapter 8, the purpose is to renew soil fertility of the swidden plot. **Terracing** (see Figure 4.24) replaces steep slopes with a series of narrow layered, level fields, providing cropland where little or none existed previously. In addition, because water moving rapidly down-slope has great erosive power, breaking the speed of flow by terracing reduces the amount of soil lost. Field trials in Nigeria indicate that cultivation on a 1% slope (a drop of 1 foot in elevation over 100 feet of horizontal distance) results in soil loss at or below the rate of soil formation; farming there on a 15% slope would totally strip a field of its soil cover in only 10 years (see "Maintaining Soil Productivity").

Farming skills have not declined in recent years. Rather, pressures on farmlands have increased with population growth and the intensification of agriculture and clearing of land for the commercial cropping that is increasingly part of the developing countries' economies. Farming has been forced higher up on steeper slopes, more forest land has been converted to cultivation, grazing and crops have been pushed farther and more intensively into semiarid areas, and existing fields have had to be worked more intensively and less carefully. Many traditional agricultural systems and areas that were ecologically stable and secure as recently as 1950, when world population stood at 2.5 billion and subsistence agriculture was the rule, are disintegrating under the pressures of more than 6 billion people and a changing global economy.

The evidence of that deterioration is found in all parts of the world (Figure 13.15). The International Food Policy Research Institute in 2000 reported that nearly 40% of the world's agricultural land is seriously degraded, though the percentages differ by region. Almost 75% of cropland in Central America shows serious degradation, as does 20% in Africa (mostly pasture), and 11% in Asia. Soil deterioration expresses itself in two ways: through decreasing yields of cultivated fields themselves and in increased stream sediment loads and downstream deposition of silt. In Guatemala, for example, some 40% of the productive capacity of the land has been lost through erosion, and several areas of the country have been abandoned because agriculture has become economically impracticable; the figure is 50% in El Salvador. In Turkey, a reported 75% of the land is affected, and 54% is severely or very severely eroded. Haiti has no high-value soil left at all. A full one-quarter of India's total land area has been significantly eroded: some 13 million hectares (32 million acres) by wind and nearly 74 million hectares (183 million acres) by water. Between 1960 and 2000, China lost over 15% of its total arable land to erosion, desertification, or conversion to nonagricultural use; some 700,000 hectares (1.7 million acres) of cultivated land annually are taken by construction. Its Huang River is the most sediment-laden of any waterway on earth; in its middle course it is about 50% silt by weight, just under the point of liquid mud. Worldwide, an estimated 6 to 7 million hectares (15–17 million acres) of existing arable land are lost to erosion each year.

Off-farm erosion evidence is provided by siltation loads carried by streams and rivers and by the downstream deposition that results. In the United States, about 3 billion tons of sediment are washed into waterways each year; off-site damage in the form of reduced reservoir capacity, fish kills, dredging costs, and the like, is estimated at over $6 billion annually. The world's rivers deliver about 24 billion tons of sediment to the oceans each year, while additional billions of tons settle along stream valleys or are deposited in reservoirs (see "Dam Trouble in the Tropics").

Agricultural soil depletion through erosion—and through salt accumulation and desertification—has been called "the quiet crisis." It continues inexorably and unfolds gradually, without the abrupt attention attracted by an earthquake or volcanic explosion. Unfortunately, silent or not, productive soil loss is a crisis of growing importance and immediacy, not just in the countries of its occurrence but—because of international markets and relief programs—throughout the world.

Water Supply and Water Quality

Solar energy and water are the indispensable ingredients of life on earth. The supply of both is essentially constant and beyond the scope of humans to increase or alter although, as we saw with aerosols and atmospheric gases, humans can affect the quality and utility of an otherwise fixed resource. Any threat of reduction in availability or lessening of quality of a material so basic to our very lives as water is certain to arouse strong emotions and deep concerns. In many parts of the world and for many competitors for limited freshwater supplies, those emotions and concerns are already real.

The problem is not with the global amount of water, but with its distribution, its availability, and its quality. The total amount of water on the earth is enormous, though only a small part of the *hydrosphere* (see p. 484) is suitable or available for use by humans, plants, or animals (Figure 13.16). And the total amount remains constant. Water is a renewable resource; the **hydrologic cycle** assures that water, no matter how often used or how much abused,

In much of the world, increasing population numbers are largely responsible for accelerated soil erosion. In the United States, economic conditions rather than population pressures have often contributed to excessive rates of soil erosion. Wind and water are blowing and washing soil off pasturelands in the Great Plains, Texas, and the Southeast. America's croplands lose almost 2 billion tons of soil per year to erosion, an average annual loss over 4 tons per acre. In some areas the average is 15 to 20 tons per acre. Of the roughly 167 million hectares (413 million acres) of land intensively cropped in the United States, over one-third are losing topsoil faster than it can be replaced naturally. In parts of Illinois and Iowa, where the topsoil was once a foot deep, less than half of it remains.

Federal tax laws and the high farmland values of the 1970s encouraged farmers to plow virgin grasslands and to tear down windbreaks to increase their cultivable land and yields. The secretary of agriculture ex-

horted farmers to plant all of their land "from fencerow to fencerow," to produce more grain for export. Land was converted from cattle grazing to corn and soybean production as livestock prices declined. When prices of both land and farm products declined in the 1980s, farmers felt impelled to produce as much as possible in order to meet their debts and make any profit at all. To maintain or increase productivity, many neglected conservation practices, plowing under marginal lands and suspending proven systems of fallowing and crop rotation.

Conservation techniques were not forgotten, of course. They were practiced by many and persistently advocated by farm organizations and soil conservation groups. Techniques to reduce erosion by holding soil in place are well known. They include contour plowing, terracing, strip-cropping and crop rotation, erecting windbreaks and constructing water diversion channels, and practicing no-till or minimum tillage farming (allowing crop residue such as cut corn stalks to remain

on the soil surface when crops are planted instead of turning them under by plowing). And farmers can be paid to idle marginal, highly erodible land.

After the mid-1980s, federal farm programs attempted to reverse some of the damage resulting from past economic pressures and farming practices. One objective has been to retire for conservation purposes some 18.6 million hectares (46 million acres) of croplands that were eroding faster than three times the natural rate of soil formation. But some 13 million hectares (32 million acres) of land annually "set-aside" by government order from 1986 and 1995 to reduce crop surpluses were released for unrestricted cropping again under the 1996 "Freedom to Farm" bill. Only by reducing the economic pressures that lead to abuse of farmlands and by continuing to practice known soil conservation techniques can the country maintain the long-term productivity of soil, the resource base upon which all depend.

Strip-cropping.

No-till farming.

will return over and over to the earth for further exploitation (Figure 13.17). Enough rain and snow fall on the continents each year to cover the earth's total land area with 83 centimeters (33 in.) of water. It has usually been reckoned that the volume of fresh water annually renewed by the hydrologic cycle would meet the needs of the growing world population. Yet, over the past 75 years, as world population has tripled—with growth par-

ticularly rapid in regions of low and variable rainfall—total water demand has increased sixfold. Even now, it is generally agreed, a little more than half of the world's available freshwater is being used each year. Based on current population growth trends alone, that fraction could well rise to 74% by 2025—and to 90% should people everywhere then consume water in the current daily amount used by the average American.

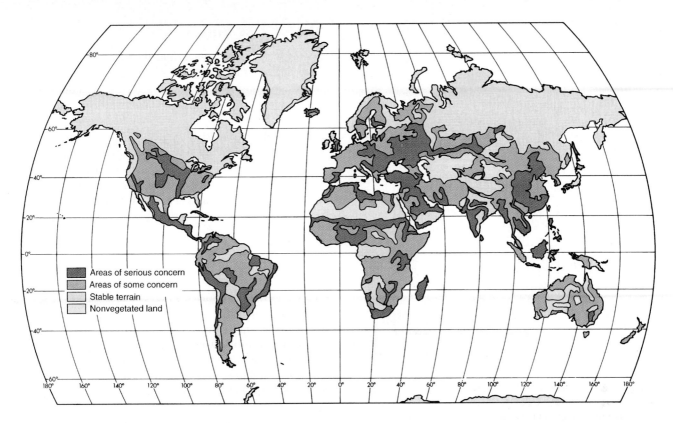

Figure 13.15 **The world pattern of soil degradation concern.** Between 1945 and 2000, nearly 2 billion hectares (almost 5 billion acres) of the world's 8.7 billion hectares (21.5 billion acres) of cropland, pastures, and forests used in agriculture—an area as large as Russia and India combined—were added to the existing total of degraded soils. Globally, about 18% of forest area, 21% of pastures, and 37% of cropland have undergone moderate to severe degradation. Water erosion accounted for 56% of that recorded deterioration, wind erosion for 28%, chemical deterioration (salinization and nutrient loss) for 12%, and physical degradation (e.g., compaction and waterlogging) for 4%.

Sources: World Resources Institute and International Soil Reference and Information Center.

In many parts of the world water supplies are inadequate and dwindling. Insufficient water for irrigation periodically endangers crops and threatens famine; permanent streams have become intermittent in flow; fresh- and saltwater lakes are shrinking; and from throughout the world come reports of rapidly falling water tables and wells that have gone dry. Reduced availability and reliability of supply are echoed in a reduced quality of the world's freshwater inventory. Increased silt loads of streams, pollution of surface and groundwater supplies, and lakes acidified and biologically dead or prematurely filled by siltation and algal growth are evidences of adverse human impact on an indispensable component of the biosphere.

Patterns of Availability

Observations about global supplies and renewal cycles of fresh water ignore the ever-present geographic reality: things are not uniformly distributed over the surface of the earth. There is no necessary relationship between the earth's pattern of freshwater availability and the distribution of consuming populations and activities. Three different world maps help us to understand why. The first, Figure 13.18, shows the spatially variable world pattern of precipitation. The second, Figure 13.4, reminds us that, as a rule, the lower the average amount of precipitation received in an area, the greater is the variability of precipitation from year to year. The recurring droughts and famines of the Sahel region of Africa are witness to the deadly impact of those expected fluctuations in areas of already low rainfall. Finally, Figure 13.19 takes account of the relationship between precipitation receipts and losses through *evapotranspiration,* the return of water from the land to the atmosphere through evaporation from soil and plants and by transpiration through plant leaves. These losses are higher in the tropics than in middle and upper latitudes, where lower rainfall amounts under cooler conditions may be more effective and useful than higher amounts received closer to the equator.

The distribution and vegetative adequacy of precipitation are givens and, except for human impact on climatic conditions, are largely independent of cultural influences. Regional water sufficiency, however, is also a function of the size of the population using the resource, its pattern of water use, and the amount of deterioration in quantity and quality the water supply experiences in the process of its use and return to the system. These are circumstances under human, not natural, control.

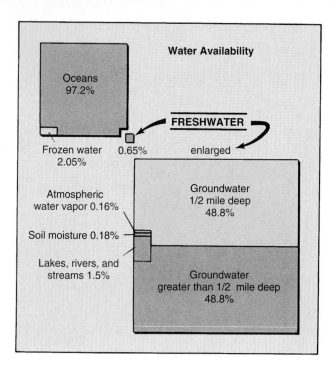

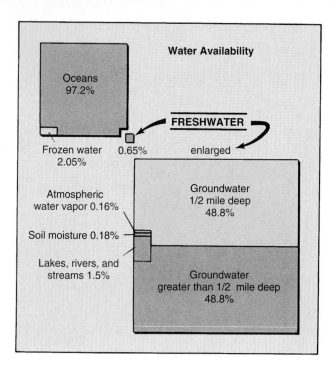

Within the figure (img_2):

Water Availability

Oceans
97.2%

FRESHWATER

Frozen water   0.65%   enlarged
2.05%

Atmospheric
water vapor 0.16%

Soil moisture 0.18%

Lakes, rivers, and
streams 1.5%

Groundwater
1/2 mile deep
48.8%

Groundwater
greater than 1/2 mile deep
48.8%

Figure 13.16 Less than 1% of the world's water supply is available for human use in freshwater lakes and rivers and from wells. An additional 2% is effectively locked in glaciers and polar ice caps.

Water Use and Abuse

For the world as a whole, irrigated agriculture accounts for nearly three-quarters (73%) of freshwater use; in the poorest countries, the proportion is 90%. The irrigation share continues to grow worldwide as irrigation farming expands by between 5 and 6 million additional hectares (12 to 15) million acres) annually. Industry uses about one-fifth (21%) of water consumption, and domestic and recreation needs account for the remainder. World figures conceal considerable regional variation.

Irrigation agriculture produces some 40% of the world's harvest from about 17% of its cropland. Unfortunately, in many instances the crops that are produced are worth less than the water itself; the difference is made up in the huge subsidies that governments everywhere offer to irrigation farming. In areas and economies as different as California's Napa Valley or Egypt's Nile Valley, farmers rarely pay over a fifth of the operating costs of public irrigation projects or any of their capital costs. Unfortunately as well, much of the water used for agriculture is lost to the regional supply through evaporation and transpiration; often less than half of the water withdrawn for irrigation is returned to streams or **aquifers** (porous, water-bearing layers of sand, gravel, and rock) for further use. Much of that returned water, moreover, is heavily charged with salts removed from irrigated soils, making it unfit for reuse.

On the other hand, most of the water used for manufacturing processes and power production is returned to streams, lakes, or aquifers, but often in a state of pollution that renders it unsuitable for alternate and subsequent uses. Industrial water use rises

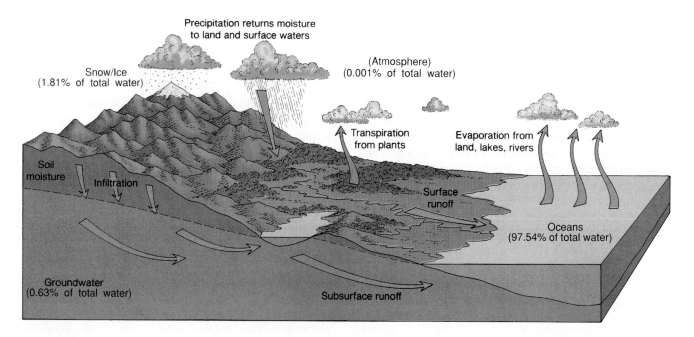

Figure 13.17 **The hydrologic cycle.** Water may change form and composition, but under natural environmental circumstances, it is marvelously purified in the recycling process and is again made available with appropriate properties and purity to the ecosystems of the earth. The sun provides energy for the evaporation of fresh and ocean water. The water is held as vapor until the air becomes supersaturated. Atmospheric moisture is returned to the earth's surface as solid or liquid precipitation to complete the cycle. Precipitation is not uniformly distributed, and moisture is not necessarily returned to areas in the same quantity as it has evaporated from them. The continents receive more water than they lose; the excess returns to the seas as surface water or groundwater. A global water balance, however, is always maintained.

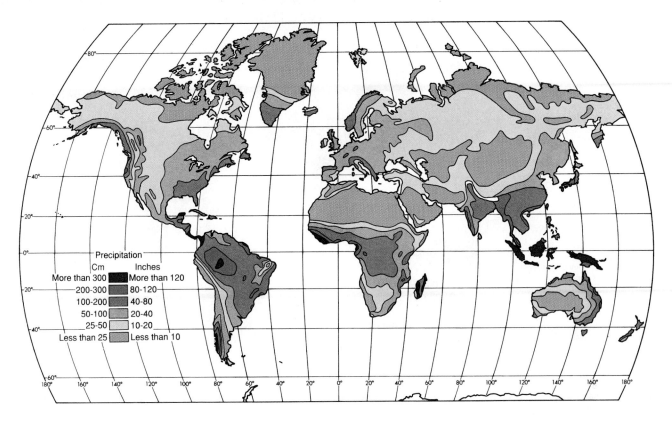

𝓕igure 13.18　**Mean annual precipitation.** Regional contrasts of precipitation receipts clearly demonstrate the truism that natural phenomena are unequally distributed over the surface of the earth. High and very high rainfall amounts are recorded in equatorial and tropical areas of Central and South America, Africa, and South and Southeast Asia. Productive agricultural regions of North America and Europe have lower moisture receipts. The world's desert regions—in North Africa, Inner Asia, the southwestern United States, and interior Australia—are clearly marked by low precipitation totals. But not all areas of low moisture receipt are arid, as Figure 13.19 makes clear.

dramatically with economic development, and in the developing countries, growing industrial demands compete directly with increasing requirements for irrigation and urban water supply.

Although municipal wastewater treatment is increasing in the most developed countries, 90% of raw sewage from urban areas in the developing world is discharged totally untreated into streams and oceans, contaminating surface water supplies, endangering drinking water sources, and destroying aquatic life. Fully 70% of total surface waters in India are polluted, in large part because only 8 of its more than 3000 sizable urban centers have full sewage treatment and no more than 200 have even partial management. Of Taiwan's 22 million people, only 600,000 are served by sewers. Hong Kong each day pours 1 million tons of untreated sewage and industrial waste into the sea. Mainland China's rivers also suffer from increasing pollution loads. More than 80% of major rivers are polluted to some degree, over 20% to such an extent that their waters cannot be used for irrigation. Four-fifths of China's urban surface water is contaminated, only six of the country's 27 largest cities have drinking water within the state standards, and the water to be impounded by the massive Three Gorges Dam project, it is predicted, will be seriously contaminated by untreated raw sewage from the dozens of cities along the new reservoir. In Malaysia, more than 40 major rivers are so

polluted that they are nearly devoid of fish and aquatic mammals. And even in developed countries of formerly communist Eastern Europe and Russia, sewage and, particularly, industrial waste seriously pollute much of the surface water supply.

When humans introduce wastes into the biosphere in kinds and amounts that the biosphere cannot neutralize or recycle, the result is **environmental pollution.** In the case of water, pollution exists when water composition has been so modified by the presence of one or more substances that either it cannot be used for a specific purpose or it is less suitable for that use than it was in its natural state. In both developed and developing countries, human pressures on freshwater supplies are now serious and pervasive concerns. If current trends of use and water abuse continue, freshwater will certainly—and soon—become a limiting factor for economic activity, food production, and maintenance of health in many parts of the world (see "A World of Water Woes"). A recent government report on global resources predicts that by 2015 nearly half the world's population—more than 3 billion people—will live in countries with insufficient water to satisfy their needs (Figure 13.20). Although a few governments have begun to face the water problem—potentially one every bit as serious as atmospheric pollution, soil erosion, deforestation, and desertification—much remains to be done.

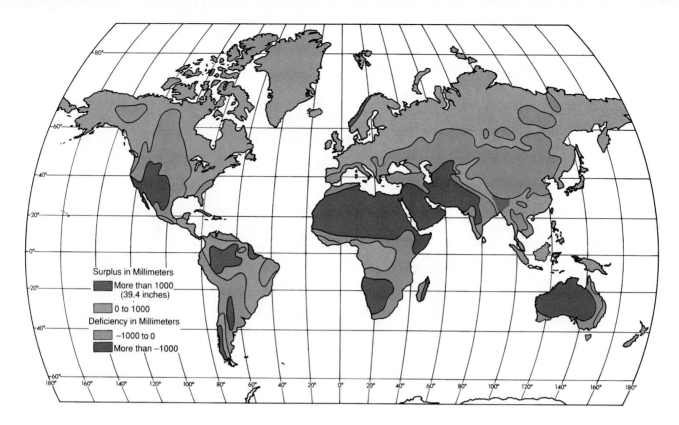

Figure 13.19 **World water supplies.** The pattern of surplus and deficit is seen in relation to the demands of the vegetation cover. Water is in surplus when precipitation is sufficient to satisfy or exceed the demands of the vegetation cover. When precipitation is lower than this potential demand, a water deficit occurs. By this measure, most of Africa (except the tropical rain forest areas of West Africa), much of the Middle East, the southwestern United States, and almost all of Australia are areas of extreme moisture deficit.

A comparison of this map with Figure 13.4 helps demonstrate the **limiting factor principle,** which notes that the single factor that is most deficient in an ecosystem is the one that determines what kind of plant and animal associations will exist there. Moisture surplus or deficit is the limiting factor that dictates whether desert, grassland, or forest will develop under natural, undisturbed conditions.

Redrawn with permission from Malin Falkenmark, "Water and Mankind—A Complex System of Mutual Interaction," Ambio 6(1977):5.

Garbage Heaps and Toxic Wastes

Humans have always managed to leave their mark on the landscapes they occupy. The search for minerals, for example, has altered whole landscapes, beginning with the pockmarks and pits marking Neolithic diggings into chalk cliffs to obtain flints or early Bronze Age excavations for tin and culminating with modern open-pit and strip-mining operations that tear minerals from the earth and create massive new landforms of depressions and rubble (Figures 13.1 and 13.21). Ancient irrigation systems still visible on the landscape document both the engineering skills and the environmental alterations of early hydraulic civilizations in the Near East, North Africa, and elsewhere. The raised fields built by the Mayas of Yucatán are still traceable 1000 years after they were abandoned, and aerial photography reveals the sites of villages and patterns of fields of medieval England.

Among the most enduring of landscape evidences of human occupance, however, are not the holes deliberately dug or the structures built but the garbage produced and discarded by all societies everywhere. Prehistoric dwelling sites are located and analyzed by their *middens,* the refuse piles containing the kitchen

wastes, broken tools, and other debris of human settlement. We have learned much about Roman and medieval European urban life and lifestyles by examination of the refuse mounds that grew as man-made hills in their vicinities. In the Near East, whole cities gradually rose on the mounds of debris accumulating under them (Figure 13.22).

Modern cultures differ from their predecessors by the volume and character of their wastes, not by their habits of discard. The greater the society's population and material wealth, the greater the amount and variety of its garbage. Developed countries of the late 20th century are increasingly discovering that their material wealth and technological advancements are submerging them in a volume and variety of wastes—solid and liquid, harmless and toxic—that threaten both their environments and their established ways of life. The United States may serve as an example of situations all too common worldwide.

Solid Wastes and Rubbish

North Americans produce rubbish and garbage at a rate of 220 million tons per year, or about 2 kilograms (4.5 pounds) per

Water covers almost three-quarters of the surface of the globe, yet "scarcity" is the word increasingly used to describe water-related concerns in both the developed and developing world. Globally, freshwater is abundant. Each year, an average of over 7000 cubic meters (some 250,000 cubic feet) per person enters rivers and underground reserves. But rainfall does not always occur when or where it is needed. Already, 80 countries with 40% of the world's population have serious water shortages that threaten to cripple agriculture and industry; 22 of them have renewable water resources of less than 1000 cubic meters (35,000 cubic feet) per person—a level generally understood to mean that water scarcity is a severe constraint on the economy and public health. Another 18 countries have less than 2000 cubic meters per capita on average, a dangerously low figure in years of rainfall shortage. Most of the water-short countries are in the Middle East, North Africa, and sub-Saharan Africa, the regions where populations (and consumption demands) are growing fastest.

In several major crop-producing regions, water use exceeds sustainable levels, threatening future food supplies. America's largest underground water reserve, stretching from west Texas northward into South Dakota, is drying up, partially depleted by more than 150,000 wells pumping water for irrigation, city supply, and industry. In parts of Texas, Oklahoma, and Kansas, the underground water table has dropped by more than 30 meters (100 feet). In some areas, the wells no longer yield enough to permit irrigation, and farmed land is decreasing; in others, water levels have fallen so far that it is uneconomical to pump it to the surface for any use.

In many agricultural districts of northern China, west and south India, and Mexico, water scarcity limits agriculture even though national supplies are adequate. In Uzbekistan and adjacent sections of Central Asia and Kazakhstan, virtually the entire flow of the area's two primary rivers—the Amu Darya and the Syr Darya—is used for often wasteful irrigation, with none left to maintain the Aral Sea or supply growing urban populations. In Poland, the draining of bogs that formerly stored rainfall, combined with unimaginable pollution of streams and groundwater, has created a water shortage as great as that of any Middle Eastern desert country. And salinity now seriously affects productivity—or prohibits farming completely—on nearly 10% of the world's irrigated lands.

Water scarcity is often a region-wide concern. More than 200 river systems draining over half the earth's land surface are shared by two or more countries. Egypt draws on the Nile for 86% of its domestic consumption, but virtually all of that water originates in eight upstream countries. Turkey, Iraq, and Syria have frequently been in dispute over the management of the Tigris and Euphrates rivers, and the downstream states fear the effect on them of Turkish impoundments and diversions. Mexico is angered at American depletion of the Colorado before it reaches the international border.

Many coastal communities face saltwater intrusions into their drinking water supplies as they draw down their underlying freshwater aquifers, while both coastal and inland cities dependent on groundwater may be seriously depleting their underground supplies. In China, 110 mostly large cities face acute water shortages; for at least 50 of them, the problem is groundwater levels dropping on average 1 to 2 meters (3 to 6 ft) each year. In Mexico City, groundwater is pumped at rates 40% faster than natural recharge; the city has responded to those withdrawals by sinking 30 feet during the 20th century. Millions of citizens of major cities throughout the world have had their water rationed as underground and surface supplies are used beyond recharge or storage capacity.

---

person per day. As populations grow, incomes rise, and consumption patterns change, the volume of disposable materials continues to expand. Relatively little residue is created in subsistence societies that move food from garden to table, and wastes from table to farm animals or compost heaps. The problem comes with urban folk who purchase packaged foods, favor plastic wrappings and containers for every commodity, and seek (and can afford) an ever-broadening array of manufactured goods, both consumer durables such as refrigerators and automobiles and many designed for single use and quick disposal.

The wastes that communities must somehow dispose of include newspapers and beer cans, toothpaste tubes and old television sets, broken stoves and rusted cars (Figure 13.23). Such ordinary household and municipal trash does not meet the usual designation of *hazardous waste:* discarded material that may pose a substantial threat to human health or the environment when improperly stored or disposed of. Much of it, however, does have a component of danger to health or to the environment. Paints and paint removers, used motor oils, pesticides and herbicides, bleaches, many kinds of plastics, and the like pose problems significantly different from apple cores and waste paper.

Landfill Disposal

The supply of open land and a free-enterprise system of waste collection and disposal led most American communities to opt for dumping urban refuse in *landfills.* In earlier periods, most of these were simply open dumps on the land, a menace to public health and an esthetic blot on the landscape. Beginning in the 1960s, more stringent federal controls began to require waste disposal in what was considered a more environmentally sound manner: the *sanitary landfill.* This involves depositing refuse in a natural depression or excavated trench, compacting it, and then covering it each day with soil to seal it (Figure 13.24). Open dumping was outlawed in 1976.

Some 75% of the country's municipal waste is disposed of by landfill. In the 1970s and 80s, there was a real fear that the available, affordable, or permitted landfill sites were rapidly disappearing and the cost of solid waste disposal would soon greatly increase. Some two-thirds of all landfills in operation in the late 1970s were filled and closed by 1990, and more than half the cities on the East Coast were without any local landfill sites in the middle 1990s. Because of changes in garbage economics during

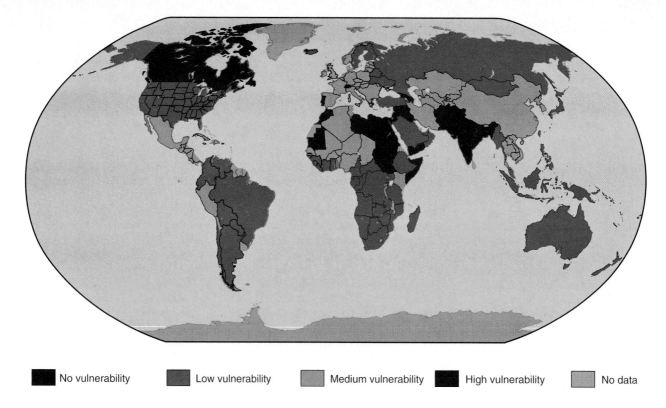

| No vulnerability | Low vulnerability | Medium vulnerability | High vulnerability | No data |

Figure 13.20 **Countries vulnerable to water shortage.** Water consumption rises with population growth, increases in the standard of living expansion of irrigation farming, and the enlarged demands of industry and municipalities that come with development. In a growing number of world areas—many where water is already scarce—limited freshwater resources severely constrain sustainable development, requiring hard choices in the allocation of water among competing users.

Source: From Environmental Science, *8th ed., by E. Enger and B. Smith. Reprinted by permission of the McGraw-Hill Companies, Inc.*

the 1990s, however, those earlier fears proved unnecessary. First, large waste management companies have built efficient mega-landfills, replacing a great many small, local, and inefficient operations, increasing disposal capacity nationwide. Second, widespread adoption of municipal recycling programs—now diverting an estimated 20% of trash away from landfills—has extended the capacity and life span of the remaining landfills. Lack of profitable (or any) market for recyclables, however, has caused some cities—New York in 2002 is an example—to suspend or curtail their recycling programs.

Over the years, of course, many filled dumps have posed a problem for the cities which gave rise to them. New York City, for example, for years placed all of its daily 14,000 tons of residential waste into the world's largest dump, Fresh Kills on Staten Island. Opened in 1947 as a 3-year "temporary" 500-acre facility, it became a malodorous 3000 acres of decomposing garbage rising some 15 stories above former ground level. Generating 140,000 cubic meters (5 million cubic feet) of methane gas annually and illegally exuding contaminated water, Fresh Kills— finally closed in 2001 at a cost of more than $1 billion—symbolized the rising tide of refuse engulfing cities and endangering the environment.

Incineration

For cities and regions faced with growing volumes of solid waste, alternatives to local landfill are few, expensive, and strongly re-

sisted. One possibility is *incineration,* a waste-to-energy option of burning refuse to produce steam or electricity that usually also involves sorting, recapturing, and recycling useful rubbish components, such as paper, glass, metals, and the like. Incinerators also produce air pollution, including highly toxic dioxin,[1] so control equipment is required. Acid gases and heavy metals are also released by waste burning. The gases add to atmospheric pollution and acid rain, although "scrubbers" and fabric filters on modern incinerators reduce emissions to very low levels; the metals contribute to the toxicity of the ash that is the inevitable product of incineration and that requires landfill disposal.

The likelihood of pollution from one or many incinerator by-products has sparked strong protest to their construction in the United States, though they have been more accepted abroad. A Supreme Court ruling of 1994 that incinerator ash was not exempt from the nation's hazardous waste law recognized the potential danger and mandated more expensive ash disposal procedures. Nonetheless, the 110 waste-to-energy incinerators operating in the United States at the end of the 1990s continued to burn about one-sixth of the country's municipal garbage (Figure 13.25). The seriousness of the dioxin and toxic ash problem,

---

[1] Any of several types of hydrocarbon compounds that are extremely toxic, persistent in the environment, and biologically magnified in the food chain. Dioxin is a frequently unavoidable trace contaminant in chemical processes and may also be formed during waste matter incineration or other combustion processes.

Figure 13.21 About 400 square kilometers (some 150 sq mi) of land surface in the United States are lost each year to the strip-mining of coal and other minerals; far more is chewed up worldwide. On flat or rolling terrain, strip-mining leaves a landscape of parallel ridges and trenches, the result of stripping away the unwanted surface material. That material—*overburden*—taken from one trench to reach the underlying mineral is placed in an adjacent one, leaving the wavelike terrain shown here. Besides altering the topography, strip-mining interrupts surface and subsurface drainage patterns, destroys vegetation, and places sterile and frequently highly acidic subsoil and rock on top of the new ground surface. Current law not always successfully requires stripped areas to be returned to their original contours.

however, has aroused concern everywhere. In Japan, where about three-quarters of municipal waste is incinerated in over 1850 municipal and 3300 private industrial incinerators, atmospheric dioxin levels triple those of the United States led the Ministry of Health in 1997 to strengthen earlier inadequate dioxin emission guidelines. Some European countries called at least temporary halts to incinerator construction while their safety was reconsidered, and increasingly landfills are refusing to take their residue.

Ocean Dumping

For coastal communities around the world the ocean has long been the preferred sink for not only municipal garbage, but for (frequently untreated) sewage, industrial waste, and all the detritus of an advanced urban society. The practice has been so common and long-standing that by the 1980s, the oceans were added to the list of great environmental concerns of the age. While the carcinogenic effect of ozone reduction had to be assumed from scientific report, the evidence of serious ocean pollution was increasingly apparent to even the most casual observer.

Along the Atlantic coast of North America from Massachusetts to Chesapeake Bay, reports of dead dolphins, raw sewage, tar balls, used syringes, vials of contaminated blood and hospital waste, diapers, plastic products in unimagined amounts and vari-

eties, and other foul refuse kept swimmers from the beach, closed coastal shellfisheries, and elicited health warnings against wading or even breathing salt spray (Figure 13.26). The Gulf of Mexico coast is also tainted and polluted by accumulations of urban garbage, litter from ships and offshore oil rigs, and the toxic effluent of petrochemical plants. Long stretches of Pacific shoreline are in similar condition.

Elsewhere, the Adriatic, Aegean, Baltic, and Irish seas and the Sea of Japan—indeed, all the world's coastal waters—are no better. Environmental surveys of the shores of the Mediterranean Sea show serious damage and pollution. Around Italy, the Mediterranean waters are cloudy with raw sewage and industrial waste, and some of the world's most beautiful beaches are fouled by garbage. The Bay of Guanabara, the grand entryway to Rio de Janeiro, Brazil, has been called a cesspool.

An international treaty to regulate ocean dumping of hazardous trash was drafted in 1972; another, the Ocean Dumping Ban to control marine disposal of wastes, was negotiated in 1988; and a "global program of action" for protection of the marine environment from land-based pollution was devised in 1995, but their effectiveness has yet to be demonstrated. In light of the length of the world's coastline, the number of countries sharing it, and the great growth of urban populations in the vicinity of the sea, serious doubt has been raised whether any international

Figure 13.22 Aerial view of Erbil, Iraq. Here and elsewhere in the Middle East, the debris of millennia of human settlement gradually raised the level of the land surface, producing *tells,* or occupation mounds. The city—one of the oldest in the world—literally was constantly rebuilt at higher elevations on the accumulation of refuse of earlier occupants. In some cases, these striking landforms may rise scores of feet above the surrounding plains.

agreement can be fully effective or enforced. Many portents—from beach litter to massive fin- and shellfish kills—suggest that the oceans' troubled waters have reached the limit of the abuse they can absorb.

Whether the solution to solid waste disposal be sought by land, by fire, or by sea, humanity's rising tide of refuse threatens to overwhelm the environments that must deal with it. The problem is present, growing, and increasingly costly to manage. Solutions are still to be found, a constant reminder for the future of the threatening impact of the environments of culture upon those of nature.

Toxic Wastes

The problems of municipal and household solid-waste management are daunting; those of treatment and disposal of hazardous and toxic wastes seem overwhelming. The definitions of the terms are imprecise, and *hazardous* and *toxic* are frequently used interchangeably, as we shall do here. More strictly defined, **toxic waste** is a relatively limited concept, referring to materials that can cause death or serious injury to humans or animals. **Hazardous waste** is a broader term referring to all wastes, including toxic ones, that pose an immediate or long-term human health risk or that endanger the environment (see definition on p. 484). The Environmental Protection Agency has classified more than 400 substances as hazardous, and currently about 10% of industrial waste materials are so categorized.

Such wastes contaminate the environment in different ways and by different routes. Because most hazardous debris is disposed of by dumping or burial on land, groundwater is most at risk of contamination. In all, some 2% of Anglo America's groundwater supply could have hazardous waste contamination. No comparable world figures exist, but in all industrial countries at least some drinking water contamination from highly toxic solvents, hydrocarbons, pesticides, trace metals, and polychlorinated biphenyls (PCBs) has been detected. Toxic waste impoundments are also a source of air pollutants through the evaporation of volatile organic compounds. Finally, careless or deliberate distribution of toxic materials outside of confinement areas can cause unexpected, but deadly, hazards. Although methods of disposal other than containment techniques have been developed—including incineration, infrared heating, and bacterial decomposition—none is fully satisfactory and none is as yet in wide use.

Radioactive Wastes

Every facility that either uses or produces radioactive materials generates at least *low-level waste,* material whose radioactivity will decay to safe levels in 100 years or less. Nuclear power plants produce about half the total low-level waste in the form of used resins, filter sludges, lubricating oils, and detergent wastes. Industries that manufacture radiopharmaceuticals, smoke alarms, and other consumer goods produce such wastes in the form of machinery parts, plastics, and organic solvents. Research establishments, universities, and hospitals also produce radioactive waste materials.

High-level waste can remain radioactive for 10,000 years and more; plutonium stays dangerously radioactive for 240,000 years. It consists primarily of spent fuel assemblies of nuclear power reactors—termed "civilian waste"—and such "military waste" as the by-products of nuclear weapons manufacture. The volume of civilian waste alone is not only great but increasing rapidly, because approximately one-third of a reactor's rods need to be disposed of every year.

By 2003, some 115,000 spent-fuel assemblies were in storage in the containment pools of America's commercial nuclear power reactors, awaiting more permanent disposition. About 6000 more are added annually. "Spent fuel" is a misleading term: the assemblies are removed from commercial reactors not because their radiation is spent, but because they have become too radioactive for further use. The assemblies will remain radioactively "hot" for thousands of years.

Unfortunately, no satisfactory method for disposing of any hazardous waste has yet been devised (see "Yucca Mountain"). Although sealing liquids with a radioactive life measured in the thousands of years within steel drums expected to last no more than 40 years seems an unlikely solution to the disposal problem, it is one that has been widely practiced. Some wastes have been sealed in protective tanks and dumped at sea, a practice that has now been banned worldwide. Much low-level radioactive waste has been placed in tanks and buried in the ground at 13 sites operated by the U.S. Department of Energy and three sites run by private firms. Millions of cubic feet of high-level military waste are temporarily stored in underground tanks at four sites: Hanford, Washington; Savannah River, South Carolina; Idaho Falls, Idaho; and West Valley, New York (Figure 13.27). Several of these storage areas have experienced leakages, with seepage of waste into the surrounding soil and groundwater.

Figure 13.23 Some of the 240 million tires Americans replace each year. Most used tires are dumped, legally or illegally. Some are retreaded, some are exported, and some are burned to generate electricity. But most remain unused and unwanted in growing dumps that remain both fire hazards and breeding grounds for insects and the diseases they carry.

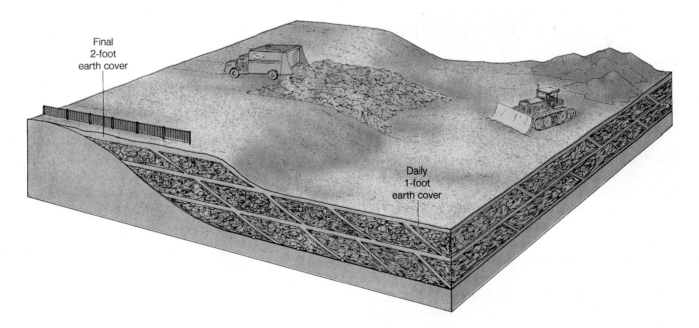

Figure 13.24 **A sanitary landfill.** Each day's deposit of refuse is compacted and isolated in a separate cell by a covering layer of soil or clay. Although far more desirable than open dumps, sanitary landfills pose environmental problems of their own, including potential groundwater contamination and seepage of methane and hydrogen sulfide, gaseous products of decomposition. By federal law, modern landfills must be lined with clay and plastic, equipped with *leachate* (chemically contaminated drainage from the landfill) collection systems to protect the groundwater, and monitored regularly for underground leaks—requirements that have increased significantly the cost of constructing and operating landfills.

Figure 13.25 This waste-to-energy incinerator at Peekskill, New York, is one of the new generation of municipal plants originally expected to convert over one-quarter of the country's municipal waste to energy by A.D. 2000. A Supreme Court ruling that the ash they produce had to be tested for hazardous toxicity and appropriately disposed in protected landfills, growing public rejection and lawsuits, and increasingly stringent controls on the amount and kind of airborne vapors they may emit have in many instances raised the operating costs of present incinerators far higher than landfill costs and altered the economic assessments of new construction. Incineration becomes more cost-effective, however, when cities are unable to dispose of their trash locally and must haul it to distant sites.

Figure 13.26 Warning signs and beaches littered with sewage, garbage, and medical debris are among the increasingly common and distressing evidences of ocean dumping of wastes.

Because low-level waste is generated by so many sources, its disposal is particularly difficult to control. Evidence indicates that much of it has been placed in landfills, often the local municipal dump, where the waste chemicals may leach through the soil and into the groundwater. By EPA estimates, the United States contains at least 25,000 legal and illegal dumps with hazardous waste; as many as 2000 are deemed potential ecological disasters.

An even less constructive response, according to increasing complaints, has been the export of radioactive materials—in common with other hazardous wastes—to willing or unwitting recipient countries with less restrictive or costly controls and its illegal and unrecorded dumping at sea—now banned by the legally binding London Convention of 1993.

Exporting Waste

Regulations, community resistance, and steeply rising costs of disposal of hazardous wastes in the developed countries encouraged producers of those unwanted commodities to seek alternate areas for their disposition. Transboundary shipments of dangerous wastes became an increasingly attractive option for producers. In total, such cross-border movement amounted to tens of thousands of shipments annually by the early 1990s, with destinations including debt-ridden Eastern European countries and impoverished developing ones outside of Europe that were willing to trade a hole in the ground for hard currency. It was a trade, however, that increasingly aroused the ire and resistance of destination countries and, ultimately, elicited international agreements among both generating and receiving countries to cease the practice.

The Organization of African Unity in 1988 adopted a resolution condemning the dumping of all foreign wastes on that continent. More broadly and under the sponsorship of the United Nations, 117 countries in March of 1989 adopted a treaty—the Basel Convention on the Control of Transboundary Movements

Geography and Public Policy

Yucca Mountain

If the U.S. government has its way, a long, low ridge in the basin-and-range region of Nevada will become America's first permanent repository for the deadly radioactive waste that nuclear power plants generate. If the opponents of the project have their way, however, Yucca Mountain will become a symbol of society's inability to solve a basic problem posed by nuclear power production: where to dispose of the used fuel. At present, no permanent disposal site for radioactive waste exists anywhere in the world.

In 1982, Congress ordered the Department of Energy (DOE) to construct a permanent repository by 1998 for the spent fuel of civilian nuclear power plants, as well as vast quantities of waste from the production of nuclear weapons. Yucca Mountain in southern Nevada was selected as the site for this high-level waste facility, which is intended to safely store wastes for 10,000 years, until

radioactive decay has rendered them less hazardous than they are today. Most of the waste would be in the form of radioactive fuel pellets sealed in metal rods; these would be encased in extremely strong glass and placed in steel canisters entombed in chambers 300 meters (1000 ft) below the Nevada desert. The steel containers would corrode in one or two centuries, after which the volcanic rock of the mountain would be responsible for containing radioactivity.

Plans now call for opening the repository in 2010, but many doubt that the Yucca Mountain facility will ever be completed and licensed. Three lines of concern have emerged. First, the storage area is vulnerable to both volcanic and earthquake activity, which could cause groundwater to well up suddenly and flood the repository. Yucca Mountain itself was formed from volcanic eruptions that occurred about 12–15 million years ago; some geologists are concerned that a new volcano could erupt within the mountain. Seven small cinder cones in the immediate area have erupted in recent times, the latest just 10,000 years ago. In addition, a number of seismic faults lie close to Yucca Mountain. One, the Ghost Dance Fault, runs right through the depository site. The epicenter of the 1992 earthquake at Little Skull Mountain was only 19 kilometers (12 mi) from the proposed dump site.

Second, rainwater percolating down through the mountain could penetrate the vaults holding the waste. Over the centuries, water could dissolve the waste itself, and the resulting toxic brew could seep down into the water table, be carried into the groundwater table under the mountain, and then beyond the boundaries of the repository.

Finally, the Yucca Mountain site lies between the Nevada Test Site, which the DOE used as a nuclear bomb testing range, and the Nellis Air Force Base Bombing and Gunnery Range. Questions have been raised about the wisdom of locating a waste repository just a few kilometers from areas subject to aerial bombardment.

In February, 2002, the Energy Department officially recommended that President Bush designate Yucca Mountain as the site for a national nuclear waste repository, a plan the President and Congress later approved. The project even then faced substantial technical, legal, and political challenges and was subject to possible engineering problems, adverse court decisions, and the vehement opposition of Nevada officials.

Questions to Consider

1. What are the advantages and disadvantages of permitting nuclear plants to operate when no system for disposing of their hazardous wastes exists?

2. Comment on the paradox that plutonium remains dangerously radioactive for 240,000 years, yet plans call for it to be stored in a repository that is intended to safely hold wastes for only 10,000 years.

3. React to the fact that even if Yucca Mountain opens, it will be too small. It is designed to accept 78,000 tons of civilian and military wastes, but the radioactive waste from nuclear power plants alone already exceeds that amount.

4. Considering the uncertainties that would attend the irreversible underground entombment of high-level waste, do you think the government should instead pursue aboveground storage in a form that would allow for the continuous monitoring and retrieval of the wastes? Why or why not?

5. High-level waste is produced at relatively few sites in the United States. Low-level waste is much more common; it is produced at more than 2000 sites in California alone. Which do you think should arouse greater public concern and discussion, high-level or low-level waste?

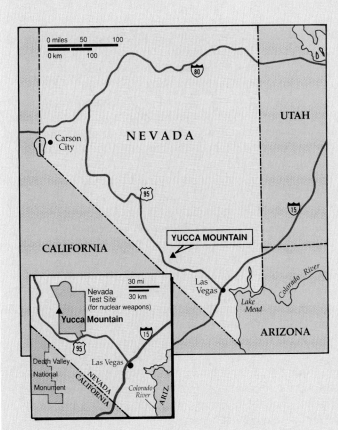

Figure 13.27 Storage tanks under construction in Hanford, Washington. Built between 1943 and 1985 to contain high-level radioactive waste, the tanks are shown before they were encased in concrete and buried underground. By the early 1990s, 66 of the 177 underground tanks were already known to be leaking. Of the approximately 55 million gallons of waste the tanks hold, about 1 million gallons of liquids have seeped into the soil, raising the fear that the radioactive waste has already reached underground water supplies and is flowing toward the Columbia River. In mid-2002, construction began on a $4 billion vitrification plant to transform the worst of the radioactive wastes into a stable, granular form of glass, eventually to be buried at Yucca Mountain.

of Hazardous Wastes and Their Disposal—aimed at regulating the international trade in wastes. That regulation was to be achieved by requiring exporters to obtain consent from receiving countries before shipping waste and by requiring both exporter and importer countries to ensure that the waste would be disposed of in an environmentally sound manner.

A still more restrictive convention was reached in March, 1994 when—with the United States dissenting—most Western industrialized countries agreed to ban the export of all poisonous or hazardous industrial wastes and residues to the developing world, the countries of Eastern Europe, and the former Soviet Union. United States' objections concerned the assumed prohibition on export of such materials as scrap metals for recycling within consenting receiving countries. Despite the agreement, a UN committee in 1998 identified the United States, Germany, Australia, Britain, and the Netherlands as continuing major toxic waste exporters. Investigating toxic wastes dumping as a violation of basic human rights, the committee reported that Africa still receives masses of developed country toxic waste in spite of its 1988 resolution. The bulk of European waste goes to the Baltic countries and to Eastern and Central Europe. Half of the United States' exports go to Latin America, and those of Britain go largely to Asia.

Prospects and Perspectives

Not surprisingly, the realities of the human impacts upon the environment that we have looked at in this chapter bring us directly back to ideas first presented in Chapter 2, at the start of our ex-

amination of the meaning of culture and the development of human geographic patterns on the surface of the earth. We noted there and see more clearly now that humans, in their increasing numbers and technical sophistication, have been able since the end of the last glaciation to alter for their own needs the physical landscapes they occupy. Humans, it is often observed, are the ecological dominant in the human–environmental equation that is the continuing focus of geographic inquiry.

That dominance reflected itself in the growing divergence of human societies as they separated themselves from common hunting-gathering origins. In creating their differing cultural solutions to common concerns of sustenance and growth, societies altered the environments they occupied. Diverse systems of exploitation of the environment were developed in and diffused from distinctive culture hearths. They were modified by the ever-expanding numbers of people occupying earth areas of differing carrying capacities and available resources. Gradually developing patterns of spatial interaction and exchange did not halt the creation of areally distinctive subsystems of culture or assure common methods of utilization of unequally distributed earth resources or environments. Sharp contrasts in levels of economic development and well-being emerged and persisted even as cultural convergence through shared technology began increasingly to unite societies throughout the world.

Each culture separately placed its imprint on the environment it occupied. In many cases—Chaco Canyon and Easter Island were our earlier examples—that imprint was ultimately destructive of the resources and environments upon which the cultures developed and depended. For human society collectively or single cultures separately, environmental damage or destruction is the unplanned consequence of the ecological dominance of humans. Our perpetual dilemma lies in the reality that what we need and want in support and supply from the environments we occupy generally exceeds in form and degree what they are able to yield in an unaltered state. To satisfy their felt needs, humans have learned to manipulate their environment. The greater those needs and the larger the populations with both needs and technical skills to satisfy them, the greater is the manipulation of the natural landscape. For as long as humans have occupied the earth the implicit but seldom addressed question has not been should we exploit and alter the environment, but how can we extract our requirements from the natural endowment without dissipating and destroying the basis of our support?

This final chapter detailing a few of the damaging pressures placed upon the environment by today's economies and cultures is not meant as a litany of despair. Rather it is a reminder of the potentially destructive ecological dominance of humans. Against the background of our now fuller understanding of human geographic patterns and interactions, this chapter is also meant to remind us yet again of the often repeated truism that everything is connected to everything else; we can never do just *one* thing. The ecological crises defined in this chapter and the human geographic patterns of interaction, contrast, and—occasionally—conflict observed in the preceding chapters show clearly and repeatedly how close and complex are the connections within the cultural world and how intimately our created environment is joined to the physical landscape we all share.

There is growing awareness of those connections, of the adverse human impacts upon the natural world, and of the unity of all cultural and physical landscapes. Climatic change, air and water pollution, soil loss and desertification, refuse contamination, and a host of other environmental consequences and problems of intensifying human use of the earth are all matters of contemporary public debate and consideration. Awareness and concern of individuals are increasingly reflected by policies of environmental protection introduced by governments and supported or enforced by international conferences, compacts, and treaties. Acceptance of the interconnectedness and indivisibility of cultural and natural environments—the human creation and the physical endowment—is now more the rule than the exception.

Our understanding of those relationships is advanced by what we have learned of the human side of the human–environmental structure. We have seen that the seemingly infinitely complex diversity of human societies, economies, and interrelations is in fact logical, explicable, and far from random or arbitrary. We now have developed both a mental map of the cultural patterns and content of areas and an appreciation of the dynamics of their creation and operation. We must have that human geographic background—that sense of spatial interaction and unity of cultural, economic, and political patterns—to understand fully the relationship between our cultural world and the physical environment on which it ultimately depends. Only with that degree of human geographic awareness can we individually participate in an informed way in preserving and improving the increasingly difficult and delicate balance between the endowment of nature and our landscapes of culture.

Summary

Cultural landscapes may buffer but cannot isolate societies from the physical environments they occupy. All human activities, from the simplest forms of agriculture to modern industry, have an impact upon the biosphere. Cumulatively, in both developed and developing countries, that impact is now evident in the form of serious and threatening environmental deterioration. The atmosphere unites us all, and its global problems of greenhouse heating, ozone depletion, and particulate pollution endanger us all. Desertification, soil erosion, and tropical deforestation may appear to be local or regional problems, but they have worldwide implications of both environmental degradation and reduced population-supporting capacity. Freshwater supplies are deteriorating in quality and decreasing in sufficiency through contamination and competition. Finally, the inevitable end product of human use of the earth—the garbage and hazardous wastes of civilization—are beginning to overwhelm both sites and technologies of disposal.

We do not end our study of human geography on a note of despondency, however. We end with the conviction that the fuller knowledge we now have of the spatial patterns and structures of human cultural, economic, and political activities will aid in our understanding of the myriad ways in which human societies are bound to the physical landscapes they occupy—and which they have so substantially modified.

Key Words

acid rain 493	environmental pollution 507	ozone 494
aquifer 506	fallowing 503	rotation 503
biome 485	greenhouse effect 488	soil 501
biosphere (ecosphere) 484	hazardous waste 512	soil erosion 499
desertification 500	hydrologic cycle 503	terracing 503
ecosystem 486	icebox effect 487	toxic waste 512
environment 484	limiting factor principle 508	

For Review

1. What does the term *environment* mean? What is the distinction between the natural and the cultural environments? Can both be part of the physical environment we occupy?

2. What is the *biosphere,* or *ecosphere?* What are its parts? How is the concept of *biome* related to that of the ecosphere?

3. Were there any evidences of human impact upon the natural environment prior to the Industrial Revolution? If so, can you provide examples? If not, can you explain why not?

4. Do we have any evidence of physical environmental change that we cannot attribute to human action? Can we be certain that environmental change we observe today is attributable to human action? How?

5. What lines of reasoning and evidence suggest that human activity is altering global climates? What kind of alteration has occurred or is expected to occur? What do the terms *greenhouse* and *icebox effect* have to do with possible climatic futures?

6. What is *desertification?* What types of areas are particularly susceptible to desertification? What kinds of land uses are associated with it? How easily can its effects be overcome or reversed?

7. What agricultural techniques have been traditionally employed to reduce or halt soil erosion? Since these are known techniques that have been practiced throughout the world, why is there a current problem of soil erosion anywhere?

8. What effects has the increasing use of fossil fuels over the past 200 years had on the environment? What is *acid rain,* and where is it a problem? What factors affect the type and degree of air pollution found at a place? What is the relationship of *ozone* to *photochemical smog?*

9. Describe the chief sources of water pollution of which you are aware. How has the supply of freshwater been affected by pollution and human use? When water is used, is it forever lost to the environment? If so, where does it go? If not, why should there be water shortages now in regions of formerly ample supply?

10. What methods do communities use to dispose of solid waste? Can *hazardous wastes* be treated in the same fashion? Since disposition of waste is a technical problem, why should there be any concern with waste disposal in modern advanced economies?

11. Suggest ways in which your study of human geography has increased your understanding of the relationship between the environments of culture and those of nature.

Focus Follow-up

1. **What are contributing causes and resulting concerns of global warming, acid rainfall, and ozone level changes?** pp. 484–495.

Following the last glaciation, relatively stable world patterns of climates and biomes persisted, broken only by occasional periods of unusual warming or cooling. Great increases in human numbers and their environmental impact over the past century resulted in apparent detectable changes in former earth system stability. Recent global warming has been attributed in significant measure to human-caused increases in greenhouse gases. Increases as well in airborne smoke, soot, and acid gases from factories and cars help produce acid precipitation that corrodes stone and metals, destroys forests, and acidifies to sterility some lakes and soils. Upper-air ozone depletion and lower-level ozone accumulation, both with serious effect on plant and animal life, are also largely attributed to humans' adverse impact on the environment.

2. **What human actions have contributed to tropical deforestation, desertification, and soil erosion? What are the consequences?** pp. 496–503.

Current rapid destruction of tropical forests reflects human intentions to expand farming and grazing areas and harvest tropical wood. Their depletion endangers or destroys the world's richest, most diversified plant and animal biome and adversely affects local, regional, and world patterns of temperature and rainfall. Their loss also diminishes a vital "carbon sink" needed to absorb excess carbon dioxide. Desertification—the expansion of areas of destroyed soil and plant cover in dry climates—results from both natural climatic fluctuations and human pressures from plowing, woody plant removal, or livestock overgrazing. Those same human actions and pressures can accelerate the normal erosional loss of soil beyond natural soil regeneration potential. Such loss reduces total and per capita area of food production, diminishing the human carrying capacity of the land.

3. **How are emerging water supply and waste disposal problems related to human numbers and impacts?** pp. 503–507.

The hydrologic cycle assures water will be continuously regenerated for further use. But growing demand for irrigation, industrial use, and individual and urban consumption means increasing lack of balance between natural water supplies and consumption demands. Pollution of those supplies by human actions further reduces water availability and utility.

4. **How are modern societies addressing the problems of solid and toxic waste disposal?** pp. 508–517.

Increasingly, all societies are becoming more dependent on modern manufacturing and packaging of industrial, commercial, and personal consumption items. The easy recycling of waste materials found in subsistence cultures is no longer possible, and humans are presented with increasing needs for sites and facilities to safely dispose of solid wastes. Sanitary landfills and incineration are employed to handle nontoxic wastes. The former demands scarce and expensive land near cities or costly export to distant locations; the latter is often opposed because of unsafe emissions and ash residue. Disposal of toxic and hazardous wastes including nuclear wastes, products of modern societies and technologies, poses problems yet to be satisfactorily and safely solved.

 Selected References

Abramovitz, Janet N. *Imperiled Waters, Impoverished Future: The Decline of Freshwater Ecosystems.* Worldwatch Paper 128. Washington, D.C.: Worldwatch Institute, 1996.

Bailey, Robert G. *Ecosystem Geography.* New York/Berlin: Springer Verlag, 1996.

Bradley, R. "1000 Years of Climate Change." *Science* 288 (May 26, 2000): 1353–1355.

Calvin, William H. "The Great Climate Flip-flop." *Atlantic Monthly* 281, no. 1 (January 1998): 47–64.

"Desertification after the UNCED, Rio 1992." *GeoJournal* 31, no. 1 (September 1993). Special issue.

Enger, Eldon D., and Bradley F. Smith. *Environmental Science: A Study of Interrelationships.* 9th ed. Boston: McGraw-Hill, 2004.

Feshbach, Murray, and Alfred Friendly, Jr. *Ecocide in the USSR.* New York: Basic Books, 1992.

Food and Agriculture Organization. *State of the World's Forests.* UN, FAO. Biennial.

Gardner, Gary. *Shrinking Fields: Cropland Loss in a World of Eight Billion.* Worldwatch Paper 131. Washington, D.C.: Worldwatch Institute, 1996.

Gleick, Peter H. *The World's Water: The Biennial Report on Freshwater Resources.* Covelo, Calif.: Island Press. Biennial.

Goudie, Andrew. *The Human Impact.* 5th ed. New York: Blackwell, 2000.

Goudie, Andrew. *Encyclopedia of Global Change: Environmental Change and Human Society.* New York: Oxford University Press, 2001.

Groombridge, Brian, and Martin D. Jenkins. *World Atlas of Biodiversity: Earth's Living Resources in the Twenty-First Century.* Berkeley: University of California Press, 2002.

Harrison, Paul, and Fred Pearce. *AAAS Atlas of Population and Environment.* Victoria D. Markham, ed. Berkeley, Calif.: American Association for the Advancement of Science and University of California Press, 2001.

Houghton, John. *Global Warming: The Complete Briefing.* 2d ed. Cambridge, England and New York: Cambridge University Press, 1997.

"Human-Dominated Ecosystems." A special report in *Science* 227, (25 July 1997): 485–525.

Lenssen, Nicholas. *Nuclear Waste: The Problem That Won't Go Away.* Worldwatch Paper 106. Washington, D.C.: Worldwatch Institute, 1991.

Lomborg, Bjorn. *The Skeptical Environmentalist: Measuring the Real State of the World.* Cambridge, Eng.: Cambridge University Press, 2001.

McGinn, Anne Platt. *Safeguarding the Health of Oceans.* Worldwatch Paper 145. Washington, D.C.: Worldwatch Institute, 1999.

Middleton, Nick, and David Thomas, eds. *World Atlas of Desertification.* 2d ed. United Nations Environmental Programme. London: Edward Arnold, 1997.

National Geographic Society. "A World Transformed" and "A Thirsty Planet." Maps and text produced by National Geographic Maps for *National Geographic Magazine.* Washington, D.C.: National Geographic Society, 2002.

Park, Chris C. *Tropical Rainforests.* London and New York: Routledge, 1992.

Pickering, Kevin T., and Lewis A. Owen. *An Introduction to Global Environmental Issues.* 2d ed. New York: Routledge, 1997.

Postel, Sandra. *Last Oasis: Facing Water Scarcity.* New York: W. W. Norton, 1992.

Postel, Sandra. "When the World's Wells Run Dry." *Worldwatch* 12, no. 5 (Sept/Oct 1999): 30–38.

Roberts, Neil. *The Changing Global Environments.* Cambridge, Mass.: Blackwell, 1993.

Sampat, Payal. *Deep Trouble: The Hidden Threat of Groundwater Pollution.* Worldwatch Paper 154. Washington, D.C.: Worldwatch Institute, 2000.

Simmons, Ian. *Changing the Face of the Earth: Culture, Environment, History.* 2d ed. Cambridge, Mass.: Blackwell, 1996.

Thomas, William, ed. *Man's Role in Changing the Face of the Earth.* Chicago: University of Chicago Press, 1956.

U.S. Global Change Research Program, National Assessment Synthesis Team. *Climate Change Impacts on the United States: The Potential Consequences of Climate Variability and Change.* New York: Cambridge University Press, 2000.

"Water: The Power, Promise, and Turmoil of North America's Fresh Water." Special edition of *National Geographic* 184, no. 5A (November 1993).

World Resources Institute/International Institute for Environment and Development. *World Resources.* Washington, D.C.: World Resources Institute. Biennial.

Worldwatch Institute. *State of the World.* New York: Norton. Annual.

Websites: The World Wide Web has a tremendous number and variety of sites pertaining to geography. Websites relevant to the subject matter of this chapter appear in the "Web Links" section of the On-line Learning Center associated with this book. Access it at **www.mhhe.com/fellmann8e**

Appendix A
Map Projections

A map projection is simply a system for displaying the curved surface of the earth on a flat sheet of paper. The definition is easy; the process is more difficult. No matter how one tries to "flatten" the earth, it can never be done in such a fashion as to show all earth details in their correct relative sizes, shapes, distances, or directions. Something is always wrong, and the cartographer's—the mapmaker's—task is to select and preserve those earth relationships important for the purpose at hand and to minimize or accept those distortions that are inevitable but unimportant.

Round Globe to Flat Map

The best way to model the earth's surface accurately, of course, would be to show it on a globe. But globes are not as convenient to use as flat maps and do not allow one to see the entire surface of the earth all at once. Nor can they show very much of the detailed content of areas. Even a very large globe of, say, 3 feet (nearly 1 meter) in diameter, compresses the physical or cultural information of some 130,000 square kilometers (about 50,000 sq mi) of earth surface into a space 2.5 centimeters (1 in.) on a side.

Geographers make two different demands on the maps they use to represent reality. One requirement is to show at one glance generalized relationships and spatial content of the entire world; the many world maps used in this and other geography textbooks and in atlases have that purpose. The other need is to show the detailed content of only portions of the earth's surface—cities, regions, countries, hemispheres—without reference to areas outside the zone of interest. Although the needs and problems of both kinds of maps differ, each starts with the same requirement: to transform a curved surface into a flat one.

If we look at the globe directly, only the front—the side facing us—is visible; the back is hidden (Figure A.1). To make a world map, we must decide on a way to flatten the globe's curved surface on the hemisphere we can see. Then we have to cut the globe map down the middle of its hidden hemisphere and place the two back quarters on their respective sides of the already visible front half. In simple terms, we have to "peel" the map from the globe and flatten it in the same way we might try to peel an orange and flatten the skin. Inevitably, the peeling and flattening process will produce a resulting map that either shows tears or breaks in the surface (Figure A.2a) or is subject to uneven stretching or shrinking to make it lie flat (Figure A.2b).

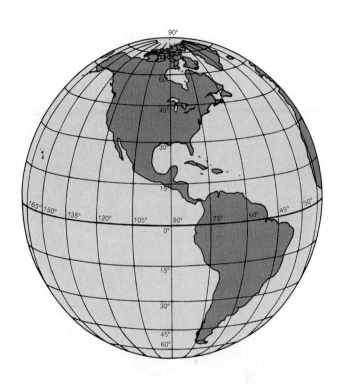

Figure A.1 An orthographic projection gives us a visually realistic view of the globe; its distortion toward the edges suggests the normal perspective appearance of a sphere viewed from a distance. Only a single hemisphere—one half of the globe—can be seen at a time, and only the central portion of that hemisphere avoids serious distortion of shape.

Projections—Geometrical and Mathematical

Of course, mapmakers do not physically engage in cutting, peeling, flattening, or stretching operations. Their task, rather, is to construct or *project* on a flat surface the network of parallels and meridians (the **graticule**) of the globe grid (see p. 20). The idea of projections is perhaps easiest visualized by thinking of a transparent globe with an imagined light source located inside. Lines of latitude and longitude (or of coastlines or any other features) drawn on that globe will cast shadows on any nearby surface. A tracing of that shadow globe grid would represent a geometrical map projection.

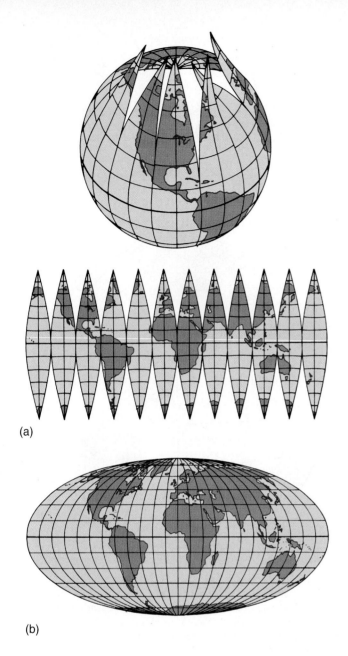

(a)

(b)

Figure A.2 (*a*) A careful "peeling" of the map from the globe yields a set of tapered "gores" which, although individually not showing much stretching or shrinking, do not collectively result in a very useful or understandable world map. (*b*) It is usually considered desirable to avoid or reduce the number of interruptions by depicting the entire global surface as a single flat circular, oval, or rectangular shape. That continuity of area, however, can be achieved only at the cost of considerable alteration of true shapes, distances, directions, or areas. Although the homolographic (Mollweide) projection shows areas correctly, it distorts shapes.

Redrawn with permission from American Congress Surveying and Mapping, Choosing a World Map. *Special Publication No. 2 of the American Cartographic Association, Bethesda, Md. Copyright 1988 American Congress on Surveying and Mapping.*

In **geometrical** (or **perspective**) **projections,** the graticule is in theory visually transferred from the globe to a geometrical figure, such as a plane, cylinder, or cone, which, in turn, can be cut and then spread out flat (or *developed*) without any stretching or tearing. The surfaces of cylinders, cones, and planes are said to be **developable surfaces**—cylinders and cones can be cut and laid flat without distortion and planes are flat at the outset (Figure A.3). In actuality, geometrical projections are constructed not by tracing shadows but by the application of geometry and the use of lines, circles, arcs, and angles drawn on paper.

The location of the theoretical light source in relation to the globe surface can cause significant variation in the projection of the graticule on the developable geometric surface. An **orthographic** projection results from placement of the light source at infinity. A **gnomonic** projection is produced when the light source is at the center of the earth. When the light is placed at the *antipode*—the point exactly opposite the point of tangency (point of contact between globe and map)—a **stereographic** projection is produced (Figure A.4).

Although a few useful and common projections are based on these simple geometric means of production, most map designs can only be derived mathematically from tables of angles and dimensions separately developed for specific projections. The objective and need for **mathematical projections** is to preserve and emphasize specific earth relationships that cannot be recorded by the perspective globe and shadow approach. The graticule of each mathematical projection is orderly and "accurate" in the sense of displaying the correct locations of lines of latitude and longitude. Each projection scheme, however, presents a different arrangement of the globe grid to minimize or eliminate some of the distortions inherent in projecting from a curved to a flat surface. Every projection represents a compromise or deviation from reality to achieve a selected purpose, but in the process of adjustment or compromise, each inevitably contains specific, accepted distortions.

Globe Properties and Map Distortions

The true properties of the global grid are detailed on p. 20. Not all of those grid realities can ever be preserved in any single projection; projections invariably distort some or all of them. The result is that all flat maps, whether geometrically or mathematically derived, also distort in different ways and to different degrees some or all of the four main properties of actual earth surface relationships: area, shape, distance, and direction.

Area

Cartographers use **equal-area,** or **equivalent,** projections when it is important for the map to show the *areas* of regions in correct or constant proportion to earth reality—as it is when the map is intended to show the actual areal extent of a phenomenon on the earth's surface. If we wish to compare the amount of land in agriculture in two different parts of the world, for example, it would

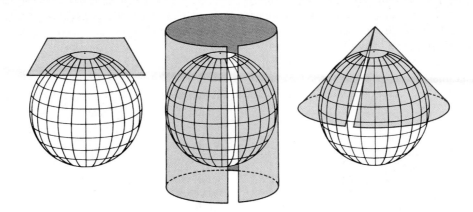

Figure A.3 The theory of *geometrical projections.* The three common geometric forms used in projections are the plane, the cylinder, and the cone.

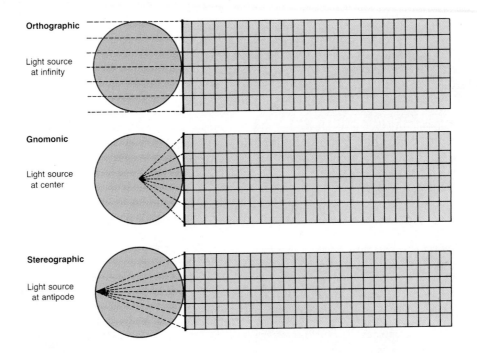

Orthographic

Light source at infinity

Gnomonic

Light source at center

Stereographic

Light source at antipode

Figure A.4 The effect of light source location on planar surface projections. Note the variations in spacing of the lines of latitude that occur when the light source is moved.

be very misleading visually to use a map that represented the same amount of surface area at two different scales.[1] To retain the needed size comparability, our chosen projection must assure that a unit area drawn anywhere on it will always represent the same number of square kilometers (or similar units) on the earth's surface. To achieve *equivalence,* any scale change that the projection imposes in one direction must be offset by compensating changes in the opposite direction. As a result, the shape of the portrayed area is inevitably distorted. A square on the earth, for

example, may become a rectangle on the map, but that rectangle has the correct area (Figure A.5). *A map that shows correct areal relationships always distorts the shapes of regions,* as Figure A.6a demonstrates.

Shape

Although no projection can reproduce correct shapes for large areas, some do accurately portray the shapes of small areas. These true-shape projections are called **conformal,** and the importance of *conformality* is that regions and features "look right" and have the correct directional relationships. They achieve these properties for small areas by assuring that lines of latitude and longitude cross each other at right angles and that the scale is the

---

[1]**Scale** is the relationship between the size of a feature or length of a line on the map and that same feature or line on the earth's surface. It may be indicated on a map as a ratio—for example, 1:1,000,000—that tells us the relationship between a unit of measure on the map and that same unit on the earth's surface. In our example, 1 centimeter of map distance equals 1 million centimeters (or 10 kilometers) of actual earth distance. See Figure 1.18.

same in all directions at any given location. Both these conditions exist on the globe but can be retained for only relatively small areas on maps. Because that is so, the shapes of large regions—continents, for example—are always different from their true earth shapes even on conformal maps. Except for maps for very small

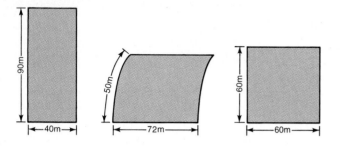

Figure A.5 These three figure are all equal in area despite their different dimensions and shapes.

areas, *a map cannot be both equivalent and conformal;* these two properties are mutually exclusive, as Figure A.6b suggests.

Distance

Distance relationships are nearly always distorted on a map, but some projections do maintain true distances in one direction or along certain selected lines. True distance relationships simply mean that the length of a straight line between two points on the map correctly represents the *great circle* distance between those points on the earth. (An arc of a great circle is the shortest distance between two points on the earth's curved surface; the equator is a great circle and all meridians of longitude are half great circles.) Projections with this property can be designed, but even on such **equidistant** maps true distance in all directions is shown only from one or two central points. Distances between all other locations are incorrect and, quite likely, greatly distorted as Figure A.6c clearly shows.

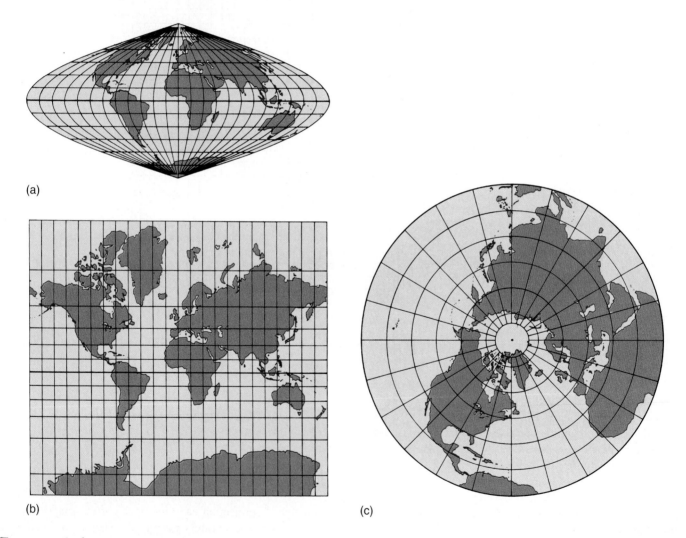

(a)

(b)

(c)

Figure A.6. Sample projections demonstrating specific map properties. (*a*) The equal-area sinusoidal projection retains everywhere the property of *equivalence.* (*b*) The mathematically derived Mercator projection is *conformal*, displaying true shapes of individual features but greatly exaggerating sizes and distorting shapes away from the equator. (*c*) A portion of an azimuthal *equidistant* projection, polar-case. Distances from the center (North Pole) to any other point are true; extension of the grid to the Southern Hemisphere would show the South Pole infinitely stretched to form the circumference of the map.

Direction

As is true of distances, directions between all points on a map cannot be shown without distortion. On **azimuthal** projections, however, true directions are shown from one central point to all other points. (An *azimuth* is the angle formed at the beginning point of a straight line, in relation to a meridian.) The azimuthal property of a projection is not exclusive but may be combined with equivalency, conformality, and equal distance. The azimuthal equal-distance ("equidistant") map shown as Figure A.6c is, as well, a true-direction map from the same North Pole origin. Another more specialized example is the gnomonic projection, displayed as Figure A.7.

Classes of Projections

Although there are many hundreds of different projections, the great majority of them can be grouped into four primary classes or families based on their origin. Each family has its own distinctive outline, set of similar properties, and pattern of distortions. Three of them are easily seen as derived from the geometric or perspective projection of the globe grid onto the developable surfaces of cylinders, cones, and planes. The fourth class is mathematically derived; its members have a variety of attributes but share a general oval design (Figure A.8).

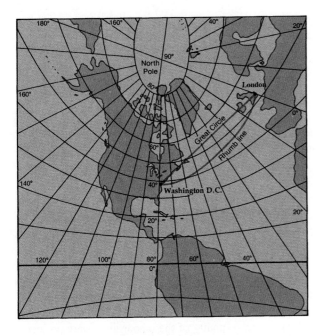

Figure A.7. A gnomonic projection centered on Washington, D.C. In this geometrical projection the light source is at the center of the globe (see Figure A.4), and the capital city marks the "standard point" where the projection plane is in contact with the globe. The rapid outward increase in graticule spacing makes it a projection impractical for more than a portion of a hemisphere. A unique property of the gnomonic projection is that it is the only projection on which all great circles appear as straight lines.

Cylindrical Projections

Cylindrical projections are developed geometrically or mathematically from a cylinder wrapped around the globe. Usually, the cylinder is tangent at the equator, which thus becomes the **standard line**—that is, transferred from the globe without distortion. The result is a globe grid network with meridians and parallels intersecting at right angles. There is no scale distortion along the standard line of tangency, but distortion increases with increasing distance away from it. The result is a rectangular world map with acceptable low-latitude representation, but with enormous areal exaggeration toward the poles.

The mathematically derived Mercator projection invented in 1569 is a special familiar but commonly misused cylindrical projection (see Figure A.6b). Its sole original purpose was to serve as a navigational chart of the world with the special advantage of showing true compass headings, or *rhumb lines,* as straight lines on the map. Its frequent use in wall or book maps gives grossly exaggerated impressions of the size of land areas away from the tropics. Equal-area alternatives to the conformal Mercator map are available, and a number of "compromise" cylindrical projections that are neither equal area nor conformal (for example, the Miller projection, Figure A.9a) are frequently used bases for world maps.

Conic Projections

Of the three developable geometric surfaces, the cone is the closest in form to one-half of a globe. **Conic projections,** therefore, are often employed to depict hemispheric or smaller parts of the earth. In the *simple conic* projection, the cone is placed tangent to the globe along a single standard parallel, with the apex of the cone located above the pole. The cone can also be made to intersect the globe along two or more lines, with a *polyconic* projection resulting; the increased number of standard lines reduces the distortion which otherwise increases away from the standard parallel. The projection of the grid on the cone yields evenly spaced

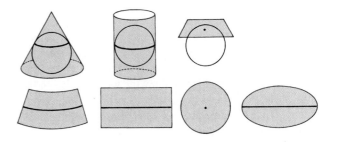

Figure A.8. Shape consistencies within families of projections. When the surface of cone, cylinder, or plane is made *tangent*—that is, comes into contact with the globe—at either a point or along a circle and then "developed," a characteristic family outline results. The tangent lines and point are indicated. A fourth common shape, the oval, may reflect a design in which the long dimension is a great circle comparable to the tangent line of the cylinder.

Redrawn with permission from American Congress on Surveying and Mapping, Choosing a World Map. *Special Publication No. 2 of the American Cartographic Association, Bethesda, Md. Copyright 1988 American Congress on Surveying and Mapping.*

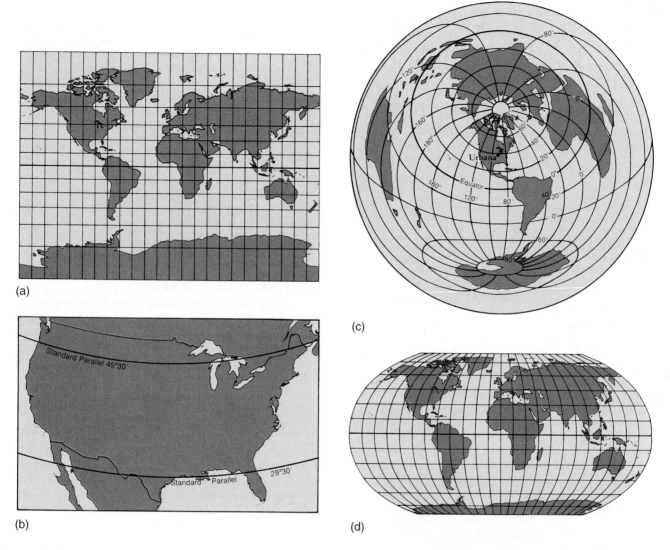

(a)

(b)

(c)

(d)

Figure A.9. Some sample members of the principal projection families. (*a*) The Miller cylindrical projection is mathematically derived. (*b*) The Albers equal-area conic projection, used for many official U.S. maps, has two standard parallels: 29 1/2° and 45 1/2°. (*c*) A planar, or azimuthal, equidistant projection centered on Urbana, Illinois. (*d*) The Robinson projection of the oval family; neither conformal nor equivalent, it was designed as a visually satisfactory world map.

straight-line meridians radiating from the pole and parallels that are arcs of circles. Although conic projections can be adjusted to minimize distortions and become either equivalent or conformal, by their nature they can never show the whole globe. In fact, they are most useful for and generally restricted to maps of midlatitude regions of greater east-west than north-south extent. The Albers equal-area projection often used for U.S. maps is a familiar example (Figure A.9b).

Planar (Azimuthal) Projections

Planar (or **azimuthal**) **projections** are constructed by placing a plane tangent to the globe at a single point. Although the plane may touch the globe anywhere the cartographer wishes, the polar case with the plane centered on either the North or the South Pole is easiest to visualize (see Figure A.6c). This equidistant projection is useful because it can be centered anywhere, facilitating the correct measurement of distances from that point to all others. When the plane is tangent at places other than the poles, the meridians and the parallels become curiously curved (Figure A.9c).

Planar maps are commonly used in atlases because they are particularly well suited for showing the arrangement of polar landmasses. Depending on the particular projection used, true shape, equal area, or some compromise between them can be depicted. The special quality of the planar gnomonic projection has already been shown in Figure A.7.

Oval or Elliptical Projections

Oval or elliptical projections have been mathematically developed usually as compromise projections designed to display the entire world in a fashion that is visually acceptable and suggestive of the curvature of the globe. In most, the equator and a cen-

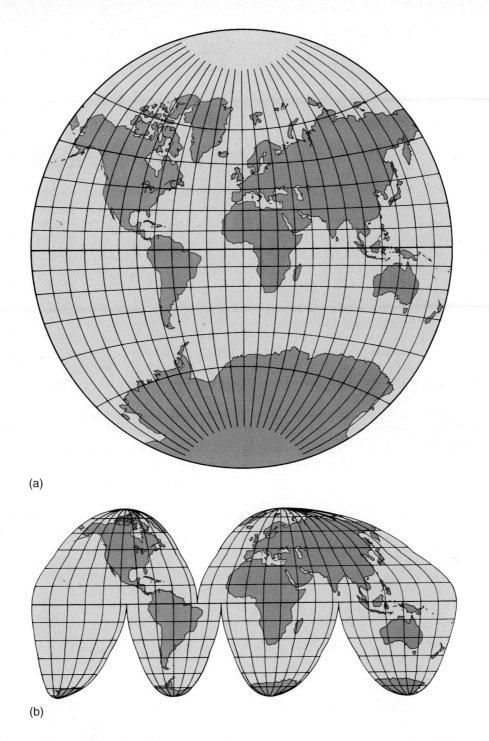

(a)

(b)

Figure A.10. (*a*) The full Van der Grinten projection; (*b*) Goode's interrupted homolosine grafts an upper latitude homolographic (Mollweide) onto a sinusoidal projection.

(b) Copyright by the Committee on Geographic Studies, University of Chicago. Used by permission.

tral meridian (usually the prime meridian) are the standard lines. They cross in the middle of the map, which thus becomes the point of no distortion. Parallels are, as a rule, parallel straight lines; meridians, except for the standard meridian, are shown as curved lines. In some instances the oval projection is a modification of one of different original shape. The world maps in this text, for example, are an oval adjustment of the circular (but not azimuthal) Van der Grinten projection (Figure A.10a), a compromise projection that achieves acceptable degrees of equivalence and conformality in lower and middle latitudes but becomes increasingly and unacceptably distorted in polar regions.

Other Projections and Manipulations

Projections can be developed mathematically to show the world or a portion of it in any shape that is desired: ovals are most common, but hearts, trapezoids, stars, and other—sometimes bizarre—forms have been devised for special purposes. One often-seen projection is the equal-area Goode's homolosine, an "interrupted" projection that is actually a product of fitting together the least distorted portions of two different projections and centering the split map along multiple standard meridians to minimize distortion of either (as desired) land or ocean surfaces (Figure A.10b).

The homolosine map clearly shows how projections may be manipulated or adjusted to achieve desired objectives. Since most projections are based on a mathematically consistent rendering of the actual globe grid, possibilities for such manipulation are nearly unlimited. Map properties to be retained, size and shape of areas to be displayed, and overall map design to be achieved may influence the cartographer's choices in reproducing the globe grid on the flat map.

Special effects and properties may also be achieved geometrically by adjusting the aspect of the projection. *Aspect* simply means the positional relationship between the globe and the developable surface on which it is visually projected. Although the fundamental distortion pattern of any given projection system onto any of the developable surfaces will remain constant, shifting of the point or line of tangency will materially alter the appearance of the graticule and of the geographical features shown on the map.

Although an infinite number of aspects are possible for any of the geometric projections, three classes of aspects are most common. Named according to the relation of the axis of the globe to the cylinder, plane, or oval projection surface, the three classes are usually called *equatorial, polar,* and *oblique*. In the equatorial, the axis of the globe parallels the orientation of the plane, cylinder, or cone; a parallel, usually the central equator, is the line of tangency. In the polar aspect, the axis of the globe is perpendicular to the orientation of the developable surface. In the oblique aspect, the axis of the globe makes an oblique angle with the orientation of the developable surface, and a complex arrangement of the graticule results.

A Cautionary Reminder

Mapmakers must be conscious of the properties of the projections they use, selecting the one that best suits their purposes. It is not ever possible to transform the globe into a flat map without distortion. But cartographers have devised hundreds of possible mathematical and geometrical projections in various modifications and aspects to display to their best advantage the variety of earth features and relationships they wish to emphasize. Some projections are highly specialized and properly restricted to a single limited purpose; others achieve a more general acceptability and utility.

If the map shows only a small area, the choice of a projection is not critical—virtually any can be used. The choice becomes more important when the area to be shown extends over a considerable longitude and latitude; then the selection of a projection clearly depends on the purpose of the map. As we have seen, Mercator or gnomonic projections are useful for navigation. If numerical data are being mapped, the relative sizes of the areas involved should be correct, and equivalence is the sought-after map property. Conformality and equal distance may be required in other instances.

While selection of an appropriate projection is the task of the cartographer, understanding the consequences of that selection and recognizing and allowing for the distortions inevitable in all flat maps are the responsibility of the map reader. When skillfully designed maps are read by knowledgeable users, clear and accurate conveyance of spatial information and earth relationships is made convenient and natural.

 *Key Words*

Selected References

American Cartographic Association, Committee on Map Projections. *Choosing a World Map: Attributes, Distortions, Classes, Aspects.* Special Publication No. 2. Falls Church, Va.: American Congress on Surveying and Mapping, 1988.

American Cartographic Association, Committee on Map Projections. *Matching the Map Projection to the Need.* Special Publication No. 3. Falls Church, Va.: American Congress on Surveying and Mapping, 1991.

American Cartographic Association, Committee on Map Projections. *Which Map Is Best? Projections for World Maps.* Special Publication No. 1. Falls Church, Va.: American Congress on Surveying and Mapping, 1986.

Brown, Lloyd. *The Story of Maps.* Boston: Little, Brown, 1949; reprint ed., New York: Dover Publications, 1977.

Campbell, John. *Map Use and Analysis.* 3d ed. Dubuque, Iowa: WCB/McGraw-Hill, 1998.

Deetz, Charles H., and Oscar S. Adams. *Elements of Map Projection.* United States Department of Commerce, Special Publication No. 68. Washington, D.C., USGPO, 1945.

Porter, Phil, and Phil Voxland. "Distortion in Maps. The Peters Projection and Other Devilments." *Focus* (Summer 1986): 22–30.

Robinson, Arthur H., et al. *Elements of Cartography.* 6th ed. New York: John Wiley & Sons, 1995.

Snyder, John P. *Map Projections—A Working Manual.* U.S. Geological Survey Professional Paper 1395. Washington, D.C.: Department of the Interior, 1987.

Snyder, John P. *Flattening the Earth: Two Thousand Years of Map Projections.* Chicago, Ill.: University of Chicago Press, 1993.

Steers, J. A. *An Introduction to the Study of Map Projections.* London: University of London Press, 1962.

Stillwell, H. Daniel. "Global Distortion: Is It Time to Retire the Mercator Projection?" *Mercator's World* 2, no. 1 (September/October 1997): 54–59.

Wilford, John Noble. "Resolution in Mapping." *National Geographic* 193, no. 2 (February 1998): 6–39.

The World Wide Web has a tremendous number and variety of sites pertaining to geography. Websites relevant to the subject matter of this appendix appear in the "Web Links" section of the On-line Learning Center associated with this book. Access it at: **www.mhhe.com/fellmann8e**

Appendix B
2003 World Population Data

	Population Mid–2003 (millions)	Births Per 1,000 Pop.	Deaths Per 1,000 Pop.	Rate of Natural Increase (%)	Projected Population in 2025 (millions)	Projected Population in 2050 (millions)	Projected Pop. Change 2003–2050 (%)	Infant Mortality Rate[a]	Total Fertility Rate[b]	Percent of Population of Age < 15/65 +		Life Expectancy at Birth (Years) Total	Percent Urban	Percent of Pop. 15–49 with HIV/AIDS End–2001	Rural % with Access to Improved Water Source	Per Capita GNI 2001 (US$)
WORLD	**6,314**	**22**	**9**	**1.3**	**7,907**	**9,198**	**46**	**55**	**2.8**	**30**	**7**	**67**	**47**	**1.2**	**71**	**—**
MORE DEVELOPED	**1,202**	**11**	**10**	**0.1**	**1,260**	**1,257**	**5**	**7**	**1.5**	**18**	**15**	**76**	**75**	**0.4**	**+**	**—**
LESS DEVELOPED	**5,112**	**24**	**8**	**1.6**	**6,647**	**7,940**	**55**	**61**	**3.1**	**33**	**5**	**65**	**40**	**1.4**	**69**	**—**
LESS DEVELOPED (excl. China)	**3,823**	**28**	**9**	**1.9**	**5,192**	**6,547**	**71**	**65**	**3.5**	**36**	**4**	**63**	**41**	**1.9**	**—**	**—**
AFRICA	**861**	**38**	**14**	**2.4**	**1,289**	**1,883**	**119**	**88**	**5.2**	**42**	**3**	**52**	**33**	**6.5**	**—**	**—**
SUB-SAHARAN AFRICA	**711**	**40**	**16**	**2.5**	**1,084**	**1,636**	**130**	**93**	**5.6**	**44**	**3**	**48**	**30**	**8.9**	**44**	**460**
NORTHERN AFRICA	**188**	**27**	**7**	**2.1**	**267**	**331**	**76**	**51**	**3.6**	**37**	**4**	**67**	**45**	**0.6**	**—**	**—**
Algeria	31.7	23	5	1.8	42.8	51.0	61	54	2.8	35	4	70	49	0.1	82	1,650
Egypt	72.1	27	6	2.1	103.2	127.4	77	44	3.5	36	4	68	43	z	96	1,530
Libya	5.5	28	4	2.4	8.3	10.8	97	30	3.7	36	4	76	86	0.2	68	—
Morocco	30.4	22	6	1.6	39.2	45.0	48	37	2.7	32	5	70	57	0.1	56	1,190
Sudan	38.1	39	10	2.8	61.3	84.2	121	70	5.5	45	2	57	27	2.6	69	340
Tunisia	9.9	17	6	1.1	11.6	12.2	23	23	2.1	29	6	73	63	z	58	2,070
Western Sahara	0.3	45	16	2.9	0.4	0.6	117	134	6.6	—	—	50	95	—	—	—
WESTERN AFRICA	**256**	**41**	**15**	**2.7**	**402**	**617**	**142**	**88**	**5.8**	**45**	**3**	**51**	**35**	**4.9**	**—**	**—**
Benin	7.0	41	14	2.7	11.8	18.0	156	89	5.6	47	2	51	40	3.6	55	380
Burkina Faso	13.2	47	19	2.8	22.5	39.5	198	105	6.5	49	3	45	15	6.5	37	220
Cape Verde	0.5	30	7	2.3	0.7	0.8	76	31	3.9	41	6	69	53	—	89	1,310
Côte d'Ivoire	17.0	37	18	1.9	24.6	34.1	101	102	5.2	46	2	43	46	9.7	72	630
Gambia	1.5	42	13	2.9	2.7	4.2	177	82	5.8	45	3	53	37	1.6	53	320
Ghana	20.5	31	10	2.1	25.4	29.8	46	56	4.2	43	3	57	37	3.0	62	290
Guinea	9.0	43	16	2.7	16.2	30.6	239	98	6.0	45	3	49	26	1.5	36	410
Guinea-Bissau	1.3	45	20	2.5	2.2	3.3	154	126	6.0	44	4	45	22	2.8	49	160
Liberia	3.3	49	17	3.1	5.5	8.8	165	141	6.6	43	3	49	45	2.8	—	140
Mali	11.6	50	20	3.0	20.0	32.5	179	126	7.0	47	3	45	26	1.7	61	230
Mauritania	2.9	44	15	2.9	5.4	8.5	190	101	6.0	44	3	54	55	0.5	40	360
Niger	12.1	55	20	3.5	25.7	51.9	330	123	8.0	50	2	45	17	1.4	56	180

530

	Population Mid–2003 (millions)	Births Per 1,000 Pop.	Deaths Per 1,000 Pop.	Rate of Natural Increase (%)	Projected Population in 2025 (millions)	Projected Population in 2050 (millions)	Projected Pop. Change 2003–2050 (%)	Infant Mortality Rate[a]	Total Fertility Rate[b]	Percent of Population of Age < 15/65 +	Life Expectancy at Birth (Years) Total	Percent Urban	Percent of Pop. 15–49 with HIV/AIDS End–2001	Rural % with Access to Improved Water Source	Per Capita GNI 2001 (US$)
WESTERN AFRICA															
(continued)															
Nigeria	133.9	41	13	2.8	206.4	307.4	130	75	5.8	44 / 3	52	36	5.8	49	290
Senegal	10.6	38	11	2.7	17.1	24.6	132	68	5.2	44 / 3	53	43	0.5	65	490
Sierra Leone	5.7	47	21	2.5	9.0	13.8	141	155	6.2	45 / 3	43	37	7.0	46	140
Togo	5.4	38	11	2.7	7.6	9.7	78	72	5.5	46 / 2	54	31	6.0	38	270
EASTERN AFRICA	**263**	**41**	**17**	**2.4**	**395**	**590**	**124**	**102**	**5.6**	**45 / 3**	**44**	**20**	**9.5**	**—**	**—**
Burundi	6.1	40	19	2.2	10.1	15.4	152	75	6.3	48 / 3	43	8	8.3	77	100
Comoros	0.6	47	12	3.5	1.1	1.8	190	86	6.8	46 / 5	56	29	0.1	95	380
Djibouti	0.7	39	19	2.0	0.8	1.1	62	117	5.9	43 / 3	43	83	11.8	100	890
Eritrea	4.4	41	13	2.8	7.0	10.5	142	48	5.9	44 / 3	54	16	2.8	42	160
Ethiopia	70.7	42	15	2.7	117.6	173.3	145	107	5.9	44 / 3	42	15	6.4	12	100
Kenya	31.6	35	15	2.0	35.3	40.2	27	66	4.4	44 / 3	46	20	15.0	42	350
Madagascar	17.0	43	13	3.0	33.0	65.5	286	85	5.8	45 / 3	55	22	0.3	31	260
Malawi	11.7	47	22	2.6	17.7	29.0	149	108	6.5	46 / 3	39	14	15.0	44	160
Mauritius	1.2	16	7	0.9	1.4	1.5	23	14.1	1.9	25 / 6	72	43	0.1	100	3,830
Mayotte	0.2	41	9	3.2	0.4	0.6	255	—	5.0	47 / 2	60	28	—	—	—
Mozambique	17.5	41	28	1.3	17.5	19.0	9	201	5.1	45 / 3	34	29	13.0	41	210
Reunion	0.8	21	5	1.6	0.9	1.0	33	27	2.5	27 / 7	75	73	—	—	—
Rwanda	8.3	40	21	1.9	11.7	17.3	108	107	5.8	43 / 3	40	5	8.9	40	220
Seychelles	0.1	18	7	1.2	0.1	0.1	11	10	2.1	29 / 8	70	63	—	—	7,050
Somalia	8.0	48	19	2.9	14.9	25.5	218	126	7.2	45 / 3	46	28	1.0	—	120
Tanzania	35.4	40	17	2.3	52.0	73.8	109	105	5.3	45 / 3	45	22	7.8	57	270
Uganda	25.3	47	17	3.0	47.3	82.5	226	88	6.9	51 / 2	44	12	5.0	47	260
Zambia	10.9	43	21	2.2	13.6	17.5	60	95	5.9	47 / 3	41	36	15.6	48	320
Zimbabwe	12.6	32	20	1.2	12.8	14.6	16	65	4.0	40 / 3	41	32	33.7	73	480
MIDDLE AFRICA	**104**	**45**	**16**	**2.9**	**184**	**305**	**193**	**104**	**6.4**	**44 / 3**	**47**	**33**	**6.3**	**—**	**—**
Angola	13.1	48	20	2.9	25.2	43.1	230	145	6.8	47 / 3	40	32	5.5	40	500
Cameroon	15.7	37	15	2.2	22.4	30.9	96	77	4.9	43 / 3	48	48	11.8	39	580
Central African Republic	3.7	38	19	1.9	4.8	6.2	68	98	5.1	44 / 3	43	39	12.9	57	260
Chad	9.3	49	16	3.2	16.7	29.2	215	103	6.6	48 / 3	49	21	3.6	26	200
Congo	3.7	44	15	2.9	6.8	10.6	186	84	6.3	46 / 3	50	41	7.2	17	640
Congo, Dem. Rep. of	56.6	47	16	3.1	104.9	181.3	220	102	6.9	43 / 4	48	29	4.9	26	80
Equatorial Guinea	0.5	38	13	2.5	0.8	1.2	143	95	4.9	43 / 4	54	37	3.4	42	700
Gabon	1.3	32	11	2.2	1.9	2.5	87	57	4.3	42 / 4	59	73	4.2	47	3,160
Sao Tome and Principe	0.2	43	8	3.5	0.3	0.5	185	50	6.1	48 / 4	65	44	—	—	280
SOUTHERN AFRICA	**50**	**24**	**14**	**1.0**	**41**	**39**	**−22**	**59**	**3.0**	**35 / 5**	**52**	**50**	**21.5**	**—**	**—**
Botswana	1.6	28	25	0.3	10	0.9	−43	60	3.6	40 / 4	37	54	38.8	90	3,100
Lesotho	1.8	33	22	1.1	2.1	2.2	24	89	4.4	43 / 5	37	17	31.0	74	530
Namibia	1.9	36	15	2.1	2.1	2.6	37	65	4.9	43 / 4	49	27	22.5	67	1,960

	Population Mid–2003 (millions)	Births Per 1,000 Pop.	Deaths Per 1,000 Pop.	Rate of Natural Increase (%)	Projected Population in 2025 (millions)	Projected Population in 2050 (millions)	Projected Pop. Change 2003–2050 (%)	Infant Mortality Rate[a]	Total Fertility Rate[b]	Percent of Population of Age < 15/65 +	Life Expectancy at Birth (Years) Total	Percent Urban	Percent of Pop. 15–49 with HIV/AIDS End–2001	Rural % with Access to Improved Water Source	Per Capita GNI 2001 (US$)
SOUTHERN AFRICA (continued)															
South Africa	44.0	23	14	0.9	35.1	32.5	–26	57	2.8	34 / 5	53	53	20.1	73	2,820
Swaziland	1.2	32	16	1.5	1.1	1.1	–2	65	5.9	42 / 3	45	25	33.4	—	1,300
NORTH AMERICA	**323**	**14**	**8**	**0.5**	**387**	**459**	**42**	**7**	**2.0**	**21 / 13**	**77**	**79**	**0.6**	**—**	**—**
Canada	31.6	11	7	0.3	36.0	36.6	16	5.3	1.5	18 / 13	79	79	0.3	+	21,930
United States	291.5	14	9	0.6	351.1	421.8	45	6.9	2.0	21 / 13	77	79	0.6	+	34,280
LATIN AMERICA AND THE CARIBBEAN	**540**	**23**	**6**	**1.7**	**690**	**789**	**46**	**29**	**2.7**	**32 / 6**	**71**	**75**	**0.7**	**66**	**3,580**
CENTRAL AMERICA	**144**	**29**	**5**	**2.4**	**192**	**230**	**60**	**27**	**3.0**	**35 / 5**	**74**	**68**	**0.5**	**—**	**—**
Belize	0.3	29	6	2.3	0.4	0.6	110	21	3.7	41 / 4	67	45	2.0	82	2,910
Costa Rica	4.2	18	4	1.4	5.6	6.3	51	10	2.1	30 / 6	79	59	0.6	92	4,060
El Salvador	6.6	29	6	2.3	9.3	12.4	86	30	3.4	38 / 5	70	58	0.6	64	2,040
Guatemala	12.4	33	7	2.6	19.8	27.2	120	41	4.4	42 / 4	66	39	1.0	88	1,680
Honduras	6.9	34	5	2.9	10.7	14.7	114	34	4.2	42 / 4	71	46	1.6	81	900
Mexico	104.9	29	5	2.4	133.8	153.2	46	25	2.8	33 / 5	75	75	0.3	69	5,530
Nicaragua	5.5	32	5	2.7	8.3	10.9	98	31	3.8	42 / 3	69	57	0.2	59	420
Panama	3.0	23	5	1.8	4.2	5.0	68	21	2.7	32 / 6	74	62	1.5	79	3,260
CARIBBEAN	**38**	**20**	**8**	**1.2**	**46**	**51**	**36**	**38**	**2.7**	**29 / 7**	**69**	**62**	**2.4**	**—**	**—**
Antigua and Barbuda	0.1	24	6	1.7	0.1	0.1	1	17	2.7	26 / 8	71	37	—	89	9,070
Bahamas	0.3	18	5	1.3	0.3	0.3	9	15.8	2.1	30 / 5	72	84	3.5	86	14,960
Barbados	0.3	15	8	0.6	0.3	0.2	–2	13.2	1.8	22 / 12	73	38	1.3	100	9,250
Cuba	11.3	12	7	0.5	11.8	11.1	–2	6	1.6	21 / 10	76	75	0.1	77	—
Dominica	0.1	17	7	1.0	0.1	0.1	17	16.1	1.9	33 / 9	73	71	—	90	3.060
Dominican Republic	8.7	25	6	1.9	11.1	13.4	54	31	3.0	35 / 5	69	61	2.7	78	2,230
Grenada	0.1	19	7	1.2	0.1	0.1	–17	17	2.1	35 / 8	71	38	—	93	3,720
Guadeloupe	0.4	18	6	1.2	0.5	0.5	5	7.6	2.1	25 / 9	78	100	—	—	—
Haiti	7.5	32	14	1.8	11.1	15.1	100	80	4.7	40 / 4	51	36	6.1	45	480
Jamaica	2.6	21	7	1.5	3.3	3.7	38	24	2.4	31 / 7	75	52	1.2	85	2,800
Martinique	0.4	15	7	0.8	0.4	0.4	4	8	1.9	24 / 10	79	95	—	—	—
Netherlands Antilles	0.2	14	6	0.7	0.2	0.2	12	13	1.8	25 / 8	76	70	—	—	—
Puerto Rico	3.9	16	7	0.8	4.2	3.9	1	10.6	1.9	24 / 11	77	71	—	—	—
St. Kitts-Nevis	0.05	21	9	1.2	0.1	0.1	35	28	2.6	31 / 9	71	43	—	—	6,880
Saint Lucia	0.2	17	6	1.1	0.2	0.2	45	13.6	2.2	31 / 5	72	30	—	—	3,970
St. Vincent and the Grenadines	0.1	18	7	1.1	0.1	0.1	–16	19.3	2.1	37 / 7	72	44	—	—	2,690
Trinidad and Tobago	1.3	13	7	0.6	1.3	1.2	–7	18.6	1.6	25 / 7	71	72	2.5	—	5,540

	Population Mid–2003 (millions)	Births Per 1,000 Pop.	Deaths Per 1,000 Pop.	Rate of Natural Increase (%)	Projected Population in 2025 (millions)	Projected Population in 2050 (millions)	Projected Pop. Change 2003–2050 (%)	Infant Mortality Rate[a]	Total Fertility Rate[b]	Percent of Population of Age < 15/65 +	Life Expectancy at Birth (Years) Total	Percent Urban	Percent of Pop. 15–49 with HIV/AIDS End–2001	Rural % with Access to Improved Water Source	Per Capita GNI 2001 (US$)
SOUTH AMERICA	**358**	**22**	**6**	**1.5**	**452**	**507**	**42**	**30**	**2.5**	**31 / 6**	**70**	**79**	**0.6**	**—**	**—**
Argentina	36.9	19	8	1.2	47.2	54.5	48	16.6	2.5	31 / 9	74	89	0.7	73	6,940
Bolivia	8.6	32	9	2.3	12.2	15.4	79	61	4.1	39 / 5	63	63	0.1	64	950
Brazil	176.5	20	7	1.3	211.2	221.4	25	33	2.2	30 / 6	69	81	0.7	53	3,070
Chile	15.8	18	6	1.2	19.5	22.2	41	10.1	2.4	26 / 7	76	87	0.3	58	4,590
Colombia	44.2	23	6	1.8	58.1	67.3	52	28	2.7	33 / 5	71	71	0.4	70	1,890
Ecuador	12.6	27	6	2.1	17.5	21.7	73	35	3.2	36 / 4	71	61	0.3	75	1,080
French Guiana	0.2	31	4	2.8	0.3	0.4	95	12	3.6	33 / 5	76	79	—	—	—
Guyana	0.8	23	9	1.4	0.7	0.5	–34	53	2.4	30 / 4	63	36	2.7	91	840
Paraguay	6.2	31	5	2.7	10.1	15.0	142	37	4.2	39 / 5	71	54	0.1	59	1,350
Peru	27.1	26	7	2.0	35.7	42.8	58	33	2.9	34 / 5	69	72	0.4	62	1,980
Suriname	0.4	23	7	1.5	0.4	0.4	–18	27	2.7	32 / 6	70	69	1.2	50	1,690
Uruguay	3.4	16	9	0.6	3.8	4.2	24	13.5	2.2	24 / 13	75	93	0.3	93	5,710
Venezuela	25.7	24	5	1.9	35.2	41.7	62	19.6	2.8	34 / 4	73	87	0.5	70	4,760
ASIA	**3,830**	**20**	**.7**	**1.3**	**4,776**	**5,353**	**40**	**54**	**2.6**	**30 / 6**	**67**	**38**	**0.4**	**—**	**—**
ASIA (excl. China)	**2,541**	**24**	**8**	**1.6**	**3,322**	**3,959**	**56**	**60**	**3.1**	**34 / 5**	**65**	**38**	**0.5**	**—**	**—**
WESTERN ASIA	**204**	**27**	**7**	**2.0**	**308**	**418**	**105**	**47**	**3.8**	**36 / 5**	**68**	**62**	**z**	**—**	**—**
Armenia	3.2	14	8	0.6	3.4	3.4	4	36	1.7	24 / 10	72	64	0.2	—	570
Azerbaijan	8.2	14	6	0.8	9.7	11.6	41	13	1.9	29 / 6	72	51	z	58	650
Bahrain	0.7	21	3	1.8	1.0	1.2	75	8	2.6	28 / 3	74	87	0.3	—	9,370
Cyprus	0.9	12	6	0.6	1.0	1.0	9	5	1.6	22 / 11	77	66	0.3	100	12,370
Georgia	4.7	9	9	–0.0	3.9	2.6	–43	15	1.1	20 / 14	77	58	z	61	590
Iraq	24.2	35	10	2.5	41.5	60.5	150	103	5.4	47 / 3	58	68	z	48	2,170
Israel	6.7	21	6	1.5	9.3	11.0	64	5.3	2.9	28 / 10	79	91	0.1	+	16,750
Jordan	5.5	29	5	2.4	8.7	11.8	115	22	3.7	40 / 5	69	79	z	84	1,750
Kuwait	2.4	18	2	1.7	4.6	7.0	192	10	4.0	26 / 2	78	100	0.1	—	18,270
Lebanon	4.2	21	7	1.4	5.2	5.7	35	33	2.4	28 / 7	73	88	0.1	100	4,010
Oman	2.6	28	4	2.5	4.4	6.3	139	16	4.1	34 / 3	73	72	0.1	30	4,940
Palestinian Territory	3.6	39	4	3.5	7.4	11.9	228	26	5.7	46 / 3	72	57	—	—	—
Qatar	0.6	20	4	1.6	0.8	0.9	43	12	3.5	26 / 1	72	91	0.1	—	12,000
Saudi Arabia	24.1	35	6	2.9	46.1	74.2	208	25	5.7	43 / 3	72	83	z	64	8,460
Syria	17.5	28	5	2.4	27.6	35.0	99	18	3.8	40 / 4	70	50	z	64	1,040
Turkey	71.2	22	7	1.5	88.9	97.5	37	39	2.5	30 / 5	69	59	z	86	2,530
United Arab Emirates	3.9	16	2	1.4	4.7	4.9	27	8	3.0	26 / 1	74	78	0.2	—	18,060
Yemen	19.4	43	10	3.3	39.6	71.1	268	75	7.0	48 / 3	60	26	0.1	68	450
SOUTH CENTRAL ASIA	**1,563**	**27**	**9**	**1.8**	**2,084**	**2,546**	**63**	**69**	**3.3**	**37 / 4**	**62**	**30**	**0.6**	**—**	**—**
Afghanistan	28.7	42	18	2.4	45.9	67.2	134	154	6.0	43 / 3	46	22	z	11	250
Bangladesh	146.7	30	8	2.2	208.3	254.6	73	66	3.6	40 / 3	59	23	z	97	360
Bhutan	0.9	34	9	2.5	1.5	2.1	117	61	4.7	39 / 5	66	16	z	60	640

	Population Mid–2003 (millions)	Births Per 1,000 Pop.	Deaths Per 1,000 Pop.	Rate of Natural Increase (%)	Projected Population in 2025 (millions)	Projected Population in 2050 (millions)	Projected Pop. Change 2003–2050 (%)	Infant Mortality Rate[a]	Total Fertility Rate[b]	Percent of Population of Age < 15/65 +	Life Expectancy at Birth (Years) Total	Percent Urban	Percent of Pop. 15–49 with HIV/AIDS End–2001	Rural % with Access to Improved Water Source	Per Capita GNI 2001 (US$)
SOUTH CENTRAL ASIA															
(continued)															
India	1,068.6	25	8	1.7	1,363.0	1,628.0	52	66	3.1	36 / 4	63	28	0.8	79	460
Iran	66.6	18	6	1.2	84.7	96.5	45	32	2.5	33 / 5	69	66	0.1	83	1,680
Kazakhstan	14.8	15	10	0.5	14.7	13.3	–10	19	1.8	29 / 7	66	56	0.1	82	1,350
Kyrgyzstan	5.0	20	7	1.3	6.4	7.1	41	23	2.4	35 / 6	69	35	z	66	280
Maldives	0.3	24	4	2.0	0.4	0.5	77	17	3.7	39 / 4	67	27	0.1	100	2,040
Nepal	25.2	34	10	2.4	37.8	50.8	102	77	4.5	41 / 4	59	11	0.5	87	250
Pakistan	149.1	37	10	2.7	249.7	348.6	134	91	4.8	42 / 4	60	34	0.1	87	420
Sri Lanka	19.3	19	6	1.3	21.7	21.4	11	13	2.0	27 / 7	72	30	z	70	880
Tajikistan	6.6	19	4	1.4	8.6	10.0	53	19	2.4	42 / 4	68	27	z	47	180
Turkmenistan	5.7	19	5	1.3	7.7	8.8	55	25	2.2	38 / 4	67	44	z	—	950
Uzbekistan	25.7	20	5	1.5	33.2	37.2	45	20	2.5	38 / 4	70	38	z	79	550
SOUTHEAST ASIA	**544**	**22**	**7**	**1.6**	**697**	**792**	**46**	**41**	**2.7**	**31 / 5**	**68**	**37**	**0.6**	**—**	**—**
Brunei	0.4	22	3	1.9	0.5	0.7	90	7	2.3	31 / 3	76	67	0.2	—	24,630
Cambodia	12.6	28	10	1.8	18.5	24.4	94	95	4.0	43 / 4	56	16	2.7	26	270
East Timor	0.8	26	13	1.3	1.2	1.4	84	129	4.1	44 / 5	49	8	—	—	—
Indonesia	220.5	22	6	1.6	281.9	315.6	43	46	2.6	31 / 5	68	40	0.1	69	690
Laos	5.6	36	13	2.3	8.5	11.3	102	104	4.9	43 / 4	54	17	0.1	29	300
Malaysia	25.1	26	4	2.1	34.3	46.6	86	11	3.3	34 / 4	73	57	0.4	94	3,330
Myanmar	49.5	25	11	1.4	59.7	64.4	30	87	3.1	33 / 5	57	27	2.0	66	220
Philippines	81.6	28	6	2.2	111.5	132.8	63	26	3.5	37 / 4	70	47	z	79	1,030
Singapore	4.2	11	4	0.7	4.8	4.4	6	2.5	1.4	21 / 7	79	100	0.2	+	21,500
Thailand	63.1	13	6	0.7	72.1	72.8	15	20	1.7	23 / 7	71	31	1.8	81	1,940
Vietnam	80.8	19	6	1.3	104.1	117.2	45	26	2.3	30 / 6	72	25	0.3	72	410
EAST ASIA	**1,519**	**13**	**7**	**0.6**	**1,688**	**1,597**	**5**	**29**	**1.7**	**22 / 8**	**72**	**45**	**0.1**	**—**	**—**
China	1,288.7	13	6	0.6	1,454.7	1,393.6	8	32	1.7	22 / 7	71	39	0.1	66	890
China, Hong Kong SAR[c]	6.8	7	5	0.2	8.4	7.5	10	2.6	0.9	16 / 11	81	100	0.1	—	25,330
China, Macao SAR[c]	0.4	7	3	0.4	0.6	0.8	76	4	0.9	22 / 7	77	99	—	—	14,580
Japan	127.5	9	8	0.1	121.1	100.6	–21	3.0	1.3	14 / 19	81	78	z	+	35,610
Korea, North	22.7	17	11	0.6	24.7	24.9	10	45	2.0	27 / 6	63	59	z	—	—
Korea, South	47.9	12	5	0.7	50.6	44.3	–8	8	1.3	21 / 8	76	79	z	71	9,460
Mongolia	2.5	18	8	1.1	3.2	3.6	45	30	2.7	36 / 5	65	57	z	30	400
Taiwan	22.6	11	6	0.5	24.4	22.1	–2	6.0	1.3	20 / 9	76	78	—	—	—
EUROPE	**727**	**10**	**12**	**–0.2**	**722**	**664**	**–9**	**8**	**1.4**	**17 / 15**	**74**	**73**	**0.4**	**—**	**—**
NORTHERN EUROPE	**95**	**11**	**10**	**0.1**	**101**	**101**	**6**	**5**	**1.6**	**19 / 16**	**78**	**83**	**0.1**	**+**	**—**
Channel Islands	0.2	11	9	0.2	0.2	0.1	–3	2.8	1.4	17 / 15	78	30	—	+	—
Denmark	5.4	12	11	0.1	5.9	5.8	8	4.9	1.7	19 / 15	77	72	0.2	+	30,600
Estonia	1.4	9	14	–0.4	1.2	0.9	–35	9	1.3	17 / 16	71	67	1.0	+	3,870
Finland	5.2	11	9	0.1	5.3	4.8	–8	3.2	1.7	18 / 15	78	62	0.1	+	23,780

	Population Mid–2003 (millions)	Births Per 1,000 Pop.	Deaths Per 1,000 Pop.	Rate of Natural Increase (%)	Projected Population in 2025 (millions)	Projected Population in 2050 (millions)	Projected Pop. Change 2003–2050 (%)	Infant Mortality Rate[a]	Total Fertility Rate[b]	Percent of Population of Age < 15/65 +	Life Expectancy at Birth (Years) Total	Percent Urban	Percent of Pop. 15–49 with HIV/AIDS End–2001	Rural % with Access to Improved Water Source	Per Capita GNI 2001 (US$)
NORTHERN EUROPE (continued)															
Iceland	0.3	14	6	0.8	0.3	0.4	31	2.7	1.9	23 / 12	80	94	0.2	+	28,880
Ireland	4.0	15	8	0.7	4.5	4.7	18	5.8	2.0	21 / 11	77	58	0.1	+	22,850
Latvia	2.3	9	14	–0.5	2.2	1.8	–24	1.1	1.2	17 / 15	71	68	0.4	+	3,230
Lithuania	3.5	9	12	–0.3	3.5	3.1	–10	8	1.2	19 / 14	72	67	0.1	+	3,350
Norway	4.6	12	10	0.3	5.1	5.6	22	3.9	1.7	20 / 15	79	74	0.1	+	35,630
Sweden	9.0	11	11	0.0	9.6	10.0	11	3.7	1.6	18 / 17	80	84	0.1	+	25,400
United Kingdom	59.2	11	10	0.1	62.9	63.7	8	5.4	1.6	19 / 16	78	90	0.1	+	25,120
WESTERN EUROPE	**185**	**11**	**10**	**0.1**	**187**	**177**	**–4**	**4**	**1.6**	**17 / 16**	**79**	**78**	**0.2**	**+**	**—**
Austria	8.2	9	9	0.0	8.4	8.2	1	4.8	1.3	16 / 16	79	54	0.2	+	23,940
Belgium	10.4	11	10	0.1	10.8	11.0	6	5.0	1.6	18 / 17	78	97	0.2	+	23,850
France	59.8	13	9	0.4	63.4	64.0	7	4.2	1.9	19 / 16	79	74	0.3	+	22,730
Germany	82.6	9	10	–0.1	78.1	67.7	–18	4.3	1.3	15 / 17	78	86	0.1	+	23,560
Liechtenstein	0.04	12	7	0.5	0.04	0.04	11	7.9	1.4	18 / 10	—	21	—	+	—
Luxembourg	0.5	12	8	0.4	0.6	0.6	31	5.9	1.7	19 / 14	78	88	0.2	+	41,770
Monaco	0.03	23	16	0.6	0.04	0.04	15	—	—	15 / 23	—	100	—	+	—
Netherlands	16.2	13	9	0.4	17.7	18.0	11	5.4	1.7	19 / 14	78	62	0.2	+	24,330
Switzerland	7.3	10	8	0.2	7.6	7.4	0	4.9	1.4	17 / 16	80	68	0.5	+	38,330
EASTERN EUROPE	**301**	**9**	**14**	**–0.5**	**285**	**247**	**–18**	**13**	**1.2**	**18 / 13**	**68**	**68**	**0.6**	**—**	**—**
Belarus	9.9	9	14	–0.5	9.4	8.5	–14	9	1.3	18 / 14	69	71	0.3	+	1,290
Bulgaria	7.5	8	14	–0.6	6.0	4.5	–40	13.8	1.2	15 / 17	72	69	z	+	1,650
Czech Republic	10.2	9	11	–0.2	10.1	9.2	–10	4.1	1.2	16 / 14	75	77	z	+	5,310
Hungary	10.1	10	13	–0.4	8.9	7.6	–25	7.2	1.3	16 / 15	72	65	0.1	98	4,830
Moldova	4.3	9	9	–0.1	4.6	4.6	8	16	1.3	22 / 10	68	46	0.2	88	400
Poland	38.6	10	9	0.0	38.6	33.9	–12	7.7	1.3	18 / 13	74	62	0.1	—	4,230
Romania	21.6	10	12	–0.3	20.6	17.1	–21	18.4	1.2	18 / 14	71	55	z	16	1,720
Russia	145.5	10	16	–0.7	136.9	119.1	–18	15	1.3	18 / 13	65	73	0.9	96	1,750
Slovakia	5.4	10	10	–0.0	5.2	4.7	–12	6.2	1.2	19 / 11	74	57	z	100	3,760
Ukraine	47.8	8	15	–0.8	45.1	38.4	–20	11	1.1	17 / 14	68	67	1.0	94	720
SOUTHERN EUROPE	**147**	**10**	**9**	**0.1**	**149**	**138**	**–6**	**6**	**1.3**	**16 / 17**	**78**	**70**	**0.4**	**—**	**—**
Albania	3.1	16	5	1.1	3.6	3.6	16	12	2.0	32 / 6	74	46	z	95	1,340
Andorra	0.1	12	4	0.8	0.1	0.1	0	4	1.3	15 / 13	—	92	—	+	1,240
Bosnia-Herzegovina	3.9	10	8	0.2	3.9	3.3	–15	8	1.3	19 / 9	72	40	z	—	1,240
Croatia	4.3	9	11	–0.2	4.4	4.3	–1	8.4	1.3	17 / 16	74	54	z	—	4,550
Greece	11.0	9	9	–0.0	10.4	9.7	–12	5.9	1.2	14 / 19	78	59	0.2	+	11,430
Italy	57.2	9	10	–0.1	57.6	52.3	–9	4.8	1.2	14 / 19	80	90	0.4	+	19,390
Macedonia	2.1	13	8	0.5	2.2	2.1	2	11.9	1.7	22 / 10	73	59	z	—	1,690
Malta	0.4	10	8	0.2	0.4	0.4	–8	3.4	1.5	19 / 13	77	91	0.1	+	9,120
Portugal	10.4	11	10	0.1	10.3	9.4	–10	5.0	1.5	16 / 16	77	48	0.5	+	10,900

	Population Mid–2003 (millions)	Births Per 1,000 Pop.	Deaths Per 1,000 Pop.	Rate of Natural Increase (%)	Projected Population in 2025 (millions)	Projected Population in 2050 (millions)	Projected Pop. Change 2003–2050 (%)	Infant Mortality Rate[a]	Total Fertility Rate[b]	Percent of Population of Age < 15/65 +	Life Expectancy at Birth (Years) Total	Percent Urban	Percent of Pop. 15–49 with HIV/AIDS End–2001	Rural % with Access to Improved Water Source	Per Capita GNI 2001 (US$)
SOUTHERN EUROPE (continued)															
San Marino	0.03	11	7	0.4	0.04	0.04	23	3.2	1.3	15 / 16	81	84	—	+	—
Serbia and Montenegro	10.7	12	11	0.2	10.7	10.2	–4	13	1.7	20 / 14	73	52	0.2	+	—
Slovenia	2.0	9	9	–0.1	2.0	1.7	–15	9.2	1.2	16 / 14	76	50	z	100	9,760
Spain	41.3	10	9	0.1	43.5	41.3	0	3.5	1.2	15 / 17	79	64	0.5	+	14,300
OCEANIA	**32**	**18**	**7**	**1.1**	**42**	**50**	**56**	**25**	**2.4**	**25 / 10**	**75**	**69**	**0.2**	**—**	**—**
Australia	19.9	13	7	0.6	25.0	29.5	48	5.1	1.7	20 / 13	80	85	0.1	+	19,900
Federated States of Micronesia	0.1	29	6	2.3	0.1	0.2	45	37	4.1	40 / 4	68	27	—	—	2,150
Fiji	0.9	25	6	1.9	1.0	1.0	15	20	3.3	35 / 3	67	46	0.1	51	2,130
French Polynesia	0.2	20	4	1.6	0.3	0.4	45	7	2.5	31 / 4	72	53	—	—	17,290
Guam	0.2	23	4	1.8	0.2	0.3	65	9.8	3.5	30 / 5	78	93	—	—	—
Kiribati	0.1	33	8	2.5	0.2	0.2	137	55	4.5	41 / 3	62	37	—	25	830
Marshall Islands	0.1	42	5	3.7	0.1	0.1	87	37	5.7	43 / 2	68	68	—	—	2,190
Nauru	0.01	28	7	2.1	0.02	0.02	92	25	3.7	41 / 2	61	100	—	—	—
New Caledonia	0.2	22	5	1.7	0.3	0.4	67	5	2.6	30 / 5	73	71	—	—	15,060
New Zealand	4.0	14	7	0.7	4.7	5.1	27	5.3	1.9	22 / 12	78	77	0.1	+	13,250
Palau	0.02	20	7	1.3	0.02	0.03	30	17	2.5	27 / 5	69	71	—	20	6,730
Papua New Guinea	5.5	33	8	2.5	8.3	11.1	102	60	4.4	39 / 4	57	15	0.7	32	580
Samoa	0.2	30	5	2.4	0.2	0.2	42	25	4.3	41 / 5	69	21	—	100	1,520
Solomon Islands	0.5	35	4	3.1	0.8	1.1	118	25	4.8	44 / 3	71	13	—	65	580
Tonga	0.1	28	7	2.1	0.1	0.1	14	22	4.0	41 / 4	68	32	—	100	1,530
Tuvalu	0.01	22	8	1.4	0.02	0.02	90	29	3.1	34 / 5	66	42	—	100	—
Vanuatu	0.2	32	6	2.7	0.3	0.4	103	45	4.4	42 / 3	67	21	—	94	1,050

A dash (—) indicates data unavailable or inapplicable.

A plus sign (+) indicates essentially 100% access to safe water.

z = Less than 0.5%.

[a]Infant deaths per 1000 live births. Rates with decimals indicate complete national registration; whole numbers are UN estimates. Rates in italics are based on less than 50 annual infant deaths and are subject to great yearly variability.

[b]Average number of children born to a woman during her lifetime.

[c]Special Administrative Region.

Urban population data are the percentage of the total population living in areas termed urban by that country.

Data for safe water supply are based on World Health Organization reports.

GNI = Gross National Income; data are from the World Bank.

Table modified from the *2003 World Population Data Sheet* of the Population Reference Bureau.

Appendix C
Anglo America Reference Map

Glossary

Terms in italics identify related glossary items.

A

absolute direction Direction with respect to cardinal east, west, north, and south reference points.

absolute distance (*syn:* geodesic distance) The shortest-path separation between two places measured on a standard unit of length (miles or kilometers, usually); also called real distance.

absolute location (*syn:* mathematical location) The exact position of an object or place stated in spatial coordinates of a grid system designed for locational purposes. In geography, the reference system is the *globe grid* of parallels of *latitude* north or south of the *equator* and of meridians of *longitude* east or west of a *prime meridian*. Absolute globe locations are cited in degrees, minutes, and (for greater precision) seconds of latitude and longitude north or south and east or west of the equatorial and prime meridian base lines.

absorbing barrier A condition that blocks the *diffusion* of an *innovation* or prevents its adoption.

accessibility The relative ease with which a destination may be reached from other locations; the relative opportunity for *spatial interaction*. May be measured in geometric, social, or economic terms.

acculturation Cultural modification or change that results when one *culture* group or individual adopts traits of a dominant or *host society*; cultural development or change through "borrowing."

acid rain *Precipitation* that is unusually acidic; created when oxides of sulfur and nitrogen change chemically as they dissolve in water vapor in the *atmosphere* and return to earth as acidic rain, snow, or fog.

activity space The area within which people move freely on their rounds of regular activity.

adaptation (1) Genetic modification making a population more fit for existence under specific environmental conditions; (2) in immigration, the term summarizes how individuals, households, and communities respond and adjust to new experiences and social and cultural surroundings.

agglomeration The spatial grouping of people or activities for mutual benefit; in *economic geography,* the concentration of productive enterprises for collective or cooperative use of *infrastructure* and sharing of labor resources and market access.

agglomeration economies (*syn:* external economies) The savings to an individual enterprise derived from locational association with a cluster of other similar economic activities, such as other factories or retail stores.

agricultural density The number of rural residents per unit of agriculturally productive land; a variant of *physiological density* that excludes urban population.

agriculture The science and practice of farming, including the cultivation of the soil and the rearing of livestock.

amalgamation theory In *ethnic geography,* the concept that multiethnic societies become a merger of the *culture traits* of their member groups.

anecumene See *nonecumene.*

animism A belief that natural objects may be the abode of dead people, spirits, or gods who occasionally give the objects the appearance of life.

antecedent boundary A *boundary* line established before the area in question is well populated.

antipode The point on the earth's surface that is diametrically opposite the observer's location.

aquaculture Production and harvesting of fish and shellfish in land-based ponds.

aquifer A porous, water-bearing layer of rock, sand, or gravel below ground level.

arable land Land that is or can be cultivated.

arithmetic density See *crude density.*

artifacts The material manifestations of *culture,* including tools, housing, systems of land use, clothing, and the like. Elements in the *technological subsystem* of culture.

artificial boundary See *geometric boundary.*

aspect In *map projections,* the positional relationship between the globe and the *developable surface* on which it is visually projected.

assimilation A two-part *behavioral* and *structural* process by which a minority population reduces or loses completely its identifying cultural characteristics and blends into the *host society.*

atmosphere The air or mixture of gases surrounding the earth.

autonomous nationalism Movement by a dissident minority intent to achieve partial or total independence of territory it occupies from the *state* within which it lies.

awareness space Locations or places about which an individual has knowledge even without visiting all of them; includes *activity space* and additional areas newly encountered or about which one acquires information.

azimuth Direction of a line defined at its starting point by its angle in relation to a *meridian.*

azimuthal projection See *planar projection.*

B

basic sector Those products or services of an *urban* economy that are exported outside the city itself, earning income for the community.

behavioral assimilation (*syn:* cultural assimilation) The process of integration into a common cultural life through acquisition of the sentiments, attitudes, and experiences of other groups.

beneficiation The enrichment of low-grade ores through concentration and other processes to reduce their waste content and increase their *transferability.*

bilingualism Describing a society's use of two *official languages.*

biomass The total dry weight of all living organisms within a unit area; plant and animal matter that can in any way be used as a source of energy.

biome A major ecological community, including plants and animals, occupying an extensive earth area.

biosphere (*syn:* ecosphere) The thin film of air, water, and earth within which we live, including the *atmosphere,* surrounding and subsurface waters, and the upper reaches of the earth's crust.

birth rate The ratio of the number of live births during one year to the total population, usually at the midpoint of the same year, expressed as the number of births per year per 1000 population.

Boserup thesis The view that population growth independently forces a conversion from extensive to intensive *subsistence* agriculture.

boundary A line separating one political unit from another; see *international boundary.*

boundary dispute See *functional dispute.*

Brandt Report Entitled *North–South: A Program for Survival,* a report of the Independent Commission on International Development Issues, published in 1980 and named for the commission chairman, Willy Brandt.

break-of-bulk point A location where goods are transferred from one type of carrier to another (e.g., from barge to railroad).

Buddhism A *universalizing religion,* primarily of eastern and central Asia, based on teachings of Siddhartha Gautama, the Buddha, that suffering is inherent in all life but can be relieved by mental and moral self-purification.

built environment That part of the *physical landscape* that represents *material culture; the* buildings, roads, bridges, and similar structures large and small of the *cultural landscape.*

carrying capacity The maximum population numbers that an area can support on a continuing basis without experiencing unacceptable deterioration; for humans, the numbers supportable by an area's known and used resources—usually agricultural ones.

cartogram A map that has been simplified to present a single idea in a diagrammatic way; the base is not normally true to scale.

caste One of the hereditary social classes in *Hinduism* that determines one's occupation and position in society.

central business district (CBD) The nucleus or "downtown" of a city, where retail stores, offices, and cultural activities are concentrated, mass transit systems converge, and land values and building densities are high.

central city That part of the *metropolitan area* contained within the boundaries of the main city around which suburbs have developed.

central place An *urban* or other settlement node whose primary function is to provide goods and services to the consuming population of its *hinterland, complementary region,* or trade area.

central place theory A deductive theory formulated by Walter *Christaller* (1893–1969) to explain the size and distribution of settlements through reference to competitive supply of goods and services to dispersed rural populations.

centrifugal force 1: In *urban geography,* economic and social forces pushing households and businesses outward from central and inner-city locations. **2:** In *political geography,* forces of disruption and dissolution threatening the unity of a *state.*

centripetal force 1: In *urban geography,* a force attracting establishments or activities to the city center. **2:** In *political geography,* forces tending to bind together the citizens of a state.

chain migration The process by which *migration* movements from a common home area to a specific destination are sustained by links of friendship or kinship between first movers and later followers.

channelized migration The tendency for *migration* to flow between areas that are socially and economically allied by past migration patterns, by economic and trade connections, or by some other affinity.

charter group In plural societies, the early arriving ethnic group that created the *first effective settlement* and established the recognized cultural norms to which other, later groups are expected to conform.

chlorofluorocarbons (CFCs) A family of synthetic chemicals that has significant commercial applications but whose emissions are contributing to the depletion of the *ozone* layer.

choropleth map A *thematic map* presenting spatial data as average values per unit area.

Christaller Walter Christaller (1893–1969), German geographer credited with developing *central place theory* (1933).

Christianity A *monotheistic, universalizing religion* based on the teachings of Jesus Christ and of the Bible as sacred scripture.

circular and cumulative causation A process through which tendencies for economic growth are self-reinforcing; an expression of the *multiplier effect,* it tends to favor major cities and *core* regions over less-advantaged *peripheral* regions.

city A multifunctional nucleated settlement with a *central business district* and both residential and nonresidential land uses.

climate A summary of weather conditions in a place or region over a period of time.

cluster migration A pattern of movement and settlement resulting from the collective action of a distinctive social or *ethnic group.*

cognitive map See *mental map.*

cohort A population group unified by a specific common characteristic, such as age, and subsequently treated as a statistical unit during their lifetimes.

collective farm In the former Soviet *planned economy,* the cooperative operation of an agricultural enterprise under state control of production and market, but without full status or support as a state enterprise.

colony In *ethnic geography,* an urban ethnic area serving as point of entry and temporary *acculturation* zone for a specific immigrant group.

commercial economy A system of production of goods and services for exchange in competitive markets where price and availability are determined by supply and demand forces.

commercial energy Commercially traded fuels, such as coal, oil, or natural gas; excluding wood, vegetable or animal wastes, or other *biomass.*

compact state A *state* whose territory is nearly circular.

comparative advantage The principle that an area produces the items for which it has the greatest ratio of advantage or the least ratio of disadvantage in comparison to other areas, assuming free trade exists.

complementarity The actual or potential relationship of two places or regions that each produce different goods or services for which the other has an effective demand, resulting in an exchange between the locales.

complementary region The area served by a *central place.*

concentration In *spatial distributions,* the clustering of a phenomenon around a central location.

concentric zone model A model describing urban land uses as a series of circular belts or rings around a core *central business district,* each ring housing a distinct type of land use.

conformality The map property of correct angles and shapes of small areas.

conformal projection A *map projection* that retains correct shapes of small areas; lines of *latitude* and *longitude* cross at right angles and *scale* **(1)** is the same in all directions at any point on the map.

Confucianism A Chinese *value system* and *ethnic religion* emphasizing ethics, social morality, tradition, and ancestor worship.

conic projection A *map projection* employing a cone placed tangent or secant to the globe as the presumed *developable surface.*

connectivity The directness of routes linking pairs of places; an indication of the degree of internal connection in a transport *network.* More generally, all of the tangible and intangible means of connection and communication between places.

consequent boundary (*syn:* ethnographic boundary) A *boundary* line that coincides with some cultural divide, such as religion or language.

conservation The wise use or preservation of natural resources so as to maintain supplies and qualities at levels sufficient to meet present and future needs.

contagious diffusion A form of *expansion diffusion* that depends on direct contact. The process of dispersion is centrifugal, strongly influenced by distance, and dependent on interaction between actual and potential adopters of the *innovation.* Its name derives from the pattern of spread of contagious diseases.

containment A guiding principle of U.S. foreign policy during the Cold War period: to prevent or restrict the expansion of the Soviet Union's influence or control beyond its then existing limits.

continental shelf A gently sloping seaward extension of the landmass found off the coasts of many continents; its outer margin is marked by a transition to the ocean depths at about 200 meters (660 feet).

conurbation A continuous, extended *urban* area formed by the growing together of several formerly separate, expanding cities.

Convention on the Law of the Sea See *United Nations Convention on the Law of the Sea.*

core area 1: In *economic geography,* a "core region," the national or world districts of concentrated economic power, wealth, innovation, and advanced technology. **2:** In *political geography,* the heartland or nucleus of a *state,* containing its most developed area, greatest wealth, densest populations, and clearest national identity.

core-periphery model A model of the spatial structure of an economic system in which underdeveloped or declining peripheral areas are defined with respect to their dependence on a dominating developed *core region.*

core region See *core area* (1).

counter migration (*syn:* return migration) The return of migrants to the regions from which they earlier emigrated.

country See *state.*

creole A *language* developed from a *pidgin* to become the native tongue of a society.

critical distance The distance beyond which cost, effort, and/or means play a determining role in the willingness of people to travel.

crop rotation The annual alteration of crops that make differential demands on or contributions to soil fertility.

crude birth rate (CBR) See *birth rate.*

crude death rate (CDR) See *death rate.*

crude density (*syn:* arithmetic density) The number of people per unit area of land.

cultural assimilation See *behavioral assimilation.*

cultural convergence The tendency for *cultures* to become more alike as they increasingly share *technology* and organizational structures in a modern world united by improved transportation and communication.

cultural divergence The likelihood or tendency for *cultures* to become increasingly dissimilar with the passage of time.

cultural ecology The study of the interactions between societies and the natural *environments* they occupy.

cultural geography A branch of *systematic geography* that focuses on culturally determined human activities, the impact of *material* and *nonmaterial* human *culture* on the environment, and the human organization of space.

cultural integration The interconnectedness of all aspects of a *culture;* no part can be altered without creating an impact on other components of the culture.

cultural lag The retention of established *culture traits* despite changing circumstances rendering them inappropriate.

cultural landscape The *natural landscape* as modified by human activities and bearing the imprint of a *culture* group or society; the *built environment.*

culture 1: A society's collective beliefs, symbols, values, forms of behavior, and social organizations, together with its tools, structures, and artifacts created according to the group's conditions of life; transmitted as a heritage to succeeding generations and undergoing adoptions, modifications, and changes in the process. **2:** A collective term for a group displaying uniform cultural characteristics.

culture complex A related set of *culture traits* descriptive of one aspect of a society's behavior or activity. Culture complexes may be as basic as those associated with food preparation, serving, and consumption or as involved as those associated with religious beliefs or business practices.

culture hearth A nuclear area within which an advanced and distinctive set of *culture traits,* ideas, and *technologies* develops and from which there is *diffusion* of those characteristics and the *cultural landscape* features they imply.

culture realm A collective of *culture regions* sharing related culture systems; a major world area having sufficient distinctiveness to be perceived as set apart from other realms in terms of cultural characteristics and complexes.

culture rebound The readoption by later generations of *culture traits* and identities associated with immigrant forebears or ancestral homelands.

culture region A *formal* or *functional region* within which common cultural characteristics prevail. It may be based on single *culture traits,* on *culture complexes,* or on political, social, or economic integration.

culture system A generalization suggesting shared, identifying traits uniting two or more *culture complexes.*

culture trait A single distinguishing feature of regular occurrence within a *culture,* such as

the use of chopsticks or the observance of a particular caste system. A single element of learned behavior.

cumulative causation See *circular and cumulative causation.*

custom The body of traditional practices, usages, and conventions that regulate social life.

cylindrical projection A *map projection* employing a cylinder wrapped around the globe as the presumed *developable surface.*

Daoism See *Taoism.*

death rate (*syn:* mortality rate) A mortality index usually calculated as the number of deaths per year per 1000 population.

deforestation The clearing of land through total removal of forest cover.

deglomeration The process of deconcentration; the location of industrial or other activities away from established *agglomerations* in response to growing costs of congestion, competition, and regulation.

deindustrialization The cumulative and sustained decline in the contribution of manufacturing to a national economy.

demographic equation A mathematical expression that summarizes the contribution of different demographic processes to the population change of a given area during a specified time period. $P_2 = P_1 + B_{1-2} - D_{1-2} + IM_{1-2} - OM_{1-2}$, where P_2 is population at time 2; P_1 is population at beginning date; B_{1-2} is the number of births between times 1 and 2; D_{1-2} is the number of deaths during that period; IM_{1-2} is the number of in-migrants and OM_{1-2} the number of out-migrants between times 1 and 2.

demographic momentum (*syn:* population momentum) The tendency for population growth to continue despite stringent family planning programs because of a relatively high concentration of people in the childbearing years.

demographic transition A model of the effect of economic development on population growth. A first stage involves stable numbers with both high *birth rates* and *death rates;* the second displays high birth rates, falling death rates, and population increases. Stage three shows reduction in population growth as birth rates decline to the level of death rates. The fourth and final stage again implies a population stable in size but with larger numbers than at the start of the transition process. An idealized summary of population history of industrializing Europe, its application to newly developing countries is questioned.

demography The scientific study of population, with particular emphasis upon quantitative aspects.

density The quantity of anything (people, buildings, animals, traffic, etc.) per unit area.

dependency ratio The number of dependents, old or young, that each 100 persons in the economically productive years must on average support.

desertification Extension of desertlike landscapes as a result of overgrazing, destruction of the forests, or other human-induced changes, usually in semiarid regions.

developable surface *Projection* surface (such as a plane, cone, or cylinder) that is or can be made flat without distortion.

development The process of growth, expansion, or realization of potential; bringing regional resources into full productive use.

devolution The transfer of certain powers from the *state* central government to separate political subdivisions within the state's territory.

dialect A *language* variant marked by vocabulary, grammar, and pronunciation differences from other variants of the same common language. When those variations are spatial or regional, they are called *geographic dialects;* when they are indicative of socioeconomic or educational levels, they are called *social dialects.*

dialect geography See *linguistic geography.*

dibble Any small hand tool or stick to make a hole for planting.

diffusion The spread or movement of a phenomenon over space or through time. The dispersion of a *culture trait* or characteristic or new ideas and practices from an origin area (e.g., *language,* plant *domestication,* new industrial *technology*). Recognized types include *relocation, expansion, contagious,* and *hierarchical* diffusion.

diffusion barrier Any condition that hinders the flow of information, the movement of people, or the spread of an *innovation.*

direction bias A statement of *movement bias* observing that among all possible directions of movement or flow, one or only a very few are favored and dominant.

dispersion In *spatial distributions,* a statement of the amount of spread of a phenomenon over area or around a central location. Dispersion in this sense represents a continuum from clustered, concentrated, or agglomerated (at one end) to dispersed or scattered (at the other).

distance bias A statement of *movement bias* observing that short journeys or interchanges are favored over more distant ones.

distance decay The declining intensity of any activity, process, or function with increasing distance from its point of origin.

domestication The successful transformation of plant or animal species from a wild state to a condition of dependency on human management, usually with distinct physical change from wild forebears.

domino theory A *geopolitics* theory made part of American *containment* (of the former Soviet Union) policy beginning in the 1950s. The theory maintained that if a single country fell under Soviet influence or control, its neighbors would likely follow, creating a ripple effect like a line of toppling dominos.

doubling time The time period required for any beginning total experiencing a compounding growth to double in size.

#

ecology The scientific study of how living creatures affect each other and what determines their distribution and abundance.

economic base The manufacturing and service activities performed by the *basic sector* of a city's labor force; functions of a city performed to satisfy demands external to the city itself and, in that performance, earning income to support the urban population.

economic geography The branch of *systematic geography* concerned with how people support themselves, with the spatial patterns of production, distribution, and consumption of goods and services, and with the areal variation of economic activities over the surface of the earth.

ecosphere See *biosphere.*

ecosystem A population of organisms existing together in a small, relatively homogeneous area (pond, forest, small island), together with the energy, air, water, soil, and chemicals upon which it depends.

ecumene That part of the earth's surface physically suitable for permanent human settlement; the permanently inhabited areas of the earth.

edge city Distinct sizeable nodal concentration of retail and office space of lower than central city densities and situated on the outer fringes of older metropolitan areas; usually localized by or near major highway intersections.

electoral geography The study of the geographical elements of the organization and results of elections.

elongated state A *state* whose territory is long and narrow.

enclave A small bit of foreign territory lying within a *state* but not under its jurisdiction.

environment Surroundings; the totality of things that in any way may affect an organism, including both physical and cultural conditions; a region characterized by a certain set of physical conditions.

environmental determinism The view that the physical *environment,* particularly *climate,* controls human action, molds human behavior, and conditions cultural development.

environmental perception The concept that people of different *cultures* will differently observe and interpret their *environment* and make different decisions about its nature, potentialities, and use.

environmental pollution See *pollution.*

epidemiologic transition The reduction of periodically high mortality rates from epidemic diseases as those diseases become essentially continual within a population that develops partial immunity to them.

equal-area (equivalent) projection A *map projection* designed so that a unit area drawn anywhere on the map always represents the same area on the earth's surface.

equator An imaginary east-west line that encircles the globe halfway between the North and South Poles.

equidistant projection A *map projection* showing true distances in all directions from one or two central points; all other distances are incorrect.

equivalence/equivalent projection In *map projections,* the characteristic that a unit area drawn on the map always represents the same area on the earth's surface, regardless of where drawn. See also *equal-area projection.*

erosion The wearing away and removal of rock and soil particles from exposed surfaces by agents such as moving water, wind, or ice.

ethnic enclave A small area occupied by a distinctive minority *culture.*

ethnic geography The study of spatial distributions and interactions of *ethnic groups* and of the cultural characteristics on which they are based.

ethnic group People sharing a distinctive *culture,* frequently based on common national origin, *religion, language,* or *race.*

ethnic island A small rural area settled by a single, distinctive *ethnic group* that placed its imprint on the landscape.

ethnicity Ethnic quality; affiliation with a group whose racial, cultural, religious, or linguistic characteristics or national origins distinguish it from a larger population within which it is found.

ethnic province A large territory, urban and rural, dominated by or closely associated with a single *ethnic group.*

ethnic religion A *religion* identified with a particular *ethnic group* and largely exclusive to it. Such a religion does not seek converts.

ethnic separatism Desired *regional autonomy* expressed by a culturally distinctive group within a larger, politically dominant *culture.*

ethnocentrism Conviction of the evident superiority of one's own *ethnic group.*

ethnographic boundary See *consequent boundary.*

European Union (EU) An economic association established in 1957 by a number of Western European countries to promote free trade among members; often called the Common Market.

evapotranspiration The return of water from the land to the *atmosphere* through evaporation from the soil surface and transpiration from plants.

exclave A portion of a *state* that is separated from the main territory and surrounded by another country.

exclusive economic zone (EEZ) As established in the *United Nations Convention on the Law of the Sea*, a zone of exploitation extending 200 nautical miles (370 km) seaward from a coastal state that has exclusive mineral and fishing rights over it.

expansion diffusion The spread of ideas, behaviors, or articles through a culture area or from one *culture* to neighboring areas through contact and exchange of information; the dispersion leaves the phenomenon intact or intensified in its area of origin.

extensive agriculture A crop or livestock system characterized by low inputs of labor per unit area of land. It may be part of either a *subsistence* or a *commercial* economy.

external economies See *agglomeration economies*.

extractive industries *Primary activities* involving the mining and quarrying of *nonrenewable* metallic and nonmetallic mineral resources.

F

fallowing The practice of allowing plowed or cultivated land to remain (rest) uncropped or only partially cropped for one or more growing seasons.

federal state A *state* with a two-tier system of government and a clear distinction between the powers vested in the central government and those residing in the governments of the component regional subdivisions.

fertility rate The average number of live births per 1000 women of childbearing age.

filtering In *urban geography*, a process whereby individuals of a lower-income group replace, in a portion of an urban area, residents who are of a higher-income group.

first effective settlement The influence that the characteristics of an early dominant settlement group exert on the later *social* and *cultural geography* of an area.

fixed cost An activity cost (as of investment in land, plant, and equipment) that must be met without regard to level of output; an input cost that is spatially constant.

folk culture The body of institutions, customs, dress, *artifacts*, collective wisdoms, and traditions of a homogeneous, isolated, largely self-sufficient, and relatively static social group.

folklore Oral traditions of a *folk culture*, including tales, fables, legends, customary observations, and moral teachings.

folkway The learned manner of thinking and feeling and a prescribed mode of conduct common to a traditional social group.

footloose A descriptive term applied to manufacturing activities for which the cost of transporting material or product is not important in determining location of production; an industry or firm showing neither *market* nor *material orientation*.

Fordism The manufacturing economy and system derived from assembly-line mass production and the mass consumption of standardized goods. Named after Henry Ford, who innovated many of its production techniques.

formal region (*syn:* uniform region, homogeneous region, structural region) A *region* distinguished by a uniformity of one or more characteristics that can serve as the basis for areal generalization and of contrast with adjacent areas.

form utility A value-increasing change in the form—and therefore in the "utility"—of a raw material or commodity.

forward-thrust capital A capital city deliberately sited in a *state's* frontier zone.

fossil fuel (*syn:* mineral fuel) Any of the fuels derived from decayed organic material converted by earth processes; especially, coal, petroleum, and natural gas, but also including tar sands and oil shales.

fragmented state A *state* whose territory contains isolated parts, separated and discontinuous.

frame In *urban geography*, that part of the *central business district* characterized by such low-intensity uses as warehouses, wholesaling, and automobile dealers.

freight rate The charge levied by a transporter for the loading, moving, and unloading of goods; includes *line-haul costs* and *terminal costs*.

friction of distance A measure of the retarding or restricting effect of distance on *spatial interaction*. Generally, the greater the distance, the greater the "friction" and the less the interaction or exchange, or the greater the cost of achieving the exchange.

frontier That portion of a country adjacent to its boundaries and fronting another political unit.

frontier zone A belt lying between two *states* or between settled and uninhabited or sparsely settled areas.

functional dispute (*syn:* boundary dispute) In *political geography*, a disagreement between neighboring *states* over policies to be applied to their common border; often induced by differing customs regulations, movement of nomadic groups, or illegal immigration or emigration.

functional region (*syn:* nodal region) A *region* differentiated by what occurs within it rather than by a homogeneity of physical or cultural phenomena; an earth area recognized as an operational unit based upon defined organizational criteria. The concept of unity is based on interaction and interdependence between different points within the area.

G

gated community A restricted access subdivision or neighborhood, often surrounded by a barrier, with entry permitted only for residents and their guests; usually totally planned in land use and design, with "residents only" limitations on public streets and parks.

gathering industries *Primary activities* involving the *subsistence* or *commercial* harvesting of *renewable* natural resources of land or water. Primitive gathering involves local collection of food and other materials of nature, both plant and animal; commercial gathering usually implies forestry and fishing industries.

GDP See *gross domestic product*.

gender In the cultural sense, a reference to socially created—not biologically based—distinctions between femininity and masculinity.

gene flow The transfer of genes of one breeding population into the gene pool of another through interbreeding.

genetic drift A chance modification of gene composition occurring in an isolated population and becoming accentuated through inbreeding.

gentrification The movement into the inner portions of American cities of middle- and upper-income people who replace low-income populations, rehabilitate the structures they occupied, and change the social character of neighborhoods.

geodesic distance See *absolute distance*.

geographic dialect (*syn:* regional dialect) See *dialect*.

geographic information system (GIS) Integrated computer programs for handling, processing, and analyzing data specifically referenced to the surface of the earth.

geometrical projection (*syn:* perspective projection; visual projection) The trace of the *graticule* shadow projected on a *developable surface* from a light source placed relative to a transparent globe.

geometric boundary (*syn:* artificial boundary) A boundary without obvious physical geographic basis; often a section of a *parallel of latitude* or a *meridian of longitude*.

geophagy The practice of eating earthy substances, usually clays.

geopolitics That branch of *political geography* treating national power, foreign policy, and international relations as influenced by geographic considerations of location, space, resources, and demography.

gerrymander To redraw voting district boundaries in such a way as to give one political party maximum electoral advantage and to reduce that of another party, to fragment voting blocks, or to achieve other nondemocratic objectives.

ghetto A forced or voluntarily segregated residential area housing a racial, ethnic, or religious minority.

GIS See *geographic information system.*

globalization A reference to the increasing interconnection of all parts of the world as the full range of social, cultural, political, and economic processes becomes international in scale and effect.

globe grid (*syn:* graticule) The set of imaginary lines of *latitude* and *longitude* that intersect at right angles to form a coordinate reference system for locating points on the surface of the earth.

GNI See *gross national income.*

gnomonic projection A *geometrical projection* produced with the light source at the center of the earth.

graphic scale A graduate line included in a map legend by means of which distances on the map may be measured in terms of ground distances.

graticule The network of meridians and parallels on the globe; the *globe grid.*

gravity model A mathematical prediction of the interaction between two bodies (places) as a function of their size and of the distance separating them. Based on Newton's law, the model states that attraction (interaction) is proportional to the product of the masses (population sizes) of two bodies (places) and inversely proportional to the square of the distance between them.

great circle Line formed by the intersection with the earth's surface of a plane passing through the center of the earth; an arc of a great circle is the shortest distance between two points on the earth's surface.

greenhouse effect Heating of the earth's surface as shortwave solar energy passes through the *atmosphere,* which is transparent to it but opaque to reradiated long-wave terrestrial energy; also, increasing the opacity of the atmosphere through addition of increased amounts of carbon dioxide and other gases that trap heat.

Green Revolution A term suggesting the great increases in food production, primarily in subtropical areas, accomplished through the introduction of very high-yielding grain crops, particularly wheat, maize, and rice.

grid system See *globe grid.*

gross domestic product (GDP) The total value of goods and services produced within the borders of a country during a specified time period, usually a calendar year.

gross national income (GNI) The total value of goods and services produced by a country per year plus net income earned abroad by its nationals; formerly called "gross national product."

gross national product (GNP) See *gross national income.*

groundwater Subsurface water that accumulates in the pores and cracks of rock and *soil.*

guest worker A foreign worker, usually male and frequently under contract, who migrates to secure permanent work in a host country without intention to settle permanently in that country; particularly, workers from North Africa and countries of eastern, southern, and southeastern Europe employed in industrialized countries of Western Europe.

H

hazardous waste Discarded solid, liquid, or gaseous material that poses a substantial threat to human health or to the *environment* when improperly disposed of or stored.

heartland theory The belief of Halford Mackinder (1861–1947) that the interior of Eurasia provided a likely base for world conquest.

hierarchical diffusion A form of *diffusion* in which spread of an *innovation* can proceed either upward or downward through a hierarchy.

hierarchical migration The tendency for individuals to move from small places to larger ones. See also *step migration.*

hierarchy of central places The steplike series of *urban* units in classes differentiated by both size and function.

high-level waste Nuclear waste with a relatively high level of radioactivity.

Hinduism An ancient and now dominant *value system* and *religion* of India, closely identified with Indian *culture* but without central creed, single doctrine, or religious organization. Dharma (customary duty and divine law) and *caste* are uniting elements.

hinterland The market area or region served by an *urban* center.

homeostatic plateau (*syn:* carrying capacity) The application of the concept of homeostasis, or relatively stable state of equilibrium, to the balance between population numbers and areal resources; the equilibrium level of population that available resources can adequately support.

horticultural farming See *truck farming.*

host society The established and dominant society within which immigrant groups seek accommodation.

human geography One of the two major divisions (the other is *physical geography*) of *systematic geography;* the spatial analysis of human populations, their *cultures,* their activities and behaviors, and their relationship with and impact on the physical landscapes they occupy.

hunter-gatherer/hunting-gathering An economic and social system based primarily or exclusively on the hunting of wild animals and the gathering of food, fiber, and other materials from uncultivated plants.

hydrologic cycle The natural system by which water is continuously circulated through the *biosphere* by evaporation, condensation, and *precipitation.*

hydrosphere All water at or near the earth's surface that is not chemically bound in rocks, including the oceans, surface waters, *groundwater,* and water held in the *atmosphere.*

I

icebox effect The tendency for certain kinds of air pollutants to lower temperatures on earth by reflecting incoming sunlight back into space and thus preventing it from reaching (and heating) the earth.

iconography In *political geography,* a term denoting the study of symbols that unite a country.

ideological subsystem The complex of ideas, beliefs, knowledge, and means of their communication that characterize a *culture.*

incinerator A facility designed to burn waste.

independent invention (*syn:* parallel invention) *Innovations* developed in two or more unconnected locations by individuals or groups acting independently. See also *multilinear evolution.*

Industrial Revolution The term applied to the rapid economic and social changes in agriculture and manufacturing that followed the introduction of the factory system to the textile industry of England in the last quarter of the 18th century.

infant mortality rate A refinement of the *death rate* to specify the ratio of deaths of infants age 1 year or less per 1000 live births.

informal economy See *informal sector.*

informal sector That part of a national economy that involves productive labor not subject to formal systems of control or payment; economic activity or individual enterprise operating without official recognition or measured by official statistics.

infrastructure The basic structure of services, installations, and facilities needed to support industrial, agricultural, and other economic development; included are transport and communications, along with water, power, and other public utilities.

innovation Introduction of new ideas, practices, or objects; usually, an alteration of *custom* or *culture* that originates within the social group itself.

insolation The solar radiation received at the earth's surface.

intensive agriculture Any agricultural system involving the application of large amounts of capital and/or labor per unit of cultivated land; this may be part of either a *subsistence* or a *commercial economy.*

interaction model See *gravity model.*

international boundary The outer limit of a *state's* claim to land or water surface, projected downward to the center of the earth and, less certainly, upward to the height the state can effectively control.

International Date Line By international agreement, the designated line where each new day begins, generally following the 180th *meridian.* The line compensates for accumulated 1-hour time changes for each 15 degrees of longitude by adding (from east to west) or subtracting (from west to east) 24 hours for travelers crossing the line.

interrupting barrier A condition that delays the rate of *diffusion* of an *innovation* or that deflects its path.

intervening opportunity The concept that closer opportunities will materially reduce the attractiveness of interaction with more distant— even slightly better—alternatives; a closer alternative source of supply between a demand point and the original source of supply.

in-transit privilege The application of a single-haul *freight rate* from origin to destination even though the shipment is halted for processing en route, after which the journey is completed.

irredentism The policy of a *state* wishing to incorporate within itself territory inhabited by people who have ethnic or linguistic links with the country but that lies within a neighboring state.

Islam A *monotheistic, universalizing* religion that includes belief in Allah as the sole deity and in Mohammed as his prophet completing the work of earlier prophets of *Judaism* and *Christianity.*

isochrone A line connecting points equidistant in travel time from a common origin.

isogloss A mapped boundary line marking the limits of a particular linguistic feature.

isoline A map line connecting points of equal value.

isotropic plain A hypothetical portion of the earth's surface assumed to be an unbounded, uniformly flat plain with uniform and unvarying distribution of population, purchasing power, transport costs, accessibility, and the like.

J-curve A curve shaped like the letter J, depicting exponential or geometric growth (1, 2, 4, 8, 16 . . .).

Judaism A *monotheistic, ethnic religion* first developed among the Hebrew people of the ancient Near East; its determining conditions include descent from Israel (Jacob), the Torah (law and scripture), and tradition.

krill A form of marine *plankton* composed of crustaceans and larvae.

L

landlocked Describing a *state* which lacks a sea coast.

land race A genetically diverse, naturally adapted, native food plant.

language The system of words, their pronunciation, and methods of combination used and mutually understood by a community of individuals.

language family A group of *languages* thought to have descended from a single, common ancestral tongue.

latitude Angular distance north or south of the *equator,* measured in degrees, minutes, and seconds. Grid lines marking latitudes are called *parallels.* The equator is 0°; the North Pole is 90° N; the South Pole is 90° S. Low latitudes are considered to fall within the tropics (23° 30′ N and 23° 30′ S); midlatitudes extend from the tropics to the Arctic and Antarctic circles (66° 30′ N and S); high latitudes occur from those circles to the North and South poles.

law of peripheral neglect The observation that a government's awareness of or concern with regional problems decreases with the square of the distance of an outlying region from the capital city.

leachate The contaminated liquid discharged from a *sanitary landfill* to either surface or subsurface land or water.

leaching The removal of soluble minerals from the upper soil horizons through the downward movement of water.

least-cost theory (*syn:* Weberian analysis) The view that the optimum location of a manufacturing establishment is at the place where the costs of transport and labor and the advantages of *agglomeration* or *deglomeration* are most favorable.

limiting factor principle The distribution of an organism or the structure of an *ecosystem* can be explained by the control exerted by the single factor (such as temperature, light, water) that is most deficient, that is, that falls below the levels required.

line-haul costs (*syn:* over-the-road costs) The costs involved in the actual physical movement of goods (or passengers); costs of

haulage (including equipment and routeway costs), excluding *terminal costs.*

lingua franca Any of various auxiliary *languages* used as common tongues among people of an area where several languages are spoken; literally, "Frankish language."

linguistic geography (*syn:* dialect geography; dialectology) The study of local variations within a speech area by mapping word choices, pronunciations, or grammatical constructions.

link A transportation or communication connection or route within a *network.*

lithosphere The earth's solid crust.

locational interdependence The circumstance under which the locational decision of a particular firm is influenced by the locations chosen by competitors.

locational triangle A simple graphic model in *Weberian analysis* to illustrate the derivation of the least-transport-cost location of an industrial establishment.

longitude Angular distance of a location in degrees, minutes, and seconds measured east or west of a designated *prime meridian* given the value of 0°. By general agreement, the *globe grid* prime meridian passes through the old observatory of Greenwich, England. Distances are measured from 0° to 180° both east and west, with 180° E and W being the same line. For much of its extent, the 180° meridian also serves as the *International Date Line.* Because of the period of the earth's axial rotation, 15 degrees of longitude are equivalent to a difference of 1 hour in local time.

long lot A farm or other property consisting of a long, narrow strip of land extending back from a river or road.

low-level waste Nuclear waste with relatively moderate levels of radioactivity.

M

malnutrition Food intake insufficient in quantity or deficient in quality to sustain life at optimal conditions of health.

Malthus Thomas R. Malthus (1766–1843). English economist, demographer, and cleric who suggested that unless self-control, war, or natural disaster checks population, it will inevitably increase faster than will the food supplies needed to sustain it. This view is known as Malthusianism. See also *neo-Malthusianism.*

map projection A systematic method of transferring the *globe grid* system from the earth's curved surface to the flat surface of a map. Projection automatically incurs error, but an attempt is usually made to preserve one or more (though never all) of the characteristics of the spherical surface: equal area, correct distance, true direction, proper shape.

map scale See *scale.*

marginal cost The additional cost of producing each successive unit of output.

mariculture Production and harvesting of fish and shellfish in fenced confinement areas along coasts and in estuaries.

market equilibrium The point of intersection of demand and supply curves of a given commodity; at equilibrium the market is cleared of the commodity.

market gardening See *truck farming.*

market orientation The tendency of an economic activity to locate close to its market; a reflection of large and variable distribution costs.

material culture The tangible, physical items produced and used by members of a specific *culture* group and reflective of their traditions, lifestyles, and technologies.

material orientation The tendency of an economic activity to locate near or at its source of raw material; this is experienced when material costs are highly variable spatially and/or represent a significant share of total costs.

mathematical location See *absolute location.*

mathematical projection The systematic rendering of the *globe grid* on a *developable surface* to achieve *graticule* characteristics not obtainable by visual means of *geometrical projection.*

maximum sustainable yield The maximum rate at which a *renewable resource* can be exploited without impairing its ability to be renewed or replenished.

Mediterranean agriculture An agricultural system based upon the mild, moist winters; hot, sunny summers; and rough terrain of the Mediterranean basin. It involves cereals as winter crops, summer tree and vine crops (olives, figs, dates, citrus and other tree fruits, and grapes), and animals (sheep and goats).

megalopolis 1: A large, sprawled *urban* complex with contained open, nonurban land, created through the spread and joining of separate *metropolitan areas;* **2:** When capitalized, the name applied to the continuous functionally urban area of coastal northeastern United States from Maine to Virginia.

mental map (*syn:* cognitive map) The maplike image of the world, country, region, city, or neighborhood a person carries in mind. The representation is therefore subjective; it includes knowledge of actual locations and spatial relationships and is colored by personal perceptions and preferences related to place.

mentifacts The central, enduring elements of a *culture* expressing its values and beliefs, including *language, religion, folklore,* artistic traditions, and the like. Elements in the *ideological subsystem* of culture.

Mercator projection A true *conformal cylindrical projection* first published in 1569, useful for navigation.

meridian A north-south line of *longitude;* on the *globe grid,* all meridians are of equal length and converge at the poles.

Mesolithic Middle Stone Age. The *culture* stage of the early postglacial period, during which earliest stages of *domestication* of animals and plants occurred, refined and specialized tools were developed, pottery was produced, and semipermanent settlements were established as climate change reduced the game-animal herds earlier followed for food.

metes-and-bounds survey A system of property description using natural features (streams, rocks, trees, etc.) to trace and define the boundaries of individual parcels.

metropolitan area In the United States, a large functionally integrated settlement area comprising one or more whole county units and usually containing several *urbanized areas;* discontinuously built up, it operates as a coherent economic whole.

microdistrict The basic neighborhood planning unit characteristic of new urban residential construction in the planned East European city under communism.

microstate (*syn:* ministate) An imprecise term for a *state* or territory small in both population and area. An informal definition accepted by the United Nations suggests a maximum of 1 million population combined with a territory of less than 700 km^2 (270 sq mi).

migration The permanent (or relatively permanent) relocation of an individual or group to a new, usually distant, place of residence and employment.

migration field The area from which a given city or place draws the majority of its in-migrants.

mineral A natural inorganic substance that has a definite chemical composition and characteristic crystal structure, hardness, and density.

mineral fuel See *fossil fuel.*

ministate See *microstate.*

model An idealized representation, abstraction, or simulation of reality. It is designed to simplify real-world complexity and eliminate extraneous phenomena in order to isolate for detailed study causal factors and interrelationships of *spatial systems.*

monoculture Agricultural system dominated by a single crop.

monolingualism A society's or country's use of only one *language* of communication for all purposes.

monotheism The belief that there is but a single God.

mortality rate See *death rate.*

movement bias Any aggregate control on or regularity of movement of people, commodities, or communication. Included are *distance bias, direction bias,* and *network bias.*

multilinear evolution A concept of independent but parallel cultural development advanced

by the anthropologist Julian Steward (1902–1972) to explain cultural similarities among widely separated peoples existing in similar environments but who could not have benefited from shared experiences, borrowed ideas, or diffused technologies. See *independent invention.*

multilingualism The common use of two or more *languages* in a society or country.

multinational corporation (MNC) A large business organization operating in a number of different national economies; the term implies a more extensive form of *transnational corporation.*

multiple-nuclei model The postulate that large cities develop by peripheral spread not from one *central business district* but from several nodes of growth, each of specialized use. The separately expanding use districts eventually coalesce at their margins.

multiplier effect The direct, indirect, and induced consequences of change in an activity. **1:** In industrial *agglomerations,* the cumulative processes by which a given change (such as a new plant opening) sets in motion a sequence of further industrial employment and *infrastructure* growth. **2:** In *urban geography,* the expected addition of *nonbasic* workers and dependents to a city's total employment and population that accompanies new *basic sector* employment.

nation A culturally distinctive group of people occupying a specific territory and bound together by a sense of unity arising from shared *ethnicity,* beliefs, and *customs.*

nationalism A sense of unity binding the people of a *state* together; devotion to the interests of a particular country or *nation;* an identification with the state and an acceptance of national goals.

nation-state A *state* whose territory is identical to that occupied by a particular *ethnic group* or *nation.*

natural boundary (*syn:* physical boundary) A *boundary* line based on recognizable physiographic features, such as mountains or rivers.

natural hazard A process or event in the physical environment that has consequences harmful to humans.

natural increase The growth of a population through excess of births over deaths, excluding the effects of immigration or emigration.

natural landscape The physical *environment* unaffected by human activities. The duration and near totality of human occupation of the earth's surface assure that little or no "natural landscape" so defined remains intact. Opposed to *cultural landscape.*

natural resource A physically occurring item that a population perceives to be necessary and useful to its maintenance and well-being.

natural selection The process resulting in the reproductive success of individuals or groups best adapted to their environment, leading to the perpetuation of their genetic qualities.

natural vegetation The plant life that would exist in an area if humans did not interfere with its development.

neocolonialism A disparaging reference to economic and political policies by which major developed countries are seen to retain or extend influence over the economies of less developed countries and peoples.

Neolithic New Stone Age. The *culture* (succeeding that of the *Mesolithic*) of the middle post-glacial period, during which polished stone tools were perfected, the economy was solely or largely based on cultivation of crops and *domestication* of animals, and the arts of spinning, weaving, smelting and metal working were developed. More formalized societies and *culture complexes* emerged as cities developed and trade routes were established.

neo-Malthusianism The advocacy of population control programs to preserve and improve general national prosperity and well-being.

net migration The difference between in-migration and out-migration of an area.

network The areal pattern of sets of places and the routes (*links*) connecting them along which movement can take place.

network bias The view that the pattern of *links* in a *network* will affect the likelihood of flows between specific *nodes*.

network cities Two or more nearby cities, potentially or actually complementary in function, that cooperate by developing transportation links and communications infrastructure joining them.

nodal region See *functional region*.

node In *network* theory, an origin, destination, or intersection in a communication network.

nomadic herding Migratory but controlled movement of livestock solely dependent on natural forage.

nonbasic sector (*syn:* service sector) Those economic activities of an urban unit that supply the resident population with goods and services and that have no "export" implication.

nonecumene (*syn:* anecumene). That portion of the earth's surface that is uninhabited or only temporarily or intermittently inhabited. See also *ecumene*.

nonmaterial culture The oral traditions, songs, and stories of a *culture* group along with its beliefs and customary behaviors.

nonrenewable resource A *natural resource* that is not replenished or replaced by natural processes or is used at a rate that exceeds its replacement rate.

North The general term applied in the *Brandt Report* to the developed countries of the Northern Hemisphere plus Australia and New Zealand.

official language A governmentally designated *language* of instruction, of government, of the courts, and other official public and private communication.

orthographic projection A *geometrical projection* that results from placing the light source at infinity.

outsourcing **1:** Producing abroad parts or products for domestic use or sale;
2: Subcontracting production or services rather than performing those activities "in house."

overpopulation A value judgment that the resources of an area are insufficient to sustain adequately its present population numbers.

over-the-road costs See *line-haul costs*.

ozone A gas molecule consisting of three atoms of oxygen (O_3) formed when diatomic oxygen (O_2) is exposed to *ultraviolet radiation*. In the upper *atmosphere* it forms a normally continuous, thin layer that blocks ultraviolet light; in the lower atmosphere it constitutes a damaging component of *photochemical smog*.

Paleolithic Old Stone Age. An early stage of human *culture* largely coinciding with the *Pleistocene* glacial period. Characterized by *hunting-gathering* economies and the use of fire and simple stone tools, especially those made from flint.

parallel invention See *independent invention*.

parallel of latitude An east-west line of *latitude* indicating distance north or south of the equator.

pattern The design or arrangement of phenomena in earth space.

peak value intersection The most accessible and costly parcel of land in the *central business district* and, therefore, in the entire *urbanized area*.

perception The acquisition of information about a place or thing through sensory means; the subjective organization and interpretation of acquired information in light of cultural attitudes and individual preferences or experiences. See *environmental perception*.

perceptual region A *region* perceived to exist by its inhabitants or the general populace. Also known as a *vernacular region* or popular region, it has reality as an element of *popular culture* or *folk culture* represented in the *mental maps* of average people.

perforated state A *state* whose territory is interrupted ("perforated") by a separate, independent state totally contained within its borders.

periodic market A market operating at a particular location (village, city, neighborhood) on one or more fixed days per week or month.

periphery/peripheral The outer regions or boundaries of an area. See also *core-periphery model*.

permeable barrier An obstacle raised by a culture group or one culture group's reluctance to accept some, but not all, innovations diffused from a related but different *culture*. Acceptance or rejection may be conditioned by religious, political, ethnic, or similar considerations of suitability or compatibility.

personal communication field An area defined by the distribution of an individual's short-range informal communications. The size and shape of the field are defined by work, recreation, school, and other regular contacts and are affected by age, sex, employment, and other personal characteristics.

personal space An invisible, usually irregular area around a person into which he or she does not willingly admit others. The sense (and extent) of personal space is a situational and cultural variable.

perspective projection See *geometrical projection*.

photochemical smog A form of polluted air produced by the interaction of hydrocarbons and oxides of nitrogen in the presence of sunlight.

physical boundary See *natural boundary*.

physical geography One of two major divisions (the other is *human geography*) of *systematic geography;* the study of the structures, processes, distributions, and change through time of the natural phenomena of the earth's surface that are significant to human life.

physical landscape The *natural landscape* plus visible elements of *material culture*.

physiological density The number of persons per unit area of cultivable land.

pidgin An auxiliary *language* derived, with reduced vocabulary and simplified structure, from other languages. Not a native tongue, it is used for limited communication among people with different languages.

place perception See *perception*.

place utility **1:** In human movement and *migration* studies, a measure of an individual's perceived satisfaction or approval of a place in its social, economic, or environmental attributes.
2: In *economic geography,* the value imparted to goods or services by *tertiary* activities that provide things needed in specific markets.

planar projection (*syn:* azimuthal projection) A *map projection* employing a plane as the presumed *developable surface*.

plankton Microscopic freely floating plant and animal organisms of lakes and oceans.

planned economy A system of production of goods and services, usually consumed or distributed by a governmental agency, in quantities, at prices, and in locations determined by governmental program.

plantation A large agricultural holding, frequently foreign owned, devoted to the production of a single export crop.

Pleistocene The geological epoch dating from 2 million to 11 thousand years ago during which four stages of continental glaciation occurred.

political geography A branch of *human geography* concerned with the spatial analysis of political phenomena.

pollution The introduction into the biosphere of materials that because of their quantity, chemical nature, or temperature have a negative impact on the *ecosystem* or that cannot be readily disposed of by natural recycling processes.

polytheism Belief in or worship of many gods.

popular culture The constantly changing mix of material and nonmaterial elements available through mass production and the mass media to an urbanized, heterogeneous, nontraditional society.

popular region See *vernacular region.*

population density A measurement of the numbers of persons per unit area of land within predetermined limits, usually political or census boundaries. See also *physiological density.*

population geography A division of *human geography* concerned with spatial variations in distribution, composition, growth, and movements of population and the relationship of those concerns with the geographic character of areas.

population momentum See *demographic momentum.*

population projection A statement of a population's future size, age, and sex composition based on the application of stated assumptions to current data.

population pyramid A bar graph in pyramid form showing the age and sex composition of a population, usually a national one.

positional dispute (*syn:* boundary dispute) In *political geography,* disagreement about the actual location of a *boundary.*

possibilism The philosophical viewpoint that the physical *environment* offers human beings a set of opportunities from which (within limits) people may choose according to their cultural needs and technological awareness. The emphasis is on a freedom of choice and action not allowed under *environmental determinism.*

postindustrial A stage of economic development in which service activities become relatively more important than goods production; professional and technical employment super-

sedes employment in agriculture and manufacturing; and level of living is defined by the quality of services and amenities rather than by the quantity of goods available.

potential model A measurement of the total interaction opportunities available under *gravity model* assumptions to a center in a multicenter system.

precipitation All moisture—solid and liquid—that falls to the earth's surface from the *atmosphere.*

predevelopment annexation The inclusion within the *central city* of nonurban peripheral areas for the purpose of securing to the city itself the benefits of their eventual development.

primary activities Those parts of the economy involved in making *natural resources* available for use or further processing; included are mining, *agriculture,* forestry, fishing and hunting, and grazing.

primate city A country's leading city, disproportionately larger and functionally more complex than any other; a city dominating an urban hierarchy composed of a base of small towns and an absence of intermediate-sized cities.

prime meridian An imaginary line passing through the Royal Observatory at Greenwich, England, serving by agreement as the 0° line of *longitude.*

private plot In the planned economies under communism, a small garden plot allotted to collective farmers and urban workers.

projection See *map projection.*

prorupt state A *state* of basically *compact* form but with one or more narrow extensions of territory.

protolanguage An assumed, reconstructed, or recorded ancestral *language.*

proved reserves That portion of a *natural resource* that has been identified and can be extracted profitably with current technology.

psychological distance The way an individual perceives distance.

pull factors Characteristics of a locale that act as attractive forces, drawing migrants from other regions.

purchasing power parity (PPP) A monetary measurement which takes account of what money actually buys in each country.

push factors Unfavorable characteristics of a locale that contribute to the dissatisfaction of its residents and impel their emigration.

quaternary activities Those parts of the economy concerned with research, with the gathering and dissemination of information, and with administration—including administration of the other economic activity levels; often considered only as a specialized subdivision of *tertiary activities.*

quinary activities A sometimes separately recognized subsection of *tertiary activity* management functions involving highest-level decision making in all types of large organizations. Also deemed the most advanced form of the *quaternary* subsector.

race A subset of human population whose members share certain distinctive, inherited biological characteristics.

rank-size rule An observed regularity in the city-size distribution of some countries. In a rank-size hierarchy, the population of any given town will be inversely proportional to its rank in the hierarchy; that is, the nth-ranked city will be $1/n$ the size of the largest city.

rate The frequency of an event's occurrence during a specified time period.

rate of natural increase *Birth rate* minus the *death rate,* suggesting the annual rate of population growth without considering *net migration.*

reapportionment The process and outcome of a reallocation of electoral seats to defined territories, such as congressional seats to states of the United States.

recycling The reuse of disposed materials after they have passed through some form of treatment (e.g., melting down glass bottles to produce new bottles).

redistricting The drawing of new electoral district boundary lines in response to changing patterns of population or changing legal requirements.

region Any earth area with distinctive and unifying physical or cultural characteristics that set it off and make it substantially different from surrounding areas. A region may be defined on the basis of its homogeneity or its functional integration as a single organizational unit. Regions and their boundaries are devices of areal generalization, intellectual concepts rather than visible landscape entities.

regional autonomy A measure of self-governance afforded a subdivision of a *state.*

regional concept The view that physical and cultural phenomena on the surface of the earth are rationally arranged by complex, diverse, but comprehensible interrelated spatial processes.

regional dialect (*syn:* geographic dialect) See *dialect.*

regional geography The study of geographic *regions;* the study of areal differentiation.

regionalism In *political geography,* group—frequently ethnic group—identification with a particular region of a *state* rather than with the state as a whole.

relational direction See *relative direction.*

relative direction (*syn:* relational direction) A culturally based locational reference, as the Far West, the Old South, or the Middle East.

relative distance A transformation of *absolute distance* into such relative measures as time or monetary costs. Such measures yield different explanations of human spatial behavior than do linear distances alone. Distances between places are constant by absolute terms, but relative distances may vary with improvements in transportation or communication technology or with different psychological perceptions of space.

relative location The position of a place or activity in relation to other places or activities. Relative location implies spatial relationships and usually suggests the relative advantages or disadvantages of a location with respect to all competing locations.

relic boundary A former *boundary* line that is still discernible and marked by some *cultural landscape* feature.

religion A personal or institutionalized system of worship and of faith in the sacred and divine.

relocation diffusion The transfer of ideas, behaviors, or articles from one place to another through the *migration* of those possessing the feature transported; also, spatial relocation in which a phenomenon leaves an area of origin as it is transported to a new location.

renewable resource A *natural resource* that is potentially inexhaustible either because it is constantly (as solar radiation) or periodically (as *biomass*) replenished as long as its use does not exceed its *maximum sustainable yield*.

replacement level The number of children per woman that will supply just enough births to replace parents and compensate for early deaths, with no allowance for *migration* effects; usually calculated at between 2.1 and 2.5 children.

representative fraction The *scale* of a map expressed as a ratio of a unit of distance on the map to distance measured in the same unit on the ground, e.g., 1:250,000.

resource See *natural resource*.

resource dispute In *political geography*, disagreement over the control or use of shared resources, such as boundary rivers or jointly claimed fishing grounds.

return migration See *counter migration*.

rhumb line A directional line that crosses each successive *meridian* at a constant angle.

rimland theory The belief of Nicholas Spykman (1894–1943) that domination of the coastal fringes of Eurasia would provide a base for world conquest.

rotation See *crop rotation*.

roundwood Timber as it is harvested, before squaring, sawing, or pulping.

S

Sahel The semiarid zone between the Sahara desert and the grassland areas to the south in West Africa; a district of recurring drought, famine, and environmental degradation and *desertification*.

salinization The process by which *soil* becomes saturated with salt, rendering the land unsuitable for *agriculture*. This occurs when land that has poor drainage is improperly irrigated.

sanitary landfill Disposal of solid wastes by spreading them in layers covered with enough soil to control odors, rodents, and flies; sited to minimize water pollution from runoff and *leachate*.

satisficing location A less-than-ideal best location, but one providing an acceptable level of utility or satisfaction.

scale 1: In cartography, the ratio between the size of area on a map and the actual size of that same area on the earth's surface. **2:** In more general terms, scale refers to the size of the area studied, from local to global.

S-curve The horizontal bending, or leveling, of an exponential or J-*curve*.

secondary activities Those parts of the economy involved in the processing of raw materials derived from *primary activities* and in altering or combining materials to produce commodities of enhanced utility and value; included are manufacturing, construction, and power generation.

sector model A description of urban land uses as wedge-shaped sectors radiating outward from the *central business district* along transportation corridors. The radial access routes attract particular uses to certain sectors, with high-status residential uses occupying the most desirable wedges.

secularism A rejection of or indifference to *religion* and religious practice.

segregation A measure of the degree to which members of a minority group are not uniformly distributed among the total population.

separatism See *ethnic separatism*.

service sector See *nonbasic sector*.

shamanism A form of *tribal religion* based on belief in a hidden world of gods, ancestral spirits, and demons responsive only to a shaman or interceding priest.

shifting cultivation (*syn:* slash-and-burn agriculture; swidden agriculture) Crop production on tropical forest clearings kept in cultivation until their quickly declining fertility is lost. Cleared plots are then abandoned and new sites are prepared.

Shinto The *polytheistic, ethnic religion* of Japan that includes reverence of deities of natural forces and veneration of the emperor as descendent of the sun-goddess.

site The *absolute location* of a place or activity described by local relief, landform, and other physical (or sometimes cultural) characteristics.

situation The *relative location* of a place or activity in relation to the physical and cultural characteristics of the larger regional or *spatial system* of which it is a part. Situation implies spatial interconnection and interdependence.

slash-and-burn cultivation See *shifting cultivation*.

social area An area identified by homogeneity of the social indices (age group, socioeconomic status, *ethnicity*) of its population.

social dialect See *dialect*.

social distance A measure of the perceived degree of social separation between individuals, *ethnic groups*, neighborhoods, or other groupings; the voluntary or enforced *segregation* of two or more distinct social groups for most activities.

social geography The branch of *cultural geography* that studies *social areas* and the social use of space, especially urban space; the study of the *spatial distribution* of social groups and of the processes underlying that distribution.

sociofacts The institutions and links between individuals and groups that unite a *culture*, including family structure and political, educational, and religious institutions. Components of the *sociological subsystem* of culture.

sociological subsystem The totality of expected and accepted patterns of interpersonal relations common to a *culture* or subculture.

soil The complex mixture of loose material including minerals, organic and inorganic compounds, living organisms, air, and water found at the earth's surface and capable of supporting plant life.

soil erosion See *erosion*.

solar energy Radiation from the sun, which is transformed into heat primarily at the earth's surface and secondarily in the *atmosphere*.

South The general term applied in the *Brandt Report* to the poor, developing countries of the world, generally (but not totally) located in the Southern Hemisphere.

space-time compression/ convergence Expressions of the extent to which improvements in transportation and communication have reduced distance barriers and permitted, for example, the instantaneous *diffusion* of ideas across space.

space-time prism A diagram of the volume of space and the length of time within which our activities are confined by constraints of our bodily needs (eating, resting) and the means of mobility at our command.

spatial Of or pertaining to space on the earth's surface. Often a synonym for *geographical* and used as an adjective to describe specific geographic concepts or processes, as *spatial interaction* or *diffusion*.

spatial diffusion See *diffusion.*

spatial distribution The arrangement of things on the earth's surface; the descriptive elements of spatial distribution are *density, dispersion,* and *pattern.*

spatial interaction The movement (e.g., of people, goods, information) between different places; an indication of interdependence between different geographic locations or areas.

spatially fixed cost An input cost in manufacturing that remains constant wherever production is located.

spatially variable cost An input cost in manufacturing that changes significantly from place to place in its amount and its relative share of total costs.

spatial margin of profitability The set of points delimiting the area within which a firm's profitable operation is possible.

spatial search The process by which individuals evaluate the alternative locations to which they might move.

spatial system The arrangement and integrated operation of phenomena produced by or responding to spatial processes on the earth's surface.

speech community A group of people having common characteristic patterns of vocabulary, word arrangement, and pronunciation.

spine In *urban geography,* a continuation of the features of the *central business district* outward along the main wide boulevard characteristic of Latin American cities.

spread effect (*syn:* trickle-down effect) The diffusion outward of the benefits of economic growth and prosperity from the power center or *core area* to poorer districts and people.

spring wheat Wheat sown in spring for ripening during the summer or autumn.

standard language A *language* substantially uniform with respect to spelling, grammar, pronunciation, and vocabulary and representing the approved community norm of the tongue.

standard line Line of contact between a projection surface and the globe; transformed from the sphere to the plane surface without distortion.

state (*syn:* country) An independent political unit occupying a defined, permanently populated territory and having full sovereign control over its internal and foreign affairs.

state farm In the former Soviet Union (and other planned economies), a government agricultural enterprise operated with paid employees.

step (stepwise) migration A *migration* in which an eventual long-distance relocation is undertaken in stages as, for example, from farm to village to small town to city. See also *hierarchical migration.*

stereographic projection A *geometrical projection* that results from placing the light source at the *antipode.*

stimulus diffusion A form of *expansion* diffusion in which a fundamental idea, though not the specific trait itself, stimulates imitative behavior within a receptive population.

structural assimilation The distribution of immigrant ethnics among the groups and social strata of a *host society,* but without their full *behavioral assimilation* into it.

subsequent boundary A *boundary* line that is established after the area in question has been settled and that considers the cultural characteristics of the bounded area.

subsistence agriculture Any of several farm economies in which most crops are grown for food nearly exclusively for local or family consumption.

subsistence economy An economic system of relatively simple technology in which people produce most or all of the goods to satisfy their own and their family's needs; little or no exchange occurs outside of the immediate or extended family.

substitution principle In industry, the tendency to substitute one factor of production for another in order to achieve optimum plant location.

suburb A functionally specialized segment of a large *urban* complex located outside the boundaries of the *central city;* usually, a relatively homogeneous residential community, separately incorporated and administered.

superimposed boundary A *boundary* line placed over and ignoring an existing cultural pattern.

supranationalism Term applied to associations created by three or more states for their mutual benefit and achievement of shared objectives.

sustained yield The practice of balancing harvesting with growth of new stocks so as to avoid depletion of the *resource* and ensure a perpetual supply.

swidden agriculture See *shifting cultivation.*

syncretism The development of a new form of *culture trait* by the fusion of two or more distinct parental elements.

systematic geography A division of geography that selects a particular aspect of the physical or cultural *environment* for detailed study of its areal differentiation and interrelationships. Branches of systematic geography are labeled according to the topic studied (e.g., recreational geography) or the related science with which the branch is associated (e.g., *economic geography*).

systems analysis An approach to the study of large systems through (**1**) segregation of the entire system into its component parts, (**2**) investigation of the interactions between system elements, and (**3**) study of inputs, outputs, flows, interactions, and boundaries within the system.

Taoism (*syn:* Daoism) A Chinese *value system* and *ethnic religion* emphasizing conformity to Tao (Way), the creative reality ordering the universe.

tapering principle A *distance decay* observation of the diminution or tapering of costs of transportation with increasing distance from the point of origin of the shipment because of the averaging of *fixed costs* over a greater number of miles of travel.

technological subsystem The complex of material objects together with the techniques of their use by means of which people carry out their productive activities.

technology The integrated system of knowledge, skills, tools, and methods developed within or used by a *culture* to successfully carry out purposeful and productive tasks.

technology gap The contrast between the *technology* available in developed *core regions* and that present in *peripheral areas* of *underdevelopment.*

technology transfer The *diffusion* to or acquisition by one *culture* or *region* of the *technology* possessed by another, usually more developed, society.

terminal costs (*syn:* fixed costs of transportation) The costs incurred, and charged, for loading and unloading freight at origin and destination points and for the paperwork involved; costs charged each shipment for terminal facility use and unrelated to distance of movement or *line-haul costs.*

terracing The practice of planting crops on steep slopes that have been converted into a series of horizontal steplike level plots (terraces).

territorial dispute (*syn:* boundary dispute; functional dispute) In *political geography,* disagreement between *states* over the control of surface area.

territoriality An individual or group attempt to identify and establish control over a clearly defined territory considered partially or wholly an exclusive domain; the behavior associated with the defense of the home territory.

territorial production complex A design in former Soviet economic planning for large regional industrial, mining, and agricultural development leading to regional self-sufficiency, diversification, and the creation of specialized production for a larger national market.

terrorism Systematic open and covert action employing fear and terror as a means of political coercion.

tertiary activities Those parts of the economy that fulfill the exchange function, that provide market availability of commodities, and that bring together consumers and providers of services; included are wholesale and retail trade, associated transportational and governmental services, and personal and professional services of all kinds.

thematic map A map depicting a specific *spatial distribution* or statistical variation of abstract objects (e.g., unemployment) in space.

Third World Originally (1950s), designating countries uncommitted to either the "First World" Western capitalist bloc or the Eastern "Second World" communist bloc; subsequently, a term applied to countries considered not yet fully developed or in a state of *underdevelopment* in economic and social terms.

threshold In *economic geography* and *central place theory,* the minimum market needed to support the supply of a product or service.

time-distance decay An influence on the rate of *expansion diffusion* of an idea, observing that the spread or acceptance of an idea is usually delayed as distance from the source of the innovation increases.

tipping point The degree of neighborhood racial or ethnic mixing that induces the former majority group to move out rapidly.

toponym A place name.

toponymy The place names of a region or, especially, the study of place names.

total fertility rate (TFR) The average number of children that would be born to each woman if during her childbearing years she bore children at the current year's rate for women that age.

town A nucleated settlement that contains a *central business district* but that is small and less functionally complex than a *city.*

toxic waste Discarded chemical substances that can cause serious illness or death.

traditional religion See *tribal religion.*

tragedy of the commons The observation that in the absence of collective control over the use of a resource available to all, it is to the advantage of all users to maximize their separate shares even though their collective pressures may diminish total yield or destroy the resource altogether.

transculturation A term describing the relatively equal exchange of cultural outlooks and ways of life between two culture groups; it suggests more extensive cross-cultural influences than does *acculturation.*

transferability Acceptable costs of a spatial exchange; the cost of moving a commodity relative to the ability of the commodity to bear that cost.

transnational corporation (TNC) A large business organization operating in at least two separate national economies; a form of *multinational corporation.*

tribal religion (syn: traditional religion) An *ethnic religion* specific to a small, localized, preindustrial culture group.

trickle-down effect See *spread effect.*

tropical rain forest Tree cover composed of tall, high-crowned evergreen deciduous species, associated with the continuously wet tropical lowlands.

truck farming (syn: horticultural farming; market gardening) The intensive production of fruits and vegetables for market rather than for processing or canning.

𝒰

ubiquitous industry A *market-oriented* industry whose establishments are distributed in direct proportion to the distribution of population.

ultraviolet (UV) radiation Electromagnetic radiation from the sun with wavelengths shorter than the violet end of visible light and longer than X-rays.

underdevelopment A level of economic and social achievement below what could be reached—given the natural and human resources of an area—were necessary capital and technology available.

underpopulation A value statement reflecting the view that an area has too few people in relation to its resources and population-supporting capacity.

unified government (syn: unigov; metro) Any of several devices federating or consolidating city governments.

uniform plain See *isotropic plain.*

uniform region See *formal region.*

unitary state A *state* in which the central government dictates the degree of local or *regional autonomy* and the nature of local governmental units; a country with few cultural conflicts and with a strong sense of national identity.

United Nations Convention on the Law of the Sea (UNCLOS) A code of maritime law approved by the United Nations in 1982 that authorizes, among other provisions, territorial waters extending 12 nautical miles (22 km) from shore and 200-nautical-mile-wide (370-km-wide) *exclusive economic zones.*

universalizing religion A *religion* that claims global truth and applicability and seeks the conversion of all humankind.

urban Characteristic of, belonging to, or related to a city or town; the opposite of rural. An agglomerated settlement whose inhabitants are primarily engaged in nonagricultural occupations.

urban geography The geographical study of cities; the branch of *human geography* concerned with the spatial aspects of **(1)** the locations, functional structures, size hierarchies, and intercity relationships of national or regional systems of cities, and **(2)** the *site,* evolution, *economic base,* internal land use, and social geographic patterns of individual cities.

urban hierarchy A ranking of cities based on their size and functional complexity.

urban influence zone An area outside of a *city* that is nevertheless affected by the city.

urbanization Transformation of a population from rural to *urban* status; the process of city formation and expansion.

urbanized area A continuously built-up *urban* landscape defined by building and population densities with no reference to the political boundaries of the city; it may contain a *central city* and many contiguous towns, *suburbs,* and unincorporated areas.

usable reserves Mineral deposits that have been identified and can be recovered at current prices and with current technology.

𝒱

value system *Mentifacts* of the *ideological subsystem* of a *culture* summarizing its common beliefs, understandings, expectations, and controls.

variable cost A cost of enterprise operation that varies either by output level or by location of the activity.

variable costs of transportation See *line-haul costs.*

verbal scale A statement of the relationship between units of measure on a map and distance on the ground, as "one inch represents one mile."

vernacular 1: The nonstandard indigenous *language* or *dialect* of a locality. 2: Of or related to indigenous arts and architecture, such as a *vernacular house.* 3: Of or related to the perceptions and understandings of the general population, such as a *vernacular region.*

vernacular house An indigenous style of building constructed of native materials to traditional plan, without formal drawings.

vernacular region A region perceived and defined by its inhabitants, usually with a popularly given or accepted nickname.

von Thünen model Model developed by Johann Heinrich von Thünen (1783–1850), German economist and landowner, to explain the forces that control the prices of agricultural commodities and how those variable prices affect patterns of agricultural land utilization.

von Thünen rings The concentric zonal pattern of agricultural land use around a single market center proposed in the *von Thünen model.*

𝒲

water table The upper limit of the saturated zone and therefore of *groundwater.*

wattle and daub A building technique featuring walls of interwoven twigs, branches, or poles (wattles) plastered (daubed) with clay and mud.

Weberian analysis See *least-cost theory.*

winter wheat Wheat sown in fall for ripening the following spring or summer.

world city One of a small number of interconnected, internationally dominant centers (e.g., New York, London, Tokyo) that together control the global systems of finance and commerce.

Z

zero population growth (ZPG) A term suggesting a population in equilibrium, fully stable in numbers with births (plus immigration) equaling deaths (plus emigration).

zoning Designating by ordinance areas in a municipality for particular types of land use.

zonal model See *concentric zone model.*

Credits

Photos

Part Openers

One: © Robert Essel /Corbis Images;
Two: © Lawrence Manning/Corbis Images;
Three: © Brian Sytnyk/Materfile;
Four: © Philip Rostron/Materfile;
Five: © Gerry Ellis/Minden Pictures.

Chapter *One*

Opener: © Daryl Benson/Masterfile; **1.1:** © Karl Weatherly/Corbis; **1.10:** © Walter Frerck/Odyssey Productions; **1.11:** NASA; **1.12:** US Geological Survey; **1.13:** Chicago Area Transportation Study, Final Report, 1959. Vol. 1, p. 44, figure 22: "Desire Lines of Internal Automobile Driver Trips."

Chapter *Two*

Opener: © Horst Klemm/Masterfile; **2.1a:** © Ian Murphy/Getty; **2.1b:** Tim McCabe, US Dept. of Agriculture Natural Resource Conservation Service; **2.2(top):** © Walter Gans/The Image Works; **2.2(bottom):** © Genzo Sugino/Tom Stack & Assoc.; **2.3:** © Kennan Ward/Corbis; **2.5:** © Charles O' Rear/ Corbis; **2.6:** © Morton Beebe/Corbis; **2.7:** © Wolfgang Kaehler/Corbis; **Box 2.1:** Jon C. Malinowski/Human Landscape Studio; **2.11:** © Aubrey Land/Valan Photos; **2.14a:** © Galen Rowell/Corbis; **2.14b:** © Bojan Brecelj/Corbis; **2.16:** © Joyce Gregory Wyles; **2.17a:** © Dave G. Houser/Hillstrom Stock Photos; **2.17b:** © R. Aguirre & G. Switkes/Amazonia Films; **2.18a:** © Cary Wolinsky/Stock Boston; **2.18b:** © John Eastcott/The Image Works; **2.18c:** © Yigal Pardo/Hillstrom Stock Photos; **2.18d:** © Paul Conklin/Photoedit; **2.19a:** © Mark Antman/The Image Works; **2.19b:** © W. Marc Bernsau/The Image Works; **2.19c:** © Owen Franken/Stock Boston; **2.24:** © Geoffrey Hiller/eStock; **2.25(all):** Jon C. Malinowski/Human Landscape Studio.

Chapter *Three*

Opener: © Peter Christopher/Masterfile; **3.1:** © Melinda Berge/Network Aspen; **3.8a:** © The McGraw-Hill Companies, Inc./Barry Barker photographer; **3.8b:** © Charles Gupton; **3.22:** © Bettmann/Corbis; **3.23:** © Sergio Moraes/Reuters; **3.27:** © Bettmann/Corbis; **Box 3.1:** © Bettmann/Corbis.

Chapter *Four*

Opener: © R. Ian Lloyd/Masterfile; **4.2:** © Herb Snitzer/Stock Boston; **Box 4.1:** © Owen Franken/Corbis; **4.18:** Library of Congress; **4.19:** Lynn Betts, USDA, Natural Resources Conservation Center; **4.24:** © Bill Cardoni; **4.25:** © William E. Ferguson; **4.29:** © Guillermo Granja/Reuters America; **4.31:** © Carl Purcell/Words and Pictures; **4.32:** © Nathan Benn/Corbis.

Chapter *Five*

Opener: © AP/Wide World Photos; **5.1:** © Susan Reisenweaver; **5.8:** © Mark Antman/The Image Works; **5.23:** © David A Burney/Visualizations; **5.26b:** © David Brownell; **5.27:** © Topham/The Observer/The Image Works; **5.29:** © Arvind Garg/Corbis; **5.30:** © Porterfield/Chickering/Photo Researchers, Inc.; **5.31:** © Fred Bruemmer/Valan Photos; **5.33:** © Wolfgang Kaehler; **5.34:** © Charlotte Kahler.

Chapter *Six*

Opener: © AP Wide World Photos; **6.1:** © Philippe Gontier/The Image Works; **6.2:** © J. Messerschmidt/ Bruce Coleman; **6.3:** © Kevin Fleming/Corbis; **6.7:** © Chromosohm Media/The Image Works; **6.14:** © Stephanie Maze/Corbis; **6 15a:** © Joseph Pierce/Valan Photos; **6.15b:** © Harold E. Green/Valan Photos; **6.18:** © Rudi Von Briel/PhotoEdit; **6.22:** © Les Stone/Sygma/Corbis; **6.23:** © Tony Freeman/PhotoEdit; **6.24:** © Susan Reisenweaver.

Chapter *Seven*

Opener: © Ted Spiegel/Corbis; **7.1:** Library of Congress; **7.2a:** © Courtesy of Jean Fellmann; **7.3:** © Farrell Grehan/Photo Researchers, Inc.; **7.6a:** © Ira Kirschenbaum/Stock Boston; **7 6b:** © Pam Hickman/Valan Photos; **7.7a, b:** © Courtesy of Professor Colin E. Thorn; **7.7c:** © Wolfgang Kaehler; **7.7d:** © Stone/Getty; **7.7e:** © Wolfgang Kaehler; **7.8a-c:** Courtesy of Professor John A. Jakle; **7.9a:** © Peter Menzel/Stock Boston; **7.9b:** Courtesy of Professor John A. Jakle; **7.9c:** © Susan Reisenweaver; **7.9d:** Courtesy of Professor John A. Jakle; **7.10:** Jon C. Malinowski/Human Landscape Studio; **7.11a:** Courtesy of Professor John A. Jakle; **7.11b:** © Susan Reisenweaver; **7.12:** © Sharon Wildman; **7.13a:** Courtesy of Professor John A. Jakle; **7.13b, 7.16a-c:** © Susan Reisenweaver; **7.16d:** © Jared Fellmann; **7.24:** © Larry Mangnio/The Image Works; **7.25:** Jon C. Malinowski/Human Landscape Studio; **7.28:** © Michael Dwyer/Stock Boston; **7.29:** © Joseph Nettis/Stock Boston.

Chapter *Eight*

Opener: © Owen Franken/Corbis Images; **8.1:** © John Maher/Coyote Crossing, Inc.; **8.3:** © Mark Gibson; **8.9:** © Wolfgang Kaehler; **8.10:** © Sean Sprague; **8.13:** © Bill Gillette/Stock Boston; **8.19:** © Jim Pickerell/Stock Boston; **8.22:** © Wolfgang Kaehler; **8.24:** Jon C. Malinowski/Human Landscape Studio; **8.25:** © David Muench/Corbis; **8.28:** © William McCloskey; **8.29:** © Cammeramann International; **8.31:** © Larry Tackett/Tom Stack & Assoc.; **8.34:** © Cammeramann International; **8.35:** © Thomas Kitchin/Valan Photos; **8.36:** © Dean Conger/Corbis; **8.37:** United Nations Photo.

Chapter *Nine*

Opener: © R. Ian Lloyd/Masterfile; **9.1:** © Carl Wolinsky/Stock Boston; **9.6:** © Rick Browne/Photo Researchers, Inc.; **9.14:** © Cammeramann International; **9.15:** © Sharon Stewart; **9.16: IBM billboard:** © Deborah Harse/Image Works; **Ford Motor Company billboard:** © Steven Harris/Newsmakers/Getty News; **Nestle billboard:** © Mark Henley/Panos Pictures; **Sony billboard:** © Mark Henley/Panos Pictures; **BP sign:** © Caroline Penn/Panos Pictures; **Nokia mobile billboard:** © Chris Stowers/Panos Pictures; **9.19:** © Peter Pearson; **9.23:** © Chris Stowers/Panos Pictures; **9.27:** © William E. Ferguson.

Chapter *Ten*

Opener: © AFP/Photo/Stephen Shaver/Corbis; **10.1:** © Dilip Mehta/Contact Press Images; **10.2a:** © Cammeramann International; **10.2b** © Jason Clay/Anthro Photo File; **10.4:** © Wolfgang Kaehler; **10.6:** © Rob Crandall/The Image Works; **10.7:** © Bettmann/Corbis; **10.12:** © Betty Press/Woodfin Camp; **10.10:** © Jim Holmes/Environmental Images; **10.14:** © AP Wide World Photos; **10.16:** © Louise Grubb/The Image Works; **10.18:** © David Hiser/Stone/Getty; **10.20:** © Liba Taylor/Corbis.

Chapter *Eleven*

Opener: © Yann Arthus-Bertrand/Corbis; **11.1:** © Elizabeth J. Leppman; **11.8a:** © William E. Ferguson; **11.8b:** Courtesy of Colin E. Thorn; **11.8c:** © Wolfgang Kaehler; **11.10a:** © Carl Purcell/Words and Pictures; **11.10b:** © Susan Reisenweaver; **11.21:** © Robert Brenner/Photoedit; **11.33:** © Marc PoKempner; **Box 11.1:** © Richard B. Levine; **Box 11.2:** © Bettmann/Corbis; **11.34:** © Carl Purcell/Words and Pictures; **11.35:** © Lester Lefkowitz/Corbis Images; **11.36:** © Thomas Kitchin/Tom Stack & Assoc.; **11.37:** © IPA/The Image Works; **11.39a:** © Aubrey Diem/Valan Photos; **11.39b:** © Eastcott/Momatuik/The Image Works; **11.40:** © Dave G. Houser/Corbis; **11.41:** AP/Wide World Photos; **11.42:** © Luiz Claudio Margio/Peter Arnold, Inc.; **11.44:** © Byron Augustin/Tom Stack & Assoc.

Chapter *Twelve*

Opener: © Gavin Hellier/Getty Images; **12.1:** Courtesy of the Pilgrim Society, Plymouth, MA; **12.2:** © Pictor/Image State; **12.11:** © Australian Information Center; **12.13:** © Fred Bavendam/Peter Arnold, Inc.; **12.14:** AP/Wide World Photos; **12.17b:** © San Diego Union Tribune/John Nelson; **12.18:** © Michael Siluk; **12.22:** © AFP/Corbis Images; **12.28:** © Cameramann International; **12.29:** © Keystone/The Image Works.

Chapter *Thirteen*

Opener: © Frans Lanting/Minden Pictures; **13.1:** Courtesy of Kennecott; **13.6:** © Rick Maiman/Stgma/Corbis; **13.7:** © Robert V. Eckert Jr./Coyote Crossing, Inc.; **13.8b:** © William E. Ferguson; **13.10a:** © Exxon Corporation and the American Petroleum Institute; **13.10b:** © Susan Reisenweaver; **Box 13.1:** NASA; **13.11:** © Martin Wendel/Peter Arnold, Inc.; **13.13:** © Jim Whitmer; **13.14:** © Wolfgang Kaehler; **Box 13.2(1):** Lynn Betts, USDA; **Box 13.2(2):** Gene Alexander/Soil Conservation Service/USDA; **13.21:** © Stephen Trimble; **13.22:** © J. Baylor Roberts/National Geographic Image Collection; **13.23:** © San Diego Union Tribune/Don Kohlbauer; **13.25:** © Mark Antman/The Image Works; **13.26:** © Roger A. Clark/Photo Researchers, Inc.; **13.27:** Illustrators Stock Photos.

Index

Page numbers in **bold** indicate key words. Page references followed by *f* and *t* refer to figures and tables, respectively.

A

Abortion, and Cairo Conference, 123
Absolute direction, **11**
Absolute distance, **12**
Absolute location, **9**
Absorbing barrier, 59
Accessibility, **15**
Acculturation, **57**–58, 152, **197**
　in America, 163
　common language and, 162
Acid rain, 487, **493**–494, 494*f,* 495*f,* 496*f*
ACLU. *See* American Civil Liberties Union
Acquired immune deficiency syndrome. *See* AIDS
Acquisitions and mergers, transnational, 332
Activity space, **72**–76, 73*f*
　definition of, 72–73
　distance limits on, 75–76, 76*f,* 77*f,* 78*f*
　time limits on, 74, 74*f*
Adaptation, **192**
Admixture of races, 192
Aerosols, effect on atmosphere, 487
Affirmative action, California Proposition 209 and, 91
Afghanistan
　demographic data, 533*t*
　ethnic strife in, 199*f*
Africa
　agriculture in, 279, 284
　AIDS in, 371, 530*t*–532*t*
　carrying capacity of land in, 129, 129*f*
　Christianity in, 172
　city characteristics in, 433–435, 435*f*
　colonial boundaries in, 445, 446*f,* 447*f*
　demographic data, 530*t*–532*t*
　desertification in, 501
　education in, 371
　energy use in, 367
　food production trends, 282*f,* 284
　fossil fuel reserves, 305*t*
　gender roles in, 381*f*
　hunter-gatherers in, 34*f,* 43, 45*f,* 47
　infant mortality rates, 109, 530*t*–532*t*
　international refugee flows in, 86, 87*f*
　Islam in, 177
　landlessness in, 366, 368*f*
　language in, 148*f*–149*f,* 151, 151*f*
　lingua francas of, 160, 160*f*
　malnutrition in, 369, 369*f,* 370*f*
　maternal mortality ratio, 111, 111*f*
　Northern
　　demographic data, 530*t*
　　environmental damage in, 41
　official languages in, 160, 161*f*
　population
　　control measures, 134
　　growth in, 107*f,* 109, 112, 132
　　by nation, 530*t*–532*t*
　poverty in, 361
　segregation in, 210
　soil degradation in, 503
　sub-Saharan
　　carrying capacity in, 129, 129*f*
　　census data for, 132
　　death rates in, 110
　　demographic data, 530*t*
　　development in, 370
　　fertility rate in, 107, 134
　　male-female ratio in, 114
　　manufacturing in, 359
　　maternal mortality ratio, 111, 111*f*
　　medical services in, 374
　　population growth in, 109, 112
　　population pyramid for, 112, 113*f*
　　poverty in, 361
　　service exports, 347, 347*t*
　　trade and, 309
　　women in, 380*f*
　toxic waste dumping in, 516
　tropical rain forest in, 496–497
　urbanization in, 130, 397*f*
　villages in, 399*f*
　Western
　　demographic data, 530*t*–531*t*
　　migration from, 92
African Americans
　arrival in U.S., 86, 193–195, 195*t*
　cultural rebound in, 217
　dispersions of, in U.S., 203–205, 203*f,* 204*f*
　folk song tradition of, 247, 247*f*
　ghettos and, 214–215, 216*f*
　as percent of population, 193*t,* 205*t*
　racial gerrymandering and, 474–475
　and segregation, in urban areas, 417
　total fertility rates for, 108
Africans, and slavery migrations, 86, 193–195, 195*t*

Afrikaans, 159
Afro-Asiatic languages, 148*f*–149*f,* 151
Age
　and activity space, 73
　of population
　　in developing nations, 135, 136*f*
　　by nation, 2003 data, 530*t*–536*t*
Age of mass consumption stage of development, 370
Agglomeration, 328
　definition of, 17
　in high tech industries, 342–343
　in manufacturing, **325**
Agglomeration economies, and industrial location, 325, **328,** 329*f*
Aging population, challenges of, 135
Agribusiness, 285
Agricultural density, 128*t,* 129
Agricultural location, model for, 286–288, 287*f,* 288*f*
Agricultural regions, of U.S., 289*f*
Agricultural Revolution, 103–104
　gender roles and, 379
　and population growth, 133*f*
Agricultural societies, classification of, 274
Agriculture
　carrying capacity and, **44, 129**–130
　in Africa, 129, 129*f*
　in Egypt, 126
　technology and, 134–135
　in China, 279–280, 282, 293–294
　commercial, 285–294
　　extensive, **288**–290
　　impact of, 283, 283*f*
　　intensive, 287, **288**
　in cultural hearths, 49
　definition of, **273**
　deforestation and, 301
　in developing nations, 274, 275*f,* 285–286
　environmental impact of, 280, 281
　folk culture annual calendar for, 244, 244*f*
　and global warming, 491, 491*f*
　grape harvesting, 268*f*
　Green Revolution, 134, **282**–285, 282*f*
　as gross domestic product share, worldwide, 276*f*
　growing season length, worldwide, 275*f*
　Irish potato famine, 270
　irrigation, costs of, 506
　land use models, 286–288, 287*f,* 288*f*

land value and, 288–289, 290*f*
Mediterranean, 291*f,* 292
origin and diffusion of, 46–47, 47*f,* 54–55
percentage of workforce engaged in, as measure of development, 366–367, 368*f*
in planned economies, 292–294
plantation, 291*f,* **292,** 292*f*
population percent employed in, 274, 275*f*
pressures on, 128, 503, 504
as primary economic activity, 273–294
principal wheat-growing areas, 290, 290*f*
production controls in, 285–286
soil preservation methods, 503, 504
special crop farming, 292, 292*f*
subsistence, 274–285
　areas practicing, globally, 277*f*
　definition of, 272–274
　environmental impact of, 280, 281
　rural settlements in, 397–400, 399*f*
terracing, and productivity, 126, 127*f*
urbanization and, 130
in U.S., farmland cost per acre, 288, 290*f*
water use in, 506
Ahimsa, 179
AIDS (acquired immune deficiency syndrome)
　in Africa, 371, 530*t*–532*t*
　cost of fighting, 373
　death toll from, 110, 121
　in developing world, 110, 376
　population growth and, 120
　spread of, 112
Air pollution, 487, **493**–494, 494*f,* 495*f,* 496*f*
　and acid rain, **493**–494, 494*f,* 495*f,* 496*f*
　toxic waste and, 512
　trash incineration and, 510
Air transportation
　advantages and disadvantages of, 324*t*
　industry location and, 323
Albers equal-area conic projection, 526, 526*f*
Alcoholic beverages, folk culture and, 245
Allah, 176
Allegiance of citizens, importance of, 456

Dioxin, and incineration of trash, 510–511
Direction
    absolute, **11**
    definition of, 11–12
    distortion of, in map projection, 525
    relative, 11–12
Direction bias
    in information flow, 80
    in spatial interaction, 71, 71*f*
Dirt, eating of, 245
Discrimination, in freight rates, 323
Disease(s)
    AIDS
        in Africa, 371, 530*t*–532*t*
        cost of fighting, 373
        death toll from, 110, 121
        in developing world, 110, 376
        population growth and, 120
        spread of, 112
    changing face of, 120, 121
    contagious diffusion of, 56*f*
    five leading killers, 121
    ongoing struggle against, 121
    and population, 119
    resurgence of, 376
    tuberculosis, 121, 376
    urban agriculture and, 280
Dispersion, **17**, 17*f*
Distance
    absolute, **12**
    distortion of, in map projection, 524
    as limitation of spatial interaction,
        74–75, 75*f*
    relative, **12**
    types of, 12
Distance bias, in spatial interaction, 71
Distance decay, 15, 68–**69**
    in human interaction, 74–75, 78*f*
    in information flow, 79
    migration and, 92
Distortion in maps, 20, 522–525
Districting
    gerrymandering and, **472**, 473*f*,
        474–475
    representation and, 471–472, 473*f*
Districts, varying, for various functions,
    472–477, 476*f*
Diversity
    in architecture, global, 234, 235*f*
    biological, in rain forests, 498–499
    of crops, diminished, 283, 283*f*
    ethnic, 191–193
        acceleration of, 194
        in American cities, 213
        in Australia, 194
        in Canada, 194
            principal ethnic groups, 196*t*
        in cities, 209–216
        colonialism and, 190
        in Europe, 194
        in Latin America, 194
        ubiquity of, 190
        in United States, 195–197
            principal ethnic groups,
                193*t*, 195*t*, 196*f*
        urban, 213
        in various nations, 194
    religious, in U.S., 144*f*
    in societies, 230
DOE. *See* United States Department of
    Energy (DOE)

Dogtrot house, 240, 241*f*
Domestication of plants and animals,
    46, 47*f*
Domino theory, **462**
Dot-com bubble, high-tech industry
    and, 341
Dot maps, 22, 24*f*
Doubling time (population), **114**–118,
    116*t*, 117*f*
Douglas, William O., 162
Dravidian language family, 148*f*–149*f*
Drink, folk preferences in, 245
Drive to maturity stage of
    development, 370
Durga, 142*f*
Dust Bowl, 500
Dutch barns. *See* Pennsylvania Dutch,
    architecture of

ℰ

Earth
    atmosphere of
        characteristics of, 484
        damage to, 487
        greenhouse effect and, 487,
            488–489, 490*f*, 491*f*, 492
        ozone layer, 487, 497
    hydrosphere of, 484, 503
    impact of humans on (*See*
        Environment, human impact
        on)
    lithosphere of, 484
Earthquakes, damage from, 81–82, 83*f*
Earth Summit (1992), 492
Easley v. *Cromartie* (2001), 475
East Asia
    ethnic religions in, 182–183
    population distribution in, 125
East Asia Manufacturing region,
    338–339, 339*f*, 340*f*
Easter Island, environmental damage in,
    41–42, 43*f*
Eastern Europe
    city characteristics in, 430–431, 430*f*
    demographic data, 535*t*
    as manufacturing center, 337–338,
        338*f*
    nationalism in, 460
    pronatalist policies in, 109
    urbanization in, 397*f*
    water pollution in, 507
Eastern Orthodox Church, 171
East Timor
    demographic data, 534*t*
    UN peacekeepers in, 464*f*
Economic activity
    categories of, 271–273, 273*f*
    cultural affects of, 267
    factors affecting, 270–271
    future of, 76
    globalization of, 267
    postindustrial, 267, 343, 345–346
    transportation as factor in, 273, 274*f*
Economic alliances, 468–470
Economic base, of a city, **402**
Economic Community of West African
    States (ECOWAS), 469
Economic development. *See*
    Development
Economic geography, **270**

Economic growth, multiplier effect
    in, **402**
Economic measures of development,
    359–371
Economic rationality assumption, 317
Economics
    as arena for global conflict, 463
    effect of religion on, 166
    supply and demand, 318, 318*f*
Economic systems, types of, 273
Economy
    of Europe, restructuring of, 273
    global
        changes in, 343
        services in, 346–349, 347*t*
        informal (underground), 360, 360*f*
        planned, **273**
            agriculture in, 292–294
            industrial location in, 331
        population pyramid and, 112–114
Economy, U.S.
    changes in, 270, 316, 317*f*
    dot-com bubble, 341
    space economy, components of,
        316–318
Ecosphere, **484**–485. *See also*
    Environment
Ecosystems, **486**
ECOWAS. *See* Economic Community
    of West African States
Ecuador
    census in, 131*f*
    demographic data, 533*t*
Ecumene, **126**
Edge cities, **420**
Edrisi, 6
*Educate American Act. See Goals
    2000: Educate American Act*
Education
    fertility rate and, 106
    as measure of development,
        371–373, 372*t*
    migration and, 92–94
    multilingual, 162
    teaching geography, as career, 8
Educational, Scientific, and Cultural
    Organization
    (UNESCO), 464
EEC. *See* European Economic
    Community
EEZ. *See* Exclusive economic zone
Effective demand, and spatial
    distribution of tertiary
    activities, 345
EFTA. *See* European Free Trade
    Association
Egypt
    carrying capacity in, 126
    demographic data, 530*t*
    infant mortality rate for, 112*f*
    population density of, 128*t*
    urbanization in, 130
Egyptian Arabic language, number of
    speakers, 146*t*
Elderly people, in developing countries,
    health care for, 376
Electoral geography, **472**
Elliptical projections, 526–527
Ellis Island, 190, 195*f*
Elongated states, **448**, 450*f*
Emerging economy, as term, 356

Employment
    in agriculture, as measure of
        development,
        366–367, 368*f*
    choice of, and space-time prisms, 75
    geographic information systems
        knowledge and, 9, 23
    high-tech industry and, 340–341
    opportunities in geography, 8–9
    tourism and, 345
    work trips, 72*f*, 75, 78*f*
    and work week, 40 hour, 418
Enclaves, **448**
Energy. *See also* Natural gas; Oil
    consumption per capita, as measure
        of development, 365–366
    developed nations and, 304–305
    developing nations and,
        365–366, 367
    mineral fuels, mining of, 304–307
    wood fuels, 300, 367
Energy Agency, 488
England. *See also* United Kingdom
    origin of place names in, 162–163
    paleolithic population of, 43
    travel preferences in, 80
English language
    as dominant official language, 154,
        155*f*, 160
    Estuary, 154
    global dialects of, 156, 157
    history of, 153–154
    in India, 152
    infection of other languages
        by, 161
    number of speakers, 146*t*
    as official language in Africa, 161*f*
    Old English dialect regions, 153*f*
    sources of additions to, 153
    standard form of, 154
    vocabulary exchange with other
        languages, 154
Environment, **484**
    aquaculture and, 299
    cultural development and, 39–40
    cultural landscape changes and, 484
    destruction of cultures by, 50
    fishing and, 296–298, 297*f*
    forestry and, 300–301
    Gross national income and, 365
    human impact on, 13, 14–15, 14*f*,
        15*f*, 40–42, 516–517 (*See
        also* Pollution)
        deforestation, 480*f*, 482*f*,
            496–499
        desertification, 499–501,
            501*f*, 502*f*
        and global warming,
            489–490, 493
        increased concern about, 481
        and increase of disease, 121
        and subsistence farming, 280, 281
        and wood as fuel, 366
    increasing damage to, 486–487
    land use and, 496–503
    mining and, 303
    as national resource, 359*f*
    perception of, and barriers to
        information, 78–81
Environmental determinism, **39**
Environmental pollution, **507**

Environmental Protection Agency, U.S.
(EPA)
    acid rain and, 494*f*
    hazardous waste and, 512
    on radioactive waste disposal, 514
    role of, 13
EPA. *See* Environmental Protection
    Agency
Epidemiologic transition, and increased
    life spans, 119–120
Equal-area projections, **522**–523,
    524*f,* 526*f*
Equatorial projections, 528
Equidistant projections, **524,** 524*f,* 526*f*
Equivalent projections, **522**–523, 524*f*
Eratosthenes, 4
Erbil, Iraq, 512*f*
Erosion, **499,** 499*f,* 501–503, 504
Erosion pavement, 500
Eskimo-Aleut language family, 147, 150*f*
Estuary English, 154
Ethiopia
    census in, 130
    demographic data, 531*t*
    maternal mortality ratio, 111
Ethnic cleansing, 192
Ethnic conflicts, 192
Ethnic culture, food and, 244–245
Ethnic diversity, 191–193
    acceleration of, 194
    in American cities, 213
    in Australia, 194
    in Canada, 194, 196*t*
    in cities, 209–216
    colonialism and, 190
    in Europe, 194
    in Latin America, 194
    ubiquity of, 190
    in United States, 195–197
        principal ethnic groups, 193*t,*
            195*t,* 196*f*
        urban, 213
    in various nations, 194
Ethnic enclaves
    characteristics of, 193, **214,** 215*f*
    urban, expansion of, 214, 215*f,* 216*f*
Ethnic geography, **190**
Ethnic groups, **190**
    desire for isolation in, 211–212
    origins of, 138
    shifting concentrations of, in U.S.,
        212–214
    as source of diversity, 230
Ethnic identity, assertion of,
    217–218, 218*f*
Ethnic islands, **201**
    in Canada, 202, 202*f*
    in U.S., 201–202, 201*f*
Ethnicity, **191**
    areal expressions of, 198–208
    as centrifugal force, 459
    concept of, **191**
    vs. culture, 191
    effect on landscape, 218–221
    employment and, 197, 197*f*
    segregation by, in urban areas,
        417, 417*f*
    suburbanization and, 419
    and territorial identity, 191–192
Ethnic neighborhoods, 209
Ethnic provinces in America, **203,** 203*f*

Ethnic regionalism, 221
Ethnic religions, **167**
    East Asian, 182–183
    Judaism, 169
Ethnic strife, 198
Ethnoburbs, 419
Ethnocentrism, **191**
Ethnographic boundaries, 454
EU. *See* European Union
Eurasia, geopolitics and, 461
Euro (currency), 468
Europe
    Central, as manufacturing center,
        336–337, 338*f*
    demographic data, 534*t*–536*t*
    Eastern
        city characteristics in,
            430–431, 430*f*
        demographic data, 535*t*
        as manufacturing center,
            337–338, 338*f*
        nationalism in, 460
        pronatalist policies in, 109
        urbanization in, 397*f*
        water pollution in, 507
    economy, restructuring of, 273
    emigration from, 124
    ethnic diversity in, 194
    fossil fuel reserves, 305*t*
    guest workers in, 85, 85*f,* 109, 190,
        190*f,* 194, 209
    in Ice Age, 43, 44*f*
    infant mortality rates, 109
    Inner Six and Outer Seven, 468, 468*f*
    Islam in, 177
    language in, 148*f*–149*f*
    network cities in, 410
    population in, 109, 126
    regional dialects in, 162
    total fertility rate in, 109
    toxic waste dumping by, 516
    urbanization in, 397*f*
    Western
        changing role of children in, 120
        city characteristics in,
            427–430, 429*f*
        demographic data, 535*t*
        guest workers in, 85, 85*f,* 109,
            190, 190*f,* 194, 209
        as manufacturing center,
            336–337, 338*f*
        population growth in, 119–120
        population pyramid for, 112, 113*f*
        secessionist movements in,
            460, 460*f*
European Coal and Steel
        Community, 468
European Economic Community
        (EEC), 468
European Free Trade Association
        (EFTA), 468
European Union (EU), **468,** 469*f*
    agricultural subsidies in, 286
    climate change agreements and, 492
    fur trade and, 301
    in international services
            industry, 347
    population in, 107
Evapotranspiration, 515
Excess vote technique, 472
Exclaves, **448**

Exclusive economic zone (EEZ), 297,
    465, 466*f*
Expansion diffusion, **55,** 55*f,* 56*f*
    of Christianity, 170
    of Islam, 177, 177*f*
    of language, 152
Exponential growth, 116–118
Extensive commercial agriculture,
    288–290
Extensive subsistence agriculture,
    276–278
External economies, and industrial
    location, **328**
Extractive industries, **294.** *See also*
        Mining and quarrying

F

Falkland Islands, population density
    of, 129
Falkland Islands War, 449, 467
Fallowing, **503**
Family, as folk culture basic unit, 243
Family status, segregation by, in urban
    areas, 416–417, 417*f*
FAO. *See* United Nations, Food and
        Agriculture Organization
Farming. *See* Agriculture
Farmland, value per acre, in U.S.,
    288, 290*f*
Fashion, history of awareness of, 251
Fast food
    local variations in, 254
    as popular culture, 252–253, 253*f*
FDI. *See* Foreign direct investment
Federal Housing Administration
        (FHA), 422
Federalism, asymmetric, 451–452
Federal states, 451
Females. *See* Women
Fencing, folk designs for, in North
        America, 241–242, 242*f*
Fertility, and overcrowding, 133
Fertility control. *See* Birth control
Fertility rate, 107–108, 108*f*
    decline in, 118, 123, 132, 135
    education and, 106
    global, 2003 data, 530*t*–536*t*
    reduction of, 121–122
Fertilizer minerals, 304
FHA. *See* Federal Housing
        Administration
Financial centers
    global, 347, 348*f*
    in U.S., 403*f*
Finland
    demographic data, 534*t*
    multilingualism in, 161
Fire, vegetation management and, 40
First effective settlement, **200**
First law of geography, 69*f*
First World, as term, 356
Fish farming, 298–299, 298*f*
Fishing
    environmental consequences of,
        296–297
    inland catch, 298
    regulation of, 297
    as resource exploitation, 295–299,
        296*f*
    total catch, 296–297, 297*f*

Five Civilized Tribes, 86*f*
Five pillars of faith, in Islam, 176, 176*f*
Five Year plans, 338
Fixed costs, of shipment, **323**
Flexible production systems, 328–329
Florida
    migration fields of, 93*f*
    tropical, as cultural region, 262*f*
Folk culture
    in Anglo America, 231
    British, 228*f*
    building traditions, 234–240, 235*f*
        diffusion of, 240–241, 241*f*
    in Canada, 231
    definition of, **225**
    gender roles in, 243
    magic in, 248
    music in, 245–248
    nonmaterial, 242–249
    as source of diversity, 230
    in U.S.
        diffusion of, 249–250, 249*f*
        diversity in, 230–251
Folk culture regions
    definition of, 231
    of Eastern U.S., 249–250, 249*f*
    passing of, 250–251
Folk life, 231
Folklore, **248**–249
Folk medicine and cures, 248
Folk music, style regions of,
    245–248, 247*f*
Folk tradition, oral, 248–249
Folkways, **249**
Food
    calorie requirements, 367–369
    ethnic, Americanization of,
        252–253, 253*f*
    fast food
        local variations in, 254
        as popular culture, 252–253, 253*f*
    fuel access and, 367
    north-south disparity in, 372*t*
    poverty and, 367–369, 369*f,* 370*f*
    preferences in, as folk trait,
        244–245
Food and Agriculture Organization
        (FAO). *See* United Nations,
        Food and Agriculture
        Organization
Food industry, growth of, 345
Food production, 282–285, 282*f. See
        also* Agriculture
Footloose industrial firms, **327,** 343
Forced migrations, 86
Ford, Henry, 328
Fordism, 328
Foreign aid, helpfulness of, 371,
    372–373
Foreign direct investment (FDI)
    and international quaternary
            services, 347
    as source of development, 371
    and transnational corporations, 331
Forest(s), destruction of, by pollution,
    493, 494, 498*f*
Forestry, 299–301, 299*f*
Formal regions, **18**
Form utility, of commodities, 271–272,
    318–319
Forward-thrust city capital, 452

traditional, **167**
world distribution of, 167–169, 167f
Religious folk music, Anglo-American, 247, 248f
Relocation diffusion, **56**
of Christianity, 171
definition of, 55f, **56**
of Hinduism, 178
of Islam, 177f
of Jews, 171f
of language, 152
Reluctant relocation, 86
Renewable resources, **294**–295
Replacement rate, 109
Representation, districting and, 471–472, 473f
Representative fraction, 21f
Research Triangle, 341
Reserves (mining), usability of, 301–302, 301f
Residential preferences, and access to information, 80, 81f
Resources, natural. *See* Natural resources
Resources disputes, **455**
Retail centers, in U.S., 403f
Retail gravitation, law of (Reilly), 70, 70f
Return migration, **90**–91, 92f
Rhine river, as boundary, 453
Rhumb lines, 525
Rice, as crop, 278–279, 280f, 281
Rimland theory, **461**
Ring highways, in Latin American cities, 436
Rio de Janeiro, 434f
Rio Earth Summit (1992), 492
Robert Taylor Homes, 423f
Robinson projection, 526f
Rockall Island, 467
Roger II, King of Sicily, 6
Roger's Book, 6
Roman Catholic Church. *See* Catholic Church
Romance languages, origins of, 146
Roman Empire
collapse of, 171, 199
environmental damage by, 41
spread of Latin language and, 152
Roosevelt, Franklin D., and New Deal, 418
Rostow, W. W., 370
Rotation of crops, **503**
Round village, 398f
Roundwood production, 300
Royalty, function of, 457
Ruhr industrial district, 337, 338f
Rural dispersal, 397, 398f
Russia. *See also* Soviet Union
agriculture in, 293, 491
AIDS in, 112
city characteristics in, 430–431
demographic data, 535t
gated communities in, 421, 431
industry in, 337–338, 338f
information direction bias in, 80
language in, 162
population pyramid for, 112, 113f
separatism and, 460
size of, 446

Russian language
number of speakers, 146t
standard form of, 154
Rustic Northeast, as cultural region, 262f
Rwanda
demographic data, 531t
refugees from, 87f

S

Saddlebag house, 240
Saharan language family, 148f–149f
St. Lawrence Valley hearth, 233, 233f
Saltbox house, 237, 237f
Salt Lake City, manufacturing in, 335, 336f
Salt Lake Oasis hearth, 233f, 234
Sanders, Beverly, 243
San Francisco, earthquake damage in, 83f
San Francisco Bay district
manufacturing in, 335, 336f
trips, by place of residence, 19f
Sanitary landfills, 509–510, 513f
Sanitation
in developing world cities, 432
improved, and mortality rates, 110, 120, 120f
as measure of development, 372t, 373–374, 375f
San people (Africa), 34f, 43, 45f, 47
Santa Barbara, California, topographical map of, 23f
Satellites, impact on information flow, 75–76, 78
Satisficing locations for industry, **327**, 327f
Saudi Arabia, demographic data, 533t
Scale, **12**–13, 20, 21f, 523
Scandinavia, pronatalist policies, 109
Schools, multilingual, 162
Schweizer barns. *See* Pennsylvania Dutch, architecture of
Scottsdale, Arizona, growth of, 426
S-curve, in population growth, **133,** 133f
Sea, International law of (U.N.), **465**–467, 466f
Sea level, global warming and, 490
Sears catalog, 251
SEATO. *See* Southeast Asia Treaty Organization
Secession, of local metropolitan areas, 476–477
Secondary economic activities, **271**–272, 271f, 272f. *See also* Manufacturing
location considerations, factors affecting, 318–333
transnational corporations and, 331
Second World, as term, 356
Sector model of urban land use, **414,** 415f, 419f
Secularism, **169**
fundamentalist reaction to, 168
Seed banks, 283
Segregation, **209**
by ethnicity and status in urban areas, 417
in European cities, 428–430
external controls in, 211

of immigrants, 193
internal controls in, 211–212
urban, worldwide, 209–210, 211
in U.S., 209–216, 210f
Self-determination, as centrifugal force, 460
Self-sufficiency, as value of folk society, 243
Senior citizens, costs associated with, 114
Separatism, **460**
Sephardim, 169–170
September 11th terrorist attacks, 90, 177
Sequoyah, 56
Serbs and Croatians, conflict between, 462
Service sector. *See also* Quaternary economic activities; Quinary economic activities; Tertiary economic activities
of city economic structure, **402**
in world economy, 346–349, 347t
Settlement
patterns of, in U.S., 220
rural, types and characteristics of, 397–400, 398f, 399f
urban, location factors in, 401–402
Shadow (informal) economy, 360, 360f
Shamanism, **167**
Shape
distortion of, in map projection, 523–524, 524f
of states, effects of, 446–448, 450f
Sharia, 166
Shaw, George Bernard, 155
Shaw v. Reno (1993), 475
Shifting cultivation, 277–278, 277f, 278f, 279
Shinto
practices and beliefs, **183,** 183f
world distribution of, 167f
Shopping malls
in Latin American cities, 436
as popular culture, 255–256, 255f
suburbanization and, 418–419
Short haul penalty, 325f
Shotgun house, 239, 240f
Siddhartha Gautama, 180
Sikhism, 166, 180
Silicon Fen, 342
Silicon Forest, 341
Silicon Glen, 342
Silicon Swamp, 341
Silicon Valley, 19f, 316, 335, 341
Silicon Valley North, 341
Simple conic projections, 525
Singapore
demographic data, 534t
economic success of, and location, 449
industry in, 314f
location of, 449
population control measures in, 102, 105, 134
population in, 357
service exports, 347
Singlish, 157
Sino-Tibetan languages, 148f–149f, 151
Site, **10,** 401–402
Situation, **10**–11, 401–402

Size
scale and, 12–13
of states
effects of, 446
ministates, 446, 449
Skeleton grid plan, 398f
Skills learned in geography, 8–9
Skylines, forces underlying creation of, 411
Slash-and-burn farming. *See* Shifting cultivation
Slater, Samuel, 362f
Slaves, arrival of Africans in America as, 193
Slavs, expansion across Siberia, 147
Slums, in Latin American cities, 436
Smart growth programs, 427
Smelting plants, 304f
Smog, 494, 498f
Smoking, origin of, 57
Snack nuts, regional variations in consumption of, 259f
Snake fence, 241–242, 242f
Soccer, popularity of, 252, 252f
Social areas of cities, 415–417, 417f
Social dialects, **156**
Social distance, **209**
Social Security, potential support ratios and, 135
Social status, segregation by, in urban areas, 416, 417f
Societies, agricultural, classification of, 274
Sociofacts, 51, 54f
Sociological subsystem, **51**–52
Software Valley, 341
Soil, **501**
degradation of, 283, 503, 505f
erosion of, **499,** 499f, 501–503, 504
preservation methods for, 503, 504
Somalia
census in, 130
demographic data, 531t
South Africa
Cape Town, 41f
demographic data, 532t
life expectancy in, 110
San people of, 34f
slums in, 375f
Zulu villages in, 399f
South America
demographic data, 533t
language in, 148f–149f
population control measures in, 134
religion in, 169
tropical rain forests in, 498
South Americans
dispersal of, in U.S., 206
as percent of Hispanics, 206t
Southeast Asia Treaty Organization (SEATO), 462
Southeastern U.S., manufacturing in, 335, 336f
Southern Backwoods and Appalachian song area, 246, 247f
Southern Coastal stream, architecture, diffusion of, 240–241, 241f
Southern Cone Community Market (MERCOSUR), 468
Southern hearths, architecture of, 238–239, 239f

Southern New England hearth, 233*f*, 234, 237–238, 237*f*
Southern Tidewater hearth, 233*f*, 234, 239
Southern U.S.
   as cultural region, 261*f*, 262*f*
   land survey techniques in, 219
South Korea
   anti-female bias in, 114
   demographic data, 534*t*
   population, 125
   population growth in, 135
   Seoul, size of, 406
South-north disparity in development, 356–357, 356*f*, 361–362, 371, 372*t*
Southwestern U.S.
   as cultural region, 261*f*, 262*f*
   Hispanic influence in, 201
Soviet Union. *See also* Russia
   agriculture in, 293, 293*f*
   city characteristics in, 430–431, 430*f*
   collapse of, 198, 199
      migration following, 85
      and nationalism, 460
      NATO and, 470
      and return to private agriculture, 293
      states created by, 445, 448*f*
   containment of, 461–462
   eastward migration in, 85
   force migrations in, 86
   fossil fuel reserves, 305*t*
   imperialism of, 461
   industry in, 331, 337–338, 338*f*
   infant mortality rates, 109
   Soviet of Nationalities, 445
   Virgin and Idle Lands program, 293*f*
Space economy, in U.S., components of, 316–318
Space-time compression, 55
Space-time path, 74, 76*f*
Space-time prisms
   definition of, **72**, 72*f*
   women and, 75
Spain
   colonialism of, 172
   demographic data, 536*t*
   dialects in, 161–162
   Islam in, 177
   population of, 109
   service exports, 347*t*
Spanish adobe house, 240
Spanish language
   number of speakers, 146*t*
   as official language in Africa, 161*f*
   standard form of, 154
Spanish Laws of the Indies (1573), 173
Spatial diffusion, **16**
   of language, 152
Spatial distribution, **16**–17, 18
Spatial interaction, **15**–16, **66**
   human behavior and, 71–84
   information flow and, 75–78
   measurement of, 68–71
   model of, 67–78
   place perception and, 78–84
Spatially fixed costs, in manufacturing, **319**, 319*f*
Spatially variable costs, in manufacturing, **319**, 319*f*

Spatial margin of profitability, **327**, 327*f*
Spatial mismatch, 422–423
Spatial monopoly, 326
Spatial science, geography as, 4
Spatial search behavior, **88**, 89*f*
Spatial system
   analysis of, **27**–29
   government entities as, 471
Special crop farming, 292, 292*f*
Special function cities, 405
Specialization, in agriculture, 285
Speech community, **154**
Spine, of urban area, 436
Sports. *See* Games
Spratly Islands, 467
Sprawl. *See* Urban sprawl
Spread effects, **358**
Spykman, Nicholas, 461, 462*f*
Sri Lanka
   ethnic strife in, 198
   reduction of death rate in, 121
Stacked gerrymandering, 472
Stage in life, and activity space, 73
Stalin, Josef, 292, 293*f*
Standard language, **154**–155
Standard line, **525**
Standard of living
   development and, 363–364, 363*t*
   and population density, 129
Starvation
   population increase and, 133
   world population and, 104
State(s)
   administration of, and citizen loyalty, 458
   boundaries of, 453–455
      natural, **453**, 453*f*
      as three-dimensional claim, 453
      types of, 453–454
   cohesiveness of, 456–459
   definition of, **443**–444
   geographic characteristics of, 446–453
   modern, evolution of, 445
   number of, 445
   super-sized, possible emergence of, 471
   unitary, 450
Stateless nations, 444*f*, 445
Statistical maps, 22
Statute of Pleading, 153
Steam power, and industrial revolution, 337
Steel industry
   material flow in, 320, 320*f*
   U.S., decline of, 316
Step migration, **88**
Stereographic projection, **522**
Sterilization programs, 134*f*
Steward, Julian, 51
Stimulus diffusion, 55–56
Stith, Mary, 251
Stone Age, 36
   humans awareness in, 84
   Mesolithic era, 44, 46
   Neolithic era, 47–49
   Paleolithic era, 42–43, 44*f*
   stages of, 44–46
Strabo, 4
Street pattern, grid-style, origin of, 17
String village, 398*f*

Strip cropping, 504, 504*f*
Stripmining, 511*f*
Structural assimilation, **197**
Stupas, 181
Style and fashion, history of awareness of, 251
Subsequent boundaries, **454**
Subsistence agriculture, 274–285
   areas practicing, globally, 277*f*
   definition of, 272–274
   environmental impact of, 280, 281
   rural settlements in, 397–400, 399*f*
Subsistence economy, **273**
Subsistence household economies, 243
Substitution principle in industry location, **327**
Suburb(s)
   in Canada, 427
   defining features of, **400**, 401*f*
Suburbanization
   effect on cities, 422–423
   in United States, 418–420, 419*f*
Sudanic language family, 148*f*–149*f*
Summer, year without a (1816), 487
Sun, and seasonal cycles, 485, 486*f*
Sunni Muslims, 167*f*, 177, 177*f*
Sunrise Strip, 342
Superimposed boundaries, **454**
Superstates, possible emergence of, 471
Supply and demand, 318, 318*f*
Supply curve, 318*f*
Supranationalism, **463**–464
Supreme Court, U.S.
   on incinerator pollution, 510
   on racial gerrymandering, 475
Survey systems
   under Ordinance of 1785, 17
   types of, 9–10
   variance by ethnicity, 219–220, 219*f*, 220*f*
Sustainable development, 359*f*
Suzerainty, 461
Swahili
   as creole, 159
   as lingua franca, 160*f*
   number of speakers, 146*t*
Swedagon pagoda, 182
Sweden
   demographic data, 535*t*
   population pyramid for, 112, 113*f*
Swidden farming. *See* Shifting cultivation
Switzerland, demographic data, 535*t*
Syncretism, **59**, 60*f*, **182**, 249
Systematic geographers, 7
Systems, as subject of human geography, 33

𝒯

Taglish, 157
Taiwan
   anti-female bias in, 114
   demographic data, 534*t*
   manufacturing in, 339, 340*f*, 341*f*
   population, 109, 125, 534*t*
Takeoff stage of development, 370, 371
Tambora, 491, 493
Tamil language, 146*t*
Tamils, 198

Tanzania
   demographic data, 531*t*
   language in, 159
Tao, 182–183
Taoism, **182**–183
Tapering principle of transportation costs, 323, 325*f*
Task Force for Child Survival program (WHO), 374, 376*f*
Taxes, agriculture and, 504
Taxpayers, cost of illegal immigrants to, in U.S., 90
Teachers, school-age population per, 371
Teaching geography, as career, 8
Technological subsystem, **51**–52, 52*f*, 56
Technology, **270**
   and carrying capacity of land, 134–135
   definition of, **362**
   diffusion of, and creation of wealth, 337, 362–363, 362*f*
   economic activity and, 270
   innovation in, 362
   national security and, 463
   spread of, 343
Technology gap, **362**
Technology transfer, **362**–363
   health and, 374
   as source of development, 371
Telegraph, impact on sports, 230
Telephone, space-cost convergence in toll charges for, 75
Tells, 512*f*
Telugu language, number of speakers, 146*t*
Teotihuacán, 48
Terminal costs, **323**
Terracing, 126, 127*f*, **503**
Territorial disputes among nations, **454**–455
Territorial identity, ethnicity and, 191–192
Territoriality of humans, **72**
Territorial sea, 465, 466*f*
Terrorism
   definition of, **462**
   as geopolitics, 462
   goals of, 462
   Islamic fundamentalism and, 462–463
   September 11th terrorist attacks, 90, 177
   state sponsors of, 463
Tertiary economic activities, 271*f*, **272**
   definition of, **343**, 343–344
   and GDP, global, 344*f*, 344*t*
   growth of, 343–344, 343*f*, 344*f*, 344*t*, 345
   locational orientation of, 344–345, 345*f*
   low-level, 345, 345*f*
Tertullian, 133
Texas City, Texas, 14*f*
Textile industry, 321
TFR. *See* Total fertility rate
Thailand
   demographic data, 534*t*
   informal economy in, 360
Thematic maps, 21–22, 24*f*

List of Maps